Brian Berkowitz · Ishai Dror
Bruno Yaron

Contaminant Geochemistry

Interactions and Transport in the Subsurface Environment

Second Edition

Springer

Brian Berkowitz
Ishai Dror
Bruno Yaron
Department of Earth and Planetary Sciences
Weizmann Institute of Science
Rehovot
Israel

ISBN 978-3-662-51777-2 ISBN 978-3-642-54777-5 (eBook)
DOI 10.1007/978-3-642-54777-5
Springer Heidelberg New York Dordrecht London

Printed on acid-free paper

Springer is part of Springer Science+Business Media (www.springer.com)

Contaminant Geochemistry

To our wives, Melanie, Tali, and Angie;
and
To our children, Noa, Sari, and Yotam;
Yaara, Maayan, and Adar; Dana
… with love and respect

Preface to the First Edition

Contaminant Geochemistry: Interactions and Transport in the Subsurface Environment combines the earth science fields of subsurface hydrology and environmental geochemistry and aims to provide a comprehensive background for students and researchers interested in protection and sustainable management of the subsurface environment. This book focuses on the upper part of the earth's crust, covering the region between the land surface and the groundwater zone; anthropogenic contamination occurs primarily in this well-defined geosystem.

Water and land are limited natural resources, and it is incumbent on humankind to manage knowledge and technology in a way that avoids, or at least minimizes, deterioration of these resources. In this context, an understanding of the interactions between subsurface components (solid, liquid, and gaseous phases) and chemical contaminants is required. Because the subsurface is an open system, contaminants may be transported, transformed, and redistributed in the subsurface under a variety of environmental influences. Contaminant interactions in the subsurface are subject to continuous changes, being affected by fluctuations in climatic conditions (particularly precipitation) and microbiological activity. Additionally, these interactions are controlled by the structure and properties of the earth materials, the molecular properties of the contaminants, and the hydrogeology of each specific location. As a consequence, a multidisciplinary approach is fundamental to understanding the governing processes.

Because our book was conceived for readers with different backgrounds, we devote Part I to revisiting aspects of classical geochemistry, focusing on the constituents of subsurface water and earth materials (Chapter 1) and selected processes related to potential interactions between subsurface liquid and solid phases with toxic chemicals (Chapter 2). Part II is an overview of potential subsurface contaminants of anthropogenic origin; properties of these chemicals are described together with their environmental hazards. Chapter 3 is devoted to inorganic chemicals, while the characteristics and hazards of organic toxic chemicals are presented in Chapter 4.

The retention of contaminants in the subsurface, controlled by properties of both chemicals and subsurface constituents as well as contaminant partitioning among the solid, aqueous, and gaseous phases, are the focus of Part III. Chapter 5 deals with the sorption, retention, and release of contaminants, while Chapters 6 and 7 examine contaminant partitioning in the aqueous phase and partitioning of volatile compounds.

Contaminant redistribution in the subsurface, as a result of transport (in dissolved form, as an immiscible-with-water phase, or adsorbed on colloids) is discussed in Part IV. These phenomena do not occur in a static domain, and contaminants are redistributed, usually by flowing water, from the land surface, through the partially saturated subsurface down to the water table, and within the fully saturated aquifer zone. After a basic presentation of water movement in the subsurface environment (Chapter 9), we focus on transport of passive contaminants (Chapter 10) and reactive contaminants (Chapter 11).

Transformation and reactions of contaminants in the subsurface are addressed in Part V. From an environmental point of view, we do not restrict the contaminant transformation to molecular changes; we also consider the effects of such changes on contaminant behavior in the subsurface. Abiotic and biologically mediated reactions of contaminants in subsurface water are discussed in Chapter 13. Abiotic transformations of contaminants at the solid–liquid interface are described in Chapter 14, while biologically mediated changes in subsurface contaminants are the subject of Chapter 15.

We used our own results and selected research findings reported in the literature to provide numerous examples of contaminant retention, redistribution, and transformation in the subsurface (Chapters 8, 12, and 16). Because a limited number of published research findings had to be selected from the vast number of available publications, the choice was very difficult. Many other research results of equal value could have been used to illustrate processes governing the fate of contaminants in the subsurface environment.

Contaminant Geochemistry was written for the use of geochemists, soil scientists, water specialists, environmental chemists, and engineers involved with understanding, preventing, controlling, and remediating subsurface contamination by chemicals of anthropogenic origin. This book also provides beginning graduate students in environmental sciences an overview of contaminant behavior in the geosystem, as a basis for their future professional development. We hope that we have succeeded in presenting the reader with a comprehensive—but not exhaustive—review of current knowledge in the field of subsurface contaminant geochemistry.

Rehovot, Israel, June 2007

Brian Berkowitz
Ishai Dror
Bruno Yaron

Preface to the Second Edition

We were pleased when the publisher asked us to prepare a second edition of "Contaminant Geochemistry: Interactions and Transport in the Subsurface Environment," as a result of the strong interest in our book published in 2008. This book focuses on the upper part of the earth's crust, covering the region between the land surface and the groundwater where anthropogenic chemical contamination primarily occurs.

In this expanded second edition, new literature has been added on contaminant fate in the soil–subsurface environment. In particular, we have included more data on the behavior of inorganic contaminants and on engineered nanomaterials, the latter comprising a group of "emerging contaminants" that may reach the soil and subsurface zones. To facilitate access to the cited literature in this second edition, we include the references after each chapter. We have also introduced corrections to small errors in the first edition.

Part VI of the book is devoted to a new perspective of contaminant geochemistry, namely irreversible changes in pristine land and subsurface systems following chemical contamination. Two new chapters were added on this topic, focusing attention on the impact of chemical contaminants on the matrix and properties of both liquid and solid phase of soil and subsurface domains. Contaminant impacts on irreversible changes occurring in groundwater are discussed in Chap. 17 and their irreversible changes on the porous medium solid phase are surveyed in Chap. 18. In contrast to the geological timescale controlling natural changes of porous media liquid and solid phases, the timescale associated with chemical pollutant-induced changes is far shorter and extends over a "human lifetime scale."

Rehovot, Israel, January 2014

Brian Berkowitz
Ishai Dror
Bruno Yaron

Contents

Part I Geochemistry Revisited: Selected Aspects

1 Characterization of the Subsurface Environment 3
1.1 Subsurface Solid Phase 4
1.1.1 Silica Minerals 4
1.1.2 Clay Minerals.................................. 6
1.1.3 Minerals Other than Silica and Clay 13
1.1.4 Organic Matter 13
1.1.5 Electrically Charged Surfaces..................... 17
1.2 Subsurface Liquid Phase 18
1.2.1 Near Solid Phase Water.......................... 19
1.2.2 Subsurface Aqueous Solutions 21
1.3 Subsurface Gaseous Phase 22
1.4 Aquifers.. 23
1.4.1 Aquifer Structure 23
1.4.2 Groundwater Composition 25
References 26

2 Selected Geochemical Processes 29
2.1 Thermodynamics and Equilibrium....................... 29
2.1.1 Enthalpy, Entropy, and the Laws of Thermodynamics.......................... 29
2.1.2 Equilibrium 31
2.1.3 Solubility, Chemical Potential, and Ion Activities...... 32
2.1.4 Kinetic Considerations and Reaction Rate Laws 34
2.2 Weathering... 39
2.2.1 Dissolution and Precipitation 39
2.2.2 Redox Processes................................ 42
2.3 Adsorption 45
2.3.1 Adsorption of Charged Ionic Compounds............ 46
2.3.2 Adsorption of Nonionic Compounds 48
2.3.3 Kinetic Considerations........................... 49
References 50

Part II Properties of Potential Contaminants: Environmental and Health Hazards

3 **Inorganic and Organometallic Compounds** 53
3.1 Nitrogen 53
3.2 Phosphorus 56
3.3 Salts 57
3.3.1 Agricultural Impacts 58
3.3.2 Impacts on Water Usage 59
3.3.3 Impacts on Infrastructure 60
3.3.4 Impacts on Biodiversity and the Environment 60
3.4 Radionuclides 61
3.5 Heavy Metals and Metalloids 63
3.5.1 Arsenic 63
3.5.2 Cadmium 65
3.5.3 Chromium 66
3.5.4 Lead 67
3.5.5 Nickel 68
3.6 Nanomaterials 70
References 72

4 **Organic Compounds** 79
4.1 Pesticides 79
4.1.1 Organochlorine Insecticides 81
4.1.2 Organophosphates 83
4.1.3 N-methyl Carbamate 83
4.1.4 Triazines 84
4.1.5 Paraquat and Diquat 84
4.2 Synthetic Halogenated Organic Substances 86
4.2.1 Chlorinated Hydrocarbons 86
4.2.2 Brominated Flame Retardants 89
4.3 Petroleum Hydrocarbons and Fuel Additives 91
4.4 Pharmaceuticals and Personal Care Products 94
4.4.1 Analgesics and Anti-inflammatory Drugs 96
4.4.2 Hormones 96
4.4.3 Antibacterial Drugs 97
4.4.4 Antiepileptic Drugs 98
4.4.5 Beta-Blockers 98
References 98

Part III Contaminant Partitioning in the Subsurface

5 Sorption, Retention, and Release of Contaminants 107
5.1 Surface Properties of Adsorbents 107
5.2 Quantifying Adsorption 109
5.2.1 Adsorption–Desorption Coefficients 109
5.2.2 Adsorption Isotherms 109
5.3 Kinetics of Adsorption 116
5.4 Adsorption of Ionic Contaminants 118
5.5 Adsorption of Nonionic Contaminants 123
5.6 Other Factors Affecting Adsorption 126
5.7 Nonadsorptive Retention of Contaminants 129
5.7.1 Contaminant Precipitation 129
5.7.2 Liquid Trapping 131
5.7.3 Particle Deposition and Trapping 132
5.8 Reversible and Irreversible Retention 135
5.8.1 Genuine Hysteresis 135
5.8.2 Apparent Hysteresis 137
5.8.3 Bound Residues 138
References 141

6 Contaminant Partitioning in the Aqueous Phase 147
6.1 Solubility Equilibrium 148
6.2 Aqueous Solubility of Organic Contaminants 150
6.3 Ligand Effects 151
6.4 Cosolvents and Surfactant Effects 153
6.5 Salting-Out Effect 156
6.6 Apparent Solubility 159
References 161

7 Partitioning of Volatile Compounds 163
7.1 Gas–Liquid Relationships 164
7.2 Volatilization from Subsurface Aqueous Solutions 165
7.3 Vapor Pressure–Volatilization Relationship 166
7.4 Volatilization of Multicomponent Contaminants 169
References 170

8 Selected Research Findings: Contaminant Partitioning 171
8.1 Partitioning Among Phases 171
8.2 Contaminant Volatilization 173
8.2.1 Inorganic Contaminants 175
8.2.2 Organic Contaminants 177
8.2.3 Mixtures of Organic Contaminants 180

8.3 Solubility and Dissolution . . . 184
8.3.1 Acidity and Alkalinity Effects . . . 185
8.3.2 Redox Processes . . . 187
8.3.3 Dissolution from Mixtures of Organic Contaminants . . . 189
8.3.4 Apparent Solubility . . . 189
8.4 Contaminant Retention in the Subsurface . . . 198
8.4.1 Contaminant Adsorption . . . 198
8.4.2 Nonadsorptive Retention . . . 220
8.5 Contaminant Release . . . 223
8.6 Bound Residues . . . 230
8.7 Contaminants in Urban Geosystems . . . 233
References . . . 238

Part IV Contaminant Transport from Land Surface to Groundwater

9 Water Flow in the Subsurface Environment . . . 247
9.1 Water Flow in Soils and the Vadose Zone . . . 247
9.2 Flow Through the Capillary Fringe . . . 250
References . . . 252

10 Transport of Passive Contaminants . . . 255
10.1 Advection, Dispersion, and Molecular Diffusion . . . 256
10.2 Preferential Transport . . . 259
10.3 Non-Fickian Transport . . . 261
References . . . 265

11 Transport of Reactive Contaminants . . . 267
11.1 Contaminant Sorption . . . 267
11.2 Colloids and Sorption on Colloids . . . 269
11.3 Dissolving and Precipitating Contaminants . . . 270
11.4 Transport of Immiscible Liquids . . . 273
11.5 Transport of Contaminants by Runoff . . . 279
11.5.1 Overland Runoff . . . 279
11.5.2 Urban Runoff . . . 281
References . . . 281

12 Selected Research Findings: Contaminant Transport . . . 285
12.1 Heavy Metals and Metalloids . . . 285
12.1.1 Heavy Metals . . . 286
12.1.2 Metalloids . . . 296
12.2 Engineered Nanoparticles (ENPs) . . . 302
12.2.1 Fate of ENPs in Aqueous Solutions . . . 302

12.2.2 Transport of ENPs in Sand and Glass Bead Columns . . . 304
12.2.3 Modeling ENP Transport in Porous Media. 312
12.2.4 Fate of ENPs in Soil Column Experiments. 312
12.2.5 ENP Transport in the Field 316
12.3 Pesticides 320
12.4 Colloid-Mediated Transport of Organic Contaminants 333
12.4.1 Surface Runoff 334
12.4.2 Transport in the Subsurface 335
12.5 Transport of Nonaqueous-Phase Liquids. 337
12.5.1 Infiltration into the Subsurface 338
12.5.2 Transport of Soluble NAPL Fractions in Aquifers. 341
References 343

Part V Transformations and Reactions of Contaminants in the Subsurface

13 Abiotic Contaminant Transformations in Subsurface Water 349
13.1 Abiotic Contaminant Transformations in Natural Subsurface Water 349
13.1.1 Factors Affecting Contaminant Transformations 349
13.1.2 Reactions in Natural Waters. 351
13.2 Selected Contaminant Transformations in Sediments and Groundwater. 362
13.2.1 pH and Hydrolysis Reactions 362
13.2.2 Redox State and Reactions. 363
References 369

14 Abiotic Transformation at the Solid–Liquid Interface 373
14.1 Catalysis. 373
14.2 Surface-Induced Transformation of Organic Contaminants 374
14.3 Surface-Induced Interactions of Inorganic Contaminants 379
References 379

15 Biologically Mediated Transformations. 381
15.1 Subsurface Microbial Populations 381
15.2 Processes Governing Contaminant Attenuation 381
15.3 Biotransformation of Organic Contaminants 383
15.4 Biotransformation of Inorganic Contaminants 389
References 394

16 Selected Research Findings: Transformations and Reactions 397
16.1 Abiotic Alteration of Contaminants 397
16.1.1 Transformation in Subsurface Water 397

16.1.2 Surface-Induced Transformation on Clays 412
16.1.3 Photochemical Reactions 416
16.1.4 Ligand Effects 422
16.1.5 Multiple-Component Contaminant Transformation 425
16.2 Biomediated Transformation of Contaminants. 437
16.2.1 Petroleum Products 437
16.2.2 Pesticides. 442
References 452

Part VI Contaminant Induced Irreversible Changes in the Soil–Subsurface Regime

17 Contaminant-Induced Irreversible Changes in Groundwater Chemistry 457
17.1 Groundwater Salinization 458
17.1.1 Salinization Under Downward Flow Conditions 458
17.1.2 Salinization Under Lateral Flow Conditions. 461
17.2 Groundwater Acidification 465
17.3 Groundwater Contamination by Inorganic Trace Elements 469
17.3.1 Radionuclides. 470
17.3.2 Heavy Metals. 471
17.3.3 Metalloids 476
17.4 Groundwater Contamination by Trace Organic Compounds 480
References 497

18 Contaminant Impacts on the Soil–Subsurface Solid Phase 501
18.1 Impact of Contaminants Originating from Agriculture 501
18.1.1 Fertilizers. 502
18.1.2 Micronutrients 505
18.1.3 Soil–Subsurface Amendments. 507
18.1.4 Pesticides. 517
18.2 Impact of Sodic Water. 528
18.2.1 Swelling 528
18.2.2 Dispersion and Flocculation 531
18.2.3 Water Transmission Properties 533
18.3 Impact of Municipal and Industrial Contaminants 537
18.3.1 Acid Rain Input 538
18.3.2 Impact of Effluents 542
18.3.3 Impact of Sludge and Compost. 546
18.3.4 Petroleum Products 557
References 562

Index 571

Part I
Geochemistry Revisited: Selected Aspects

We conceived this book on contaminant geochemistry not only for researchers in hydrology, soil science, and geochemistry but also for advanced undergraduate and graduate students in environmental sciences and various specialists working on environmental pollution problems. As a consequence, we consider it useful and necessary to include an introductory material that presents a short characterization of porous media together with selected geochemical processes. Part I characterizes subsurface components and the geochemical pathways and processes related to potential contamination. Keeping in mind that contaminant geochemistry covers a diverse range of natural and human-induced processes, we have selected only those that are most relevant to our field of interest.

The first chapter presents the porous medium solid, liquid, and gaseous phases in an environment controlled generally by fluctuations of rainy and dry periods, reflected by a variety of saturated and partially saturated conditions. A description of mineral and organic components in porous media is followed by a discussion of the electrically charged surface properties that affect the near-solid phase water. The composition of the subsurface water solution as affected by the surrounding solid phase, by the natural biological environment, and possibly by human factors also is discussed. The gaseous composition of porous media, which affects the chemistry of the liquid phase, is another part of the subsurface environment. Consideration of the aquifer environment, characterized by a saturated regime, includes a brief discussion of groundwater geology and composition.

Selected geochemical processes that relate to the behavior of contaminants in the subsurface are described in the second chapter. We focus on thermodynamic considerations and equilibrium processes in the subsurface, accounting for interactions among solid, liquid, and gaseous phases. Kinetic considerations in defining the solubility of organics and minerals, as well as knowledge of chemical potentials, ion activities, and reaction rate laws, are also included. The weathering of subsurface solid phases, which occurs as a result of interactions with the liquid phase, is a natural geological process that may be accelerated by anthropogenic influences. Dissolution-precipitation and redox mechanisms are other important processes affecting subsurface weathering that are included. The solid phase is a potential contaminant adsorbent, and we therefore consider adsorption as a main geochemical pathway, relevant in defining the retention, release, and persistence of pollutants in the subsurface.

Chapter 1
Characterization of the Subsurface Environment

The zone between land surface and the water table, which forms the upper boundary of the groundwater region, is known as the *vadose zone.* This zone is mostly unsaturated—or more precisely, partially saturated—but it may contain a saturated fraction in the vicinity of the water table due to fluctuations in water levels or capillary rise above the water table. The near-surface layer of this zone—the soil—is generally partially saturated, although it can exhibit periods of full saturation. Soil acts as a buffer that controls the flow of water among atmosphere, land, and sea and functions as a sink for anthropogenic contaminants.

Soil forms from the disintegration or decomposition of parent rock material, due to weathering processes and decomposition of organic materials. The mineral fraction of soil is formed by physical weathering of rock, resulting from thermal expansion of minerals along crystallographic axes or expansion of water in rock fissures. Stable weathering products from mineral decomposition may consist of ionic or neutral organic and inorganic components, solid organic residues, and newly formed silicates, oxides, hydroxides, carbonates, sulfides, and other solids (Greenland and Hayes 1978). The organic fraction of soil is an heterogeneous mixture of products resulting from microbial and chemical transformation of nonliving organic residues, plant roots, and living microorganisms.

Natural subsurface zones of the earth are characterized by heterogeneous assemblies of materials, forming a porous medium where solid, liquid, and gaseous phases are present. The open boundaries between the solid, liquid, and gaseous phases, as well as the activity of microorganisms, create conditions suitable for developing processes of chemical and biological origin, which dynamically affect the properties of subsurface environments. The porosity of the near surface (soil layer) is controlled mainly by the association of mineral and organic parts, with solid particles tending to be molded into aggregates, or peds, by biologically induced exudates, plant roots, fungi, soil microorganisms, and animals. The porosity of this region may be changed either by a shrink–swell phenomenon under wetting and drying–freezing conditions or by a redistribution of the colloidal fraction under water transport from the land surface to the water table. Over time, activity between the solid and liquid phases also can affect the stability of the porous medium itself and subsequently change the physicochemical properties of the solid phase.

B. Berkowitz et al., *Contaminant Geochemistry*,
DOI: 10.1007/978-3-642-54777-5_1, © Springer-Verlag Berlin Heidelberg 2014

1.1 Subsurface Solid Phase

The solid phase of the subsurface is a porous medium composed of a mixture of inorganic and organic natural materials in various stages of development. The surface area and the surface (chemical) properties of the solid phase are major factors that control the behavior of chemicals.

Based on their origin, minerals are grouped into two broad classes: primary minerals and secondary minerals. Primary minerals have not been altered chemically since the time of their crystallization from molten lava and their subsequent deposition. This group includes quartz (SiO_2), feldspars (e.g., $(Na,K)AlSi_3O_8$; i.e., aluminosilicates containing varying amounts of calcium, potassium, or sodium), micas (e.g., muscovite: $KAl_2(AlSi_3O_{10})(OH)_2$ and/or chlorites: $(Mg,Fe)_3(Si,Al)_4O_{10}(OH)_2$–$(Mg,Fe)_3(OH)_6$, which belong to the phyllosilicate mineral group), amphibole (e.g., hornblende: $(Ca,Na)_{2,3}(Mg,Fe,Al)_5Si_6(Si,Al)_2O_{22}(OH)_2$, i.e., magnesium–iron silicates, often with traces of calcium, aluminum, sodium, titanium, and other elements), pyroxenes (e.g., augite: $(Ca,Na)(Mg,Fe,Al)(Si,Al)_2O_6$, i.e., silicate minerals rich in calcium, iron, and magnesium and commonly found in basalt), olivine (magnesium–iron silicate $(Mg,Fe)_2SiO_4$) and rutile (TiO_2). Secondary minerals originate from the decomposition of primary minerals and subsequent reprecipitation into new, chemically distinct minerals. This group contains many minerals, including kaolinite, smectite, vermiculite, chlorite, imogolite, gibbsite, goethite, hematite, birnessite, calcite, and gypsum. Layer aluminosilicates are the dominant minerals formed in most temperate region soils. These layer silicates are composed of various arrangements of silicon/oxygen sheets in tetrahedral coordination and aluminum/oxygen sheets in octahedral coordination.

Primary minerals with low surface area (e.g., silica minerals) and low reactivity mainly affect the physical transport of water, dissolved chemicals, colloids, immiscible (in water) liquids, and vapors. Secondary minerals generally have high surface area (e.g., clay minerals) and high reactivity that affect the transport of chemicals, their retention and release onto and from the solid phase, and their surface-induced transformations. The solid phase also can indirectly induce the degradation of chemical compounds, through its effects on the water–air ratio in the system and, thus, on microbiological activity.

1.1.1 Silica Minerals

Silica minerals are a primary mineral classified as tectosilicates, characterized by repeating SiO_4 units in a frame-like structure. Quartz, one of the most abundant minerals on earth, often comprises up to 95 % of all sand and silt fractions. It therefore is representative of the structure and properties of silica minerals.

Wilding et al. (1977) showed that the silica tetrahedron in quartz is almost symmetrical and has a Si–O distance of 0.16 nm. They noted that the structure of

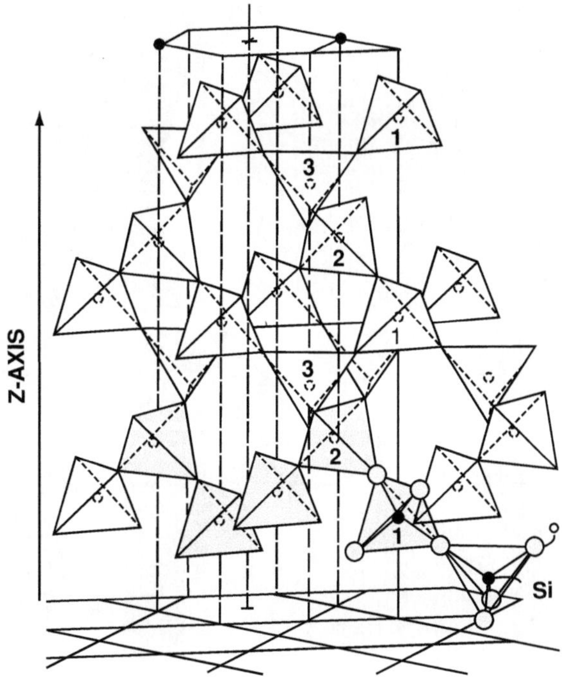

Fig. 1.1 The structure of quartz (Dress et al. 1989)

quartz can be visualized as a spiral network of silica tetrahedra around the *z*-axis. From Fig. 1.1, it is seen that each tetrahedron is repeated in the network by a rotation of 120° and a translation of c/3.

Quartz yields a characteristic X-ray pattern with well-defined peaks exhibiting a d-spacing ranging between 0.426 and 1.182 nm (Dress et al. 1989). Quartz is distinguished from other silica polymorphs by a distinctive infrared (IR) absorption band at 692 cm^{-1}, with two strong doublets at 798 and 780 cm^{-1} and at 395 and 370 cm^{-1} (Chester and Green 1968).

Quartz in the subsurface usually is altered by in situ chemical and physical weathering. Quartz appears as an anhydrous grain losing its prismatic form and containing trace elements other than silicon and oxygen. Aluminum is the major potential contaminant of the quartz mineral, but other trace elements such as Ti, Fe, Na, Li, K, Mg, Ca, and H (OH) also are present (Dennen 1966). Quartz grains are in general rounded or have an angular morphology due to physical attrition. A cleavage mechanism leads to the formation of flat grains when the quartz particles are <100 μm (Krinsley and Smalley 1973). Scanning electron micrographs of quartz grains are presented in Fig. 1.2.

The dense packing of the crystal structure and the high activation energy required to alter the Si–O–Si bond contribute to the high stability of quartz (Stober 1967). Quartz in the subsurface includes chemically precipitated forms commonly associated with carbonates or carbonate-cemented sandstones (Dapples 1979). In

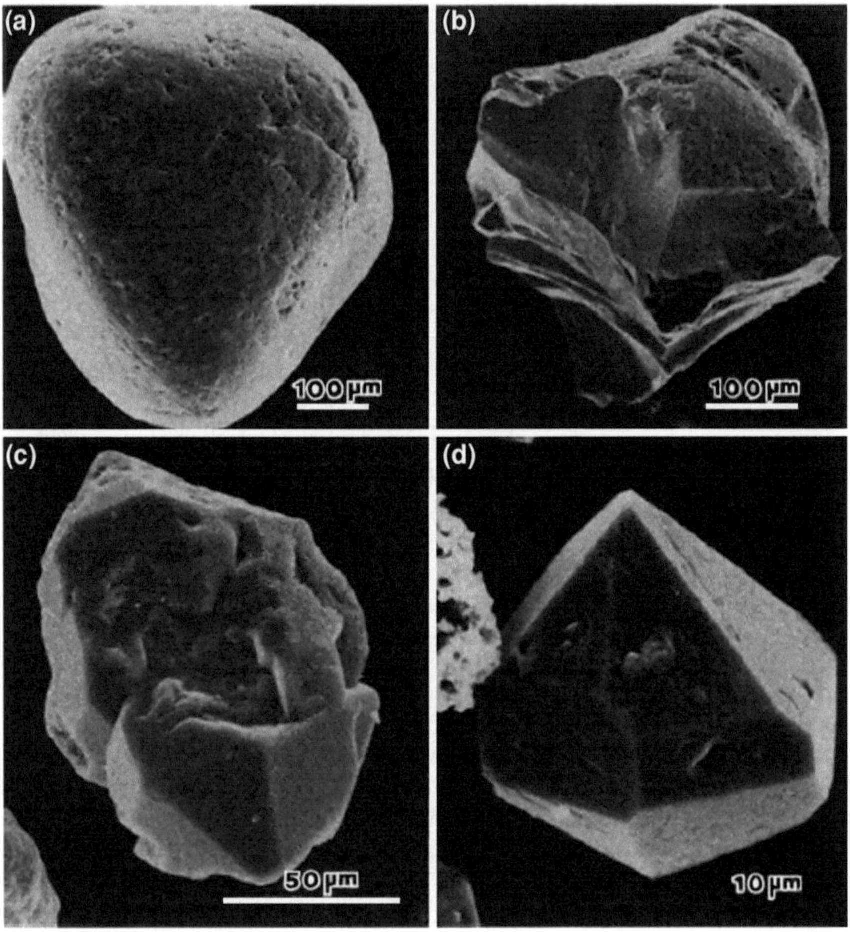

Fig. 1.2 Electron scanning microscope picture of various quartz grains (Dress et al. 1989)

soil, quartz generally is found in sand and silt fractions, the amount of silica mineral being determined by the parent material and the degree of weathering.

1.1.2 Clay Minerals

Clay minerals are natural nanomaterials that constitute the smallest particles in the subsurface. They are defined as the fraction of particles smaller than a nominal diameter of 2 µm. Many clay minerals have layer structures in which the atoms within a layer are strongly bound to each other, with the binding between layers being weaker. As a consequence, each layer can behave as an independent structural unit. Several layer arrangements exist, but they do not differ greatly in

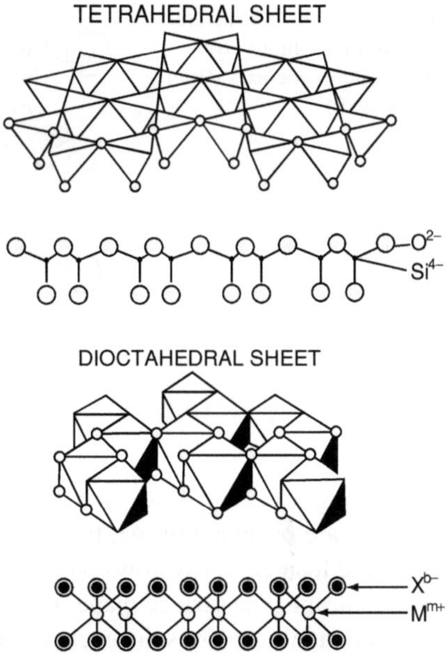

Fig. 1.3 Typical three-layered phyllosilicate structures (Sposito 1984)

their free energy. The most common inorganic structural units in clay minerals present in the subsurface are the silica tetrahedron SiO_4 and octahedral complex MX_{6m-6b} composed of a metal unit (M^{m+}) and six anions (X^{b-}). Figure 1.3 shows sheet structures formed by polymerization of these two structural units.

The architecture of a silicate layer results from SiO_2 coordination in which each SiO_2 unit shares oxygen atoms with three neighboring SiO_4 groups, thus forming rings containing six Si and six O atoms. Each ring joins the neighboring ring through shared oxygen atoms. An additional structural element in layered silicate is an octahedral sheet that contains cations in MO_6 coordination between the two planes of oxygen atoms.

As a function of their structural properties, clays interact differently with organic and inorganic contaminants. Two major groups of clay minerals are selected for discussion here: (a) kaolinite, with a 1:1 layered structured aluminosilicate and a surface area ranging from 6 to 39 m^2/g (Schofield and Samson 1954); and (b) smectites with a 2:1 silicate layer and a total surface area of about 800 m^2/g (Borchardt 1989).

1.1.2.1 Kaolinite

Kaolinite crystals in the subsurface are submicron sized and exhibit a platelike morphology. They usually are found mixed with other layered structured minerals. In a comprehensive review, Dixon (1989) summarizes the structural properties of kaolinite. This mineral is composed of tetrahedral and octahedral sheets

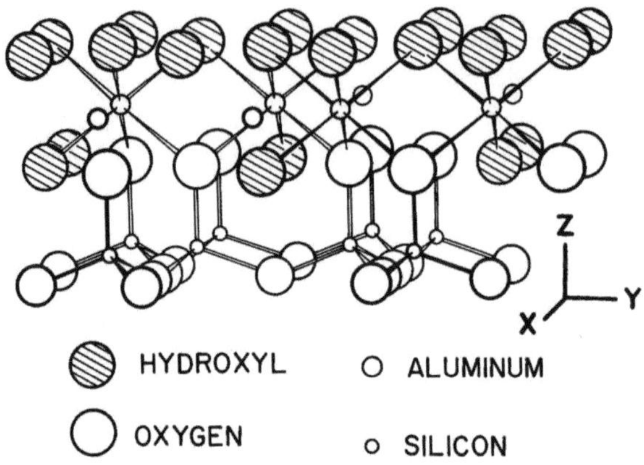

Fig. 1.4 Idealized structural diagram of kaolinite layer viewed along one axis (from Dixon 1989)

constituting a 0.7 nm layer in a triclinic unit cell. Two-thirds of the octahedral positions are occupied by Al; the tetrahedral positions are occupied by Si and Al, which are located in two rows parallel to the *x*-axis. Every third row of octahedral sites is vacant. The surface plane of octahedral anions and a third layer of the inner plane containing anions in each 0.7 nm layer is built up by hydroxyl (OH^-) groups. The surface hydroxyls bond through their hydrogens to the oxygen sheet of the bordering layer. The idealized structural diagram of a kaolinite layer, as suggested by Brindley and MacEwan (1953) and modified by Dixon (1989), is shown in Fig. 1.4.

Giese (1982) showed that the kaolinite structure exhibits covalent sharing of H, which leads to conceptualization of a kaolinite structure with two inner surface hydroxyls perpendicular to the layer, bonded to the O of an adjacent layer. As a result of polarization of the Si–O and O–H groups, the surface oxygen and proton plane become negatively or positively charged. Electrostatic attraction, in addition to van der Waals forces, contributes to the stacking of kaolinite unit layers. The negative charge of kaolinite was proven by demonstrating that the surface retains Na^+ under acidic conditions (Schofield and Samson 1953). The exchangeable cations are bound only to the tetrahedral basal plane of kaolinite (Weiss and Russow 1963). Kaolinite exhibits a very low cation exchange capacity, in general lower than 1 $cmol_c\ kg^{-1}$ (centimoles per kg) at pH = 7 (Lim et al. 1980). Schofield and Samson (1953) also show that the positive charge on kaolinite occurs on the edges of the plates, which become positive by acceptance of H^+ in the acid pH range.

1.1.2.2 Smectites

Smectites are clay minerals with an expanding nature, a negative charge, and a large total surface area. These properties are of major importance in controlling the fate of chemicals in the subsurface, by affecting their retention, transport, and persistence.

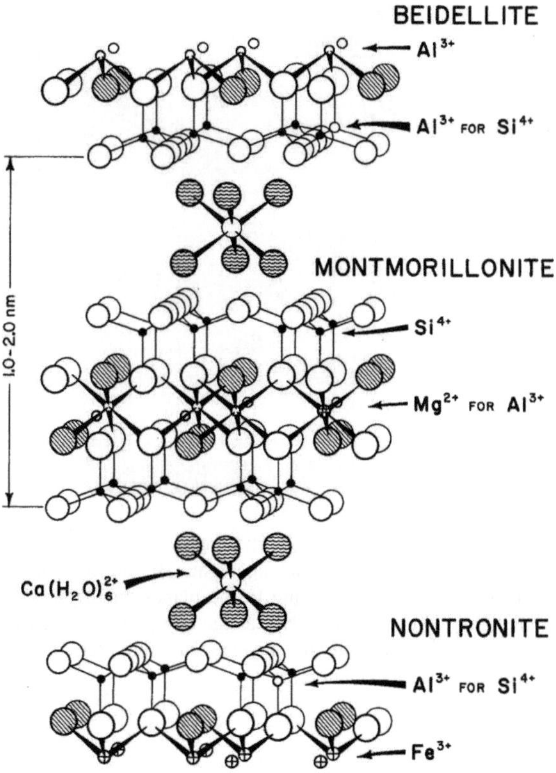

Fig. 1.5 The crystal structure of smectite, illustrating beidelite, montmorillonite, and nontronite (Borchard 1989, Brindley and MacEwan 1953)

Of the naturally existing smectites, montmorillonite is a clay of major interest in the subsurface environment. Figure 1.5 shows the crystal structure of montmorillonite, compared to those of beidellite and nontronite, and their possible substitutions. Montmorillonite has an octahedral sheet that shares oxygen atoms between two tetrahedral sheets. Cationic substitution may occur in the octahedral or tetrahedral sheets.

Smectites are classified according to differences in properties and chemical composition. For example, a typical formula for montmorillonite is $Si_4Al_{1.5}Mg_{0.5}$ with a cation exchange capacity of 135.5 $cmol_c/kg$, while a typical formula for beidellite is $Si_{3.5}Al_{0.5}Al_2$ with a cation exchange capacity of 135.2 $cmol_c/kg$. Note that montmorillonites contain significant amounts of tetrahedral Al and octahedral Fe. Layer separation in smectite depends both on the interlayer cation and the amount of water associated with the cation. The interlayer cations are replaced when the clay is wetted with an electrolyte solution, and this affects the interlayer spacing. The hydration water of the exchangeable cation forms the first layer, and an additional water layer is held with less energy (Barshad 1960). Changes in the hydration status of smectites, as a result of an increase in ambient temperature, are determined by differential thermal analysis (DTA). Smectites lose water that

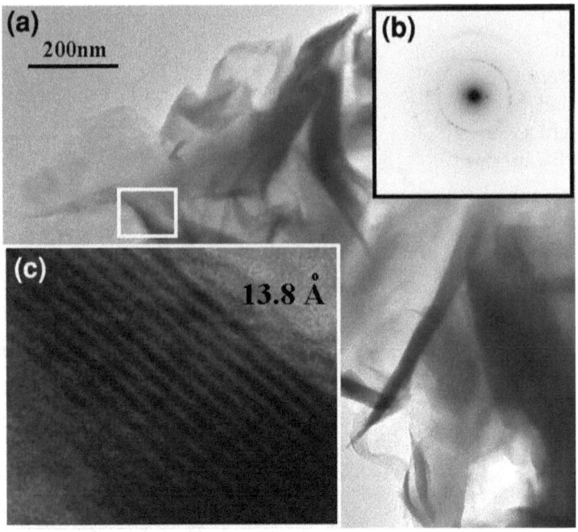

Fig. 1.6 TEM images of Na-montmorillonite particles: **a** low-magnification TEM image showing a typical crystalline aggregate; **b** the corresponding SAED pattern: note the 001 diffraction *spots* corresponding to the planes parallel to the e-beam; the calculated spacing is 13.83 Å; **c** high-resolution TEM image of a particle attached to the edge of the aggregate [*white frame* in (**a**)] exhibiting the layered structure; the measured lattice spacing (13.8 Å) is in agreement with calculations from the SAED in (**b**). Reprinted with permission from Yaron-Marcovich et al. (2005)

originates from three different sources: pore water lost below 110 °C, adsorbed water lost below 300 °C, and OH water at a temperature above 300 °C.

Structural patterns of smectites may be obtained by X-ray diffraction (XRD), high-resolution transmission electron microscopy (HRTEM), and Fourier transform infrared spectroscopy (FTIR). Borchardt (1989) reported on the effect of saturating cations on the basal spacing of smectites (a mixture of montmorillonite and beidellite from a Californian soil) using XRD measurements. All magnesium-saturated smectites give a peak corresponding to a d-spacing of 1.5 nm at 54 % relative humidity, while potassium-saturated smectites give a peak corresponding to a d-spacing of 1.25 nm at the same moisture content. These differences are explained by variations in the hydration status of the saturating ions. When smectites are kept at 0 % relative humidity after 110 °C heating, they yield a peak corresponding to a d-spacing of 1.0 nm. Simultaneous morphological observation by HRTEM and chemical analysis by AEM (EDS) techniques provide a comprehensive, nanoscale level understanding of smectite interlayer configuration and composition.

TEM images (Yaron-Marcovich et al. 2005) show the morphology of a sodium-montmorillonite along with the corresponding selected area electron diffraction (SAED) pattern (Fig. 1.6). At lower magnification, the presence of crystalline aggregates composed of relatively small, thin, flakelike, and pointed nanoscale

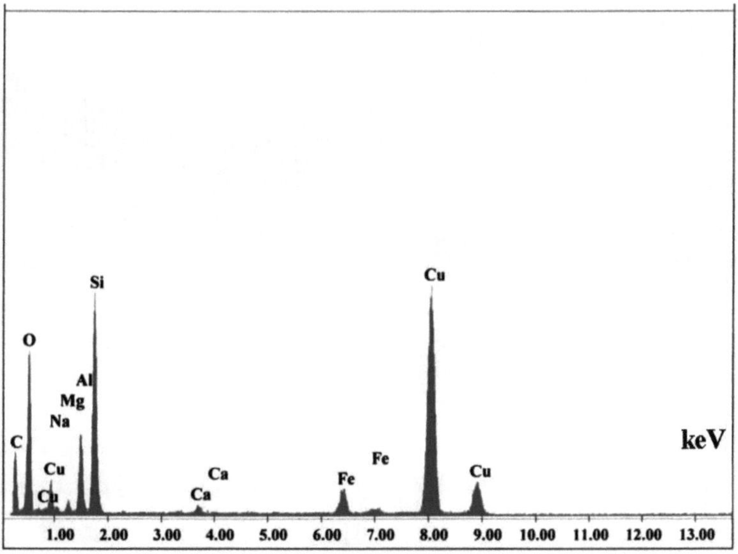

Fig. 1.7 Representative EDS spectra of Na-montmorillonite. Cu peaks arise from the TEM grid. Reprinted with permission from Yaron-Marcovich et al. (2005)

silicate particles (about 10–50 nm wide and 50–400 nm long) with no obvious orientation are visible. Closer inspection of the microstructure clearly discloses a layered structure with layers occurring as well-defined stacks separated by a regular van der Waals gap along the *c*-axis. The interlayer spacing measured from HRTEM images was 0.138 nm; this result (compared to 0.149 nm obtained by XRD) may be explained by the influence of interlayer hydration on the d-spacing of montmorillonite. The interlayer compositional variation of the same clay sample determined by X-ray energy-dispersive spectroscopy (EDS) is shown in Fig. 1.7. It may be observed that the nanoparticles are of similar composition, containing Na, K, Al, Mg, and O. The presence of copper, iron, and carbon signals can be ignored, as they arise from the analytical technique.

Infrared analysis of smectites provides considerable information on their structural properties. Silicate minerals have strong Si–O bands near 600 and 1,000 cm^{-1}, which can be affected by substitution of silicon atoms by Al atoms. The OH bending vibrations produce the absorption characteristics of the octahedral sheet. When Al is present, the absorption is near 920 cm^{-1}, and when only Fe is present, absorption is near 820 cm^{-1}. A mixture of Al, Fe, and Mg leads to intermediate values. Hydroxyl stretching vibrations absorb in the 3,000–3,800 cm^{-1} region. Farmer and Russell (1967) show that this band appears to broaden as a result of Al substitution for Si in the tetrahedral sheet and, at the same time, reflect the type of saturating cation and the hydration status in the interlayer space. The surface oxygen atoms are weak electron donors and form weak hydrogen bonds. As a consequence and as a result of their association with exchangeable cations, water molecules on the smectite

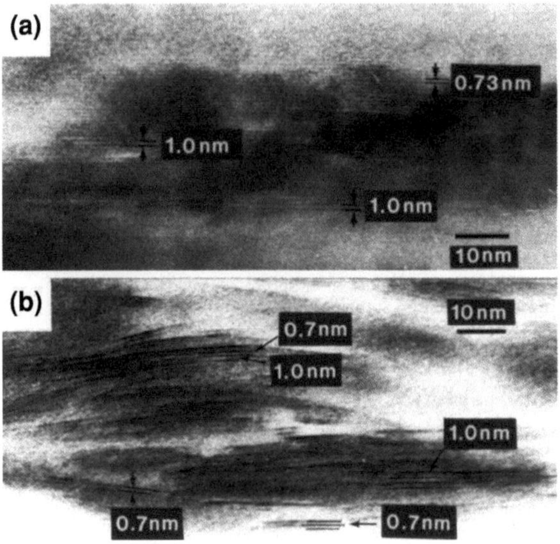

Fig. 1.8 High-resolution electron micrographs of thin sections showing electron optical fringes **a** indicative of inclusion of mica in kaolinite, and **b** interstratification of kaolinite and other layer silicates. *Fringes* shown indicate the spacings of basal planes viewed from the edge (Dixon 1989)

surface are more acidic than the interlayer water (Borchardt 1989). Hydrogen migration from exchangeable positions into the tetrahedral sheet may occur. Heating of Al- and H-saturated montmorillonite results in fading out of the 1,700 cm^{-1} band as a result of proton migration (Yariv and Heller-Kallai 1973).

Of particular interest for chemical transport into a predominantly smectite medium is the shrink–swell property of the clay material. The swelling properties of smectites are explained by two concepts. The first one, developed by Sposito (1973), shows that smectite swelling is caused by the hydration and mobility of the cations, which in turn balance the negative charge of the layer silicates. The second concept, presented by Low (1981), emphasizes the direct interaction of water molecules with the silicate surface. Both viewpoints fit the common observation that smectite swells in a high-hydration environment and at low electrolyte concentrations and shrinks when water is lost and salt is added to the bulk solution.

A mixture of intercalating clays is generally found in the subsurface. Interstratification of kaolinite and smectite has been reported in some cases (e.g., Schultz et al. 1971; Lee et al. 1975a, b; Yerima et al. 1985). This fact is reflected in an XRD spacing of about 0.8 nm (Dixon 1989) or in HRTEM micrographs (Fig. 1.8), which show the inclusion of mica (1:2 layer) and other layer silicates in kaolinite (Lee et al. 1975a, b).

Environmentally induced processes occurring in the subsurface (e.g., leaching, acidification) may induce weathering processes and structural changes in natural clays. For example, in soils with low organic matter, moderate leaching and a pH of about 5 causes smectites to be transformed into pedogenetic chlorites (Barnhisel and Bertsch 1989). Laboratory studies proved a direct transformation of smectites into kaolinite during intense weathering. In natural environmental conditions, the transformation of smectite to kaolinite, in the presence of iron oxide, may occur under enhancement of drainage conditions during landscape evolution. In the case of sedimentary environments (Morgan et al. 1979) and well-drained red–black soils (Herbillon et al. 1981), the transformation may proceed through an intermediate step consisting of interstratified kaolinite–smectite.

1.1.3 Minerals Other than Silica and Clay

In addition to silica and clay minerals, the subsurface contains a variety of minerals (e.g., oxides, carbonates), which may react with organic and inorganic contaminants. Gilkes (1990), summarizing the properties of the metal oxides in earth materials, states that iron oxides (e.g., hematite α-Fe_2O_3, magnetite β-Fe_2O_3, goethite α-FeOOH, and lepidocrocite β-FeOOH) are common constituents with crystals that vary greatly in size, shape, and surface morphology. The surface of iron oxides in the subsurface environment often is hydroxylated either structurally or through hydration of the Fe atoms. Crystals of the aluminum oxides that commonly occur (e.g., gibbsite, boehmite) are small but often larger than the associated iron oxides. Other oxide minerals are less abundant than Fe and Al oxides, but because of their very small crystal sizes and large surface areas, they may affect very significantly the geochemical properties of the subsurface and their interaction with chemicals. For example, the various Mn oxides that can be found in the subsurface can occur as very small (about 10 nm), structurally disordered crystals. Similarly titanium oxides (rutile, anatase, TiO_2) and even corundum, a rare pyrogenic mineral, occur within the clay fraction as approximately 30 nm crystals. The ability of Fe and certain other metal ions to undergo redox reactions further increases the role of metal oxides in the activity of the solid phase.

Other major components found in the subsurface include significant quantities of relatively high surface area, soluble calcium carbonate ($CaCO_3$), and calcium sulfate ($CaSO_4$). It is difficult to estimate the contribution of amorphous materials (e.g., allophone or imogolite) to the surface activity of earth materials. Amorphous materials often coat crystal minerals, which may further affect interaction of these minerals with contaminants.

1.1.4 Organic Matter

Soil organic matter is a general term for the nonliving portion of the organic fraction in the soil layer. It is a mixture of products resulting from microbial transformation of organic residues and includes organic compounds originating from undecayed plant and animal tissues, their partial decomposition products, and the near-surface biomass. Major components of the soil organic matter and their definition are presented in Table 1.1.

Table 1.1 Definition and characterization of soil organic matter (Stevenson 1994)

Term	Definition
Litter	Macroorganic matter (e.g., plant residues) that lies on the soil surface
Light fraction	Undecayed plant and animal tissues and their partial decomposition products that occur within the soil proper and can be recovered by flotation with a liquid of high density
Soil biomass	Organic matter present as live microbial tissue
Humus	Total of the organic compounds in soil, exclusive of undecayed plant and animal tissues, their "partial decomposition" products, and the soil biomass
Soil organic matter	Same as humus
Humic substances	A series of relatively high-molecular-weight, yellow- to black-colored substances formed by secondary synthesis reactions. The term is used as a generic name to describe the colored material or its fractions obtained on the basis of solubility characteristics. These materials are distinctive to the soil (or sediment) environment in that they are dissimilar to the biopolymers of microorganisms and higher plants (including lignin)
Nonhumic substances	Compounds belonging to known classes of biochemistry, such as amino acids, carbohydrates, fats, waxes, resins, and organic acids. Humus probably contains most, if not all, of the biochemical compounds synthesized by living organisms
Humin	The alkali-insoluble fraction of soil organic matter or humus
Humic acid	The dark-colored organic material that can be extracted from soil by dilute alkali and other reagents and is insoluble in dilute acid
Hymatomelanic acid	Alcohol-soluble portion of humic acid
Fulvic acid fraction	Fraction of soil organic matter that is soluble in both alkali and acid
Generic fulvic acid	Pigmented material in the fulvic acid fraction

Soil organic matter is found wherever organic matter is decomposed, mainly in the near surface. However, soil organic matter may also be transported as suspended particles into deeper layers of the vadose zone or via surface- and groundwater-forming sediments. Although these components form a minor part of the total solid phase, they are of major importance in defining the surface properties of the solid phase and have a great impact on the chemical behavior.

Organic matter extracted from earth materials usually is fractionated on the basis of solubility characteristics. The fractions commonly obtained include humic acid (soluble in alkaline solution, insoluble in acidic solution), fulvic acid (soluble in aqueous media at any pH), hymatomelamic acid (alcohol-soluble part of humic acid), and humin (insoluble in alkaline solutions). This operational fractionation is based in part on the classical definition by Aiken et al. (1985). It should be noted, however, that this fractionation of soil organic matter does not lead to a pure compound; each named fraction consists of a very complicated, heterogeneous mixture of organic substances. Hayes and Malcom (2001) emphasize that biomolecules, which are not part of humic substances, also may precipitate at a pH of 1 or 2 with the humic acids. Furthermore, the more polar compounds may precipitate with fulvic acids.

Dark-colored pigments extracted from earth materials result from multiple reactions, the major pathway being through condensation reactions involving polyphenols and quinones. According to Stevenson (1994), polyphenols derived from lignin are synthesized by microorganisms and enzymatically converted to quinines, which subsequently undergo self-condensation or combine with amino compounds to form N-containing polymers. The number of molecules involved in this process as well as the number of ways in which they combine is almost unlimited, explaining the heterogeneity of humic materials.

The major atoms composing humic materials are C (50–60%) and O (30–35%). Fulvic acid has lower carbon and higher oxygen contents. The percentage of hydrogen (H) and nitrogen (N) varies between 2 and 6 % and that of sulfur (S) varies from 0 to 2 %. The various fractions of humic substances obtained on the basis of solubility characteristics are part of a heterogeneous mixture of organic molecules, which originate from different earth materials and locations and might range in molecular weight from several hundred to several hundred thousand. The average molecular weight range for humic acids is on the order 10,000–50,000, and a typical fulvic acid has a molecular weight in the range of 500–7,000. The humic fraction in the near surface represents a colloidal complex, including long-chain molecules or two- or three-dimensional cross-linked molecules whose size and shape in solution are controlled by the pH and the presence of neutral salts. Under neutral or slightly alkaline conditions, these molecules are in an expanded state, as a result of the repulsion of the charged acidic groups, whereas at a low pH and high salt concentration, molecular aggregation occurs due to charge reduction. These large organic molecules may exhibit hydrophobic properties, which govern their interaction with nonionic solutes.

Numerous structures have been proposed for humic substances in the past, based mainly on speculative or "in vogue" research results over the years. A basic structure for modeling humic substances was the preliminary concept of a two-dimensional representation of humic acid (Schulten and Schnitzer 1993, 1997). Recent models were developed on the basis of analytical and spectroscopic data associated with computer calculations (e.g., Bailey et al. 2001; Jansen et al. 1996). Schulten (2001) improved the initial model by trapping biological substances, such as sugars and peptides, thus developing different models for terrestrial humic acids, soil organic matter, and dissolved organic matter. Because organic matter is such an important component of soil, development of models to describe subsurface behavior now incorporates chemical interaction studies, molecular mechanism calculations, structural modeling, and geometry optimization.

An example of a structural model of organic matter is presented in Fig. 1.9, reproducing Schulten's (2001) model for a terrestrial humic acid tetramer in open form (Fig. 1.9a) and of a tetramer formed by a trimer trapping an additional monomer (Fig. 1.9b). In these models, molecular mechanics calculations were used for geometry optimization and determination of total potential energy, bond,

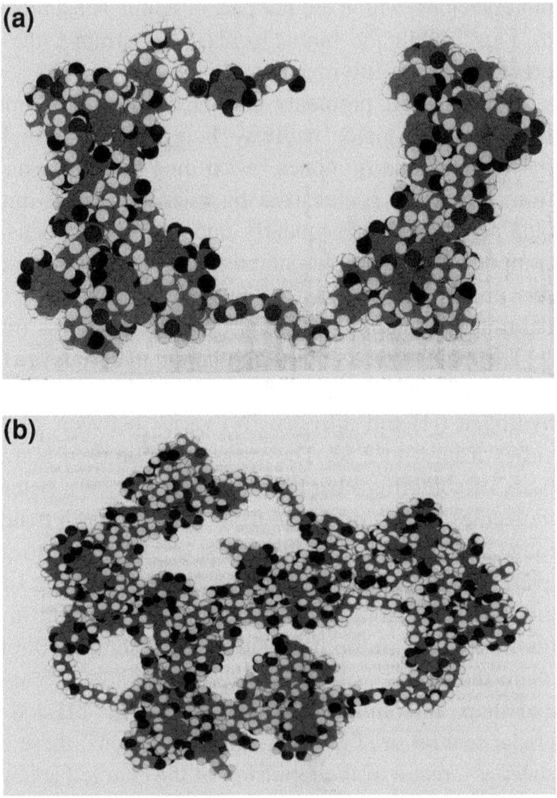

Fig. 1.9 Terrestrial humic acid model **a** tetramer open form, and **b** trimer trapping an additional monomer (Schulten 2001)

angle, dihedral, van der Waals, stretch bend and electrostatic energy derivatives, and tentatively association energy. Atomic and molecular properties were calculated at the nanometer level. Schulten (2001) compared the covalently bound tetramer humic acid, which is geometrically optimized in a widely open molecular structure (Fig. 1.9a), to the tetramer design formed by trapping a monomer in a trimer humic acid structure (Fig. 1.9b). Both molecules have the same number of atoms, molecular weight, and elemental composition. Surface areas, volumes, and density are quite similar in the wide-open and trapped tetramer forms. The only substantial difference is that the accessible surface area is much larger in the open (tetramer) system. Natural organic matter is considered by Schaumann (2006) as a three-dimensional polymer-like, entangled phase of macro- and micromolecules having a pH-dependent charge. The molecular size range is defined by the natural organic matter form (dissolved or solid) and age (young or old) and is affected by various environmental factors.

These types of models, while incomplete, are steps toward the formulation of composite models, which depend on future availability of compositional data. Moreover, these structural models are an important aid in understanding the interactions between anthropogenic chemicals and terrestrial organic matter. However, due to the heterogeneity of humic substances in the environment, provision of an exact, general structure does not seem feasible. Comprehensive discussion on effects of soil–sediment organic matter heterogeneity on sorption and desorption of organic contaminants is given by Huang et al. (2003).

A reevaluation of molecular structure of humic substances based on data obtained primarily from nuclear magnetic resonance spectroscopy, X-ray absorption near-edge structure spectroscopy, electrospray ionization mass spectrometry, and pyrolysis studies was presented by Sutton and Sposito (2005). The authors consider that "humic substances are collections of diverse, relatively low molecular mass components forming dynamic associations stabilized by hydrophobic interactions and hydrogen bonds. These associations are capable of organizing into micellar structures in suitable aqueous environments. Humic components display contrasting molecular motional behavior and may be spatially segregated on a scale of nanometers. Within this new structural context, these components comprise any molecules intimately associated with a humic substance." Sutton and Sposito (2005) conclude by stating that biomolecules bound strongly within humic fractions are, by definition, humic components.

1.1.5 Electrically Charged Surfaces

The electrically charged surface of the solid phase is characterized by a net (positive or negative) charge on the solid surface that is in contact with the liquid or gaseous phase. These charged surfaces usually are faced by one or more layers of counterions having a net charge separate from the surface charge. The adsorption of charged solutes onto a solid phase surface is subject to both chemical binding forces and an electric field at the interface, and it is controlled by an electrochemical system. Considerable differences exist between the surface properties of original minerals constituting the bulk solid phase and surface properties of organic and inorganic colloids.

Some of the functional groups (e.g., OH) on the clay surface exhibit electrical charges. The magnitude of the electrical charge, as well as its character, is controlled by the properties of the surfaces to which the functional groups are bound and by the composition of the surrounding liquid. Sposito (1984) classified the surface charge density of soils as follows:

- *Intrinsic surface charge density,* defined by the number of Coulombs per square meter bound by surface functional groups, either because of isomorphic substitutions or because of dissociation/protonation reactions.
- *Structural surface charge density,* defined as the number of Coulombs per square meter, as a result of isomorphic substitutions in soil minerals.
- *Proton surface charge density,* defined as the difference between the number of moles of complexed proton charges and of complexed hydroxyl charges per unit mass of colloids.

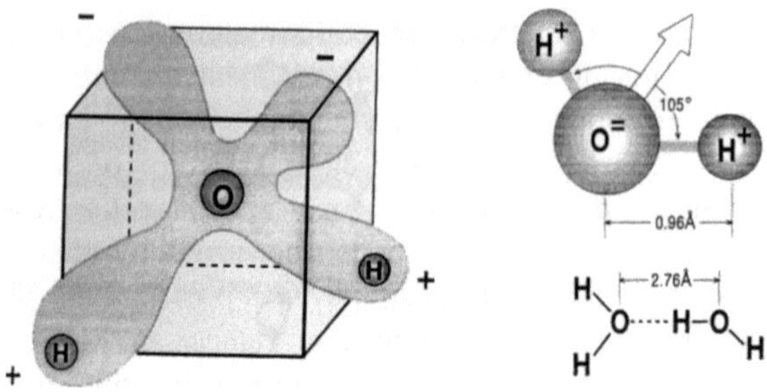

Fig. 1.10 Water as a solvent: electron cloud depiction for a water molecule and the structure of the angular water molecule and hydrogen bond (Horne 1969)

1.2 Subsurface Liquid Phase

Within the subsurface zone, two liquid phase regions can be defined. One region, containing water near the solid surfaces, is considered the most important surface reaction zone. This "near solid phase water," which is affected by the solid phase properties, controls the diffusion of the mobile fraction of the solute adsorbed on the solid phase. The second region constitutes the "free" water zone, which governs liquid and chemical flow in the porous medium.

The composition and reactivity of the liquid phase (known as the *soil solution)* is defined by the quality of the incoming water and affected by fluxes of matter and energy originating from the vicinity of the solid phase, microbiological activity, and the gas phase. To understand the properties of the subsurface liquid phase, it is first necessary to consider the structure of the water molecule.

The H_2O molecule has a dipolar character with a high negative charge density near the oxygen atom and a high positive charge density near the protons. Figure 1.10 depicts the electron cloud of the angular water molecule resulting from the hybridization of electrons, to yield two bonding orbitals between the O and the two H atoms. This specific character strongly influences the interaction of water with the solid and air phases in the subsurface.

The local structure of water has been compared to ordinary hexagonal ice structures and calculated spectra. Synchrotron X-ray measurements have led to contrasting opinions regarding the H-bond coordination environment in liquid water. Wernet et al. (2004) used this technique, together with X-ray Raman scattering, to probe the molecular arrangement in the first coordination shell of liquid water. Most molecules in liquid water are in two hydrogen-bonded configurations with one strong donor and one strong acceptor hydrogen bond, in contrast to the four hydrogen-bonded tetrahedral structure specific to ice. Heating water to 90 °C causes about 10 % of the molecules to change from tetrahedral environments to two hydrogen-bonded configuration.

Water is a dynamic liquid, where H-bonds are continuously broken and reformed. Wernet et al. (2004) show that water, probed on the subfemtosecond time scale, consists mainly of structures with two strong H-bonds, implying that most molecules are arranged in strongly H-bonded chains, or in rings embedded in a disordered cluster network connected mainly by weak H-bonds. These results are consistent with neutron and X-ray diffraction data and confirm the theoretical model of Weinhold (1998); this model considers an asymmetric H-bonding pattern in agreement with general quantum mechanical principles underlying H-bonding.

The subsurface liquid phase generally is an open system, and its composition is a result of dynamic transformation of dissolved constituents in various chemical species over a range of reaction time scales. At any particular time, the liquid phase is an electrolyte solution, potentially containing a broad spectrum of inorganic and organic ions and nonionized molecules. The presently accepted description of the energy characteristics of the liquid phase is based on the concept of matrix and osmotic potentials. The matrix potential is due to the attraction of water to the solid matrix, while the osmotic potential is due to the presence of solute in the subsurface water.

The composition of the subsurface liquid phase varies over time, mainly due to recharge with rainwater, irrigation water, or the fluctuation of the water table (groundwater) level. In addition, disposal and discharge of wastes and effluents and the application of agrochemicals are key sources of pollution into the subsurface, contributing significant changes to the composition of the liquid phase.

1.2.1 Near Solid Phase Water

Water molecules that are oriented preferentially with the polar axis perpendicular to the solid surface, in the vicinity of a solid surface, are considered *near solid phase water*. When the net surface charge of the polar phase is negative, the hydration occurs through one hydrogen of water forming a hydrogen bond with specific atoms at the boundary of the polar surface, in such a way that the second hydrogen still can form a hydrogen bond with another water molecule outside the primary hydration level. In contact with a nonpolar solid, water molecules are oriented such that the positive hydrogen points into the bulk solution (Yariv and Cross 1979).

A water molecule exhibits a series of spatial arrangements with great irregularity as it moves through a bulk liquid phase. Unlike a solid, which has a well-defined structure, the liquid phase has instantaneous structures (I type) comprising molecules in highly irregular arrangements. Sposito (1984) showed that, with lengthening of the time scale, two additional type structures are defined: vibrationally averaged (V type) and diffusionally averaged (D type). These structures highlight the fact that the concept of molecular structure in liquid water is a dynamic one.

Stillinger (1980) conceives of liquid water as consisting of macroscopically connected random networks of hydrogen bonds with frequent strained and broken bonds that continually undergo topological reformation. The water properties arise from competition among relatively bulky means of connecting molecules into local patterns, characterized by strong bonds and nearly tetrahedral angles and more compact arrangements characterized by more strain and bond linkage. The presence of an electrolyte introduces a localized perturbation of the "tetrahedral configuration." This perturbation derives from several sequences. Near the ion, water molecules are dominated by a dense electromagnetic field, resulting in the formation of the primary solvation shell. In the next zone, called the secondary solvation shell, water molecules interact weakly with the ion. An example of such behavior of an electrolyte solution near a clay surface was discussed by Sposito (1984). It was shown that the primary solvation shell of a monovalent cation contains between three and six water molecules that exchange relatively rapidly with the surrounding bulk liquid. A secondary solvation shell, if it exists, is very weakly developed. The primary solvation shell of a bivalent cation contains between six and eight water molecules, which exchange rapidly with the surrounding bulk liquid. A secondary solvation shell containing about 15 water molecules develops as the cation concentration decreases, and it also moves with the cation as a unit.

The configuration of near solid phase water can be altered in close proximity to the phyllosilicate. The siloxane surface influences the character of the water due to the nature of their charge distribution and the complexes formed between the cation and the surface functional groups. Both the type of charge and degree of charge localization, as well the valence and size of the complexed cations, control the characteristics of the water molecules near the surface.

Clay minerals with their own surface properties affect the near-surface water in different ways. The adsorbed water in the case of kaolinite consists only of water molecules ("pure" water), whereas water adsorbed on a smectite-type mineral is an aqueous solution, due to the presence of exchangeable cations on the 2:1 layer silicate. Sposito (1989) noted the generally accepted description that the spatial extent of adsorbed water on a phyllosilicate surface is about 1 nm (two to three layers of water molecules) from the basal plane of the clay mineral.

The interlayer water of clay minerals is structurally different from bulk liquid water or water in aqueous solution (Sposito and Prost 1982), and chemical reactions in this region are affected by a perturbed water structure. Another ability of the interlayer water is to diffuse widely over surface oxygens (e.g., smectites); a strong hydrogen bonding to the layer is not essential. The same patterns also follow when the cation is large and univalent. Under these conditions, a much less polar liquid can completely replace the interlayer water (Farmer 1978). Water is retained on oxides and hydroxides of aluminum and iron through the hydroxyl groups, but it can also involve oxide bridges and water coordinated with structural cations.

Retention of water on organic surfaces also occurs at a molecular level. Farmer (1978) stated that the principal polar sites where water adsorption occurs are likely to include carboxyl groups, such as phenolic and alcoholic groups, oxides, amines, aldehydes, and esters. Ionized carboxyilic groups and their associated cations are likely to have the greatest affinity for aqueous solutes. Recall that, in addition to polar sites, organic surfaces exhibit important hydrophobic regions, which are involved largely in the retention of organic contaminants.

1.2.2 Subsurface Aqueous Solutions

The *aqueous solution* here refers to free water in the subsurface having a composition affected by the interaction between the incoming water and the solid and gaseous phases. This composition is achieved under a dynamic equilibrium with natural processes and may be disturbed by anthropogenic activities. The chemical composition of the subsurface aqueous solution at a given time is the end product of all the reactions to which the liquid water has been exposed.

The thermodynamic properties of subsurface aqueous solutions are expressed in terms of a single-species solution activity coefficient for each molecular constituent. Its composition, however, should be considered on the basis of molecular speciation in the aqueous solution, which in turn is related to biological uptake exchange reactions and transport through the subsurface.

Yaron et al. (1996) summarize the characteristics of aqueous solutions as follows:

- *Acidity–alkalinity* of the solution, measured as pH, is affected by the quality of the incoming water (generally acidic for rain, neutral or alkaline for irrigation and effluents) and buffered by the environmental system.
- *Salinity,* or total salt concentration, is usually expressed in terms of total dissolved solids (TDS) or as the electrical conductivity (EC) of the solution. The major fractions of anions are composed of Cl^-, SO_4^{2-}, and NO_3^- and the common cations are Ca^{2+}, Mg^{2+}, Na^+, and K^+. The composition of the subsurface solution varies between the composition of water entering the system and that of the solution in equilibrium with the solid phase.
- *Trace elements* of natural or anthropogenic origin may enter in the composition of the subsurface aqueous solution. Alkali and cationic materials, transition metals, nonmetals, and heavy metals are inorganic trace elements potentially found in the composition of the subsurface solution. Adsorption is the most significant mechanism for distributing trace elements between the solid and liquid phases in the subsurface.
- *Organic ligands* found in solution cause complexation of inorganic trace elements, which influences the equilibrium status between solid and aqueous phases and affects their concentration in the subsurface solution. The presence of organic trace compounds of natural or anthropogenic origin in the aqueous phase is controlled by the nature and properties of subsurface colloids, the chemical and physicochemical characteristics of the organic molecules, and the nature of the environmental system.

The volume of solution in the subsurface, under partially saturated conditions, varies with the physical properties of the medium. In the soil layer, the composition of the aqueous solution fluctuates as a result of evapotranspiration or addition by rain or irrigation water to the system. Changes in the solution concentration and composition, as well as the rate of change, are controlled by the buffer properties of the solid phase. Because of the diversity in the physicochemical properties of the solid phase, as well as changes in the amount of water in the subsurface as result of natural and human influences, it is difficult to make generalizations concerning the chemical composition of the subsurface aqueous solution.

1.3 Subsurface Gaseous Phase

The volume of the subsurface gas phase is controlled by the medium porosity and moisture content, or in other words, by the ratio between the gas and water phases occupied in pores. From the point of view of chemical component behavior, the subsurface gas phase can assist the movement of organic molecules in the vapor phase or chemicals dissolved in the water; it also can affect microbiological activity and consequently define chemical persistence mainly in a near-surface environment. Gaseous transport through pores makes the subsurface gas phase an important channel for subsurface pollution, particularly by volatile toxic chemicals. From the gaseous phase, chemicals might be adsorbed on solid surfaces or dissolved in subsurface water. On the other hand, the transport of water as vapor into pores might lead to the formation of a water layer that coats potentially available sites for nonpolar gaseous pollutants, thus reducing pollutant fixation on the solid phase. The presence of gaseous phases in the region near the water table (capillary fringe) may be of particular importance for geochemical transformations of pollutants.

The subsurface gas phase is composed mostly of CO_2, N_2, and O_2, which are the major gases in the atmosphere. Gases arising from microbiological activity, such as nitrogen oxides, may be present at any time, but because of their high reactivity with subsurface components and their susceptibility to microbiological activity, they usually are transitory (Paul and Clark 1989). In general, in well-aerated soil, the amount (by volume) of O_2 is around 20 % and that of CO_2 is between 1 and 2 %. In clay-dominated material, with a high water content, the CO_2 concentration may reach values as high as 10 %.

The composition of the subsurface gas phase may change as a result of gas dissolution into the liquid phase. The solubility of gases in water depends on the type of gas, temperature, salt concentration, and the partial pressure of the gases in the atmosphere. The most soluble gases are those that become ionized in water (CO_2, NH_3, H_2S), while O_2 and N_2 are much less soluble (Table 1.2).

The CO_2 concentration in the subsurface may be different in small and large pores and varies as a function of the aerobic or anaerobic activity of the microbial population. Paul and Clark (1989) showed that a change from aerobic to anaerobic metabolism occurs at an O_2 concentration of less than 1 % (by volume). The overall aeration of the soil layer is not as important as that of individual aggregates. Calculations show that water-saturated aggregates larger than 3 mm in radius have no O_2 in their center (Harris 1981). This means that aerobic and

Table 1.2 Diffusion constants of CO_2, O_2, and N_2 in air and water and their solubility in water at 20 °C. Reprinted from Paul and Clark (1989)

	Diffusion constant (cm^2/s)		Solubility in H_2O (mL/L)
	Air	Water	
CO_2	0.161	0.177×10^{-4}	8.878
O_2	0.205	0.180×10^{-4}	0.031
N_2	0.205	0.164×10^{-4}	0.015

anaerobic zones may coexist in a porous medium even under partially saturated conditions.

1.4 Aquifers

Aquifers form the region below the vadose zone, where the solid phase is in contact with a flowing groundwater phase and a local chemical equilibrium between the solid and aqueous phases has a tendency to be reached. An aquifer is defined as a saturated geological unit that can transmit significant quantities of water under ordinary hydraulic gradients. Within the context of aquifers, an *aquitard* is defined as a less permeable geological unit that has the potential to store water in significant quantities. In addition to geological controls of an aquifer, groundwater composition is affected by the properties of recharge water, which originates from the initial incoming water constituents. The properties of the solid matrix (porous medium) were discussed in Sect. 1.1.

1.4.1 Aquifer Structure

The lithology, stratigraphy, and structure of a geological system control the nature and distribution of aquifers. Lithology includes the mineral composition, grain size, and grain aggregation of the sediments or rocks; stratigraphy depicts the relation among the lenses, beds, and formations of geological sediments; and structural features describe the geometry of the geological system resulting from deformation.

According to their origins, aquifer materials may be classified as deposits of various origins and sedimentary or metamorphic rocks. Aquifers of fluvial origin are characterized by alluvial deposits composed of particles of gravel, silt, and clay of various sizes that are not bound or hardened by mineral cement, pressure, or temperature. The topography controls the deposition of sediments and their spatial redistribution as a function of their textural properties. Aeolian deposits consisting of sand or silt are more homogeneous than fluvial deposits. Glacial deposits,

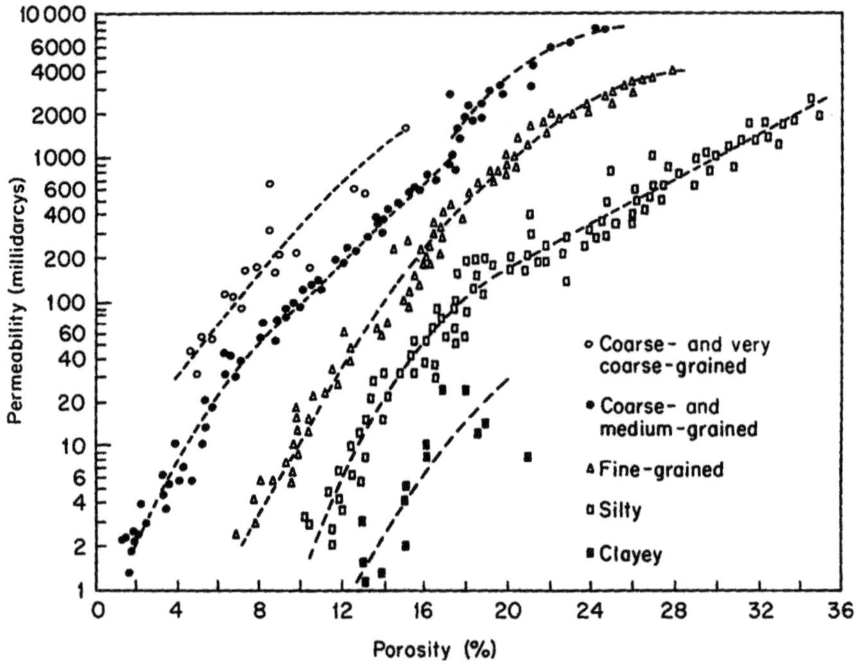

Fig. 1.11 Relation between porosity and permeability for various grain size categories (Chilingar 1963)

including glacial till, glaciofluvial, and glaciolacustrine sediment forms, are the principal components of aquifers. The type and origin of sediments affect their permeability, which in turn controls the water transmission potential of aquifers.

Sedimentary rocks, consisting mostly of sandstones and carbonates, are bodies of major hydrological significance. Sandstone makes up about 25 % of the sedimentary rocks, originating from environmental depositions (e.g., floodplain, deltaic, and marine shoreline), which are characterized by cementing materials like quartz, calcite, and clays in various stages of alteration (Freeze and Cherry 1979). The porosity of sandstone in extreme cases can be as low as 1 %, as compared to sands that reach only as low as 30 %. The relationship between porosity and permeability for various grain sizes is presented in Fig. 1.11. It was observed that an increase in porosity of a few percent generally corresponds to a large increase in permeability.

Carbonate rocks consist mostly of calcite and dolomite with minor amounts of clay. The porosity of carbonate rocks ranges from 20 to 50 %, but in contrast to sandstone, it tends to decrease with depth. Often, carbonate rocks are fractured, providing a permeability that is much greater than the primary one. In some cases, initial small-scale fractures in calcite and dolomite are enlarged by dissolution during groundwater flow, leading to an increase in rock permeability with time.

Table 1.3 Range of hydrologic properties of lower Yakima basalt flows and interbeds (Freeze and Cherry 1979)

	Hydraulic conductivity (m/s)	Porosity (%)
Dense basalt	10^{-11}–10^{-8}	0.1–1
Vesicular basalt	10^{-9}–10^{-8}	5
Fractured, weathered, or brecciated basalt	10^{-9}–10^{-5}	10
Interbeds	10^{-8}–10^{-5}	20

Igneous and metamorphic rocks, which exhibit a low porosity generally smaller than 2 %, are characterized by a minute permeability. As result of changes in stress conditions occurring during geological periods, wide fractures are formed, which lead to a substantial increase in the permeability of igneous and crystalline meta-morphic rocks. Dissolution of siliceous rocks increases the widths of fractures and consequently increases their permeability. However, the permeability of crystalline rocks usually decreases with depth. Volcanic rocks, for example, are formed by solidification of magma, and their hydraulic properties are determined by the cooling regime of magma reflected in an anisotropic permeability. The variation of properties in a basalt is presented in Table 1.3.

1.4.2 Groundwater Composition

The groundwater composition is controlled by chemical and biochemical interactions with the geological materials through which water flows as well as by the chemistry of incoming water. Because inorganic components dissolved in groundwater mainly are in an ionic form, groundwater may be considered an electrolyte solution with a conductance ranging from tens of microsiemens (a value close to that of rainwater) to hundreds of thousands of microsiemens (for brines in sedimentary basins). Dissolved inorganic constituents are grouped in three classes: major constituents, with concentrations greater than 5 mg/L; minor constituents, with a concentration between 0.01 and 5.0 mg/L; and trace constituents, having a concentration of less than 0.01 mg/L. Table 1.4 gives examples of constituents included in these groups.

Organic constituents may be found in groundwater in dissolved forms, associated with ligands, or adsorbed on colloidal materials. Natural products, such as humic and fulvic acids, have a low aqueous solubility but may serve as ligands for inorganic trace components. The majority of organic pollutants—including petroleum hydrocarbons, solvents, and toxic organic chemicals of industrial origin known as NAPLs (nonaqueous phase liquids)—exhibit a limited solubility in aqueous solutions, with usually very low concentrations. However, it should be recognized that even low solubility limits (and the corresponding concentration) usually are several orders of magnitude higher than the maximal allowable concentrations for potable water. As a consequence (NAPLs), as a mixture or as

Table 1.4 Dissolved inorganic substances in groundwater (Freeze and Cherry 1979)

Category	Components
Major constituents (>5 mg/L)	Bicarbonate, calcium, carbonic acid, chloride, magnesium, silicon, sodium, sulfate
Minor constituents (0.01–10.0 mg/L)	Boron, carbonate, fluoride, iron, nitrate, potassium, strontium
Selected trace constituents (<0.1 mg/L)	Aluminum, arsenic, barium, bromide, cadmium, chromium, cobalt, copper, gold, iodide, lead, lithium, manganese, molybdenum, nickel, phosphate, radium, selenium, silver, tin, titanium, uranium, vanadium, zinc, zirconium

individual components, can be found in groundwater in concentrations exceeding concentrations stipulated by environmental protection agencies. In addition, inorganic polymers, various natural ligands (e.g., bio-exudates), or surfactants also may serve as mediators that shuttle organic pollutants toward groundwater via the vadose zone.

Groundwater may contain dissolved gases as a result of exposure to the surface environment prior to water infiltration, contact with the subsurface gaseous phase, and gas produced biologically below the water table. The most important dissolved gas in groundwater is CO_2.

References

Aiken GR, McKnight DM, Wershaw EL, MacCarthy P (1985) An introduction to humic substances in soil, sediment and water. In: Aiken GR et al (eds) humic substances in soil, sediment and water. Wiley, New York, pp 1–9

Bailey GW, Akim LG, Shevcenko SM (2001) Predicting chemical reactivity of humic substances for minerals and xenobiotics: use of computational chemistry, scanning probe microscopy and virtual reality. In: Clapp CE et al (eds) Humic substances and chemical contaminants. Soil Science Society of America, Madison, pp 40–72

Barnhisel RL, Bertsch PM (1989) Chlorites and hydroxy-interlayered vermiculite and smectite. Minerals in Soils. Soil Science Society of America, Madison, pp 730–789

Barshad I (1960) Thermodynamics of water adsorption and desorption on montmorillonite. Clays Clay Miner 8:84–101

Borchardt G (1989) Smectites. In: Dixon JB, Weeds SB (eds) Minerals in Soil. Soil Science Society of America, Madison, pp 675–728

Brindley GW, MacEwan DMC (1953) Structural aspects of the mineralogy of clays and related silicates. In: Green AT, Stewart GH (eds) Ceramics: a symposium. The British Ceramic Society, Stoke-on-Trent, pp 15–59

Chester R, Green RN (1968) The infrared determination of quartz in sediments and sedimentary rocks. Chem Geol 3:199–212

Chilingar GV (1963) Relationship between porosity, permeability and grain size distribution of sands and sandstones. In: Proceedings of International Sedimentol Congress, Amsterdam

Dapples EC (1979) Diagenesis in sandstones. In: Larsen G, Chilingar GV (eds) Diagenesis in sediments and sedimentary rocks. Elsevier, New York, pp 99–141

Dennen WH (1966) Stoichiometric substitution in natural quartz. Geochim Cosmochim Acta 30:1235–1241
Dixon JB (1989) Kaolin and serpentine group minerals. In: Dixon JB, Weed SB (eds) Minerals in Soils. Soil Science Society of America, Madison, pp 468–527
Dress LR, Wilding LP, Smeck NE, Senkayi AL (1989) Silica in soils: quartz and disordered silica polymorphs. In: Dixon JB, Weed SB (eds) Minerals in soil environment, 2nd edn. Soil Science Society of America, Madison, WI
Farmer VC (1978) Water on partial surfaces In: Greenland DJ, Hayes MHB (eds) The chemistry of soil constituents. Wiley, New York, pp 405–449
Farmer VC, Russel JD (1967) Infrared absorption spectrometry in clay studies. Clays Clay Miner 15:121–142
Freeze RA, Cherry JA (1979) Groundwater. Prentice-Hall Englewood Cliffs, NJ
Giese RF Jr (1982) Theoretical studies of the kaolin minerals: electrostatic calculations. Bull Mineral 105:417–424
Gilkes RJ (1990) Mineralogical insights into soil productivity: an anatomical perspective. In: Proceedings of 14th Congress of Soil Science, Kyoto, Japan. Transaction Plenary Papers pp 63–75
Greenland DG, Hayes MHB (1978) The chemistry of soil constituents. Wiley, New York
Harris RF (1981) Effect of water potential on microbial growth and activity. In: Par J et al (eds) Water potential relations in soil microbiology. Soil Science Society of America, Madison
Hayes MHB, Malcom RL (2001) Considerations of compositions and aspects of the structure of humic substances. In: Clapp CE, Hayes MHB, Senesi N, Bloom PR Jardine PM (eds) Humic substances and chemical contaminants. Soil Science Society of America, Madison, pp 1–39
Herbillon AJ, Frankart R, Vielvoye L (1981) An occurrence of interstratified kaolinite-smectite minerals in a red-black soil top sequence. Clay Miner 16:195–201
Horne RA (1969) Marine chemistry. Wiley, New York
Huang W, Peng P, Yu Z, Fu J (2003) Effects of organic matter heterogeneity on sorption and desorption of organic contaminants by soil and sediments. Appl Geochem 18:955–972
Jansen S, Malaty AM, Nabara S, Johnson E, Ghabbour E, Davies G, Varnum JM (1996) Structural modeling in humic acids. Mater Sci Eng C4:175–179
Krinsley DH, Smalley IJ (1973) Shape and nature of small sedimentary quartz particles. Science 180:127–129
Lee SY, Jackson ML, Brown JL (1975a) Micaceous occlusions in kaolinite observed by ultramicotomy and high resolution electron microscopy. Clays Clay Miner 23:125–129
Lee SY, Jackson ML, Brown JL (1975b) Micaceous vermiculite, glauconite and mixed-layered kaolinite-montmorillonite by ultramicotomy and high resolution electron microscopy. Proc Soil Sci Soc Amer 39:793–800
Lim CH, Jackson ML, Koons RD, Helmke PA (1980) Kaolins: sources of differences in cation exchange capacity and cesium retention. Clays Clay Miner 28:223–229
Low PF (1981) The swelling of clay: III. Dissociation of exchangeable cation. Soil Sci Soc Amer J 45:1074–1078
Morgan DJ, Highley DE, Bland DJ (1979) A montmorillonite, kaolinite association in the Lower Cretaceous of south-west England. In: Mortland M, Farmer VC (eds) Proceedings of International Clay Conference, Pergamon Press, Oxford, 1978, pp 301–310
Paul FA, Clark FE (1989) Soil microbiology and biochemistry. Academic Press, New York 234 pp. Copyright 1989 with permission of Elsevier
Schaumann GE (2006) Soil organic matter beyond molecular structure: amorphous nature and physical aging. J Plant Nutr Soil Sci 169:157–167
Schofield RK, Samson HR (1953) The deflocculation of kaolinite suspensions and the accompanying change-over from positive to negative chloride adsorption. Clay Miner Bull 2:45–51
Schofield RK, Samson HR (1954) Flocculation of kaolinite due to the attraction of opposite charged crystal faces. Discuss Faraday Soc 18:135–145

Schulten HR (2001) Models of humic structures: association of humic acids and organic matter in soils and water. In: Clapp CE et al. (eds) Humic substances and chemical contaminants. Soil Science Society of America, Madison, pp 73–88
Schulten HR, Schnitzer M (1993) A state of the art structural concept for humic substances. Naturwissenschaften 80:29–30
Schulten HR, Schnitzer M (1997) Chemical model structures for soil organic matter and soils. Soil Sci 162:115–130
Schultz IG, Shepard AO, Blackmon PD, Starkey HC (1971) Mixed layer kaolinite-montmorillonite from the Yucatan Peninsula, Mexico. Clays Clay Miner 19:137–150
Sposito G (1973) Volume changes in swelling clays. Soil Sci 115:315–320
Sposito G (1984) The surface chemistry of soils. Oxford University Press, New York
Sposito G (1989) The chemistry of soils. Oxford University Press, New York
Sposito G, Prost R (1982) Structure of water adsorption on smectites. Chem Rev 82:553–573
Stevenson FJ (1994) Humus Chemistry, 2nd edn. Wiley, New York
Stillinger FH (1980) Water revisited. Science 209:451–453
Stober W (1967) Formation of silicic acid in aqueous suspensions of different silica modification. In: Goulded RF (ed) Equilibrium concepts in natural water systems. Adv Chem Ser 67:161–172
Sutton R, Sposito G (2005) Molecular structure in soil humic substances: the new view. Environ Sci Technol 39:9009–9015
Weinhold F (1998) Quantum cluster equilibrium theory of liquids: Illustrative application to water. J Chem Phys 109:373–384
Weiss A, Russow J (1963) Uber die Lage der austauschbaren Kationen bei Kaolinit. Proceedings of International Clay Conference, Stockhom, vol 1, pp 203–213
Wernet P, Nordlund D, Bergmann U, Cavalleri M, Odelius M, Ogasawara H, Naslund LA, Hirsch TK, Ojamae L, Glatzel P, Pettersson LGM, Nilsson A (2004) The structure of the first coordination shell in liquid water. Science 304:995–999
Wilding LP, Smeck NE, Dress LR (1977) Silica in soils. In: Dixon JB, Weed SB (eds) Minerals in soil environment. Soil Science Society of America, Madison, pp 471–553
Yariv S, Cross H (1979) Geochemistry of colloid systems. Springer, Heidelberg
Yariv S, Heller-Kallai L (1973) IR evidence for migration of protons in H and organo-montmorillonite. Clays Clay Miner 21:199–200
Yaron B, Calvet R, Prost R (1996) Soil pollution: processes and dynamics. Springer, Heidelberg
Yaron-Marcovich D, Chen Y, Nir S, Prost R (2005) High resolution electron microscopy (HRTEM) structural studies of organo-clay nanocomposites. Environ Sci Technol 39:1231–1239 (Copyright 2005 American Chemical Society)
Yerima BPK, Calhoun FG, Senkayi AL, Dixon JB (1985) Occurrence of inter-stratified kaolinite-smectite in El Salvador vertisols. Soil Sci Soc Amer J 49:462–466

Chapter 2
Selected Geochemical Processes

In physical chemistry, the world is divided in two parts: the system, containing the portion of the world of particular interest, and the *surroundings*, comprising the region outside the system (Atkins and de Paula 2002). A geochemical system is an open system that may be studied within two basic frameworks:

1. *Thermodynamic system*, which is a state of equilibrium where environmental parameters, such as pressure and temperature, are imposed on the bulk composition of the system. This approach is used to predict the system stability and the impact of changing environmental conditions.
2. *Kinetic system*, wherein the pathways along the system are moving toward some state of local equilibrium, which in turn determines the rate of change along the pathway. In the context of a kinetic approach, which is relevant to geochemical processes, dissolution–precipitation, exchange-adsorption, oxidation–reduction, vaporization, and formation of new phases are discussed here.

2.1 Thermodynamics and Equilibrium

When a thermodynamic approach is used to describe geochemical phenomena in the subsurface, it is necessary to define the solids, liquids, gases, and soluble species that exist at equilibrium.

2.1.1 Enthalpy, Entropy, and the Laws of Thermodynamics

In thermodynamics, the total energy of a system is given by the sum of the total kinetic and potential energies of the molecules in the system.

The *first law of thermodynamics* is the application of the conservation of energy principle. In geochemistry, the first law considers that the change in internal

B. Berkowitz et al., *Contaminant Geochemistry*,
DOI: 10.1007/978-3-642-54777-5_2,

energy (dU) is equal to the heat added to the system (dq) plus the work (dw) done on the system:

$$dU = dq + dw. \tag{2.1}$$

The first law of thermodynamics applied to an adiabatic system may be expressed as the work done on a system by an adiabatic process, which is equal to the increase in its internal energy, and a function of the state of the system.

In any natural system, the system boundary is prescribed as a nonadiabatic (insulated) wall that allows the passage of heat to and from the surrounding system. The change in internal energy is not equal to the heat supplied when the system is open and free to change its volume. Under this condition, part of the energy supplied is returned to the surroundings, and the heat supplied at constant pressure is equal to the *enthalpy* (H) of the system, which is defined as

$$H = U + PV, \tag{2.2}$$

where U is the internal energy, P is the pressure, and V is its volume. Because U, P, and V are state functions, the enthalpy also is a state function.

The *second law of thermodynamics*, as formulated by Kelvin, states that "no process is possible in which the sole result is the absorption of heat from a reservoir and its complete conversion into work;" in other words, in any natural process involving a transfer of energy, some energy is converted irreversibly into heat that cannot be involved in further exchange. The second law of thermodynamics, therefore, is a recognition of spontaneous and nonspontaneous processes and the fact that natural processes have a sense of direction.

The second law of thermodynamics can be expressed in terms of another state function, the *entropy* (S). The thermodynamics definition considers the change in entropy dS that occurs as a result of a physical or chemical change and is based on the expression

$$dS = dq_{rev}/T, \tag{2.3}$$

in which dq is an infinitesimal amount of heat gained by a body at temperature T in a reversible (rev) process. A reversible change in thermodynamics is a change that can be reversed by an infinitesimal modification of a variable. Because real processes are never completely reversible, the entropy is a measure of the degree in which a system has lost heat and therefore part of its capacity to do work. In general, a local decrease in entropy is possible but must be accompanied by an increase in entropy in the surrounding system. In real geochemical (open) systems, a change in entropy must be equal to (in the case of reversible change) or greater than zero:

$$dS \geq dq/T. \tag{2.4}$$

A *fundamental equation* combines the first and second laws of thermodynamics and, in this manner, addresses the behavior of matter. For a reversible change in a closed system of constant composition and without nonexpansion work, one can write

$$dw_{rev} = -PdV. \tag{2.5}$$

Then, for a reversible change in a closed system, substitution of Eqs. (2.5) and (2.3) (written for dq_{rev}) into Eq. (2.1) yields

$$dU = TdS - PdV. \tag{2.6}$$

Equation (2.6) is called a *fundamental equation* and because dU is an exact differential, its value is independent of path. Hence, Eq. (2.6) applies to any change—reversible or irreversible—of a closed system that does no additional work (Atkins and de Paula 2002).

Given that, in the subsurface, we are dealing with an open system, the fundamental equation may be applied only when the macroscopic system is decoupled in isolated, well-defined systems. As an example, we can consider that an adiabatic "zone" of the subsurface solid phase is in contact with an aqueous solution through a rigid barrier, surrounded by an insulating wall.

Gibbs free energy (G) is probably the most frequently used quantity in thermodynamics; it measures spontaneity of a reaction or energy available to do work in a system. Free energy is a state function because it is defined formally only in terms of the state functions enthalpy and entropy and the state variable temperature. The Gibbs free energy is defined as

$$G = H - TS. \tag{2.7}$$

At constant temperature and pressure, chemical reactions are spontaneous in the direction of decreasing Gibbs free energy. Some reactions are spontaneous because they give off energy in the form of heat ($\Delta H < 0$). Other reactions are spontaneous because they lead to an increase in the disorder of the system ($\Delta S > 0$). Calculations of ΔH and ΔS can be used to probe the driving force behind a particular reaction.

2.1.2 Equilibrium

A system is in equilibrium when all acting influences are canceled by others, resulting in a stable, balanced, or unchanging system in relation to its surroundings. In the subsurface, equilibrium can be defined in terms of thermal, chemical, or mechanical equilibrium. Usually, in the subsurface, changes occur slowly over geological time scales, so that a state of equilibrium is never reached. However, in the subsurface, when mainly anthropogenic-induced effects are involved, various changes do occur over "observable" time scales, ranging from seconds to years. Under these conditions, an *apparent equilibrium* may be reached. The concept of equilibrium in subsurface systems is discussed in classical books of geochemistry, soil science, or aquatic chemistry (e.g., Stumm and Morgan 1995; Sposito 1981), and the reader is directed to such sources for ample information on this topic.

In a closed system when temperature and pressure are constant, the sum of chemical potentials of all components is fixed; in contrast, in an open system, the chemical potential of all components is influenced by both the thermodynamic parameters of the phases and various parameters outside the system. The main relationship among phases, components, and physical conditions is given by the phase rule.

The *phase rule* states that when equilibrium conditions are sustained, a minimum number of intensive properties of the (subsurface) system can be used to calculate its remaining properties. An *intensive property* is a property that is independent of the amount of substance in the domain. Examples of intensive properties include temperature (T), pressure (P), density (ρ), and chemical potential (μ), which is a relative measure of the potential energy of a chemical compound. The phase rule specifies the minimum number of intensive properties that must be determined to obtain a comprehensive thermodynamic depiction of a system.

A *phase* is defined as a state of matter that is uniform throughout in terms of its chemical composition and physical state; in other words, a phase may be considered a pure substance or a mixture of pure substances wherein intensive properties do not vary with position. Accordingly, a gaseous mixture is a single phase, and a mixture of completely miscible liquids yields a single liquid phase; in contrast, a mixture of several solids remains as a system with multiple solid phases. A phase rule, therefore, states that, if a limited number of macroscopic properties are known, it is possible to predict additional properties.

A system is *homogeneous* when the intensive properties are not a function of position, while a system is *heterogeneous* when the composition of a given mixture varies as a function of position. For example, the subsurface liquid phase usually comprises an aqueous solution incorporating a number of solutes; in contaminated subsurface environments, nonaqueous phase liquids also may be present. The air phase of the subsurface includes gases with various partial pressures, and the solid phases comprise a mixture of minerals and organic compounds.

A *phase diagram* describes how a system reacts to changing conditions of pressure and temperature and consists of a field in which only one phase is stable, separated by boundary curves along which a combination of phases coexist in equilibrium.

2.1.3 Solubility, Chemical Potential, and Ion Activities

The chemical potential of species i, μ_i, is expressed in terms of the Gibbs free energy added to a system at constant T and P, as well as relative to the mole fraction of each added increment of i. When adding an incremental number of molecules of i, free energy is introduced in the form of internal energies of i as well as by the interaction of i with other molecules in the system. As i increases, the composition of the mixture changes and consequently μ_i changes as a function of the amount in moles (n) of i:

$$\mu_i = \left[\frac{\partial G}{\partial n_i}\right]_{T,P,n_{j\neq 1}}. \tag{2.8}$$

Because $\mu_i = G_i = H_i - TS_i$, the chemical potential can be used to assess the tendency of components i to be transferred to another system or transformed within a system; in other words, matter flows spontaneously from a region of high chemical potential to a region of low chemical potential, just as mass flows from a position of high gravitational potential to a position of low gravitational potential. The chemical potential, therefore, can be used to determine whether or not a system is in equilibrium: at equilibrium, the chemical potential of each substance is the same in all phases appearing in the system.

An *ideal solution* can be defined as a solution in which the chemical potential of each species is given by the expression

$$\mu_i = \mu_i^0(P,T) + RT\ln x_i, \tag{2.9}$$

where R is the gas constant (=0.001987 kcal/mol/K), T is the temperature, and x_i is the mole fraction of species i. The chemical potential of a pure species i, $\mu_i^0(P,T)$, is a measure of the activity of compound i in its standard state, that is, pure organic liquid at the same pressure (P) and temperature (T). The term $\mu_i^0(P,T)$ is referred to as the *standard state chemical potential*. From Eq. (2.9), it is seen that the chemical potential of a species in an ideal solution is lower than the chemical potential of the pure component.

Usually, only very dilute solutions can be considered ideal. In most aqueous solutions, ions are stabilized because they are solvated by water molecules. As the ionic strength is increased, ions interact with each other. Thus, when calculating the chemical potential of species i, a term that takes into account the deviation from ideal conditions is added. This term is called an *excess term* and can be either positive or negative. The term usually is written as $RT\ln\gamma_i$, where γ_i is the activity coefficient of component i. The complete expression for the chemical potential of species i then becomes

$$\mu_i = \mu_i^0(P,T) + RT\ln x_i + RT\ln\gamma_i = \mu_i^0(P,T) + RT\ln(x_i\gamma_i). \tag{2.10}$$

As mentioned previously, in this expression, $\mu_i^0(P,T)$ is the chemical potential of a pure species i. For a pure species i, $x_i = 1$, and consequently, from Eq. (2.10), $\gamma_i = 1$, too.

The expression $x_i\gamma_i$ is referred to as the *activity* of the species, a_i, and is a measure of how active a compound is in a given state compared to its standard state (e.g., the pure liquid at the same T and P).

For aqueous solutions of salts, $\mu_i^0(P,T)$ represents the chemical potential of pure ions. This chemical potential cannot be measured experimentally. Instead of using this hypothetical standard state, the activity coefficients of ions often are normalized by introducing the "asymmetrical activity coefficient," γ_i^* defined as

$$\gamma_i^* = \frac{\gamma_i}{\gamma_i^\infty}, \tag{2.11}$$

where γ_i^∞ is the activity coefficient of species i at infinite dilution. If the chemical potential of species i is expressed in terms of γ_i^*, we obtain the expression

$$\begin{aligned} \mu_i = \mu_i^0 + RT\ln(x_i\gamma_i^*\gamma_i^\infty) &= \mu_i^0 + RT\ln\gamma_i^\infty + RT\ln(x_i\gamma_i^*) \\ &= \mu_i^* + RT\ln(x_i\gamma_i^*). \end{aligned} \tag{2.12}$$

The standard state chemical potential, $\mu_i^* = \mu_i^0 + RT\ln\gamma_i^\infty$, has the advantage that it can be measured experimentally.

In the *molality concentration scale*, the molality m_i of solute i is the amount of solute i per kg of solvent. If the solvent is water (subscript w), the following relation between mole fraction and molality of solute i can be derived:

$$m_i = \frac{n_i}{n_w M_w} \Rightarrow n_i = m_i n_w M_w \tag{2.13}$$

$$x_i = \frac{n_i}{\sum\limits_{\text{ions}} n_i + n_w} = \frac{m_i n_w M_w}{\sum\limits_{\text{ions}} n_i + n_w} = m_i M_w x_w \tag{2.14}$$

where n is the number of moles and M_w is the molecular mass of water (kg/mol). Using this relation, the chemical potential of ion i can be expressed as a function of the molality and the molal activity coefficient γ_i^∇:

$$\begin{aligned} \mu_i = \mu_i^0 + RT\ln\left(m_i M_w x_w \gamma_i^* \gamma_i^\infty\right) &= \mu_i^0 + RT\ln\left(m_w m^0 \gamma_i^\infty\right) + RT\ln\left(\frac{m_i}{m^0}\gamma_i^* x_w\right) \\ &= \mu_i^\nabla + RT\ln\left(\frac{m_i}{m^0}\gamma_i^\nabla\right) \end{aligned} \tag{2.15}$$

where the term $m^0 = 1$ mol/kg has been included to make the expression dimensionless. The standard state chemical potential is $\mu_i^\nabla = \mu_i^0 + RT\ln M_w m^0 \gamma_i^\infty$ when the molality concentration scale is used. The molal activity coefficient is related to the asymmetrical mole fraction activity coefficient by $\gamma_i^\nabla = \gamma_i^* x_w$, where x_w is the mole fraction of water.

2.1.4 Kinetic Considerations and Reaction Rate Laws

Thermodynamic considerations provide a basic approach for predicting what may or may not happen in a given system. On a practical level, large and growing numbers of chemical species are contained in thermodynamic databases. Given the increasing ease of data retrieval and extrapolation and the availability of free energy minimization algorithms, this information is accessible and useful, allowing application of thermodynamics to a wide variety of geochemical systems.

Despite this usefulness, thermodynamic considerations have limitations, and these most often are apparent in environmental systems at lower temperatures, in biological systems, and in the description of reactions at phase boundaries. Thermodynamics applies to chemical processes among large numbers (i.e., Avogadro's number) of molecules and deals with overall reactions among a set of chemical species. Strictly speaking, equilibrium thermodynamics provide no information about how a chemical system reached its current state.

The earth's subsurface is not at complete thermodynamic equilibrium, but parts of the system and many species are observed to be at local equilibrium or, at least, at a "dynamic" steady state. For example, the release of a toxic contaminant into a groundwater reservoir can be viewed as a perturbation of the local equilibrium, and we can ask questions such as, What reactions will occur? How long will they take? and Over what spatial scale will they occur? Addressing these questions leads to a need to identify actual chemical species and reaction processes and consider both the thermodynamics and kinetics of reactions.

For any chemical reaction, whether inorganic or organic, we must choose which kinetic species to include in the elementary reactions that make up the overall process; ideally, molecular or chemical information is available to guide this choice. In general, for an elementary (irreversible) reaction among species A and B, to give species C and D, in relative amounts a, b, c, and d, respectively,

$$aA + bB \rightarrow cC + dD, \tag{2.16}$$

the rate of the reaction is

$$\text{Rate} = -\frac{1}{a}\frac{\mathrm{d}[A]}{\mathrm{d}t} = -\frac{1}{b}\frac{\mathrm{d}[B]}{\mathrm{d}t} = \frac{1}{c}\frac{\mathrm{d}[C]}{\mathrm{d}t} = \frac{1}{d}\frac{\mathrm{d}[D]}{\mathrm{d}t} = k[A]^a[B]^b, \tag{2.17}$$

where [ ] denotes concentration, $a + b$ is the order of the reaction, and the coefficient k is the rate constant for the reaction.

In a *first-order reaction*, the rate-determining step involves a transformation where one reactant reacts to give one product, that is, $A \rightarrow B$. In first-order reactions, there is an exponential decrease in the reactant concentration, so that at any given time, the transformation rate is dependent on the corresponding concentration of the reactant at the same time. This can be expressed in the following way:

$$\text{Rate} = \frac{\mathrm{d}[A]}{\mathrm{d}t} = -k[A], \tag{2.18}$$

$$[A]_t = [A]_0 \mathrm{e}^{-kt}, \tag{2.19}$$

where here k is the first-order rate constant with dimension [time^{-1}] and the subscripts t and 0 denote time and initial time, respectively. Plotting $\ln[A]_t/[A]_0$

against time yields a linear relation with a slope equal to $-k$ that intersects the origin.

In *second-order reactions*, the rate-determining step involves a transformation where two reactants interact to give one product. The simplest case of such a reaction is $2A \rightarrow B$, and in such a case, we can write

$$\frac{\mathrm{d}[A]}{\mathrm{d}t} = -k[A]^2, \tag{2.20}$$

$$[A]_t = \frac{[A]_0}{1 + kt[A]_0}, \tag{2.21}$$

where here k is the second-order rate constant with dimensions [mol^{-1} volume time^{-1}]. As a consequence, the decay period depends on the initial concentration $[A]_0$. This result has implications for environmental pollutants that decompose by second-order reactions; in such cases, pollutants may persist for longer times at low concentrations, compared to first-order reactions, because decay times become longer as the concentration decreases.

When the reaction rate is not dependent on the reactant concentration, the reaction is *zero order*:

$$\frac{\mathrm{d}[A]}{\mathrm{d}t} = k. \tag{2.22}$$

In nature, the reaction rate depends on the reactant concentration. However, practically speaking, when a reactant exists at a very high concentration, it is essentially unchanged due to the reaction, and the reaction is called *pseudo-zero order*.

In geochemistry, interest is focused on *open systems*, in which mass is added or removed over observable time periods. A simple example of such a case is that of steady fluid flow in a system, with constant inflow of species A. Then, for constant recharge of species A at concentration $[A]_0$ (and discharge at concentration A), at rate k^*, with loss A in reaction with species B (recall Eq. 2.16), we have

$$\frac{\mathrm{d}[A]}{\mathrm{d}t} = -k[A] + k^*[A]_0 - k^*[A] \tag{2.23}$$

so that

$$\frac{\mathrm{d}[A]}{\mathrm{d}t} = -k[A] + k^*([A]_0 - [A]) = -(k + k^*)[A] + k^*[A]_0. \tag{2.24}$$

The solution of this first-order differential equation is

$$[A] = \frac{k^* + k\mathrm{e}^{-(k+k^*)t}}{k^* + k}[A]_0. \tag{2.25}$$

When $k^* \ll k$, that is, the recharge is very slow, we return to a simple first-order reaction. When the recharge is very fast, compared to the reaction rate, that is, when $k^* \gg k$, then $[A] \cong [A]_0$ and the concentration of A remains constant in the system during the period of observation.

Many reactions use the *Arrhenius parameter*; for example, plotting ln k against $1/T$ gives a linear relationship that can be written in the form

$$\ln k = \ln A - \frac{Ea}{RT}, \tag{2.26}$$

where Ea is the activation energy, and ln A corresponds to the interception point when $1/T = 0$; A is called the *pre-exponential factor* or *frequency factor*. The higher the Ea, the stronger the dependence of the rate constant on temperature. The *activation energy Ea* is the minimum kinetic energy that reactants must have to form a product. The pre-exponential factor is a measure of the rate at which transformation of a reactant to products occurs, irrespective of their energy.

An excellent example combining thermodynamics, kinetics, and equilibrium considerations was presented by O'Day (1999), who considered the precipitation reaction of solid lead carbonate, in the form of the mineral cerussite ($PbCO_{3(s)}$) according to the reaction

$$Pb^{2+} + CO_3^{2-} \rightleftharpoons PbCO_{3(s)}. \tag{2.27}$$

The equilibrium constant is governed by a function of temperature and pressure and can be expressed using the standard state equilibrium constant (K_{eq}) and the change in standard free energy of the reaction:

$$K_{eq} = \frac{a_{PbCO_{3(s)}}}{a_{Pb^{2+}} a_{CO_3{}^{2-}}} \tag{2.28}$$

$$\Delta G_r^0 = -RT \ln K_{eq}, \tag{2.29}$$

where a is the activity of each species, and ΔG_t^0 denotes the change in free energy of a reaction at standard state. By convention, the activities of chemically pure solids are set equal to 1. The description of cerussite precipitation given by Eq. (2.27) is an example of a thermodynamic equilibrium. However, when examining the reaction system at the molecular level, before equilibrium is achieved, the reaction is much more complicated. For example, aqueous species take part in pH-dependent reactions that determine the form of carbon in solution and thus affect the precipitation of cerussite:

$$CO_{2(aq)} + H_2O \rightleftharpoons H_2CO_3 \tag{2.30}$$

$$H_2CO_3 \rightleftharpoons HCO_3^- \tag{2.31}$$

$$H^+ + HCO_3^- \rightleftharpoons H^+ + CO_3^{2-}. \tag{2.32}$$

Thus, at the molecular level, the reactions that actually take place and their associated chemical pathways can be completely different from the cerussite precipitation scheme presented in Eq. (2.27).

In addition to a number of solution species and a solid, species adsorbed to and desorbed from the surface can be included and described using elementary reactions. As discussed previously, to determine the reaction order, rate-determining steps, or other kinetic parameters, one must choose the kinetic species to be included in the elementary reactions that make up the overall process. Ideally, molecular or chemical information is available to guide this choice. Therefore, the set of kinetic species for any overall chemical reaction is generally larger (and usually more complicated) than the set of equilibrium species. The only constraints on overall reactions are that they proceed in the same direction as ΔG_r^0 and that the overall reaction rate approaches zero at equilibrium. This connection between the rate of an overall reaction and the driving force supplied by thermodynamics can be expressed by including a free energy term in a rate equation.

The change in free energy of a reaction not at equilibrium (ΔG_r) is given by

$$\Delta G_r = RT \ln \frac{Q}{KE}, \tag{2.33}$$

where R, T, and K_{eq} have the same meaning as in Eq. (2.29) and Q is the measured ion activity product for the reaction. By comparing the activity product of a species observed in the system with the expected concentration product at equilibrium (K_{eq}), the ratio Q/K_{eq} provides a measure of how far from equilibrium the reaction is and in which direction it is going (note that $Q = K_{eq}$, and therefore $\Delta G_r = 0$, at equilibrium). In general, rate equations derived from experiments have many terms that account explicitly for variables such as concentrations of species in solution, pH, ionic strength, possible catalytic or inhibitory effects, and different forms of expression for $f(\Delta G_r)$. Rate expressions should be able to account for the difference in reaction rate far from and close to equilibrium. However, it cannot be assumed that the reaction mechanism is the same for both dissolution and precipitation. In the context of environmental systems, the first, and sometimes the most difficult, task is determining the species and stoichiometry of reactions that govern the fate of the elements of interest, then deciding whether they can be treated as equilibrium reactions for the time scale of interest. Reactions that occur at surfaces and the molecular species involved are inherently difficult to characterize because their concentrations usually are lower than those of bulk species and they often are transitory.

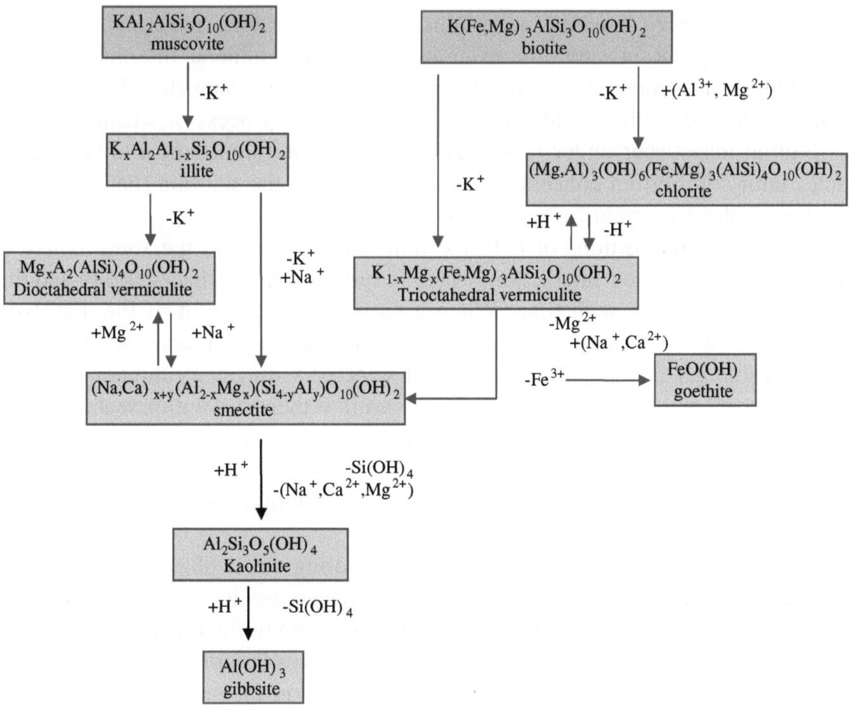

Fig. 2.1 Possible chemical weathering pathways of muscovite and biotite

2.2 Weathering

Weathering of subsurface solid phases occurs as a result of their direct interaction with liquid phases, which may also in turn be affected by the gaseous environment. Examples of weathering processes include reactions that convert primary minerals such as quartz and clays into metal oxides and metal hydroxides.

The major chemical weathering agent in the subsurface is water, which can act as either a weak acid or a base. Oxygen can oxidize organic hydrocarbons and a variety of metals that include Fe^{2+} and S^{2-}. Carbon dioxide can be transformed to function as an inorganic acid (e.g., HCO_3^-) or as an organic acid (e.g., HCOOH), and the conjugate bases are often strong ligands that complex metals.

A possible chemical weathering process of two primary minerals, muscovite and biotite, and their various mineral products is presented in Fig. 2.1.

2.2.1 *Dissolution and Precipitation*

Dissolution and precipitation in the subsurface are controlled by the properties of the solid phases, by the chemistry of infiltrating water, by the presence of a gas phase, and by environmental conditions (e.g., temperature, pressure, microbiological activity). Rainwater, for example, may affect mineral dissolution paths differently than groundwater, due to different solution chemistry. When water comes in contact with a solid surface, a simultaneous process of weathering and dissolution may occur under favorable conditions. Dissolution of a mineral continues until equilibrium concentrations are reached in the solution (between solid and liquid phases) or until all the minerals are consumed.

The initial compositions of both the infiltrating water and the solid materials may change due to their interaction, which in turn may affect the solubility and the pathway of processes with time. When a particular component of the dissolved solution reaches a concentration greater than its solubility, a precipitation process occurs. Table 2.1 includes the solubility of selected sedimentary minerals in pure water at 25 °C and total pressure of 1 bar, as well as their dissolution reactions. All of the minerals listed in Table 2.1 dissolve, so that the products of the mineral dissolution reactions are dissolved species. Figure 2.2 shows the example of gypsum precipitation with its increasing concentration in an NaCl aqueous solution.

The presence of organic ligands in the infiltrating water phase may cause the complexation of some minerals, leading to an increase in their solubility. For example, oxalate anions enhance the solubility of ZnO in the range of pH = 6–8, and EDTA (ethylenediamine tetraacetic acid) dissolves hydrous ferric oxide up to a pH = 9 (Stumm and Morgan 1995). Fulvic acids in the soil layer may act as chelating agents, contributing to an increase in solubility of the minerals and, as a consequence, enhancing their mobility. Laboratory experiments performed by Bennett et al. (1988) showed that organic acids can greatly enhance the dissolution process, especially where reactions take place in an open, flow-through (non-equilibrium) system.

Carbon dioxide has a dominant effect on the dissolution of carbonate minerals, such as calcite and dolomite (Table 2.1). If a carbonate mineral dissolves in water that is equilibrated with a constant source of CO_2, then the concentration of dissolved carbonic acid remains constant and high. However, when calcite dissolution is accompanied by consumption of carbonic acid and a continuous source of CO_2 is not maintained, the reaction proceeds further to achieve equilibrium.

The subsurface generally is an open system. The presence of CO_2 and other gases in the atmosphere affects the partial pressure of gas constituents in the subsurface. For example, carbonate mineral dissolution in a system open to atmospheric CO_2 does not achieve equilibrium. However, higher local subsurface CO_2 concentrations can originate from biological activity and other oxidation processes.

The rate of chemical weathering of minerals in the subsurface depends on a number of factors, including mineralogy, temperature, flow rate, surface area, the presence of ligands and CO_2, and H^+ concentrations in the subsurface water (Stumm et al. 1985). Figure 2.3 shows the rate-limiting steps in mineral dissolution consisting of (a) transport of solute away from the dissolved crystal or

Table 2.1 Dissociation reactions and solubility of selected minerals that dissolve congruently in water (Freeze and Cherry 1979)

Mineral	Dissociation reaction	Solubility at pH = 7 (mg/L or g/m^3)
Gibbsite	$Al_2O_3 \cdot 2H_2O + H_2O = 2Al_3^+ + 6OH^-$	0.001
Quartz	$SiO_2 + 2H_2O = Si(OH)_4$	12
Hydroxylapatite	$Ca_5OH(PO_4)_3 = 5Ca^{2+} + 3PO_4{}^{3-} + OH^-$	30
Amorphous silica	$SiO_2 + 2H_2O = Si(OH)_4$	120
Fluorite	$CaF_2 = Ca^{2+} + 2F^-$	160
Dolomite	$CaMg(CO_3)_2 = Ca^{2+} + Mg^{2+} + 2CO_3{}^{2-}$	90–480
Calcite	$CaCO_3 = Ca^{2+} + CO_3{}^{2-}$	100–500
Gypsum	$CaSO_4 \cdot 2H_2O = Ca^{2+} + SO_4{}^{2-} + 2H_2O$	2100
Sylvite	$KCl = K^+ + Cl^-$	264,000
Epsomite	$MgSO_47H_2O = Mg^{2+} + SO_4{}^{2-} + 7H_2O$	267,000
Mirabillite	$Na_2SO_4 \cdot 10H_2O = 2Na^+ + SO_4{}^{2-} + 10H_2O$	280,000
Halite	$NaCl = Na^+ + Cl^-$	360,000

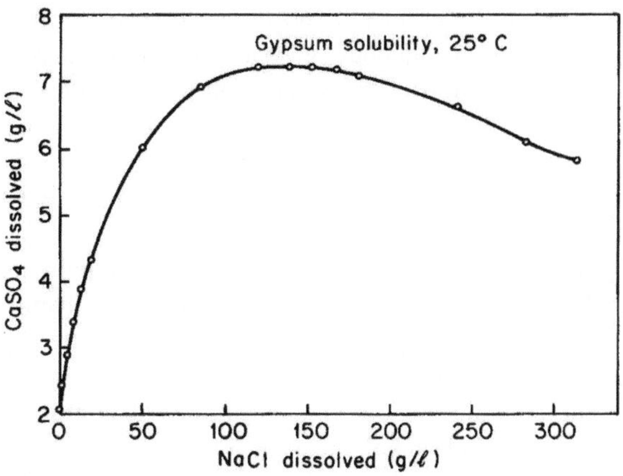

Fig. 2.2 Solubility of gypsum in aqueous solutions with an increasing NaCl content (Shternina 1960)

transport-controlled kinetics, (b) surface reaction-controlled kinetics where ions or molecules are detached from the crystal surface, and (c) a combination of transport and surface reaction-controlled kinetics.

Based on the results of Berner (1978, 1983), Sparks (1988) showed that in transport-controlled kinetics, the dissolution ions are detached very rapidly and accumulate to form a saturated solution adjacent to the surface. In surface reaction-controlled kinetics, ion detachments are slow, and ion accumulation at the crystal surface is equivalent to the surrounding solution concentration. The third type of rate-limiting mechanism for dissolution of minerals occurs when the surface detachment is fast, and the surface concentration is greater than the solution concentration.

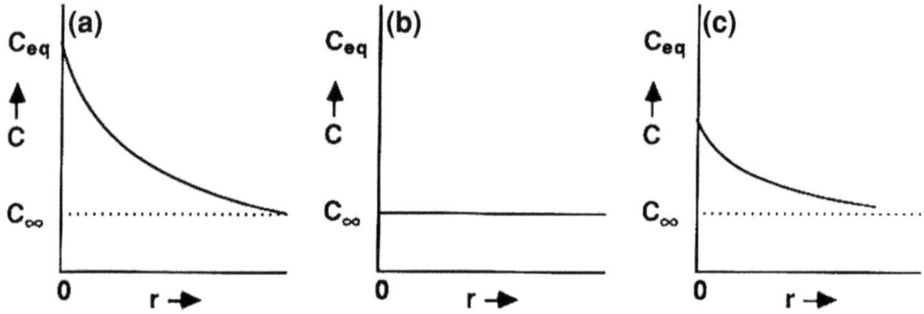

Fig. 2.3 Rate-limiting steps in mineral dissolution: **a** transport-controlled, **b** surface reaction-controlled, and **c** mixed transport and surface reaction control. Concentration (C) versus distance (r) from a crystal surface for three rate-controlling processes, where C_{eq} is the saturation concentration and C_∞ is the concentration in an infinitely diluted solution. Reprinted from Sparks (1988). Copyright 2005 with permission of Elsevier

2.2.2 Redox Processes

Redox processes in the subsurface involve the transfer of electrons among the constituents of aqueous, gaseous, and solid phases. As a result, changes in the oxidation state of the reactants and products occur. *Oxidation* is the half reaction where electrons are lost or removed from a species, while *reduction* is the half reaction where electrons are gained or added to a species. Accordingly, in a redox reaction, an oxidation agent (or oxidant) is an electron acceptor, and a reducing agent (or reductant) is an electron donor.

All oxidation reactions are coupled to reduction reactions. In many cases, redox reactions can also involve or be affected by changes in the surrounding environment, such as changes in the pH or temperature (i.e., endothermic or exothermic reactions). Many elements in the subsurface can exist in various oxidation states; some examples include elements like carbon, nitrogen, oxygen, sulfur, iron, cobalt, vanadium, and nickel.

The great importance of redox reactions and redox potentials in biogeochemical processes and their effect on the subsurface system is well established. In general, the redox potential pE (or Eh) and pH are considered the principal variables controlling geochemical reactions in geological and aquatic environments; many of the processes that are discussed in the next chapters involve redox reactions. The redox potential, Eh, is also commonly given as pE, the measure of electron activity analogous to pH. By convention, half reactions are written in terms of reduction; for example,

$$aA + bB + e^- \rightleftharpoons cC + dD. \tag{2.34}$$

In aqueous solutions, the general form of a redox reaction equation is given as

$$mA_{\mathrm{ox}} + nH^{+} + ne^{-} \rightleftharpoons pA_{\mathrm{red}} + s\mathrm{H_2O_{(1)}} \quad (2.35)$$

where A is a chemical species in any phase and ox and red denote oxidized and reduced states, respectively. The parameters m, n, p, and s represent stoichiometric coefficients, and H^+ and e^- refer to protons and electrons, respectively, in aqueous solution. The equilibrium constant that corresponds to the half reaction in Eq. (2.35) is

$$K_{\mathrm{eq}} = \frac{[A_{\mathrm{red}}]^p[\mathrm{H_2O}]^s}{[A_{\mathrm{ox}}]^m[\mathrm{H^+}]^n[e^-]^n}. \quad (2.36)$$

An example of such a reaction is the reduction of iron,

$$\mathrm{Fe(OH)_{3(s)}} + 3\mathrm{H^+} + \mathrm{e^-} \rightleftharpoons \mathrm{Fe^{2+}} + 3\mathrm{H_2O_{(1)}} \quad (2.37)$$

with $\log K_{\mathrm{eq}} = 15.87$.

The voltage of the general reaction presented in Eq. (2.34) is given by the Nernst equation:

$$\mathrm{Eh(volts)} = \mathrm{E}^\circ + \frac{RT}{nF}\ln\frac{[A]^a[B]^b}{[C]^c[D]^d} \quad (2.38)$$

where R is the gas constant (=0.001987 kcal/mol/K), T is the temperature (in °K), F is the Faraday constant (23.061 kcal/mol/volt). Using 2.303 $\log_{10}$ instead of ln, then we obtain the term (2.303 RT/F), which is called the *Nernst factor*. This factor is equal to 0.05916 volts at 25 °C, so that Eq. (2.38) becomes

$$\mathrm{Eh(volts)} = \mathrm{E}^\circ + \frac{0.05916}{n}\log\frac{[A]^a[B]^b}{[C]^c[D]^d} \quad (2.39)$$

In a redox reaction, the energy released in a reaction due to movement of charged particles gives rise to a potential difference. The maximum potential difference is called the *electromotive force* (EMF), E, and the maximum electric work, W, is the product of charge q in Coulombs (C), and the potential ΔE in volts or EMF:

$$W = q\Delta E. \quad (2.40)$$

Note that the EMF (or ΔE) is determined by the nature of the reactants and electrolytes, not by the size of the system or amounts of material in it. The change in Gibbs free energy, ΔG, is the negative value of maximum electric work,

$$\Delta G = -W = -q\Delta E. \quad (2.41)$$

A redox reaction equation requires well-defined amounts of reactants and products. The number (n) of electrons in such a reaction equation is related to the amount of charge transferred when the reaction is completed. Because each mole of electron has a charge of 96,485 coulombs (known as the Faraday constant, F), $q = nF$, so that

$$\Delta G = -nF\Delta E. \tag{2.42}$$

The free electron activity, pE, which indicates the redox intensity in a system, is defined as

$$\mathrm{pE} = -\log[e^-] \tag{2.43}$$

Based on the pE value, redox environments are classified as follows: (a) pE > 7 indicates an oxic environment (b) at pE values between 2 and 7, the environment is considered suboxic, and (c) pE < 2 indicates an environment considered anoxic. The occurrence of redox reactions in the subsurface environment is limited by the decomposition and reduction of water:

- The upper bound is defined by the decomposition of water

$$\begin{gathered}
{}^1\!/\!{}_2\mathrm{H_2O} \leftrightarrow {}^1\!/\!{}_4\mathrm{O_2(g)} + \mathrm{H^+} + e^- \\
K_{eq} = -20.78, \mathrm{E}^\circ = 1.22\mathrm{V} \\
(\mathrm{P_{O_2}} = 0.21\ \mathrm{atm}, \mathrm{E}^\circ = 1.22\mathrm{V}[\mathrm{H_2O}] \approx 1) \\
\mathrm{pE} = -\log e^- = \log K_{\mathrm{eq}} - \mathrm{pH} + {}^1\!/\!{}_4 \log[\mathrm{O_2}] \\
\mathrm{Eh(volts)} = \mathrm{E}^\circ + 0.059 \log \frac{[\mathrm{O_2}]^{0.25}[\mathrm{H^+}]}{[\mathrm{H_2O}]} \\
\mathrm{pE} = 20.61 - \mathrm{pH}; \quad \mathrm{Eh} = 1.22 - 0.059\ \mathrm{pH}
\end{gathered}$$

- The lower bound is defined by the reduction in water:

$$\begin{gathered}
2\mathrm{H_2O} + 2e^- \leftrightarrow \mathrm{H_2} + 2\mathrm{OH^-}\ \text{or}\ \mathrm{H^+} + e^- = {}^1\!/\!{}_2\mathrm{H_2} \\
(\mathrm{P_{H_2}} = 1\ \mathrm{atm},\ \mathrm{E}^\circ = 0.0\,\mathrm{V}\ \text{by definition}) \\
\mathrm{pE} = -\log[e^-] = \log\ \mathrm{K_{eq}} - \mathrm{pH} + {}^1\!/\!{}_2 \log[\mathrm{H_2}] \\
\mathrm{Eh(volts)} = \mathrm{E}^\circ + 0.059 \log \frac{[\mathrm{H_2}]^{0.5}}{[\mathrm{H^+}]} \\
\mathrm{pE} = -\mathrm{pH};\ \mathrm{Eh} = -0.059\,\mathrm{pH}
\end{gathered}$$

Redox diagrams are used to express the stability of dissolved species and minerals. An example diagram is presented in Fig. 2.4, where the redox potentials of various types of aqueous systems are shown as a function of pH. It can be seen that at acidic pH, a mine water system has a very high oxidation potential (Eh > 500 mv). In contrast, groundwater at natural to basic pH shows a reduction capability as low as −500 mv and even lower. As mentioned previously, redox reactions are affected by the environmental conditions prevailing in the system and thus the presence of acid functional groups (e.g., humic acids) on porous material interfaces, for example, can affect redox activity in the subsurface.

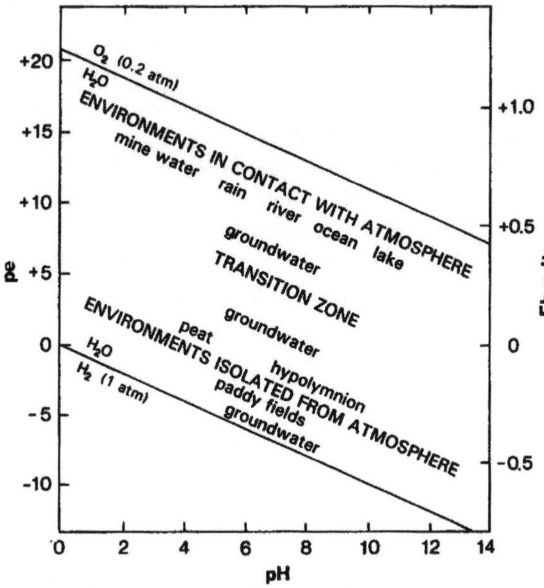

Fig. 2.4 The stability of water and the ranges of pE and pH conditions in natural environments (Appelo and Postma 1993)

In many cases, redox reactions that are favorable from a thermodynamic point of view may not actually take place: Sometimes, the activation energy barriers for such reactions are too high to allow fast transformation, according to the preferred thermodynamic considerations. For example, the complete oxidation of any organic molecule to carbon dioxide and water is thermodynamically favorable. However, such oxidation is not favorable kinetically, which implies that organic molecules—including all forms of living species—are not oxidized immediately; this fact explains the ability to sustain life. The reason for this difference between kinetic and thermodynamic considerations, for redox reactions, is partly because redox reactions are relatively slow compared to other reactions and partly due to the fact that in many cases, reactions are poorly coupled because of slow species diffusion from one microenvironment to another. Therefore, many redox reactions are dependent on catalytic processes.

2.3 Adsorption

Adsorption is the net accumulation of matter on the solid phase at the interface with an aqueous solution or gaseous phase. In this process, the solid surface is the *adsorbent* and the matter that accumulates is the *adsorbate*. *Adsorption* also may be defined as the excess concentration of a chemical at the subsurface solid interface compared to that in the bulk solution, or the gaseous phase, regardless of the nature of the interface region or the interaction between the adsorbate and the solid surface that causes the excess. Surface adsorption is due to interactions between electrical charges, or nonionized functional groups, on mineral and organic constituents.

Adsorption removes a compound from the bulk phase and thus affects its behavior in the subsurface environment. Due to some hysteresis effects, sometimes reflected in formation of bound residues, the release of compounds from the solid phase to the liquid or gaseous phase does not always reach the amount of adsorbate retained on solid surfaces.

When measured adsorption data are plotted against the concentration value of the adsorbate at equilibrium, the resulting graph is called an *adsorption isotherm.* The mathematical description of isotherms invariably involves adsorption models described by Langmuir, Freundlich, or Brauner, Emmet and Teller (known as the BET model). Discussion of these models is given in Part III, as conditions relevant to chemical-subsurface interactions are examined.

2.3.1 Adsorption of Charged Ionic Compounds

Adsorption of charged ionic compounds on the surface of a solid phase is subject to a combination of chemical binding forces and electric fields. The solid phase has a net charge that, in contact with liquid or gaseous phases, is faced by one or more layers of counter- or co-ions having a net charge equal to and separated from the surface charge. Electrical neutrality on the colloidal surface requires that an equal amount of charge of the opposite sign must accumulate in the liquid phase near the charged surface. For a negatively charged surface, this means that positively charged cations are electrostatically attracted to the charged surface. Anions are repelled by such a surface with diffusion forces acting in an opposite direction, such that there is a deficit of anions near the surface. The overall pattern, known as a *diffuse double layer* (DDL), is described by the Gouy–Chapman theory (see Sect. 5.4). This theory assumes that the exchangeable cations exist as point charges, the colloid surfaces are planar and infinite in extent, and the surface charge is distributed uniformly over the entire colloid surface. Stern (1924) and Grahame (1947) refined this theory, showing that the counter-ions are unlikely to approach the surface more closely than the ionic radii of anions and the hydrated radii of cations. Detailed, critical presentations of diffuse double-layer theory applied to earth materials can be found in Sposito (1981, 1989), Sparks (1988), and Bolt et al. (1991).

Cation exchange and selectivity are processes involving the cationic concentration in solution, the cation dimensions, and the configuration of the exchange sites. For cationic molecules, the retention properties follow the relation

$$\frac{A_{\text{aq}}}{B_{\text{aq}}} = K_k \frac{A_{\text{ads}}}{B_{\text{ads}}}, \tag{2.44}$$

where K_k is a selectivity coefficient that expresses the inequality of the activity ratios of the cationic molecules, A and B, in solution (aq) and adsorbed (ads). Kerr (1928) was the first to propose this equation, and K_k is therefore called the *Kerr coefficient.*

When cations of differing charge are involved, the exchange equation (Eq. 2.44) is modified. Gapon (1933) proposed that in the case of an exchange between mono- and bivalent cations, the expression should be

$$\left(\frac{M^{+}}{M^{2+}}\right)_{\text{ads}} = K_G \left(\frac{M^{+}}{\left(\frac{M^{2+}}{2}\right)^{1/2}}\right)_{\text{aq}}, \tag{2.45}$$

where the solute concentration is measured in terms of activity and the adsorption is measured in terms of equivalents, M denotes metal ion, and K_G is the Gapon coefficient.

The exchange properties of negatively charged surfaces do not affect different cations that have the same valence. This is due to differences in size and polarization among the cations themselves, the structural characteristics of the surfaces, and the differences in the surface charge distribution. The preference of the minerals in the subsurface for monovalent cations decreases according to the order $Cs > Rb > K > NH_4 > Na > Li$, which is known as the *lyotropic series*. In this case, a greater attraction of the surfaces for less-hydrated cations may be observed. This mechanism is relevant to minerals having isomorphic substitution in the tetrahedral sheet of the layer (recall Sect. 1.1) and is less evident in the case of divalent alkaline earth cations and trivalent cations. In the case of organomineral complexes, the pattern of cation selectivity is modified considerably (Greenland and Hayes 1981).

The selectivity of the exchange between two cations is specific to saline subsurface systems, described by the sodium adsorption ratio (SAR). Based on Na^{+} content in soil solutions, the US Salinity Laboratory (US Salinity Laboratory 1954) derived the equation

$$\text{SAR} = \frac{[Na^{+}]}{\sqrt{\frac{[Ca^{2+}]+[Mg^{2+}]}{2}}} \tag{2.46}$$

where the solution concentration is in μmol/L.

Metal cation adsorption processes include exchange, Coulombic, and site-specific adsorption. Heavy metal cations exhibit exchange reactions with negatively charged surfaces of clay minerals. Cationic adsorption is affected by the pH and in an acid environment ($pH < 5.5$), and some heavy metals do not compete with Ca^{2+} (a ubiquitous constituent in the subsurface) for mineral adsorption sites. At a higher pH, heavy metal adsorption increases abruptly and becomes irreversible.

Negative adsorption occurs when a charged solid surface faces an ion in an aqueous suspension and the ion is repelled from the surface by Coulomb forces. The Coulomb repulsion produces a region in the aqueous solution that is depleted of the anion and an equivalent region far from the surface that is relatively enriched. Sposito (1984) characterized this macroscopic phenomenon through the definition of the relative surface excess of an anion in a suspension, by

$$\Gamma_i^{(\mathrm{w})} = \frac{n_i - M_{\mathrm{w}} m_i}{S} \tag{2.47}$$

where n_i is total moles of ion i in the suspension per unit mass of solid, M_{w} is the molecular mass of water in the suspension per unit mass of solid, m_i is the molality of ion i in the supernatant solution, and S is the specific surface area of the suspended solid. This $\Gamma_i^{(\mathrm{w})}$ is the excess moles of ion (per unit area of suspended solid) relative to an aqueous solution containing M_{w} kilograms of water and the ion of molality, m_i.

If an anion approaches a charged surface, it is subject to attraction by positively charged surface sites on the surface or repulsion by negatively charged ones. Because clay materials in the subsurface normally are negatively charged, anions tend to be repelled from mineral surfaces. Negative adsorption of anions is affected by the anion charge, concentration, pH, the presence of other anions, and the nature and charge of the surface.

2.3.2 *Adsorption of Nonionic Compounds*

Adsorption of nonionic compounds on subsurface solid phases is subject to a series of mechanisms such as protonation, water bridging, cation bridging, ligand exchange, hydrogen bonding, and van der Waals interactions. Hasset and Banwart (1989) consider that the sorption of nonpolar organics by soils is due to enthalpy-related and entropy-related adsorption forces.

Enthalpy-related adsorption forces include the following processes:

1. Hydrogen bonding, which refers to the electrostatic interaction between a hydrogen atom covalently bound to one electronegative atom and to another electronegative atom or group of atoms in a molecule. The hydrogen atom may be regarded as a bridge between electronegative atoms. In general, this bonding is conceptualized as a dipole phenomenon, where the hydrogen bond exhibits an asymmetrical distribution of the first electron of the H atom induced by various electronegative atoms.
2. Ligand exchange processes, involving replacement of one or more ligands by the adsorbing species and, in some cases, can be considered a condensation reaction.
3. Protonation mechanisms, including a Coulomb electrostatic force resulting from charged surfaces. Due to the surface acidity, solutes having proton-selective organic functional groups can be adsorbed through a protonation reaction.
4. Pi (π) bonds, occurring as a result of overlapping of π orbitals when they are perpendicular to aromatic rings.

5. London–van der Waals forces, which are multipole interactions produced by correlation between fluctuation-induced multipole moments in two nearly uncharged polar molecules. These forces also include dispersion forces that arise from the correlation between the movement of electrons in one molecule and those of neighboring molecules. The van der Waals dispersion interaction between two molecules is generally very weak, but when many groups of atoms in a polymeric structure act simultaneously, the van der Waals components are additive.
6. Chemisorption, denoting the situation when an actual chemical bond is formed between the molecules and surface atoms.
7. Atoms that are rearranged, forming new compounds at the demand of the unsatisfied valences of the surface atoms.

Entropy-related adsorption, known as *hydrophobic sorption*, involves the partitioning of nonpolar organics from a polar aqueous phase onto hydrophobic surfaces, where they are retained by dispersion forces. The major feature of hydrophobic sorption is the weak interaction between the solute and the solvent. The entropy change is due largely to the destruction of the cavity occupied by the solute in the solvent and the destruction of the structured water shell surrounding the solvated organic.

2.3.3 Kinetic Considerations

A perspective based on kinetics leads to a better understanding of the adsorption mechanism of both ionic and nonionic compounds. Boyd et al. (1947) stated that the ion exchange process is diffusion controlled and the reaction rate is limited by mass transfer phenomena that are either film diffusion (FD) or particle diffusion (PD) controlled. Sparks (1988) and Pignatello (1989) provide a comprehensive overview on this topic.

In the case of subsurface cation exchange, charge compensation cations are held in the solid phase within crystals in interlayer positions, structural holes, or surface cleavages and faults of the crystals as well as on the external surfaces of clay minerals. Cations held on external surfaces are immediately accessible to the aqueous phase. On reaching this phase, they move by diffusion to regions of smaller concentration, the diffusion being affected by the tortuosity of the porous medium. An additional restriction affecting the rate of exchange is given by the fact that the arrival rate of incoming cations to the exchange sites is much slower than the release rate of the outgoing cations. The characteristic period of ion exchange in the subsurface ranges from a few seconds to days, being due to various constituents of the solid phase and the properties of the adsorbate.

In the case of nonionic compounds, the driving forces for adsorption consist of entropy changes and weak enthalpic (bonding) forces. The sorption of these compounds is characterized by an initial rapid rate followed by a much slower approach to an apparent equilibrium. The faster rate is associated with diffusion on the surface, while slower reactions have been related to particle diffusion into micropores.

References

Appelo CAJ, Postma D (1993) Geochemistry, groundwater and pollution. Balkema, Rotterdam
Atkins P, de Paula J (2002) Physical chemistry, 7th edn. Oxford University Press, Oxford
Bennett PC, Melcer ME, Siegel DI, Hassett JP (1988) The dissolution of quartz in dilute aqueous solutions of organic acids at 25 °C. Geochim Cosmochim Acta 52:1521–1530
Berner RA (1978) Rate control of mineral dissolution under earth surface conditions. Am J Sci 78:1235–1252
Berner RA (1983) Kinetics of weathering and diagenesis. Mineral 8:111–134
Bolt GH, De Boodt MF, Hayes MHB, McBride MB (1991) Interactions at the soil colloid-soil solution interface. Kluwer, Dordrecht
Boyd GE, Adamson AW, Myers LS (1947) The exchange adsorption of ions from aqueous solutions by organic zeolites: II. Kinetics. J Am Chem Soc 69:2836–2848
Freeze RA, Cherry JA (1979) Groundwater. Prentice-Hall Englewood Cliffs, NJ
Gapon EN (1933) Theory of exchange adsorption in soils. J Gen Chim 3:144–163
Grahame DC (1947) The electrical double layer and the theory of electrocapillarity. Chem Rev 41:441–501
Greenland DG, Hayes MHB (1981) The chemistry of soil constituents. Wiley, New York
Hasett IJ, Banwart WL (1989) The sorption of nonpolar organics by soils and sediments. In: Sawhney BL, Brown K (eds) Reactions and movement of organic chemicals in soils. Soil Science Society of America, Madison, pp 31–45
Kerr HW (1928) The identification and comparison of soil aluminosilicate active base exchange and soil acidity. Soil Sci 26:385–398
O'Day PA (1999) Molecular environmental geochemistry. Rev Geophy 37:249–274
Pignatello JJ (1989) Sorption dynamics of organic compounds in soils and sediments. In: Sawhney BL, Brown K (eds) Reactions and movement of organic chemicals in soils. Soil Science Society of America, Madison, pp 45–80
Shternina EB (1960) Solubility of gypsum in aqueous solutions of salts. Intern Geol Rev 1:605–616
Sparks DL (1988) Kinetics of soil chemical processes. Academic Press, New York
Sposito G (1981) The thermodynamics of soil solutions. Oxford University Press, Oxford
Sposito G (1984) The surface chemistry of soils. Oxford University Press, New York
Sposito G (1989) The chemistry of soils. Oxford University Press, New York
Stern O (1924) Zur theorie der electrolytischen doppelschicht. Z Electrochem 30:508–516
Stumm W, Furrer G, Wieland E, Zinder B (1985) The effects of complex-forming ligands on the dissolution of oxides and aluminosilicates. In: Drever JI (ed) The chemistry of weathering. Reidel, Dordrecht, pp 55–74
Stumm W, Morgan JJ (1995) Aquatic chemistry, 3rd edn. Wiley, New York
United States Salinity Laboratory (1954) Diagnosis and improvement of saline and alkali soils. Handbook 60. U.S. Department of Agriculture, Washington, DC

Part II
Properties of Potential Contaminants: Environmental and Health Hazards

In Part II, we discuss the potential sources, chemical properties, and toxicity of several major groups of contaminants found in the subsurface environment. Usually, the release of contaminants to the environment originates from anthropogenic processes. Even when the contaminants are naturally occurring species, we often find that human intervention or changes in natural conditions are involved in the development of pollution. Furthermore, many contaminants are relatively persistent and therefore may be found in the subsurface environment long after their actual release.

The massive industrial development that has improved quality of life and affected the world over the last two centuries has also had a profound impact on the amounts and types of compounds released into the subsurface. During this period, many thousands of new materials have been produced, used, stored, and transported. The amounts of these substances produced, used, and subsequently discarded have also increased exponentially, leading to the release of huge amounts of contaminants into the environment. In parallel, the understanding that many compounds may be toxic or hazardous to ecological systems, in general, and to humans in particular, has gradually evolved during the last 50 years. This understanding is dependent largely on complex analytical capabilities: environmental samples contain very low concentration of target substance(s), and such samples often contain large amounts of interfering compounds.

Part II divides contaminants into two groups: inorganic substances and organic compounds. The following chapters touch only a small portion of the potential contaminants that belong to each group; we chose representative materials that provide a broad view of the subject.

Finally, it should be mentioned that, because other aspects of chemical interactions and transport of such substances in the subsurface environment are discussed in other parts of this book, we focus here on the contamination potential of these compounds.

Chapter 3
Inorganic and Organometallic Compounds

3.1 Nitrogen

Nitrogen is a key building block in all life forms; it is an essential element in many fundamental cell components, such as proteins, DNA, RNA, and vitamins, as well as in hormones and enzymes. Nitrogen is an extremely versatile element, existing in both inorganic and organic forms as well as in many oxidation states. Nitrogen gas constitutes almost 80 % of the atmosphere and is therefore present almost everywhere. However, nitrogen gas is not available for use by most organisms. For plants and animals to be able to use nitrogen, N_2 gas must first be converted to a more chemically available form, such as ammonium (NH_4^+), nitrate (NO_3^-), or organic nitrogen (e.g., urea). Furthermore, higher organisms (e.g., animals) cannot use these simple forms of nitrogen and require even more complex forms, such as amino acids and nucleic acids.

To sustain the food needs of the world's growing population, it is essential to maintain modern agriculture, which in turn is dependent on a constant supply of nitrogen fertilizers. Currently, the main technology for the "fixation" of atmospheric nitrogen is the *Haber–Bosch method,* which was developed in Germany between 1908 and 1910 and is now the primary process in the production of over 99 % of nitrogen fertilizer materials (Alan 2004). Nitrogen fertilizers have been produced and applied in very large amounts to field crops around the world over several decades; in the years 1980, 1990, and 2000, for example, the world "fixed" N production by the Haber–Bosch process, in amounts that increased from 59,290, to 76,320, to 85,130 KT, respectively.

Fixed nitrogen in the subsurface results from many processes and is applied in many forms and materials. Fixation by rhizobium–legume combinations or free-living organisms adds N directly to crop or soil organic forms. Animal manure, sewage sludge, crop residues, and roots add organic nitrogen materials, which are then mineralized to give ammonium, which is further transformed to nitrate. Because plants often cannot utilize all the nitrogen applied to agricultural fields, some is left in the soil, which subsequently leaches into the groundwater (Pratt and

B. Berkowitz et al., *Contaminant Geochemistry*,
DOI: 10.1007/978-3-642-54777-5_3,

Jury 1984). In addition, not all the applied nitrogen enters the soil; some is washed off fields in the form of runoff and flows into surface waters, such as streams and rivers.

Although various nitrogen forms are present in subsurface environments, environmental risks usually are linked to nitrate concentrations. Nitrates can have adverse effects on humans, animals, and plants. High nitrate concentrations (i.e., more than a few tenths of mg/L) may promote eutrophication in surface waters. Over 30 years ago, nitrate concentration criteria for irrigation water were suggested by Ayers and Westcott (1976), differentiating "no problems" (<5 mg/L), "increasing problems" (5–30 mg/L), and "severe problems" (>30 mg/L) for the production of sensitive crops.

The major sources of nitrate for human intake are food (e.g., vegetables and meat) and water, although polluted air, cigarette smoking, and certain medications also contribute to nitrate ingestion. Specifically, the contribution from nitrogen-containing fertilizers to high levels of nitrate in food and drinking water has been identified as an environmental health concern.

In general, exposure to nitrate is not of particular interest with respect to human health. However, about 5 % of the total dietary nitrate can be reduced endogenously (within the human body) to nitrite through bacterial and other reactions; and nitrite can be further reduced to N-nitroso compounds (NOCs). Methemoglobinemia ("blue baby syndrome") in infants up to six months old, various cancers, and birth defects have been listed as possibly being associated to exposure to elevated nitrate levels in drinking water. A maximum (allowable) contaminant level (MCL) of 10 ppm for nitrates in drinking water was established by the US Environmental Protection Agency (EPA) on the basis of concerns about methemoglobinemia. This syndrome is expressed as an inability of the blood to effectively transport oxygen and carbon dioxide and may result in acute distress to the system; in severe cases, it can cause a bluish tinge in the skin color, hence the term *blue baby*. The World Health Organization (WHO 1985) refers to several cases of nitrite poisoning due to use of drinking water with high concentrations of nitrates, in the range of 44–88 ppm. Other health outcomes usually are sporadic and debatable. However, in some cases, exposure to nitrate concentrations in drinking water has been associated with an increased incidence of hyperthyroidism (goiter) (Seffner 1995; van Maanen et al. 1994). Also, an increased risk of central nervous system malformations in infants whose mothers consumed drinking water with high nitrate levels has been reported (Dorsch et al. 1984; Arbuckle et al. 1988). A different study suggested a link between consumption of water with high nitrate levels and risk of delivering malformed children, relative to an area where nitrate concentration in the water supply was <5 ppm (Scragg et al. 1982). A significant increase in the mean number of chromatid/chromosome breaks in children exposed to nitrate concentrations that exceeded 70.5 mg/L was detected, indicating that chronic elevated concentrations of nitrate in drinking water are capable of inducing cytogenetic effects (Tsezou et al. 1996). Another study suggested that consumption of drinking water, especially well water, with high nitrate levels can imply a genotoxic risk for humans as indicated by increased hypoxanthine

Table 3.1 Results of various correlation studies between nitrate intake and cancer in humans

	Induced cancer effect	No effect	Protective effect
Brain cancer	Barrett et al. (1998)	Steindorf et al. (1994), Muelleret al. (2001)	
Central nervous system tumors	Barrett et al. (1998)		
Gastric cancer		Barrett et al. (1998), van Loon et al. (1998)	
Esophageal cancer		Barrett et al. (1998)	
Non-Hodgkin's lymphoma	Weisenburger (1993), Ward et al. (1996)		
Lymphatic cancer		Jensen (1982)	
Hematopoietic (blood or bone marrow) cancer		Jensen (1982)	
Uterine cancer	Jensen (1982), Thouez et al. (1981)		Weyer et al. (2001)
Ovarian cancer		Jensen (1982), Thouez et al. (1981)	
Bladder cancer	Weyer et al. (2001)		Ward et al. (2003)
Rectal cancer			Weyer et al. (2001)

phosphoribosyltransferase gene (HPRT) variant frequencies and by endogenous formation of carcinogenic NOCs from nitrate-derived nitrite (van Maanen et al. 1996).

NOCs are known carcinogens (Odashima 1980; Lijinsky and Epstein 1970; Magee and Barnes 1967; Tricker and Preussmann 1991) and have been found to induce cancer in a variety of organs in more than 40 animal species, including higher primates (Bogovski and Bogovski 1981). However, and despite the fact that NOCs are produced from nitrate in the human body, only mixed results linking nitrates and their derivatives to various cancers are reported in ecological studies. For example, stomach, esophagus, nasopharynx, urinary bladder, and prostate cancer have been reported to be linked to nitrate intake (e.g., Cantor 1997; Eicholzer and Gutzwiller 1990). However, other ecological studies comparing populations that were exposed to varying nitrate concentrations in drinking water did not find any correlation to gastric and esophageal cancer rates (Barrett et al. 1998) but found increased incidence of adult brain and central nervous system tumors in regions with high nitrate concentrations; this latter effect was not observed by others (Steindorf et al. 1994; Mueller et al. 2001). Table 3.1 summarizes several examples of the mixed results that correlate nitrate intake to cancer.

3.2 Phosphorus

Phosphorus is one of the inorganic macronutrients in all known forms of life. Inorganic phosphorus in the form of phosphate (PO_4^{3-}) plays a major role in vital biological molecules, such as DNA and RNA. Living cells also utilize phosphate to transport cellular energy via adenosine triphosphate (ATP). Phospholipids are the main structural components of all cellular membranes. Calcium phosphate salts are used by animals to stiffen their bones, and adequate P supplies are necessary for seed and root formation.

The principal phosphorus forms in the subsurface are Ca-phosphate, adsorbed phosphates, occluded phosphates, and organic phosphate (Lindsay 1979; Mengel 1985). The principal phosphate mineral is apatite (or fluorapatite), $Ca_{10}(PO_4)_6(OH,F)_2$. Secondary minerals include silica, silicates, and carbonates, generally as calcite or dolomite. Usually, these minerals are utilized as a raw material source for the production of phosphorus fertilizers. Many soils in their natural state are low in readily available P. For example, typical levels of phosphorus in subsurface solutions of unfertilized soils range from 0.001 to 0.1 mg/L P (Hook 1983). Therefore, intensive agricultural activity requires fertilization to achieve maximum possible yields. Application of phosphorus to soil as fertilizer, manure, or effluents results in an immediate rise in the concentration of utilizable phosphorus. It was once thought that P was completely immobile in soil, and therefore, farmers were encouraged to increase phosphate fertilizer application without fear that P applied in excess of crop requirements would be lost from the soil profile.

Phosphorus loss from soils occurs mainly through crop removal, runoff, and leaching. The uptake of phosphorus by agricultural crops, for example, varies greatly (i.e., 10–100 kg/ha/yr). While subsurface pathways can be significant in P transfer to water, especially in soils with low P retention properties or significant preferential flow pathways, it is reasonably well established that in most watersheds, P export occurs mainly by overland flow. Soils that have been used heavily for agricultural crops are often deficient in phosphorus, as are acidic sandy and granitic soils. In landscaped urban soils, however, phosphorus is rarely deficient, and the misapplication of this element can have serious repercussions on plants, the soil environment, and adjoining watersheds.

In ecological terms, phosphorus often is a limiting nutrient in many environments. However, the result of phosphate overfertilization is leaf chlorosis. Phosphorus is known to compete with iron and manganese uptake by roots, and deficiencies in these two metal micronutrients cause interveinal yellowing. Moreover, it has been demonstrated experimentally that high levels of phosphorus are detrimental to mycorrhizal health and lower the rate of mycorrhizal infection of root systems (Manna et al. 2006). This mutually beneficial relationship between the fungus and the plant roots allows the plant to more effectively explore the soil environment and extract needed nutrients. Often in aquatic systems, an excess of phosphorus may cause eutrophication. Eutrophication has been linked to many aspects of water quality degradation, including fish kills, loss of biodiversity and recreational uses of waters, and the onset of harmful algal blooms that can pose a threat to human health (Burkholder et al. 1999). Phosphorus losses from agricultural production systems are known to contribute to accelerated eutrophication of natural waters (Sims et al. 1998; Toor et al. 2003). This is especially true in areas with intensive animal farming, where repeated manure applications have led to excessive accumulation of P in soils. Substantial evidence exists to show that higher P concentrations in soils can result in increased P losses to natural waters (Sharpley and Tunney 2000; Sims et al. 1998).

Organic compounds of phosphorus form a wide class of materials, some of which are extremely toxic. Fluorophosphate esters are among the most potent neurotoxins known. Many organophosphorus compounds are employed for their toxicity to certain organisms as pesticides (e.g., herbicides, insecticides, fungicides) and developed as nerve agents in weapons. In contrast, most inorganic phosphates are relatively nontoxic and essential nutrients.

3.3 Salts

Salts, in contrast to the other main groups presented in this chapter, are natural species that have always been in the environment in large amounts. However, natural processes, along with the extensive use of salts, overpumping of fresh groundwater, and changes in land use are altering the delicate natural balance and causing salinization of soils and water. The salinization process can be divided into natural and anthropogenic mechanisms; in the latter case, these mechanisms can be due to direct application of salts and saline solutions to the environment and indirect salinization by changes in the natural equilibrium. Examples of direct salinization include the distribution of salts on roads and use of treated wastewater, which usually contains large concentrations of salts. Examples of indirect salinization include overpumping of groundwater, which in turn may lead to seawater intrusion into coastal aquifers, and changing land use, which replaces natural vegetation.

Unrelated to its origin, increases in salt concentrations carry a variety of implications, which include altering important ecological niches, degradation of water quality to a level where it may become unpotable or unusable for agriculture and industrial uses, loss of soil fertility and change in its structure, and degradation of infrastructure utilities such as roads, pipelines, and buildings. Problems related to salinization have been known for many decades and extensive literature is available (e.g., Eaton 1942; Richards 1954; Bresler et al. 1982; Vengosh 2003, and references within these publications). In spite of the significant attention devoted to the subject over many years, salts still are considered one of the major groups of contaminants because of their devastating effects on the environment and the ever-growing demand for fertile soil and fresh water. The impact of salinization and its sources will be discussed here on the basis of its effects on four major areas of interest: agriculture, water supply, degradation of infrastructure, and biodiversity ecology.

3.3.1 Agricultural Impacts

Agriculture activities are one of the major generators of salinity but also one of the first to suffer its damage. The salinity of agricultural return flow is derived from two principal sources: salinity of the irrigation water and the effects of added agrochemicals (e.g., fertilizers, boron compounds) or animal waste. Drainage water is sometimes saline and may be accompanied by elevated concentrations of selenium, boron, arsenic, and mercury (Hern and Feltz 1998; Beltran 1999; Causape et al. 2004; Farber et al. 2004).

Land use changes from natural vegetation to agricultural crops, and application of irrigation water, add salts to the system (Cao et al. 2004; Vaze et al. 2004; Peck and Hatton 2003). Approximately 60 % of supplied irrigation water is consumed by growing crops, but the salts remain in the residual solution, as they are not consumed by evaporation and transpiration (Vengosh 2003). As a consequence, adequate drainage is a key factor that determines soil salinization: in arid and semiarid areas, the natural salinity of soil is high; therefore, flushing with irrigation water enhances the dissolution of stored salts.

Salinity contributes to significant losses of productivity in agricultural land and may take some land entirely out of production. Plant roots absorb water from the soil through the process of osmosis. Osmosis moves water from areas of lower salt concentration to areas of higher concentration. Once the salt concentration gradient is reduced, transport of water to plants is slowed; and if extreme conditions of salinity prevail, the osmosis may stop or even change direction, causing plant dehydration.

When high salinity is associated with high sodium content, exchangeable sodium may replace exchangeable calcium on the soil clays. The increased level of adsorbed Na^+ causes soil to become dispersed, with significantly reduced porosity and permeability (Yaalon and Yaron 1966; Panayiotopoulos et al. 2004; Bethune and Batey 2002). As a result of this interaction the soil becomes impermeable. Breakdown in soil structure, together with the associated loss of plant cover, results in greater exposure of the soil to erosion. Sheet, rill, gully, and wind erosion are common effects of salinity (e.g., Funakawa et al. 2000; Spoor 1998). This also is a major problem for drainage of soil water and salt flush in the partially saturated zone.

Land degradation by salinization, which decreases soil fertility, is a significant component of desertification processes. The World Bank states that soil salinization caused by inappropriate irrigation practices affects ~60 Mha, or 24 % of all irrigated land worldwide. In Africa, salinization accounts for 50 % of irrigated land (Vengosh 2003). Increasing soil salinization also is occurring in the Middle East, Australia, China, America, and central Asia (Kang et al. 2004; Dehaan and Taylor 2002; Kotb et al. 2000; Salama et al. 1999; Spoor 1998; Hern and Feltz 1998).

Soil salinization is the first stage of environmental destruction caused by salinity and is related to river and lake salinization. For example, the diversion of the Amu Darya and Syr Darya Rivers caused significant desiccation of the Aral Sea, but it also caused salinization of associated agricultural land (e.g., Funakawa et al. 2000; Spoor 1998). In Australia, soil salinization is the most severe environmental problem on the continent, causing dramatic changes in landscape, industry, and the future of farmland (e.g., Dehaan and Taylor 2002).

3.3.2 Impacts on Water Usage

The salt content of groundwater, surface water, and soils is a major factor in determining their benefit to the community, the economy, and the environment. In many areas around the globe, increasing demand for water has created tremendous pressures on water resources, which resulted in lowering water levels and increasing salinization.

Both natural and anthropogenic changes result in surface water salinization, and in many cases, a combination of the two processes enhances the phenomenon. Natural processes include discharge of saline groundwater to surface water and the dissolution and transport of salts accumulated in the partially saturated zone (James et al. 1996; Kolodny et al. 1999). Additionally, saline feed streams can result from runoff water that dissolved salts during transport, which then are drained to surface water. These natural processes are more pronounced in semiarid and arid zones, where high evaporation rates often exceed transport rates. Anthropogenic changes also affect the water–salt balance, especially in more heavily populated areas. Some examples of such anthropogenic interventions include

1. Diversion of rivers or constriction of dams, which causes decreases in flow and available water, amounts for natural washing of salts (e.g., Funakawa et al. 2000; Spoor 1998).
2. Directed deposition of saline solutions by disposal of effluents, drainage of saline solutions (containing, e.g., agricultural waste), or migration of road deicing salts after dissolution and transport of the resulting brine (e.g., Cardona et al. 2004; Thunqvist 2004).
3. Land use changes that eliminate natural vegetation and increase drainage and recharge rates (e.g., Gordon et al. 2003; Leaney et al. 2003).

Groundwater salinization is caused mainly by saltwater intrusion and recharge of saline solution. Natural processes involved in these phenomena include advection and diffusion of saline fluids entrapped in aquitards connected to an aquifer, dissolution of soluble salts (such as gypsum and halite minerals) within the aquifer, intrusion of underlying or adjacent saline groundwater, and flow from adjacent or underlying aquifers. Population growth (especially near coastal zones, where approximately two-thirds of the world's population lives) and dense urbanization increasingly cause groundwater to become the most important source of freshwater. Furthermore, large water volumes from domestic industrial and agricultural use are recharged to the subsurface as saline wastewater effluents. Rising water demand leads to extensive (over)pumping of fresh groundwater. This, in turn, reduces or reverses the natural offshore flow that otherwise counters the invasion of saline water. In many places throughout the world, coastal water supplies have been rendered unsuitable for domestic and agricultural purposes by such exploitation (e.g., Khan and Sapna 2004; Ergil 2000; Bear et al. 1999). The Ghyben-Herzberg relation, for example, suggests that under hydrostatic conditions, the weight of a unit column of fresh water extending from the water table to the interface is balanced by a unit column of saltwater extending from sea level to the same depth as the point on the interface. This analysis assumes hydrostatic conditions in a homogeneous, unconfined coastal aquifer. According to this relation, if the water table in an unconfined coastal aquifer is lowered by 1 m, the saltwater interface will rise 40 m. Therefore, rising sea levels, which may occur by global warming, will result in seawater intrusion; this may damage many coastal aquifers that currently supply major quantities of fresh water (e.g., Sherif and Singh 1999).

3.3.3 Impacts on Infrastructure

Salinization also has consequences on infrastructure maintenance and durability (e.g., Wilson 2001). Some of the effects include:

- Damage to houses, buildings, and other structures caused by the deterioration of brick, mortar, and concrete, resulting from saline water crystallizing in brickwork (e.g., Cole and Ganther 1996).
- Corrosion of metal buried in the ground or set in structural concrete (e.g., Cole and Ganther 1996).
- Shifting or sinking of foundations, which may result in structural cracking, damage, or collapse. Damage to heritage buildings may be of particular concern, and land values may be degraded.
- Salt damage to roads and highways including the breakdown of concrete, bitumen, and asphalt with associated pot holing, cracking, and crumbling of the road base.
- Damage to underground pipes, cables, and other infrastructure due to the breakdown of unprotected metal, cement, and other materials.
- Poorer water quality for industrial and domestic washing purposes, resulting in an increased need for water conditioners and detergents, as well as the deterioration of vegetated parks and gardens.

3.3.4 Impacts on Biodiversity and the Environment

Salinization has profound effects on many ecological niches and may cause distortion of biodiversity in many delicate zones. Fluctuating water table levels and increasing salinity have serious impacts on native vegetation, in the same way they do for crops and pastures (e.g., Cramer and Hobbs 2002). Remnant vegetation may be threatened, together with a variety of animal species and their habitats (e.g., Hannam et al. 2003). Changes in vegetation often are the first signs of increasing salinity. These may involve the appearance of a salt-tolerant plant species or the reduced growth of existing species. The salinity rise of a lake may lead to a dramatic change in the ecological system and species composition, which in turn modifies the food chain and thus the bioaccumulation of toxic elements such as selenium (e.g., Funakawa et al. 2000; Spoor 1998; Cramer and Hobbs 2002). Wetland salinization is a good example of such processes. More than one-fourth of the world's wetlands have been lost due to salinization (Vengosh 2003). The destruction of wetlands has reduced the diverse assembly of millions of waterfowl and shorebirds. Moreover, salinization is likely to harm the reproductive capacity of birds and place further stress on their diversity. Wetland destruction has been particularly devastating in arid zones. Coastal wetland salinization is caused by a significant decrease in river discharge, diversion of freshwater for irrigation networks, and upstream seawater intrusion under the influence of tides.

3.4 Radionuclides

Radionuclides are present in the environment both as naturally occurring species and as anthropogenic substances. Emission of ionizing radiation during the decay of active atoms is the main contamination route for this group of pollutants: This radiation can disrupt atoms, creating positive ions and negative electrons, and cause biological harm. Exposure to large radiation doses can be fatal to humans, while lower doses cause mainly elevated cancer risks (Stannard 1973; Zhu and Shaw 2000; Shaw 2005). Ionizing radiation may be emitted in the process of natural decay of some unstable nuclei or following excitation of atoms and their nuclei in nuclear reactors, cyclotrons, X-ray machines, or other instruments.

The exposure to ionizing radiation from natural sources is continuous and unavoidable. For most individuals, this exposure exceeds that from all human-made sources combined (UNSCEAR 2000). The two main contributors to natural radiation exposures are high-energy cosmic ray particles incident on the earth's atmosphere and radioactive nuclides that originate in the earth's crust and are present everywhere in the environment, including the human body itself.

Naturally occurring radionuclides of terrestrial origin (also called *primordial radionuclides*) are present in various degrees in all media in the environment. Only those radionuclides with half-lives comparable to the age of the earth and their decay products exist in significant quantities in these materials. Irradiation of the human body from external sources is mainly by gamma radiation from radionuclides in the ^{238}U and ^{232}Th series and from ^{40}K. These radionuclides are present in the body and irradiate the various organs with alpha and beta particles, as well as gamma rays. Some other terrestrial radionuclides, including those of the ^{235}U series, ^{87}Rb, ^{138}La, ^{147}Sm, and ^{176}Lu, exist in nature but at such low levels that their contributions to the dose in humans are small.

Exposure from terrestrial radionuclides present at trace levels in all soils is specific and relates to the types of rock from which the soils originate. Higher radiation levels are associated with igneous rocks, such as granite, and lower levels with sedimentary rocks. There are exceptions, however, as some shale and phosphate rocks have relatively high contents of radionuclides. Radon and its short-lived decay products in the atmosphere are the most important contributors to human exposure from natural sources.

The main human-made contribution to the exposure of the world's population has come from the testing of nuclear weapons in the atmosphere, from 1945 to 1980 (UNSCEAR 2000). Each nuclear test resulted in unrestrained release into the environment of substantial quantities of radioactive materials, which were dispersed widely in the atmosphere and deposited everywhere on the earth's surface. Underground testing caused exposures beyond the test sites only if radioactive gases leaked or were vented. Most underground tests had much lower yields than atmospheric tests, and it usually was possible to contain the debris. Underground tests were conducted at the rate of 50 or more per year from 1962 to 1990. During the time when nuclear weapon arsenals were being expanded, especially in the earlier years (1945–1960), radionuclide releases exposed local populations downwind or downstream of nuclear installations. At the time, there was little recognition of exposure potentials, and monitoring of releases was limited. More recent controls on the military fuel cycle have now diminished exposures to very low levels.

Several industries process or utilize large volumes of raw materials containing natural radionuclides. Discharges from these industrial plants to air and water and the use of by-products and waste materials may contribute to enhanced exposure of the general public. A list of radionuclides responsible for most environmental concerns is given in Table 3.2. Estimated maximum exposures arise from phosphoric acid production, mineral sand processing industries, and coal-fired power stations. Except in the case of accidents or at sites where wastes have accumulated, causing localized areas to be contaminated to significant levels, no other practices result in important exposures of radionuclides released into the environment. Estimates of releases of isotopes produced and used in industrial and medical applications are being reviewed, but these seem to be associated with rather insignificant levels of exposure.

When accidents occur, environmental contamination and exposures may become significant. The accident at the Chernobyl nuclear power plant was a notable example. Exposures were highest in local areas surrounding the reactor, but low-level exposures could be identified for the European region and the entire northern hemisphere. In the first year following the accident, the highest regionally averaged annual doses in Europe, outside the former Union of Soviet Socialist Republics, were less than 50 % of the natural background dose. Subsequent exposures decreased rapidly.

Table 3.2 Characteristics of major radionuclides that occur in soil

Isotope	Half-life (yr)	Principal radiation	Main occurrence
^{14}C	5.7×10^3	β	Natural and nuclear reactor
^{40}K	1.3×10^9	β	Natural
^{90}Sr	28	β	Nuclear reactor
^{134}Cs	2	β, γ	Nuclear reactor
^{137}Cs	30	β, γ	Nuclear reactor
^{239}Pu	2.4×10^4	α, X-rays	X-rays, nuclear reactor

Reprinted from Zhu and Shaw (2000).

3.5 Heavy Metals and Metalloids

The term *heavy metal* refers to any metallic chemical element that has a relatively high density (usually specific density of more than 5 g/mL) and is toxic or poisonous at low concentrations. Examples of heavy metals include arsenic (As), cadmium (Cd), chromium (Cr), mercury (Hg), lead (Pb), and thallium (Tl). The sources, uses, and environmental effects of several exemplary specific metals are discussed briefly here.

Heavy metals are natural components of the earth's crust. They cannot be degraded or destroyed. To a small extent, they generally enter human and animal bodies via food, drinking water, and air. However, exposure to increasingly higher amounts of pollutants from this group usually is due to technological progress that, in many cases, is linked to the ability to extract and process metals. Therefore, already several thousands of years ago, polluted zones were identified, and some effects of metal poisoning were known during the period of development of methods for use of metals (Nariagu 1996; Jarup 2003; Maskall and Thornton 1998).

In trace amounts, some heavy metals (e.g., copper, nickel, selenium, zinc) are essential to all organisms, to accomplish specific catalytic functions. However, at levels exceeding these requirements, all metals can disturb the metabolism by binding nonspecifically to biomolecules and inflicting oxidative damage, due to their ability to catalyze redox reactions. This can result in the deactivation of essential enzymatic reactions, damage to cellular structures (especially membranes), and DNA modification (mutagenesis). In humans, exposure to high levels of metals can cause acute toxicity symptoms, while long-term exposure to lower levels can trigger allergies and even cancers.

3.5.1 Arsenic

Arsenic is historically the poison of choice for many murders, both in fiction and reality (e.g., Christie 1924; CNN 1998). The element is considered a metalloid (having both metallic and nonmetallic properties) and is widely distributed in the earth's crust. Arsenic occurs in trace quantities in all rock, soil, water, and air (WHO 2001). Under reducing conditions, arsenite (As^{III}) is the dominant form, while arsenate (As^{V}) generally is the stable form in oxygenated environments. Arsenic salts exhibit a wide range of solubilities, depending on pH and the ionic environment.

The average subsurface abundance of arsenic is 5–10 mg/kg (Han et al. 2003), and it is present in more than 200 mineral species. Approximately 60 % of natural arsenic minerals are arsenates, 20 % sulfides and sulfosalts, and the rest are arsenides, aresnites, oxides, alloys, and polymorphs of elemental arsenic. Inorganic arsenic of geological origin is found in groundwater used as drinking water in several parts of the world, especially in the Bengal Basin. Organic arsenic compounds are found mainly in marine organisms.

Mining, smelting of nonferrous metals, and burning of fossil fuels are the major industrial processes that contribute to anthropogenic arsenic contamination of air, water, and soil. Use of chromated copper arsenate (CCA) for wood preservation is still in widespread use in many countries and was used heavily during the latter half of the twentieth century as a structural and outdoor building component, where there was a risk of rot or insect infestation in untreated timber. Lead arsenate has been used, well into the twentieth century, as a pesticide on fruit trees, and copper arsenate has even been recorded in the nineteenth century as a coloring agent in sweets.

Arsines released from microbial sources in soils or sediments undergo oxidation in the air, reconverting the arsenic to nonvolatile forms, which settle back to the ground. Dissolved forms of arsenic in the water column include arsenate, arsenite, methylarsonic acid (MMA), and dimethylarsinic acid (DMA). Some arsenic species have an affinity for clay mineral surfaces and organic matter, and this can affect their environmental behavior. There is a potential for arsenic release when there is fluctuation in Eh, pH, soluble arsenic concentration, and sediment organic content. Many arsenic compounds tend to adsorb to soils, and leaching usually results in transportation over only short distances in soil. Three major modes of arsenic biotransformation have been found to occur in the environment: redox transformation between arsenite and arsenate, reduction and methylation of arsenic, and biosynthesis of organo-arsenic compounds. Arsenic levels in groundwater average about 1–2 μg/L except in areas with volcanic rock and sulfide mineral deposits, where arsenic levels can reach up to 3 mg/L. Naturally elevated levels of arsenic in soils may be associated with geological substrata such as sulfide ores. Anthropogenically contaminated soils can have concentrations of arsenic up to several grams per 100 mL. Nonoccupational human exposure to arsenic in the environment is primarily through the ingestion of food and water. Contaminated soils, such as mine tailings, also are a potential source of arsenic exposure. The daily intake of total arsenic from food and beverages generally is between 20 and 300 μg/day.

Soluble inorganic arsenic is acutely toxic, and ingestion of large doses leads to gastrointestinal symptoms, disturbances of cardiovascular and nervous system functions, and eventually death. In survivors, bone marrow depression, hemolysis, hepatomegaly, melanosis, polyneuropathy, and encephalopathy may be observed.

Long-term exposure to arsenic in drinking water is causally related to increased risks of cancer in the skin, lungs, bladder, and kidney, as well as other skin changes such as hyperkeratosis and pigmentation changes. Conclusions on the causality of the relationship between arsenic exposure and other health effects are less clear. The evidence is strongest for hypertension and cardiovascular disease, suggestive for diabetes and reproductive effects, and weak for cerebrovascular disease, long-term neurological effects, and cancer at sites other than lung, bladder, kidney, and skin. Studies on laboratory animals and in vitro systems generally suggest that inorganic arsenicals are considered more toxic than organic arsenicals, and within these two classes, the trivalent forms are more toxic than the pentavalent forms, at least at high doses.

3.5.2 Cadmium

Cadmium is found naturally deep in the subsurface in zinc, lead, and copper ores, in coal, shales, and other fossil fuels; it also is released during volcanic activity. These deposits can serve as sources to ground and surface waters, especially when in contact with soft, acidic waters. Chloride, nitrate, and sulfate salts of cadmium are soluble, and sorption to soils is pH dependent (increasing with alkalinity). Cadmium found in association with carbonate minerals, precipitated as stable solid compounds, or coprecipitated with hydrous iron oxides is less likely to be mobilized by resuspension of sediments or biological activity. Cadmium absorbed to mineral surfaces (e.g., clay) or organic materials is more easily bioaccumulated or released in a dissolved state when sediments are disturbed, such as during flooding.

Roughly 15,000 tons of cadmium is produced worldwide each year (McMurray and Tainer 2003). It is produced as an inevitable by-product of zinc, lead, and copper refining and smelting, because these combined metals occur naturally within the raw ore. The most significant uses of cadmium are metal plating and coating for corrosion protection in alloys and electronic compounds and in nickel/cadmium batteries. Additional uses for cadmium are in pigments and stabilizers for PVC. Cadmium also is present as an impurity in several products, including phosphate fertilizers, detergents, and refined petroleum products (EPA 2005; Jarup 2003). Industrial contamination of topsoil is likely the major source of human exposure, via uptake into food plants and tobacco (Hayes 1997).

Cadmium is biopersistent and, once absorbed by an organism, remains resident for many years (with a biological half-life of 10–30 yr) although it is eventually excreted (McMurray and Tainer 2003). Acute exposures may cause several health effects in humans, including nausea, vomiting, diarrhea, muscle cramps, salivation, sensory disturbances, liver injury, convulsions, shock, and renal failure. Chronic effects may include a potential to cause kidney, liver, bone, and blood damage from long-term exposure at levels above the MCL (EPA 2005). Cadmium compounds are classified as human carcinogens by several regulatory agencies. The most convincing data that cadmium is carcinogenic in humans come from studies indicating that occupational cadmium exposure is associated with lung cancer. Cadmium exposure has also been linked to human prostate and renal cancer, although this linkage is weaker than for lung cancer. Other target sites of cadmium carcinogenesis in humans, such as the liver, pancreas, and stomach, are considered equivocal. In animals, cadmium effectively induces cancers at multiple sites and by various routes (Waalkes and Rehm 1994; Waalkes 2000, 2003; Hertz-Picciotto and Hu 1995; Satarug et al. 2003; McMurray and Tainer 2003; Hayes 1997). Cigarette smoking is the major source of cadmium exposure, causing an average blood cadmium level four to five times higher than for nonsmokers (Jarup et al. 1998). In general, for the nonsmoking population, the major exposure pathway is through food, via the addition of cadmium to agricultural soil from various sources. Additional exposure to humans arises through the presence of cadmium in ambient air and drinking water.

3.5.3 Chromium

Chromium toxicity is different for each of its species (especially Cr^{VI} and Cr^{III}). It is, therefore, regulated separately for each oxidation state, unlike all other heavy metals, which are regulated on the basis of their total concentration (Kimbrough et al. 1999). Chromium (III) occurs naturally in the environment and is an essential nutrient. Chromium (VI) and chromium (0) are generally produced by industrial processes. Chromium exists in small quantities throughout the environment. Chromite ore ($FeCr_2O_4$) is the most important commercial ore and usually is associated with ultramafic and serpentine rocks. Chromium also is associated with other ore bodies (e.g., uranium and phosphorites) and may be found in tailings and other wastes from these mining operations. Acid mine drainage can make chromium available to the environment. In air, chromium compounds are present mostly as fine dust particles, which eventually settle over land and water. Chromium can attach strongly to soil, and only a small amount can dissolve in water and move deeper in the soil to underground water.

Chromium and its compounds are used in refractories, drilling muds, electroplating cleaning agents, catalytic manufacture, and in the production of chromic acid and specialty chemicals. The metal chromium, which is the chromium (0) form, is used for making steel. Chromium (VI) and chromium (III) are used for chrome plating, dyes and pigments, leather tanning, and wood preserving. Chromium enters the air, water, and soil mostly in the chromium (III) and chromium (VI) forms. The key to these uses is that under typical environmental and biological conditions of pH and oxidation–reduction potential, the most stable form of chromium is the trivalent oxide. This form has very low solubility and low reactivity, resulting in low mobility in the environment and low toxicity in living organisms. However, the stable and generally nontoxic trivalent form of chromium can be transformed (oxidized) in the environment to chromate (CrO_4) and dichromate (Cr_2O_7) anions, which are mobile and toxic. Hexavalent chromium anions are the predominant form of dissolved chromium in waters that are alkaline and mildly oxidizing. While hexavalent chromium contamination is generally associated with industrial activity, it can occur naturally.

Chromium and certain chromium compounds are classified as substances known to be carcinogenic. EPA classifies chromium as a "de minimis" carcinogen, meaning that the minimum amount of the chemical set by the US Occupational Safety and Health Administration (OSHA) is considered to be carcinogenic. Chromium compounds vary greatly in their toxic and carcinogenic effects. Trivalent chromium compounds are considerably less toxic than the hexavalent compounds and are neither irritating nor corrosive.

Breathing high levels of chromium (VI) can cause irritation to the nose, including nosebleeds, ulcers, and holes in the nasal septum. Ingesting large amounts of chromium (VI) can cause stomach upset and ulcers, convulsions, kidney and liver damage, and even death. Skin contact with certain chromium (VI) compounds can cause skin ulcers. Some people are extremely sensitive to chromium (VI) or chromium (III). Allergic reactions consisting of severe redness and swelling of the skin have been noted. Several studies have shown that chromium (VI) compounds can increase the risk of lung cancer. Animal studies also have shown an increased risk of cancer. The World Health Organization determined that chromium (VI) is a human carcinogen. The US Department of Health and Human Services (DHHS) determined that certain chromium (VI) compounds are known to cause cancer in humans.

3.5.4 Lead

Lead is a bluish-white lustrous metal. It is very soft, highly malleable, ductile, and a relatively poor conductor of electricity. It is very resistant to corrosion but tarnishes on exposure to air. It is one of the oldest metals used by humans and has been used widely since 5000 BC; lead was known to the ancient Egyptians and Babylonians. The Romans used lead for pipes and in solder. The element has four naturally occurring stable isotopes, three of which result from the decay of naturally occurring radioactive elements (thorium and uranium). Although lead is seldom found uncombined in nature, its compounds are widely distributed throughout the world, principally in the ores galena (a lead sulfide ore), cerussite ($PbCO_3$), and anglesite ($PbSO_4$).

To date, the single most important commercial use of lead is in the manufacture of lead-acid storage batteries. However, for most of the twentieth century, the most important environmental source of Pb was gasoline combustion. It is also used in alloys, such as fusible metals, antifriction metals, and solder. Lead foil is made with lead alloys. Lead is used for covering cables and as a lining for laboratory sinks, tanks, and the "chambers" in the lead-chamber process for the manufacture of sulfuric acid. It is used extensively in plumbing. Because it has excellent vibration-dampening characteristics, lead is often used to support heavy machinery.

Contamination of lead in groundwater originates from the dissolution of lead from soil and the earth's crust. Lead particulate from the combustion of leaded gasoline and ore smelting can contaminate local surface water by surface runoff. Lead itself is only of minor content in the earth's crust. A wide distribution of lead in sedimentary rock and soils has been reported, with an average lead content of 10 mg/kg (10 ppm) in topsoil, and a range of 7–12.5 ppm in sedimentary rock (EPA 1992). In soils, lead generally is present in the form of carbonates and hydroxide complexes. Strong absorption by soil and complexation by humus can further limit lead concentrations in surface waters and groundwater.

Lead poisoning is an environmental and public health hazard of global proportions. Children and adults in virtually every region of the world are being exposed to unsafe levels of lead in the environment. In fact, children are exposed to lead from different sources, such as paint, gasoline, and solder, and through different pathways such as air, food, water, dust, and soil. Poisoning can occur from a single, high-level exposure or the cumulative effect of repeated high- or low-level exposures.

The adverse health effects of lead are indicated by blood levels, which cause changes in blood pressure at 10 μg/dL, to severe retardation and even death at very high blood levels of 100 μg/dL. For example, lead interferes with synthesis of red blood cells and causes anemia, kidney damage, impaired reproduction function, interference with vitamin D metabolism, and delayed neurological and physical development. For adult men, high blood lead can cause elevated blood pressure, hypertension, strokes, and heart attack. Pregnant women exposed to lead are at risk of complications in their pregnancies, shorter gestational period, and damage to the fetus.

3.5.5 Nickel

Nickel is a silver-white, lustrous, hard, malleable, ductile, ferromagnetic metal that is relatively resistant to corrosion and is a fair conductor of heat and electricity. Nickel is a ubiquitous trace metal that occurs in soil, water, air, and in the biosphere. The average content in the earth's crust is about 0.008 %. Nickel ore deposits are accumulations of nickel sulfide minerals (mostly pentlandite) and laterites. Nickel exists in five major forms: elemental nickel and its alloys; inorganic, water-soluble compounds (e.g., nickel chloride, nickel sulfate, and nickel nitrate); inorganic, water-insoluble compounds (e.g., nickel carbonate, nickel sulfide, and nickel oxide); organic, water-insoluble compounds; and nickel carbonyl $Ni(CO)_4$.

Nickel is used mostly for the production of stainless steel and other nickel alloys with high-corrosion and temperature resistance. Nickel alloys and nickel platings are used in vehicles, processing machinery, armaments, tools, electrical equipment, household appliances, and coins. Nickel compounds also are used as catalysts, pigments, and in batteries.

Nickel, which is emitted into the environment from both natural and human-made sources, is circulated throughout all environmental compartments by means of chemical and physical processes and is biologically transported by living organisms. Nickel enters groundwater and surface waters from erosion and dissolution of rocks and soils, as well as from biological cycles, atmospheric fallout, industrial processes, and waste disposal. Nickel leached from dump sites can contribute to nickel contamination of an aquifer, with potential ecotoxicity. Acid rain has a tendency to mobilize nickel from soil and increase nickel concentrations in groundwater, leading eventually to increased uptake and possible toxicity in microorganisms, plants, and animals.

Depending on the soil type, nickel may exhibit a high mobility within the soil profile, finally reaching groundwater, rivers, and lakes. Terrestrial plants take up nickel from soil, primarily via the roots. The amount of nickel uptake from soil depends on various geochemical and physical parameters, including soil type, soil pH and humidity, the organic matter content of the soil, and the concentration of extractable nickel. Drinking water generally contains less than 10 μg/L of nickel, but nickel may occasionally be released from plumbing fittings, resulting in concentrations of up to 500 μg/L.

In terms of human health, nickel carbonyl is the most acutely toxic nickel compound. The effects of acute nickel carbonyl poisoning include frontal headache, vertigo, nausea, vomiting, insomnia, and irritability, followed by pulmonary symptoms similar to those of viral pneumonia. Pathological pulmonary lesions include hemorrhage, edema, and cellular derangement. The liver, kidneys, adrenal glands, spleen, and brain also are affected. Cases of nickel poisoning have been reported in patients dialyzed with nickel-contaminated dialysate and in electroplaters who accidentally ingested water contaminated by nickel sulfate and nickel chloride.

Chronic effects such as rhinitis, sinusitis, nasal septal perforations, and asthma have been reported in nickel refinery and nickel-plating workers. Some authors reported pulmonary changes with fibrosis in workers inhaling nickel dust. In addition, nasal dysplasia has been reported in nickel refinery workers. Nickel contact hypersensitivity has been documented extensively in both the general population and in a number of occupations in which workers were exposed to soluble nickel compounds. Very high risks of lung and nasal cancer have been reported in nickel refinery workers employed in the high-temperature roasting of sulfide ores, involving substantial exposure to nickel subsulfide, oxide, and perhaps sulfate. Similar risks have been reported in processes involving exposure to soluble nickel (electrolysis, copper sulfate extraction, hydrometallurgy), often combined with some nickel oxide exposure, but with low nickel subsulfide exposure.

3.6 Nanomaterials

Engineered nanomaterials are usually described as inorganic materials of high uniformity, with at least one critical dimension below 100 nm. This group of substances has received considerable attention recently as the basis for the next "scientific revolution." In this framework, nanomaterials are expected to be incorporated in a wide range of applications, from medicine and cosmetics to new construction materials and industrial processes, and to many more applications in all areas of technological development. Based on current estimates, nanotechnology is projected to become a trillion-dollar market by 2015. As a consequence, these materials are expected to be spread around the globe rapidly due to massive production and use.

Nanomaterials hold promise for elegant solutions to numerous environmental concerns, from implementation of green chemistry processes for industrial and agrochemical uses (e.g., Mohanty et al. 2003; McKenzie and Hutchison 2004) to production of novel materials for treatment of various contaminants (e.g., Dror et al. 2005; Nurmi et al. 2005; Nagaveni et al. 2004; Kuhn et al. 2003). However, these possibilities also bring new threats that must be considered and monitored. Nanomaterials magnify and stimulate properties that, at larger scales, are in many cases, of minor importance. Below 100 nm—the upper limit of the nanomaterial range—the surface area to mass ratio and the proportion of total number of atoms at the surface of a structure are large enough that surface properties become very significant. This can alter the chemical reactivity, thermal and electrical conductivity, and tensile strength of known substances (Owen and Depledge 2005). Additional physicochemical properties of engineered nanomaterials include specific chemical composition (e.g., purity, crystallinity, electronic properties), surface structure (e.g., surface reactivity, surface groups, inorganic or organic coatings), solubility, shape, and aggregation (Nel et al. 2006). Furthermore, when the size of a structure is 10–30 nm, it approaches certain physical length scales, such as the electron mean free path and the electron wavelength, resulting in quantum-size effects that alter the electronic structure of the particle (Nurmi et al. 2005; Brus 1986; Wang and Herron 1991). These changes modify optical, magnetic, and electrical behavior.

The desired size-related properties of nanomaterials also raise major concerns: the same characteristics that make these substances so appealing may have negative health and environmental impacts. Recent studies (e.g., Owen and Depledge 2005; Royal Society and Royal Academy of Engineering 2004; Balbus et al. 2005; Colvin 2003) state that the data collected to date are inadequate to provide full risk assessments, but substantial basis for concern exists and should be further investigated. Only a handful of ecotoxicological studies on nanosized materials have been published so far. However, the collected data are alarming.

In a recent review, Nel et al. (2006) discuss the toxic potential of nanomaterials. For example, Hote et al. (2004) state that nanoparticles entering the liver can induce oxidative stress locally. It was also found that uptake of polymeric

Table 3.3 Nanomaterial effects as the basis for pathophysiology and toxicity

Experimental NM effects	Possible pathophysiological outcomes
ROS generation[a]	Protein, DNA and membrane injury[a], oxidative stress[b]
Oxidative stress[a]	Phase II enzyme induction, inflammation[b], mitochondrial perturbation[a]
Mitochondrial perturbation[a]	Inner membrane damage[a], permeability transition (PT) pore opening[a], energy failure[a], apoptosis[a], apo-necrosis, cytotoxicity
Inflammation[a]	Tissue infiltration with inflammatory cells[b], fibrosis[b], granulomas[b], atherogenesis[b], acute phase protein expression (e.g., C-reactive protein)
Uptake by reticulo-endothelial system[a]	Asymptomatic sequestration and storage in liver[a], spleen, lymph nodes[b], possible organ enlargement and dysfunction
Protein denaturation, degradation[a]	Loss of enzyme activity[a], autoantigenicity
Nuclear uptake[a]	DNA damage, nucleoprotein clumping[a], autoantigens
Uptake in neuronal tissue[a]	Brain and peripheral nervous system injury
Perturbation of phagocytic function[a], "particle overload," mediator release[a]	Chronic inflammation[b], fibrosis[b], granulomas[b], interference in clearance of infectious agents[b]
Endothelial dysfunction, effects on blood clotting[a]	Atherogenesis[a], thrombosis[a], stroke, myocardial infarction
Generation of neoantigens, breakdown in immune tolerance	Autoimmunity, adjuvant effects
Altered cell cycle regulation	Proliferation, cell cycle arrest, senescence
DNA damage	Mutagenesis, metaplasia, carcinogenesis

[a] Effects supported by limited experimental evidence; [b] Effects supported by limited clinical evidence. From Nel et al. (2006). Reprinted with permission from AAAS

nanoparticles by Kupffer cells in the liver induces modifications in hepatocyte antioxidants systems, probably due to the production of radical oxygen species (Fernandez-Urrusuno et al. 1997). Exposure of laboratory-grown human liver and skin cells for 48 h to solutions containing 20 ppb of buckyballs demonstrated that the solution could kill half the cells (Goho 2004). Green and Howman (2004) report the observation of DNA damage in plasmid nicking assays with water-soluble CdSe/ZnS quantum dots (cadmium selenide capped with a shell of zinc sulfide, complete with biotin surface functionality). Studies involving pulmonary exposure of carbon nanotubes in rodents (Lam et al. 2004; Warheit et al. 2004) suggest that lung histopathological responses, including inflammation and granuloma formation, may be significant. Nanoparticles can enter cells and cross the blood–brain barrier where they may have unexpected health effects (Guzman et al. 2006). The US EPA attributed 60,000 deaths per year to the inhalation of atmospheric nanoparticles; and there is evidence for direct transfer into the brain (Oberdörster et al. 2004; Raloff 2003). Table 3.3 summarizes the current knowledge of the impact of nanomaterials in relation to pathophysiology and toxicity.

At this point, there is very limited knowledge about the transport and fluxes of nanoparticles in the natural environment. Colvin (2003) suggested that if engineered nanomaterial applications develop as projected, increasing concentrations of nanomaterials in groundwater and soil may present the most significant exposure avenues for environmental risk. Release to the environment, either by accident or by approved and regulated discharge of effluents and waste water, may result in direct exposure of humans to nanoparticles via ingestion or inhalation of airborne (Brigger et al. 2002; Dowling 2004; Moore 2002, 2006; Warheit 2004) or water aerosols, skin contact, and direct ingestion of contaminated drinking water or particles adsorbed on vegetables or other foodstuffs (Daughton 2004; Howard 2004; Moore 2006).

In addition, direct passage across fish gills and other external surface epithelia may constitute a route of toxic nanoparticle contamination. A limited investigation of fish (Oberdörster 2004) has indicated that C_{60}-fullerenes may be internalized by these routes. At the cellular level, uptake of nanoparticles into biological systems may be facilitated by the caveolar and endocytotic systems in cells, but knowledge of the pathological consequences, if any, is currently very limited (Panyam and Labhasetwar 2003; Pelkmans and Helenius 2002; Reiman et al. 2004). Also, indirect exposure could arise from ingestion of organisms such as fish and shellfish (i.e., mollusks and crustaceans) as part of the human diet. Surface sediment- and filter-feeding mollusks are prime candidates for uptake of manufactured nanoparticles from environmental releases, if it transpires that some of these nanomaterials associate with natural particulates, because these mollusks already are known to accumulate suspended particle and sediment-associated conventional pollutants (Galloway et al. 2002; Livingstone 2001).

References

Alan WT (2004) Fertilizers. In: Kirk-Othmer encyclopedia of chemical technology. John Wiley. doi:10.1002/0471238961.0605182008150606.a01.pub2

Arbuckle TE, Sherman GJ, Corey PN, Walters D, Lo B (1988) Water nitrates and CNS birth defects: a population-based case–control study. Arch Environ Health 43:162–167

Ayers RS, Westcott DW (1976) Water quality for agriculture. Irrigation and drainage paper 29, FAO Rome

Balbus J, Denison R, Florini K, Walsh S (2005) Getting nanotechnology right the first time. Issues Sci Technol (Summer):65–71

Barrett JH, Parslow RC, McKinney PA, Law GR, Forman D (1998) Nitrate in drinking water and the incidence of gastric, esophageal, and brain cancer in Yorkshire, England. Cancer Causes Control 9:153–159

Bear J, Cheng A, Sorek S, Ouazar D, Herrera I (eds) (1999) Seawater intrusion in coastal aquifers concepts, methods and practices. In: Theory and applications of transport in porous media series. Springer Heidelberg 14:640

Beltran JM (1999) Irrigation with saline water: benefits and environmental impact. Agric Water Manag 40:183–194

Bethune MG, Batey TJ (2002) Impact on soil hydraulic properties resulting from irrigating saline-sodic soils with low salinity water. Austral J Exp Agri 42:273–279

Bogovski P, Bogovski S (1981) Animal species in which N-nitroso compounds induce cancer. Int J Cancer 27:471–474

Bresler E, McNeal BL, Carter DL (1982) Saline and sodic soils. Advanced series in agricultural sciences 10. Springer, Berlin

Brigger I, Dubernet C, Couvreur P (2002) Nanoparticles in cancer therapy and diagnosis. Adv Drug Deliv Rev 54:631–651

Brus L (1986) Electronic wave functions in semiconductor clusters: experiment and theory. J Phys Chem 90:2555–2560

Burkholder JM, Mallin MA, Glasgow HB (1999) Fish kills, bottom-water hypoxia, and the toxic Pfiesteria complex in the Neuse River and Estuary. Mar Ecol Prog Ser 179:301–310

Cantor KP (1997) Drinking water and cancer. Cancer Causes Control 8:292–308

Cao ZH, Huang JF, Zhang CS, Li AF (2004) Soil quality evolution after land use change from paddy soil to vegetable land. Environ Geochem Health 26:97–103

Cardona A, Carrillo-Rivera JJ, Huizar-Alvarez R, Graniel-Castro E (2004) Salinization in coastal aquifers of arid zones: an example from Santo Domingo, Baja California Sur, Mexico. Environ Geol 45:350–366

Causape J, Quilez D, Aragues R (2004) Assessment of irrigation and environmental quality at the hydrological basin level—II. Salt and nitrate loads in irrigation return flows. Agric Water Man 70:211–228

Christie A (1924) The mysterious affair at styles. New York: bartleby.com

CNN (1998) Florida's "Black Widow" executed. Available at http://www.cnn.com/US/9803/30/black.widow/

Cole IS, Ganther W (1996) A preliminary investigation into airborne salinity adjacent to and within the envelope of Australian houses. Constr Build Mat 10:203–207

Colvin VL (2003) The potential environmental impact of engineered nanomaterials. Nat Biotechnol 21:1166–1170

Cramer VA, Hobbs RJ (2002) Ecological consequences of altered hydrological regimes in fragmented ecosystems in Southern Australia: impacts and possible management responses. Austral Ecol 27:546–564

Daughton CG (2004) Non-regulated water contaminants: emerging research. Environ Impact Assess Rev 24:711–732

Dehaan RL, Taylor GR (2002) Field-derived spectra of salinized soils and vegetation as indicators of irrigation-induced soil salinization. Remote Sens Environ 80:406–417

Dorsch MM, Scragg RK, McMichael AJ, Baghurst PA, Dyer KF (1984) Congenital malformations and maternal drinking water supply in rural South Australia: a case–control study. Am J Epidemiol 119:473–486

Dowling A (2004) Development of nanotechnologies. Mater Today 7:30–35

Dror I, Baram D, Berkowitz B (2005) Use of nanosized catalysts for transformation of chloroorganic pollutants. Environ Sci Technol 39:1283–1290

Eaton FM (1942) Toxicity and accumulation of chloride and sulfate salts in plants. J Agr Res 64:357–399

Eicholzer M, Gutzwiller F (1990) Dietary nitrates, nitrites and N-nitroso compounds and cancer risk: a review of the epidemiologic evidence. Nutr Rev 56:95–105

EPA (1992) U.S. environmental protection agency. Lead poisoning and your children (800-B-92-0002). Office of pollution prevention and toxics. Washington, DC

EPA (2005) http://www.epa.gov/OGWDW/dwh/t-ioc/cadmium.html

Ergil ME (2000) The salinization problem of the Guzelyurt aquifer. Cyprus Water Res 34:1201–1214

Farber E, Vengosh A, Gavrieli I, Marie A, Bullen TD, Mayer B, Holtzman R, Segal M, Shavit U (2004) The origin and mechanisms of salinization of the lower Jordan river. Geochim Cosmochim Acta 68:1989–2006

Fernandez-Urrusuno R, Fattal E, Feger J, Couvreur P, Therond P (1997) Evaluation of hepatic antioxidant systems after intravenous administration of polymeric nanoparticles. Biomaterials 18:511–517

Funakawa S, Suzuki R, Karbozova E, Kosaki T, Ishida N (2000) Salt-affected soils under rice-based irrigation agriculture in Southern Kazakhstan. Geoderma 97:61–85
Galloway TS, Sanger RC, Smith KL, Fillmann G, Readman JW, Ford TE, Depledge MH (2002) Rapid assessment of marine pollution using multiple biomarkers and chemical immunoassays. Environ Sci Technol 36:2219–2226
Goho A (2004) Sci News Online 166:14
Gordon L, Dunlop M, Foran B (2003) Land cover change and water vapour flows: learning from Australia. Philos Trans Roy Soc Lond Ser B-Biol Sci 358:1973–1984
Green M, Howman E (2004) Semiconductor quantum dots and free radical induced DNA nicking. Chem Comm 121:121–123
Guzman KAD, Taylor MR, Banfield JF (2006) Environmental risk of nanotechnology: national nanotechnology initiative funding 2000–2004. Environ Sci Technol 40:1401–1407
Han FX, Su Y, Monts DL, Plodinec MJ, Banin A, Triplett GE (2003) Assessment of global industrial-age anthropogenic arsenic contamination. Naturwissenschaften 90:395–401
Hannam KM, Oring LW, Herzog MP (2003) Impacts of salinity on growth and behavior of American avocet chicks. Waterbirds 26:119–125
Hayes RB (1997) The carcinogenicity of metals in humans. Cancer Cause Control 8:371–385
Hern J, Feltz HR (1998) Effects of irrigation on the environment of selected areas of the Western United States and implications to world population growth and food production. J Environ Manage 52:353–360
Hertz-Picciotto I, Hu SW (1995) Contribution of cadmium in cigarettes to lung cancer: an evaluation of risk assessment methodologies. Lung Cancer 12:116
Hook JE (1983) Movement of phosphates and nitrogen in soils following application of municipal wastewater. In: Nielson DW, Elrik DE, Tanji KK (eds) Chemical mobility and reactivity in soil system. Soil Sci Soc Am Spec Publ 11:241–255
Hote PHM, Hohlfeld IB, Salata O (2004) Nanoparticles—known and unknown health risks. J Nanobiotechnol 2:12–27
Howard CV (2004) Small particles—big problems. Int Lab News 34:28–29
James WP, Chakka KB, Mascianglioli PA (1996) Control of natural brine springs in Brazos River basin 1 Recovery System. Water Resour Bull 32:475–484
Jarup L (2003) Hazards of heavy metal contamination. Br Med Bull 68:167–182
Jarup L, Berglund M, Elinder CG, Nordberg G, Vahter M (1998) Health effects of cadmium exposure—a review of literature and a risk estimate. Scand J Work Environ Health 24:1–51
Jensen OM (1982) Nitrate in drinking water and cancer in Northern Jutland, Denmark, with special reference to stomach cancer. Ecotoxicol Environ Saf 9:258–267
Kang SZ, Su XL, Tong L, Shi PZ, Yang XY, Abe YK, Du TS, Shen QL, Zhang JH (2004) The impacts of human activities on the water-land environment of the Shiyang river basin, an arid region in Northwest China. Hydrol Sci J-J Des Sci Hydrol 49:413–427
Khan AB, Sapna PV (2004) Groundwater salination in different aquifers—a case study. J Geo Soc India 63:183–190
Kimbrough ED, Cohen Y, Winer AM, Creelman L, Mabuni C (1999) A critical assessment of chromium in the environment. Crit Rev Environ Sci and Technol 29:1–46
Kolodny Y, Katz A, Starinsky A, Moise T, Simon E (1999) Chemical tracing of salinity sources in lake Kinneret (Sea of Galilee), Israel. Limnol Ocean 44:1035–1044
Kotb THS, Watanabe T, Ogino Y, Tanji KK (2000) Soil salinization in the Nile delta and related policy issues in Egypt. Agric Water Man 43:239–261
Kuhn KP, Chaberny IF, Massholder K, Stifkler M, Benz VW, Sonntag HG, Erdinger L (2003) Disinfection of surfaces by photocatalytic oxidation with titanium dioxide and UVA light. Chemosphere 53:71–77
Lam CW, James JT, McCluskey R, Hunter RL (2004) Pulmonary toxicity of single-wall carbon nanotubules in mice 7 and 90 days after intratracheal instillation. Toxicol Sci 77:126–134
Leaney FW, Herczeg AL, Walker GR (2003) Salinization of a fresh palaeo-ground water resource by enhanced recharge. Ground Water 41:84–92
Lijinsky W, Epstein SS (1970) Nitrosamines as environmental carcinogens. Nature 225:21–23

Lindsay WL (1979) Chemical equilibria in soils. Wiley, New York
Livingstone DR (2001) Contaminant-stimulated reactive oxygen species production and oxidative damage in aquatic organisms. Mar Pollut Bull 42:656–666
Magee PN, Barnes JM (1967) Carcinogenic nitroso compounds. Adv Cancer Res 10:163–246
Manna MC, Rao AS, Ganguly TK (2006) Effect of fertilizer P and farmyard manure on bioavailable P as influenced by rhizosphere microbial activities in soybean-wheat rotation. J Sustain Agric 29:149–166
Maskall JE, Thornton I (1998) Chemical partitioning of heavy metals in soils, clays and rocks at historical lead smelting sites. Water Air Soil Poll 108:391–409
McKenzie LC, Hutchison JE (2004) Green nanoscience. Chimica Oggi-Chem Today 22:30–33
McMurray CT, Tainer J (2003) Cancer, cadmium and genome integrity. Nat Genet 34:239–241
Mengel K (1985) Dynamics and availability of major nutrients in soils. Adv Soil Sci 2:67–134
Mohanty AK, Drzal LT, Misra M (2003) Nano reinforcements of bio-based polymers—the hope and the reality. Polym Mater Sci Eng 88:60–61
Moore MN (2002) Biocomplexity: the post-genome challenge in ecotoxicology. Aquat Toxicol 59:1–15
Moore MN (2006) Do nanoparticles present ecotoxicological risks for the health of aquatic environment? Environ Inter 32:967–976
Mueller BA, Newton K, Holly EA, Preston-Martin S (2001) Residential water source and the risk of childhood brain tumors. Environ Health Perspec 109:551–556
Nagaveni K, Silalingam G, Hegde MS, Madras G (2004) Photocatalytic degradation of organic compounds over combustion-synthesized nano TiO_2. Environ Sci Technol 38:1600–1604
Nariagu JO (1996) History of global metal pollution. Science 272:223–224
Nel A, Xia T, Madler L, Li N (2006) Toxic potential of materials at the nanolevel. Science 311:622–627
Nurmi JT, Tratnyek PG, Sarathy V, Baer DR, Amonette JE, Pecher K, Wang CM, Linehan JC, Matson DW, Penn RL, Driessen MD (2005) Characterization and properties of metallic iron nanoparticles: spectroscopy, electrochemistry, and kinetics. Environ Sci Technol 39:1221–1230
Oberdorster E (2004) Manufactured nanomaterials (fullerenes, C60) induce oxidative stress in the brain of juvenile largemouth bass. Environ Health Pers 112:1058–1062
Oberdörster G, Sharp Z, Atudorei V, Elder A, Gelein R, Kreyling W, Cox C (2004) Translocation of inhaled ultrafine particles to the brain of rats. Inhal Toxicol 16:437–445
Odashima S (1980) Overview: N-nitroso compounds as carcinogens for experimental animals and man. Oncology 37:282–286
Owen R, Depledge M (2005) Nanotechnology and the environment: risks and rewards. Mar Poll Bull 50:609–612
Panayiotopoulos KP, Barbayiannis N, Papatolios K (2004) Influence of electrolyte concentration, sodium adsorption ratio, and mechanical disturbance on dispersed clay particle size and critical flocculation concentration in alfisols. Comm Soil Sci Plant Anal 35:1415–1434
Panyam J, Labhasetwar V (2003) Biodegradable nanoparticles for drug and gene delivery to cells and tissues. Adv Drug Deliv Rev 55:329–347
Peck AJ, Hatton T (2003) Salinity and the discharge of salts from catchments in Australia. J Hydrol 272:191–202
Pelkmans L, Helenius A (2002) Endocytosis via caveolae. Traffic 3:311–320
Pratt PF, Jury WA (1984) Pollution of unsaturated zone with nitrate. In: Yaron B, Dagan G, Goldshid J (eds) Pollutants in porous media. Springer, Berlin, pp 52–67
Raloff J (2003) Air sickness: how microscopic dust particles cause subtle but serious harm. Sci News 164:1–11
Reiman J, Oberle V, Zuhorn IS, Hoekstra D (2004) Size-dependant internalization of particles via the pathways of clathrin- and caveolae-mediated endocytosis. Biochem J 377:159–169
Richards LA (ed) (1954) Agriculture handbook no. 60, diagnosis and improvement of saline and alkali soils. United States Department of Agriculture, Washington, DC

Royal Society and Royal Academy of Engineering (2004) Nanoscience and nanotechnologies: opportunities and uncertainties. RS policy document 19/04. The Royal Society, London, p 113
Salama RB, Otto CJ, Fitzpatrick RW (1999) Contributions of groundwater conditions to soil and water salinization. Hydrogeol J 7:46–64
Satarug S, Baker JR, Urbenjapol S, Haswell-Elkins M, Reilly PEB, Williams DJ, Moore MR (2003) A global perspective on cadmium pollution and toxicity in non-occupationally exposed population. Toxicol Lett 137:65–83
Scragg RK, Dorsch MM, McMichael AJ, Baghurst PA (1982) Birth defects and household water supply: epidemiological studies in the Mount Gambier region of South Australia. Med J Australia 2:577–579
Seffner W (1995) Natural water contents and endemic goiter—a review. Zantralblatt fur Hygiene und Umweltmedizin 196:381–398
Sharpley H, Tunney A (2000) Phosphorus research strategies to meet agricultural and environmental challenges of the 21st century. J Environ Qual 29:176–181
Shaw G (2005) Applying radioecology in a world of multiple contaminant. J Environ Radio 81:117–130
Sherif MM, Singh VP (1999) Effect of climate change on sea water intrusion in coastal aquifers. Hydrol Proc 13:1277–1287
Sims JT, Simard RR, Joern BC (1998) Phosphorus loss in agricultural drainage—historical perspective and current research. J Environ Qual 27:277–293
Spoor M (1998) The Aral Sea basin crisis: transition and environment in former Soviet Central Asia. Dev Change 29:409–435
Stannard JN (1973) Toxicology of radionuclides. Ann Rev Pharmacol 13:325–357
Steindorf K, Schlehofer B, Becher H, Hornig G, Wahrendorf J (1994) Nitrate in drinking water: a case–control study on primary brain tumours with an embedded drinking water survey in Germany. Int J Epidemiol 23:451–457
Thouez JP, Beauchamp Y, Simard A (1981) Cancer and the physicochemical quality of drinking water in Quebec. Soc Sci Med 15D:213–223
Thunqvist EL (2004) Regional increase of mean chloride concentration in water due to the application of deicing salt. Sci Total Environ 325:29–37
Toor GS, Condron LM, Di HJ, Cameron KC, Cade-Menun BJ (2003) Characterization of organic phosphorus in leachate from a grassland soil. Soil Bio Biochem 35:1317–1323
Tricker AR, Preussmann R (1991) Carcinogenic N-nitrosamines in the diet: occurrence, mechanisms and carcinogenic potential. Mutat Res 259:277–289
Tsezou A, Kitsiou-Tzeli S, Galla A, Gourgiotis D, Papageorgiou J, Mitrou S, Molybdas PA, Sinaniotis C (1996) High nitrate content in drinking water: cytogenetic effects in exposed children. Arch Environ Health 51:458–461
UNSCEAR (2000) Annex C exposures to the public from man-made sources of radiation. Available at http://www.unscear.org/pdffiles/annexc.pdf
van Loon AJM, Botterweck AAM, Goldbohm RA, Brants HAM, van Klaveren JD, van den Brandt PA (1998) Intake of nitrate and nitrite and the risk of gastric cancer: a prospective cohort study. Br J Cancer 7:129–135
van Maanen JM, van Dijk A, Mulder K, de Baets MH, Menheere PC, van der Heide D, Mertens PL, Kleinjans JC (1994) Consumption of drinking water with high nitrate levels causes hypertrophy of the thyroid. Toxicol Lett 72:365–374
van Maanen JM, Welle IJ, Hageman G, Dallinga JW, Mertens PL, Kleinjans JC (1996) Nitrate contamination of drinking water: relationship with HPRT variant frequency in lymphocyte DNA and urinary excretion of N-nitrosamines. Environ Health Perspec 104:522–528
Vaze J, Barnett P, Beale G, Dawes W, Evans R, Tuteja NK, Murphy B, Geeves G, Miller M (2004) Modelling the effects of land-use change on water and salt delivery from a catchment affected by dryland salinity in South-East Australia. Hydrol Proc 18:1613–1637
Vengosh A (2003). Salinization and saline environments. In: Treatise on geochemistry. Elsevier 9:333–365
Waalkes MP (2000) Cadmium carcinogenesis in review. J Inorg Biochem 79:241–244

Waalkes MP (2003) Cadmium carcinogenesis. Mutat Res 533:107–120

Waalkes MP, Rehm S (1994) Cadmium and prostate cancer. J Toxicol Environ Health 43:251–269

Wang Y, Herron N (1991) Nanometer-sized semiconductor clusters: material synthesis, quantum size effects, and photophysical properties. J Phys Chem 95:525–532

Ward MH, Mark SD, Cantor KP, Weisenburger DD, Correa-Villasenore A, Zahm SH (1996) Drinking water and the risk of non-Hodgkin's lymphoma. Epidemiology 7:465–471

Ward MH, Cantor KP, Riley D, Merkle S, Lynch CF (2003) Nitrate in public water supplies and risk of bladder cancer. Epidemiology 14:183–190

Warheit DB (2004) Nanoparticles: health impacts? Mater Today 7:32–35

Warheit DB, Laurence BR, Reed KL, Roach DH, Reynolds GAM, Webb TR (2004) Comparative pulmonary toxicity assessment of single-wall carbon nanotubes in rats. Toxicol Sci 77:117–125

Weisenburger D (1993) Potential health consequences of ground-water contamination of nitrates in Nebraska. Nebr Med J 78:7–10

Weyer PJ, Cerhan JR, Kross BC, Hallberg GR, Kantamneni J, Breuer G, Jones MP, Zheng W, Lynch CF (2001) Municipal drinking water nitrate level and cancer risk in older women: the Iowa Women's Health Study. Epidemiology 11:327–338

WHO (1985) World Health Organization. Environmental health criteria 52, toluene. World Health Organization, Geneva

WHO (2001) Arsenic and arsenic compounds. Environmental health criteria 224. the International Programme on Chemical Safety (IPCS). Available at http://www.inchem.org/documents/ehc/ehc/ehc224.htm

Wilson SM (2001) Assessing the cost of dryland salinity to non-agricultural stakeholders in selected Victorian and NSW catchments: a methodology report. Available at www.ndsp.gov.au

Yaalon DH, Yaron B (1966) Framework for man made soil changes—outline of metapedogenesis. Soil Sci 102:272–276

Zhu YG, Shaw G (2000) Soil contamination with radionuclides and potential remediation. Chemosphere 41:121–128

Chapter 4
Organic Compounds

4.1 Pesticides

Pesticides are substances, or mixtures of substances, intended to prevent, destroy, repel, or mitigate any pest. They may be chemical substances, biological agents (such as viruses or bacteria), antimicrobial and disinfectant agents, or other devices. The term *pest* includes insects, plant pathogens, weeds, mollusks, birds, mammals, fish, nematodes (roundworms), and microbes that compete with humans for food, destroy property, spread or are a vector for disease, or are a nuisance (EPA 2006). Pests have attacked and destroyed crops, and pesticides have been developed, as long as agriculture has been practiced. Some historical examples include selection of seed from resistant plants in Neolithic times ($\sim$7000 BC; Ordish 2007), sulfur dusting by the Sumerians ($\sim$2500 BC), and over 800 recipes in the Ebers Papyrus, the oldest known medical document (dated around 1550 BC), which describes recognizable substances that were used as poisons and pesticides. More recently, during the fifteenth century, arsenic, mercury, and lead were used to fight pests. The first book to deal with pests in a scientific way was John Curtis's *Farm Insects*, published in 1860, but massive production and application of pesticides only began around World War II. The major development at that time was the discovery of the insecticidal properties of DDT by P. H. Muller in 1942, who received the Nobel Prize (medicine) for his discovery in 1948.

From this point on, many millions of tons of active ingredients have been released intentionally each year to the environment, spreading around the globe. An illustration of the amounts of active-ingredient pesticides applied is shown in Fig. 4.1, where total annual amounts used in the USA surpass billions of pounds for the period from 1982 to 2001.

In contrast to all other groups of contaminants mentioned in Part II, pesticides are released to the environment in staggering quantities, even though they are designed to suppress the normal biological growth of different pests. Pesticides are formulated specifically to be toxic to living organisms, and as such, they are usually hazardous to humans. In fact, most pesticides used today are acutely toxic

B. Berkowitz et al., *Contaminant Geochemistry*,
DOI: 10.1007/978-3-642-54777-5_4,

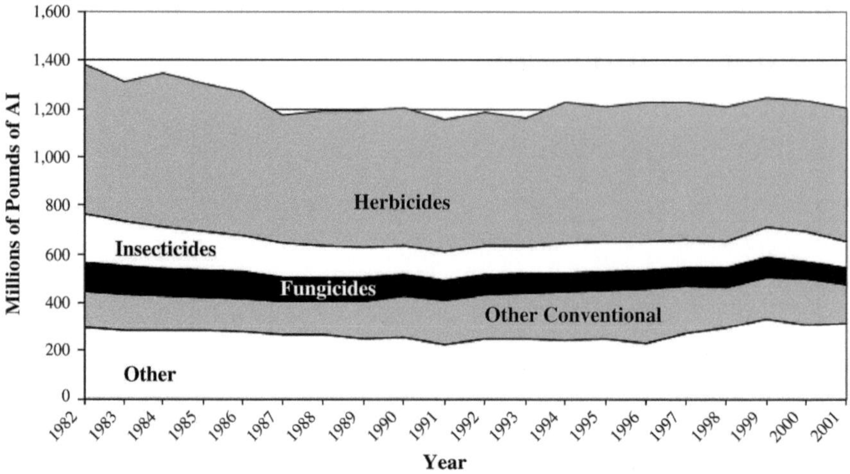

Fig. 4.1 Annual amount of pesticide active ingredients used in the USA, by pesticide type, from 1982 to 2001; estimates are for all market sectors (Kiely et al. 2004)

to humans, causing poisonings and deaths every year. Annual poisoning from pesticides, in the USA alone, is estimated to range between 10,000 and 40,000 diagnosed illnesses and injuries among agricultural workers (Blondell 1997). Chronic health effects from pesticides also have been reported, including neurological effects, reproductive problems, interference with infant development, and cancer. The effect of exposure to various pesticides can induce, on the longer term, shortened attention span and reduced coordination (Rosenstock et al. 1991); a more drastic outcome of such exposure includes increased risk of early onset of Parkinson's disease (Butterfield et al. 1993; Gorell et al. 1998). Moreover, from animal studies, it appears that developing brains during the cell division stage (i.e., during early infancy) are more susceptible to destructive impact of some pesticides, which may lead to long-term abnormal behavior (Chanda and Pope 1996; Eriksson 1996).

Increased risk of various cancers (e.g., lymphatics, blood, stomach, prostate, testes, brain, and soft tissues) was reported upon exposure to pesticides (Zahm and Blair 1993; Hayes et al. 1995; Zahm et al. 1997). Children's cancer including brain tumors, leukemia, non-Hodgkin's lymphoma, sarcoma, and Wilms' tumor due to direct or parental exposure to pesticides also has been determined (Fear et al. 1998; Kristensen et al. 1996; Pogoda and Preston-Martin 1997; Sharpe et al. 1995; Buckley et al. 1994). Often, the risk of cancer due to exposure to pesticides has been higher in children than in adults (Zahm and Ward 1998). Moreover, birth defects, including limb-reduction defects, have been linked to exposures to pesticides (Restrepo et al. 1990; Schwartz and LoGerfo 1988; Lin et al. 1994), as has a higher-than-normal risk of stillborn births (Pastore et al. 1997).

Usually, only small fractions of applied pesticides reach the target organisms, the majority of the chemicals being distributed to the air, soil, and water. On such release, many nontarget organisms can be affected severely. Pesticide residues and their degradation products are toxic to many components of ecological systems, either by being lethal to certain living forms or by changing environmental conditions that in turn alter and usually decrease biodiversity in these systems. In some cases, pesticides can have synergistic effects with other contaminants, thus increasing overall toxicity. Exposure routes of pesticides include direct intake of pesticide residues through, for example, digestion of pesticides adsorbed on crops or dissolved in water and indirect paths through intake of pesticides or their degradation products that have been concentrated in food chains and intake of other environmental mediators.

Official lists of pesticides contain over 1,500 substances (Wood 2006). Pesticides are divided into several major groups (e.g., herbicides, fungicides, insecticides), which are subdivided into chemical or other classes (e.g., chloroacetanilide herbicides or auxins). These compounds may be arranged according to their toxicity (WHO 2005), chemical structure, and/or activity. Here, we briefly discuss only a few of the more common groups of pesticides; the interested reader should consult readily available references (e.g., Barbash 2003; Matthews 2006; Milne 1995; Briggs 1992) as a primary source of more detailed information.

4.1.1 Organochlorine Insecticides

Organochlorine insecticides (e.g., DDT, aldrin, dieldrin, heptachlor, mirex, chlordecone, and chlordane) were used commonly in the past, but many have been removed from the market due to their negative health and environmental effects and their persistence. However, insecticides of this group are in some cases still used as active ingredients in various pest control products, such as gamma-hexachlorocyclohexane (lindane). Lindane also is used as the active agent in the medicine Kwell®, used for human ectoparasitic disease, although it has been associated with acute neurological toxicity either from ingestion or in persons treated for scabies or lice. The general chemical structures of some of the organochlorine insecticides are given in Fig. 4.2. These compounds are characterized by cyclic structures, a relatively large number of chlorine atoms on the molecule, and low volatility. As a result, they usually are resistant to natural degradation processes and thus stable for very long periods after release to the environment.

Organochlorines are absorbed in the body through ingestion, inhalation, and across the skin. These substances tend to concentrate in fatty tissues following exposure. The chief acute toxic action of organochlorine pesticides is on the nervous system, where these compounds induce a hyperexcitable state in the brain (Joy 1985; Reigart and Roberts 1999). This effect is manifested mainly as convulsions, sometimes limited to myoclonic jerking, but often expressed as violent seizures. Other less severe signs of neurological toxicity, such as paresthesias, tremor, ataxia, and hyperreflexia, also are characteristic of acute organochlorine poisoning. Convulsions may cause death by interfering with pulmonary gas

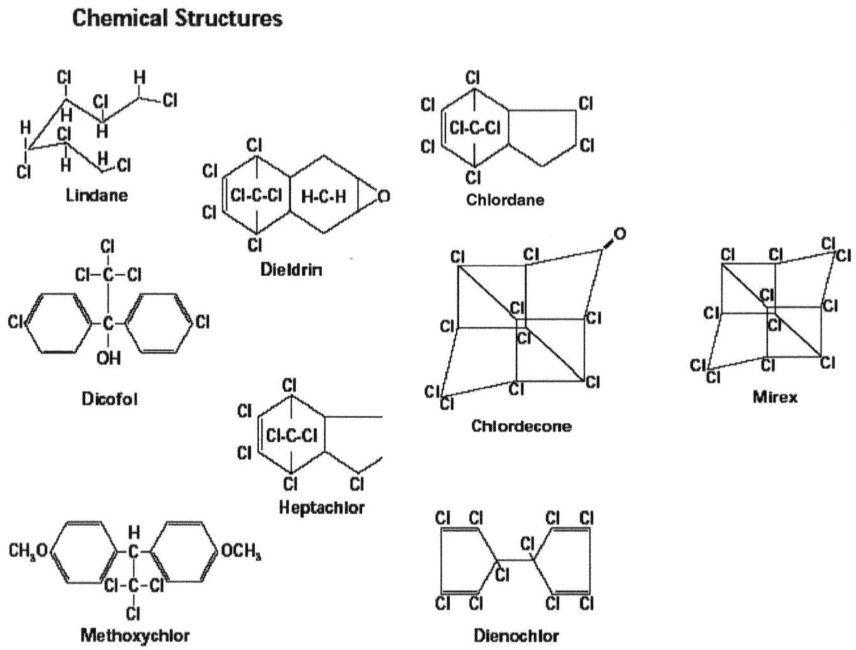

Fig. 4.2 General chemical structures of some organochlorine insecticides

exchange and generating severe metabolic acidosis. High tissue concentrations of organochlorines increase myocardial irritability, predisposing to cardiac arrhythmia. Human absorption of organochlorine sufficient to cause enzyme induction is likely to occur only as a result of prolonged intensive exposure.

There has been considerable interest recently in the interaction of organochlorines with endocrine receptors, particularly estrogen and androgen receptors. In vitro studies and animal experimentation support the view that the function of the endocrine system may be altered by these interactions. This in turn may alter the reproductive development and success of animals and humans. The International Association for Research on Cancer evaluated organochlorine insecticides as being either possibly carcinogenic to humans (DDT, chlordane, heptachlor, toxaphene) or not classifiable as to carcinogenicity (aldrin, dieldrin, lindane) (IARC 1987, 1991, 2001). Inconclusive epidemiological studies linked organochlorine insecticides to increased risks of soft tissue sarcoma, non-Hodgkin's lymphoma, leukemia, and prostate, lung, pancreas, and breast cancers (Purdue et al. 2006). Due to evidence of their toxicity and carcinogenic potential, some organochlorines have been banned or restricted for use. Several organochlorine pesticides, including DDT, methoxychlor, endosulfan, and dicofol, mimic estrogen (Gillette et al. 1994; Cummings 1997). Lindane, which is sometimes used to treat head lice in children, acts as an antiestrogen and is toxic to the nervous system (Cooper et al. 1989).

4.1.2 Organophosphates

The general chemical structure of organophosphate pesticides is shown in Fig. 4.3. The functional group R usually is either ethyl or methyl. Pesticides with double-bonded sulfur moieties are organothiophosphates but are converted to organophosphates in the liver. Phosphonate contains an alkyl (R–) in place of one alkoxy group (RO). The "leaving group" is the principal metabolite for a specific identification.

Acute pesticide poisonings frequently involve organophosphate pesticides; these pesticides were originally derived from chemical warfare agents developed during World War II. Some common organophosphates in use today include chlorpyrifos, diazinon, azinphos-methyl, malathion, and methyl-parathion, all of which apparently share a common mechanism of cholinesterase inhibition and cause similar health effects. Organophosphates poison insects and mammals primarily by phosphorylation of the acetylcholinesterase enzyme (AChE) at nerve endings. The result is a loss of available AChE so that the affected organ becomes overstimulated by the excess acetylcholine (ACh, the impulse-transmitting substance) in the nerve ending. The enzyme is critical to normal control of nerve impulse transmission from nerve fibers to smooth and skeletal muscle cells, glandular cells, and autonomic ganglia, as well as within the central nervous system.

4.1.3 N-methyl Carbamate

The general chemical structure of *N*-methyl carbamate is shown in Fig. 4.4. Common *N*-methyl carbamates in use today include aldicarb, carbofuran, methiocarb, oxamyl, and carbaryl. *N*-methyl carbamates share with organophosphates the capacity to inhibit cholinesterase enzymes and, therefore, share similar symptomatology during acute and chronic exposure.

The *N*-methyl carbamate esters cause reversible carbamylation of the acetylcholinesterase enzyme, allowing accumulation of acetylcholine, the neuromediator substance, at parasympathetic neuroeffector junctions (muscarinic effects), at skeletal muscle myoneural junctions and autonomic ganglia (nicotinic effects), and in the brain (CNS effects). The carbamyl–acetylcholinesterase combination dissociates more readily than the phosphoryl–acetylcholinesterase complex produced by organophosphate compounds. This property has several important consequences:

1. It tends to limit the duration of *N*-methyl carbamate poisoning.
2. It accounts for the greater span between symptom-producing and lethal doses than most organophosphate compounds.
3. It frequently invalidates the measurement of blood cholinesterase activity as a diagnostic index of poisoning.

N-methyl carbamates are absorbed by inhalation and ingestion, and somewhat by skin penetration, although the last tends to be a less toxic route.

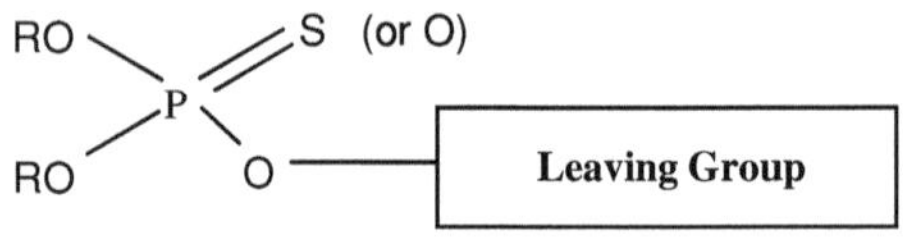

Fig. 4.3 General structure of organophosphate pesticides

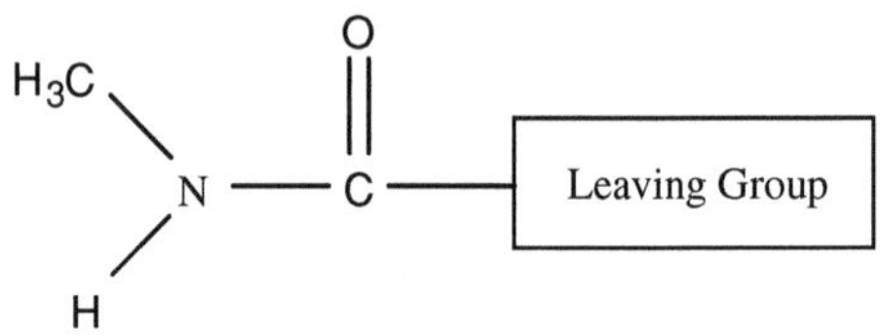

Fig. 4.4 General structure of *N*-methyl carbamate

4.1.4 Triazines

Triazine pesticides and their metabolites are a group of closely related herbicides used widely on agricultural and nonagricultural sites; they are inhibitors of electron transport in photosynthesis. As a family, their chemical structures are heterocyclic, composed of carbon and nitrogen in their rings. Most, except for metribuzin, are symmetrical with their altering carbon and nitrogen atoms. Herbicide members in this family include atrazine, hexazinone, metribuzin, prometon, prometryn, simazine, and their degradates. Atrazine is used widely in corn production and is estimated to have been the most often-used pesticide in the USA during the late 1990s. Its toxic effects may include disruption of ovarian function, generation of mammary (breast) tumors in animals, and interference with the binding of steroid hormones and the breakdown pathway of estrogen (Bradlow et al. 1995; Cooper et al. 1996; Danzo 1997). Some uses of atrazine are classified as restricted because of groundwater and surface water concerns. Many of the triazines show acute and chronic toxicities at low concentrations (Letterman 1999; Montgomery 1993), and they generally are known or suspected to be carcinogenic, mutagenic, and/or teratogenic (Newman 1995; Letterman 1999; Montgomery 1993; C&EN 2000, 2002). Recent evidence (Reeder et al. 1998; Renner 2002; Tavera-Mendoza et al. 2002a, b) implicated specific triazines and/or their degradation products as endocrine disruptors and teratogens in amphibians.

4.1.5 Paraquat and Diquat

Paraquat (1,1′-dimethyl, 4,4′-bipyridyl) is a nonselective contact herbicide. It is used almost exclusively as a dichloride salt and usually is formulated to contain surfactants. Both its herbicidal and toxicological properties are dependent on the ability of the parent cation to undergo a single-electron addition, to form a free radical that reacts with molecular oxygen to reform the cation and concomitantly

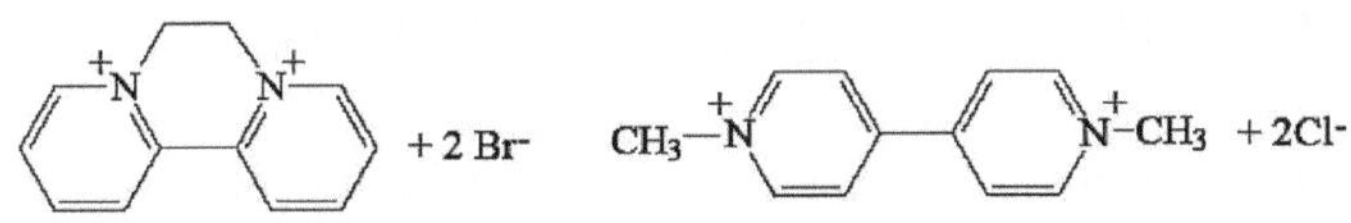

Fig. 4.5 General chemical structures of paraquat and diquat

produce a superoxide anion. This oxygen radical may directly or indirectly cause cell death. Diquat, 1,1′-ethylene-2,2′-dipyridylium, is a charged quaternary ammonium compound often found as the dibromide salt. The structure of diquat dibromide and that of the closely related herbicide paraquat is shown in Fig. 4.5.

Paraquat and diquat are nonselective contact herbicides that are relatively widely used and highly toxic. Paraquat has life-threatening effects on the gastro-intestinal (GI) tract, kidney, liver, heart, and other organs. The LD50 (lethal dose to 50 % of the population) in humans is approximately 3–5 mg/kg. The lung is the primary target organ of paraquat, and pulmonary effects represent the most lethal and least treatable manifestation of toxicity (Pond 1990; Giulivi et al. 1995). However, toxicity from inhalation is rare. Both types I and II pneumatocytes appear to selectively accumulate paraquat. Biotransformation of paraquat in these cells results in free-radical production with resulting lipid peroxidation and cell injury (Pond 1990; Giulivi et al. 1995; Honore et al. 1994). There is a progressive decline in arterial oxygen tension and CO_2 diffusion capacity. Such a severe impairment of gas exchange causes progressive proliferation of fibrous connective tissue in the alveoli and eventual death from asphyxia and tissue anoxia (Harsany et al. 1987). Local skin damage includes contact dermatitis. Prolonged contact produces erythema, blistering, abrasion and ulceration, and fingernail changes (Tungsanga et al. 1983; Vale et al. 1987).

Following ingestion of the substance, the GI tract is the site of initial or phase I toxicity to the mucosal surfaces. This toxicity is manifested by swelling, edema, and painful ulceration of the mouth, pharynx, esophagus, stomach, and intestine. With higher levels, other GI toxicity includes centrizonal hepatocellular injury, which can cause elevated bilirubin, and hepatocellular enzyme levels such as AST, ALT, and LDH.

Diquat poisoning is much less common than paraquat poisoning, so that human reports and animal experimental data for diquat poisoning are less extensive than for paraquat. However, diquat has severe toxic effects on the central nervous system (CNS) that are not typical of paraquat poisoning (Vanholder et al. 1981; Olson 1994).

4.2 Synthetic Halogenated Organic Substances

Halogenated hydrocarbons constitute a widely used family of products. Some of the more common halogenated hydrocarbons are brominated flame retardants and the chlorinated derivatives of methane, ethane, and benzene, which are used mainly as solvents and chemical intermediates. Broad production and use of these compounds began in the early 1900s when chlorinated solvents replaced other flammable substances in a variety of industrial process. These compounds became popular because the progressive halogenation of a hydrocarbon molecule yields a succession of liquids or solids of increasing density, viscosity, and improved solubility for a large number of inorganic and organic materials. Other physical properties such as flammability, specific heat, dielectric constant, and water solubility decrease with increasing halogen content (Marshall 2003).

In addition to use as dry-cleaning and degreasing solvents, many of the halogenated organic solvents have been used in adhesives, pharmaceuticals, and textile processing; as extraction solvents, paint solvents, and coating solvents; and as feedstocks for production of other chemicals. The widespread use and subsequent disposal of chlorinated solvents has led to their being among the most commonly found contaminants at hazardous waste disposal sites. In general, these compounds are considered persistent in the environment, having long half-lives in soil, air, and water in comparison with other, nonhalogenated hydrocarbons. The health effects of these compounds have been studied extensively, as a result of concerns raised about their toxicity and their carcinogenic nature. Due to the large diversity of this group, only a few major ubiquitous substances are discussed further.

4.2.1 Chlorinated Hydrocarbons

Chlorinated derivatives of methane include methyl chloride, methylene chloride, chloroform, carbon tetrachloride, and several chlorofluorohydrocarbons (CFCs). We discuss carbon tetrachloride (CT) as a representative example of this group. CT was originally prepared in 1839 and was one of the first organic chemicals to be produced on a large scale by the end of the nineteenth century and beginning of the twentieth century. CT is the most toxic of the chloromethanes and the most unstable on thermal oxidation (Holbrook 2000).

In the past, the main uses of CT were for dry-cleaning, fabric-spotting, and fire extinguisher fluids; as a grain fumigant; and as a solvent in various chemical processes (DeShon 1979). Until recently, CT was used as a solvent for the recovery of tin in tin-plating waste, for metal degreasing, in the manufacture of semiconductors, as a petrol additive and a refrigerant, as a catalyst in the production of polymers, and as a chemical intermediate in the production of fluorocarbons and some pesticides (HSDB 1995).

Acute inhalation and oral exposure to high levels of CT have been observed primarily to damage the liver (swollen, tender liver; changes in enzyme levels; and jaundice) and kidneys (nephritis, nephrosis, proteinuria) of humans. Depression of the CNS also has been reported. Symptoms of acute exposure in humans include headache, weakness, lethargy, nausea, and vomiting (EPA 2000a). Occasional reports have noted the occurrence of liver cancer in workers exposed to CT by inhalation; however, the data are not sufficient to establish a cause-and-effect relationship. Liver tumors have developed in rats and mice exposed to CT, by experimentally placing the chemical in their stomachs (ATSDR 1994; IARC 1972, 1982). The EPA has classified CT as a Group B2, probable human carcinogen (EPA 2000a).

Chlorinated ethanes and ethylenes comprise ethyl chloride, ethylene dichloride (EDC) (1,2-dichloroethane), vinyl chloride, trichloroethylene (TCE), perchloroethylene (PCE, also called tetrachloroethylene), and several CFCs. Some of the major uses of these compounds are as degreasing agents, dry-cleaning solvents, building blocks for manufacturing of polymers (e.g., PVC, ethyl cellulose), and raw material for the production of tetraethyl lead and CFCs. We discuss EDC, TCE, and PCE as examples of this group.

EDC is used primarily in the production of vinyl chloride monomer (HSDB 2000). It also is an intermediate in the manufacture of trichloroethane and fluorocarbons and used as a solvent. In the past, EDC was used as a gasoline additive and a soil fumigant. The reported toxicological effects on exposure of workers to levels of 10–37 ppm were nausea, vomiting, dizziness, and unspecified blood changes (Brzozowski et al. 1954). In other studies, adverse CNS and liver effects were reported in workers occupationally exposed to concentrations of 16 ppm EDC (Kozik 1957) and less than 25 ppm (Rosenbaum 1947). EDC is reasonably anticipated to be a human carcinogen based on experiments on animals (IARC 1987). When administered by gavage, 1,2-dichloroethane increased the incidence of hepatocellular carcinomas in male mice, mammary gland adenocarcinomas and endometrial stromal neoplasms of the uterus in female mice, and lung adenomas in mice of both sexes. Furthermore, gavage administration of 1,2-dichloroethane increased the incidence of squamous cell carcinomas of the forestomach, subcutaneous fibromas, and hemangiosarcomas in male rats and mammary gland adenocarcinomas in female rats. No adequate data were available to evaluate the carcinogenicity of 1,2-dichloroethane in humans (IARC 1987).

The first documented synthesis of TCE was in 1864, and by the early 1900s, a manufacturing process was initiated, becoming a full industrial process by the 1920s (Mertens 2000). The main use of TCE is metal degreasing (over 90 % of production and consumption). TCE also was used extensively for dry cleaning and, in the past, as an extraction solvent for natural fats and oils for food, cosmetic, and drug production (e.g., extraction of palm, coconut, and soybean oils; decaffeination of coffee; isolation of spice oleoresins) (Doherty 2000a; Linak et al. 1990). Additional applications of TCE are as components in adhesive and paint stripping formulations, as a low-temperature heat-transfer medium, as a nonflammable solvent carrier in industrial paint systems, and as a solvent base for metal phosphatizing systems. TCE is used in the textile industry as a carrier solvent for spotting fluids and as a solvent in waterless preparation dying and finishing operations (Mertens 2000; Doherty 2000a).

TCE is now a common contaminant at hazardous waste sites and many federal facilities in the USA. TCE has been identified in at least 1,500 hazardous waste sites regulated under Superfund or the Resource Conservation and Recovery Act (EPA 2005). TCE can enter surface waters via direct discharges and groundwater through leaching from disposal operations and Superfund sites; the maximum contaminant level for TCE in drinking water is 5 ppb. TCE can be released to indoor air from use of consumer products that contain it, vapor intrusion through underground walls and floors, and volatilization from the water supply.

On acute exposure, TCE is considered toxic, primarily because of its anesthetic effect on the central nervous system. Exposure to high vapor concentrations is likely to cause headache, vertigo, tremors, nausea and vomiting, fatigue, intoxication, unconsciousness, and even death. Ingestion of large amounts of TCE may cause liver damage, kidney malfunction, cardiac arrhythmia, and coma (Mertens 2000; EPA 2000b). TCE is anticipated to be a human carcinogen, based on limited studies on humans and evidence from studies of animals (NTP 2002). Studies have found that occupational exposures to TCE are associated with excess in liver cancer, non-Hodgkin's lymphoma, prostate cancer, and multiple myeloma, with the strongest evidence for the first three cancers (Wartenberg et al. 2000).

PCE was first prepared in 1821, but industrial production of PCE reportedly began in the first decade of the twentieth century (Gerhartz 1986); significant use began only about 100 years after its discovery (Doherty 2000b). The main use of PCE is in the dry-cleaning industry. It is also used as a feedstock for chlorofluorocarbon production, for metal cleaning, as a transformer insulating fluid, in chemical masking formulations, and as a process solvent for desulfurizing coal (Hickman 2000).

Overexposure to tetrachloroethylene by inhalation affects the CNS and the liver. Dizziness, headache, confusion, nausea, and eye and mucous tissue irritation occur during prolonged exposure to vapor concentrations of 200 ppm (Rowe et al. 1952). These effects are intensified and include lack of coordination and drunkenness at concentrations in excess of 600 ppm. At concentrations in excess of 1,000 ppm, anesthetic and respiratory depression effects can cause unconsciousness and death (Hickman 2000).

PCE inhalation may affect the CNS and the liver. At higher concentrations, the effects become more pronounced, and at high concentrations, PCE was used as an anesthetic substance, which also can cause depression, difficulty in speaking and walking, respiratory system damage, unconsciousness, and death (Hickman 2000). The International Agency for Research on Cancer determined that PCE probably is carcinogenic to humans. Results of animal studies, conducted with amounts much higher than those to which most people are exposed, show that tetrachloroethylene can cause liver and kidney damage and liver and kidney cancers, even though the relevance to people is unclear (ATSDR 2006a).

Chlorinated aromatics, including monochlorobenzene (MCB), *o*-dichlorobenzene (*o*-DCB), and *p*-dichlorobenzene (*p*-DCB), are the major chlorinated aromatic species produced on an industrial scale. MCB is used as both a chemical intermediate and a solvent. As an intermediate, it is used to produce chloronitrobenzene, pesticides, and pharmaceutical products. In solvent applications, MCB is used in the manufacture of isocyanates. Its high solvency allows it to be used with many types of resins, adhesives, and coatings. The *o*-DCB is used primarily for organic synthesis, especially in the production of 3,4-dichloroaniline herbicides. Like MCB, it can be used as a solvent, especially in the production of isocyanates. It is also used in motor oil and paint formulations. The *p*-DCB is used as a moth repellent and for the control of mildew and fungi. It also is used for odor control. It is a chemical intermediate for the manufacture of pharmaceuticals and other organic chemicals.

4.2.2 Brominated Flame Retardants

The term *brominated flame retardant* (BFR) incorporates more than 175 different types of substances, which form the largest class of flame retardants; other classes are phosphorus-containing, nitrogen-containing, and inorganic flame retardants (Birnbaum and Sttaskal 2004). The major BFR substances in use today (depicted in Fig. 4.6) are tetrabromobisphenol A (TBBPA), hexabromocyclododecane (HBCD), and mixtures of polybrominated diphenyl ethers (PBDEs) (namely, decabromodiphenyl ether (DBDE), octabromodiphenyl ether (OBDE), and pentabromodiphenyl ether (pentaBDE)).

BFRs have been added to various products (e.g., electrical appliances, building materials, vehicle parts, textiles, furnishings) since the 1960s, in growing rates (15-fold from the mid-1960s to 2003; DePierre 2003). BFRs usually are classified as semivolatile and hydrophobic, but these properties vary due to the large diversity of this group of compounds.

BFRs tend to accumulate in organic-rich media, such as soils and sediments, and lipid-rich biotic tissues and are expected to biomagnify in food chains (DePierre 2003). Two incidents in the 1970s brought attention to the toxic potential of BFR. The first incident was in a farm in Michigan in 1974, where polybrominated biphenyls were mixed accidentally with animal feed. As a result, individuals living on affected farms and consumers of contaminated farm products were exposed to these compounds for months before the mistake was discovered. The outcomes of the contamination were loss of livestock and long-term impact on the health of farm families (Birnbaum and Sttaskal 2004; Dunckel 1975). The second case involved tris(2,3-dibromopropyl)phosphate (tris-BP), which is a mutagen and causes cancer and sterility in animals; it was found to be absorbed from fabric by people (Blum et al. 1978). These two BFRs were phased out, as a consequence (Birnbaum and Sttaskal 2004).

The potential toxicity of these compounds is considered here for the PBDE group and HBCD, which are among the most ubiquitous BFRs to date. PBDEs

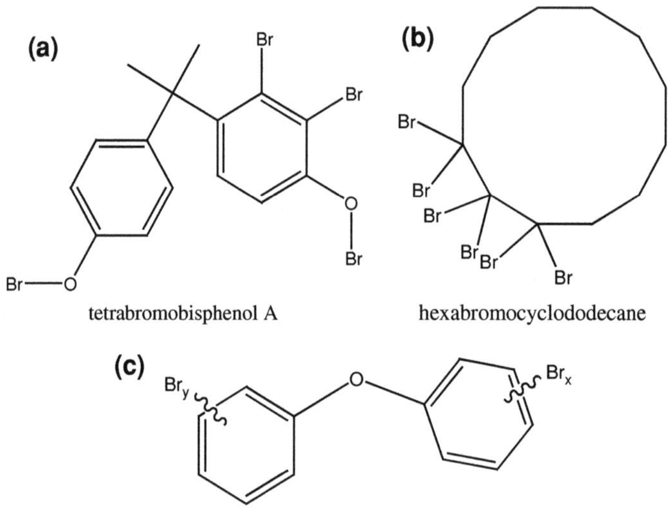

Fig. 4.6 Chemical structures of **a** tetrabromobisphenol A (*TBBPA*), **b** hexabromocyclododecane (*HBCD*), and **c** polybrominated diphenyl ethers (*PBDEs*)

have been associated with a wide variety of toxic effects, affecting (1) thyroid hormone balance, which can cause hypothyroidism and tumors; (2) the central nervous system, which may manifest abnormalities in development dysfunction; (3) hepatic functions, which may cause increased activities of a number of enzymes, including cytochrome P-450, reduction in vitamin A levels, and tumors; (4) disturbances to the estrogen balance; and (5) in utero development, which can cause increased embryo mortality and delayed skeletal formation (e.g., DePierre 2003; EPA 2000c; EU 1997; Hooper and McDonald 2000; Meerts et al. 2000). Exposure to PBDEs disrupts the thyroid hormone both in humans (Bahn et al. 1980) and in animals (Hallgren and Darnerud 1998, 2002; Hallgren et al. 2001). It should be further noted that such effects on the thyroid hormone and its regulatory functions may cause brain developmental abnormalities, especially in children exposed in utero or through breast-feeding. PBDEs were found to bind the Ah receptor in experiments on rats, which in turn regulate several enzymes, including the cytochrome P-450 system (Hooper and McDonald 2000). Due to similarities in their chemical structure and physical properties with other toxic compounds such as PCBs, dioxins, and several pesticides such as DDT, PBDEs are suspected of sharing some toxicological properties as well (DePierre 2003).

HBCD distribution in the environment and its effects on humans were discussed in a review by Covaci et al. (2006). HBCD was reported to be capable of inducing cancer by a nonmutagenic mechanism (Helleday et al. 1999; Yamada-Okabe et al. 2005). Similar to the BPDEs, HBCD is considered capable of disrupting the thyroid hormone system (Yamada-Okabe et al. 2005). Following neonatal exposure experiments in rats, developmental neurotoxic effects can be induced, such as aberrations in spontaneous behavior, learning, and memory function (Eriksson et al. 2002). HBCDs can also alter the normal uptake of neurotransmitters in rat brains (Mariussen and Fonnum 2003).

4.3 Petroleum Hydrocarbons and Fuel Additives

Petroleum hydrocarbons (PHs) constitute a group of compounds characterized by complex mixtures of hydrocarbons. Overall, hundreds to thousands of individual compounds can be found in the different mixtures, although information about their physical and chemical properties is available only for approximately 250 of these compounds, and substantial toxicological data exist for just a small fraction (~ 10 %) of the identified substances (Vorhees et al. 1999). Often these compounds are referred to in the literature as light nonaqueous-phase liquids (LNAPLs), when they exist as a separate phase. In terms of volumes of contaminants released to the environment, the contamination of land surface, partially saturated zone, and groundwater by PHs generally is one of the most serious. This is due to the staggering amounts of PHs used mainly as energy sources for electricity, transportation, and heating around the world. Leaking underground and aboveground storage tanks, improper disposal of petroleum wastes, and accidental spills are major routes of soil and groundwater contamination by petroleum products (Nadim et al. 2000).

Upon release to the environment, the bulk phase migrates downward by gravity. As the NAPL moves through the partially saturated zone, a fraction of the PH is retained by capillary forces as residual globules in the soil pores, thereby depleting the contiguous PH mass until movement ceases. If sufficient PH is released, it will migrate until it encounters a physical barrier (e.g., low-permeability stratum) or is affected by buoyancy forces near the water table. Additionally, PH vapors migrate in the porous matrix creating a larger impact zone. Once the capillary fringe is reached, the PH may move laterally as a continuous, free-phase layer along the upper boundary of the water-saturated zone, due to gravity and capillary forces. On contact with water in the saturated or partially saturated zone, dissolution of compounds from the PH mixture begins. A schematic description of PH distribution patterns in the subsurface is given in Fig. 4.7.

There are different approaches to estimating the toxicity of various PHs. One method is to examine the known individual compounds in each PH fraction, based on the data collected for a limited number of compounds and assuming that the known materials are representative of the entire mixture. A second method is to divide the mixture into several fractions that contain substances with similar chemical and physical properties, which therefore are considered to have comparable toxicity. A third approach is to consider the entire mixture. The actual content of each mixture depends mainly on the origin of the PH and the distillate fractions.

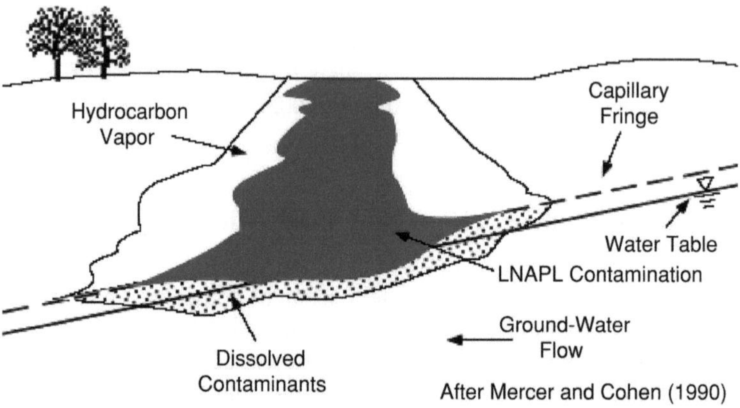

Fig. 4.7 Simplified conceptual model for light nonaqueous-phase liquid (*LNAPL*) release and migration. Reprinted from Mercer and Cohen (1990). Copyright 1990 with permission from Elsevier

The direct exposure pathways of humans to PH following a leak are described in Fig. 4.8. Once released to the surface and subsurface environment, PHs can reach humans directly as vapors, solutes, or adsorbed on particles.

Here, we briefly discuss several individual compounds that are common constituents of different PH mixtures, the main groups being (1) small aromatic compounds, mostly benzene derivatives (e.g., benzene, toluene, ethylbenzene, and xylenes), which are considered slightly soluble (150–1,800 mg/L); (2) branched and linear aliphatics (e.g., *n*-dodecane and *n*-heptane), which are characterized by relatively low water solubility; and (3) polar hydrocarbons and petroleum additives (e.g., methyl tertiary-butyl ether (MTBE) and alcohols), which are highly soluble. The weight percentage of three selected compounds in various commercial petroleum products is given in Table 4.1.

Benzene is important from an environmental point of view, as it is an important component of various petroleum products. Its weight fraction ranges from practically zero for the heavy distillates to 3.5 % for gasoline, as shown in Table 4.1. Its solubility is 1,780 mg/L, and it is very volatile (vapor pressure 100 torr at 26.1 °C). The acute (short-term) effects of benzene toxicity include dizziness, headache, nausea, vomiting, and drowsiness; with higher levels of benzene toxicity come the threat of convulsions, coma, and death. The long-term or chronic results of benzene toxicity include reproductive damage, chromosomal aberrations, immunodeficiencies, and several types of leukemia (ATSDR 2006b).

Xylene belongs to the group of small aromatic compounds with relatively higher solubility, like benzene. Exposure to toluene causes CNS depression (Faust 1994). Short-term exposure effects include fatigue, confusion, lack of coordination, impaired reaction time, perception, and motor control and function (NTP 1990). Exposure to high concentrations results in narcosis and death (WHO 1985). Prolonged abuse of toluene or solvent mixtures containing toluene has led to permanent

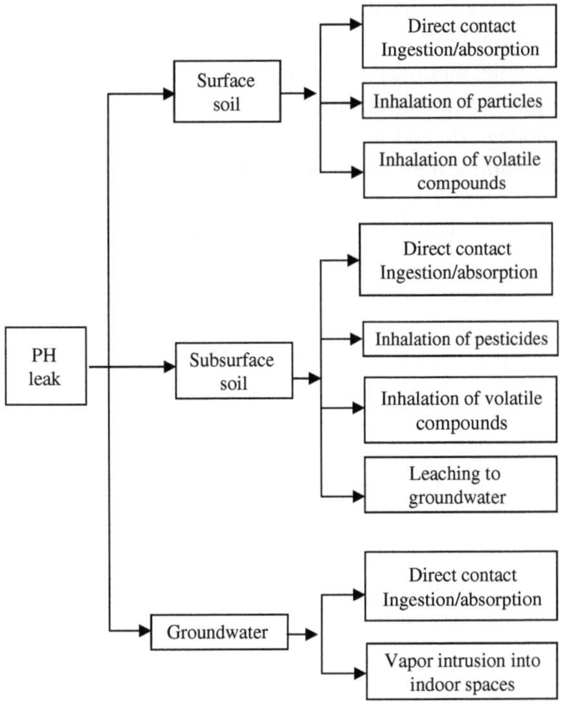

Fig. 4.8 Pathways of direct human exposure to petroleum hydrocarbons following release to the surface and subsurface environment (Vorhees et al. 1999)

Table 4.1 Fraction range (in wt.%) of three hydrocarbons in selected commercial petroleum products, based on data from Gustafson et al. (1997)

	Crude oil	Diesel	Fuel oil #2	Gasoline	JP-4
Benzene	0.04–0.4	0.003–0.10	<0.125	0.12–3.50	0.5
Toluene	0.09–2.5	0.007–0.70	0.025–0.110	2.73–21.80	1.33
n-Octane	0.9–1.9	0.1	0.1	0.36–1.43	3.8

CNS effects. Hepatomegaly and impaired liver and kidney function have been reported in some humans chronically exposed to toluene (Greenburg et al. 1942).

Hexane represents the aliphatic group. Most of the aliphatic compounds are branched, but the same trend of low solubility that decreases with increasing C number is typical of all substances in this group. Due to their very low solubility, these compounds hardly partition to water and migrate mainly as vapors, as a separate phase, or adsorbed on particulate matter. In very high concentrations (thousands of ppm), hexane is a lethal narcotic to humans (HCN 2005). High-level exposure affects several enzyme functions, which lead to increased liver weight. No data on octane toxicity are available, and it is considered nontoxic.

MTBE is an octane enhancer; that is, it promotes more complete burning of gasoline and thereby reduces carbon monoxide and ozone levels. MTBE is very soluble, and once released, it moves through soil and into water more rapidly than other chemical compounds present in gasoline. In groundwater, it is slow to biodegrade and more persistent than other gasoline-related compounds. Exposure to vapors may result in several health effects including dizziness, nausea, sore eyes, and respiratory irritation (McCarthy and Tiemann 2001). The US EPA concluded that MTBE poses a potential for carcinogenicity to humans at high doses; however, because of uncertainties and limitations in available data, the EPA has been unable to reliably estimate the risk at low exposure levels (EPA 1997). Based on this, several wells contaminated with MTBE have been closed, but because MTBE is an additive, its effect should be considered against other alternatives, which usually are more problematic.

In general, because of the combination of solubility and toxicity characteristics, aromatic compounds are the major group of PH contaminants in groundwater. However, due to the large amounts of PH released to the environment and lack of information, much more research is needed to understand the behavior and toxicity of these complex mixtures and their potential effect on the subsurface environment.

4.4 Pharmaceuticals and Personal Care Products

A large class of chemicals gaining attention in recent years comprises pharmaceutical, veterinary, and illicit ("recreational") drugs and the ingredients in cosmetics, food supplements, and other personal care products, together with their respective metabolites and transformation products; they are collectively referred to as *pharmaceuticals and personal care products* (PPCPs). PPCPs are used in large amounts throughout the world, and some studies demonstrated their occurrence in aquatic environments in Austria, Brazil, Canada, Croatia, Denmark, England, Germany, Greece, Italy, Spain, Switzerland, the Netherlands, and the USA (Heberer et al. 2002; Daughton and Ternes 1999; Erickson 2002).

Most PPCPs are disposed of or discharged into the environment on a continuous basis via domestic and industrial sewage systems and wet weather runoff. In many instances, untreated sewage is discharged into receiving waters (e.g., flood overload events, domestic "straight piping," or sewage waters lacking municipal treatment). A scheme of possible pathways for the occurrence of PPCP residues in aquatic environments is depicted in Fig. 4.9. The bioactive ingredients are first subjected to metabolism by the dosed user; the excreted metabolites and unaltered parent compounds then can be subjected to further transformations in sewage treatment facilities. The literature shows, however, that many of these compounds survive biodegradation, eventually being discharged into receiving waters. Many of these PPCPs and their metabolites are ubiquitous and display persistence in, and bioconcentration from, surface waters. Additionally, by way of continuous

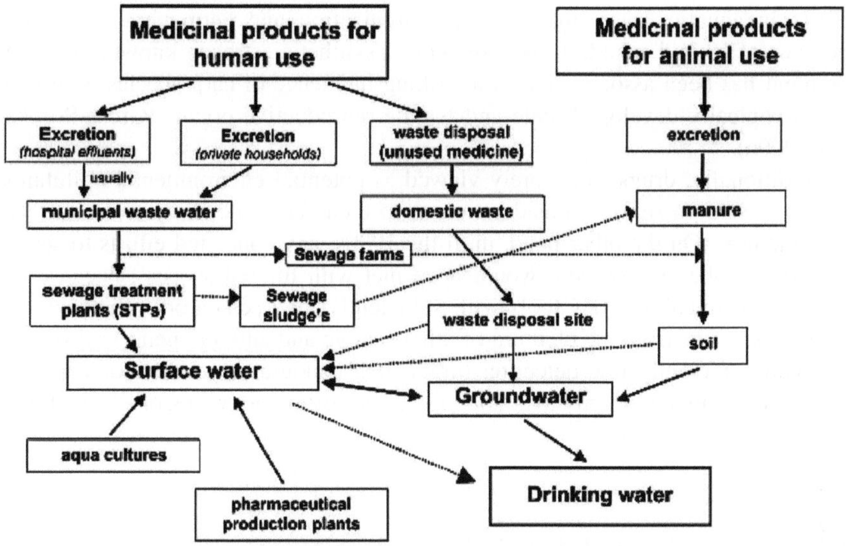

Fig. 4.9 Scheme showing possible sources and pathways for the occurrence of pharmaceutical residues in the aquatic environment. Reprinted from Heberer T (2002). Copyright 2002 with permission from Elsevier

infusion into the aquatic environment, those PPCPs that might have low persistence can display the same exposure potential as truly persistent pollutants, because their transformation and removal rates can be compensated by their replacement rates. While the concentration of individual drugs in the aquatic environment often is low (subparts per billion or subnanomolar, often referred to as *micropollutants*), the presence of numerous drugs sharing a specific mode of action could lead to significant effects through additive exposures.

Many PPCPs are used on a daily basis for very long periods, sometimes a good portion of the user's lifetime. Although drugs are usually designed with a specific mode of action in mind, they also can have numerous effects on nontarget, or as yet unknown receptors, and possibly cause side effects in the target organism. Furthermore and of equal importance, nontarget organisms can have receptors, or receptor tissue distributions, that do not exist in the target organisms, and therefore, unexpected effects can result from unintentional exposure. Often PPCPs are released to the environment in low concentrations for long periods, which in turn may cause genetic selection of the more resistant pathogens that can reduce the effectiveness of current medications.

Some studies (Migliore et al. 1995) demonstrated that drugs alter the normal post-germinative development of plants and the growth of roots, hypocotyls, and leaves. This effect becomes more important with time, so it is more evident in structures produced later. In other cases, drugs such as natural and synthetic estrogens that reach the environment have been shown to produce deleterious effects in aquatic organisms, such as feminization and hermaphroditism. The presence of ethinyl estradiol, the most potent synthetic estrogen known, in a river sediment has been associated with a striking incidence of carp species with both macroscopically developed male and female reproductive organs (Gross-Sorokin et al. 2006).

Traditionally, drugs were rarely viewed as potential environmental pollutants; there was seldom serious consideration as to their fates once they were excreted from the user. On the other hand, until the 1990s, any concerted efforts to search for drugs in the environment would have met with limited success, because the requisite chemical analysis tools with sufficiently high separatory efficiencies to resolve the drugs from the plethora of other (native and anthropogenic) substances and with sufficiently low detection limits (i.e., nanograms per liter or parts per trillion) were not commonly available. Examples of major groups of PPCPs found in the environment follow.

4.4.1 Analgesics and Anti-inflammatory Drugs

This group refers mainly to drugs used primarily as painkillers, although they may also have anti-inflammatory and antipyretic properties. Drugs in this group are sold in large quantities by prescription and even larger amounts without prescription, as so-called over-the-counter (OTC) drugs. Acetaminophen (paracetamol) and acetylsalicylic acid (ASA, aspirin) are the two most popular painkillers sold as OTC drugs. In Germany, the total quantities of ASA sold per year have been estimated at >500 tons (Heberer et al. 2002; Ternes 2001).

Other examples of analgesics or their metabolites that have been found in the environment include 4-aminoantipyrine, aminophenazone, codeine, fenoprofen, hydrocodone, indomethacin, ketoprofen, mefenamic acid, naproxen, propyphenazone, diclofenac, ibuprofen, phenazone, gentisic acid and *N*-methyl phenacetin. Many studies have identified these compounds in various locations around the world and in different water resources (e.g., Heberer et al. 1997, 2001a, b; Ternes 2001; Stumpf et al. 1999; Ahrer et al. 2001; Sedlak and Pinkston 2001; Holm et al. 1995; Ahel and Jelicic 2001; Sacher et al. 2001).

4.4.2 Hormones

Synthetic steroids, especially estrogenic drugs, are used extensively in estrogen-replacement therapy and oral contraceptives, in veterinary medicine for growth enhancement, and in athletic performance enhancement. In general, large portions of these endocrine disruptors, used by humans as well as for stimulating beef, poultry, and fish production, are excreted unchanged in feces and urine. Synthetic steroids have been found in the environment in very low concentrations (usually less than 5 ng/L). Furthermore, the physicochemical properties (lipophilicity) of such hormones are expected to allow removal via sorption processes in sewage treatment facilities or adsorption to subsurface soil. However, even the low concentrations found in different water bodies (e.g., sewage, surface water, and groundwater) (Daughton and Ternes 1999; Heberer et al. 2002) may pose a threat for ecosystems. For example, exposure of wild male fish to only 0.1 ng/L of xenoestrogens may provoke feminization in some species.

4.4.3 Antibacterial Drugs

Antibacterial drugs (i.e., antibiotic and bacteriostatic drugs) have received considerable attention because of their heavy use and their potential hazardous effect on ecosystems. Antibiotics used to treat infections are an invaluable tool, and their introduction has revolutionized the treatment of infectious diseases. Because of their widespread use, it is not surprising that antibiotics have been found in liquid waste at animal feedlots and spread into many surface water and groundwater supplies. In general, large portions of antibiotics used by humans, as well as for beef and poultry production, are excreted unchanged in feces and urine. With increasingly wide use of antibiotics, resistant strains of bacteria are replacing antibiotic-susceptible bacteria. Furthermore, resistant bacteria in one environment may not be confined to that specific environment and can be carried over distances of thousands of kilometers by wind, water, animals, food, or people. And, most important, antibiotic-resistant organisms that develop in animals, fruits, or vegetables can be passed to humans through the food chain and environment. All these factors have had the effect of changing the balance between antibiotic-susceptible and antibiotic-resistant bacteria in ecosystems, both locally and globally.

Macrolide antibiotics (clarithromycin, dehydroerythromycin (a metabolite of erythromycin), roxithromycin, lincomycin, sulfonamides (sulfamethoxazole, sulfadimethoxine, sulfamethazine, sulfathiazole), fluoroquinolones (ciprofloxacin, norfloxacin, enrofloxacin), chloramphenicol, tylosin, and trimethoprim) have been found up to low μg/L levels in sewage and surface water samples. Sacher et al. (2001) reported the occurrence of sulfamethoxazole (up to 410 ng/L) and dehydroerythromycin (up to 49 ng/L) in groundwater samples in Baden-Wurttemberg, Germany. Sulfamethoxazole and sulfamethazine have also been detected at low concentrations in several groundwater samples in the USA and Germany (e.g. Hartig et al. 1999). Holm et al. (1995) found residues of different sulfonamides at high concentrations in groundwater samples collected down gradient of a landfill in Grindsted, Denmark.

4.4.4 Antiepileptic Drugs

Treatment of seizures by antiepileptic drugs began in 1850, and since then, a variety of medications have been applied. The main groups of antiepileptic drugs include sodium channel blockers, calcium current inhibitors, gamma-aminobutyric acid (GABA) enhancers, glutamate blockers, carbonic anhydrase inhibitors, hormones, and drugs with unknown mechanisms of action. One of the widespread antiepileptic drugs, carbamazepine, has been detected frequently in municipal sewage and surface water samples (Heberer et al. 2001a; Ahrer et al. 2001). Various field studies have shown that carbamazepine (Heberer et al. 2001b) and primidone (Heberer et al. 2001b) are not attenuated during riverbank infiltration. Both compounds have been detected in shallow wells and water supply wells of a transect built to study the behavior of drugs during riverbank filtration (Heberer et al. 2001b). This also explains why carbamazepine has been detected in a number of groundwater samples at a maximum concentration up to 1.1 μg/L (Seiler et al. 1999; Sacher et al. 2001; Ternes 2001) and in drinking water at a concentration of 30 ng/L (Ternes 2001).

4.4.5 Beta-Blockers

Beta-blockers are medications that reduce the workload of the heart and lower blood pressure. They are commonly prescribed to relieve angina (a type of chest pain, pressure, or discomfort) or treat heart failure. They also are prescribed for people who have high blood pressure (hypertension). Several beta-blockers (metoprolol, propranolol, betaxolol, bisoprolol, and nadolol) have been detected in municipal sewage effluents up to the μg/L level (Ternes 1998) and in groundwater samples (Sacher et al. 2001).

References

Ahel M, Jelicic I (2001) Phenazone analgesics in soil and groundwater below a municipal solid waste landfill. In: Daughton CG, Jones-Lepp T (eds) Pharmaceuticals and personal care products in the environment: scientific and regulatory issues. American Chemical Society, Washington DC, pp 100–115

Ahrer W, Scherwenk E, Buchberger W (2001) Determination of drug residues in water by the combination of liquid chromatography or capillary electrophoresis with electrospray mass spectrometry. J Chrom A 910:69–78

ATSDR (1994) Agency for toxic substances and disease registry. Toxicological profile for carbon tetrachloride (update). Public Health Service, U.S. Department of Health and Human Services, Atlanta, GA

ATSDR (2006a) Agency for toxic substances and disease registry. Evaluation of tetrachloroethylene vapor intrusion into buildings located above a contaminated aquifer. Available at http://www.atsdr.cdc.gov/HAC/pha/SchlageLockCompany/SchlageLockCompanyHC113006.pdf

ATSDR (2006b) Agency for toxic substances and disease registry. Medical management guidelines for benzene—agency for toxic substances and disease registry. Available at http://www.atsdr.cdc.gov/mhmi/mmg3.html#bookmark02

Bahn AK, Mills JL, Synder PJ, Gann PH, Houten L, Bialik O, Hollmann L, Utiger RD (1980) Hypothyroidism in workers exposed to polybrominated biphenyls. N Engl J Med 302:31–33

Barbash JE (2003) The geochemistry of pesticides. In: Sherwood Lollar B (ed) Treatise on geochemistry, vol 9. Elsevier, Oxford pp 541–577

Birnbaum LS, Sttaskal DF (2004) Brominated flame retardants: cause for concern? Environ Health Perspect 112:9–17

Blondell J (1997) Epidemiology of pesticide poisonings in the United States, with special reference to occupational cases. Occ Med State Art Rev 12:209–220

Blum A, Gold MD, Ames BN, Jones FR, Hett EA, Dougherty RC, Horning EC, Dzidic I, Carroll DI, Stillwell RN, Thenot JP (1978) Children absorb tris-BP flame retardant from sleepwear: urine contains the mutagenic metabolite, 2,3-dibromopropanol. Science 201:1020–1023

Bradlow HL, Davis DL, Lin G, Sepkovic D, Tiwari R (1995) Effects of pesticides on the ratio of 16 alpha/2-hydroxyestrone: a biologic marker of breast cancer risk. Environ Health Perspect 103:147–150

Briggs SA (1992) Basic guide to pesticides: their characteristics and hazards. CRC Press, Boca Raton

Brzozowski J, Czajka J, Dutkiewicz T, Kesy I, Wojcik J (1954) Work hygiene and the health condition of workers occupied in combating the Leptinotarsa decemlineata with HCH and dichloroethane. Med Pr 5:89–98

Buckley JD, Buckley CM, Ruccione K, Sather HN, Waskerwitz MJ, Woods WG, Robison LL (1994) Epidemiological characteristics of childhood acute lymphocytic leukemia. Analysis by immunophenotype. The childrens cancer group. Leukemia 8:856–864

Butterfield PG, Valanis BG, Spencer PS, Lindeman CA, Nutt JG (1993) Environmental antecedents of young-onset Parkinson's disease. Neurology 43:1150–1158

C&EN (2000) Atrazine is not likely human carcinogen. Government Concentrates C&EN, vol 78(51)

C&EN (2002) Call for investigation of Syngenta. Government Concentrates C&EN, vol 80(23)

Chanda SM, Pope CN (1996) Neurochemical and neurobehavioral effects of repeated gestational exposure to chlorpyrifos in maternal and developing rats. Pharmacol Biochem Behav 53:771–776

Cooper RL, Chadwick R, Rehnberg G, Goldman J (1989) Effect of lindane on hormonal control of reproductive function in the female rat. Tox Appl Pharm 99:384–394

Cooper RL, Stoker TE, Goldman JM, Parrish MB, Tyrey L (1996) Effect of atrazine on ovarian function in the rat. Reprod Toxicol 10:257–264

Covaci AN, Gerecke ASC, Low RJ, Voorspoels S, Kohler M, Heeb N, Leslie H, Allchin C, Deboer J (2006) Hexabromocyclododecanes (HBCDs) in the environment and humans: a review. Environ Sci Technol 40:3679–3688

Cummings AM (1997) Methoxychlor as a model for environmental estrogens. Crit Rev Toxicol 27:367–379

Danzo BJ (1997) Environmental xenobiotics may disrupt normal endocrine function by interfering with the binding of physiological ligands to steroid receptors and binding proteins. Environ Health Perspect 105:294–301

Daughton CG, Ternes TA (1999) Pharmaceuticals and personal care products in the environment: agents of subtle change. Environ Health Perspect 107:907–942

DePierre JW (2003) Mammalian toxicity of organic compounds of bromine and iodine. Handb Environ Chem 3:205–251. doi:10.1007/b11992

DeShon ND (1979) Carbon tetrachloride. In: Grayson M, Eckroth D (eds) Kirk-Othmer encyclopedia of chemical technology, 3rd edn, vol 5. Wiley, New York, pp 704–714 [as cited by Santodonato J. 1985. Monograph on human exposure to chemicals in the workplace: Carbon tetrachloride; PB86-143377; SRC-TR-84-1123]

Doherty RE (2000a) A history of the production and use of carbon tetrachloride, tetrachloroethylene, trichloroethylene and 1,1,1-trichloroethane in the United States: part 2. Trichloroethylene and 1,1,1-trichloroethane. J Environ Forensics 1:83–93

Doherty RE (2000b) A history of the production and use of carbon tetrachloride, tetrachloroethylene, trichloroethylene and 1,1,1-trichloroethane in the United States: part 1. Historical background; carbon tetrachloride and tetrachloroethylene. J Environ Forensics 1:69–81

Dunckel AE (1975) An updating on the polybrominated biphenyl disaster in Michigan. J Am Vet Med Assoc 167:838–841

EPA (1997) Environmental protection agency. Drinking water advisory: consumer acceptability advice and health effects analysis on methyl tertiary-butyl ether (MTBE). Available at EPA-822-F-97-009, pp 1–2. http://www.epa.gov/OST/drinking/mtbe.html

EPA (2000a) Environmental protection agency. Carbon tetrachloride—hazard summary. Available at http://www.epa.gov/ttn/atw/hlthef/carbonte.html

EPA (2000b) Trichloroethylene hazard summary. Available at http://www.epa.gov/ttn/atw/hlthef/tri-ethy.html

EPA (2000c) Environmental protection agency. Integrated risk information system (IRIS) data for brominated diphenyl ethers. Available at http://www.epa.gov/iris

EPA (2005) Environmental protection agency. Trichloroethylene (TCE) health risk assessment: overview. Available at http://oaspub.epa.gov/eims/xmlreport.display?deid=119268&z_chk=31804

EPA (2006) Environmental protection agency. About pesticides. Available at http://www.epa.gov/pesticides/

Erickson BE (2002) Analyzing the ignored environmental contaminants. Environ Sci Tech 36:140A–145A

Eriksson P (1996) Developmental neurotoxicology in the neonate-effects of pesticides and polychlorinated organic substances. Arch Toxicol Suppl 18:81–88

Eriksson P, Viberg H, Fischer C, Wallin M, Fredriksson A (2002) A comparison of developmental neurotoxic effects of hexabromo cyclododecane, 2,2′,4,4′,5,5′-hexabromodiphenylether (PBDE 153) and 2,2′,4,4′,5,5′-hexachlorobiphenyl (PCB 153). Organohalogen Comp 57:389–392

EU (1997) European union. Draft risk assessment biphenylether, octabromo derivative, human health

Faust RA (1994) Toxicity summary for toluene, oak ridge reservation environmental restoration program. Available at http://rais.ornl.gov/tox/profiles/toluene_f_V1.shtml

Fear NT, Roman E, Reeves G, Pannett B (1998) Childhood cancer and paternal employment in agriculture: the role of pesticides. Br J Cancer 77:825–829

Gerhartz W (ed) (1986) Ullman's encyclopedia of industrial chemistry, 5th edn. Weinheim, New York

Gillette LJ Jr, Gross TS, Masson GR, Matter JM, Percival HF, Woodward AR (1994) Developmental abnormalities of the gonad and abnormal sex hormone concentrations in juvenile alligators from contaminated and control lakes in Florida. Environ Health Perspect 102:680–688

Giulivi C, Lavagno CC, Lucesoli F, Bermudez MJN, Boveris A (1995) Lung damage in paraquat poisoning and hyperbaric oxygen exposure: superoxide-mediated inhibition of phospholipase A2. Free Radic Biol Med 18:203–213

Gorell JM, Johnson CC, Rybicki BA, Peterson EL, Richardson RJ (1998) The risk of Parkinson's disease with exposure to pesticides, farming, well water, and rural living. Neurology 50:1346–1350

Greenburg L, Mayers MR, Heimann H, Moskowitz S (1942) The effects of exposure to toluene in industry. J Am Med Assoc 118:573–578

Gross-Sorokin MY, Roast SD, Brighty GC (2006) Assessment of feminization of male fish in English rivers by the Environment Agency of England and Wales. Environ Health Perspect 147–151

Gustafson JB, Tell JG, Orem D (1997) Selection of representative TPH fractions based on fate and transport considerations total petroleum hydrocarbon criteria. Working group series, vol 3. Amherst Scientific Publishers, Amherst

Hallgren S, Darnerud PO (1998) Effects of polybrominated diphenyl ethers (PBDEs), polychlorinated biphenyls (PCBs) and chlorinated paraffins (CPs) on thyroid hormone levels and enzyme activities in rats. Organohalogen Comp 35:391–394

Hallgren S, Darnerud PO (2002) Polybrominated diphenyl ethers (PBDEs), polychlorinated biphenyls (PCBs) and chlorinated paraffins (CPs) in rats—testing interactions and mechanisms for thyroid hormone effects. Toxicology 177:227–243

Hallgren S, Sinjari T, Hakansson H, Darnerud PO (2001) Effects of polybrominated diphenyl ethers (PBDEs) and polychlorinated biphenyls (PCBs) on thyroid hormone and vitamin A levels in rats and mice. Arch Toxicol 75:200–208

Harsany L, Nemeth A, Lang A (1987) Paraquat (Gramoxone) poisoning in south-west Hungary, 1984. Am J Forensic Med Pathol 8:131–134

Hartig C, Storm T, Jekel M (1999) Detection and identification of sulphonamide drugs in municipal wastewater by liquid chromatography coupled with electrospray ionisation tandem mass spectrometry. J Chromatogr A 854:163–173

Hayes HM, Tarone RE, Cantor KP (1995) On the association between canine malignant lymphoma and opportunity for exposure to 2,4-dichlorophenoxyacetic acid. Environ Res 70:119–125

HCN (2005) Health council of the Netherlands: committee on updating of occupational exposure limits. Octane; health-based reassessment of administrative occupational exposure limits. Health Council of the Netherlands, The Hague. Available at 2000/15OSH/156. http://www.gr.nl/pdf.php?ID=1315&p=1

Heberer T (2002) Occurrence, fate, and removal of pharmaceutical residues in the aquatic environment: a review of recent research data. Toxicol Lett 131:5–17

Heberer T, Dünnbier U, Reilich C, Stan HJ (1997) Detection of drugs and drug metabolites in groundwater samples of a drinking water treatment plant. Fresen Environ Bull 6:438–443

Heberer T, Fuhrmann B, Schmidt-Baumler K, Tsipi D, Koutsouba V, Hiskia A (2001a) Occurrence of pharmaceutical residues in sewage, river, ground and drinking water in Greece and Germany. In: Daughton CG, Jones-Lepp T (eds), Pharmaceuticals and personal care products in the environment: scientific and regulatory issues. Symposium series 791, American Chemical Society, Washington DC, pp 70–83

Heberer T, Verstraeten IM, Meyer MT, Mechlinski A, Reddersen K (2001b) Occurrence and fate of pharmaceuticals during bank filtration—preliminary results from investigations in Germany and the United States. Water Resour Update 120:4–17

Heberer T, Reddersen K, Mechlinski A (2002) From municipal sewage to drinking water: fate and removal of pharmaceutical residues in the aquatic environment in urban areas. Water Sci Technol 46:81–88

Helleday T, Tuominen KL, Bergman A, Jenssen D (1999) Brominated flame retardants induce intragenic recombination in mammalian cells. Mutat Res 439:137–147

Hickman JC (2000) Tetrachloroethylene. In: Kirk-Othmer encyclopedia of chemical technology. Available at http://www.mrw.interscience.wiley.com/emrw/9780471238966/kirk/article/tetrhick.a01/current/pdf

Holbrook MT (2000) Carbon tetrachloride. In: Kirk-Othmer encyclopedia of chemical technology. Wiley, New York. Available at http://www.mrw.interscience.wiley.com/emrw/9780471238966/kirk/article/carbholb.a01/current/pdf

Holm JV, Rugge K, Bjerg PL, Christensen TH (1995) Occurrence and distribution of pharmaceutical organic compounds in the groundwater downgradient of a landfill (Grindsted, Denmark). Environ Sci Technol 29:1415–1420

Honore P, Hantson P, Fauville JP, Peeters A, Mahieu P (1994) Paraquat poisoning: state of the art. Acta Clin Belg 49:220–228

Hooper K, McDonald TA (2000) The PBDEs: an emerging environmental challenge and another reason for breast-milk monitoring programs. Environ Health Perspect 108:387–392

HSDB (1995) Hazardous substances data bank. National Library of Medicine, Bethesda, MD (CD-ROM version). Micromedex, Denver, CO

HSDB (2000) Hazardous substances data bank. National Library of Medicine, Bethesda, MD. Available through toxicology data network at http://toxnet.nlm.nih.gov

IARC (1972) International agency for research on cancer. IARC monographs on the evaluation of the carcinogenic risk of chemicals to man, vol 1. World Health Organization, Lyon

IARC (1982) International agency for research on cancer. IARC monographs on the evaluation of the carcinogenic risk of chemicals to humans: chemicals, industrial processes and industries associated with cancer in humans, supplement 4. World Health Organization, Lyon

IARC (1987) International agency for research on cancer. Overall evaluations of carcinogenicity: an updating of IARC monographs volumes 1 to 42. Monographs on the evaluation of carcinogenic risk to humans, supplement 7

IARC (1991) International agency for research on cancer. Occupational exposures in insecticide application, and some pesticides. Monographs on the evaluation of carcinogenic risk to humans, vol 53. IARC, Lyon

IARC (2001) International agency for research on cancer. Some thyrotropic agents. Monographs on the evaluation of carcinogenic risk to humans, vol 79. IARC, Lyon

Joy RM (1985) The effects of neurotoxicants on kindling and kindled seizures. Fund Appl Toxicol 5:41–65

Kiely T, Donaldson D, Grube A (2004) Pesticides industry sales and usage, 2000 and 2001. Market estimates biological and economic analysis division office of pesticide programs office of prevention, pesticides, and toxic substances. U.S. Environmental Protection Agency, Washington DC

Kozik I (1957) Problems of occupational hygiene in the use of dichloroethane in the aviation industry. Gig Tr Prof Zabol 1:31–38

Kristensen P, Andersen A, Irgens LM, Bye AS, Sundheim L (1996) Cancer in offspring of parents engaged in agricultural activities in Norway: incidence and risk factors in the farm environment. Int J Cancer 65:39–50

Letterman RD (ed) (1999) Water quality and treatment: a handbook of community water supplies, 5th edn. McGraw-Hill, New York Amer. Water Works Assoc

Lin S, Marshall EG, Davidson GK (1994) Potential parental exposure to pesticides and limb reduction defects. Scand J Work Environ Health 20:166–179

Linak E, Lutz HJ, Nakamura E (1990) C_2 chlorinated solvents. In: Linak E, Lutz HJ, Nakamura E (eds) C_2 chlorinated solvents, chemical economics handbook. Stanford Research Institute, Menlo Park, pp 632.30000a–632.3001z

Mariussen E, Fonnum F (2003) The effect of brominated flame retardants on neurotransmitter uptake into rat brain synaptosomes and vesicles. Neurochem Int 43:533–542

Marshall KA (2003) Chlorocarbons and chlorohydrocarbons, survey. In: Kirk-Othmer encyclopedia of chemical technology, vol 6. Wiley, New York, pp 226–253. Available at http://www.mrw.interscience.wiley.com/emrw/9780471238966/search/firstpage

Matthews GA (2006) Pesticides: health, safety and the environment. Blackwell Publishing, Oxford

McCarthy JE, Tiemann M (2001) MTBE in gasoline: clean air and drinking water issues CRS report to the congress. Available at http://www.ncseonline.org/NLE/CRSreports/air/air-26.cfm#_1_4

Meerts IA, van Zanden JJ, Luijks EA, van Leeuwen-Bol I, Marsh G, Jakobsson E, Bergman A, Brouwer A (2000) Potent competitive interaction of some brominated flame retardants and related compounds with human transthyretin in vitro. Toxicol Sci 56:95–104

Mercer JW, Cohen RM (1990) A review of immiscible fluids in the subsurface: properties, models, characterization, and remediation. J Contam Hydrol 6:107–163

Mertens JA (2000) Trichloroethylene. In: Kirk-Othmer encyclopedia of chemical technology. Wiley, New York. Available at http://www.mrw.interscience.wiley.com/emrw/9780471238966/kirk/article/tricmert.a01/current/pdf

Migliore L, Brambilla G, Cozzolino S, Gaudio L (1995) Effect on plants of sulphadimethoxine used in intensive farming (*Panicum miliaceum, Pisum sativum* and *Zea mays*). Agric Ecosyst Environ 52:103–110

Milne WGA (1995) Handbook of pesticides. CRC Press, Boca Raton

Montgomery JH (1993) Agrochemicals desk reference: environmental data. Lewis Publishers, Chelsea

Nadim F, Hoag GE, Liu SL, Carley RJ, Zack P (2000) Detection and remediation of soil and aquifer systems contaminated with petroleum products: an overview. J Pet Sci Eng 26:169–178

Newman A (1995) Atrazine found to cause chromosomal breaks. Environ Sci Technol 29:450A

NTP (1990) National toxicology program. Toxicology and carcinogenesis studies of toluene (CAS No. 108-88-3) in F344/N rats and $B6C3F_1$ mice (inhalation studies). Technical report series no. 371. U.S. Department of Health and Human Services, Public Health Service, National Institutes of Health, Research Triangle Park, NC

NTP (2002) National toxicology program. Tenth report on carcinogens. Public Health Service, U.S. Department of Health and Human Services, Research Triangle Park, NC. Available at http://ehp.niehs.nih.gov/roc/tenth/intro.pdf

Olson KR (1994) Paraquat and diquat. In: Olson KR et al (eds) Poisoning and drug overdose, 2nd edn. Appelton and Lange, Norwalk CT, pp 245–246

Ordish G (2007) History of agriculture: beginnings of pest control. In: Encyclopedia Britannica. Available at http://www.britannica.com/eb/article-10711/history-of-agriculture

Pastore LM, Hertz-Picciotto I, Beaumont JJ (1997) Risk of stillbirth from occupational and residential exposures. Occup Environ Med 54:511–518

Pogoda JM, Preston-Martin S (1997) Household pesticides and risk of pediatric brain tumors. Environ Health Perspect 105:1214–1220

Pond SM (1990) Manifestations and management of paraquat poisoning. Med J 2:256–259

Purdue MP, Hoppin JA, Blairl A, Dosemeci M, Alavanja MCR (2006) Occupational exposure to organochlorine insecticides and cancer incidence in the agricultural health study. Int J Cancer 120:642–649

Reeder AL, Foley GL, Nichols DK, Hansen LG, Wikoff B, Faeh S, Eisold J, Wheeler MB, Warner R, Murphy JE, Beasley VR (1998) Forms and prevalence of intersexuality and effects on environmental contaminants on sexuality of cricket frogs. Environ Health Perspect 106:261–266

Reigart JR, Roberts JR (1999) Recognition and management of pesticide poisonings, 5th edn. Pesticide Programs, U.S. Environmental Protection Agency. Available at http://www.epa.gov/pesticides/safety/healthcare

Renner R (2002) Is sludge safe? Environ Sci Technol 36:46A

Restrepo M, Munoz N, Day N, Parra JE, Hernandez C, Blettner M, Giraldo A (1990) Birth defects among children born to a population occupationally exposed to pesticides in Colombia. Scand J Work Environ Health 16:239–246

Rosenbaum ND (1947) Ethylene dichloride as an industrial poison. Gig Sanit 12:17–21

Rosenstock L, Keifer M, Daniell W, Mcconnell R, Claypoole K (1991) Chronic central nervous system effects of acute organophosphate intoxication. Lancet 338:223–227

Rowe VK, McCollister DD, Spencer HC, Adams EM, Irish DD (1952) Vapor toxicity of tetrachloroethylene for laboratory animals and human subjects. AMA Arch Ind Hyg Occup Med 5:566–579

Sacher F, Lange FTh, Brauch H-J, Blankenhorn I (2001) Pharmaceuticals in groundwaters. Analytical methods and results of a monitoring program in Baden-Wurttemberg, Germany. J Chromatogr A 938:199–210

Schwartz DA, LoGerfo JP (1988) Congenital limb reduction defects in the agricultural setting. Am J Public Health 78:654–658

Sedlak DL, Pinkston KE (2001) Factors affecting the concentrations of pharmaceuticals released to the aquatic environment. Water Resour Update 56–64

Seiler RL, Zaugg SD, Thomas JM, Howcroft DL (1999) Caffeine and pharmaceuticals as indicators of waste water contamination in wells. Ground Water 37:405–410

Sharpe CR, Franco EL, de Camargo B, Lopes LF, Barreto JH, Johnsson RR, Mauad MA (1995) Parental exposures to pesticides and risk of Wilms' tumor in Brazil. Am J Epidemiol 141:210–217

Stumpf M, Ternes TA, Wilken RD, Rodrigues SV, Baumann W (1999) Polar drug residues in sewage and natural waters in the state of Rio de Janeiro, Brazil. Sci Total Environ 225:135–141

Tavera-Mendoza L, Ruby S, Brousseau P, Fournier M, Cyr D, Marcogliese D (2002a) Response of the amphibian tadpole (*Xenopus laevis*) to atrazine during sexual differentiation of the testis. Environ Toxicol Chem 21:527–531

Tavera-Mendoza L, Ruby S, Brousseau P, Fournier M, Cyr D, Marcogliese D (2002b) Response of the amphibian tadpole *Xenopus laevis* to atrazine during sexual differentiation of the ovary. Environ Toxicol Chem 21:1264–1267

Ternes TA (1998) Occurrence of drugs in German sewage treatment plants and rivers. Water Res 32:3245–3260

Ternes TA (2001) Pharmaceuticals and metabolites as contaminants of the aquatic environment. In: Daughton CG, Jones-Lepp T (eds) Pharmaceuticals and personal care products in the environment: scientific and regulatory issues. Symposium series 791, American Chemical Society, Washington DC, pp 39–54

Tungsanga K, Chusilp S, Israsena S, Chusilp S, Sitprija V (1983) Paraquat poisoning: evidence of systemic toxicity after dermal exposure. Postgrad Med J 59:338–339

Vale JA, Meredith TJ, Buckley BM (1987) Paraquat poisoning: clinical features and immediate general management. Hum Toxicol 6:41–47

Vanholder R, Colardyn F, DeReuck J, Praet M, Lameire N, Ringoir S (1981) Diquat intoxication: report of two cases and review of the literature. Am J Med 70:1267–1271

Vorhees DJ, Weisman WH, Gustafson JB (1999) Human health risk-based evaluation of petroleum release sites: implementing the working group approach. In: Total petroleum hydrocarbon criteria working group. Amherst Scientific Publishers, Amherst

Wartenberg D, Reyner D, Scott CS (2000) Trichloroethylene and cancer: epidemiologic evidence. Environ Health Perspect 108:161–176

WHO (1985) Health hazards from nitrates in drinking water. World Health Organization, Geneva

WHO (2005) The World Health Organization recommended classification of pesticides by hazard. Available at http://www.who.int/ipcs/publications/pesticides_hazard_rev_3.pdf

Wood A (2006) Compendium of pesticide common names—classified lists of pesticides. Available at http://www.alanwood.net/pesticides/class_pesticides.html

Yamada-Okabe T, Sakai H, Kashima Y, Yamada-Okabe H (2005) Modulation at a cellular level of the thyroid hormone receptor mediated gene expression by 1,2,5,6,9,10-hexabromocyclododecane (HBCD), 4,4′-diiodobiphenyl (DIB), and nitrofen (NIP). Toxicol Lett 55:127–133

Zahm SH, Blair A (1993) Cancer among migrant and seasonal farmworkers: an epidemiologic review and research agenda. Am J Ind Med 24:753–766

Zahm SH, Ward MH (1998) Pesticides and childhood cancer. Environ Health Perspect 106:893–908

Zahm SH, Ward MH, Blair A (1997) Pesticides and cancer. Occup Med 12:269–289

Part III
Contaminant Partitioning in the Subsurface

Contaminants may reach the subsurface in a gaseous phase, dissolved in water, as an immiscible liquid, or as suspended particles. Contaminant partitioning in the subsurface is controlled by the physicochemical properties and the porosity of the earth materials, the composition of the subsurface water, as well as the properties of the contaminants themselves. While the physicochemical and mineralogical characteristics of the subsurface solid phase define the retention capacity of contaminants, the porosity and aggregation status determine the potential volume of liquid and air that are accessible for contaminant redistribution among the subsurface phases. Environmental factors, such as temperature and water content in the subsurface prior to contamination, also affect the pollution pattern.

Under natural conditions, subsurface contaminants can be composed of single organic or inorganic compounds or mixtures thereof. These compounds have different properties, so they react differently, even if they reach the subsurface simultaneously. Therefore, knowledge of subsurface partitioning among individual components in a contaminant mixture is of major importance.

Chapter 5 discusses contaminant adsorption on geosorbents and includes a short description of the surface properties of adsorbents and the methodology for quantifying adsorption. The chapter continues with a presentation of adsorption of various types of toxic chemicals on the subsurface solid phase. In addition to physicochemical adsorption, contaminants can be retained in the subsurface by precipitation, deposition, and trapping. These topics, as well as hysteresis phenomena and formation of bound residues, are discussed.

Contaminant partitioning is described in the following chapters. Chapter 6 examines aqueous solubility equilibria as affected by environmental factors and the apparent solubility of toxic chemicals in the presence of natural and industrial ligands, cosolvents, and electrolytes. Volatilization of contaminants from the water phase is discussed in Chap. 7. A significant portion of Part III is devoted to the presentation of numerous examples, selected from literature, that illustrate contaminant partitioning among the subsurface solid, aqueous, and gaseous phases.

Chapter 8 reports experimental results on the partitioning of toxic chemicals under various environmental conditions, as individual compounds or contaminant mixtures, when adsorbed or retained on mineral or organic solid phases, dissolved in subsurface water, or volatilized into the surface or subsurface atmosphere.

Chapter 5
Sorption, Retention, and Release of Contaminants

Contaminant retention on geosorbents is controlled by their physicochemical properties, their structural pattern as well as the properties of the contaminants themselves. The properties of these adsorbents control their capacity to retain release contaminants in the subsurface environment.

Contaminants may be adsorbed on the solid phase or on suspended particles in the liquid phase. Environmental factors, such as temperature, pH, and water content in the subsurface prior to contamination, also affect the nature of contaminant adsorption. Other physical processes of retention include precipitation, deposition, and trapping. Under natural conditions, pollutants often consist of more than a single contaminant, comprising a mixture of organic and inorganic toxic compounds. Each of these compounds can react differently with the existing minerals and chemicals in the subsurface.

5.1 Surface Properties of Adsorbents

Clay minerals, oxides, and humic substances are the major natural subsurface adsorbents of contaminants. Under natural conditions, when humic substances are present, humate–mineral complexes are formed with surface properties different from those of their constituents. Natural clays may serve also as a basic material for engineering novel organo-clay products with an increased adsorption capacity, which can be used for various reclamation purposes.

Clay minerals are characterized by a high surface charge and a very small particle size. A detailed presentation of two types of layered silicate clay (kaolinite and smectite) is given in Chap. 1.

Clay minerals have a permanent negative charge due to isomorphous substitutions or vacancies in their structure. This charge can vary from zero to >200 $cmol_c$/kg (centimoles/kg) and must be balanced by cations (counter-ions) at or near the mineral surface (Table 5.1), which greatly affect the interfacial properties. Low counter-ion charge, low electrolyte concentration, or high dielectric constant of the solvent lead to an increase in interparticle electrostatic repulsion

B. Berkowitz et al., *Contaminant Geochemistry*,
DOI: 10.1007/978-3-642-54777-5_5,

Table 5.1 Chemical composition and charge characteristics of selected layer silicates (McBride 1994)

Mineral	Chemical structure	Structure	Charge per half unit cell		Structural charge ($cmol_c/kg$)
			Tetrahedral	Octahedral	
Montmorillonite	$Ca_{0.165}Si_4(Al_{1.67}Mg_{0.33})O_{10}(OH)_2$	2:1 dioctahedral	0	−0.33	92
Beidelite	$Ca_{0.25}(Si_{3.5}Al_{0.5})Al_2O_{10}(OH)_2$	2:1 dioctahedral	−0.5	0	135
Talc	$Si_4Mg_3O_{10}(OH)_2$	2:1 trioctahedral	0	0	0
Vermiculite	$Mg_{0.31}(Si_{3.15}Al_{0.85})(Mg_{2.69}Fe^{3+}_{0.23}Fe^{2+}_{0.08})O_{10}(OH)_{10}$	2:1 trioctahedral	−0.85	0.23	157
Kaolinite	$Si_2Al_2O_5(OH)_4$	1:1 dioctahedral	0	0	0
Serpentine	$Si_2Mg_3O_5(OH)_4$	1:1 trioctahedral	0	0	0

forces, which in turn stabilize colloidal suspensions. An opposite situation supports interparticle association in negatively charged colloids and induces flocculation. These behaviors validate the diffuse double-layer model, which assumes that the layer-silicate surface can be treated as a structurally featureless plane with an evenly distributed negative charge (van Olphen 1967).

Oxides and hydroxides of Al, Fe, Mn, and Si may exist in the subsurface mainly as a mixture (known also as a *solid solution*) rather than as pure mineral phases. They are considered amphoteric materials, characterized by no permanent surface charge. Their cation and anion exchange capacities reflect adsorption of potential-determining ions such as H^+ and OH^-. Different surfaces have a diverse affinity for H^+ and OH^- ions and thus exhibit various points of zero charge (PZC). Details of various models for variable charge minerals may be found in the extensive review of McBride (1989).

Humic substances, including both humic and fulvic acids, are the main subsurface organic components capable of adsorbing contaminants. The functional groups on humic materials control the cation exchange capacity (CEC) and the complexation of metals. In the case of humic substances, for example, the CEC generally is calculated to be at least one electric charge (i.e., ionized group) per square nanometer (Oades 1989) although in some cases it may range from 0.3 to 1.3 (Greenland and Mott 1978). Because humic substances are polydisperse and characterized by diverse chemistry, it is difficult to obtain a well-defined understanding of their capacity for ion exchange and metal complexation.

Organo-mineral association in the subsurface is a natural process controlled by a range of bonding mechanisms, and therefore, it is practically impossible to separate one from other. The resulting organo-mineral complex has surface properties different from the original components. For example, hydrophilic clay surfaces may become hydrophobic.

5.2 Quantifying Adsorption

Quantifying adsorption of contaminants from gaseous or liquid phases onto the solid phase should be considered valid only when an equilibrium state has been achieved, under controlled environmental conditions. Determination of contaminant adsorption on surfaces, that is, interpretation of adsorption isotherms and the resulting coefficients, helps to quantify and predict the extent of adsorption. The accuracy of the measurements is important in relation to the heterogeneity of geosorbents at a particular site. The spatial variability of the solid phase is not confined only to field conditions; variability is present at all scales, and its effects are apparent even in well-controlled laboratory-scale experiments.

5.2.1 Adsorption–Desorption Coefficients

Adsorption–desorption coefficients are determined by various experimental techniques related to the status of a contaminant (solute or gas) under static or continuous conditions. Solute adsorption–desorption is determined mainly by batch or column equilibration procedures. A comprehensive description of various experimental techniques for determining the kinetics of soil chemical processes, including adsorption–desorption, may be found in the book by Sparks (1989) and in many papers (e.g., Nielsen and Biggar 1961; Bowman 1979; Boyd and King 1984; Peterson et al. 1988; Podoll et al. 1989; Abdul et al. 1990; Brusseau et al. 1990; Hermosin and Carnejo 1992; Farrell and Reinhard 1994; Schrap et al. 1994; Petersen et al. 1995).

Application of analytical techniques from molecular geochemistry can be used to study reactions at the molecular level. Such studies can elucidate the partitioning and interactions of contaminant species in aqueous, solid, and gas phases. While spectroscopic methods provide information on chemical reactions on the contaminant–solid interface, other techniques may provide additional spatial information at an atomic level. In an extensive review on molecular geochemistry, O’Day (1999) summarizes common analytical methods (Table 5.2) and discusses their benefits in understanding contaminant–solid interactions at the molecular level.

5.2.2 Adsorption Isotherms

The sorption process generally is studied by plotting the equilibrium concentration of a compound on the adsorbent, as a function of equilibrium concentration in the gas or solution at a given temperature. Adsorption isotherms are graphs obtained

Table 5.2 Summary of selected analytical methods for molecular environmental geochemistry

Analytical method	Type of energy	
	Source	Signal
Absorption, emission, and relaxation spectroscopies, IR and FTIR	Infrared radiation	Transmitted infrared radiation
Synchrotron X-ray absorption spectroscopy (XAS), X-ray absorption near-edge spectroscopy, extended X-ray absorption fine structure	Synchrotron X rays	Transmitted or fluorescent X rays; electron yield
Synchrotron microanalysis, X-ray diffraction	Synchrotron X rays	Fluorescent X rays
EELS (also called PEELS)	Electrons	Electrons
XPS and Auger spectroscopy	X rays	Electrons
Resonance spectroscopies		
NMR	Radio waves (+ magnetic field)	Radio waves
ESR (also called EPR)	Microwaves (+ magnetic field)	Microwaves
Scattering and ablation		
X-ray scattering (small angle, SAXS; wide angle, WAXS)	X rays (synchrotron or laboratory)	Scattered X rays
SIMS	Charged ion beam	Atomic mass
LA-ICP-MS	Laser	Atomic mass
Microscopies		
STM	Tunneling electrons	Electronic perturbations
AFM (also called SFM)	Electronic force	Force perturbation
HR-TEM and STEM	Electrons	Transmitted or secondary electrons
SEM/EM with EDS or WDS chemical analysis	Electrons	Secondary, or backscattered electrons

AAS Atomic absorption spectroscopy, *AFM* atomic force microscopy (also known as SFM), *CT* computerized tomography, *EDS* energy-dispersive spectrometry. *EELS* electron energy loss spectroscopy, *EM* electron microscopy, *EPR* electron paramagnetic resonance (also known as ESR), *ESR* electron spin resonance (also known as EPR), *EXAFS* extended X-ray absorption fine structure, *FTIR* Fourier transform infrared, *HR-TEM* high-resolution transmission electron microscopy, *ICP-AES* inductively coupled plasma atomic emission spectrometry, *ICP-MS* inductively coupled plasma mass spectrometry. Reproduced by permission of American Geophysical Union. O'Day (1999). Copyright 1999 American Geophysical Union

by plotting measured adsorption data against the concentration value of the adsorbate. Several mechanisms may be involved in the retention of contaminants on adsorbents, and therefore, several adsorption isotherms with different shapes may exist. Giles et al. (1960) related the shape of the adsorption isotherms to the adsorption mechanism for a solute–solvent adsorbent system as follows:

- The *S-curve* isotherm exhibits an initial slope that increases with the concentration of a substance in the solution. This suggests that the relative affinity of

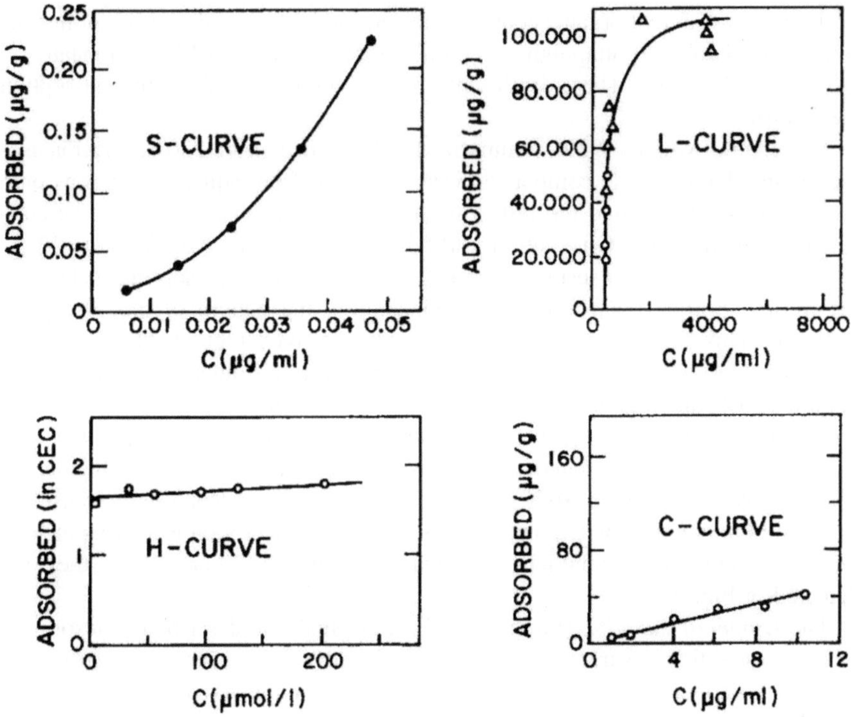

Fig. 5.1 Examples of adsorption isotherms. *S-type* aldrin on oven-dry kaolinite from aqueous solution. *L-type* parathion on oven-dry attapulgite from hexane solution. *H-type* methylene blue at pH = 6 on montmorillonite from aqueous solution. *C-type* parathion on clay soil from hexane solution (Yaron et al. 1996)

the adsorbent for the solute at low concentration is less than the affinity of the solid surface for the solvent.

- The *L-curve* isotherm is characterized by an initial slope that does not increase with the concentration of the substance in the solution. This behavior corresponds to high relative affinity of the adsorbent at low concentration and a decrease in the free adsorbing surface.
- The *H-curve* isotherm is characterized by a linear increase that remains independent of the solute concentration in the solution (i.e., constant partitioning of the solute between the solvent and the adsorbing surface). This behavior indicates a high affinity of the solid phase for the solvent.
- The *C-curve* isotherm is similar to the *H*-curve, being characterized by a linear increase, but also passing through the origin. This behavior may be due to a proportional increase in the adsorbing surface as well as to surface accessibility.

Based on their molecular properties as well as the properties of the solvent, each inorganic or organic contaminant exhibits an adsorption isotherm that corresponds to one of the isotherm classifications just described. Figure 5.1 illustrates these isotherms for different organic contaminants, adsorbed from either water or hexane solution on kaolinite, attapulgite, montmorillonite, and a red Mediterranean soil (Yaron et al. 1996). These isotherms may be used to deduce the adsorption mechanism.

Weber and Miller (1989) summarized published data of 230 adsorption isotherms in which organic compounds were adsorbed from aqueous solutions onto various soils. They found the following distribution of behaviors using the classification defined by Giles et al. (1960): $S = 16\ \%$, $L = 64\ \%$, $H = 12\ \%$, $C = 8\ \%$. Based on this result, it can be concluded that the adsorbing material is the most important controlling factor in defining the pattern of the adsorption isotherm. This fact is confirmed by the work of Weber et al. (1992) for the herbicide fluoridone, which exhibited the *S*-type sorption isotherm on soil with low organic matter and high montmorillonite content, and the *L*-type sorption isotherm on a soil with moderate organic matter content and mixed mineralogy.

In addition to these characterizations of adsorption curves, mathematical descriptions of adsorption isotherms, based on physical models, often are used to study solid interactions with contaminants. The main adsorption isotherms include those of Langmuir, Freundlich, and Brunauer–Emmet–Teller (BET); they are depicted in Fig. 5.2.

The *Langmuir equation* (Eq. 5.1), derived originally to describe the adsorption of gases on solids, assumes that the adsorbed entity is attached to the surface at specific, homogeneous, localized sites, forming a monolayer. It is also assumed that the heat of adsorption is constant over the entire monolayer, that there is no lateral interaction between adsorbed species, that equilibrium is reached, and that the energy of adsorption is independent of temperature:

$$\frac{x}{m} = \frac{KCb}{1 + KC} = \frac{Kb}{\frac{1}{C} + K}, \tag{5.1}$$

where x is the amount of adsorbed chemical, m is the mass of adsorbent, C is the equilibrium concentration, K is a constant related to the bonding strength, and b is the maximum amount of adsorbate that can be adsorbed.

The best way to determine the parameter values is to plot the distribution coefficient (K_d), which is the ratio between the amount adsorbed per unit mass of adsorbent (x/m) and the concentration in solution (C):

$$K_d = \frac{x/m}{C}. \tag{5.2}$$

Multiplying Eq. (5.1) by $1/C + K$ and substituting into Eq. (5.2) gives a linear equation for K_d, expressed as

$$K_d = Kb - K\frac{x}{m}. \tag{5.3}$$

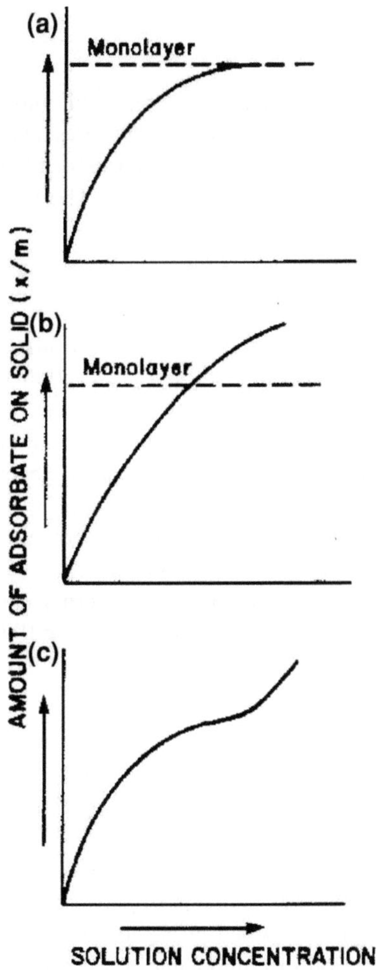

Fig. 5.2 Typical adsorption isotherms described by **a** Langmuir, **b** Freundlich, and **c** BET equations (Yaron et al. 1996)

If a straight line is obtained when K_d is plotted against x/m at low concentrations, the Langmuir equation is applicable. However, due to the restrictive assumptions, Langmuir isotherms usually are of minor importance in heterogeneous media such as the subsurface environment. Corrections therefore have been introduced to the Langmuir equation to overcome the problems of heterogeneous sites, coupled adsorption–desorption reactions, and adsorption of inorganic and organic trace elements on geosorbents.

The *Freundlich equation* was derived empirically, based on the logarithmic decrease in adsorption energy with increasing coverage of the adsorbent surface. Freundlich found that adsorption data for many dilute solutions could be fit by the expression

$$\frac{x}{m} = KC^{1/n}, \tag{5.4}$$

where K and n are empirical constants, and the other terms are as defined previously. The value of $1/n$ represents a joint measure of both the relative magnitude and diversity of energies associated with a particular sorption process (Karickhoff 1981; Weber et al. 1992). The linear form of the Freundlich equation is

$$\log\frac{x}{m} = \frac{1}{n}\log C + \log K. \tag{5.5}$$

The Freundlich equation can also be derived theoretically by assuming that the decrease in energy of adsorption with increasing surface coverage is due to the surface heterogeneity (Fripiat et al. 1971).

The main limitation of the Freundlich equation is that it does not predict a maximum adsorption capacity, because linear adsorption generally occurs at very low solute concentration and low loading of the sorbent. However, in spite of this limitation, the Freundlich equation is used widely for describing contaminant adsorption on geosorbents.

Composite linear isotherms express the natural conditions of heterogeneity specific to geosorbents (Lafleur 1979; McCarty et al. 1981; Karickhoff 1984). The relative equation expressing composite conditions of geosorbents may be of the type

$$q = \sum_{i=1}^{m} x_i q_i = \left(\sum_{i=1}^{m} x_i K_{di}\right) C = K_d C, \tag{5.6}$$

where q is the total solute mass sorbed per unit mass of bulk solid at equilibrium, x_i is the mass fraction of geosorbent constituting the reaction region or component i, q_i is the sorbed-phase concentration at equilibrium expressed per unit mass of that region or component, K_{di} is the partition coefficient for a reaction expressed per unit mass of component i, and K_d is the mass-averaged partition coefficient.

When one or more of the component elements of sorption is governed by a nonlinear relationship between the solution and the sorbed phase, the composite isotherm deviates from linearity. In these cases, modifications to the Freundlich isotherm have been developed (e.g., Lambert 1967; Weber et al. 1992) to express these conditions.

Overlapping patterns of some Langmuir-type sorption processes, which can occur at different sites of a complex sorbent (such as a geosorbent) and show different interaction energies, may be quantified by a Freundlich-type isotherm. A meaningful thermodynamic interpretation of this equation has been developed by Wauchope and Koskinen (1983), using a fugacity approach, with a proposed standard state for a sorbed organic contaminant (herbicide). This interpretation was based on the assumption that the organic fraction of the geosorbent forms a homogeneous solid comprising many components, which is known as a *solid solution*. The fugacity approach for the interpretation of environmental behavior of a chemical contaminant is described in detail by Mackay and Paterson (1981).

The *BET equation* describes the phenomenon of multilayer adsorption, which is characteristic of physical or van der Waals interactions. In the case of gas adsorption, for example, multilayer adsorption merges directly into capillary condensation when the vapor pressure approaches its saturation value and often proceeds with no apparent limit. The BET equation has the form

$$\frac{P}{V(P_0 - P)} = \frac{1}{V_m C_h} + \frac{(C_h - 1)P}{V_m C_h P_0} \tag{5.7}$$

where P is the equilibrium pressure at which a volume V of gas is adsorbed, P_0 is the saturation pressure of the gas, V_m is the volume of gas corresponding to an adsorbed monomolecular layer, and C_h is a constant related to the heat of adsorption of the gas on the solid in question. If a plot of $P/(P_0 - P)$ against P/P_0 results in a straight line, the effective surface area of the solid can be calculated after V_m has been determined, either from the slope of the line $(C_h - 1)/V_m C_h$ or from the intercept $1/V_m C_h$.

It is interesting to note the effect of laboratory-scale variability on the nonlinear sorption behavior of contaminants in a porous medium, composed of various particles that are characterized individually by randomly distributed sorptive capacities and selectivity coefficients. A discrepancy is observed between the results obtained for an individual particle and for an ensemble of particles. As the variability in underlying sorptive properties increases, the Langmuir isotherm ceases to describe the behavior of the aggregates of individual particles, under either static or dynamic conditions. Assessment of the pollution hazard from parameters obtained in the laboratory therefore should consider the variability among the individual particles making up the analyzed geosorbent sample. Schwarzenbach et al. (2003) introduced the concept of "the complex nature of the distribution coefficient," showing that this parameter may lump together many chemical species. The solute–geosorbent exchange should describe an appropriate equilibrium expression that incorporates properties of the various geosorbent components. The resulting K_d parameter is weighted by the availability of sorbent properties in the total solid phase of the sample measured. Despite this limitation, the distribution coefficient gives an effective representation of the solute–geosorbent relationship with regard to contaminant adsorption–desorption behavior.

In conclusion, the different shapes of isotherms describing equilibrium distributions of a contaminant, between geosorbents and aqueous or gaseous phases, depend on the sorption mechanism involved and the associated sorption energy. At low contaminant concentration, all models reduce to essentially linear correlation. At higher contaminant concentration, when sorption isotherms deviate from linearity, an appropriate isotherm model should be used to describe the retention process.

5.3 Kinetics of Adsorption

Adsorption kinetics involve a time-dependent process that describes the rate of adsorption of chemical contaminants on the solid phase. The "standard" chemical meaning of *kinetics* usually covers the study of the rate of reactions and molecular processes when transport is not a limiting factor; however, this definition is not applicable to subsurface conditions. In the "real" subsurface environment, many kinetic processes are a blend of chemical- and transport-controlled kinetics.

Understanding the kinetics of contaminant adsorption on the subsurface solid phase requires knowledge of both the *differential rate law*, explaining the reaction system, and the *apparent rate law*, which includes both chemical kinetics and transport-controlled processes. By studying the rates of chemical processes in the subsurface, we can predict the time necessary to reach equilibrium or quasi-state equilibrium and understand the reaction mechanism. The interested reader can find detailed explanations of subsurface kinetic processes in Sparks (1989) and Pignatello (1989).

The *mechanistic rate law* is not applicable to processes in the subsurface, if we assume only that chemically controlled kinetics occur and neglect the transport kinetics. Instead, *apparent rate laws*, which comprise both chemical- and transport-controlled processes, are the proper tool to describe reaction kinetics on subsurface soil constituents. Apparent rate laws indicate that diffusion and other microscopic transport phenomena, as well as the structure of the subsurface and the flow rate, affect the kinetic behavior.

Based on these rate laws, various equations have been developed to describe kinetics of soil chemical processes. As a function of the adsorbent and adsorbate properties, the equations describe mainly first-order, second-order, or zero-order reactions. For example, Sparks and Jardine (1984) studied the kinetics of potassium adsorption on kaolinite, montmorillonite (a smectite mineral), and vermiculite (Fig. 5.3), finding that a single-order reaction describes the data for kaolinite and smectite, while two first-order reactions describe adsorption on vermiculite.

The *Elovich equation* was developed to determine the kinetics of heterogeneous chemisorption of gases on solid surfaces. This equation assumes a heterogeneous distribution of adsorption energies, where the energy of activation (E) increases linearly with surface coverage (Rao et al. 1989). A simplified Elovich equation used to study the rate of soil chemical processes is given by

$$q = \frac{1}{Y}\ln(XY) + \frac{1}{Y}\ln(t + t_0), \tag{5.8}$$

where q is the amount sorbed at time t, X and Y are constants, and t_0 is an integration constant. An application of Eq. (5.8) for the case of PO_4 sorption on soils is shown in Fig. 5.4. In this particular case, a linear relationship is observed. Chien and Clayton (1980) found that the Elovich equation was best based on the highest values of the simple correlation coefficient. Polysopoulos et al. (1986),

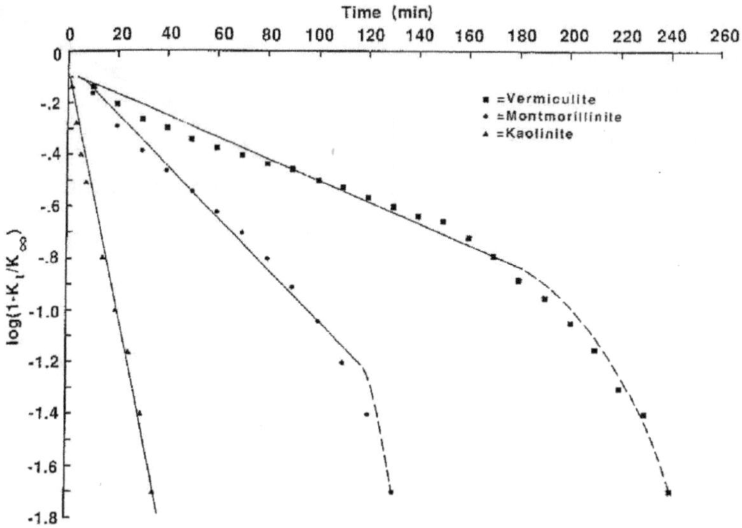

Fig. 5.3 First-order plots of potassium adsorption on clay, where K_t is the quantity of potassium adsorbed at time t, and K_∞ is the quantity of potassium adsorbed at equilibrium (Sparks and Jardine 1984)

however, show that the pre- and post-Elovichian sections, in many cases, are not observed, which leads to the incorrect conclusion that the entire rate process may be explained by one single kinetic law.

Sparks (1989) discusses the application of various kinetic equations to earth materials based on the analysis of a large number of reported studies. Even though different equations describe rate data satisfactorily, Sparks (1989) uses linear regression analysis to show that no single equation best describes every study.

According to the Arrhenius law, the rate of reaction is correlated linearly to the increase in temperature, with the rate constant k given by

$$k = Ae^{-E/RT}, \tag{5.9}$$

where A is a frequency factor, E is the energy of activation, R is the universal gas constant, and T is the absolute temperature. A low activation energy usually indicates a diffusion-controlled process, while higher activation energy indicates chemical-reaction-controlled processes (Sparks and Huang 1985; Sparks 1986). Data on the effect of temperature on the rate of potassium release from potassium-bearing minerals were presented by Huang et al. (1968) and are reproduced in Table 5.3. Huang et al. (1968) showed that a 10 K rise in temperature during the reaction period resulted in a twofold to threefold increase in the rate constant.

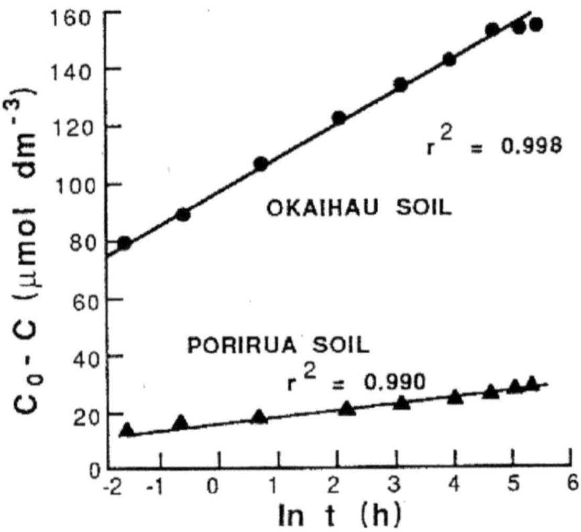

Fig. 5.4 Plot of Elovich equation for phosphate (PO_4) sorption on two soils, where C_0 is the initial concentration added at time zero, and C is the concentration in the soil solution at time t (Chien and Clayton 1980)

Table 5.3 Apparent rate constants for the release of potassium from potassium minerals, as a function of temperature (Huang et al. 1968)

Mineral	Temperature	
	301 K	311 K
Biotite	1.46×10^{-2}	3.09×10^{-4}
Phlogopite	9.01×10^{-4}	2.44×10^{-4}
Muscovite	1.39×10^{-4}	4.15×10^{-4}
Microcline	7.67×10^{-5}	2.63×10^{-4}

5.4 Adsorption of Ionic Contaminants

Chapter 2 mentioned that the adsorption of charged ionic compounds on the solid phase is a result of a combination of chemical binding forces and electric fields at the interface. Here, we extend the discussion on this topic, focusing mainly on aspects relevant to behavior of ionic contaminants in the subsurface environment.

Electrical neutrality on the solid surface requires that an equal amount of positive and negative charge accumulates in the liquid phase near the surface. If the surface is negatively charged, positively charged cations are electrostatically attracted to the surface. Simultaneously the cations are drawn back toward the equilibrating solution; as a result, a diffuse layer is formed and the concentration of cations increases toward the surface. On the other hand, ions of the same sign (anions) are repelled by the surface with diffusion forces acting in an opposite direction. The overall pattern is known as a *diffuse double layer* (DDL). The existence of a DDL was developed theoretically by Gouy and Chapman about 100 years ago and is an integral part of *electric double layer theory.*

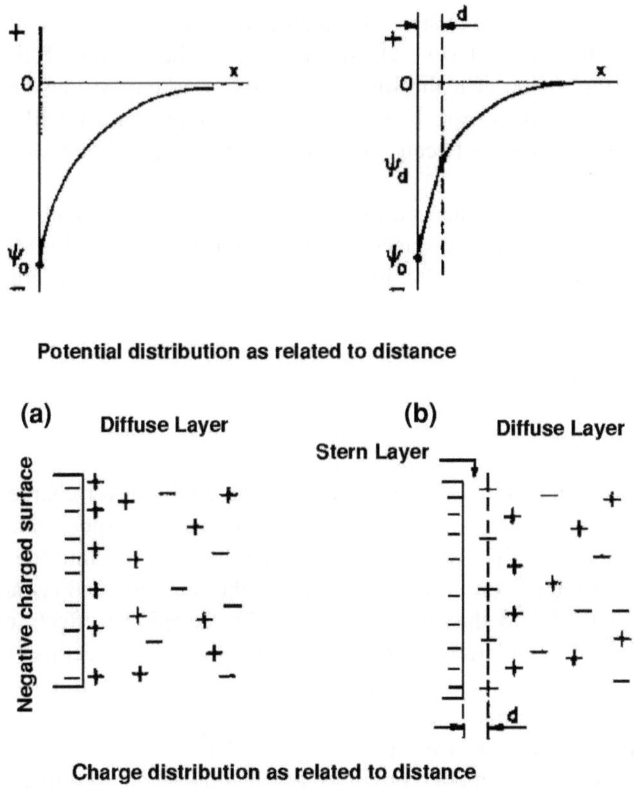

Fig. 5.5 Distribution of electrical charges and potentials in a double layer according to **a** Gouy–Chapman model and **b** Stern model, where ψ_0 and ψ_d are surface and Stern potentials, respectively, and d is the thickness of the Stern layer

The Gouy–Chapman model assumes (1) the exchangeable cations exist as point charges, (2) colloid surfaces are planar and infinite in extent, and (3) surface charge is distributed uniformly over the entire colloid surface. Even though this assumption does not correspond to the subsurface environment, it works well for the clay colloid component of the subsurface, a fact that may be explained by mutual cancelation of other interferences. Stern (1924) and Grahame (1947) refined the Gouy–Chapman model by recognizing that counter-ions are unlikely to approach the surface more closely than the ionic radii of the anions and the hydrated radii of the cations.

The Gouy–Chapman model assumes that the charge is spread uniformly over the surface, with the overall charge allocation in solution consisting of a non-uniform distribution of point charges. The solvent is treated as a continuous medium influencing the double layer only through its dielectric constant, which is assumed independent of its position in the double layer. Moreover, it is assumed that ions and surfaces are involved only in electrostatic interactions. The derivation is for a flat surface, infinite in size. The double-layer theory applies equally well to rounded or spherical surfaces (Overbeek 1952). The model of Stern (1924) assumes that the region near the surface consists of a layer of ions known as the *Stern layer* and a diffuse ion layer. A schematic representation illustrating the fundamental differences between the Gouy–Chapman and Stern models is presented in Fig. 5.5.

The surface charge is balanced by the charge in solution, which is distributed between the Stern layer at a distance d from the surface and a diffuse layer having an ionic Boltzmann-type distribution. The total charge σ is therefore due to the charge in the two layers:

$$\sigma = (\sigma_1 + \sigma_2), \tag{5.10}$$

where σ_1 is the Stern layer, and σ_2 is the diffuse layer charge.

A development of the DDL theory also considers the interactions between the two flat layers of the Gouy–Chapman model. The double-layer charge is affected only slightly when the distance between the two plates is large. Grahame (1947) suggests that specifically adsorbable anions may be adsorbed into the Stern layer when they lose their hydration water, whereas the hydrated cations are attracted only electrostatically to the surface. Bolt (1955) added the effects of ion size, dielectric saturation, polarization energy, and coulombic interactions of the ions, as well as short-range repulsion of ions into the Gouy–Chapman model. Note that the simple Gouy–Chapman model gives fairly reliable results for colloids with a constant charge density not exceeding 0.2–0.3 C/m^2.

The Gouy–Chapman model provides an invaluable answer to a number of processes occurring in the subsurface system, by explaining the exchange capacity concept for the range of surface charge densities normally encountered in clays (Bolt et al. 1991). In general, double-layer theory explains the processes occurring in the contaminant-subsurface system when the pollutants have a charge opposite to that of the surface. By studying molecular dynamics simulations of the electrical double layer on smectite surfaces contacting concentrated mixed electrolyte (Na–$CaCl_2$) solutions, Bourg and Sposito (2011) defined three ion adsorption planes: 0, β, and d with different qualifications. The locations of β- and d-planes are independent of ionic strength or ion type, and "indifferent electrolyte" ions can occupy all three planes.

CEC and *selectivity* are among the most important processes that control the fate of charged (ionic) contaminants in the subsurface. These processes involve the cationic concentration in solution and the cation dimensions, as well as the configuration of exchange sites on the interface. The Gapon relation approaches the process as an exchange of equivalents of electric charges, where the solute concentration is measured in terms of activity and the adsorption on an equivalent basis.

Negatively charged surfaces having the same exchange properties do not necessarily interact in the same manner with different cations having the same

Table 5.4 Adsorption of heavy metals on goethite as a function of pH

Metal	pH							
	4.7	5.2	5.5	5.9	6.4	7.2	7.5	8.0
Cu	17	55	75	90				
Pb		43	56	75				
Zn				13	22	68		
Cd					23	44	53	
Co						39	54	78

Data expressed as percent of initial amount of metallic cation solution. Reprinted from Quirk and Posner (1975).

valence. This is caused by the differential sizes and polarizability of the cations, by the structural properties of the adsorbent surfaces, and by the differences in the surface charge distribution. For example, ammonium ions (NH_4^+) are sorbed preferably over anhydrous H^+ or Na^+ in 1:2 clay minerals, because they may form NH-oxygen links in the hexagonal holes of Si-O sheets, and they may also link to adjacent oxygen planes in the interlayer space by an OH–N–HO bond. Selectivity of the divalent alkaline earth cations is less pronounced. Trivalent cations, such as Al^{3+}, coordinated octahedrally to water molecules, link more strongly than hydrated Ca^{2+} ions.

Cation selectivity on organic matter is related mainly to the disposition of the acidic groups in the adsorbent. Multivalent cations adsorb preferentially over monovalent cations, and transition metals adsorb preferentially over strong basic metals. Organo-mineral complexes exhibit a CEC smaller than the sum of the CECs of the components. This phenomenon is reflected in the pattern of cation selectivity (Greenland and Hayes 1981). Two aspects should be considered in the cation exchange process: the number of exchange sites occupied by the cation investigated and the selectivity of the cation relative to the concentration of the exchanging cation.

Heavy metal cations participate in exchange reactions with negatively charged surfaces of clay minerals, with Coulombic and specific adsorption being the processes involved in the exchange. Metal cationic adsorption is affected by pH. At low pH values (<5.5), some heavy metals do not compete with alkali metals (e.g., Ca^{2+}) for the mineral adsorption site. At higher pH values, heavy metal adsorption increases greatly; an example of heavy metal adsorption on goethite as a function of pH is given in Table 5.4.

Cationic organic contaminants often compete with mineral ions for the same adsorption site. At low pH, organic cationic molecules are adsorbed more strongly on earth materials than on mineral ions of a similar valence. At moderate pH values, however, mineral ions are favored over organic cations. In general, the charge density of the adsorbing surface is a determining factor in adsorption of cationic organic molecules, but their adsorption also is affected by the molecular configuration (Mortland 1970). Organic molecules may be adsorbed by clays via a cationic adsorption mechanism, but this process depends on the acidity of the medium.

Early work by Boyd et al. (1947), performed on zeolites, showed that the ion exchange process is diffusion controlled and the reaction rate is limited by mass transfer phenomena that are either film-diffusion (FD) or particle-diffusion (PD) dependent. Under natural conditions, the charge compensation cations are held on a representative subsurface solid phase as follows: within crystals in interlayer positions (mica and smectites), in structural holes (feldspars), or on surfaces in cleavages and faults of the crystals and on external surfaces of clays, clay minerals, and organic matter.

Cations held on external surfaces are immediately accessible to (subsurface) water. Once removed from the solid phase, they move to a region of reduced concentration. This movement is controlled by diffusion, and the diffusion coefficient (D) can be calculated using the equation described by Nye and Tinker (1977):

$$D = D_e \theta f \frac{dC_{\mathrm{solution}}}{dC_{\mathrm{ss}}} \tag{5.11}$$

where D_e is the diffusion coefficient in water, θ is the water content in the subsurface solid phase, f is the "impedance" factor related to the tortuosity, C_{ss} is the cation concentration in the subsurface (solid and water phases) expressed as mass/volume, and C_{solution} is the cation concentration in solution.

Cations held on the external surfaces of clays exhibit relatively rapid diffusion but are subject to an additional limitation. Because the arrival rate of ingoing cations at the exchange site is much slower than the release rate of the outgoing cations, the rate-determining step is the influx of exchanging cations to negatively charged sites. Sparks (1986) defined the following concurrent processes that take place during Na^+ and K^+ exchange in vermiculite: (1) diffusion of Na^+ with Cl^- through the solution film that surrounds the particle (FD), (2) diffusion of Na^+ ions through a hydrated interlayer space and chemical reaction leading to exchange of Na^+ in the particle (PD), (3) chemical reaction leading to exchange of Na^+ by K^+ ions on the particle surface (CR), (4) diffusion of displaced K^+ ions through the hydrated interlayer space of the particle (PD), and (5) diffusion of displaced K^+ ions with Cl^- through the solution film away from the particle (FD).

To enable a chemical reaction, the exchange ions must be transported to the active sites of the particles. The film of water adhering to and surrounding the particle, as well as the hydrated interlayer space within the particle, are zones of low contaminant concentration that are being depleted constantly by ion adsorption to the sites. The decrease in concentration of contaminant ions in these interfacial zones is then compensated for by ion diffusion from the bulk solution. Ion exchange occurs when a driving force, such as a chemical potential gradient, is maintained between solid and solution or when access to sites is kept free by the use of a hydrated and less preferred cation for exchange.

The properties of organic and inorganic constituents in the subsurface solid phase, as well as the properties of the contaminants (e.g., ion charge and radius), define the time span of ion exchange, which may range from a few seconds to days (Yaron et al. 1996). The slowly exchangeable cations are situated on exchange sites in interlayer spaces of the minerals (e.g., smectites) or in cages and channels of organic matter, and exchange and move into solution by diffusive flux.

In general, when a charged solid surface faces an ion of similar charge in an aqueous suspension, the ion is repelled from the surface by Coulomb forces. The Coulomb repulsion produces a region in the aqueous solution that is relatively depleted of the anion and an equivalent region far from the surface that is relatively enriched.

Anionic negative adsorption may occur in the subsurface when negatively charged clay minerals repel anions from the mineral surface. If, for example, a dilute neutral solution of KCl is added to dry clay, the Cl^- equilibrium concentration in the bulk solution will be greater than the Cl^- concentration in the solution originally added to the clay. Anionic negative adsorption is affected by the anion charge, concentration, pH, the presence of other anions, and the nature and charge of the surface. Negative adsorption may decrease as the subsurface pH decreases and when anions can be adsorbed by positively charged surfaces. The larger negative charge of the surface results in a greater anion negative adsorption. Acidic organic contaminants in their anionic form are expected to be repelled by negatively charged clay surfaces.

5.5 Adsorption of Nonionic Contaminants

The sorption of a nonpolar organic contaminant on a solid phase is derived by enthalpy- and entropy-related forces. Hassett and Banwart (1989) suggested that sorption occurs when the free energy of the reaction is negative due to enthalpy or entropy. The enthalpy is primarily a function of the changes in the bonding between the adsorbing surface and the sorbate (solute) and between the solvent (water) and the solute. The entropy is related to the increase or decrease in the order of the system on sorption.

The forces that control adsorption of nonionic contaminants on the solid phase were summarized by Yaron et al. (1996) in terms of enthalpy and entropy adsorption forces. These are discussed next.

Enthalpy-Related Adsorption Forces

Hydrogen bonding refers to the electrostatic interaction between a hydrogen atom covalently bound to one electronegative atom (e.g., oxygen) and another electronegative atom or group of atoms in a neighboring molecule. The hydrogen atom may be regarded as a bridge between electronegative atoms; this bonding is conceived of as an induced dipole phenomenon. The H bond generally is considered as the asymmetrical distribution of the first electron of the H atom induced by various electronegative atoms.

Ligand exchange processes involve replacement of one or more ligands by the adsorbing species. In some instances, the ligand exchange process can be regarded as a condensation reaction (e.g., between a carboxyl group and a hydroxyl aluminum surface). Under some conditions, ligand exchange reactions are very likely to be involved when humic substances interact with a clay material.

The *protonation mechanism* includes Coulomb electrostatic forces resulting from charged surfaces. The development of surface acidity by the solid phase of the subsurface offers the possibility that solutes having proton-selective organic functional groups can be adsorbed through a protonation reaction.

The π *bonds* occur as a result of the overlapping of π orbitals when they are perpendicular to aromatic rings. This mechanism can be used to explain the bonding of alkenes, alkylenes, and aromatic compounds to subsurface organic matter.

London-van der Waals forces generally are multipole (dipole–dipole or dipole-induced dipole) interactions produced by a correlation between fluctuating induced multipole (principal dipole) moments in two nearly uncharged polar molecules. Even though the time-averaged, induced multipole in each molecule is zero, the correlation between the two induced moments does not average to zero. As a result, an attractive interaction between the two is produced at very small molecular distances.

The van der Waals forces also include dispersion forces that arise from correlations between the movement of electrons in one molecule and those of neighboring molecules. Under such conditions, even a molecule with no permanent dipole moment forms an instantaneous dipole as a result of fluctuations in the arrangements of its electron cloud. This instantaneous dipole polarizes the charge of another molecule to give a second-induced dipole, resulting in a mutual dipole–dipole attraction. All molecules are subject to attraction by dispersion forces whether or not more specific interactions between ions or dipoles occur. Although the momentary dipoles and induced dipoles constantly change positions, the net result is a weak attraction. When many groups of atoms in a polymeric structure interact simultaneously, the van der Waals components are additive.

Chemisorption denotes the situation in which an actual chemical bond is formed between the molecules and the surface atoms. A molecule undergoing chemisorption may lose its identity as the atoms are rearranged, forming new compounds that better satisfy the valences of the surface atoms. The enthalpy of chemisorption is much greater than that of physical adsorption. The basis of much catalytic activity at surfaces is that chemisorption may organize molecules into forms that can readily undergo reactions. It often is difficult to distinguish between chemisorption and physical sorption, because a chemisorbed layer may have a physically sorbed layer deposited above it.

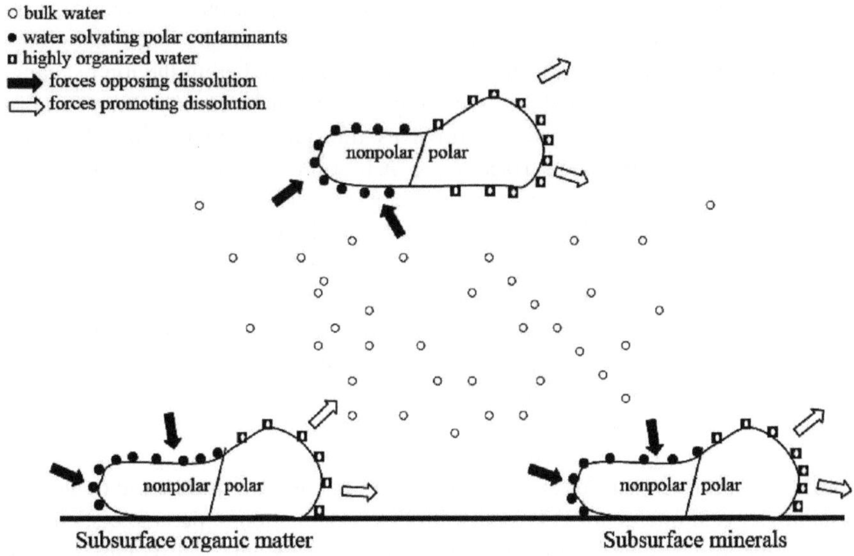

Fig. 5.6 Forces affecting sorption of nonpolar organic contaminants

Entropy-Related Adsorption Force

Entropy-related adsorption, denoted *hydrophobic sorption* (or *solvophobic interaction*), is the partitioning of nonpolar organics out of the polar aqueous phase onto hydrophobic surfaces. Figure 5.6 shows a schematic model of forces that contribute to the sorption of hydrophobic organics, relevant to the subsurface environment.

A major feature of hydrophobic sorption is the weak interaction between the solute and the solvent. The primary force in hydrophobic sorption appears to be the large entropy change resulting from the removal of a solute from solution. The entropy change is due largely to the destruction of the cavity occupied by the solute in the solvent and the destruction of the structured water shell surrounding the solvated organic. Hydrophobic interfaces may be found mainly on organic matter and on organically coated minerals.

Hydrophobic sorption, being an entropy-driven process, provides the major contribution to sorption of hydrophobic contaminants on subsurface solid phases. When a hydrophobic organic compound is adsorbed on a solid phase, the partitioning of the compound and its adsorption by the surface directly from the water phase should be considered. These processes occur in partially saturated and saturated subsurface regimes, where water is likely to be the wetting phase. In such cases, the wetting phase completely or partially coats the solid-phase surface, thus increasing the retention capacity because the wetting phase serves as an additional sink.

Rao et al. (1989) suggested that at least four mechanisms of adsorption should be considered for hydrophobic organic compounds. The first mechanism involves the sorption of the neutral molecular species from the aqueous phase, which is similar to hydrophobic sorption. The second mechanism of interest comprises the specific interactions of a dissociated (ionic) species with various functional groups on the sorbent surface. Several models developed for predicting the ion exchange of inorganic ions may be used for predicting this type of sorption. A third sorption mechanism, molecular ion pairing, involves transfer of organic ions from the aqueous phase to the organic surface phase. A fourth mechanism covers transfer of organic ions from the aqueous phase to the organic surface, while the counter-ions remain in the electric double layer of the aqueous phase. The relative contribution of each of these mechanisms depends on (1) the extent of compound dissociation as a function of the acid dissociation constant, pK_a, and solution pH; (2) the ionic charge status of the solid interface as a function of the pH and of the point of zero charge; and (3) the ionic strength and composition of the water phase.

Often, contaminants reach the subsurface as complex mixtures, and therefore, an understanding of the adsorption process under these more complicated conditions is required. Under a waste disposal site, where organic or organo-metal complexes are involved, for example, sorption may involve multiphase (water and organic) solvent interactions. To deal with this combination of parameters, one can use the theoretical approach of Rao et al. (1985). This approach is based on the predominance of solvophobic interactions for predicting sorption of hydrophobic organic chemicals from mixed solvents. With increasing volume fractions of a completely miscible organic solvent in a binary mixed solvent, the hydrophobic organic solvent sorption coefficient decreases exponentially because the solubility and sorption coefficient are inversely related.

Further in-depth discussions of nonionic pollutant adsorption on subsurface components can be found in the classical review of Mortland (1970) or in the reviews of Calvet (1989), Hassett and Banwart (1989), Hayes and Mingelgrin (1991), Delle Site (2001), as well as in a number of books (such as those by Theng 1974; Greenland and Hayes 1981; Saltzman and Yaron 1986; Yaron et al. 1996; and Schwarzenbach et al. 2003).

5.6 Other Factors Affecting Adsorption

Independent of the molecular properties of contaminants, the subsurface solid-phase constituents are a major factor that control the adsorption process. Both the mineral and organic components of the solid phases interact differentially with ionic and nonionic pollutants, and in all cases, environmental factors, such as temperature, subsurface water content, and chemistry, affect the mechanism, extent, and rate of contaminant adsorption.

The structural properties of the subsurface clay fraction are a controlling factor in defining the rate and extent of the ion exchange process. In the case of kaolinite, for example, the tetrahedral layers of adjacent clay sheets are held tightly by H bonds and only planar external surface sites are available for exchange. In contrast, under adequate hydration conditions, smectites are able to swell, allowing a rapid passage of ions into the interlayer space. Vermiculite is characterized by a more structured interlayer space because the region between layers of silicate is selective for certain types of cations like K^+ and NH_4^+ (Sparks and Huang 1985). Cation exchange also is affected by the particle size of the mineral fraction. For example, it was reported (Kennedy and Brown 1965) that, of the total Ca–Na content of a sand layer, 90 % is composed of particles of 0.12–0.20 mm and only 10 % contains a 0.20–0.50 mm sand fraction. Similar behavior was observed on silt materials where the exchange rates (Ba–K) on medium and coarse silt diminish with increasing particle size.

The organic fraction composition may influence the exchange capacity. A key contribution to the exchange capacity of humus is given by the carboxyl and phenolic hydroxyl functional groups. Under appropriate pH conditions, uranic acids in polysaccharides or carboxy-terminal structures in peptides can contribute to the negative charge and CEC of the soil organic matter. The basic amino acids lysine, arginine, and histidine are positively charged at pH = 6; the amino terminal groups in peptides and polypeptides can be expected to be the principal contributors to positive charges in subsurface organic materials, in an appropriate pH environment (Talibuden 1981). Based on a study of pH dependence of organic contaminant sorption on soil organic matter, Bronner and Goss (2011) suggest that protonation/deprotonation of carboxylic groups in humic matter has no significant influence on the sorption process.

The CEC of the organo-mineral complexes is less than the sum of each of the separate organic and mineral components. The CEC decrease may be explained by changes occurring in the humus configuration following coating of the mineral surface. A significant elucidation of the relative contributions of mineral and organic colloids to the adsorption of organic contaminants was made through studies with separated fractions and well-defined model materials (Gaillardon et al. 1977; Kang and Xing 2005; Celis et al. 1996). A different approach was to study and compare adsorption before and after organic matter removal (Saltzman et al. 1972) to assess the relative importance of soil minerals in parathion uptake. Although the removal of organic matter from soil by oxidation with hydrogen peroxide (a commonly used, strong oxidation agent) could affect the properties of an adsorbent, the results obtained may provide qualitative information about its role and properties in the contaminant retention process. The reported results showed that parathion has a greater affinity for organic adsorptive surfaces than for mineral ones. The important finding from this approach suggests that adsorption is dependent on the type of association between organic and mineral colloids, which determines the nature and the magnitude of the adsorptive surfaces. Although the importance of organic matter has been well established, the properties of organic colloids relevant to the adsorption of contaminants remain to be characterized thoroughly. The available information suggests that these properties could be related to the ratio among humic acid, fulvic acid, and humin, as well as the presence of active groups (e.g., carboxyl, hydroxyl, carbonyl, methoxy), high CEC, and surface area.

The main properties affecting adsorptive capacity of clay are considered to be the available surface and the CEC, as well as the nature of the saturating cation, the hydration status, and the surface acidity. Although amorphous oxides and hydroxides of iron, aluminum, and silica can adsorb organic molecules, only limited information exists in this direction. It is known, however, that Al and Fe hydroxides can adsorb organic contaminants, and therefore, their presence leads to an increase in the adsorption capacity of montmorillonite (Terce and Calvet 1977). For example, after removing Al and Fe oxides from soil particles, the adsorptive capacity of soils for atrazine (an organic herbicide) decreased significantly and the adsorption kinetics were affected (Huang et al. 1984).

The adsorption of contaminants on geosorbents also is affected by climatic conditions reflected in the subsurface temperature and moisture status. Calvet (1984) showed how the soil moisture content may affect adsorption of contaminants originating from agricultural practices. The moisture content determines the accessibility of the adsorption sites, and water affects the surface properties of the adsorbent. The competition for adsorption sites between water and, say, insecticides may explain this behavior. Preferential adsorption of the more polar water molecules by soil hinders insecticide adsorption at high moisture content; reduced competition is found at low moisture content, leading to an increase in adsorption. A negative relationship between pesticide adsorption and soil moisture content has been known for a long time (e.g., Ashton and Sheets 1959; Yaron and Saltzman 1978).

It is important to examine the effect of moisture content on the surface properties of clays and organic matter in relation to the adsorption of organic contaminants. In general, it is accepted that water molecules are attracted to clay surfaces mainly by the exchangeable cations, forming hydration shells. Adsorbed water provides adsorption sites for organic contaminants. An important feature of water associated with clay surfaces is its increased dissociation, giving the surface a slightly acidic character. A negative relationship usually exists between the surface acidity of clays and their water content.

Adsorption usually increases as the temperature decreases, while desorption is favored by temperature increases. The temperature may indirectly affect adsorption by its effect on organic–water interactions. The complex relationship among adsorbent, adsorbate, and solvent as affected by temperature was described by Mills and Biggar (1969) for the case of an organic insecticide. The adsorption of lindane (1,2,3,4,5,6-hexachlorocyclohexane) and its beta-isomer by a peat (high organic content), a clay soil, a Ca-bentonite, and silica gel decreased as the temperature of the system increased. The authors suggested that this adsorption–temperature relationship reflects not only the influence of energy on the adsorption process but also the change in solubility of the adsorbate. They considered that the change in activity in solution with temperature is related to the change in reduced concentration, which is the ratio between the actual concentration of the solute at a given temperature and its solubility at the same temperature. Adsorption isotherms obtained by using the reduced concentration, in contrast to normal adsorption isotherms, showed an increase in adsorption with increasing temperature. This finding suggests that the heat involved in the adsorption process mainly affects solute solubility. Similar results emphasizing the significant influence of temperature on adsorption through its solubility effect were reported by Yamane and Green (1972) for atrazine and by Yaron and Saltzman (1978) for parathion.

Strong, sometimes irreversible retention of organic contaminants on hydrated humic substances can be explained by penetration and trapping in the internal structure. Burchill et al. (1981) showed that hydrated exchangeable cations and some dissociated functional groups, as well as water held by various polar groups of the humic substances, provide adsorption sites for organic contaminants. At low moisture content, the hydrophobic portions of the organic matter structure may bind hydrophobic nonionic organic contaminants. Pignatello (2012) showed that in some soils, it is possible that there are (qualitatively) many different kinds of natural organic matter (NOM) microdomains, which form microstructures that play an important role in selectivity, kinetics and reversibility of organic contaminant sorption. The physical microstructure of the NOM responds to changes in environmental conditions, such as concentration of sorbing chemicals, temperature, and moisture content, thus developing a dynamic system that evolves with time. Pignatello (2012) considers that NOM apparently has a "memory" of its exposure history to environmental stresses, which "challenges the historical paradigm of NOM as a passive sorbent with immutable structure."

5.7 Nonadsorptive Retention of Contaminants

Nonadsorptive (physical) retention of chemicals in the subsurface has received less attention, despite the fact that significant quantities of contaminants can be retained by processes other than purely adsorptive ones.

5.7.1 Contaminant Precipitation

Contaminant precipitation involves accumulation of a substance to form a new bulk solid phase. Sposito (1984) noted that both adsorption and precipitation imply a loss of material from the aqueous phase, but adsorption is inherently two-dimensional (occurring on the solid-phase surface), while precipitation is inherently three-dimensional (occurring within pores and along solid-phase boundaries). The chemical bonds that develop due to formation of the solid phase in both cases can be very similar. Moreover, mixtures of precipitates can result in heterogeneous solids with one component restricted to a thin outer layer, because of poor diffusion. Precipitate formation takes place when solubility limits are reached and occurs on a microscale between and within aggregates that constitute the subsurface solid phase. In the presence of lamellar charged particles with

Table 5.5 Values of pK_a for major heavy metal contaminants (de Boodt 1991)

Heavy metal	pK_a
Cd^{2+}	8.7
Co^{2+}	8.9
Cr^{2+}	6.5
Cu^{2+}	6.7
Mn^{2+}	10.6
Ni^{2+}	8.9
Pb^{2+}	7.3
Zn^{2+}	7.6
Fe^{3+}	3.0

impurities, precipitation of cationic pollutants, for example, might occur even at concentrations below saturation (with respect to the theoretical solubility coefficient of the solvent).

Considering that heavy and transition metals may reach subsurface water as hydrated cations at neutral pH, they may behave as acids, due to formation of a hydration shell surrounding the cation. The "acidity" of hydrated cations depends on the acid dissociation constant (pK_a) values. The lower the pK_a value of the metal, the lower the pH at which precipitates are formed. Values of pK_a for major heavy metals are presented in Table 5.5.

There is a relationship between the solubility of a metal in water, the amount of precipitates formed, and the pH. Formation of a solid precipitate is expressed according to the equation

$$A_aB_{b(\text{solid})} \rightleftharpoons aA^{+b}_{(\text{aq})} + bB^{-a}_{(\text{aq})}, \tag{5.12}$$

where A is a metal and B is a ligand, which precipitate to form the solid $A_aB_{b(s)}$, and a and b are stoichiometric coefficients subject to the constraint of electroneutrality.

The ion activity product (IAP) is a measure of the activity of ions present in the solvent. By definition, the activity of a mineral phase (if present) is unity. Thus, the amount of precipitate does not affect the reaction between the solid and the solvent. When the IAP is much smaller than equilibrium values, there is no precipitation, and because the activity denominator generally is equal to 1, the IAP is given by

$$\text{IAP} = [A^+]_{\text{measured}}[B^-]_{\text{measured}} \tag{5.13}$$

when equilibrium is reached, solubility product constants are used to describe saturated solutions of ionic compounds of relatively low solubility. When the ion concentration in solution reaches saturation, equilibrium between the solid and dissolved ions is established.

The equilibrium constant is given by the product of the concentration of ions present in a saturated solution of ionic compounds,

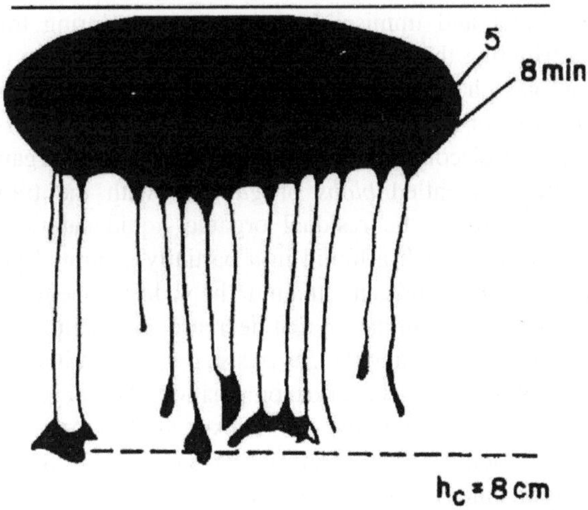

Fig. 5.7 Illustration of a water-immiscible liquid trapped in the vadose zone immediately after a spill. The *dashed line* represents the water table region (Schwille 1984)

$$K_{sp} = [A^{b+}][B^{a-}]^{b}, \tag{5.14}$$

where K_{sp} is the solubility product, or the equilibrium constant, between an ionic solid and its saturated solution. When IAP < K_{sp}, the solution is below saturation and minerals dissolve on contact with the solution. When IAP > K_{sp}, the solution is "supersaturated" and precipitation occurs.

The IAP can be larger than the corresponding solubility product constant for the solid if the active shell of the solid is of radius <1 μm. This behavior may be explained by the fact that the surface energy of these very small particles contributes to the Gibbs energy of the precipitate, increasing the activity relative to that in the standard state, where the interfacial energy component is negligible. Additional precipitate formation processes may occur in the presence of nucleating agents.

It should be noted that, in the natural subsurface solid phase, differentiation between adsorption and precipitation can be very difficult, because the new solid phase may precipitate homogeneously onto the surface of an existing solid phase. Weathering may provide host surfaces for the more stable phase into which they transform chemically.

5.7.2 Liquid Trapping

Trapping is an important form of nonsorptive retention of contaminants in the subsurface. Trapping may occur, for example, when spills of water-immiscible fluid compounds (e.g., petroleum products) leave residual ganglia or bulb configurations in the subsurface.

Water and immiscible fluids interact during transport through pore spaces, distributing themselves according to the properties of the liquids and of the solid and gas phases. Above the retention capacity, subsurface pore geometry permits the flow of nonwetting fluids, leaving behind clusters of water-immiscible liquids that are disconnected from the main body of organic liquid. These clusters are sometimes called *blobs* or *ganglia*, with the trapped immiscible fluid being referred to as the residual organic liquid saturation. Figure 5.7 illustrates the retention of such a liquid in a partially saturated porous medium. Thus, trapped immiscible liquid can remain in the vadose zone for an indefinite time, serving as a source of contamination that decreases in magnitude as a result of processes such as volatilization into the gas phase, dissolution in the leaching water, or chemically or biologically induced decomposition. The degree of porous medium saturation by an immiscible liquid can be expressed in terms of the utilization of pore space by the liquid and air phases (van Dam 1967; Schwille 1984) or as the organic liquid content in volume units.

If the organic liquid saturation is measured as the volume of organic liquid per unit void volume, measured over a representative volume of the porous medium, then S_o, the fraction of pore space occupied by the organic liquid is

$$S_o = V_{\text{organic liquid}}/V_{\text{voids}}, \tag{5.15}$$

where the subscript *o* indicates the organic liquid. The residual saturation at which the organic liquid becomes discontinuous and immobile is then

$$S_{or} = V_{\text{discontinuous organic liquid}}/V_{\text{voids}}, \tag{5.16}$$

where the subscript *or* indicates residual organic liquid. In the saturated zone, the water saturation (S_w) is given by $S_w = 1.0 - S_o$.

The extent of trapping is determined primarily by the physical properties of the vadose zone. If the organic liquids are characterized by a low vapor pressure and a low solubility in water, they remain trapped in the partially saturated zone. In this particular case, the porous medium behaves like an inert material and the behavior of the organic liquids depends only on their own properties, with no interaction between the liquid and the solid phases.

5.7.3 Particle Deposition and Trapping

Retention of suspended particles in porous media occurs by straining (trapping), physicochemical filtration (deposition), and detachment. Depending on the size of the suspended particle, a number of mechanisms may be responsible for physicochemical filtration: (1) *gravitational sedimentation*, where the gravitational forces acting on the particle cause it to settle onto a sediment grain (collector), (2) *interception*, where the particle size and trajectory are such that it encounters the collector grain while flowing past, and (3) *Brownian diffusion*, where the particle is brought into contact with a collector due to its Brownian motion (Yao et al. 1971; Elimelech et al. 1995). Geometric models (Sakthivadivel 1966, 1969) suggest that straining could have a significant influence when the ratio of the particle diameter to the median grain diameter of the porous medium is greater than 0.05. Similarly, a limiting ratio of 0.154 for predicting straining of particles in constrictions has been proposed (Herzig et al. 1970). However, recent experimental evidence suggests that straining could be important at much smaller particle to grain size ratios (Bradford et al. 2003). Mobilization (detachment) of deposited particles also is a key process governing colloid transport and fate. Mobilization can take place following drastic changes in pore water chemistry and when the hydrodynamic forces overcome the adhesive forces between particles and the medium grains (Amirtharajah and Raveendran 1993).

Deposition and trapping of contaminants on colloidal materials and other suspended particles may occur during their transport through the vadose zone and thus create an additional route for pollutant distribution in the subsurface. Below hazardous waste sites, for example, an unexpected transport process of cationic radionuclides (e.g., Pu, Am) or various heavy metals has been observed, which can be explained only by colloid-facilitated transport (McCarthy and Zachara 1989; Penrose et al. 1990; Ryan and Elimelech 1996). Laboratory experiments testing colloid-facilitated redistribution in the partially saturated zone confirmed that colloids can accelerate the transport of cationic and anionic metals (e.g., Vilks et al. 1993) or toxic organic chemicals (e.g., Vinten et al. 1983). Colloidal materials involved in the process of enhanced redistribution of contaminants in the subsurface include inorganic matter like clay minerals, oxides and carbonate particles, with sizes in the range of 10 nm to a few micrometers, as well as organic colloids like humic substances and microbial exudates.

Vinten et al. (1983) demonstrated that the vertical retention of contaminated suspended particles in soils is controlled by the soil porosity and the pore size distribution. Figure 5.8 illustrates the fate of a colloidal suspension in contaminated water during transport through soil. Three distinct steps in which contaminant mass transfer may occur can be defined: (1) contaminant adsorption on the porous matrix as the contaminant suspension passes through subsurface zones, (2) contaminant desorption from suspended solid phases, and (3) deposition of contaminated particles as the suspension passes through the soil.

The suspended solid particle size and the volume of effluent also must be considered in examining deposition in the subsurface. For example, under leaching of a waste disposal site or following irrigation with sewage effluent, the coarse fraction of suspended solids is retained in the upper layer, while the finer colloidal fraction is more mobile, and its transport is controlled by the porosity of the subsurface solid phase.

Particle deposition from aqueous suspensions onto stationary surfaces is a dynamic phenomenon characterized by a transient or time-dependent rate of deposition. The deposition of contaminated suspended particles is affected by the

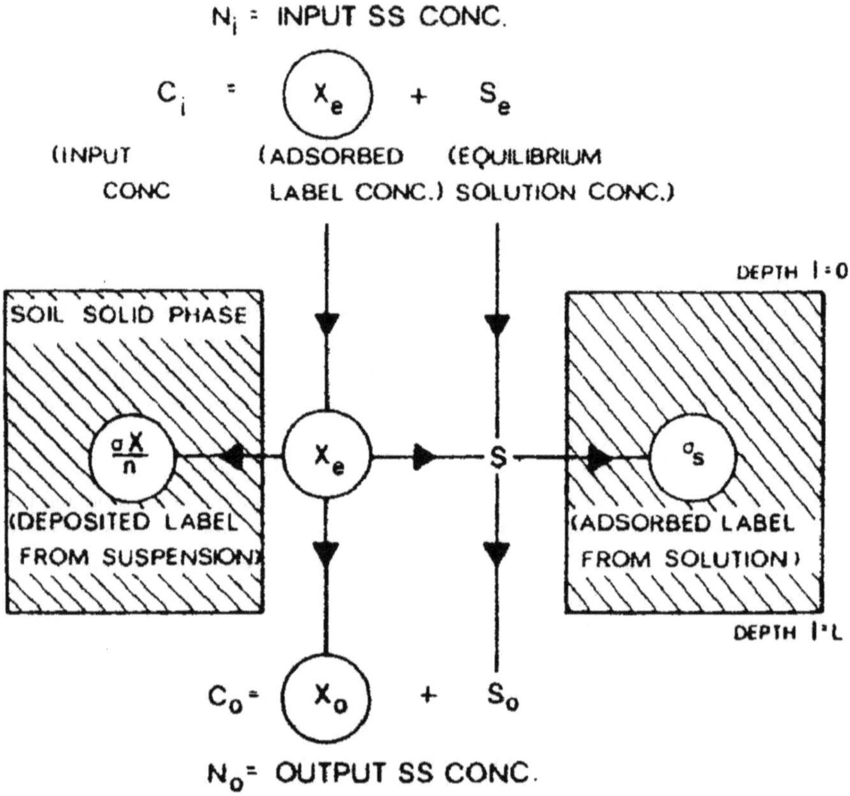

Fig. 5.8 Diagrammatic representation of transport of a labeled suspension through soil; *SS* denotes suspended solids (Vinten et al. 1983)

nature of the surrounding porous medium. A declining deposition rate is observed when particle–particle interactions are repulsive, so that the potential deposition zone becomes progressively occluded as particles accumulate; this leads to a blocking phenomenon.

Ryan and Elimelech (1996) note that conventional filtration theories are applicable only to the initial stage, when mineral grains are devoid of retained particles. As particles deposit onto mineral surface sites with charge characteristics favorable for deposition, the particle deposition rate progressively declines due to the blocking phenomenon. The heterogeneity of the subsurface makes application of deposition models very difficult, and therefore, they usually are relevant only for well-defined materials. Amitay-Rosen et al. (2005) used magnetic resonance imaging to demonstrate spatial and temporal deposition and trapping patterns of colloids in porous media.

5.8 Reversible and Irreversible Retention

Reversible and irreversible retention of contaminants on the subsurface solid phase is a major process in determining pollutant concentrations and controlling their redistribution from the land surface to groundwater. After being retained in the solid, contaminants may be released into the subsurface liquid phase, displaced as water-immiscible liquids, or transported into the subsurface gaseous phase or from the near surface into the atmosphere. The form and the rate of release are governed by the properties of both contaminant and solid phase, as well as by the subsurface environmental conditions. We consider here contaminants adsorbed on the solid phase.

Release through reversible retention can be assessed on the basis of physico-chemical and biological processes. In the case of the former, release is caused by a change in the properties of the fluid surrounding the retaining solid phase. Lowering of the pollutant concentration in the liquid phase, for example, may cause a change in the established equilibrium, and as a consequence, enhanced transfer of the adsorbed compound to the liquid phase occurs. Also, contaminants that previously entered living organisms by an uptake process may subsequently be released. In many cases, the release isotherms do not coincide with the retention isotherms, indicating the phenomenon of hysteresis. This means that not all adsorbed molecules can be transferred back into the solution phase. In the subsurface, where a multicomponent solid phase is present, and where phenomena other than adsorption–desorption may occur, it is better to use the term *retention hysteresis* rather than *adsorption hysteresis*. Retention hysteresis may vary according to the nature of the contaminant and the solid phase, the site and sample history (e.g., wetting-dry cycles), and the experimental procedure used.

Genuine (true) and apparent hysteresis may be considered to explain contaminant release from the subsurface solid phase. Genuine hysteresis assumes that observed data are real and the equilibrium results can be explained on the basis of well-identified phenomena. Apparent hysteresis results from an experimental artifact due, for example, to a failure to reach retention or release equilibrium.

5.8.1 Genuine Hysteresis

Genuine hysteresis is considered when contaminant release results only from desorption. Experimental data can be interpreted in terms of genuine desorption only when the system is at equilibrium and released molecules are those adsorbed onto the solid-phase surface. Molecules brought back into the solution as result of dissolution, diffusion out of the solid matrix, or biotic/abiotic transformation cannot be considered desorbed molecules. In the subsurface, it is almost impossible to distinguish between desorbed molecules and molecules that were not subjected to adsorption and desorption.

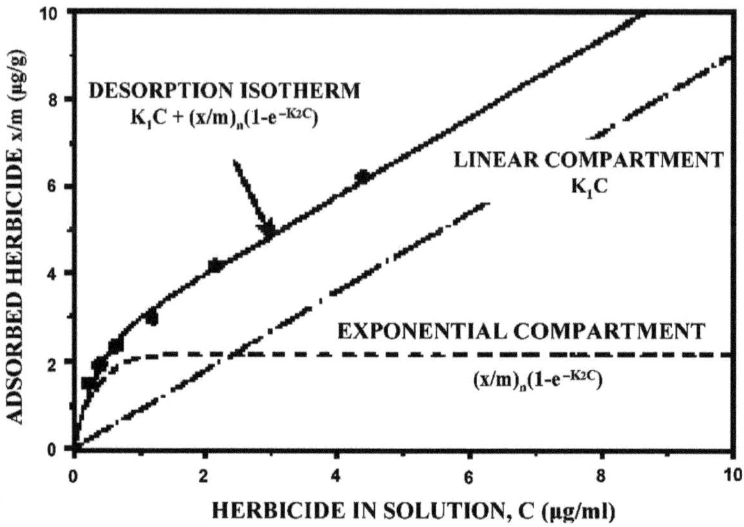

Fig. 5.9 Splitting of the desorption isotherm of dimefuron in the presence of 0.01 M $CaCl_2$ into two other isotherms, corresponding to the two-compartment (*linear*, *exponential*) model of desorption isotherms (Barriuso et al. 1992b)

Desorption isotherm may differ from adsorption isotherms for systems that are not at equilibrium, because the desorption rate is lower than the adsorption rate. Theoretical treatments by Ponec et al. (1974), for gas adsorption, and Giles et al. (1974), for solute adsorption, indicate that the activation energy for adsorption is zero or near zero. Under these conditions, these authors showed that the activation energy for desorption is greater than that for adsorption; consequently, the rate of desorption is lower than adsorption. This behavior pattern also is valid when adsorption is accompanied by dissociation of the adsorbed molecules (Ponec et al. 1974). Adsorbed molecules may be classified according to two categories: molecules retained through physical interactions and able to desorb and molecules that interact strongly with the solid matrix and therefore are released slowly or not at all (Barriuso et al. 1992a, b). Because different mechanisms are involved in the adsorption–desorption process, different types of desorption isotherms can be observed. In one case, desorption is described by a linear isotherm; in another case, the release is described by an exponential function for equilibrium concentrations in solution. Figure 5.9 gives an example of a combination of such correlations for the release of the herbicide dimefuron adsorbed on a clay loam soil.

Several other explanations have been put advanced to explain retention hysteresis, including (1) surface precipitation of metallic cations whose hydroxides, phosphates, or carbonates are sparingly soluble; (2) chemical reactions with solid surfaces, including organic surfaces, which form complexes with metallic cations; and (3) incorporation into the subsurface organic matter through chemical reactions and biochemical transformation. For the case described by Fig. 5.9 or explanations (1) and (2), the contaminant release always exhibits a hysteresis pattern. When dealing with chemical reactions and biochemical transformations (explanation 3), contaminant retention, in some situations, may reach total irreversibility.

5.8.2 Apparent Hysteresis

Apparent hysteresis occurs mainly when complete equilibrium is not reached. Diffusion into the solid matrix or into micropores of aggregates is considered a main cause of apparent hysteresis. In a transitory state, sorption occurs concurrently with desorption and the concentration of contaminant in the liquid phase is erroneously low because some fraction is associated with sorption.

Apparent hysteresis also may be caused by other phenomena. During the consecutive extractions and dilution steps used as a common technique in desorption studies, *weathering* of the sorbent may occur, resulting in a possible increase of contaminant sorption and decrease in its release. *Degradation* of the contaminant induced by physicochemical or biological factors, or a volatilization process leading to a decreased contaminant concentration in solution, are additional factors affecting a true hysteresis result.

The *moisture status* of the subsurface solid phase also may lead to an apparent hysteresis in the adsorption–desorption process. It is known that clay materials, mainly smectites, and humic substances can hydrate and swell or dehydrate and shrink. The physicochemical state of many molecules sorbed in wet conditions may be modified on drying, making the substances more difficult to desorb. The retention of sorbate molecules during drying or slow swelling of organic surfaces may be the reason for the decrease in their desorption. On rewetting, when molecules are sorbed at polar surface sites, they orient their hydrophobic part toward the solution phase, reducing considerably the access of water, and thus slowing down the swelling and desorption (Mingelgrin and Gerstl 1993).

Drying of the subsurface solid phase can cause an increase in the rate of desorption. If penetration of a sorbate toward inner surfaces does not reach its equilibrium by the time drying commences, a fraction of the sorbate may remain localized at more accessible outer surfaces in an amount greater than that corresponding to the equilibrium level. Under these conditions, the drying of the system may increase the rate of desorption during successive rewetting.

The *history of the surface* is an additional factor affecting the release of contaminants adsorbed on solid phases into the liquid or gaseous phase. For example, the effect of drying on contaminant desorption is influenced by the time allowed for its transport into the aqueous phase. In sorbing systems, like sediments that are permanently wet, the history of the system determines the fate of sorbed molecules (Pignatello 1989).

Release methodology may lead to incorrect desorption parameters, which in turn may be (erroneously) interpreted as hysteresis. For example, Hodges and

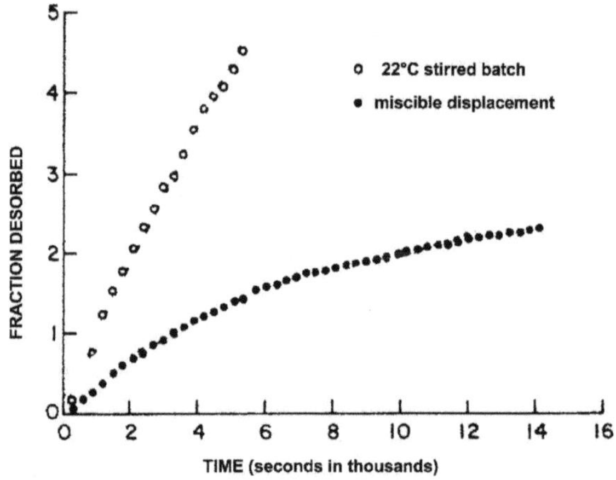

Fig. 5.10 Effect of measurement technique on sulfate desorption from soils (modified after Hodges and Johnson 1987)

Johnson (1987) used two experimental techniques (rapidly stirred batch and miscible displacement with slow flow rate) to study sulfate desorption and found that, in the stirred batch experiments, the desorption readings were less than those obtained by the miscible displacement technique (Fig. 5.10). Even within the miscible displacement technique, the time of leaching was found to have a major effect. The estimated (by extrapolation) time required for complete desorption was 10–20 times greater than for adsorption.

Bowman and Sans (1985) compared two methods (dilution and consecutive desorption) for measuring the desorption of selected synthetic organic pesticides from organo-clay systems. Note that dilution of suspensions may increase the accessibility of an adsorbing surface, so this method is not strictly comparable with the classical method. In all cases studied, only minimal hysteresis in the desorption isotherm was obtained using the dilution method, whereas almost all systems investigated with the consecutive desorption method exhibited considerably larger hysteresis. Rao and Davidson (1980) also suggested that, in the case of pesticides, the centrifugation–resuspension step is in some way responsible for the hysteresis effect, explained by the fact that partially irreversible compaction of the adsorbent during centrifugation greatly increases the time required for desorption.

5.8.3 Bound Residues

The term *bound residue* was adopted by the International Union of Pure and Applied Chemistry (IUPAC) in 1984. According to this definition and that of the European Commission (adopted in 1991), nonextractable residues in soil are chemical species, originating from pesticides, which are not extracted by methods that do not significantly change the chemical nature of the residue. Fuhr et al. (1998) expanded the meaning of *bound residues* to the “compounds in soil, plant

Table 5.6 Examples of bound pesticide residues in soils (Calderbank 1989)

Structural type	Bound residues (% of applied)	Parent detected
Herbicides		
Anilides and ureas	34–90	No
Bipyridyliums	10–90	Yes
Nitroanilines	7–85	No
Phenoxy	28	No
Phosphonate (glyphosphate)	12–95	Yes
Triazines	47–57	?
Insecticides		
Carbamates	32–70	Yes
Organochlorines	7–25	?
Organophosphates	18–80	Yes
Pyrethoids	3–23	No
Fungicides		
Chlorophenols	45–90	Yes?
Nitroaromatic (dinocap)	60–90	Yes?

or animals which persist in the matrix form of the parent substance or its metabolites after extraction." Gevao et al. (2000) include the proviso that bound residues do not include metabolites that are indistinguishable from naturally occurring compounds. Expanding this definition for the subsurface environment, we can state that *bound residues may comprise all toxic chemical species of anthropogenic origin (parent and metabolites) associated with the subsurface solid phase that cannot be separated by current extraction technology.*

Bound residues were first mentioned in the literature by Bailey and White (1964), in relation to pesticide extraction from soils. Over the years, many experiments have shown that the extraction of pesticides from soils is never complete, even when using solvents for which the molecules are highly soluble. Wanner et al. (2005) showed by analysis of ^{14}C-labeled molecules that the fungicide dithianon in soil exhibits bound residues of $\approx$63 % of the applied amount after 64 days. Calderbank (1989) showed that, for a large number of organic agrochemicals, up to 90 % of the applied radioactive-labeled substances become unextractable (Table 5.6). Calderbank (1989) examined a series of experimental data and noticed that different amounts of parent products become irreversibly retained as a function of their molecular structure (Table 5.6). Moreover, it was observed that the extractability decreases with aging, probably because the phenomena responsible for hysteresis become more efficient with increasing residence in soil.

The environmental significance of bound residues must be considered in relation to NOM (Barraclough et al. 2005). Contaminants entering the subsurface contain many functional groups similar to those of NOM and thus become involved in many of the same biological and chemical transformations. If, with aging, the bound residues become indistinguishable from subsurface organic matter, no environmental risk occurs. In contrast, however, if over time, the bound residues exhibit properties different than those of NOM, compounds having a toxic character become a contamination risk for the subsurface.

Bound residues of anthropogenic origin, found in the soil layer, may be compared with those of natural organic molecules released from plant and animal debris and utilized as a source of energy by microbial populations. Parent molecules and their metabolites may interact in the subsurface with the organic matter and then be desorbed and develop further by long-time contact. In this process, known as *aging*, molecules become more tightly bound or entrapped into organic matter or clay fractions of the solid phase. Barraclough et al. (2005) noted that the mass balance of xenobiotics in the subsurface exhibits the same variation as that seen with natural products, in terms of their partitioning between evolved CO_2 and their incorporation into the humic and fulvic substances. For example, more than 80 % of the labeled carbon from a number of xenobiotic compounds was still in the soil several years following their application (Burauel and Fuhr 2000). In a different case, 73 % of the carbon originating from a labeled phenanthrene was recovered as CO_2 after only 82 days (Richnow et al. 2000). Note that the data on carbon evolution may show the rate of incorporation of the labeled carbon from xenobiotics into the subsurface solid phase, but such studies alone do not give information on the bonding of parent compounds and their metabolites on molecular levels. The type of interaction, however, is an important factor determining both the likelihood and rate of release and the form in which the molecules are mobilized.

The mechanism of bound residue formation is better understood today due to the use of advanced extraction, analytic, and mainly spectroscopic techniques (e.g., electron spin resonance, ESR; nuclear magnetic resonance, NMR; Fourier transform infrared spectroscopy), methods that are applied without changing the chemical nature of the residues.

Physical entrapment following intraorganic matter diffusion (IOMD) or interparticle diffusion in clay minerals is another potential explanation for the formation of bound residues. Diffusion out of the solid phase may account partly for hysteresis, particularly for molecules that diffused into the organic aggregates. Entrapping in humic polymer aggregates, suggested by Kahn (1982) and further examined by Wershaw (1986) and Kan et al. (2000), is a possible explanation for hysteresis of substances compatible with the structure of humic substances. The rapid desorption phase is a result of an entrapped pool of readily desorbed material, and the slow phase is controlled by an entrapped or irreversible compartment inside the most hydrophobic part of humic aggregates.

To calculate the release through diffusion of an entrapped residue, Barraclough et al. (2005) considered the size of organic matter particles (effective radius 10^{-7} to 10^{-9} cm) and the effective diffusion coefficient of small organic molecules in a sorbing medium ($D \approx 10^{-9}$ cm^2/s). The time for 50 % of the material in a sphere to diffuse out is given by

$$t_{1/2} = 0.03r^2/D, \tag{5.17}$$

where r is the effective radius (cm) and $t_{1/2}$ is in seconds (Helfferich 1962).

For these entrapped contaminant "spheres," the diffusion is rapid, on the order of seconds rather than days. Kan et al. (2000) suggested a diffusion model for xenobiotics with a slow desorption phase, with a half-life of years rather than seconds, assuming that diffusion is hindered by the NOM matrix and occurs when the dimensions of diffusing molecules approach those of the pores. Under these conditions, hindrance from the wall becomes significant (Renkin 1954) and the drag factor F can be expressed as

$$F = 1 - 2.09(r_m/r_p) + 2.14(r_m/r_p)^3 - 0.95(r_m/r_p)^5 + \cdots, \tag{5.18}$$

where r_m and r_p are the radii of the molecule and the pore, respectively. To extend the diffusion half-life from seconds to years would require a drag factor of around 10^{-8}, in the case where no interaction occurs between the diffusing molecules and the entrapping matrix.

Another process leading to irreversible retention involves chemical binding of contaminant molecules to organic matter (Bollag and Loll 1983). Fulvic and humic acids are the compounds commonly involved in such binding. If binding on organic matter matrix involves physical entrapment, van der Waals forces, or charge transfer, significant release occurs only as a result of matrix-induced degradation by microorganisms or plant enzymes. The reactions involved appear to be the same as those responsible for humic substance formation. Phenol and aromatic amines may bind to organic matter by oxidative coupling, while substituted urea and triazines may not (Bollag et al. 1992). Binding of toxic organic molecules on an organic matter matrix can take place also during humic substance formation by polymerization processes.

References

Abdul AS, Gibson TL, Rai DN (1990) Use of humic acid solution to remove organic contaminants from hydrogeologic systems. Environ Sci Technol 24:328–333

Amirtharajah A, Raveendran P (1993) Detachment of colloids from sediments and sand grains. Coll Surfaces A 73:211–227

Amitay-Rosen T, Cortis A, Berkowitz B (2005) Magnetic resonance imaging and quantitative analysis of particle deposition in porous media. Environ Sci Technol 39:7208–7216

Ashton FM, Sheets TJ (1959) The relationship of soil adsorption of EPTC to oats injury in various soil types. Weeds 7:88–90

Bailey GW, White JL (1964) Review of adsorption and desorption of organic pesticides by soil colloids with implications concerning pesticide bioaviailability. Agri Food Chem 12:324–382

Barraclough D, Kearney T, Croxford A (2005) Bound residues: environmental solution or future problem? Environ Poll 133:85–90

Barriuso E, Baer U, Calvet R (1992a) Dissolved organic matter and adsorption-desorption of difemuron, atrazine and carbetamide by soils. J Environ Qual 21:359–367
Barriuso E, Koskinen WC, Sorenson B (1992b) Modification of atrazine desorption during field incubation experiments. Sci Total Environ 123(124):333–344
Bollag JM, Loll MJ (1983) Incorporation of xenobiotics in soil humus. Experentia 39:1221–1225
Bollag JM, Myers CJ, Minard RD (1992) Biological and chemical interactions of pesticides with soil organic matter. Sci Total Environ 123(124):205–217
Bolt GH (1955) Ion adsorption by clays. Soil Sci 79:267–278
Bolt GH, De Boodt MF, Hayes MF, McBride MB (eds) (1991) Interactions at the soil colloid-solution interface. NATO ASI Series—Applied Science—Series F, vol 190. Kluwer, Dordrecht
Bourg CI, Sposito G (2011) Molecular dynamics simulations of the electrical double layer on smectite surfaces contacting concentrated mixed electrolyte (Nacl–Ca Cl2) solutions. J Colloid Interface Sci 360:201–715
Bowman BT (1979) Method of repeated additions for generating pesticide adsorption-desorption isotherm data. Can J Soil Sci 59:435–437
Bowman BT, Sans WW (1985) Partitioning behavior of insecticide in soil-water system: desorption hysteresis effect. J Environ Qual 14:270–273
Boyd SA, King R (1984) Adsorption of labile organic-compounds by soil. Soil Sci 137:115–119
Boyd GE, Adamson AW, Mayers LS Jr (1947) The exchange adsorption of anions from aqueous solution by organic zeolites. J Am Chem Soc 69:2836–2848
Bradford SA, Simunek J, Bettahar M, van Genuchten MT, Yates SR (2003) Modelling colloid attachment, straining and exclusion in saturated porous media. Environ Sci Technol 37:2242–2250
Bronner G, Goss KU (2011) Sorption of organic chemicals to soil organic matter: influence of soil variability and pH dependence. Env Sci Technol 45:1307–1312
Brusseau ML, Jessup RE, Rao PSC (1990) Sorption kinetics of organic chemicals: evaluation of gas purge and miscible displacement techniques. Environ Sci Technol 24:727–735
Burauel P, Fuhr F (2000) Formation and long term fate of non-extractable residues in outdoor lysimetr studies. Environ Pollution 108:45–52
Burchill SM, Hayes MHB, Greenland DJ (1981) Adsorption. In: Greenland DJ, Hayes MHB (eds) Chemistry of soil processes. Wiley, New York, pp 224–400
Calderbank A (1989) The occurrence and significance of bound pesticide residues in soil. Rev Environ Contam Toxicol 108:69–103
Calvet R (1984) Behavior of pesticides in unsaturated zone. Adsorption and transport phenomenon. In: Yaron B, Dagan G, Goldschmid J (eds) Pollutants in porous media. Springer, Heidelberg, pp 143–151
Calvet R (1989) Analyse du concept de biodiponibilite d'une substance dans le sol. Sci Sol 26:183–201
Celis R, Cox L, Hermosin MC, Cornejo J (1996) Retention of metamitron by model and natural particulate matter. Intern J Environ Anal Chem 65:245–260
Chien SH, Clayton WR (1980) Application of Elovich equation to the kinetics of phosphate release and sorption in soils. Soil Sci Am J 44:265–268
de Boodt MF (1991) Application of the sorption theory to eliminate heavy metals from waste water and contaminated soils. In: Bolt GH, de Boodt MF, Hayes MHB, McBride MB (eds) Interactions at the soil colloid-soil solution interface NATO ASI Series E Applied Sciences, vol 190. Kluwer, Dordrecht, pp 291–321
Delle Site A (2001) Factors affecting sorption of organic compounds in natural sorbent-water systems and sorption coefficients for selected pesticides: a review. J Phys Chem Ref Data 30:187–439
Elimelech M, Gregory J, Jia X, Williams RA (1995) Particle deposition and aggregation: measurement, modelling, and simulation. Butterworth-Heinemann, Oxford
Farrell J, Reinhard M (1994) Desorption of halogenated organics from model solids, sediments and soils under unsaturated conditions. Environ Sci Technol 28:53–72

Fripiat J, Chaussidon J, Jelli A (1971) Chimi-physique des phenomenes de surface. Masson, Paris
Fuhr F, Ophoff H, Burauel P, Wanner U, Haider K (1998) Modification of definition of bound residues. In: Fuhr F, Phoff H (eds) Pesticide bound residues in soils. Wiley-VCH, Weinheim, pp 175–176
Gaillardon P, Calvet R, Terce M (1977) Adsorption and desorption de la terbutryne par une montmorillonite-Ca et des acides humiques seules ou en melanges. Weed Res 17:41–48
Gevao B, Semple KT, Jones KC (2000) Bound pesticide residues in soils: a review. Environ Pollut 108:3–14
Giles CH, MacEwan TH, Nakhwa SN, Smith D (1960) Studies in adsorption, part XI. A system of classification of solution adsorption isotherms and its use in diagnosis and adsorptive mechanism and measurements of specific area of solids. Chem Soc J 3:3973–3993
Giles CH, Smith D, Huitson A (1974) A general treatment and classification of the solute adsorption isotherms. J Colloid Interface Sci 47:755–765
Grahame DC (1947) The electrical double layer and the theory of electro capillarity. Chem Rev 41:441–449
Greenland DJ, Hayes MHB (eds) (1981) The chemistry of soil processes. Wiley, Chichester
Greenland DJ, Mott CJB (1978) Surfaces of soil particles. In: Greenland DJ, Hayes MHB (eds) The chemistry of soil processes. Wiley, Chichester, pp 321–354
Hassett IJ, Banwart WL (1989) The sorption of non polar organics by soil and sediments. In: Sawhney BL, Brown K (eds) Reactions and movement of organic chemicals in soils, vol 22. Soil Science Society of America Special Publication, Madison, pp 31–45
Hayes MHB, Mingelgrin U (1991) Interaction between small organic chemicals and soil colloidal constituents. In: Bolt GH, De Boodt MF, Hayes MF, McBride MB (eds) Interactions at the soil colloid-solution interface. NATO ASI Series—Applied Science—Series F vol 190, Kluwer, Dordrecht, pp 324–401
Helfferich F (1962) Ion exchange. McGraw-Hill, New York
Hermosin MC, Carnejo J (1992) Removing 2,4-d from water by organo-clays. Chemosphere 24:1493–1503
Herzig JP, Leclerc DM, LeGoff P (1970) Flow of suspensions through porous media—Application to deep filtration. Ind Eng Chem 62:8–35
Hodges SC, Johnson GC (1987) Kinetics of sulfate adsorption and desorption by Cecil soils using miscible displacement. Soil Sci Soc Am J 51:323–327
Huang PM, Crossan LS, Rennie DA (1968) Chemical dynamics of K release from potassium minerals common in Soils. Trans 9th Int Congr Soil Sci 2:705–712
Huang PM, Grover R, McKercher RB (1984) Components and particle size fractions involved in atrazine adsorption by soils. Soil Sci 138:220–224
Kahn SU (1982) Bound pesticides residues in soil and plant. Residue Rev 84:1–25
Kan AT, Chen W, Tomson MB (2000) Desorption kinetics from neutral hydrophobic organic compounds from field contaminated sediment. Environ Pollut 108:81–89
Kang SH, Xing BS (2005) Phenanthrene sorption to sequentially extracted soil humic acids and humans. Environ Sci Technol 39:134–140
Karickhoff SW (1981) Semi-empirical estimation of sorption of hydrophobic pollutants on natural sediments and soils. Chemosphere 10:833–846
Karickhoff SW (1984) Organic pollutant sorption in aquatic systems. J Hydraul Eng 110:707–735
Kennedy VC, Brown TC (1965) Experiments with a sodium ion electrode as a mean to studying cation exchange rate. Clays Clay Miner 13:351–352
Lafleur KS (1979) Sorption of pesticides by model soils and agronomic soils: rates and equilibria. Soil Sci 127:94–101
Lambert SM (1967) Functional relationship between sorption in soil and chemical structure. J Agr Food Chem 15:572–576
Mackay D, Paterson S (1981) Calculating fugacity. Environ Sci Technol 15:1006–1614
McBride MB (1989) Reactions controlling heavy metals solubility in soils. Adv Soil Sci 10:1–47
McBride MB (1994) Environmental chemistry of soils. Oxford University Press, Oxford

McCarthy JF, Zachara JM (1989) Subsurface transport of contaminants. Environ Sci Technol 23:496–502

McCarty PL, Reinhard M, Rittman BE (1981) Trace organics in ground water. Environ Sci Technol 15:40–51

Mills AC, Biggar JW (1969) Solubility-temperature effect on the adsorption of gamma and beta—BHC from aqueous and hexane solutions by soils materials. Soil Sci Am Soc Am Proc 33:210–216

Mingelgrin U, Gerstl Z (1993) A unified approach to the interactions or small molecules with macrospecies. In: Beck AJ, Jones KC, Hayes MHB, Mingelgrin U (eds) Organic substances in soil and water. Royal Society of Chemistry, Cambridge, pp 102–128

Mortland MM (1970) Clay-organic complexes and interactions. Adv Agron 22:75–117

Nielsen DR, Biggar JW (1961) Miscible displacement in soils. I. Experimental information. Soil Sci Soc Am Proc 25:1–5

Nye PH, Tinker PB (1977) Solute movement in the soil-root system. Blackwell, Oxford

O'Day PA (1999) Molecular environmental geochemistry. Rev Geophys 37:249–274

Oades JM (1989) An introduction to organic matter in mineral soils. In: Dixon JB, Weed SB (eds) Minerals in soil environment. Soil Science Society of America, Madison, pp 89–153

Overbeek JT (1952) Electrochemistry of the double layer. Colloid Sci 29:119–123

Penrose WR, Polzer WL, Essington EH, Nelson DM, Orlandini KA (1990) Mobility of plutonium and americium through a shallow aquifer in a semiarid region. Environ Sci Technol 24:228–234

Petersen LW, Moldrup P, El-Farhan YH, Jacobsen OH, Yamaguchi Y, Rolston DE (1995) The effect of moisture and soil texture on the adsorption of organic vapors. J Environ Qual 24:752–759

Peterson MS, Lion LW, Shoemaker CA (1988) Influence of vapor phase sorption and diffusion on the fate of trichloroethylene in an unsaturated aquifer system. Environ Sci Technol 22:571–578

Pignatello JJ (1989) Sorption dynamics of organic compounds in soils and sediments. In: Sawhney BL, Brown K (eds) Reactions and movement of organic chemicals in soils, vol 22. Soil Science Society of America Special Publication, Madison, pp 45–81

Pignatello JJ (2012) Dynamic interactions of natural organic matter and organic compounds. J Soils Sediments 12:1241–1256

Podoll RT, Irwin KC, Parish HJ (1989) Dynamic studies of naphthalene sorption on soil from aqueous-solution. Chemosphere 18:2399–2412

Polysopoulos NA, Keramidas VZ, Pavlatou A (1986) On the limitation of the simplified Elovich equation in describing the kinetics of phosphate sorption and release from soils. J Soil Sci 37:81–87

Ponec V, Knor Z, Cerny S (1974) Adsorption on solids. Butterworth, London

Quirk JP, Posner AM (1975) Trace element adsorption by soil minerals. In: Nicholas DJ, Egan AR (eds) Trace elements in soil plant animal system. Academic, New York, pp 95–107

Rao PSC, Davidson JM (1980) Estimation of pesticide retention and transformation parameters required in nonpoint source pollution models. In: Overcash MR, Davidson JM (eds) Environmental impact of nonpoint source pollution. Ann Arbor Science, Ann Arbor, pp 23–67

Rao PSC, Horsnby AG, Kilcrease DP, Nkedi-Kizza P (1985) Sorption and transport of hydrophobic organic chemicals in aqueous and mixed solvent system: Model development and preliminary evaluation. J Environ Qual 14:376–383

Rao PSC, Lee LS, Nkedi-Kizza P, Yalkowsky SH (1989) Sorption and transport of organic pollutants at waste disposal sites. In: Gerstl Z, Chen Y, Mingelgrin U, Yaron B (eds) Toxic organic chemicals in porous media. Springer, Heidelberg, pp 176–193

Renkin EM (1954) Filtration, diffusion and molecular sieving through porous cellulose membranes. J Gen Phys 38:224–243

Richnow HH, Annweiler E, Koning M, Luth JC, Stegmann R, Garms C, Franke W, Michaelis W (2000) Tracing the transformation of labeled [1-^{13}C] phenanthrene in a soil bioreactor. Environ Pollut 108:91–101

Ryan JN, Elimelech M (1996) Colloid mobilization and transport in groundwater. Colloids Surf A 107:1–56
Sakthivadivel R (1966) Theory and mechanism of filtration of non-colloidal fines through a porous medium. Rep. HEL 15–5, Hydraulic Engineering Laboratory, University of California, Berkeley
Sakthivadivel R (1969) Clogging of a granular porous medium by sediment. Rep. HEL 15–7, Hydraulic Engineering Laboratory, University of California, Berkeley
Saltzman S, Yaron B (eds) (1986) Pesticides in soils. Van Nostrand Reinhold, New York
Saltzman S, Kliger L, Yaron B (1972) Adsorption-desorption of parathion as affected by soil organic matter. J Agric Food Chem 20:1224–1227
Schrap SM, De Vries PJ, Opperhuizen A (1994) Experimental problems in determining sorption coefficients of organic-chemicals—An example for chlorobenzenes. Chemosphere 28:931–945
Schwarzenbach RP, Gschwend PM, Imboden DM (2003) Environmental organic chemistry, 2nd edn. Wiley-Interscience, New York
Schwille F (1984) Migration of organic fluids immiscible with water in the unsaturated zone phenomenon. In: Yaron B, Dagan G, Goldschmid J (eds) Pollutants in porous media. Springer, Heidelberg, pp 143–151
Sparks DL (ed) (1986) Soil physical chemistry. CRC Press, Boca Raton
Sparks DL (1989) Kinetics of soil processes. Academic, San Diego
Sparks DL, Huang PM (1985) Physical chemistry of soil potassium. In: Munson RE (ed) Potassium in agriculture. ASA, Madison, pp 201–276
Sparks DL, Jardine PM (1984) Comparison of kinetic equations to describe K–Ca exchange in pure and mixed systems. Soil Sci 138:115–122
Sposito G (1984) The surface chemistry of soils. Oxford University Press, Oxford
Stern O (1924) Zur theorie der elecktrolytischen doppelschicht. Z Electrochem 30:508–516
Talibuden O (1981) Cation exchange in soils. In: Greenland DJ, Hayes MHB (eds) The chemistry of soil processes. Wiley, Chichester, pp 115–178
Terce M, Calvet R (1977) Some observations on the role of Al and Fe and their hydroxides in the adsorption of herbicides by montmorillonite. Sonderdruck Z Planzen—kr Pfanzenschutz Sonderheft VIII
Theng BKG (1974) The chemistry of clay-organic reactions. Adam Hilger, London
van Dam J (1967) The migration of hydrocarbon in a water bearing stratum. In: Heppe P (ed) The joint problems of oil and water industries. Institute of Petroleum, London, pp 56–96
van Olphen H (1967) An introduction to clay colloid chemistry. Wiley, New York
Vilks P, Cramer JJ, Bachinski DB, Doern DC, Miller AG (1993) Studies of colloids and suspended particles, Cigar Lake uranium deposit, Saskatchewan, Canada. Appl Geochem 8:605–616
Vinten AJ, Mingelgrin U, Yaron B (1983) The effect of suspended solids in waste water on soil hydraulic conductivity I. Suspended solid labeling method II. Vertical distribution of suspended solids. Soil Sci Soc Am J 47:402–412
Wanner U, Fuhr F, deGraaf AA, Burauel P (2005) Characterization of non-extractable residues of the fungicide dithianon in soil using C-13/C-14-labelling. Environ Poll 133:35–41
Wauchope RD, Koskinen WC (1983) Adsorption-desorption equilibria of herbicides in soil: a thermodynamic perspective. Weed Sci 31:504–512
Weber JB, Miller CT (1989) Organic chemical movement over and through soil. In: Sawhney BL,Brown K (eds) Reactions and movement of organic chemicals in soils, vol 22. Soil Science Society of America Special Publication, Madison, Wisconsin, pp 305–335
Weber WJ Jr, McGinley PM, Katz LE (1992) A distributed reactivity model for sorption by soils and sediments 1. Conceptual basis and equilibrium assessments. Environ Sci Technol 26:1955–1962
Wershaw RL (1986) A new model for humic materials and their interactions with hydrophobic organic chemicals in soil-water or in sediment-water systems. J Contam Hydrol 1:29–45

Yamane VK, Green RE (1972) Adsorption of ametryne and atrazine on an oxisol, montmorillonite and charcoal in relation to pH and solubility effects. Soil Sci Soc Am Proc 36:58–64
Yao KM, Habibian MT, O'Melia CR (1971) Water and waste water filtration: concepts and applications. Environ Sci Technol 5:1105–1112
Yaron B, Saltzman S (1978) Soil-parathion surface interactions. Residue Rev 69:1–34
Yaron B, Calvet R, Prost R (1996) Soil pollution: processes and dynamics. Springer, Heidelberg

Chapter 6
Contaminant Partitioning in the Aqueous Phase

One way that contaminants are retained in the subsurface is in the form of a dissolved fraction in the subsurface aqueous solution. As described in Chap. 1, the subsurface aqueous phase includes "retained water," near the solid surface, and "free" water. If the "retained water" has an apparently static character, the subsurface "free" water is in a continuous feedback system with any incoming source of water. The amount and composition of incoming water are controlled by natural or human-induced factors. Contaminants may reach the subsurface liquid phase directly from a polluted gaseous phase, from point and nonpoint contamination sources on the land surface, from already polluted groundwater, or from the release of toxic compounds adsorbed on suspended particles. Moreover, disposal of an aqueous liquid that contains an amount of contaminant greater than its solubility in water may lead to the formation of a type of emulsion containing very small droplets. Under such conditions, one must deal with "apparent" solubility, which is greater than "handbook" contaminant solubility values.

It is understood that contaminant solubility in an aqueous solution may be affected by environmental factors, such as ambient pressure, temperature, and composition of the aqueous solution. However, reference data usually found in the literature are related to "pure" water and a conventionally accepted temperature of 25 °C. These are considered standard conditions for a standard state of the chemicals. Any deviation from standard conditions might be explained by defining the effect of each isolated factor on the amount and rate of chemical solubility.

The solubility of most inorganic and organic contaminants in water increases with temperature. Changes in the "real" concentration of a solute during changes in ambient temperature should be considered when dealing with the partitioning of pollutants among subsurface phases. Temperature has a direct effect on chemical solubility, but it also has an indirect effect on various reactions occurring in the subsurface. Moreover, the seasonal variations in temperature might affect the solubility of toxic chemicals in subsurface solutions; an observed solubility equilibrium therefore only reflects the solubility of a compound at a given time and ambient temperature.

B. Berkowitz et al., *Contaminant Geochemistry*,
DOI: 10.1007/978-3-642-54777-5_6,

6.1 Solubility Equilibrium

The solubility equilibrium, subject to natural processes in the subsurface matrix, was examined in Chap. 2. The process of contaminant dissolution is affected by the molecular properties of the compound, the composition of the aqueous solution, and the ambient temperature. Here, we focus our discussion on pollutant behavior.

The subsurface system contains a dynamic and complex array of inorganic and organic constituents. The fate of pollutants in the aqueous phase therefore is governed by a variety of reactions, including acid–base equilibria, oxidation–reduction equilibria, complexation with organic and inorganic ligands, precipitation and dissolution, ion exchange, and adsorption. The rate at which these reactions occur, together with the rate of volatilization and biologically or chemically induced degradation, controls contaminant concentrations in subsurface water. Figure 6.1 shows the relationships among these reactions.

Solubility equilibrium is the final state to be reached by a chemical and the subsurface aqueous phase under specific environmental conditions. Equilibrium provides a valuable reference point for characterizing chemical reactions. Equilibrium constants can be expressed on a concentration basis (K^c), on an activity basis (K^o), or as mixed constants (K^m) in which all parameters are given in terms of concentration, except for H^+, OH^-, and e^- (electron) which are given as activities.

Next, we provide a short description of the reactions involving transfer of protons and electrons that affect the solubility equilibria.

Acid–base equilibria are described by a group of reactions covering the transfer of protons, in which the proton donor is an acid and the acceptor is a base, acid $\rightleftharpoons$ base + proton, with the equilibrium constant given by

$$K_A^\circ = [\mathrm{H}^+][\mathrm{base}]/[\mathrm{acid}] \tag{6.1}$$

where K_A^o is an acidity constant, and concentrations are denoted by []. From Eq. 6.1, it follows that, for a given ratio of activities of a particular acid and its conjugate base, the proton activity has a fixed value. In a system with a complex of two bases and two acids, the relation becomes $\mathrm{acid}_1 + \mathrm{base}_2 \rightleftharpoons \mathrm{base}_1 + \mathrm{acid}_2$ and the corresponding equilibrium becomes

$$K_{1,2}^\circ = K_{A1}^\circ / K_{A2}^\circ = [\mathrm{base}_1][\mathrm{acid}_2]/[\mathrm{acid}_1][\mathrm{base}_2], \tag{6.2}$$

where K_{A1}^o and K_{A2}^o denote the acidity constants of the two acids.

Water is the ever-present proton acceptor in the subsurface. During the dissociation of an acid in subsurface water, H_3O^+ is one of the dissociation products and the acid strength is a measurable parameter. In a dilute solution, the activity of the hydrated protons equals that of H_3O^+ and the pH value characterizes the H-ion activity. Substituting for pH in Eq. 6.1, we obtain

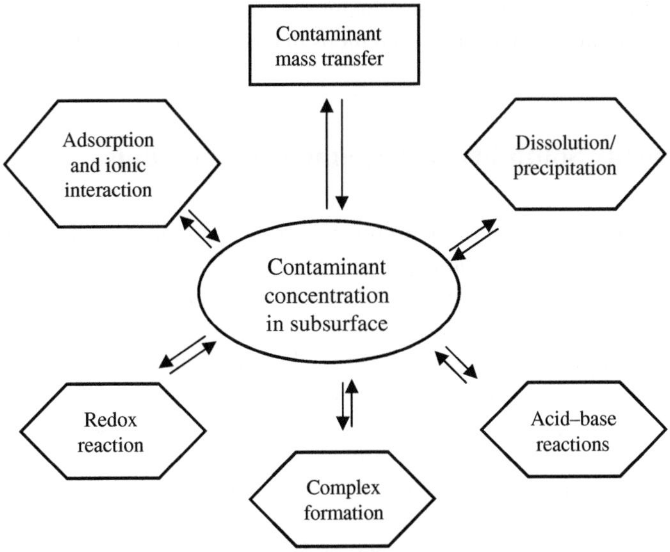

Fig. 6.1 Major processes controlling the fate of contaminants in subsurface water

$$pH - pK_A = \log([\text{base}]/[\text{acid}]). \tag{6.3}$$

Oxidation–reduction equilibria exhibit a conceptual analogy to acid–base equilibria. Similar to the approach of acids and bases acting as proton donors and proton acceptors, reducing and oxidization agents are electron donors and electron acceptors, respectively (recall Sect. 2.2.2). The redox reaction between m moles of an oxidant A_{ox} and n moles of a reductant B_{red} can be written as

$$mA_{ox} + nB_{red} \rightleftharpoons nA_{red} + mB_{ox}, \tag{6.4}$$

which is equivalent to the proton exchange equation.

Every redox reaction includes a reduction half-reaction and an oxidation half-reaction. A reduction half-reaction involving water, in which a chemical species accepts electrons, may be written in the form

$$aA_{ox} + bH^+ + e^- \rightleftharpoons zA_{red} + gH_2O, \tag{6.5}$$

where A represents a chemical species in any phase and the subscripts ox and red denote its oxidized states. The parameters a, b, z, and g are stoichiometric coefficients, while H^+ and e^- denote the proton and the electron in the water.

The potential for redox reactions to take place in subsurface aqueous solutions always exists, but these reactions do not necessarily occur, even though the redox reaction may be favored thermodynamically. When the reaction is slow, in some cases, thermodynamic equilibrium can be achieved only in the presence of a catalysis process. In the subsurface, mainly in the near surface (soil layer) region, conditions may favor microbial populations that act as a catalyst in redox reactions.

6.2 Aqueous Solubility of Organic Contaminants

Aqueous solubility of organic contaminants from a gaseous, liquid, or solid source is of major interest in contaminant geochemistry. To define the solubilities and activity coefficients of organic gases, the partial pressure of a compound in the gas phase at equilibrium above a liquid solution is considered identical to the fugacity (see Sect. 7.3) of the compound in the solution. Figure 6.2 conceptualizes the fugacity of a compound in a nonideal liquid mixture (e.g., water solution) where the gas and liquid phases are in equilibrium. Fugacity is a measure of chemical potential that indicates the tendency of a substance to move from one phase to another or from one site to another.

It is known that water is an associated liquid, in which molecules are hydrogen-bonded forming packets of a number of H_2O molecules. In this situation, the organic solutes are not in direct contact with individual water molecules. During the transfer of organic contaminant from its pure liquid phase into a pure aqueous phase, a number of water molecules surround each organic molecule. The water molecules adjacent to the organic solute are different than the bulk water molecules, because of the presence of hydrogen bonds (Schwarzenbach et al. 2003). Meng and Kollman (1996) suggest that water surrounding a nonpolar (organic) molecule maintains but does not enhance its H-bonding network. At ambient temperature, water molecules adjacent to a nonpolar organic molecule lose only a very small proportion of their total hydrogen bonds and, consequently, are able to host a nonpolar solute of limited size (Blokzijl and Engberts 1993).

Organic compounds exhibit, as a function of their molecular properties, a spectrum of solubility values starting from almost zero (for compounds completely immiscible with water, mainly nonpolar organic compounds) and reaching a large value for polar organic compounds. Because of their very low solubility in water, nonpolar organic compounds often are referred to as *nonaqueous phase liquids* (NAPLs). When an NAPL comes into contact with a water source, some organic molecules leave the organic phase and dissolve into the water, while some water molecules enter the organic liquid. When the flux of molecules into and out of the organic phase reaches an equilibrium, the amount of organic matter in the water is constant.

The dissolution of composite nonaqueous phase liquids (CNAPLs) in water is of major importance when considering the fate of organic contaminants in the subsurface. In spite of their limited aqueous solubility, these organic contaminants generally are toxic even in relatively minute amounts. Based on a laboratory experiment examining the dissolution in water of a mixture of seven structurally similar polycyclic aromatic hydrocarbons in toluene, Weber et al. (1998) concluded that variations from predictions using Raoult's law by up to a factor of 2

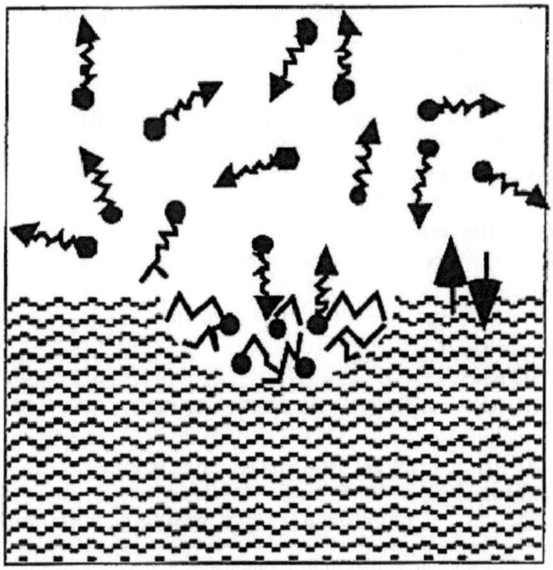

Fig. 6.2 Conceptualization of the fugacity of a compound in a nonideal liquid mixture when gas and liquid phases are in equilibrium (Schwarzenbach et al. 2003)

may be expected in activity coefficients for components deviating significantly from an average mixture. Moreover, the single-impedance mass transfer coefficient determined for one component can be representative of others with similar free liquid diffusivity but is limited when a CNAPL is incompletely characterized, with uncertainties in bulk property values and mole fraction composition.

6.3 Ligand Effects

A molecule or an ion present in a solution in various forms is called a *chemical species* or, more concisely, a *species*. Any combination of cations with molecules or anions that contain free pairs of electrons is called a *complex formation* and might be electrostatic, covalent, or a mixture of both. Ligands are the molecules coordinated around a central cation atom. If a molecule contains more than one ligand, the atoms form a multidentate complex; such complexes are called *chelates*. In a broad classification of chemical reactions, formation of these complexes involves a coordination reaction, similar to acid–base and precipitation–dissolution reactions. Coordination reactions of cations in the subsurface aqueous solution are exchange reactions, wherein the coordinated water molecules are exchanged for some preferred ligands. On the basis of Bjerrum's ion association model, Stumm and Morgan (1996) distinguished two types of complex species: ion pairs formed by two ions that approach a critical distance and are no longer electrostatically effective, and complexes resulting from the formation of largely covalent bonds between a metal ion and an electron donating ligand.

Table 6.1 Hard and soft Lewis acids and bases in solution (Sposito 1981)

Lewis acids
Hard acids
H^+, Li^+, Na^+, K^+ (Rb^+, Cs^+), Mg^{2+}, Ca^{2+}, Sr^{2+} (Ba^{2+}), Ti^{4+}, Zr^{4+}, Cr^{3+}, Cr^{6+}, MoO^{3+}, Mn^{2+}, Mn^{3+}, Fe^{3+}, Co^{3+}, Al^{3+}, Si^{4+}, CO_2
Borderline acids
Fe^{2+}, Co^{2+}, Ni^{2+}, Cu^{2+}, Zn^{2+} (Pb^{2+})
Borderline acids
Fe^{2+}, Co^{2+}, Ni^{2+}, Cu^{2+}, Zn^{2+} (Pb^{2+})
Soft acids
Cu^+, Ag^+, Au^+, Cd^{2+}, Hg^+, Hg^{2+}, CH_3Hg^+; pi-acceptors such as quinones; bulk metals
Lewis bases
Hard bases
NH_3, RNH_2, H_2O, OH^-, O^{2-}, ROH, CH_3COO^-, $CO_3{}^{2-}$, $NO_3{}^-$, $PO_4{}^{3-}$, $SO_4{}^{2-}$, F^-
Borderline bases
$C_6H_5NH_2$, C_2H_5N, N_2, $NO_2{}^-$, $SO_3{}^{2-}$, Br^- (Cl^-)
Soft bases
C_2H_4, C_6H_6, R_3P $(RO)_3P$, R_3As, R_2S, RSH, $S_2O_3{}^-$, S^{2-}, I^-

R = organic molecular unit. ( ) indicates a tendency to softness

The principle of hard and soft Lewis acids and bases, proposed by Pearson (1963), is useful to describe these reactions. A Lewis acid is any chemical species that employs an empty electronic orbital available for reaction, while a Lewis base is any chemical species that employs a doubly occupied electronic orbital in a reaction. Lewis acids and bases can be neutral molecules, simple or complex ions, or neutral or charged macromolecules. The proton and all metal cations of interest in subsurface aqueous solutions are Lewis acids. Lewis bases include H_2, O_2, oxyanions, and organic N, S, and P electron donors. A list of selected hard and soft Lewis acids and bases found in soil solutions is presented in Table 6.1.

An *inner-sphere* complex is formed between Lewis acids and bases, while an *outer-sphere* complex involves a water molecule interposed between the acid and the base. A hard Lewis acid is a molecular unit of small size, high oxidation state, high electronegativity, and low polarizability; whereas a soft Lewis acid is a molecular unit of relatively large size, characterized by low oxidation state, low electronegativity, and high polarizability. Based on this characterization, hard bases prefer to complex hard acids, and soft bases prefer to complex soft acids, under similar conditions of acid–base strength.

The presence of outside ligands in the subsurface aqueous solution leads to an increase in the solubility of coordinating ions. A complex with any ligand, L_n, or its protonated form, H_kL_n, has a total solubility expressed by

$$A \rightleftharpoons [A]_{\text{free}} + \sum\nolimits_i [A_m H_k L_n (\text{OH})]_i, \tag{6.6}$$

where A represents the ion to be coordinated with different ligand types L_n and all values of m, n, or $k \geq 0$ must be considered in the summation.

Free metal ion	Inorganic complexes	Organic complexes	Colloids large polymers	Surface bound	Solid bulk phase, lattice
$Cu^{2+}_{(aq)}$	$CuCO_3$ $CuOH^+$ $Cu(CO_3)_2$ $Cu(OH)_2$	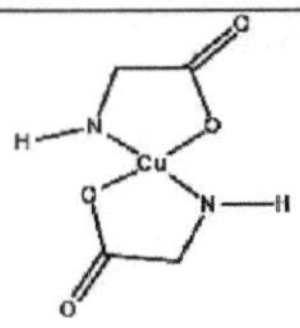 Fulvate	Inorganic Organic	Fe—OCu 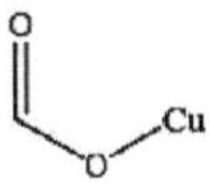	CuO $Cu_2(OH)_2CO_3$ Solid sollution

Fig. 6.3 Forms of occurrence of metal species in natural waters (Stumm and Morgan 1996)

Humic and fulvic acids are the main natural ligands acting in the subsurface aqueous solution. An example of a metal species that may occur in natural waters as a result of potential inorganic and organic ligands is presented in Fig. 6.3. It is difficult, however, to differentiate experimentally between dissolved and colloid substances in subsurface aqueous solutions, and this should be considered when considering the solubility of toxic trace elements in the subsurface system.

6.4 Cosolvents and Surfactant Effects

Many contaminants, such as pesticides and pharmaceuticals, reach the subsurface formulated as mixtures with dispersing agents (surfactants). Such formulations increase the aqueous solubility of the active compounds, and these surfactants form nearly ideal solutions with the aqueous phase.

Addition of a cosolvent is an alternative mechanism to increase contaminant solubility in an aqueous solution. When a contaminant with low solubility enters an aqueous solution containing a cosolvent (e.g., acetone), the logarithm of its solubility is nearly a linear function of the mole fraction composition of the cosolvent (Hartley and Graham-Bryce 1980). The amount of contaminant that can dissolve in a mixture of two equal amounts of different solvents, within an aqueous phase, is much smaller than the amount that can dissolve solely by the more powerful solvent. In the case of a powerful organic solvent miscible with water, a more nearly linear slope for the log solubility versus solvent composition relationship is obtained if the composition is plotted as volume fraction rather than mole fraction.

Yalkowsky and Roseman (1981) and Rubino and Yalkowsky (1987) suggest the following equations for relating solubility of a nonpolar solute (S_m) in a binary mixture of an organic solvent and water to that in pure water (S_w):

$$\log S_m = \log S_w + \sigma_s f_s$$

with

$$\sigma_s = (A \ \log K_{ow} + B), \tag{6.7}$$

where f_s is the volume fraction of cosolvent in binary mixed solvent, A and B are empirical constants dependent on cosolvent properties, and K_{ow} is the octanol–water partition coefficient of the solute. The solute mixture of interest could comprise, for example, two hydrophobic organic chemicals or one hydrophobic and one ionizable organic chemical.

Crystalline salts of many organic acids and bases often have a maximum solubility in a mixture of water and water-miscible solvents. The ionic part of such a molecule requires a strongly polar solvent, such as water, to initiate dissociation. A mixture of water-miscible solvents hydrates and dissociates the ionic fraction of pollutants at a higher concentration than would either solvent alone. Therefore, from a practical point of view, the deliberate use of a water-soluble solvent as a cosolvent in the formulation of toxic organic chemicals can lead to an increased solubility of hydrophobic organic contaminants in the aqueous phase and, consequently, to a potential increase in their transport from land surface to groundwater.

When industrial effluents or other waste materials, constituting a mixture of contaminants, are disposed of on the land surface, situations arise in which completely water-miscible organic solvents (CMOS) change the solvation properties of the aqueous phase. A similar situation may be encountered when, during remediation procedures, a mixture of water and water-miscible solvents is used to "wash" a contaminated soil (Li et al. 1996). A large number of organic solvents with various molecular properties (e.g., methanol, propanol, ethanol, acetone, dioxane, acetonitrile, dimethylformamide, glycerol) are included in the CMOS group. Each has a different effect on the activity coefficient and thus the solubility and partitioning behavior of an organic contaminant in a water/CMOS mixture.

Table 6.2 presents data showing the effect of various CMOS on the activity coefficient or mole fraction solubility of naphthalene, for two different solvent/water ratios. To examine the cosolvent effect, Schwarzenbach et al. (2003) compare the Hildebrand solubility parameter (defined as the square root of the ratio of the enthalpy of vaporization and the molar volume of the liquid), which is a measure of the cohesive forces of the molecule in pure solvent.

Schwarzenbach et al. (2003) note that, qualitatively, the more "waterlike" solvents (e.g., glycerol, ethylene glycol, methanol) have a much smaller impact on the activity coefficient of an organic solute than organic solvents for which hydrogen bonding is important but not the overall dominating factor. The CMOS are relatively small molecules with strong H-acceptor or H-donor properties. In an aqueous mixture, CMOS are able to break the hydrogen bonds between the water molecules and thus form a new H-bond solvent. In this case, the properties of the water–cosolvent solution change as a function of the nature and the relative

Table 6.2 The effect of various CMOS on the activity coefficient, γ or mole fraction solubility of naphthalene, for two different solvent–water ratios, $f_{v,\mathrm{solv}}$ (Schwarzenbach et al. 2003)

Cosolvent	Structure	Solubility parameter $(\mathrm{MPa})^{1/2}$	Naphthalene $\gamma_{iw}^{sat}/\gamma_{il}^{sat} = x_{il}^{sat}/x_{iw}^{sat}$	
			$f_{v,\mathrm{solv}} = 0.2$ (σ_i^c)	$f_{v,\mathrm{solv}} = 0.4$
Glycerol	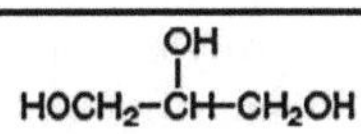	36.2	2.5 (2.0)	5.5
Ethylene glycol	$HOCH_2–CH_2OH$	34.9	3 (2.4)	9
Methanol	CH_3OH	29.7	3.5 (2.7)	14
Dimethylsulfoxide (DMSO)	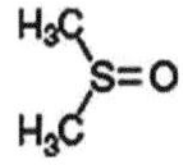	26.7	5.5 (3.7)	3.6
Ethanol	H_3CCH_2OH	26.1	7 (4.2)	48
Propanol	$H_3CCH_2CH_2OH$	24.9	17 (6.2)	180
Acetonitrile	$H_3C–C{=}N$	24.8	14 (5.7)	140
Dimethylformamide	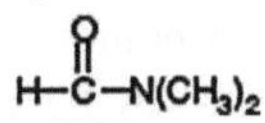	24.8	15 (5.9)	130
1,4-Dioxane	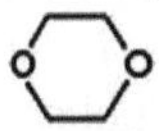	20.7	14 (5.7)	180
Acetone	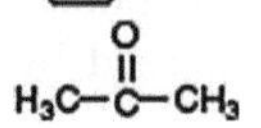	19.7	20 (6.5)	270

amount of the cosolvent. In general, cosolvents can completely change the solvation properties of subsurface water; the solubility of hydrophobic organic contaminants increases exponentially as the cosolvent fraction increases. The extent of solubility enhancement is controlled by the types of cosolvent and solute.

An increase in solubilization of nonpolar organic chemicals in water is obtained when surfactants are present in the water solution, the solvating strength of surfactants being much greater than that of simple cosolvents. This situation may occur, for example, during effluent disposal on land surface, sewage water irrigation, or point disposal of municipal wastes. The structure of a surfactant solution, above a rather well-defined critical micelle concentration (denoted CMC), is that of an ultrafine emulsion. The surfactant molecules are aggregated forming a cluster of 20–200 units or more. The hydrophobic tails are oriented to the interior of the cluster, and the hydrophilic heads are oriented to the exterior in contact with the water phase. The total solvent strength is determined by the number of micelles, their size, and their structure, but it is not proportional to surfactant concentration.

This type of dissolving action has been called *solubilization,* despite the fact that it does not correspond to the mechanism involved. Hartley and Graham-Bryce (1980) showed that the solubility of a crystalline solute of low water solubility cannot increase continuously with the expansion of the micelle. The limitation comes from the fact that the micelle cannot increase indefinitely in size without modification of its structure and properties, due to reorientation of surfactant molecules. There is a disadvantage, energetic or entropic, in indefinite expansion, which sets a limit to the solubility.

Cosolvents and salts mix completely with water to form homogeneous solutions, but with different effects. Cosolvents decrease the polarity of water and reduce the ability of an aqueous system to "squeeze out" nonpolar solutes, resulting in an increase in the solubility of nonelectrolytes. On the other hand, salts decrease the solubility on nonelectrolytes by increasing the polarity of water, thereby increasing the ability of the aqueous system to "squeeze out" the nonpolar solutes.

6.5 Salting-Out Effect

Effects of electrolytes on the solubility of organic compounds in aqueous solutions were established empirically more than 100 years ago by Setschenow (1889). He found that the presence of dissolved inorganic salts in an aqueous solution decreases the aqueous solubility of nonpolar and weak polar organic compounds. This effect, known as the *salting-out effect,* is expressed by the empirical Setschenow formula

$$\log(\gamma/\gamma_0) = \log(S^0/S) = k_s C_s, \tag{6.8}$$

where γ and γ_0 are the activity coefficients of the organic solute in salt solution and in water, respectively; S^0 and S are solubilities of the solute in water and in salt solution, respectively; k_s (L/mol) is the salting-out constant; and C_s (mol/L) is the molar concentration of the salt solution. In general, the dissolution process requires overcoming water–water interactions that allow formation and occupation of a cavity (Turner 2003; Schwarzenbach et al. 2003). In the presence of dissolved salt, solutions become "salted out" or "squeezed out" due to higher organization and compressibility of the water molecules when bound up in hydration spheres (Millero 1996). In some cases, the presence of salts leads to increases in solubility of organic compounds; this behavior is known as the *salting-in effect,* and the value of the Setschenow constant in this case becomes negative. Several examples of "salting in" and "salting out" are given in Table 6.3.

The salting-out constant expresses the potential of a salt or a mixture of salts to change the solubility of a given nonpolar compound. On contact between organic and aqueous phases, some of the organic molecules dissolve in the water, while some water molecules enter the organic phase. The values of the activity coefficients or relative solubilities of nonelectrolytes in electrolyte solutions have been examined over the years, mainly for saline concentrations close to seawater salinity (Randall and Failey 1927; Xie et al. 1997; Millero 2000; Dror et al. 2000a, 2000b). Because it is virtually impossible to quantify (experimentally) the contribution of individual ions, salting-out constants are available only for combined

Table 6.3 Effects of salts on the solubility of selected organic compounds expressed as Setschnow constant (L/mol). Reprinted from Xie WH, Shiu WY, Mackay DA (1997) review of the effect of salts on the solubility of organic compounds in seawater. Marine Envir Res 44:429–444. Copyright 1997 with permission of Elsevier

Salt	Benzene	Toluene	o-Xylene	m-Xylene	p-Xylene	Naphthalene
Salting out						
NaCl	0.195	0.225	0.227	0.248	0.251	0.260
KCl	0.166	0.206	0.205	0.222	0.217	0.204
$NaNO_3$	0.119	0.144	0.141	0.165	0.146	0.131
$MgSO_4$	0.488	0.457	0.491	0.531	0.491	0.516
$BaCl_2$	0.344	0.376	0.393	0.412	0.407	0.401
CH_3COONa	0.165	0.209	0.206	0.237	0.208	0.21
Salting in						
$(C_2H_5)_4NBr$	−0.56	−0.323	−0.414	−0.402	−0.413	−0.11
$(CH_3)_4NBr$	−0.24	−0.163	−0.198	−0.178	−0.2	−0.5

salts. Salting-out constants for selected organic compounds and aqueous electrolyte solutions at 25 °C are presented in Table 6.3.

Dror et al. (2000a, 2000b) report an experiment dealing with the effect of type and concentration of electrolytes, in an artificial soil aqueous solution, on the solubility of kerosene, a petroleum product containing more than 100 hydrocarbons. At increasing concentrations in water of NaCl or $CaCl_2$, from 0.2 to 2.0 M, a linear decrease in kerosene solubility occurred. It was also observed that the decrease in kerosene dissolution in the artificial soil solution is controlled by the electrolyte concentration but is not influenced by the sodium adsorption ratio (SAR; defined in Eq. 2.46). As noted previous, the SAR is an important factor in the relationship between soil solutions and soils. The salting-out effect is consistent with published data (Xie et al. 1997; Schwarzenbach et al. 2003).

It is interesting to note that smaller ions (e.g., Na^+, Mg^{2+}, Ca^{2+}, Cl^-) form hydration shells larger than bigger ions, which tend to bind water molecules only very weakly. In a simple way, the salting out of nonpolar and weakly polar compounds was explained by Schwarzenbach et al. (2003) by imagining that the dissolved ions compete successfully with the organic compound for solvent molecules. The freedom of some water molecules to solvate an organic molecule depends on the type and concentration of salts.

Under natural conditions, we usually deal with the solubility of nonelectrolytes in *mixed electrolyte* solutions. This aspect has not been thoroughly studied to date. Early studies of Randall and Failey (1927) used ionic strength fractions for modeling salting coefficients in mixed electrolyte solutions. No changes were observed in the Setschenow constant values when the electrolyte concentration was held constant, but NaCl ions were replaced with $CaCl_2$, showing that the ionic composition did not affect the dissolution in water of NAPL compounds. The solubility of naphthalene in some electrolyte mixtures was determined by Gordon and Thorne (1967a, b) using the simple mixing rule

$$\log(S^0/S) = kM_T = \sum_i N_i k_i M_T, \tag{6.9}$$

where M_T is the total molar concentration of the salts and N_i is the mole fraction of salt i ($M_T = \sum_i M_i$.); S^0 and S are solubilities of the organic compound measured in water and in salt solution, respectively; and k is the salting-out constant. This calculated solubility was validated experimentally for seawater. Millero (2000) used the Pitzer parameters for ions, which are related to the interaction of an ion with a nonelectrolyte, to estimate the activity coefficients and the solubility of nonelectrolytes in a water solution containing an electrolyte mixture.

A *salting-in* effect is promoted when there is a simultaneous presence of large organic molecules (e.g., tetramethyl-ammonium) and electrolytes in a water solution. In contrast to the salting-out process, salting in leads to an increase in organic compound solubility or a decrease in activity coefficient (Table 6.3). Almeida et al. (1983) observed a similar salting-in effect for very polar compounds, which may interact strongly with certain ions.

Despite the possibility of a salting-in route in some particular cases, the salting-out effect—and the decrease in organic compound solubility—in a saline environment is the main process controlling organic contaminant solubility in subsurface water solutions. Whitehouse (1984) studied the effect of salinity (from 0 to 36.7 %) on the aqueous solubility of polycyclic aromatic hydrocarbons (PAHs) and found that phenanthrene, anthracene, 2-methylanthracene, 2-ethylantracene, and benzo(a)pyrene experienced salting out, while only 1,2-benzanthracene exhibits a salting-in effect. Where salting out was observed, fairly large changes in salinity were required to cause significant changes in the solubility; such was not the case for the compound exhibiting a salting-in effect.

A simple correlation was determined for estimating the Setschenow constants for a variety of organic solutes in seawater, which yields an overall reduction in solubility by a factor 1.36 (Xie et al. 1997). The hydrophobicity of organic solutes increases by this factor, but the salting-out effect must be quantified when comparing the behavior of specific organic contaminants in freshwater and in subsurface aqueous solutions.

Under certain environmental conditions, when subsurface water contains potential organic ligands, trace metals also may be subject to salting out following their organic complexation. Turner et al. (2002) demonstrated that sediment–water partitioning of Hg(II) in estuaries characterized by high salinity may be controlled by a salting-out process. This partitioning applies to the aqueous solubility of nonelectrolytes and suggests that the organic complexes of mercury are removed from the aqueous phase via a coupled sorption–salting-out mechanism. Evidence of salting out of other metal complexes (Cd, Cr, Cu, Ni, Pb, Zn) also was reported (Turner et al. 2002). In highly contaminated, organic-rich estuaries, an increase in the sediment–water distribution of metal complexes was observed as salinity increased, except for Cd. Such behavior contradicts conventional speciation and partitioning processes and may be explained only by a salting-out effect. Turner et al. (2002) propose a mechanism by which trace metals are complexed and

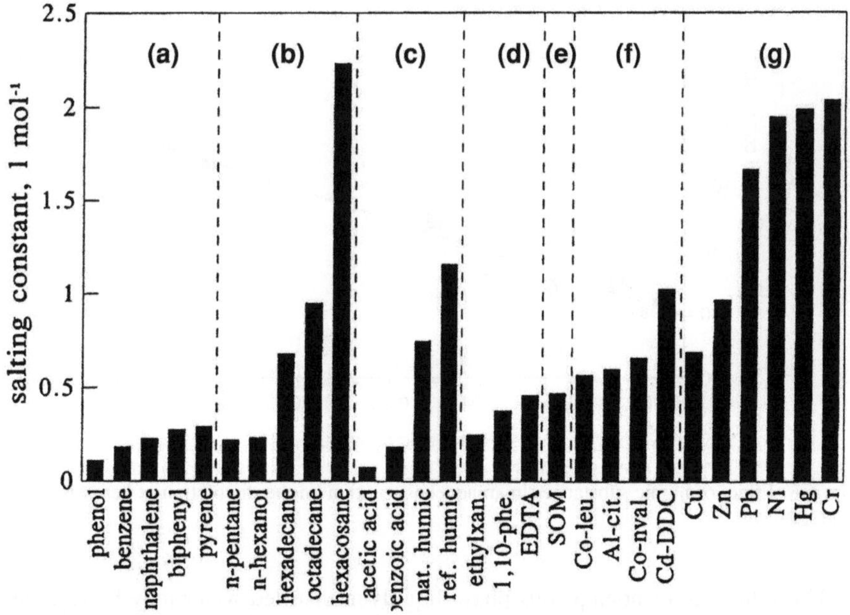

Fig. 6.4 Calculated, estimated, or apparent salting-out constants for various chemicals: **a** selected aromatic compounds, **b** selected aliphatic compounds, **c** natural or surrogate ligands, **d** anthropogenic ligands, **e** sediment organic matter (SOM), **f** transition metal complexes, **g** trace metal complexes in the Mersey Estuary. Reprinted with permission from Turner A, Martino M, Le Roux SM (2002) Trace metal distribution coefficients in the Mersey Estuary UK: Evidence for salting out of metal complexes. Environ Sci Technol 36:4578–4584. Copyright 2002 American Chemical Society

subsequently neutralize organic ligands, with the resulting neutral assemblage possibly being salted out via electrostriction. This behavior has significant implications for reactivity, availability, and transport of organic contaminants in saline water bodies rich in potential organic ligands. Calculated, estimated, or apparent salting constants for various chemicals are presented in Fig. 6.4.

6.6 Apparent Solubility

Changes in the pH of subsurface aqueous solutions may lead to an apparent increase or decrease in the solubility of organic contaminants. The pH effect depends on the structure of the contaminant. If the contaminant is sensitive to acid–base reactions, then pH is the governing factor in defining the aqueous solubility. The ionized form of a contaminant has a much higher solubility than the neutral form. However, the apparent solubility comprises both the ionized and the neutral forms, even though the intrinsic solubility of the neutral form is not affected.

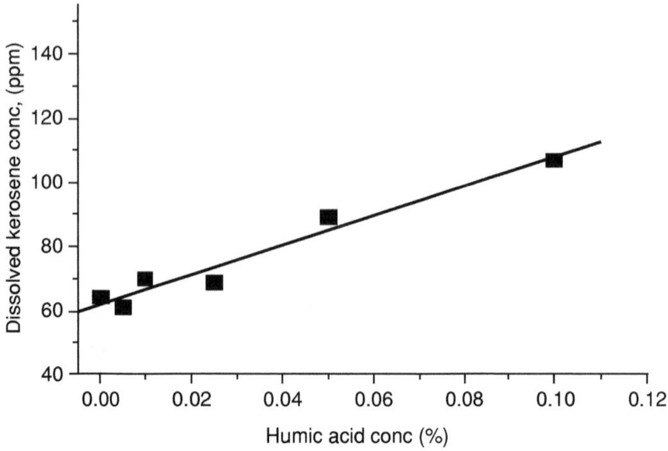

Fig. 6.5 Effect of humic substances in an aqueous solution on the dissolution of kerosene (Dror et al. 2000a)

The solubility of nonaqueous phase liquids in distilled water may be regarded as an unambiguous physical constant, whereas the apparent enhanced solubility might be due to the presence of naturally occurring organic macromolecules in subsurface solutions. Such macromolecules, including humic and fulvic substances, and negatively charged polysaccharides may increase the dissolution of organic contaminants in the subsurface aqueous solution. All the organic molecules that are constituents in subsurface organic matter and may be dissolved in the subsurface water are referred to as dissolved organic matter (DOM). Solubility in water in the presence of DOM is given by the relation

$$C_{\text{sat,DOM}} = C_{\text{sat}}(1 + [\text{DOM}]K_{\text{DOM}}), \tag{6.10}$$

where $C_{\text{sat,DOM}}$ and C_{sat} are the saturated concentrations of the organic compound measured in the presence and absence of DOM, respectively; [DOM] denotes concentration of DOM in water (kg/L); and K_{DOM} is the DOM–water partition coefficient. Note that the intrinsic solubility of the compound is not affected.

As mentioned previously, the physical state of a solute is susceptible to modifications by interaction with cosolvents. In principle, a cosolvent can enhance solute solubility by changing the solvency of the medium, by direct solute interaction, by adsorption, or by partitioning (Chiou et al. 1986). In a batch experiment testing the effect of humic acid on kerosene dissolution in an aqueous solution, Dror et al. (2000a) found a linear correlation between the amount of humic acid and the amount of kerosene that dissolved (Fig. 6.5).

The enhancement of kerosene dissolution occurs even at low humic acid content in the aqueous solution. In view of the fact that humic substances are relatively high molecular weight species containing nonpolar organic moieties, Chiou et al. (1986) assumed that a partition-like interaction between a solute of very low solubility in aqueous solution and a "microscopic organic environment" of dissolved humic molecules can explain solute solubility enhancement.

When a NAPL reaches the subsurface, it may be subject to mechanical forces that lead to the formation of a mixed NAPL–water micro-/nano-emulsion characterized by the presence of micro- and nano-droplets of organic compounds. These micro- and nano-emulsions are transparent or translucent systems, kinetically (nano-) or thermodynamically (micro-) stable, and display an apparent increase in aqueous solubility as compared to the intrinsic solubility of the NAPL itself (Tadros 2004). The very small droplet size (50–200 nm in the case of a nano-emulsion) causes a large reduction in the force of gravity, enabling the system to remain dispersed and preventing a coalescence process. In a laboratory experiment mimicking the contamination of a coastal freshwater aquifer by hydrocarbon-polluted seawater, Dror et al. (2003) report hydrocarbon concentrations in the source saltwater that are higher than the intrinsic solubilities of the contaminants, due to the formation of droplets. Contaminant concentrations in the adjacent freshwater also were higher than the intrinsic aqueous solubilities. The experiments also demonstrated the stability of hydrocarbon–water emulsions as they are transported through a porous sand layer.

References

Almeida MB, Alvarez AM, de Miguel EM, del Hoyo ES (1983) Setschenow coefficients for naphtols by distribution method. Can J Chem 61:244–248

Blokzijl W, Engberts BFN (1993) Hydrophobic effects—opinions and facts. Angew Chem Int Ed 32:1544–1579

Chiou CT, Malcom RL, Brinton TI, Kile DE (1986) Water solubility enhancement of some organic pollutants and pesticides by dissolved humic and fulvic acids. Environ Sci Technol 20:502–508

Dror I, Gerstl Z, Prost R, Yaron B (2000a) Behavior of neat and enriched volatile petroleum hydrocarbons mixture in the subsurface during leaching. Land Contam Reclam 8:341–348

Dror I, Gerstl Z, Braester C, Rubin H, Yaron B (2000b) In situ effect of amendments on the dynamics of kerosene dissipation in soil subsurface. Land Contam Reclam 8:349–356

Dror I, Amitay T, Yaron B, Berkowitz B (2003) A "salt-pump" mechanism for induced intrusion of organic contaminants from marine sources into coastal aquifers. Science 300:950

Gordon JE, Thorne RL (1967a) Salt effect on the activity coefficient of naphthalene in mixed aqueous electrolyte solutions salts 1. Mixture of two electrolytes. J Phys Chem 71:4390–4399

Gordon JE, Thorne RL (1967b) Salt effect on the activity coefficient of naphthalene in mixed aqueous electrolyte solutions salts 2. Artificial and natural seawater. Geochim Cosmochim Acta 31:2433–2443

Hartley GS, Graham-Bryce IJ (1980) Physical principles of pesticide behavior, vol 1. Academic Press, London

Li A, Andren AW, Yalkowsky SH (1996) Choosing a cosolvent: solubilization of naphthalene and cosolvent properties. Environ Toxicol Chem 15:2233–2239

Meng EC, Kollmann PA (1996) Molecular dynamics studies of the properties of water around simple organic solutes. J Phys Chem 100:11460–11470

Millero FJ (2000) The activity coefficients of non-electrolytes in seawater. Mar Chem 70:5–22

Millero FJ (1996) Chemical oceanography, 2nd edn. CRC Press, Boca Raton, Florida

Pearson RG (1963) Hard and soft acids and bases. J Am Chem Soc 85:3533–3539
Randall M, Failey CF (1927) The activity coefficient of the undissociated part of weak electrolytes. Chem Rev 4:117–128
Rubino JT, Yalkowsky SH (1987) Co-solvency and co-solvent polarity. J Pharmacol Res 4:220–230
Schwarzenbach RP, Gschwend PM, Imboden DM (2003) Environmental organic chemistry, 2nd edn. Wiley-Interscience, New York
Setschenow J (1889) Uber die constitution der salzlosungen auf grund ihres verhaltens zu kholensaure. Z Phys Chem Vierter 1:117–124
Sposito G (1981) The thermodynamics of soil solutions. Clarendon Press, Oxford
Stumm W, Morgan JJ (1996) Aquatic chemistry, 3rd edn. Wiley, New York
Tadros T (2004) Application of rheology for assessment and prediction of the long-term physical stability of emulsions. Adv Coll Interface Sci 108:227–258
Turner A (2003) Salting out of chemicals in estuaries: Implication for contaminant partitioning and modeling. Sci Total Environ 315:599–612
Turner A, Martino M, Le Roux SM (2002) Trace metal distribution coefficients in the Mersey Estuary UK: evidence for salting out of metal complexes. Environ Sci Technol 36:4578–4584
Weber WJ, Mukherji S, Peters CA (1998) Aqueous dissolution of constituents of composite non-aqueous phase liquid contaminants. In: Rubin H, Nakis N, Carberry J (eds) Soil and aquifer pollution. Springer, Heidelberg, pp 123–135
Whitehouse BG (1984) The effect of temperature and salinity on the aqueous solubility of polynuclear aromatic hydrocarbons. Mar Chem 14:319–332
Xie WH, Shiu WY, Mackay DA (1997) Review of the effect of salts on the solubility of organic compounds in seawater. Marine Envir Res 44:429–444
Yalkowsky SH, Roseman TJ (1981) Solubilization of drugs by cosolvents. In: Yalkowsky SH (ed) Techniques of solubilization of drugs. Marcel Dekker, New York

Chapter 7
Partitioning of Volatile Compounds

Volatilization (also referred to as vaporization or evaporation) is the conversion of a chemical from the solid or liquid phase to a gas or vapor phase. The partitioning of a volatile compound in the subsurface environment comprises two distinct patterns: volatilization of contaminant molecules (from the liquid, solid, or adsorbed phase) and dispersion of the resulting vapors in the subsurface gas phase or the overlying atmosphere by diffusive and turbulent mixing. Even though the two processes are fundamentally different and controlled by different chemical and environmental factors, they are not wholly independent under natural conditions: only by integrating their effects can volatilization be characterized.

The volatilization process changes the contaminant from a solid or liquid state, where the molecules are held together by intermolecular forces, into a vapor phase. The molar heats of fusion (ΔH_f), volatilization (ΔH_v), and sublimation (ΔH_s) are related according to the Born-Haber cycle by

$$\Delta H_s = \Delta H_f + \Delta H_v. \tag{7.1}$$

Even at low temperature, some molecules may overcome the energy barrier of the cohesive forces and escape from the solid or liquid state into the gaseous phase.

The *vapor dispersion process* is described mathematically as a vapor flux (I) through any plane at a height z and is expressed by

$$I = K_z \left(\frac{\mathrm{d}p}{\mathrm{d}z} \right)_z, \tag{7.2}$$

where K_z is the transfer coefficient and $\mathrm{d}p/\mathrm{d}z$ is the vapor pressure gradient at height z. Air flow just above the land surface may be turbulent and can be described by *eddy diffusion.* In the boundary layer along the soil surface and in the subsurface matrix, vapor dispersion may be described by a molecular diffusion process.

Taylor and Spencer (1990) pointed out that a laminar layer can be regarded as the limiting distance above the soil surface to which the smallest eddies of the overland turbulent flow can penetrate. Therefore, above this layer, transport takes

B. Berkowitz et al., *Contaminant Geochemistry*,
DOI: 10.1007/978-3-642-54777-5_7,

place through eddy diffusion, and the corresponding vapor flux can be described by Eq. 7.2, where K_z is then the eddy diffusion coefficient (Taylor and Spencer 1990). The height of the turbulent zone, within the atmospheric boundary layer, is orders of magnitude greater than that of the laminar flow layer, and dispersion of contaminant vapors in the turbulent zone is relatively rapid.

The concept of separate regions of dilution by molecular diffusion and turbulent mixing is of major importance in understanding the exchange of gases at land surface and for the identification of physical factors that impede the dispersion process under various microclimates.

7.1 Gas–Liquid Relationships

Gas–liquid relationships, in the geochemical sense, should be considered liquid–solid–gas interactions in the subsurface. The subsurface gas phase is composed of a mixture of gases with various properties, usually found in the free pore spaces of the solid phase. Processes involved in the gas–liquid and gas–solid interface interactions are controlled by factors such as vapor pressure–volatilization, adsorption, solubility, pressure, and temperature. The solubility of a "pure" gas in a closed system containing water reaches an equilibrium concentration at a constant pressure and temperature. A gas–liquid equilibrium may be described by a partition coefficient, relative volatilization, and Henry's law.

The *partition* or *distribution coefficient* between a gas and a liquid is constant at a given temperature and pressure. The *relative volatility* is used in defining the equilibrium between a volatile liquid mixture and the atmosphere. The partition coefficient expresses the relative volatility of a species A distributed between a vapor phase ($A1$) and a liquid phase ($A2$). *Henry's law* applies to the distribution of dilute solutions of chemicals in a gas, liquid, or solid at a specific ambient condition. Equilibrium is defined by

$$P_{A1} = H_A x_A, \tag{7.3}$$

where P_{A1} is the partial pressure of the chemical $A1$ in the gas phase, H_A is Henry's law constant for species A, and x_A is the mole fraction of chemical A in solution. Under these conditions, H_A has dimensions of pressure.

The Henry's law constant, as a function of the activity coefficient of $A2$ in water, γ_{A2}, is

$$H_A = \frac{\gamma_{A2} P^0_{A1}}{P} \tag{7.4}$$

where P and P^0_{A1} denote the total pressure and partial pressure of $A1$, respectively. Henry's law constants and relative volatility values for the most common gases, as well as their solubility in pure water, can be found in any chemical handbook.

However, not all existing data obtained under laboratory conditions can be used in a subsurface environment, where instead of pure water, an electrolyte solution that may contain surface active organic compounds exists. As a consequence, the validation of Henry's law for specific subsurface conditions must be carried out in specially designed experiments that mimic as closely as possible the natural environment.

7.2 Volatilization from Subsurface Aqueous Solutions

Water evaporation and contaminant volatilization from subsurface aqueous solutions are two companion processes that affect contaminant partitioning between the liquid and gaseous phases. Temperature-induced evaporation may affect the concentration of the natural constituents of the subsurface water and thus affect contaminant dissolution in this water.

Water evaporation occurs when the vapor pressure of the water at the surface, which is temperature dependent, is greater than the water pressure in the subsurface, which is dependent on relative humidity and temperature. The isothermal evaporation process is described by Stumm and Morgan (1996) via a "reaction progress model," in which the effects of the initial reaction path are based on the concept of partial equilibrium. Stumm and Morgan (1996) describe partial equilibrium as a state in which a system is in equilibrium with respect to one reaction but out of equilibrium with respect to others. As an example, Stumm and Morgan (1996) indicate (Fig. 7.1) that water with a "negative residual alkalinity" (i.e., $2[Ca^{2+}] > [HCO_3^-]$) tends to increase its calcium hardness as a result of water evaporation into the atmosphere, and consequently its alkalinity and pH decrease. Eventually, after excessive evaporation, the water may reach saturation, becoming a Na^+, SO_4^{2-}, or Cl^- brine.

From Fig. 7.1, it may be observed that residual alkalinity initially increases with water evaporation, concomitant with a pH increase, and that negative residual alkalinity and pH decrease. Because evaporation leads to a change in the quality of water, the water properties as a solvent for organic contaminants are also changed. This fact should be considered when dealing with contaminant partitioning between phases.

The molecular weight of water vapor (MW $=$ 18) is less than that of air (MW $\approx$ 29). As such, the diffusion of water vapor into the surrounding atmosphere, which consists of a mixture of water and air, leads to a buoyant force with upward macroscopic movement. The natural evaporation phenomenon is not only the effect of heat transfer but also a buoyancy-induced motion. The system is at steady state when the vapor pressure of the water at the surface is less than that in the air above, and the resulting condensation is governed by the slow process of molecular diffusion and laminar flow.

Volatilization of contaminants from subsurface aqueous solutions into the subsurface gas phase or the (above ground) atmosphere is controlled by the vapor pressure. Compounds with high vapor pressure tend to accumulate in the gas

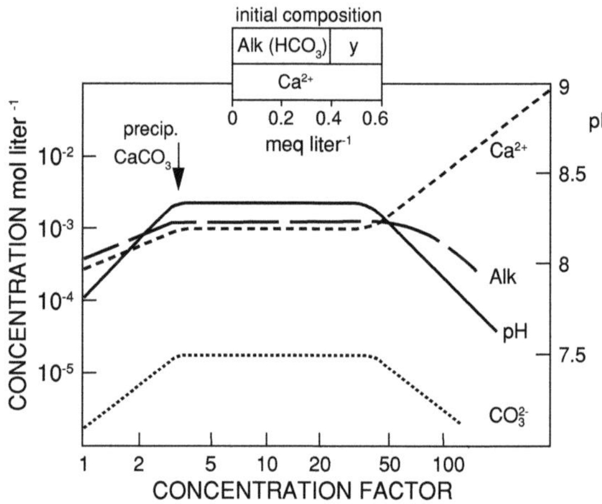

Fig. 7.1 Isothermal evaporation (25 °C) of natural water leading to $CaCO_3$ precipitation $2[Ca^{2+}] > [Alk]$ (with $pCO_2 = 10^{-3.5}$ atm). The concentration factor is the degree by which the water has been reduced by volume compared to the initial solution (Stumm and Morgan 1996)

phase, which may be considered a kind of compound "solubility" in the atmosphere. Partitioning between the liquid and the gas phases is described by Henry's law and is expressed as

$$\frac{[A_{(\mathrm{aq})}]}{p_A} = K_H \tag{7.5}$$

where $[A_{(\mathrm{aq})}]$ is the concentration of contaminant A in the aqueous solution, p_A is the vapor pressure of A (atm), and K_H (sometimes also denoted H) is the Henry's constant. The physicochemical significance of Henry's law is reflected by the fact that there is a linear relationship between the activity of a volatile species in the liquid phase and its activity in the gas phase. Table 7.1 shows the vapor pressure and Henry's constant for selected organic compounds at 25 °C.

The rate of mass transfer of a substance across a water–gas boundary is controlled by the diffusion film model as well. Gas transfer from a water source is faster than from a solid source, and the chemical does not undergo a chemical reaction during the transfer process. Under these conditions, the interface concentration may be interpreted in terms of the Henry's constant (K_H) which indicates whether the controlling resistance is in the liquid or the gas film. When $K_H < 5$, a water film is the controlling factor, while a gas film controls the behavior when $K_H > 500$.

7.3 Vapor Pressure–Volatilization Relationship

If a chemical is placed in an empty vessel that is greater in volume than the chemical itself, a portion of the chemical volatilizes to fill the remaining free space of the vessel with vapors. The pressure in the vessel at equilibrium is affected only by the

Table 7.1 Vapor pressure and Henry's constant for various organic compounds at 25 °C (Stumm and Morgan 1996)

Compound A	Molecular weight	Water solubility $[A^\circ_{(aq)}]$(M)	Vapor pressure $p_A{}^\circ$ (atm)	Concentration in gas phase $(A^\circ)_g = p_A{}^\circ/RT$ (mol/L)	$H^{sat} = K_H^{sat}RT = \frac{[A^\circ_{(aq)}]}{(A^\circ)_g}$
Alkanes					
n-Octane (C_8H_{18})	114	5.8×10^{-6}	1.8×10^{-2}	7.5×10^{-4}	7.7×10^{-3}
1-Hexane (C_6H_{12})	84	7.0×10^{-4}	2.5×10^{-1}	1.0×10^{-2}	7×10^{-2}
Aromatic substances					
Benzene	78	2.3×10^{-2}	1.2×10^{-1}	5.1×10^{-3}	4.5
Toluene	92	5.6×10^{-3}	3.7×10^{-2}	1.5×10^{-3}	3.7
Naphthalene	128	2.6×10^{-4}	1.0×10^{-4}	4.3×10^{-6}	6.1×10^{1}
Biphenyl	154	4.9×10^{-5}	7.5×10^{-5}	3.1×10^{-6}	1.6×10^{1}
Pesticides					
p,p'-DDT ($C_{14}H_9Cl_5$)	355	1.4×10^{-8}	1.3×10^{-10}	5.4×10^{-12}	2.6×10^{3}
Lindane	291	2.6×10^{-5}	8.3×10^{-8}	3.4×10^{-9}	7.6×10^{3}
Dieldrin	381	5.8×10^{-7}	6.6×10^{-9}	2.7×10^{-10}	2.1×10^{3}
Polychlorinated biphenyls (PCBs)					
2,2′4,4′-CBP ($C_{12}H_6Cl_4$)	292	2.0×10^{-8}	6.2×10^{-9}	2.5×10^{-10}	79
2,2′3,3′4,4′-CBP ($C_{12}H_4Cl_6$)	361	1.9×10^{-9}	2.2×10^{-10}	9.2×10^{-12}	2.1×10^{2}
Other					
Dimethylsulfide (C_2H_6S)	62	3.5×10^{-1}	6.3×10^{-1}	2.6×10^{-2}	1.4×10^{1}
Mercury (Hg^o)	201	1.5×10^{7}	1.7×10^{-6}	$7 \times 10-8$	2.1
Water	18	(55.5)	3.1×10^{-2}	1.3×10^{-3}	(4.3×10^{4})

temperature and is independent of the vessel volume. The pressure that develops, called *vapor pressure,* characterizes any chemical in the liquid or solid state.

When the temperature increases, the proportion of molecules with energy in excess of the cohesive energy also increases, and an excess vapor pressure is observed. The Clausius–Clapeyron equation describes the variation of vapor pressure with temperature as follows:

$$\mathrm{d}(\ln p)/\mathrm{d}T = \Delta H_\mathrm{v}/RT^2 \tag{7.6}$$

where p is the vapor pressure, ΔH_v is the heat of volatilization, R is the universal gas constant, and T is temperature (K).

Because the vapor pressure of chemicals is a key factor in controlling their dissipation within the subsurface, and from the subsurface to the atmosphere, accurate estimation of this value is required. Comprehensive reviews on this subject are given by Plimmer (1976) and Glotfelty and Schomburg (1989). For contaminants with low vapor pressure that reach the subsurface as a result of a nonpoint disposal (e.g., pesticides used in agricultural practices), their vapor pressure is sufficiently low to be below detection limits, which may explain some discrepancies in the reported results.

Partitioning between phases (corresponding to the maximum of entropy) can be expressed by equating the chemical potentials in the respective phases (Mackay 1979). In partitioning between phases, matter flows from high chemical potential to low chemical potential, and at equilibrium, the chemical potential is the same in both phases. Due to difficulties in measuring the chemical potential, Mackay (1979) reintroduced the much simpler concept of fugacity, which can be used quite simply to express distribution of organic contaminants among the various phases of the subsurface. It is important to stress that high fugacity denotes a greater tendency for a chemical to escape from a phase. In fact, the role of fugacity in mass diffusion is similar to the role of temperature in heat diffusion: Mass always flows from high to low fugacity, just as heat flows from high to low temperature. Fugacity is linearly related to concentration, given as

$$C_i = f_i Z_i \tag{7.7}$$

where i denotes the ith phase, C is the concentration (mol/V), f is fugacity, and Z is the fugacity capacity of the phase (Mackay 1979). A phase with large Z accepts a large amount of chemical, just as a substance with high heat capacity stores much heat. Each phase has a Z value for each chemical at each temperature; if the Z values are known, the equilibrium distribution between the phases can be determined.

The maximum vapor pressure value that can be established at the inner surface of the laminar flow layer in air moving over a soil surface is reached only when the surface is covered uniformly by the contaminant. This situation is almost impossible to find in the subsurface, where a chemical is partially adsorbed on the subsurface solid or dissolved in the subsurface water, which reduces the vapor pressure below the equilibrium value of the pure compound.

7.4 Volatilization of Multicomponent Contaminants

In general, contaminants reach the subsurface not as a single component but as a mixture of various components. Each of the chemicals in the mixture has its own physicochemical characteristics and therefore each behaves differently in its partitioning into the subsurface gas phase. Petroleum products, for example, contain hydrocarbons whose vapor pressures range over several orders of magnitude. The composition of a volatile organic mixture in the subsurface changes with time, due to differential volatilization of the components into the gas phase. Also, different rates of volatilization create concentration gradients among compounds in the liquid, which leads to mass diffusion. This mechanism may affect the physical properties of the liquid and their redistribution in the subsurface.

Woodrow et al. (1986) assumed that the mixture of components (W_L) is ideal, so that the volatilization of each component (W_i) proceeds by a first-order process that is described by

$$W_i = W_L^0 \exp^{-k_i t} \tag{7.8}$$

where k_i is the volatilization rate for component i, and W_L^0 denotes initial state. In such mixtures, the more volatile components volatilize faster, causing a decrease in the volatilization rate for the mixture as its "total" vapor pressure decreases.

It is known that, in a water phase, immiscible liquids such as gasoline or other petroleum products may form multicomponent droplets of various forms and sizes, under dispersive conditions. These droplets are transported by convection and diffusion, which contributes to the contamination of fresh water systems. However, during droplet transport, more volatile substances partition to the gas phase at the droplet surface, leaving less volatile material that volatilizes more slowly. More volatile material still exists in the droplet interiors, and it tends to diffuse toward the surface because of concentration gradients created by prior volatilization. Different components in a droplet have different volatilization rates, which may vary significantly during droplet transport, and as a result, the contamination of fresh water is affected accordingly.

Nye et al. (1994) studied the volatilization of single- and multicomponent liquids through soil columns, emphasizing differences between the two cases. For a single liquid, three stages were identified: (1) soil sorbs the gas as it diffuses upward through the column; (2) steady-state conditions are reached in which the vapor escapes to the atmosphere at a constant rate, after diffusing upward through the column under a constant concentration gradient; and (3) the soil reaches a stage of depletion when the gas desorbs from the soil and diffuses to the atmosphere. When a mixture of components characterized by various vapor pressures volatilizes, similar stages are observed, but important variations may exist. In the accumulation stage, each component from the mixture is sorbed according to its volatility, diffusivity, and the competitive adsorption isotherms. As a consequence, no true steady state is reached because the composition of the source liquid changes continuously and the amount of contaminant adsorbed on the soil surface is adjusted accordingly. If the supply of volatile liquid is large compared to the total adsorption capacity of the soil, this adjustment is slow and evaporation approaches an apparent steady state. In the depletion stage, the soil already enriched in less volatile components loses the volatile components slowly. The soil still contains a small proportion of the more volatile components and does not lose them completely until all components have disappeared.

References

Glotfelty DE, Schomburg CJ (1989) Volatilization of pesticides from soils. In: Shawney BL, Brown K (eds) Reactions and movement of organic chemicals in soils. Soil Science Society of America, Madison, pp 181–209 (Special Publicaion)

Mackay D (1979) Finding fugacity feasible. Environ Sci Technol 13:1218–1223

Nye PH, Yaron B, Galin T, Gerstl Z (1994) Volatilization of multicomponent liquid through dry soils: Testing a model. Soil Sci Soc Am J 58:269–278

Plimmer JR (1976) Volatility. In: Kearny PC, Kaufmann DD (eds) Herbicides: Chemistry, degradation and mode of action, vol 2, 2nd edn. Marcel Dekker, New York, pp 891–934

Stumm W, Morgan JJ (1996) Aquatic chemistry, 3rd edn. Wiley, New York

Taylor AW, Spencer WF (1990) Volatilization and vapor transport processes. In: Cheng HH (ed) Pesticides in the soil environment. Soil Science Society America, Madison, pp 213–369

Woodrow JE, Selber JN, Yong-Hwe K (1986) Measured and calculated evaporation losses of two petroleum hydrocarbon herbicide mixtures under laboratory and field conditions. Environ Sci Technol 20:783–786

Chapter 8
Selected Research Findings: Contaminant Partitioning

In the previous chapters of Part III, we discussed various aspects that affect the partitioning of contaminants of anthropogenic origin among the liquid, solid, and gaseous phases in saturated and partially saturated porous media. The partitioning process controls the fate of contaminants in the subsurface, defining their redistribution and transformation. This chapter presents selected research findings that illustrate various aspects of contaminant partitioning in the subsurface. We stress that, although many examples are included here, this chapter does not cover the entire spectrum of contaminant partitioning phenomena that occur in the subsurface.

8.1 Partitioning Among Phases

Under natural conditions, contaminants often reach the earth's surface as a mixture of (potentially) toxic chemicals, having a range of physicochemical properties that affect their partitioning among the gaseous, liquid, and solid phases. As a consequence, contaminant retention properties in the subsurface are highly diverse. Contaminants may reach the subsurface from the air, water, or land surface.

As an introductory example, we consider gasoline (a common and universally used petrochemical product comprising a mixture of about 200 hydrocarbons with different properties) to illustrate contaminant redistribution among phases in relation to their environmental behavior. The recent model of Foster et al. (2005) provides an excellent approach for assessing differential volatilization, dissolution, and retention of various gasoline hydrocarbons in the environment. Based mainly on their volumetric composition, relevant properties (vapor pressure, water solubility, octanol/water partition coefficient K_{ow}), environmental/human hazardous aspects (e.g., toxicity), and persistence (in water, soil, sediments), the authors grouped gasoline hydrocarbons into 24 blocks with a density increasing from 0.564 to 0.837 g/mL. Hydrocarbons were grouped primarily into structural classification including alkanes and alkenes with normal, branched, or cyclic structure; aromatics with one or two rings; and the number of carbon atoms in the

B. Berkowitz et al., *Contaminant Geochemistry*,
DOI: 10.1007/978-3-642-54777-5_8,

Table 8.1 Component groups for gasoline based on C—number and main chemical composition

Structural class	Number of carbon atoms (C)									
	3	4	5	6	7	8	9	10	11	12
n-alkanes	1		2	3	4			7		
Isoalkanes			5		6					
n-alkenes		8		9		20				
Isoalkenes				10						
Cyclic alkanes			11		13		14			
Cyclic alkenes				15						
Monoaromatics				16	17	19	21			22
Diaromatics								23	24	
Cyclohexane				12						
Ethylbenzene						18				

Reprinted with permission from Foster et al. (2005). Copyright 2005 American Chemical Society

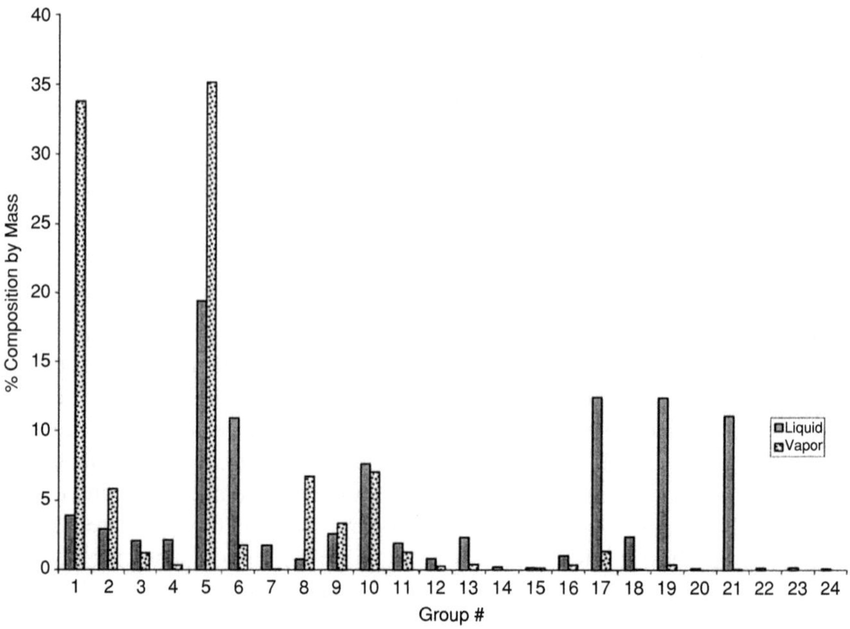

Fig. 8.1 Liquid and vapor gasoline compositions. Reprinted with permission from Foster et al. (2005). Copyright 2005 American Chemical Society

compounds, ranging from 3 to 12. The physicochemical properties are summarized in Table 8.1, and the compositional differences between each compartment and those of the original gasoline mixture are given in Fig. 8.1. These data provide an overview of the multiple properties of a contaminant mixture; these properties control its partitioning among the subsurface phases.

The mode of entry to the subsurface environment affects partitioning among subsurface components. Considering that the composition of gasoline depends on the nature of the emission process, Foster et al. (2005) developed a model scenario where a known amount of gasoline emission was divided equally among air (gas), water, and soil. The composition of gasoline in the air phase was assumed to be in the form of a fugitive vapor from liquid gasoline, while emissions to soil or to water were considered as liquid gasoline. Figure 8.2 exhibits a four by four set of charts displaying the relative gasoline composition by mass in air, water, soil, and sediment as a result of three modes of entry and the total simultaneous emission. For each mode of entry, the percentage of gasoline by mass and the concentration in each of the 24 compartments was calculated and the contribution of the mode of entry estimated.

When the emission of gasoline is mainly into the air phase, more than 99 % of the contaminant remains in the air. In this case, the gasoline emission consists predominantly of alkanes (70 % of the total gasoline composition) with a lower number of carbon atoms. When gasoline emission is mainly to water, 85 % remains in the aqueous phase, including groups 5, 6, 7, 19, and 21 (see Fig. 8.1), about 8 % partitions into sediment, and less than 1 % into soil. The mode of entry only into soil leads to retention of about 91 %. From Fig. 8.2, we see that simultaneous entry of a total emission of gasoline leads to a partitioning of 52 % in soil, 31 % in water, 14 % in air, and 3 % in sediments. This scenario and calculation provide a representative overview of contaminant partitioning among the gaseous, aqueous, and solid phases, as affected by the properties of the contaminants, the surrounding phase, and their mode of entry into the environment.

8.2 Contaminant Volatilization

Contaminant volatilization from subsurface solid and aqueous phases may lead, on the one hand, to pollution of the atmosphere and, on the other hand, to contamination (by vapor transport) of the vadose zone and groundwater. Potential volatility of a contaminant is related to its inherent vapor pressure, but actual vaporization rates depend on the environmental conditions and other factors that control behavior of chemicals at the solid–gas–water interface. For surface deposits, the actual rate of loss, or the proportionality constant relating vapor pressure to volatilization rates, depends on external conditions (such as turbulence, surface roughness, and wind speed) that affect movement away from the evaporating surface. Close to the evaporating surface, there is relatively little movement of air and the vaporized substance is transported from the surface through the stagnant air layer only by molecular diffusion. The rate of contaminant volatilization from the subsurface is a function of the equilibrium distribution between the gas, water, and solid phases, as related to vapor pressure solubility and adsorption, as well as of the rate of contaminant movement to the soil surface.

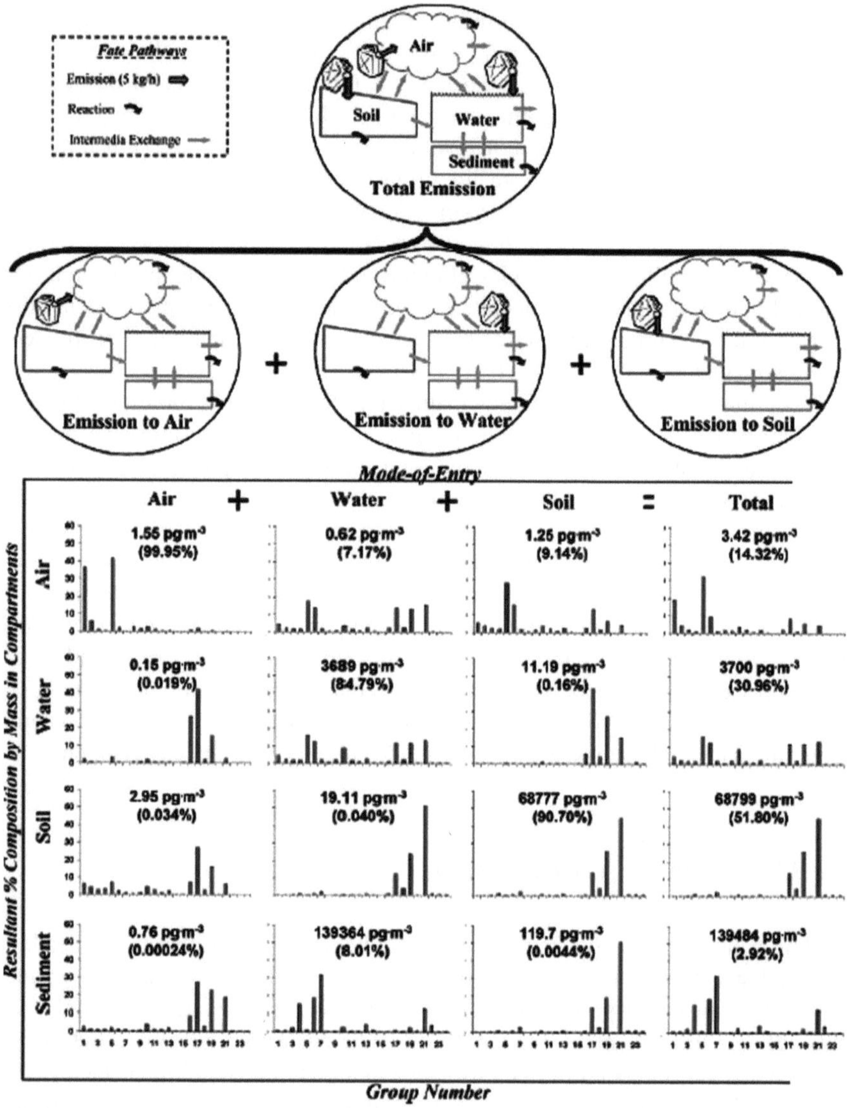

Fig. 8.2 Gasoline composition in environmental compartments as a result of emissions to air, water, soil, and sediment, individually at an arbitrary rate of 5 kg/ha, and simultaneously. Reprinted with permission from Foster et al. (2005). Copyright 2005 American Chemical Society

To illustrate these phenomena, we discuss relevant examples of volatile chemical products that originate mainly from agricultural and industrial–municipal practices. The rates and extent of volatilization of these chemicals are presented in relation to subsurface environmental conditions.

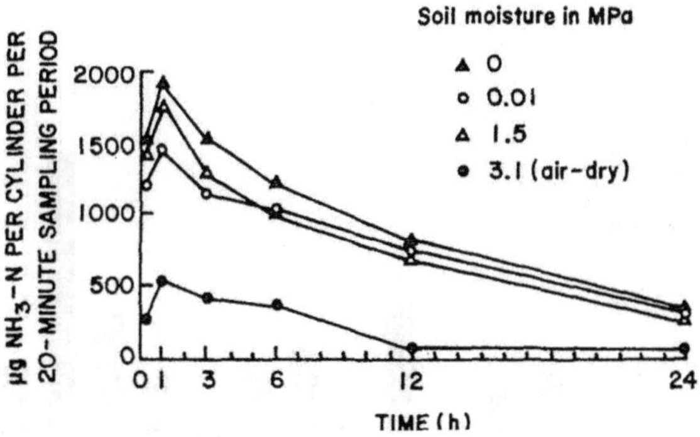

Fig. 8.3 Ammonia-N volatilization as a function of time from sewage sludge applied to soils at 0, 0.01, and 1.5 MPa, and air-dry initial moisture levels (Donovan and Logan 1983)

8.2.1 Inorganic Contaminants

Ammonia volatilization illustrates the behavior of inorganic chemicals in the subsurface under aerobic or anaerobic conditions. It is recognized that ammonia volatilization is affected by the time and depth of release, pH, temperature, and moisture content as well as by the cation exchange capacity (CEC).

The relative concentrations of NH_3 and NH_4^+ in an aqueous solution are pH dependent, in accordance with the following reaction equilibrium:

$$NH_4^+ \rightleftharpoons NH_{3(aq)} + H^+ \tag{8.1}$$

$$[NH_{3(aq)}][H^+]/[NH_4^+] = K = 10^{-9.5} \tag{8.2}$$

$$\log\left([NH_{3(aq)}]/[NH_4^+]\right) = -9.5 + \text{pH}, \tag{8.3}$$

where $NH_{3(aq)}$ is NH_3 in solution. According to these equations, the concentrations of $NH_{3(aq)}$ at pH 5, 7, and 9 are 0.0036, 0.36, and 36 %, respectively, of the total ammonia-N in solution.

Loss of NH_3 from calcareous subsurface media is considerably greater than from noncalcareous media and involves production of $(NH_4)_2CO_3$ or NH_4HCO_3. The fate of $(NH_4)_2SO_4$ added to calcareous material is expressed by the reaction between $(NH_4)_2SO_4$ and $CaCO_3$, and as a result, volatile ammonia is produced according to the equation

$$CaCO_3 + (NH_4)_2SO_4 \rightleftharpoons 2NH_3 + CO_2 + H_2O + CaSO_{4\,(solid)}. \tag{8.4}$$

A similar pattern of ammonia formation and volatilization occurs when sewage effluent is disposed on the land surface or in sludge-enriched soils. Donovan and Logan (1983) showed that ammonia loss increases with increasing pH and temperature and is affected by the type of sludge added to the land surface. Loss from air-dry soil is much lower than from a soil at moisture tension <1.5 MPa, the kinetics of volatilization with time being depicted in Fig. 8.3. Donovan and Logan

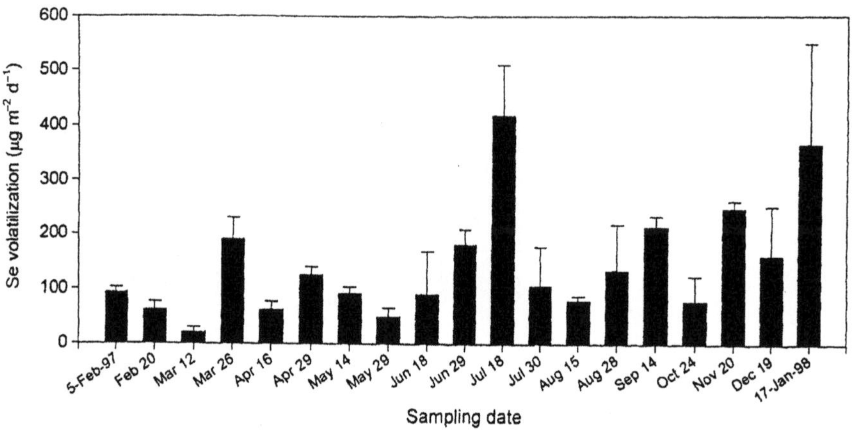

Fig. 8.4 Rates of selenium volatilization from soil versus time (Lin et al. 2000)

(1983) also report that NH_3 volatilization losses are greater from lime-stabilized sludge with pH = 12 than from aerobic or anaerobic sludge. A linear relationship between volatilization decrease and temperature decrease also characterizes NH_3 behavior in the sludge-amended lands.

A special case is given by ammonia volatilization from flooded land surfaces, which involves a more complex pathway. This is because the kinetics and extent of the volatilization are affected by water quality, type of land, and biological and environmental factors. In this particular case, the rate of NH_3 volatilization is mainly a function of ammonia concentration in the flooding water (Jayaweera and Mikkelsen 1991).

Nitrite formation may lead to nitrous oxide (N_2O) emission. An example of such a process under reclaimed effluent disposal on the land surface is reported by Master et al. (2004). Irrigating a grumosol (<60 % clay content) with fresh and reclaimed effluent water, it was found that, under effluent irrigation, the amount of N_2O emissions was double the amount emitted under freshwater treatment, at 60 % w/w. The N_2O emission from effluent-treated bulk soil was more than double the amount formed from large aggregates.

Plant-mediated volatilization of inorganic compounds is an accepted method for reclaiming selenium-contaminated lands. Volatile Se is formed mostly from dimethyl selenide, evolving from the land surface. For example, dimethyl selenide produced more than 90 % of the total volatile Se in the San Joachim Valley, California (Karlson and Frankenberger 1988). Lin et al. (2000) report on Se volatilization from a *Salicornia bigelovii* field which occurs as a result of Se methylation by plants. In their experiment, Lin et al. (2000) found that biological volatilization removed 62 mg Se m/yr, which accounted for 6.5 % of the total annual Se input to the *S. bigelovii* field. The fluctuation of Se volatilization rates during a 12-month study period is presented in Fig. 8.4. Linear regression analysis

Table 8.2 Saturation vapor pressures and densities of selected pesticides (modified after Taylor and Spencer 1990)

Pesticide	Molecular weight (g)	Temperature (°C)	Vapor pressure (MPa)	Vapor density (μg/L)
Alachlor	270	25	2.9	3.2×10^{-1}
Atrazine	216	25	9.0×10^{-2}	8.0×10^{-3}
Bromacil	261	25	2.9×10^{-2}	3.0×10^{-3}
DDT	354	25	4.5×10^{-2}	6.0×10^{-3}
Dieldrin	381	25	6.8×10^{-1}	1.0×10^{-1}
Diuron	233	25	2.1×10^{-2}	2.0×10^{-3}
Lindane	291	25	8.6	1.0
Malathion	330	20	1.3	1.8×10^{-1}
Parathion	291	25	1.3	1.5×10^{-1}
Trifluralin	335	20	1.5×10^{1}	2.0

of all 57 measurements, conducted during the study period, yielded the relationship $y = 51.96 \pm 9.42x$ ($r^2 = 0.252$), where x denotes the time during the year and y is the rate of Se volatilization. The regression shows a general increase in Se volatilization from spring to winter.

8.2.2 Organic Contaminants

Organic contaminants can be released to the surface in different ways, and contamination can be classified as point source and nonpoint source (or diffuse source). As an example of a *nonpoint source*, we discuss the case of pesticides applied during agricultural activity over large areas; an example of *point source* contamination is given by the behavior of petroleum products that reach the subsurface as a result of leakage (or a spill) from pipes or from a gas station.

Pesticides are characterized by a range of (saturation) vapor pressure and densities (Table 8.2); they therefore evaporate from the land surface in different patterns. Peck and Hornbuckle (2005) studied atmospheric concentrations of currently used pesticides in Iowa (United States) during the years 2000–2002. The average detected concentrations of five heavily used herbicides were 0.52 ng/m^3 for trifluralin, 4.6 ng/m^3 for acetoclor, 2.3 ng/m^3 for metolaclor, 1.7 ng/m^3 for pendimethalin, and 1.2 ng/m^3 for atrazine. The survey considered about 45 organic pesticides (herbicides and insecticides) used in the field; only seven of them were not detected in the air phase. A similar study was performed on herbicides used on the Canadian prairies (Waite et al. 2004), focusing on five main products: bromoxynil, dicamba, diclofop, MCPA, and trifluralin. Figure 8.5, which depicts the atmospheric concentration of trifluralin in three dugouts (Waite et al. 2004), clearly shows the seasonal variation of herbicide presence. In general, none of the herbicides was detected continuously throughout the sampling period.

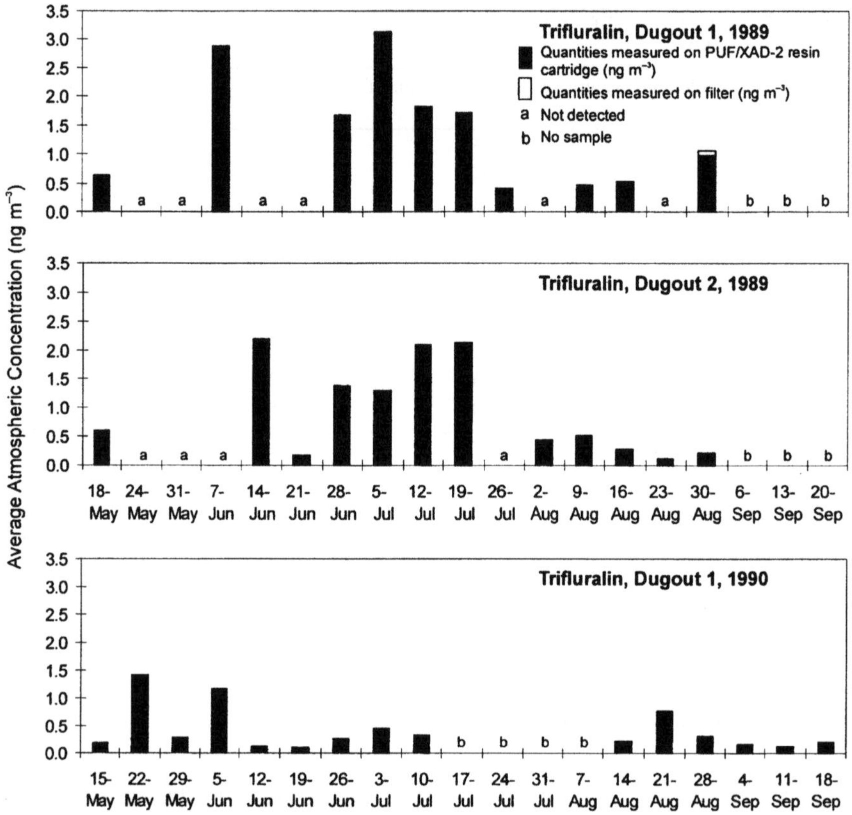

Fig. 8.5 Atmospheric concentrations of trifluralin in three dugouts on the Canadian Prairies during 1989–1990 (Waite et al. 2004)

Because each herbicide may degrade during volatilization, it is interesting to compare the cumulative volatilization of the parent contaminant and its metabolites. This behavior was studied in a wind-tunnel experiment by Wolters et al. (2003) for a mixture of parathion, terbuthylazine, and fenpropimorph, as well as for the metabolites fenpropimorph acid and desethyl-terbuthylazine. Figure 8.6 shows the volatilization dynamics of these pesticides and their metabolites when the products were applied initially on the surface of Gleyic Cambisol ($\sim$73 % clay, 23 % silt, 4 % clay). The volatilization, however, is controlled not only by the properties of the molecules but also by the properties of the subsurface composition, moisture content, and environmental factors.

Spencer and Cliath (1969, 1973) studied the effect of organic matter (OM) and clay content on vapor density of various pesticides. In general, they found that subsurface OM content and partition coefficients are of primary importance in describing the rate of volatilization for compounds having a high affinity for OM.

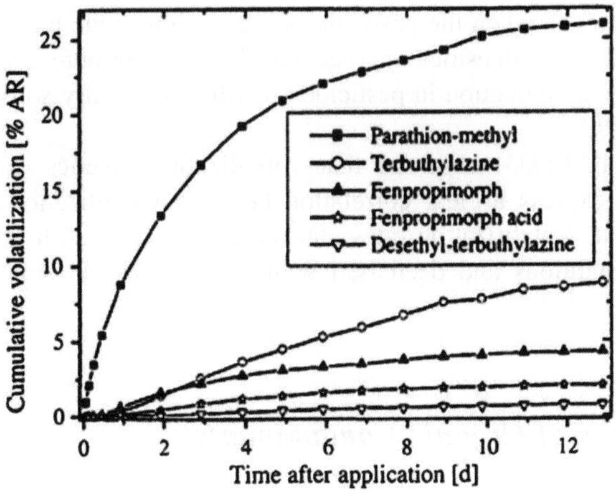

Fig. 8.6 Cumulative volatilization of ^{14}C-labeled pesticides and metabolites after soil surface application on gleyic cambisol (Wolters et al. 2003)

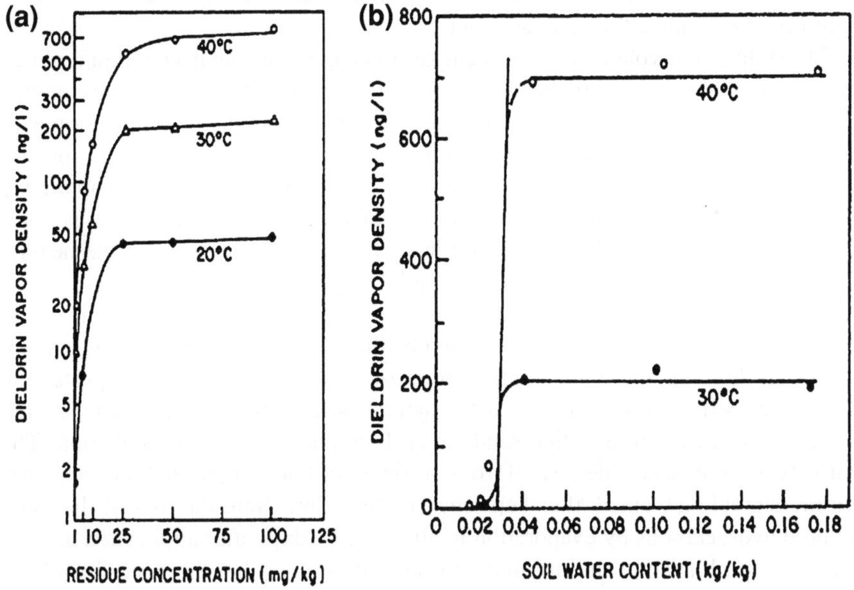

Fig. 8.7 Effect of **a** temperature and **b** soil moisture content on dieldrin vapor density (Taylor and Spencer 1990)

Temperature and moisture content are other important factors that control volatilization of organic contaminants in the subsurface. Spencer and Cliath (1969, 1973) showed that a temperature increase from 20 to 40 °C led to an increase in dieldrin vapor density from 45 to 700 ng/L (Fig. 8.7a). It also may be observed that a reduction in the soil moisture content caused a large reduction in the dieldrin vapor densities, even when the pesticide concentration in the moist soil was high enough to yield vapor densities approaching those of the pure compound. These results explain why reduction in pesticide volatilization in dry soils was observed over many years.

Wolters et al. (2003) observed that volatilization kinetics of the fungicide fenpropimorph express a clear correlation between volatilization rates and soil moisture content. Volatilization rates reached a maximum 24 h after application under moist conditions and decreased with the decrease in soil moisture over following days.

8.2.3 Mixtures of Organic Contaminants

Volatilization from mixtures of organic contaminants brings about changes in both the physical and the chemical properties of the residual liquid. We consider data on kerosene volatilization, as summarized in Yaron et al. (1998). Kerosene is an industrial petroleum product composed of more than 100 hydrocarbons, which may become a subsurface contaminant.

The differential volatilization of neat kerosene components from a liquid phase, directly into the atmosphere during volatilization up to 50 % (w/w), is presented in Fig. 8.8. Ten kerosene components were selected, and their composition was depicted as a function of gas chromatograph peak size (%), which is linearly related to their concentration. It may be seen that the lighter fractions evaporate at the beginning of the volatilization process. Increasing evaporation causes additional components to volatilize, which leads to a relative increase in the heavier fractions of kerosene in the remaining liquid.

In the subsurface, kerosene volatilization is controlled by the physical and chemical properties of the solid phase and by the water content. Porosity is a major factor in defining the volatilization process. Galin et al. (1990) reported an experiment where neat kerosene at the saturation retention value was recovered from coarse, medium, and fine sands after 1, 5, and 14 days of incubation. The porosity of the sands decreased from coarse to fine. Figure 8.9 presents gas chromatographs obtained after kerosene volatilization. Note the loss of the more volatile hydrocarbons by evaporation in all sands 14 days after application and the lack of resemblance to the original kerosene. It is clear that the pore size of the sands affected the chemical composition of the remaining kerosene. For example, the C_9–C_{12} fractions disappeared completely 14 days after their application, except for the saturated fine sand case, where 5 % of the initially applied C_{12} remained.

The effect of aggregation of the subsurface solid phase on kerosene volatilization was studied by Fine and Yaron (1993), who compared the rate of aggregation in two size fractions of a vertisol soil: the <1 mm fraction and 2 mm aggregates. The total porosity of these two fractions was similar (53 and 55 % of the total volume, respectively). Differences in aggregation are reflected in the air

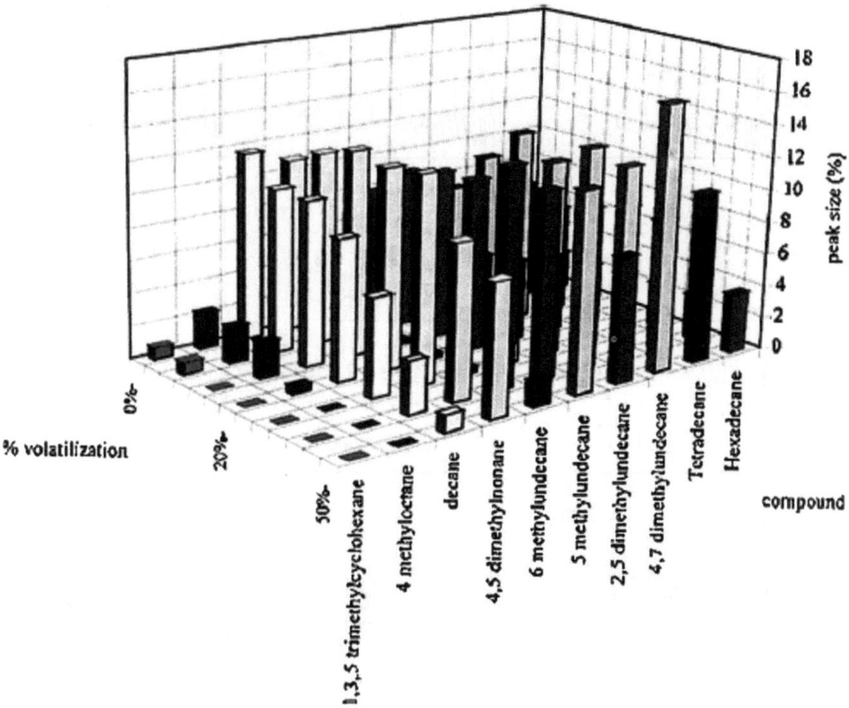

Fig. 8.8 Major remaining components of kerosene during the volatilization process (Yaron et al. 1998)

permeability; that is, their respective values were 0.0812 ± 0.009 and 0.145 ± 0.011 cm^2. Figure 8.10 presents the volatilization of kerosene as affected by the soil aggregation, when the initial amount applied was equivalent to the retention capacity. The more permeable fraction releases kerosene faster and thus enhances volatilization.

These examples indicate that aggregation and pore size distribution parameters affect volatilization of petroleum products from a contaminated subsurface. Fine and Yaron (1993) report that kerosene volatilization depends on the type of soil. Tests on four soils with a clay content increasing from 0.3 to 74.4 %, and OM content ranging from 0.01 to 5.1 % (Fig. 8.11), showed significant variations in rate and amount of volatilization over an 18-day period. This effect was attributed not only to the soil composition but also to the soil porosity and aggregation status.

Changes in the chemical composition of residual kerosene, resulting from volatilization of the light fractions, cause changes in the physical properties of the remaining product. Table 8.3 shows the effect of the differential volatilization on kerosene viscosity, surface tension, and density. When 20, 40, and 60 % of the initial amount of kerosene was removed by the transfer of light fractions to the atmosphere, the viscosity of the remaining kerosene was affected strongly. Negligible effects on liquid density and on surface tension were observed.

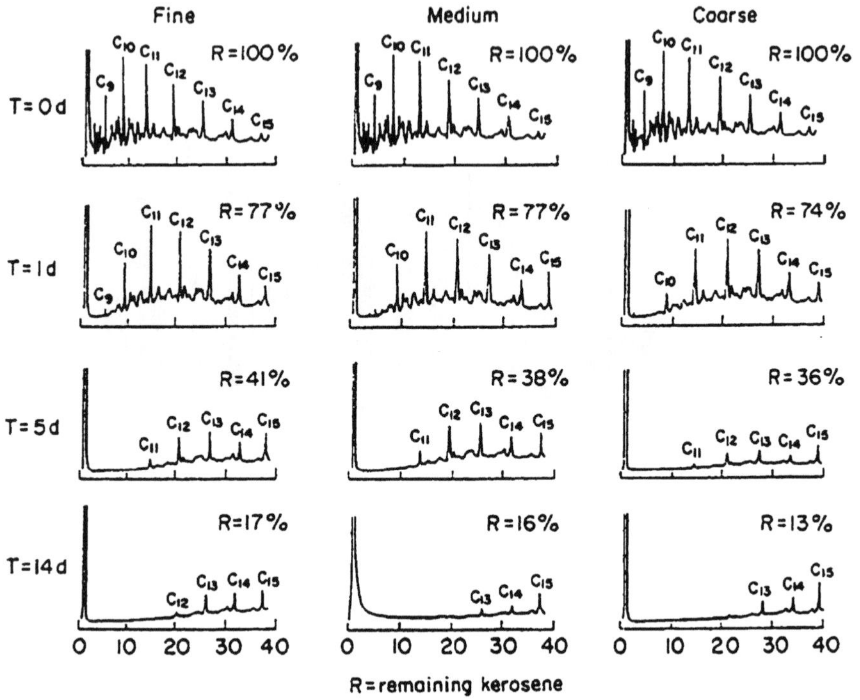

Fig. 8.9 Effect of porosity on composition of kerosene during 14 days of volatilization from fine, medium, and coarse sand, as seen from gas chromatograph analyses. Reprinted from Galin et al. (1990). Copyright 1990 with permission of Elsevier

Fig. 8.10 Volatilization of kerosene from vertisol as affected by aggregate size. Reprinted from Fine and Yaron (1993). Copyright 1994 with permission of Elsevier

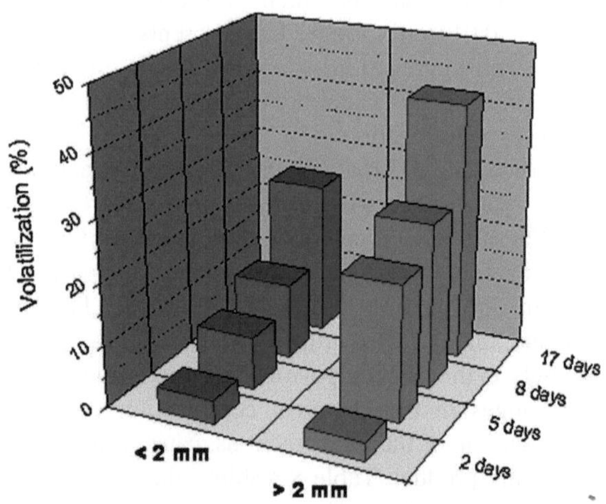

Differential volatilization, as affected by properties of the subsurface and its environmental conditions, leads to changes in residual composition of the hydrocarbon mixture. Jarsjo et al. (1994) investigated volatilization of kerosene

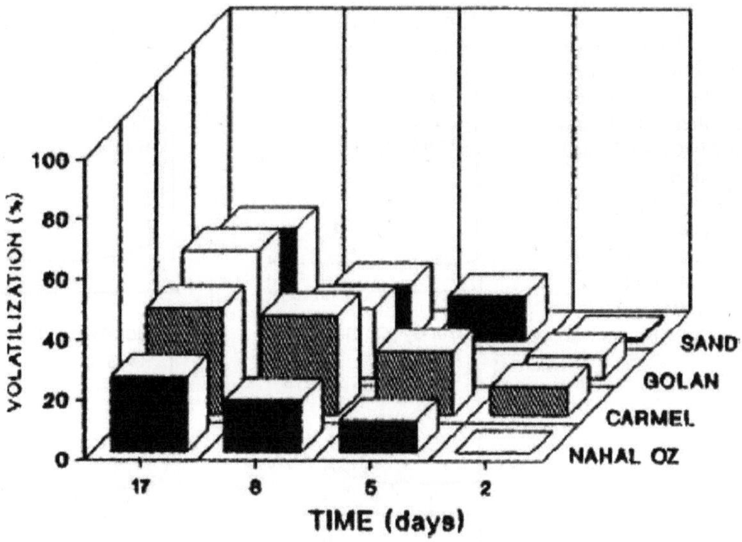

Fig. 8.11 Volatilization of kerosene as affected by soil type: sand (dune), chromoxeret (Golan), pelloxeret (Carmel), and calcic haploxeralf (Nahal Oz). Reprinted from Fine and Yaron (1993). Copyright 1994 with permission of Elsevier

Table 8.3 Effect of volatilization on physical properties of residual kerosene

Amount volatilized (%)	Density (g/cm^3)	Surface tension (1/Nm)	Viscosity (Pa s × 10^{-3})
0	0.805	2.75	1.32
20	0.810	2.78	1.48
40	0.818	2.80	1.78
60	0.819	2.78	1.96

Reprinted from Galin et al. (1990). Copyright 1990 with permission of Elsevier

from glacial and post-glacial soils. The composition of the residual kerosene mixture recovered from sand, sandy loam, and peat materials, after volatilization at 5 and 27 °C, for 7 and 30 days, respectively, is shown in Fig. 8.12. From this figure, we see that all the lighter components disappeared from the sand and the sandy loam and the residual kerosene percentages are similar regardless of temperature. The relative concentration of the light fraction, C_{10}–C_{12}, diminished with time in all soils, while those of the heavy fractions, C_{14} and C_{15}, increased.

Volatilization of an organic mixture of contaminants, distributed vertically in the subsurface, may induce not only a decrease in the component concentrations but also an enrichment of the deeper layers during the volatilization process. Figure 8.13 shows the actual content of three representative hydrocarbons—m-xylene (C_8), n-decane (C_{10}), and hexadcane (C_{16})—which originated from the applied kerosene found along a 20-cm soil column, 18 days after application on

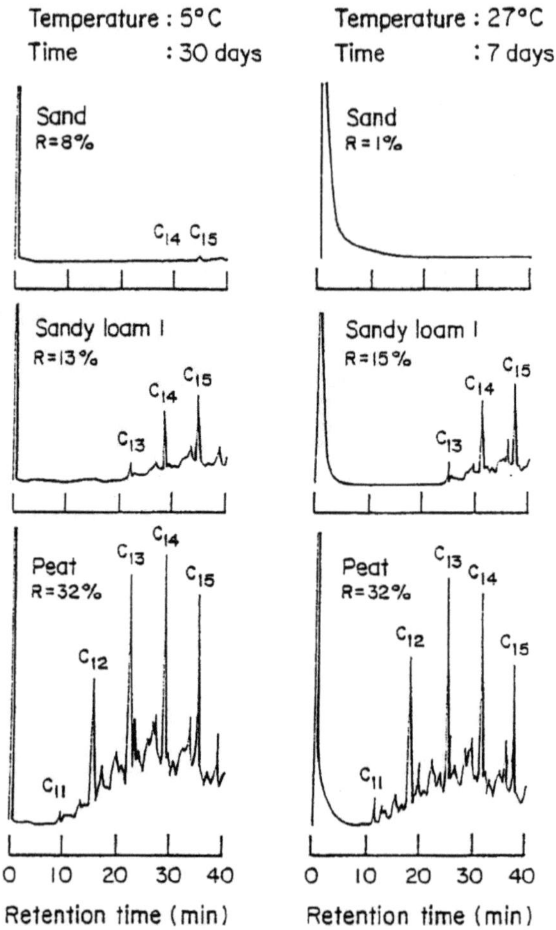

Fig. 8.12 Gas chromatographs of residual kerosene recovered from glacial and post-glacial earth materials after volatilization at 5 and 27 °C, for 30 and 7 days, respectively. *R* denotes the total remaining kerosene (percentage of initial amount). Reprinted from Jarsjo et al. (1994). Copyright 1994 with permission of Elsevier

dry soil. Roughly 30 % of the initial content of m-xylene still remained in the soil after 18 days. Furthermore, the content of m-xylene increased somewhat after the third day; a similar trend was found for the n-decane distribution. Hexadecane was partially removed from deeper layers and redistributed near the soil surface.

8.3 Solubility and Dissolution

The solubility of contaminants in subsurface water is controlled by (1) the molecular properties of the contaminant, (2) the porous media solid-phase composition, and (3) the chemistry of the aqueous solution. The presence of potential cosolvents or other chemicals in water also affects contaminant solubility. A number of relevant examples selected from the literature are presented here to illustrate various solubility and dissolution processes.

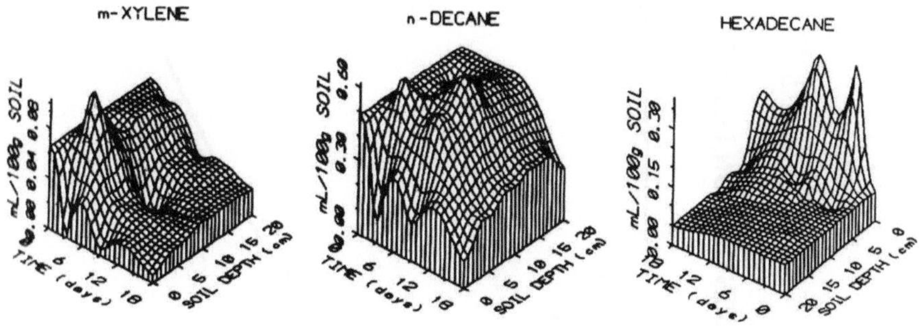

Fig. 8.13 Concentration of selected petroleum hydrocarbons (mL/100 g soil) during volatilization of kerosene from air-dry vertisol. Reprinted from Fine and Yaron (1993). Copyright 1994 with permission of Elsevier

8.3.1 Acidity and Alkalinity Effects

Hydrogen ion regulation in subsurface water is provided by numerous homogeneous or heterogeneous buffer systems. A buffer system is characterized by a range of pH within which it is efficient, relative to its acid- or base-neutralizing capacity. In a natural environment, the pH of the subsurface water does not generally correspond to an equilibrium state; this results in instability of pH values, which is reflected in short-term fluctuations due to seasonal variations and anthropogenic processes. The effects of low and high pH on dissolution of a number of earth minerals are discussed here.

Aluminum dissolution under acid rain is the outcome of an anthropogenically induced effect on the subsurface acid–base equilibrium. As a result of acidic atmospheric pollutants (e.g., HCl, HNO_3, H_2SO_4), rainwater becomes acidic; and when it reaches the subsurface, where buffering by existing bases is missing, the acidity of the subsurface water increases. Soil acidification usually is a long-term process. For example, Johnston et al. (1986) report a decrease from pH 7.2 to pH 4.2 after 100 years at the Rothamsted agricultural experimental station (United Kingdom).

Dissolution of Al, which has devastating effects on soil biological populations, is the main consequence of acidic atmospheric deposition in the forest environment in large areas of northern and central Europe and North America (Mulder et al. 1987). In a study of acid effects on sandy soils, van Grunsven et al. (1992) observed that the logarithm of Al dissolution rates in individual earth samples follows an inverse linear relationship with pH. Hargrove (1986) investigated Al dissociation from mica–OM complexes, observing that Al concentrations in water solutions reach minimum values in a pH range of about 4–5 (Fig. 8.14).

Trace element dissolution from soils surrounding abandoned waste incinerators, as a combined effect of pH and speciation, was reported by Bang and Hesterberg (2004), who determined the concentration of dissolved metals eluted from soil samples as a function of pH. In parallel, the speciations of Cu and Zn were analyzed using X-ray absorption and near-edge structure spectroscopy analysis

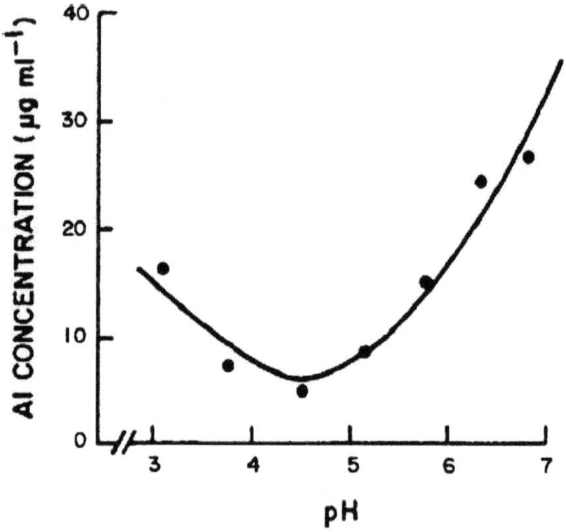

Fig. 8.14 Effect of pH on dissolution from Al–mica–organo matter complexes (Hargrove 1986)

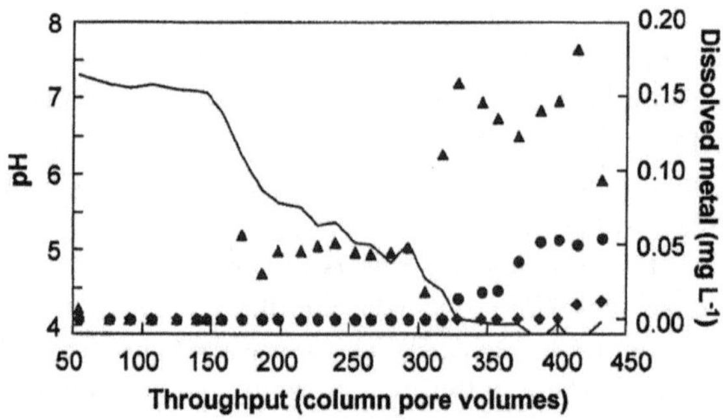

Fig. 8.15 Typical trends of dissolved Cu (*black circle*), Pb (*black diamond*), and Zn (*black up-pointing triangle*) concentrations and pH (–) in effluent solutions collected during a flow column experiment on soil samples from the Bogue site (USA), as a function of the throughput volume of acidified, 0.01 M $CaCl_2$ solution (Bang and Hesterberg 2004)

(XANES). Typical trends for dissolution of Cu, Pb, and Zn, as affected by pH, are presented in Fig. 8.15. The maximum dissolved concentration of these compounds occurred at the lowest pH. Dissolved Cu, Pb, and Zn concentrations increased as pH declined, compensating for the Na^+. This phenomenon occurs below pH threshold levels of pH ~4 for Cu and Pb and pH ~5 for Zn. The Cu and Zn results were consistent with synchrotron (XANES) spectroscopy data, suggesting that Cu was bound mainly to soil OM, while Zn was associated with Al- and Fe-oxide-type minerals.

Another key process involves gypsum dissolution; gypsum is found in the subsurface as a natural constituent or as an added reclamation material. In cases

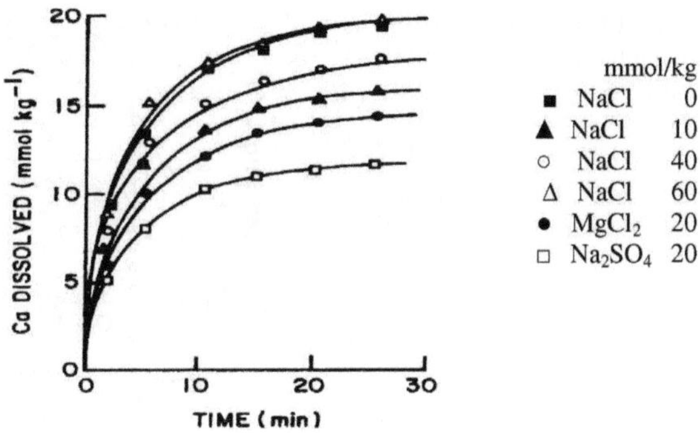

Fig. 8.16 Kinetics of Ca dissolution from a gypsum sample in various salt solutions (Gobran and Miyamoto 1985)

where human actions enhance subsurface alkalinity, as a result of irrigation with alkali water or poor management of irrigation and drainage, the distribution of gypsum with depth may be changed by gypsum dissolution. The effective solubility of gypsum depends only on the subsurface water chemistry and the exchange-phase composition of the solid matrix. The dissolution exchange reaction is

$$2NaX + CaSO_4 \rightleftharpoons CaX_2 + 2Na^+ + SO_4^{2-} \tag{8.5}$$

where X is one equivalent of exchanger and indicates that cation exchange removes Ca^{2+} from subsurface water and allows additional gypsum dissolution. Disposal of alkali water or alkali effluents on land brings an increase in the exchangeable Na^+ in the solid phase. Under such conditions, the dissolution of $CaSO_4$ in the subsurface increases to compensate for the Na^+ that exchanges with Ca^{2+}.

Effective gypsum solubility is enhanced through ion-pair formation and electrolyte effects. The amount of water required for dissolution of gypsum in an alkaline subsurface is likely to be much less than commonly inferred for gypsum solubility in water. When the alkalinity is associated with an increase in the total salt content, gypsum dissolution is affected by the presence of electrolytes. An example is given in Fig. 8.16, which shows dissolution kinetics of gypsum samples from Egypt in NaCl and $MgCl_2$ solutions (Gobran and Miyamoto 1985). It appears that the dissolved amounts at equilibrium are related to the ionic composition of the aqueous solution. We see that the dissolution in the subsurface water apparently increases with increasing salt content, as expected from ionic strength considerations.

8.3.2 Redox Processes

Redox processes affect contaminant solubility and may result from fluctuating saturation and drying processes in the subsurface due to natural or anthropogenic

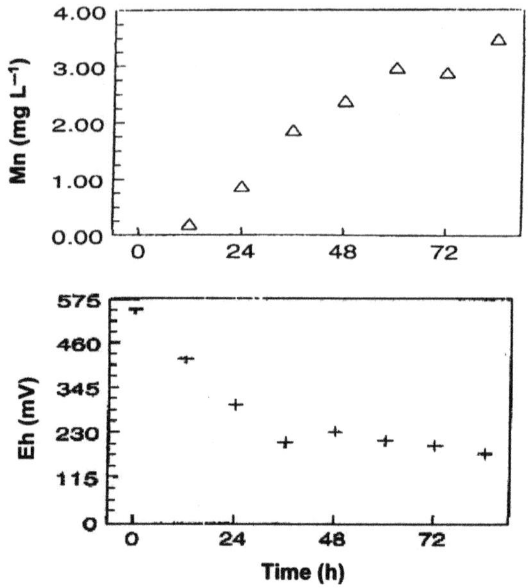

Fig. 8.17 Changes with time in soluble Mn (*increment*) and Eh (+). (Green et al. 2003)

factors. Reduction and oxidation processes also may occur simultaneously in a partially saturated soil matrix at the aggregate level: reduction into the solid-phase aggregate and oxidation at the aggregate surface. The reduction–oxidation process may be enhanced by biological activity or by chemical catalysis, which affects heavy metal solubility.

Examples of the ability of Mn oxides to oxidize metals directly or to catalyze metal oxidation have been reported widely (e.g., Lindsay 1979; McBride 1989, 1994; Stumm and Morgan 1996; Xiang and Banin 1996; Charlatchka and Cambier 2000). As an example, we consider a case study by Green et al. (2003) on solubilization of manganese and other trace metals from soils irrigated previously with water affected by acid mine runoff. The experiment was designed to simulate waterlogged irrigated soils, saturated for short-term periods of up to 5 days. Increases in soluble Mn were observed, which are associated with a decrease in redox potential (Eh; see Sect. 2.2.2). Dissolved Mn concentrations in the soil water solutions, studied after 84 h of saturation, exceeded US EPA drinking water quality standards of 0.05 mg/L. Changes in soluble Mn and in Eh as obtained in one replicate of the experiment are presented in Fig. 8.17.

As seen in Fig. 8.17, Mn was released into solution after the Eh decreased below approximately 450 mV. Once the critical Eh (~450–500 mV) needed for dissolution of Mn was reached, time became the limiting factor in determining soluble Mn concentration. The soluble Mn concentration continued to increase throughout the duration of the experiment, suggesting that equilibrium conditions were not reached. Concentrations of Zn and Ni also increased following soil reduction. Lack of significant correlation for Cu was explained by the formation of complexes with dissolved organic matter (DOM). Increases in the soluble concentration of Zn and Ni at values greater than those expected from the dissolution of metal-bearing Mn oxides are due to exchangeable trace elements. These elements can be released due to changes in the electrolyte concentration, for example, as a result of the reduction process.

8.3.3 Dissolution from Mixtures of Organic Contaminants

Dissolution of volatile organic mixtures was observed by studying the behavior of kerosene. Dror et al. (2002) compared the gas chromatographs of neat kerosene to those of the fraction of kerosene dissolved in an aqueous electrolyte solution of 0.01 M NaCl, at an ambient temperature of 22 °C (Fig. 8.18). Aliphatic and branched aliphatic hydrocarbons in the range C_9–C_{16} constitute the major group of components in neat kerosene, with only a minor set of aromatic components. In aqueous electrolyte solutions, on the other hand, aromatic compounds, especially branched benzene and naphthalene, make up the majority of compounds. The C_9–C_{16} aliphatic and branched aliphatic components do not appear in the gas chromatograph of the aqueous electrolyte solution. The reason for the difference between the neat and aqueous dissolved kerosene is the wide range of aqueous solubility typical of the different kerosene components. Aromatic compounds usually are much more soluble than aliphatic components. Therefore, aromatic components, which are minor constituents of neat kerosene, dissolve in a water solution much more readily than the major group of aliphatic compounds.

Gasoline commonly contains appreciable amounts of aniline, phenol, and their alkyl substituted homologs as well as many other polar compounds. Schmidt et al. (2002) measured the fuel–water partitioning coefficients, K_{fw}, of several polar compounds during a batch experiment. The K_{fw} values for the investigated phenols, anilines, benzotriazoles, and S-heterocycles ranged from 0.2 to 1,700; these values are up to three orders of magnitude lower than the K_{fw} of benzene. The nonaqueous-phase liquid (NAPL)–water partitioning of anilines and phenols depends strongly on the structure of the compound as well as on the pH of the gasoline composition. Due to their polarity, anilines and phenols associated with the gasoline, after a spill on the land surface, may reach groundwater ahead of the benzene and may be used as a marker to predict groundwater contamination.

8.3.4 Apparent Solubility

The "apparent solubility" of contaminants can be defined when the theoretical solubility deviates from the initial value, as a result of a number of factors encountered in the subsurface environment. Waste and sludge disposal sites usually contain a mixture of salts, toxic trace metals, and organic contaminants. When the disposal site is leached by rain or irrigation water, a solution containing soil

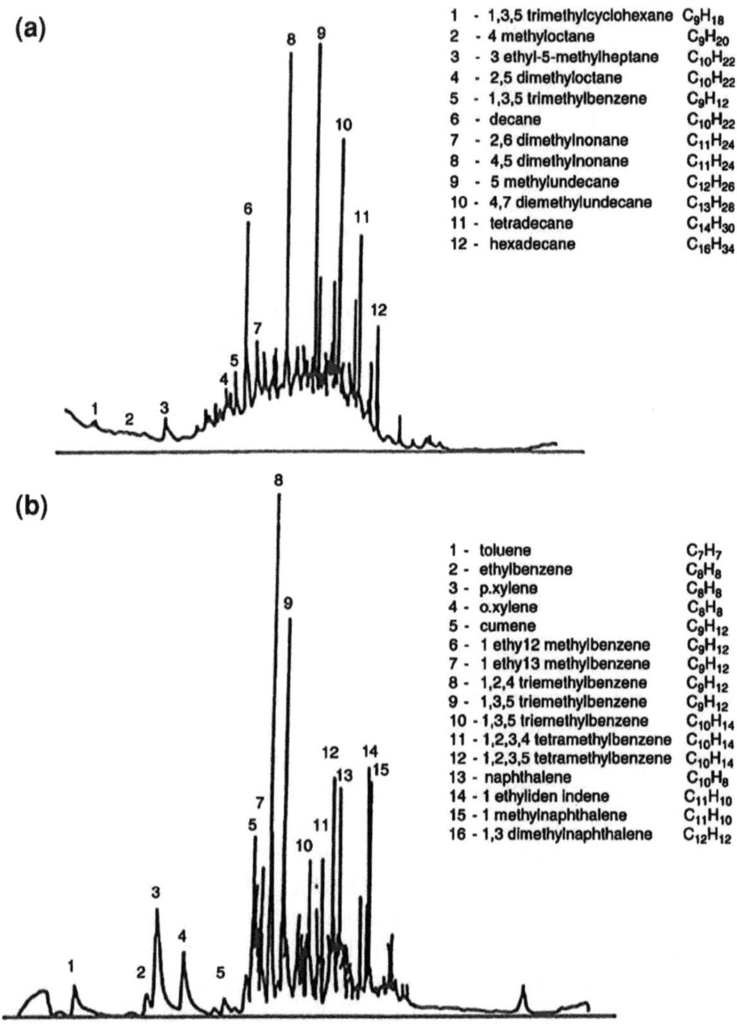

Fig. 8.18 Gas chromatographs of **a** neat kerosene and **b** kerosene dissolved in aqueous solution. Reprinted from Dror et al. (2002). Copyright 2002 with permission of Elsevier

and dissolved or suspended waste and sludge is obtained. The solubility of the contaminants in the water solution is affected by the presence of potential organic ligands and surfactants, electrolytes, and cosolvents; therefore, measurements represent an "apparent solubility" that differs from the theoretical solubility.

Surfactants are major compounds that reach the subsurface alone or accompanying other contaminants. Their effect depends highly on the solution chemistry. For example, Park and Bielefeldt (2003) report the partitioning of Tergitol, a nonionic surfactant, and pentachlorophenol (PCP), from NAPL to an aqueous

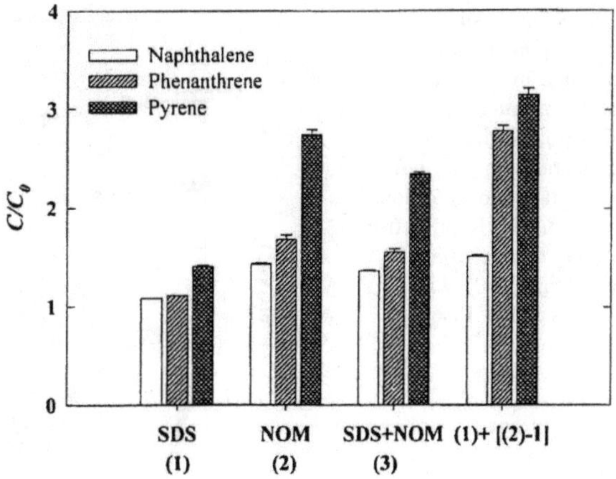

Fig. 8.19 Solubility of polycyclic aromatic hydrocarbons (*PAH*) in sodium dodecyl sulfate (*SDS*, 10 mg/L) and Suwannee River natural organic matter (*NOM*, 10 mg/L) solution. *C* denotes increased solubility in SDS and NOM solutions, and C_0 is the solubility in water (Cho et al. 2002)

solution. Enhanced PCP dissolution into water from the NAPL was achieved at aqueous Tergitol concentrations >200 mg/L. Surfactant addition of 1,200 mg/L increased the aqueous PCP concentration by 14-fold at pH 5 and by twofold to threefold at pH 7. This result is explained by the ionizable nature of PCP and the effect of pH and ionic strength on equilibrium partitioning. The significant response at the lower pH may be explained by the greater hydrophobicity of PCP molecules at a low pH. To the ionizable nature of PCP and the effect of pH on the equilibrium partitioning, one must also add the effect of ionic strength. These results can be used to improve the use of surfactants for remediation of subsurface sites contaminated by NAPLs.

A combined effect of natural organic matter (NOM) and surfactants on the apparent solubility of polycyclic aromatic hydrocarbons (PAHs) is reported in the paper of Cho et al. (2002). Kinetic studies were conducted to compare solubilization of hydrophobic contaminants such as naphthalene, phenanthrene, and pyrene into distilled water and aqueous solutions containing NOM and sodium dodecyl sulfate (SDS) surfactant. The results obtained after 72-h equilibration are reproduced in Fig. 8.19. The apparent solubility of the three contaminants was higher in SDS and NOM solutions than the solubility of these compounds in distilled water. When a combined SDS–NOM aqueous solution was used, the apparent solubility of naphthalene, phenanthrene, and pyrene was lower than in the NOM aqueous solution.

Properties of surfactant and cosolvent additives affect the rate of apparent solubilization of organic contaminants in aqueous solutions and may serve as a tool in remediation of subsurface water polluted by NAPLs. Cosolvents (synthetic or natural) are organic solutes present in sufficient quantities in the subsurface water to render the aqueous phase more hydrophobic. Surfactants allow NAPLs to partition into the micelle core, when present at levels above the critical micelle concentration, causing an apparent solubility increase. An extensive discussion of surfactant- and cosolvent-induced dissolution of NAPLs in porous media is given

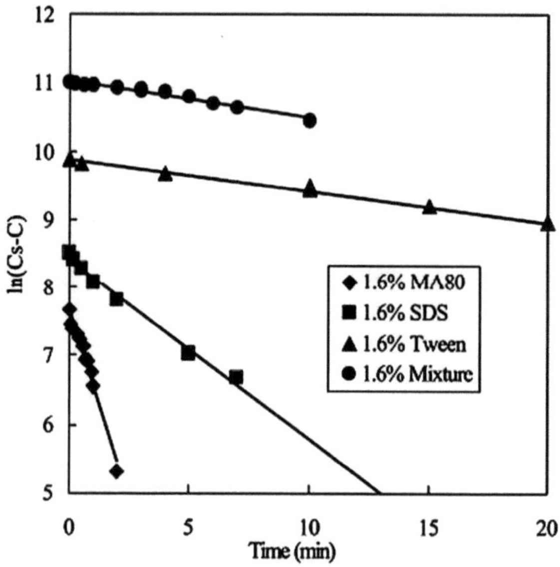

Fig. 8.20 Solubilization rate of trichloroethylene as affected by surfactant type and concentration; *Cs* and *C* denote surfactant concentration and initial surfactant concentration, respectively. Reprinted from Zhong et al. (2003). Copyright 2003 with permission of Elsevier

by Khachikian and Harmon (2000). Here, we discuss selected examples to illustrate this phenomenon.

Zhong et al. (2003) studied the apparent solubility of trichloroethylene in aqueous solutions, where the experimental variables were surfactant type and cosolvent concentration. The surfactants used in the experiment were sodium dihexyl sulfosuccinate (MA-80), SDS, polyoxyethylene 20 (POE 20), sorbitan monooleate (Tween 80), and a mixture of Surfonic PE-2597 and Witconol NP-100. Isopropanol was used as the alcohol cosolvent. Figure 8.20 shows the results of a batch experiment studying the effects of type and concentration of surfactant on solubilization of trichloroethylene in aqueous solutions. A correlation between surfactant chain length and solubilization rate may explain this behavior. However, the solubilization rate constants decrease with surfactant concentration. Addition of the cosolvent isopropanol to MA-80 increased the solubility of isopropanol at each surfactant concentration but did not demonstrate any particular trend in solubilization rate of isopropanol for the other surfactants tested. In the case of anionic surfactants (MA-80 and SDS), the solubility and solubilization rate increase with increasing electrolyte concentration for all surfactant concentrations.

Speciation of transition metals by "natural" organic substances that behave as complexing ligands may occur in the subsurface following waste and sludge disposal. As a result, metal solubility increases, favoring metal mobility with depth. Haitzer et al. (2002) report binding of mercury(II) to DOM as a function of the Hg–DOM concentration ratio. Humic and fulvic acids, which were isolated from water originating from a site in the Florida Everglades, were used as DOM binding material. Comparing the Hg binding capacity of natural DOM to a chelate (ethylenediamine tetraacetic acid, EDTA), the natural DOM ligand was observed to have a limited number of strong Hg binding sites. Haitzer et al. (2002) suggest that the binding of Hg to DOM under natural conditions (very low Hg–DOM ratios) is controlled by a small functional fraction of DOM molecules containing a thiol functional group. Therefore, Hg–DOM distribution coefficients used in modeling biogeochemical behavior of Hg in natural systems must be determined at low Hg–DOM ratios.

Contaminants bound to colloids also may lead to an increase in the apparent solubility of the compounds. Most colloidal phases are effective sorbents of low-solubility contaminants, due to their large surface area. For example, Fig. 8.21 depicts the solubilization of *p*-nitrophenol into hydrophobic microdomains, which defines the trace metal level in the groundwater of a coastal watershed (Sanudo-Wilhelmy et al. 2002). The authors emphasize that the (heavy) metals contained in the colloidal size fraction in some instances may reach more than 50 % of what is considered "dissolved" metal; this should be considered to properly understand the cycling of metals and carbon in the subsurface water.

Babiarz et al. (2001) examined total mercury (Hg) and methylmercury (Me–Hg) concentrations in the colloidal phase of 15 freshwaters from the upper Midwest and southern United States. On average, Hg and Me–Hg forms were distributed evenly between the particulate (0.4 μm), colloidal, and dissolved (10 kDa) phases. The amount of Hg in the colloidal phase decreased with increasing specific electric conductance. Furthermore, experiments on freshwater with artificially elevated electric conductance suggest that Hg and Me–Hg may partition to different subfractions of colloidal material. The two colloidal Hg phases act differently with the same type of adsorbent. For example, the colloidal phase Hg correlates poorly with organic carbon (OC), but a strong correlation between Me–Hg and OC was observed.

Once reaching a water system, the components of a crude oil or a petroleum hydrocarbon are "truly dissolved" at a molecular level or "apparently soluble" at a colloidal level when droplets characterized by radii of tens to hundreds of microns are formed. The apparent solubility of PAHs from oil in an aquatic system is reported by Sterling et al. (2003), who consider that the colloidal concentration of a given hydrocarbon contaminant in aqueous phase, C, is described by the equation

$$C = \phi X / v \tag{8.6}$$

where ϕ denotes the volume fraction of oil emulsion in water (vol. emulsion/vol. water), X is a (dimensionless) chemical mole fraction in the organic phase, and v is average molar volume of component in the petroleum product.

Figure 8.22 highlights the fraction of naphthalenes present in water due to colloidal entrainment. As the compound molecular weight increases, the relative fraction of the compound in the colloidal phase increases. This occurs regardless of the level of entrainment. Sterling et al. (2003) suggest that this behavior has

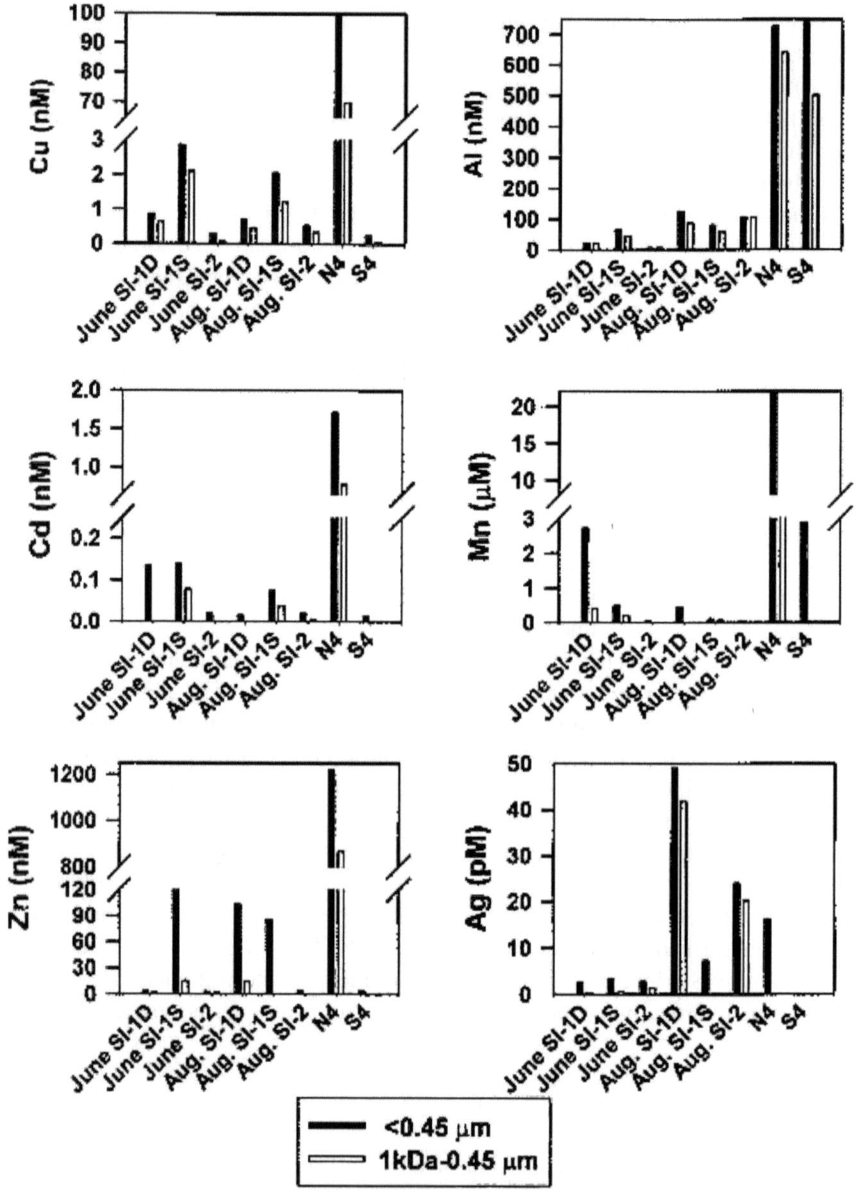

Fig. 8.21 Dissolved and colloidal concentrations of metals measured in the groundwater of Shelter Island and in the Peconics (North and South Forks of Long Island). Reprinted with permission from Sanudo-Wilhelmy et al. (2002). Copyright 2002 American Chemical Society

significant implications for the fate and transport of PAHs in natural aquatic systems, where the majority of PAH transport is due to water-entrained oil droplets.

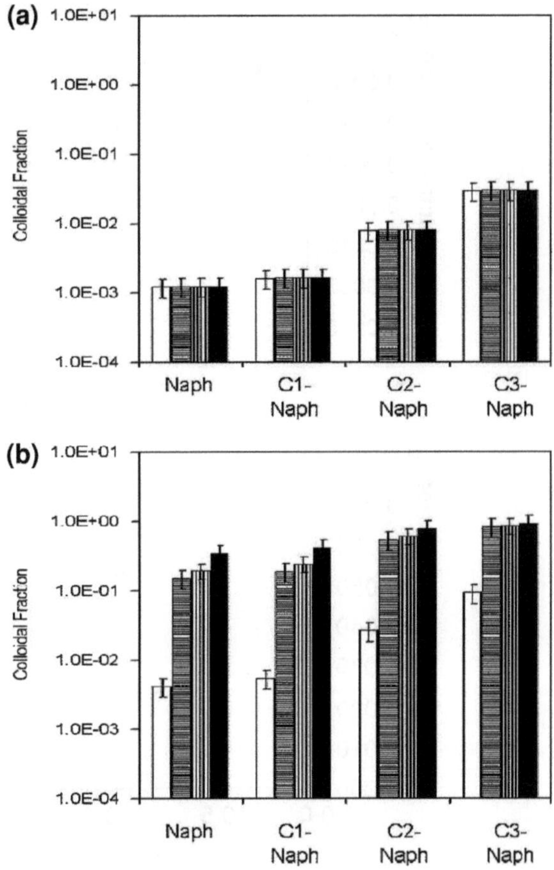

Fig. 8.22 Fraction of naphthalene concentration due to colloidal entrainment at **a** $G_m = 5\ s^{-1}$ and **b** $G_m = 20\ s^{-1}$, for naphthalene and naphthalene compounds containing 1, 2, or 3 C as side chains, where G_m is the mean shear rate. Reprinted with permission from Sterling et al. (2003). Copyright 2003 American Chemical Society

The salting-out effect (see Sect. 6.5) may lead to lower solubility of organic or organo-metallic contaminants in saline waters compared to those obtained in pure water. The solubility decreases with an increase in salt concentration in water. An extreme example may be found in Sorensen et al. (2002), who examined solubility of gaseous methane in pure and saline water. Referring to their experimental results, saline solutions of NaCl (up to 2.5 molality and ~11 wt%) were used with $CaCl_2$ (up to 2 molality and ~20 wt%). Figure 8.23 shows the decrease in solubility in saline water in the gas–water–salt system. Because methane dissolution in NaCl and $CaCl_2$ saline water occurred at different pressure and temperature, the results cannot be compared directly. However, the decreasing trend is obvious in both cases.

The electrolyte concentration in an aqueous solvent may affect dissolution of petroleum products, which are composed of a mixture of hydrocarbons. Dror et al. (2000) considered increasing concentrations of NaCl in water, up to a value approaching the salinity of seawater, and showed that kerosene dissolution decreases as the electrolyte concentration in water increases, according to the

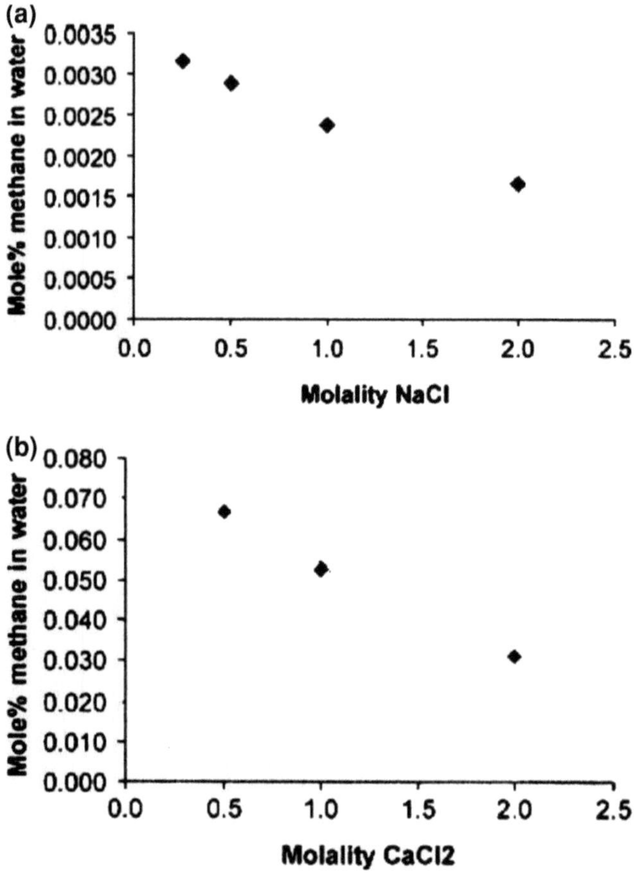

Fig. 8.23 Solubility of methane **a** in NaCl at 1 atm and 283.15 °K and **b** in $CaCl_2$ solutions at 37.4 atm and 298.15 °K. Reprinted from Sorensen et al. (2002). Copyright 2002 with permission of Elsevier

salting-out effect (Fig. 8.24). More data on effects of salts on the solubility of a large number of organic compounds are presented in Table 6.3.

The partitioning behavior of alkylphenols in crude oil–brine subsurface systems was reported by Bennett and Larter (1997). Partition coefficients were measured in the laboratory for simulated environmental conditions, from the near to the deep subsurface, as a function of pressure (25–340 bar), temperature (25–150 °C), and water salinity (0–100,000 mg/L sodium chloride) for a variety of oils. Alkylphenol partition coefficients between crude oil and brines decreased with increasing temperature, increased with water salinity and concentration of nonhydrocarbon compounds in the crude oil, and showed little change with varying pressure. The results of Bennett and Larter (1997) clearly show that, with increasing salt addition, the partition coefficient of alkylphenols increases, indicating increased phenol preference for the petroleum phase at higher brine salinity.

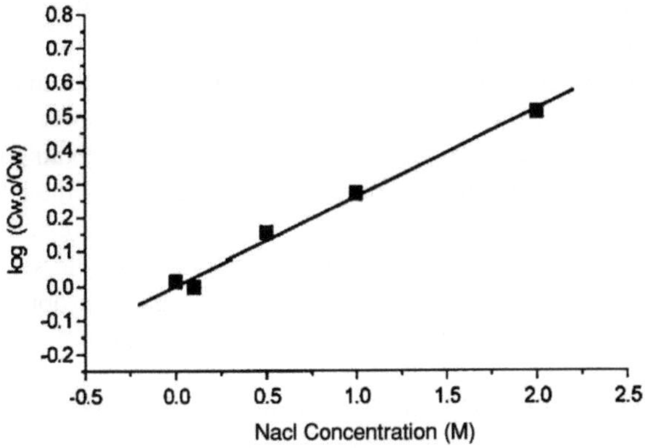

Fig. 8.24 Decrease in kerosene solubility in a NaCl aqueous solutions of various concentration as result of the salting-out effect (Dror et al. 2000). $C_{w,0}$ and C_w denote concentrations at initial maximum solubility and at various salinities

Table 8.4 K_{salt} values for three tested drugs

Solute	Experimental K_{salt}	$\log K_{ow}$	Calculated[a] K_{salt}
Phenytoin	0.191	2.08	0.198
Theophylline	0.100	−0.06	0.115
Cytosine	−0.005	−1.65	0.053

[a] Based on $K_{salt} = 0.039 \log K_{ow} + 0.117$ (Ni et al. 2000)

Pharmaceuticals and personal care products represent potential contaminants; we consider it appropriate to include an example of the salting-out effect on drug solubility in aqueous solutions. Ni et al. (2000) examined the relation between the Setschenow constant (K_{salt}) of a nonelectrolyte in a NaCl solution and the logarithm of its octanol–water partition coefficient, log K_{ow}. The relation $K_{salt} = A \log K_{ow} + B$, where A and B are constants, was tested for 15 compounds including a number of drugs. The values of A (=0.039) and B (=0.117) were determined empirically from the literature data for 62 organic compounds. This linear relationship provides a simple and accurate method to predict the salting-out effect on organic compound solubility and was used to determine the K_{salt} for the drugs phenytoin, theophylline, and cytosine; the results are presented in Table 8.4.

When a NAPL is "mixed" with saline water, the coupled effects of salting out and droplet formation can occur. Despite an expected decrease in concentration due to the salting-out effect, a higher total organic concentration may be found. This phenomenon is explained by formation of a pseudo "oil-in-water" emulsion, with droplets of various sizes leading to increased apparent solubilities that may be up to several times greater than those known theoretically. This phenomenon also contradicts the salting-out effect. These droplets, which are not visible to the naked eye, are observed by optical microscope and detected by gas chromatography analysis.

Examples of NAPL droplet formation in water are described by Yaron-Marcovich et al. (2007). Solutions of benzene, toluene, xylene, trichloroethylene, and a mixture of them were prepared in excess in freshwater and saltwater, and solution stability was examined. High organic concentrations were found to remain stable in both freshwater and saltwater. In saltwater, for example, toluene and xylene concentrations remained as high as 14 and 26 times their theoretical solubilities, respectively, over a period of six days, while in freshwater, their concentrations remained 8 and 30 times their solubilities over the same period. This phenomenon is attributed to the presence of stable organic droplets. An image of organic droplets in saline solution captured by optical microscope is presented in Fig. 8.25.

A similar effect was observed in experiments reported by Dror et al. (2003) in a saline solution simulating seawater. They found that the concentration of a mixture of organic contaminants may be much higher than expected from theoretical considerations, accounting for the salting-out effect (Table 8.5). This is because the total concentration of organic mixture in natural saline (or sea) water is given not only by the dissolved contaminant and electrolyte concentration but also by the presence of mechanically dispersed organic droplets formed under various environmental conditions. This apparent solubility and concentration is called the *carrying capacity of the aqueous solution* (CCAS). It may be observed from Table 8.5 that, in saline water, the low-density hydrocarbons (benzene and toluene) have a CCAS value much greater than the CCAS in freshwater.

8.4 Contaminant Retention in the Subsurface

As mentioned previously, the retention of contaminants on geosorbents may occur by surface adsorption on or into the colloid fraction of the solid phase and by physical retention as liquid ganglia or as precipitates into the porous media. The type of retention is defined by the properties of the solid phase and the contaminants as well as by the composition of the subsurface water solution and the ambient temperature.

8.4.1 Contaminant Adsorption

Contaminant adsorption includes retention on the porous medium solid phase, as a result of cation exchange processes, and surface retention of neutral molecules, due to van der Waals forces.

Sodic water reaching the land surface through irrigation or disposal of sewage effluent leads to soil deterioration, caused by the exchange process between Na^+ from the water and clay-saturating cations. The sodium–calcium exchange process leads to retention of Na^+ in the solid phase, affecting not only the earth material

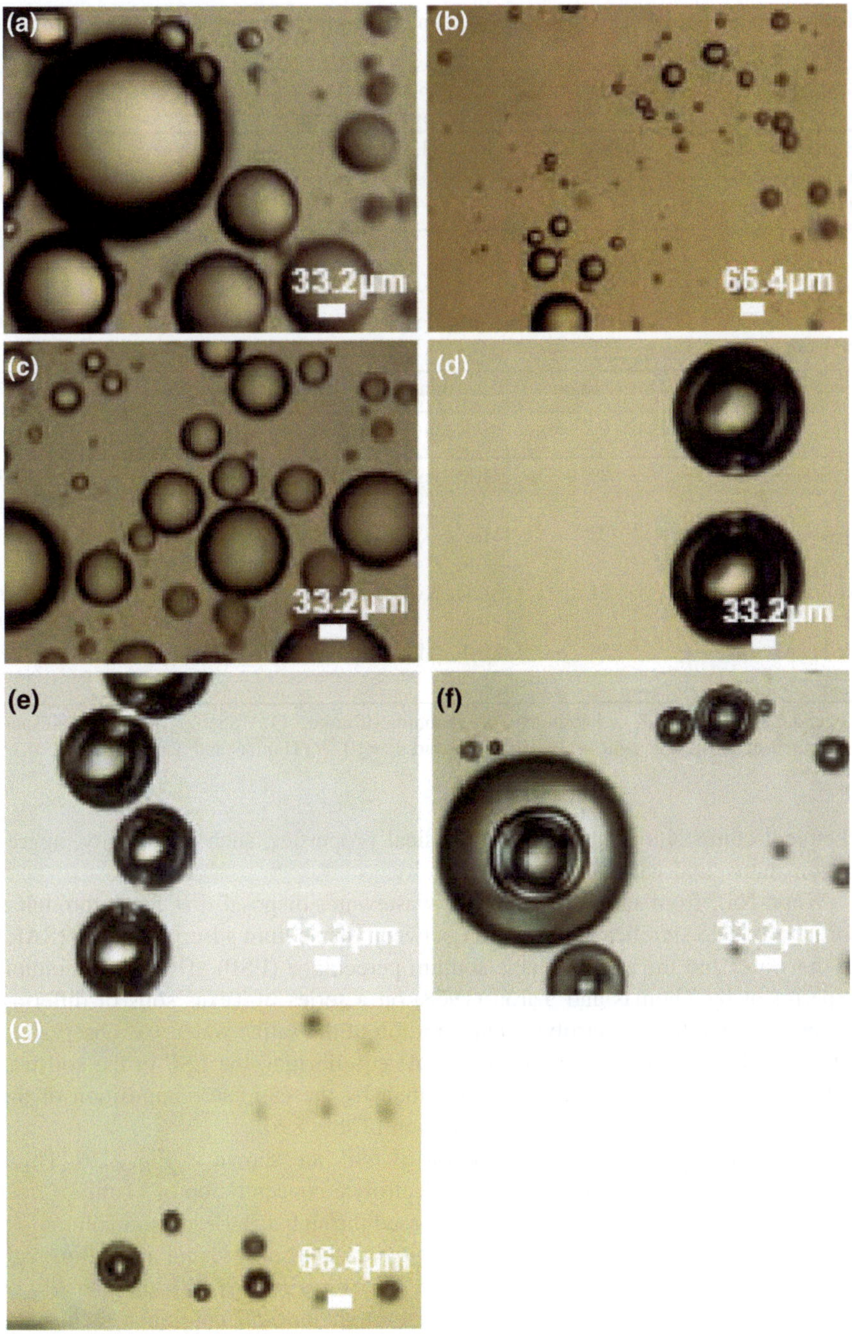

Fig. 8.25 Organic compound droplets in saline (30 g/L NaCl) water as captured by optical micrographs: **a** benzene; **b** toluene; **c** xylene; **d** TCE; **e** benzene and TCE; **f** xylene and TCE; and **g** organic compound mixture. Reprinted from Yaron-Marcovich et al. (2007). Copyright 2007 with permission of Elsevier

Table 8.5 Carrying capacity of aqueous solution (CCAS) for hydrocarbons as a result of droplet formation (Dror et al. 2003)

Compound	Distilled water solubility (mg/L)	Seawater solubility (mg/L)	Freshwater CCAS (mg/L)	Saline water CCAS (mg/L)
Benzene (C_6H_6)	1,740	1,391	1,940	3,340
Toluene (C_7H_8)	535	379	1,345	7,461
Trichloroethylene (C_2HCl_3)	1,100	769	1,520	1,306

Table 8.6 Effect of SAR and Na^+ concentration in water on the ESP of soils with different mineralogy

Location	Clay distribution		Clay mineralogy		SAR	Na^+ concentration, mEq/L	ESP
	<0.2 μm	0.2–2 μm	<0.2 μm	0.2–2 μm			
Burleson	71	29	$M_1Mi_3K_3$	$M_2MiK_2Q_2$	28	11.0	18.9
						33.0	20.4
Houston black	78	22	M_1	$M_2K_2Mi_2Q_2$	28	11.0	16.0
						33.0	16.0
Miller	49	51	$M_1Mi_2K_3$	$M_2Mi_2K_2Q_2F_3$	28	11.0	25.8
						33.0	37.9
Pullman	41	59	Mi_2K_2 M_2	$Mi_2K_2Q_2F_3$	28	11.0	31.7
						33.0	32.5

Legend: Mi = mica, K = kaolinite, M = montmorillonite, Q = quartz, and F = feldspar. Estimated quantities: 1: >40 %; 2: 10–40 %; and 3: <10 % (Thomas and Yaron 1968)

chemical composition but also soil physical properties, such as porosity, aggregation status, and water infiltration capacity.

When Na^+ from saline irrigation or wastewater disposal first flows through a soil, there is no significant correlation between the sodium adsorption ratio (SAR) of the water and the exchangeable sodium percentage (ESP) of soil. In a column experiment by Thomas and Yaron (1968) on a series of Texas soils of differing mineralogy, the total electrolyte concentration of the saline water was observed to influence the rate of sodium adsorption. At equilibrium, the ESP in the soil was influenced more by the soil mineralogy than by the cationic composition of the water and the total electrolyte concentration (Table 8.6).

Characteristic data for the adsorption of Na^+ are shown in Fig. 8.26. Three synthetic aqueous solutions with a total electrolyte concentration of 11 mEq/L and SAR values of 7.5, 14.0, and 28.0 were passed through a Burleson soil column. At equilibrium, a solution with SAR = 7.5 gave an ESP of 8.1, and a solution with SAR = 28 exhibited an ESP of 18.0. We see from Fig. 8.26 that the quantity of aqueous solution that had to be passed through the soil column to achieve a constant Na^+ content in the entire profile increased with SAR. In other words, for a given applied volume of solution, greater SAR values led to larger depths where a constant ESP was achieved.

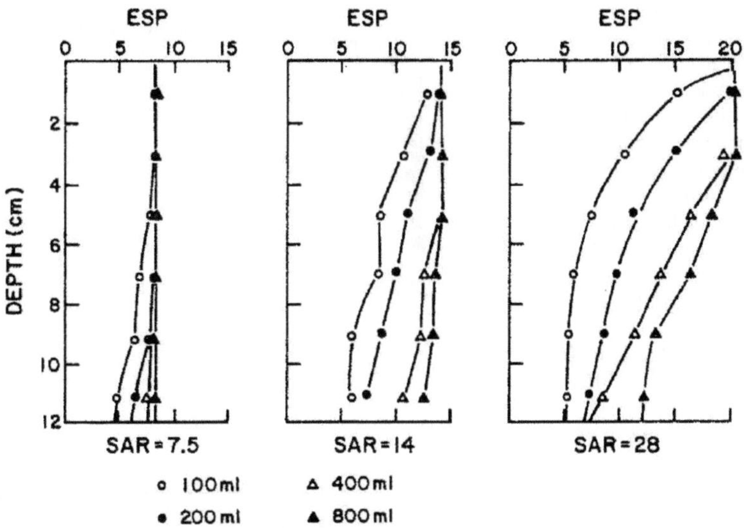

Fig. 8.26 Exchangeable sodium percentage (*ESP*) along a Burleson soil column, as a function of the sodium adsorption ratio (*SAR*) of irrigation water. The values were obtained by percolating the soil columns with sodic water (total electrolyte concentration of 11 mEq/L). Each curve corresponds to a given applied volume of solution (Thomas and Yaron 1968)

Heavy metal sorption on soil constituents may occur by one of the two mechanisms: (1) in a two-step mechanism where a rapid chemical reaction takes place on the surface of soil mineral followed by a long-time process recognized as a film diffusion route; (2) a three-step mechanism including (a) rapid adsorption of heavy metals on the external surface of the soil constituents, (b) delayed move of the heavy metal from the external to internal sites by solid-state film diffusion, and (c) binding and fixation of heavy metal at positions in the internal phase of the soil/mineral particle.

Sparks (2005) illustrated these adsorption mechanisms by comparing kinetics of some heavy metal sorption on a soil and various soil minerals. In this study, Ni, Pb, and As sorption on kaolinite, pyrophylite, gibbsite, montmorillonite, and Matapeake soil was compared as shown in Fig. 8.27. The sorption process was controlled by both sorbent and sorbate properties. An example of the three-step mechanism was demonstrated for As(V) adsorption on ferrihydrite where a great part of applied contaminant was sorbed within the first 5 min followed by slow sorption during ~200 h (Fig. 8.27c).

The desorption of heavy metals may exhibit an hysteretic pathway, the hysteresis being controlled by residence (contact) time with the sorbent. An example of the residence time effect on desorption of Pb sorbed on Matapeake soil is given in Table 8.7. By extrapolating the data of Strawn and Sparks (2000), we can assume that Pb adsorption under specific environmental conditions may become irreversible on a lifetime scale.

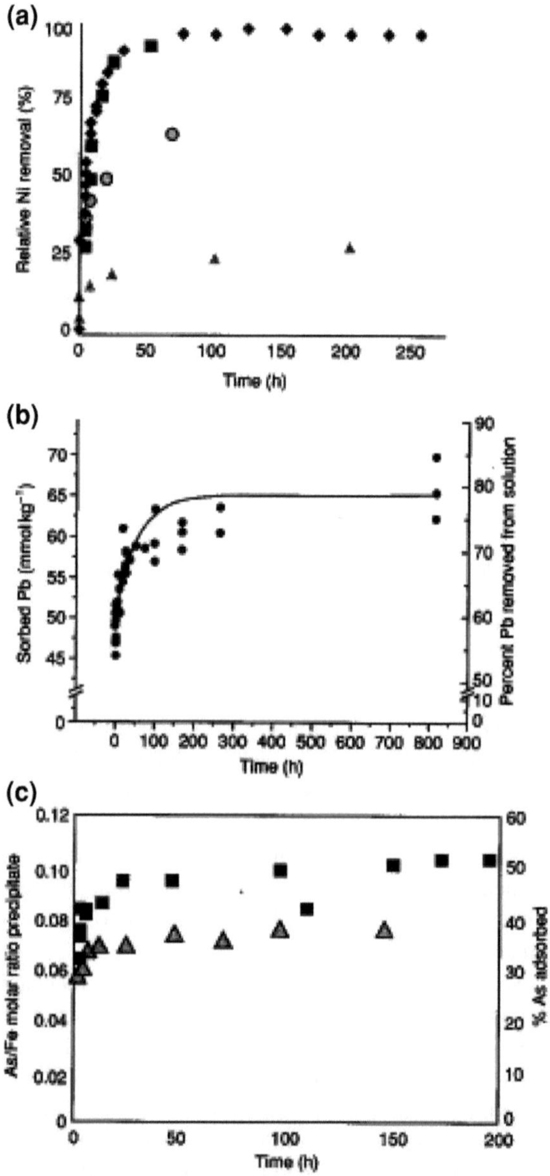

Fig. 8.27 Kinetics of metal and oxyanions on soil minerals and soil. **a** Kinetics of Ni sorption (%) on pyrophylite (*black diamond*), kaolinite (*black square*), gibbsite (*black up-pointing triangle*), and montmorillonite (*open circle*) from a 3 $mmol^{-1}$ Ni solution, an ionic strength $I = 0.1$ mol^{-1} $NaNO_3$ and a pH of 7.5; **b** kinetics of sorption on a Matapeake soil from a 12.25 $mmol^{-1}$ Pb solution, an ionic strength $I = 0.05$ mol^{-1} and a pH of 5.5; **c** kinetics of As(v) sorption on ferrihydrite at pH 8.0 and 9.0. As(v) total 1×10^{-4} mol^{-1}; Fe (III) total 5×10^{-4} mol^{-1} (*black square*) adsorption at pH 8.0 (*black up-pointing triangle*) adsorption at pH 9.0 (Sparks 2005)

Sparks (2005) notes that in addition to diffusion processes, the formation of metal hydroxide surface precipitates on phyllosilicates, metal oxides, and soils, together with residence time effects, can greatly affect the release of heavy metal and the hysteresis pattern. Aging of the metal hydroxide surface precipitate affects metal release, enhancing metal stability on soil minerals. With time, heavy metal sorption on soil minerals may develop a chain of processes leading to soil

Table 8.7 Effect of residence time on Pb desorption from a Matapeake soil (Strawn and Sparks 2000)

Residence time (days)	Sorbed Pb (mmol/kg)	Desorbed Pb (mmol/kg)	Percentage Pb desorbed (mmol/kg)
1	54.9	27.9	50.8
10	60	28.7	47
32	66.1	30.5	46.1

Table 8.8 Influence of the resident cation on the adsorption of diquat and paraquat by homo-ionic-exchanged clays (Hays and Mingelgrin 1991) indicates data for incomplete cation exchange

Clay	Resident inorganic cation	CEC (μEq/g)	Organic adsorption (μEq/g)		Enthalpy change ΔH (kJ/mol)	
			Diquat	Paraquat	Diquat	Paraquat
Montmorillonite	Na^+	999	950	970	−32	−48
	K^+	850	920	940	−31	−51
	Ca^{2+}	950	940	950	−36	−51
Vermiculite	Na^+	1,440	1,210	1,130	26	42
	K^+		270	230	1.6	−0.9
	Ca^{2+}	1130	(870)	(630)	(−9.2)	(−5.2)
Illite	Na^+	230	240	230	−13	−22
	K^+	ND	220	220	−6.8	−17
	Ca^{2+}	240	230	220	−16	−24
Kaolinite	Na^+	30	36	36	0	−11
	K^+		35	34	1.3	−4.9
	Ca^+	60	36	36	−18	−25

solid-phase transformation. Xu et al. (2008) confirm that Ni^{2+} adsorption on montmorillonite and fulvic–humic momtmorillonite complexes may be irreversible and that the adsorption–desorption pattern is not affected by the presence of humic substances.

Pesticides are used widely in plant protection practices, by dispersal on agricultural lands, and generally are cationic or nonionic organic compounds. After application, these products may volatilize, being transported in the groundwater as solutes or retained on the porous medium solid phase. Pesticide retention is controlled by the molecular properties of the compound, the solid-phase constituents, and environmental factors. Diquat and paraquat organo-cationic herbicides exhibit sorption isotherms of the *L* type according to the Giles et al. (1960) classification (Sect. 5.2.2), with the data fitting well to the Langmuir equation. Kookana and Aylmore (1993) noted that the sorption capacities of Australian soils studied for diquat ranged from very high in clay soil (146,400 μmol/kg) to very low in sand (1,765 μmol/kg). The clay soil showed the highest value of the Langmuir coefficient, indicating a high bonding energy. Table 8.8 shows the adsorption of these cationic pesticides by four types of homoionic-exchanged clays: montmorillonite,

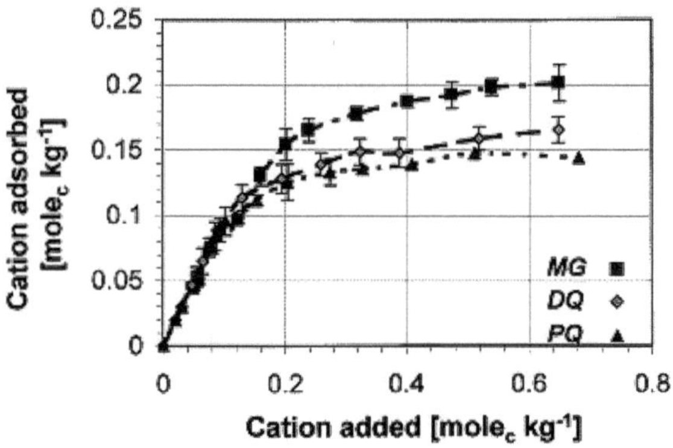

Fig. 8.28 Adsorption isotherms of methyl green (*MG*), diquat (*DQ*), and paraquat (*PQ*) on sepiolite. Reprinted from Rytwo et al. (2002). Copyright 2002 with permission of Elsevier

vermiculite, illite, and kaolinite. We see from these data that the cations can have a significant influence on the extent and energy of the adsorption processes. Exchange of the resident cations by both of the organo-cations was essentially complete in the case of kaolinite, montmorillonite, and illite clays, although there was a slight preference for paraquat over diquat in homoionic montmorillonite. The affinities were reversed in the case of adsorption by Na^+-vermiculite and illite clays.

Competitive adsorption between the organo-cationic herbicides diquat and paraquat and salts or a monovalent organic compound also was considered by Kookana and Aylmore (1993). An increase in the salt concentration of the soil solution from 0.005 to 0.05 M $CaCl_2$ resulted in decreases in sorption capacities for the studied herbicides.

The effect of NaCl concentration on the rate of paraquat adsorption on activated clays is reported by Tsai et al. (2003). The rate constant increases with an increase in salts in the aqueous paraquat solution: from 0.046 (g/mg/min) at a NaCl concentration of 0.05 M, to 0.059 (g/mg/min) at a solution concentration of 2.50 M NaCl. Studying the effect of various alkali metals ions on paraquat adsorption capacity on activated clay surfaces, Tsai et al. (2003) found the order Li^+ (32.79 mg/g) > Na^+ (31.25 mg/g) > K^+ (24.75 mg/g). Moreover, the rate constant appears to be inversely proportional to the adsorption capacity.

Competitive adsorption on sepiolite clay of a monovalent dye (e.g., methyl green or methyl blue) and of the divalent organo-cationic herbicides diquat and paraquat was studied by Rytwo et al. (2002). To evaluate a possible competitive adsorption between the two organic compounds, separate aqueous solutions of each cation were used and adsorption isotherms were obtained. Figure 8.28 shows the amount of diquat, paraquat, and methyl green adsorbed on sepiolite as a function of total added divalent cation. It may be observed that, when the added amounts were lower than the CEC of the sepiolite (0.14 mol$_c$/kg), all cations were completely adsorbed.

Rytwo et al. (2002) also found that, when the added amounts were higher than the CEC, the adsorbed amounts of methyl green and diquat increased up to 140

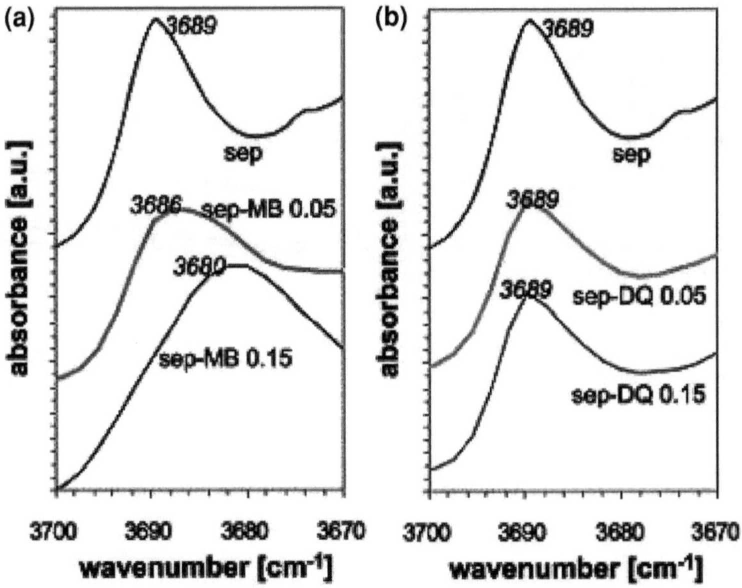

Fig. 8.29 IR spectra of the Si–OH vibration in **a** pure sepiolite (sep), sepiolite with 0.05 and 0.15 mol_c methylene blue (MB)/kg clay (sep-MB 0.05 and sep-MB 0.15, respectively) and **b** pure sepiolite (sep), sepiolite with 0.05 and 0.15 mol_c diquat/kg clay (sep-DQ 0.05 and sep-DQ 0.15, respectively). Reprinted from Rytwo et al. (2002). Copyright 2002 with permission of Elsevier

and 115 % of CEC, respectively, while paraquat adsorption reached only a value close to the CEC level. These behaviors were examined using Fourier transform infrared (FTIR) measurements. Pure sepiolite exhibits two clear peaks at 789 and 764 cm^{-1}. Adsorption of monovalent organic cations such as methylene blue (MB) leads to a considerable deformation of the peaks at high adsorption loads, but the peaks are not affected when the adsorption is smaller than one-third of the CEC. A divalent cation compound such as diquat or paraquat does not affect the O–H doublet at any adsorbed load. When both diquat and MB are coadsorbed up to CEC, the peak deformation appears but is ascribed to the binding of the monovalent organic cation to the neutral sites. Similar effects are observed for the main Si–OH vibration at 3,700 cm^{-1}. The effects of adsorption of MB and diquat on sepiolite IR spectra are presented in Fig. 8.29. With increasing amounts of adsorbed MB, the peak shift is observed (Fig. 8.29a); no shift is observed when diquat is adsorbed (Fig. 8.29b).

Rytwo et al. (2002) show that diquat and paraquat adsorb on neutral sites of sepiolite; the authors speculate that this might be a general pattern for organic cation contaminants interacting with sepiolite.

Competitive adsorption on clay and soil surfaces between a heavy metal contaminant (Cu) and a cationic herbicide (chlordimeform) was reported by Maqueda et al. (1998) and Undabeytia et al. (2002). In the presence of herbicides, Cu

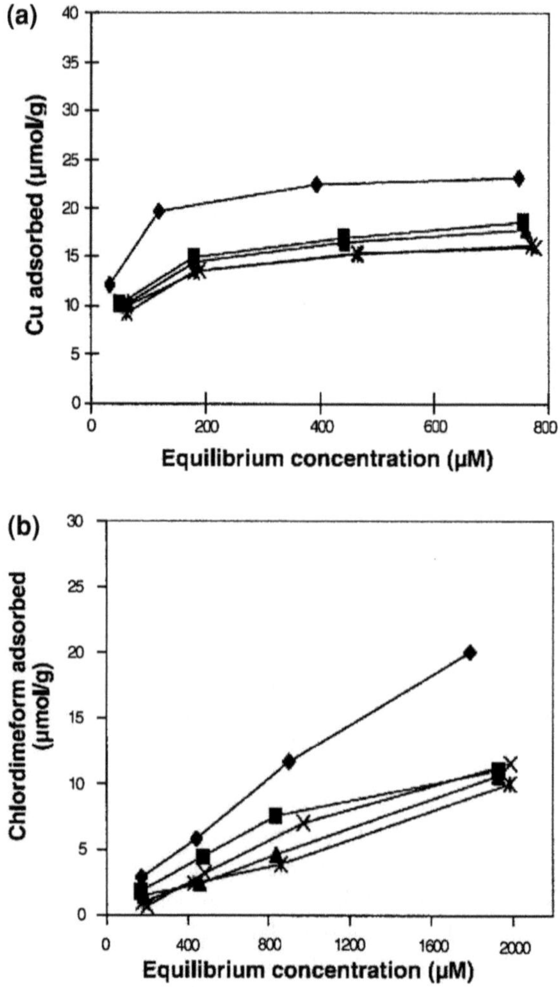

Fig. 8.30 Adsorption isotherms of chlordimeform on Cu-treated soil **a** at several chlordimeform initial concentrations (mM): 0 (*black diamond*), 0.2 (*black square*), 0.5 (*black up-pointing triangle*), 1.0 (×), and 2.0 (*black star*) and **b** at several Cu concentrations (mM): 0 (*black diamond*), 0.16 (*black square*), 0.31 (*black up-pointing triangle*), 0.63 (×), and 0.94 (*black star*) (Undabeytia et al. 2002)

adsorption on clay minerals decreases due to competition for the same adsorption sites. Preadsorption of Cu on soil surfaces leads to a decrease in chlordimeform adsorption, also as a result of competition for the adsorption sites. This behavior is shown in Fig. 8.30, including adsorption isotherms of chlordimeform at several Cu concentrations and for Cu adsorption with increasing chlordimeform values on an Ultic Haploxeralf soil derived from shales of Devonian-age parent material. The soil adsorbent is characterized by 1.9 % OM and a CEC of 9.8 mol_c/kg.

Considerable attention has been given to the adsorption of diquat and paraquat by humic acids. On the basis of data from Burns et al. (1973), Hayes and Mingelgrin (1991) explained how the exchangeable cations on humic macromolecules can influence the adsorption of paraquat. Adsorption decreased according to the order Ca^{2+}-humate > H^{+}-humic acid > Na^{+}-humate. Adsorption by the Na^{+}-

humate approached the CEC of the polyelectrolyte. This can be explained, in part, by the preference of the ion exchanger for mineral ions of higher valence. However, the effects of macromolecules are more important in this instance than consideration of valence. Adsorbed organo-cations bridge negative charges within strands and between strands, and the interstrand binding causes the macromolecules to shrink; water is then excluded from the matrix and precipitation of the paraquat–humate complex may occur. This behavior was observed for Na^+-smectite uptake of paraquat by Na^+-humate, approaching the CEC value of the polyelectrolyte. Adsorption by the Ca^{2+}- and H^+-exchanged humic acids was significantly less than the CEC of the preparation. Comparisons of the isotherms in Fig. 8.31 indicate a high affinity for adsorption of paraquat by the H^+-humic acid in water and by the Ca^{2+}-humate. These isotherms are representative of adsorption occurring on readily available sites or close to exteriors of macromolecular structures. The presence of $CaCl_2$ reduces the adsorption, due to the competition from the excess Ca^{2+} ions in solution.

In a review on the effects of OM heterogeneity on sorption of organic contaminants, Huang et al. (2003) emphasized the possible presence of matured kerogen and black carbon or soot particles as a fraction of soils and sediments. For example, analyzing the naturally occurring total organic matter (TOC) in a sediment, Song et al. (2002) found 8 % humic acid, 52 % black carbon, and 40 % kerogen materials. The molecular structures of kerogen and black carbon are very different from those of humic acids, so their interaction with organic contaminants in the subsurface are different. Kerogen, for example, has a three-dimensional structure with aromatic nuclei cross-linked by aliphatic chainlike bridges. The nuclei appear to be formed mainly from clusters (about 100 nm in diameter) of 24 or more parallel aromatic sheets, separated by gaps or voids of 30–40 nm, and therefore are capable of trapping small hydrophobic organics. The aromatic sheets contain up to 10 condensed aromatic homocyclic or heterocyclic rings. Bridges are linear or branched aliphatic chains or O- or S-containing functional groups (Engel and Macko 1993). Absorption capacities of TOC components in a pond sediment vary, for example, with a lower sorption capacity for phenanthrene compared to that calculated only for humic acid (Xiao et al. 2004). The lower sorption capacity is explained by the presence of particulate kerogen and black carbon associated with the sediment; because these surfaces (in soil/sediment aggregates) are coated with metal oxides and hydroxides, the TOC components are not fully accessible by phenanthrene.

The charge characteristics of many pesticides are pH dependent. Some anionic species are formed through dissociation of protons, and cationic compounds may be formed by the uptake of protons. Compounds with carboxylic acid groups are characteristic of anionizable compounds, although phenolic groups rise to anionic species in alkaline subsurface conditions. The s-triazine family of herbicides applied to soil is considered the classical representative of the cationizable species. Several compounds containing carboxylic acid groups enter the subsurface when sprayed on the vegetation canopy or directly onto the land surface. These act as neutral molecules when pH values are well below their pK_a values but become

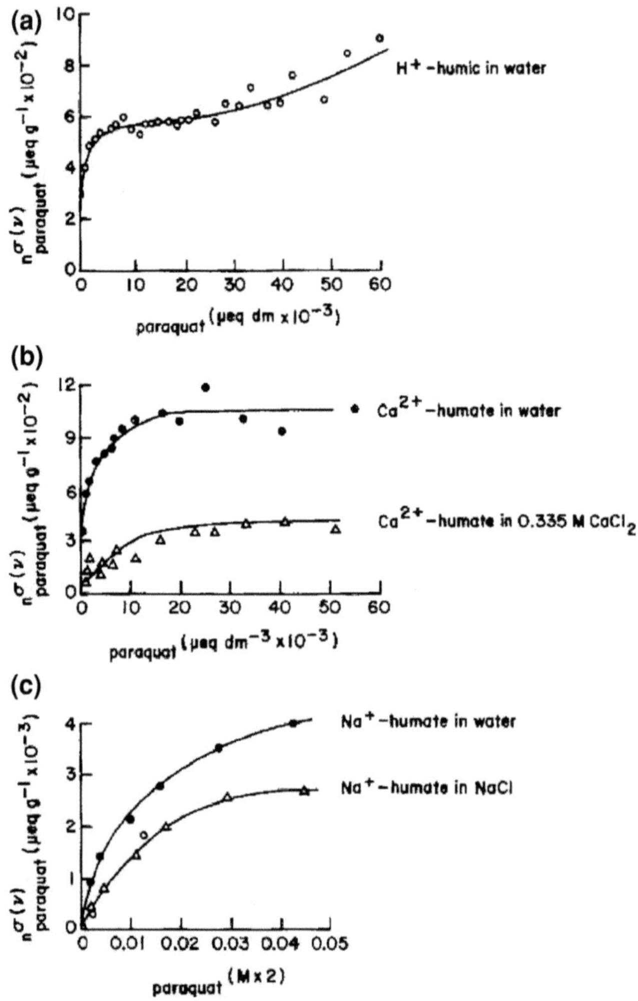

Fig. 8.31 Isotherms for the adsorption of paraquat **a** by H^+-humic acid, **b** by Ca^{2+}-humate in water and 0.335 M $CaCl_2$ solution, and **c** by Na^+-humate solution in water and dilute NaCl solutions (Burns et al. 1973)

increasingly anionic as pK_a is reached and exceeded. Table 8.9 shows the adsorption from aqueous solution of a series of compounds from the s-triazine family onto Na^+-smectites, as affected by the solution pH. Hayes and Mingelgrin (1991) emphasize that, under low pH conditions, prometone adsorption can take place through association of triazine with carboxyl groups in the humic substances, followed by electrostatic binding to the conjugate carboxylate base of the proton donor.

Table 8.9 Adsorption from aqueous solutions of s-triazine compounds by Na^+-montmorillonite (Hayes and Mingelgrin 1991)

Compound	pK_a	Water solubility (ppm)	Adsorption (μmol/g)			
			pH 2	pH 3	pH 4	pH 5
Atrazine	1.68	33	275	200	115	70
Atratone	4.20	1,654	410	450	475	400
Ametryne	3.12	193	520	610	650	560
Propazine	1.85	9	150	110	20	20
Hydroxypropazine	5.20	310	150	220	240	245
Prometone	4.30	750	290	380	400	350
Prometryn	3.08	48	460	490	490	415

Sposito et al. (1996) analyzed a broad set of data and suggested two general complexation mechanisms to characterize the retention of s-triazines on humic acids: Proton transfer is favored for humic acids of high acidic functional groups and for s-triazine with low basicity; electron transfer mechanisms are favored for humic acids of low acidic functional group content and for s-triazines of high basicity.

The effect of initial pH (3–11) on paraquat intake rate by activated clay adsorption also is reported by Tsai et al. (2003). As the pH increased, the amount of adsorbed cationic paraquat increased in response to the increasing number of negatively charged sites, which are available due to the loss of H^+ from the surfaces. The surface of activated clay at pH > 3.0 exhibited negative charges due mainly to the variable charge from pH-dependent surface hydroxyl sites. With the increase in pH, the negative charges increased, which in turn increased paraquat sorption. The distribution coefficient, K_d, decreased as pH decreased, for example, from 3,521 (L/kg) at pH 11.0 to 981 (L/kg) at pH 3.0. In fact, the rates of adsorption for all of the studied pH values decreased with time, until they gradually approached a plateau.

Barriuso et al. (1992) analyzed the relationship between soil properties and adsorption behavior of cationic, anionic, and neutral herbicides. Atrazine, terbutryn, and 2,4-D from the triazine herbicides family were considered. The study included 58 soils, covering a wide range of pH values, OC contents, and mineralogical compositions. For this range of soils and pH, atrazine was in a neutral form, 2,4-D in an anionic form, and terbutryn in either neutral or cationic form. Based on the adsorption measurements, the authors concluded that (1) K_d for atrazine is strongly related to the OC content but not correlated to the soil pH; (2) K_d for terbutryn is less correlated to the OC content and is correlated to the soil pH; and (3) K_d for 2,4-D is not correlated to the OC content, but it is inversely correlated to the soil pH.

Because pesticides in the subsurface can reach layers with different properties, as they are redistributed with depth, it is interesting to examine their adsorption on earth materials during vertical transport. Dror et al. (1999) report adsorption of two s-triazines (atrazine and terbuthylazine), as they are redistributed in a 120-cm-deep

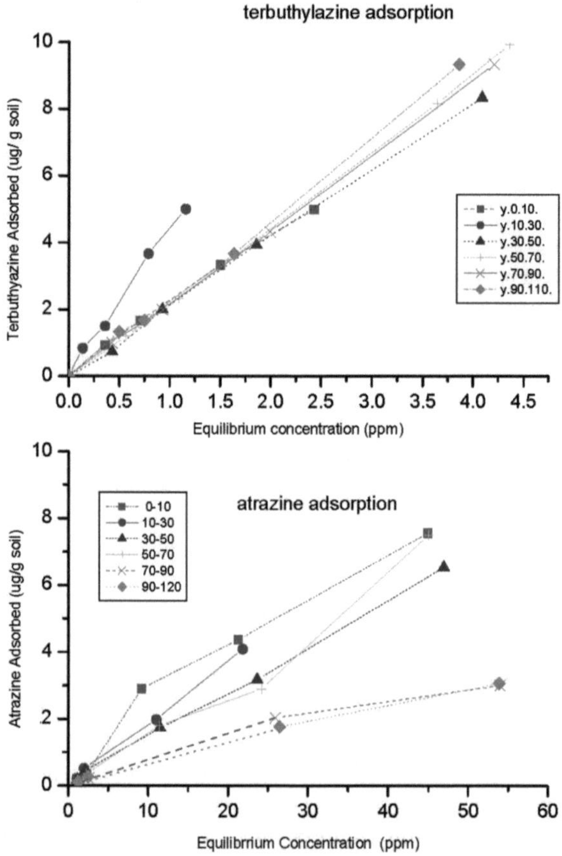

Fig. 8.32 Adsorption isotherms of terbuthylazine and atrazine at different depths along the soil profile, as affected by the *vertical* variability of soil properties (Dror et al. 1999)

soil profile, following surface application and subsequent leaching by irrigation or rainfall. Figure 8.32 shows adsorption isotherms of these two herbicides as affected by the vertical variability of the soils. The adsorption isotherms are linear along the soil profile. The K_d coefficients of atrazine range from 0.10 to 0.21 mL/g, while those of terbuthylazine range from 1.9 to 2.9 mL/g. The vertical variability of K_d, quantified in terms of the mean and standard deviation (SD), is $K_\mathrm{d} = 0.15$ with an SD of 0.04 for atrazine and $K_d = 2.17$ with an SD of 0.36 for terbuthylazine. The effect of soil OM on herbicide adsorption is given by the distribution coefficient on OC, K_oc, obtained from the relation $K_\mathrm{oc} = K_d/\mathrm{OC} \times 100$, assuming that OC = OM/1.7. The vertical spatial variability of K_oc is similar to those of K_d.

Organophosphorus pesticides are used as an example of the adsorption of nonpolar (nonionic) or slightly polar toxic chemicals. The phosphoric acid ester group has a general formula $(RO)_2PO(OX)$, where R is an alkyl group and X is a leaving group (see Sect. 4.1.2). In contact with clay surfaces or other earth materials, such organic nonpolar molecules are retained on the surface. These

Table 8.10 Percent of adsorption from aqueous solution of organophosphorus pesticides by clays (Yaron 1978)

Insecticides	Solution equilibrium concentration (ppm)	Clays		
		Ca^{2+}-montmorillonite (%)	Ca^{2+}-kaolinite (%)	Ca^{2+}-attapulgite (%)
Parathion ($C_{10}H_{14}NO_5PS$)	6.5	73	14	87
Pirimiphos-methyl ($C_{11}H_{20}N_3O_3PS$)	16.0	94	30	
Pirimiphos-methyl ($C_{13}H_{24}N_3O_3PS$)	16.0	92	75	
Menazon ($C_6H_{12}N_5O_2PS_2$)	16.0	16	5	

esters are stable at neutral or acidic pH but susceptible to hydrolysis in the presence of alkali compounds, where the P–O–X ester bond breaks down. The rate of hydrolysis is related to the nature of the constituent X, the presence of catalytic agents, pH, and temperature. Because the solubilities of organophosphorus compounds in water are low, their adsorption from water solutions also is low. Table 8.10 shows the percentage of some organophosphorus compounds adsorbed from aqueous solution onto clay surfaces. The maximum adsorption capacity of clays for organophosphorus pesticides is achieved only when the chemical reaction occurs in an appropriate organic solvent. The maximum adsorption capacity of parathion, for example, on a monoionic clay surface from a hexane solution is 10 % of the initial concentration on montmorillonite, 8 % on attapulgite, and less than 0.5 % on kaolinite.

The amount of adsorbed chemical is controlled by both properties of the chemical and of the clay material. The clay-saturating cation is a major factor affecting the adsorption of the organophosphorus pesticide. The adsorption isotherm of parathion from an aqueous solution onto montmorillonite saturated with various cations (Fig. 8.33) shows that the sorption sequence ($Al^{3+} > Na^{+} > Ca^{2+}$) is not in agreement with any of the ionic series based on ionic properties. This shows that, in parathion-montmorillonite interactions in aqueous suspension, such factors as clay dispersion, steric effects, and hydration shells are dominant in the sorption process. In general, organophosphorus adsorption on clays is described by the Freundlich equation, and the K_d values for parathion sorption are 3 for Ca^{2+}-kaolinite, 125 for Ca^{2+}-montmorillonite, and 145 for Ca^{2+}-attapulgite.

The hydration status of the clay or earth material may affect the adsorption capacity of nonpolar (or slightly polar) toxic chemicals. Continuing with parathion as a case study, Fig. 8.34 shows the increase adsorbed parathion on attapulgite from a hexane solution, as the adsorbed water on the clay surface decreases. This behavior may be explained by the competition for adsorption sites between the polar water and the slightly polar parathion. Possibly, however, the reduction in adsorption due to the presence of water is caused by the increased time required for parathion molecules to diffuse through the water film to the adsorption sites.

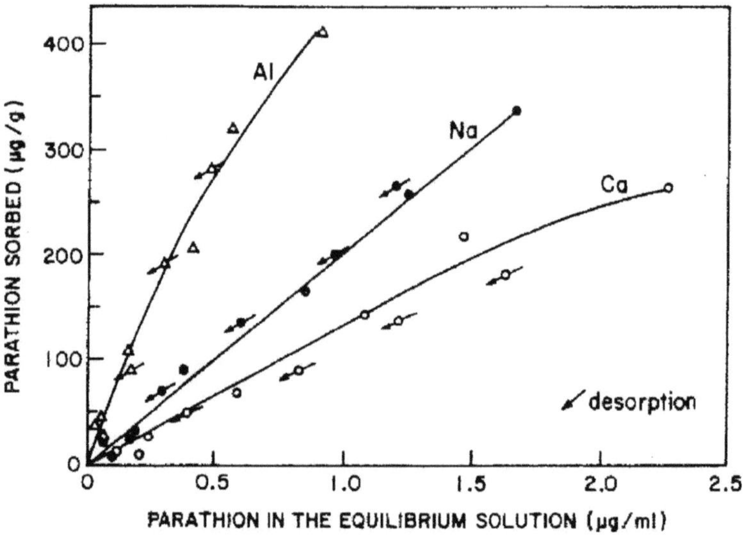

Fig. 8.33 Adsorption from aqueous solutions of parathion onto montmorillonite saturated with various cations (Yaron and Saltzman 1978)

Fig. 8.34 Adsorption of parathion by attapulgite from hexane solution, as affected by initial hydration status of mineral; RH = relative humidity (Gerstl and Yaron 1981)

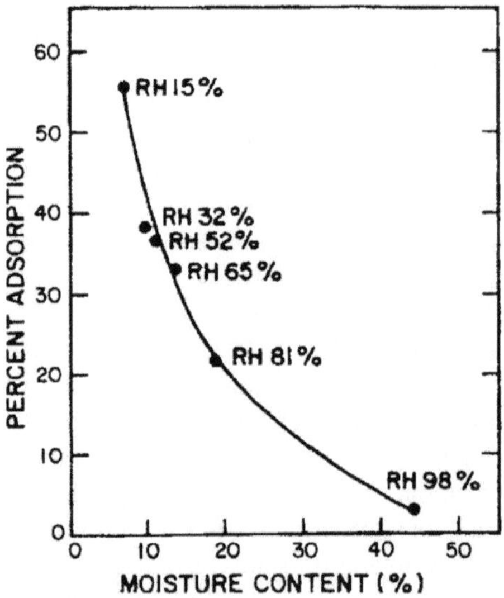

In a dry attapulgite–parathion–hexane system, parathion molecules compete effectively with nonpolar hexane molecules for the adsorption sites. In partially hydrated systems, parathion molecules cannot replace the strongly adsorbed water molecules, so that parathion adsorption occurs only on water-free surfaces and a decrease in adsorption per total surface area may be observed. Infrared studies lead

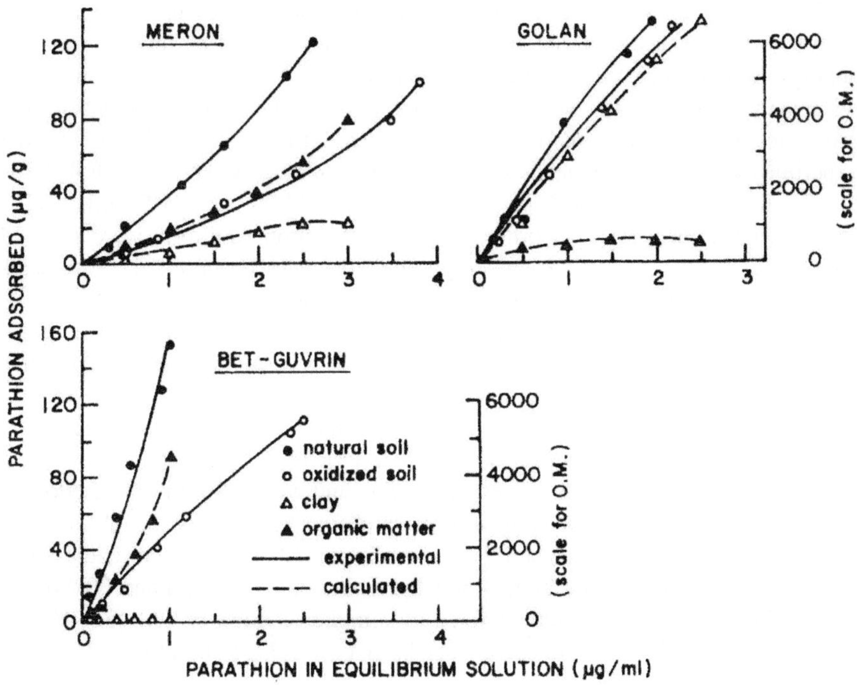

Fig. 8.35 Parathion adsorption from aqueous solutions by three soils, before and after oxidation, as well as on clay and organic matter. Reprinted with permission from Saltzman et al. (1972). Copyright 1972 American Chemical Society

to an understanding of the adsorption mechanism of organophosphorus compounds on clay surfaces. For example, IR spectra indicate that, on both montmorillonite and attapulgite, parathion was sorbed by clay coordinates through water molecules to the metallic cations. When clay–parathion complexes are dehydrated, parathion becomes coordinated directly, and the type of cation determines the structure of the complex. The main interaction is through the oxygen atoms of the nitro group. Interactions through the P=S group have also been observed, especially for complexes saturated by polyvalent cations (Saltzman and Yariv 1976; Prost et al. 1977).

Subsurface adsorption of nonpolar or slightly polar toxic chemicals also was found to be related to the solid-phase OM and mineralogy. Saltzman et al. (1972) studied the importance of mineral and organic surfaces in parathion adsorption on semiarid soils characterized by low OM content and different mineralogy. They found that parathion adsorption depends on the type of association with organic and mineral colloids. Following removal of OM by soil treatment with hydrogen peroxide, the adsorptive affinity of mineral soils decreases, mainly due to a decrease in the OM content and not due to other soil modifications that may occur during OM oxidation. Figure 8.35 presents the isotherms of parathion adsorption on three semiarid soils, before and after oxidation, as well as on clay and OM.

Steroid hormones form a group of pollutants that includes natural hormones such as estradiol, testosterone, and their metabolites as well as several synthetic analogs. Steroid hormones used as growth promoters have already been found in water and sediments (Lai et al. 2000; Thorpe et al. 2003), and their adsorption properties on earth materials have been considered. Lee et al. (2003) report batch experiments where simultaneous sorption of three hormones (17-β-estradiol, 17-α-ethyl estradiol (EED), and testosterone) on four Midwestern US soils and a freshwater sediment was performed. Apparent sorption equilibria were reached within a few hours. Sorption isotherms generally were linear for the chemicals studied; on one of these soils (Drummer soil), K_d ranged from 23.4 to 83.2 L/kg and $\log K_{oc}$ ranged from 2.91 to 3.46. The distribution coefficients on OC, K_{oc}, indicate that hydrophobic partitioning was consistent with the aqueous solubilities and octanol–water partition coefficients.

Endocrine disruptor compounds (EDCs), which are hormone and hormone-like substances, may enter the subsurface through agricultural practices or urban and industrial effluents, or through improper sludge and waste disposal. Loffredo and Senesi (2006) reported adsorption–desorption of four EDCs—bisphenol A (BPA), octyphenol (OP), 17-α-EED, and 17-β-estradiol (17ED)—onto two acidic soils, with the adsorptive material being collected from depths of 0–30 and 30–90 cm. The surface samples were characterized by a higher content of OM (6.3–16.0 g/kg) than those collected from the deeper layer (1.8–3.3 g/kg) and a clay content ranging from 1.5 to 26 %. Adsorption of EDCs on the soils was relatively fast, occurring mainly during the first hours of contact. Over the concentration range tested, no limiting adsorption was observed for BPA, OP, or EED, whereas adsorption of 17ED reached a saturation level. In general, EDC adsorption in the upper layer is greater than on the samples collected from a 30–90 cm depth. The type of adsorption isotherm (e.g., *C* or *L* type) is controlled by both the characteristics of the EDCs and the properties of the earth material. Based on their experimental results, Loffredo and Senesi (2006) suggested that, in addition to the OC content, EDC adsorption is affected by the nature of the OC and its association with other colloidal fractions (such as clay materials).

EDC desorption rates and amounts define their distribution in the subsurface. The adsorption of OP and EED is mostly pseudo-irreversible, with a partial desorption that occurs very slowly; OP and EED thus are likely to accumulate in the upper soil layers. In contrast, BPA adsorption is reversible; desorption occurs quickly and is completed after a few steps of leaching. As a consequence, BPA released to the soil environment is more likely to be transported to deeper layers, leading eventually to groundwater contamination.

Inorganic trace element retention on earth materials is illustrated in several case studies, where selected contaminants (e.g., fluoride, cesium, Hg, lead, and cadmium) interact with rocks, clays, soils, and sediments under different environmental conditions (e.g., pH, presence of organic ligands, and salinity).

Fluoride in minor concentrations is beneficial for animals and humans, but it becomes toxic when ingested in excessive amounts. Bar-Yosef et al. (1989) investigated adsorption kinetics and isotherms of K^+-montmorillonite and a series

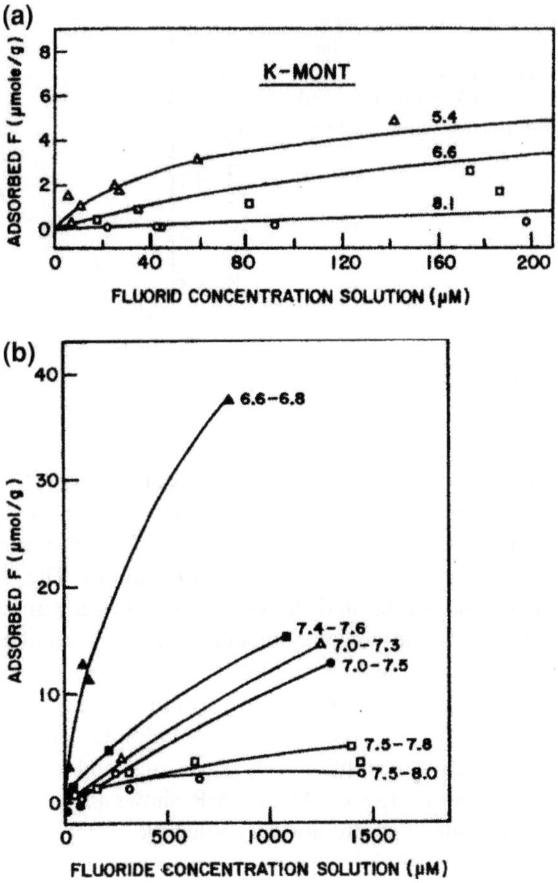

Fig. 8.36 Fluoride adsorption from aqueous suspensions as affected by pH **a** on K^+-montmorillonite and **b** on six soils (Bar-Yosef et al. 1989)

of soils (clay, 4–61 %; OM, 2–7 %), as affected by solution pH. The fluoride adsorption isotherms are shown in Fig. 8.36. In all cases, fluoride adsorption is a function of its concentration at the selected pH values. Soil pH is correlated inversely to the maximum number of fluoride adsorption sites, and this correlation stems from the effect of pH on the charge density of the clay edges. No significant correlation was found between the OM content and the maximum fluoride adsorption. Fluoride partitioning between solid and liquid phases of neutral and high pH is in accord with the earlier results of Flühler et al. (1982). The time needed to attain quasi-equilibrium in fluoride sorption reactions is satisfactorily described by a Langmuir adsorption isotherm model.

Cesium is a radionuclide found in radioactive wastes intended for storage in underground repositories; it is used as a standard marker for highly radioactive and long-lived materials. Argillaceous or clay media are considered potential earth materials for inclusion as barriers in radioactive waste repositories, due to their favorable properties for confinement: low permeability, high retention capacity,

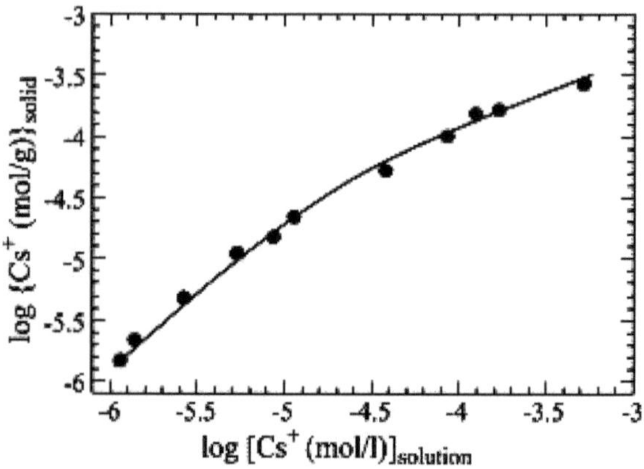

Fig. 8.37 Cs^+ adsorption isotherm on Na^+-loess clay. Reprinted from Bergaoui et al. (2005). Copyright 2005 with permission of Elsevier

and chemical buffering effect. Bergaoui et al. (2005) studied cesium adsorption on a Na^+-loess clay, where batch experiments were analyzed by X-ray diffraction and infrared and far-infrared measurements. The adsorption isotherm (Fig. 8.37) shows that loess clay is selective for cesium cations. The raw material contained a large amount of quartz, and the clay material was a mixture of kaolinite and an interstratified illite–smectite mineral; as a result, equilibrium Cs^+ adsorption data are not consistent with a single-site Langmuir model. Cesium adsorption on this particular soil clay occurs by cation exchange on sites with various cesium affinities. At low concentration, far-infrared spectroscopy shows the presence of very selective adsorption sites that correspond to internal collapsed layers. At high concentration, 133 Cs MAS-NMR shows that cesium essentially is adsorbed to external sites that are not very selective.

Variation of Cs^+ adsorption with depth, as a function of changes in clay content, is reported by Melkior et al. (2005); the study aimed to test the efficiency of a host rock for radionuclide confinement. Mudrock samples were collected from Callovo-Oxfordian layers in Bure (France), at depths between −422 and −478 m. The total clay content increases with depth by a factor of two to three between the measured depths. Figure 8.38 depicts the K_d of Cs^+ as a function of its concentration in solution at equilibrium.

Soil and sediment samples collected from Fukushima prefecture were investigated for Cs content using a sequential extraction (EXAFS) procedure by Qin et al. (2012). The investigations revealed that 94 % of ^{137}Cs was retained in the interlayer of the phyllosilicate mineral (Fig. 8.39). The $k^3\chi(k)$ spectra of vermiculite, soil, and sediments showed significantly different oscillations in frequency and amplitude compared to those for the CsCl solution, suggesting that inner-sphere complexes were formed on contaminated vermiculate and soil–sediment materials were formed on vermiculite. However, the Cs–soil/sediment complexes were smaller than those formed on vermiculite. The authors attribute these differences to the OM content of the soil–sediment samples.

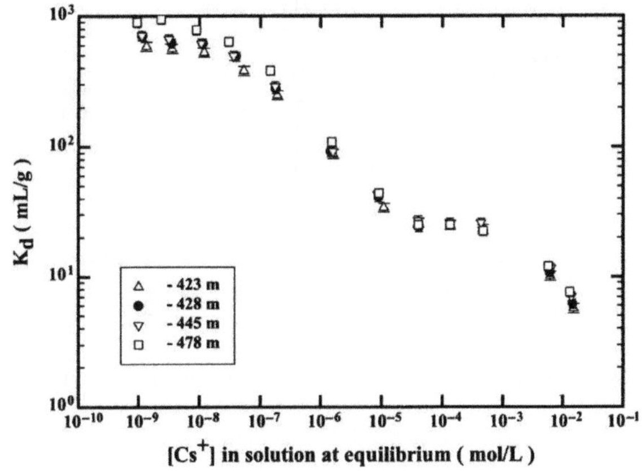

Fig. 8.38 Cesium sorption on Bure mudrock samples: K_d as a function of concentration in solution at equilibrium, at four different depths (Melkior et al. 2005)

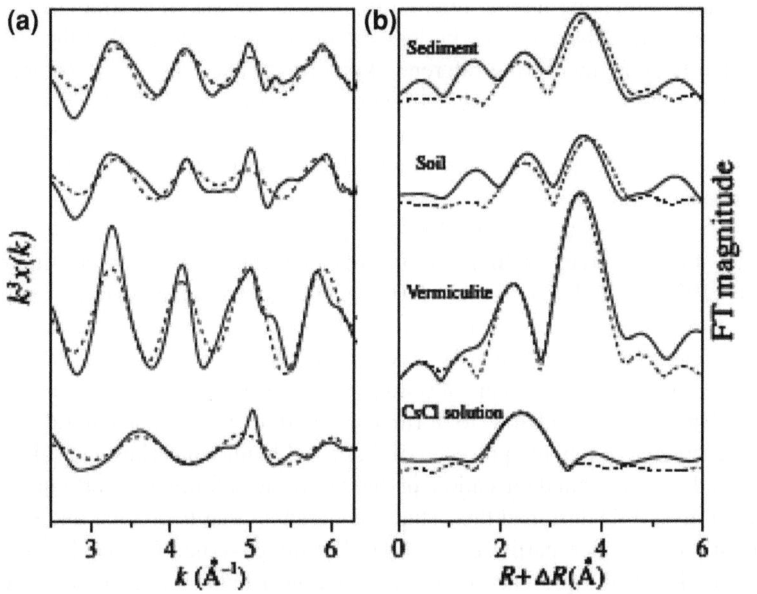

Fig. 8.39 Adsorption of cesium on vermiculite compared to CsCl solution. Cesium L_{III}-edge EXAFS spectra of reference material, Cs-adsorbed samples. **a** k^3-weighted $\chi(k)$ spectra and **b** RSF of (**a**) (phase shift not corrected). *Solid lines* are spectra obtained by experiments, and *dotted lines* are fitted spectra by curve-fitting analysis (Qin et al. 2012)

Buerge-Weirich et al. (2002) examined copper and nickel adsorption on a goethite surface in the presence of organic ligands, in natural groundwater samples from an infiltration site of the river Glatt at Glattfelden (Switzerland). Figure 8.40 exhibits Cu and Ni adsorption isotherms at pH 7.35, in the presence of

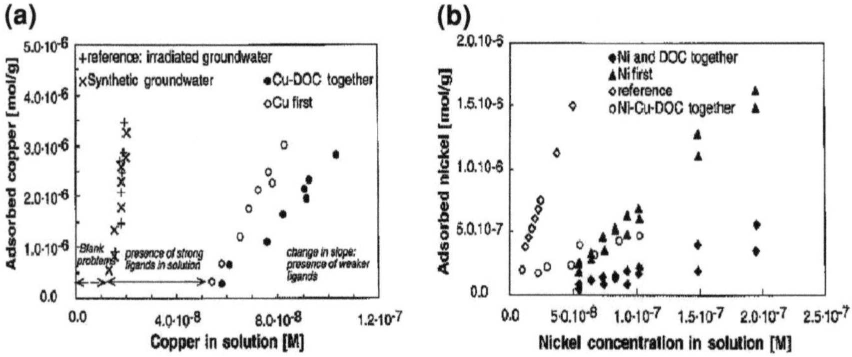

Fig. 8.40 Experimental results for **a** Cu and **b** Ni adsorption isotherms on goethite, at pH 7.35, in the presence of groundwater containing natural organic ligands, as compared to irradiated natural groundwater and synthetic groundwater. Reprinted with permission from Buerge-Weirich et al. (2002). Copyright 2002 American Chemical Society

groundwater containing natural organic ligands, as compared to irradiated natural and synthetic groundwater. Clearly, less Cu and Ni were adsorbed in the presence of organic ligands than in the reference systems. In both cases, the isotherms do not intercept the x-axis at the origin but at a value near 5×10^{-8} M. This behavior indicates the presence of strong complexing ligands in solution. When these ligands were saturated, the amount of adsorbed metal increased linearly with total metal concentration.

Dissolved inorganic trace metal (e.g., Hg, Ag, Cd, Ni, Zn) speciation and the tendency to form stable soluble complexes in saline water was considered by Turner (1996) and Turner et al. (2001, 2002), who developed an empirical equation to describe the direct relationship between the sediment–water distribution coefficient (K_d) and salinity. We focus on Hg sediment–water partitioning as a result of a coupled speciation and a salting-out process, as reported by Turner et al. (2001). Hg has the tendency to form chloro-complexes. In a natural aqueous system characterized by the presence of dissolved humic materials and salts, Hg is complexed by the abundant cation or has a strong affinity for particulate OM. Turner et al. (2001) designed their study recognizing that the nature and extent of sediment–saline water partitioning of Hg(II) are governed by the hydrophobic characteristics in the presence of OM, including a tendency to be salted out of solution by seawater ions. They found that the increase in K_d with increasing salinity is at least partially the result of an increase in the proportion of the relatively hydrophobic and lipophilic $HgCl_2$ complex, which is subject to salting out.

Distribution coefficients (K_d) and (K_{oc}) defining ^{203}Hg(II) sorption to three estuarine sediments are plotted against salinity in Fig. 8.41. Despite different Hg(II) concentrations and K_{oc} values in each location, a common exponential increase is evident. Salting out did not occur when Hg(II) is only in an ionic form, so that an increase in Hg(II) dissolution in estuarine water may be explained by the

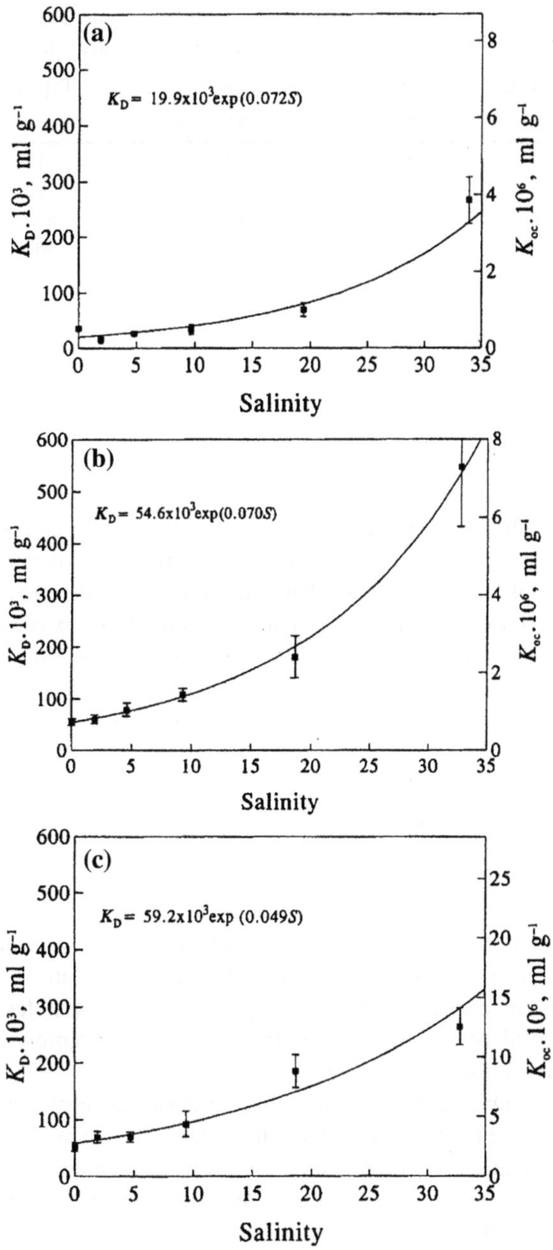

Fig. 8.41 Distribution coefficients (K_d and K_{oc}) defining the sorption of inorganic Hg(II) to estuarine particles versus salinity (S, in g/L) in the **a** Beaulieu, **b** Mersey, and **c** Plym estuaries. Reprinted with permission from Turner et al. (2001). Copyright 2001 American Chemical Society

formation of organically complexed Hg(II), due to the presence of dissolved or particulate OM. In particular, Hg(II) tends to form uncharged complexes with Cl^- that are covalent, nonpolar, and lipophilic. An increase in dissolved Hg(II) with increasing salinity may be explained by an increase in the proportion of the

Table 8.11 Range of selected trace element concentrations in dry, digested sewage sludge, and in comparison with element concentrations in median sludges and in soils (Chaney 1989)

Element	Reported range (mg/kg)		Typical median sludge	Typical soil
	Minimum	Maximum		
As	1.1	230	10	
Cd	1	3,410	10	0.1
Cu	84	17,000	800	15
Cr	10	99,000	500	25
F	80	33,500	260	200
Hg	0.6	56	6	
Ni	2	5,300	80	25
Pb	13	26,000	250	15
Se	1.7	17.2	5	
Zn	101	49,000	1,700	50

relatively hydrophobic and lipophilic $HgCl_2$ complex, which is more prone to salting out than other, more hydrophilic, complexes. Based on this study, Turner et al. (2001) consider that neutral methylated forms of Hg(II) behave in a similar fashion because they have a strong affinity for particulate OM and are known to be salted out from aqueous solution. Extending this finding to other trace metals, Turner et al. (2002) suggest that such contaminants are complexed by and subsequently neutralize organic ligands and the resulting neutral assemblages are salted out, possibly by electrostriction.

8.4.2 Nonadsorptive Retention

Nonadsorptive retention of contaminants may occur when chemicals reach the subsurface as a separate liquid phase or are adsorbed on suspended particles or organic residues. Contaminated suspended particles originating from sludge disposal or polluted runoff, for example, can represent a substantial hazard to the subsurface environment.

Table 8.11 shows the concentration range of potential toxic trace elements in US sewage sludges, as summarized by Chaney (1989). In this table, data on maximum concentration of toxic trace elements in dry, digested sewage sludges are compared to concentrations of the elements in median sludges and in soils. The subsurface contamination that may result from uncontrolled disposal on land surfaces is evident. In a laboratory study carried out by Vinten et al. (1983), the distribution of suspended particles from a sewage effluent on various types of soils was studied. If, for example, heavy metals in the concentration range of the values shown in Table 8.11 are adsorbed on transported sludge, a highly concentrated aggregate of polluted solid material is formed. Vinten et al. (1983) examined the effect of subsurface properties on the vertical redistribution of deposited solids and

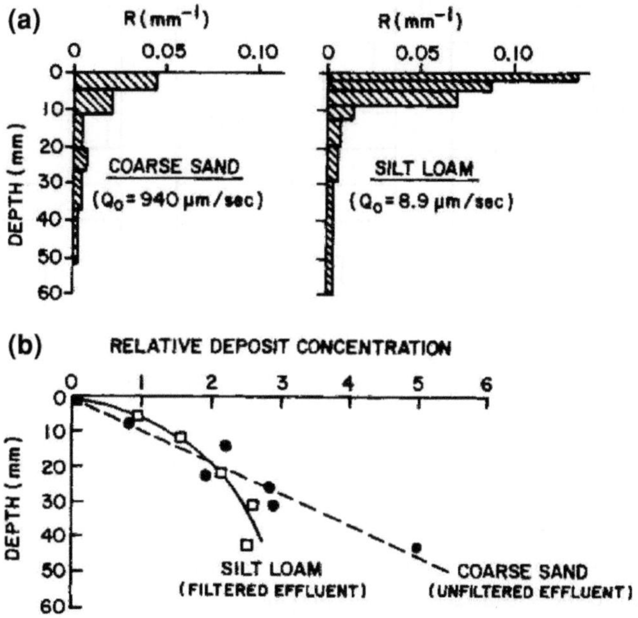

Fig. 8.42 Retention (R, defined as deposited solids per unit length of section/total added suspended solids, mm^{-1}) of suspended solids in the soil subsurface: **a** distribution of deposited solids in coarse sand and silt loam and **b** relative deposition (defined as—$\ln(\sigma/\sigma_i)$, where σ denotes initial amount of applied suspended solids and σ_i denotes measured amount of deposited solid mass per unit length of section at each depth) of suspended solids in silt loam and coarse sand leached by filtered and unfiltered effluents (Vinten et al. 1983)

on the resulting change in flow rate (Fig. 8.42). In the experiment, coarse sand was leached with unfiltered effluent, and silt loam was leached with filtered effluent. Passing unfiltered effluent through the coarse sand resulted in retention of 43 % of the total suspended solids, and the flow rate declined by less than 25 %. The relatively low retention is caused by the coarseness of the porous medium; finer suspended solids remain highly mobile. In a heavier silt loam, a very large relative reduction in flow rate occurred in the silt loam; after 400 mm of filtered effluent were added, the flow rate declined to 20 % of its initial value. A very small fraction of suspended solids was present in the leachate; most of the deposit was in the top 10 mm of soil.

Nonadsorptive retention of contaminants can also be beneficial. For example, oil droplets in the subsurface are effective in developing a reactive layer or decreasing the permeability of a sandy porous medium. Coulibaly and Borden (2004) describe laboratory and field studies where edible oils were successfully injected into the subsurface, as part of an in situ permeable reactive barrier. The oil used in the experiment was injected in the subsurface either as a NAPL or as an oil-in-water emulsion. The oil-in-water emulsion can be distributed through sands

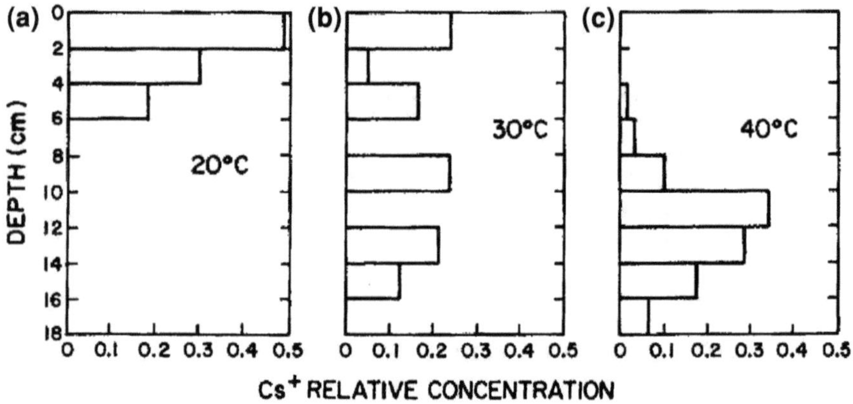

Fig. 8.43 Simulated Chernobyl Cs^+ distribution in the soil ecosystem in soil columns when the microbial decomposition of the organic material is enhanced by an increase in ambient temperature **a** 20 °C, **b** 30 °C, **c** 40 °C (Tengen et al. 1991)

without excessive pressure buildup, contrary to NAPL injection, which requires introduction to the subsurface by high pressure.

Subsurface organisms and organic residues also may affect vertical migration of contaminants. In a laboratory experiment by Tengen et al. (1991), the influence of microbial activity on the migration of Cs^+ and the effect of OM residue on Cs^+ retention were illustrated. These experiments were performed to understand the consequences of the Chernobyl accident in 1986. Figure 8.43 shows the distribution of Cs^+ in soil columns leached at temperatures of 20, 30, and 40 °C. Based on these results, Tengen et al. (1991) note the dominant role of the turnover of nondecomposed soil OM on Cs^+ retention in the subsurface: After deposition on land surface, the contaminant is fixed mainly by nondecomposed organic material and subsequently redistributed with depth as a result of microbial decomposition. This process is induced by an increase in temperature and the dissolution of organic C compounds in the infiltrating water. While the organic material is not decomposed, the Cs^+ is retained in the top layer of the subsurface. By decomposition of OM, the Cs^+ is complexed and transported to deeper layers.

The residual content of immiscible liquids can be defined by the amount of NAPL remaining in the subsurface when pore geometry permits NAPL flow greater than the retention capacity. In an outdoor pilot experiment, Fine and Yaron (1993) studied the effect of soil constituents and soil moisture contents on the retention of kerosene in the subsurface. This retention is termed the kerosene residual content (KRC). Ten soils were studied, with a broad spectrum of clay and OM contents, together with four soil moisture contents corresponding to oven-dried, air-dried, 33 kPa tension and 199 kPa tension. The KRC of the soils, studied as a function of clay content and moisture content, is presented in Fig. 8.44. The KRC of oven-dried soils ranged from 3.5 to 18.1 mL/100 g. It is affected by the

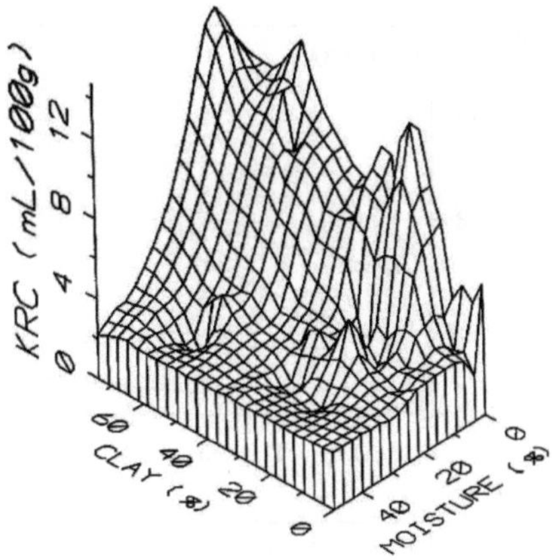

Fig. 8.44 Kerosene residual content (KRC) of soils as a function of clay and moisture contents. Reprinted from Fine and Yaron (1993). Copyright 1994 with permission of Elsevier

clay content of the soils, with KRC values of clayey soils being 1.5 times greater than for less clayey ones.

The OM content also contributes to the KRC. The relationship between clay, OM, and moisture contents and KRC is described by

$$\mathrm{KRC} = 0.13[\%\mathrm{clay}] + 1.48[\%\mathrm{OM}] - 0.32[\%\mathrm{moisture}] + 4.31. \tag{8.7}$$

KRC exhibits a linear relationship toward the combination of clay and OM contents and an inverse correlation to moisture content. The range of values of the three independent variables was clay, 0.3–74 %; OM, 0–5.2 %; and moisture content, 0–49 %. The crucial effect of moisture content on KRC of the soils was demonstrated by comparing the fine clay material to dune sand. The respective KRCs were 14.8 and 3.2 mL/100 g for air-dried materials and 2.17 and 1.84 mL/100 g for moist materials at 33 % pK_a tension. Hayden et al. (1997) investigated the influence of NOM on the residual content of another petroleum product, gasoline, and found that higher organic content leads to a higher residual saturation under air-dried environmental conditions. Contamination with gasoline when the soils are water saturated leads to a virtually identical gasoline residual content value, despite the differences in the soil OM content.

8.5 Contaminant Release

Contaminant release can result from a change in physicochemical characteristics of the liquid phase surrounding the retaining solid phase. Such a release usually is obtained by lowering the contaminant concentration in the solution, which previously reached equilibrium with the solid phase. Changes in the liquid phase concentration can occur as a result of physicochemical or biologically induced processes. Moreover, the release of trace elements can occur following a decrease

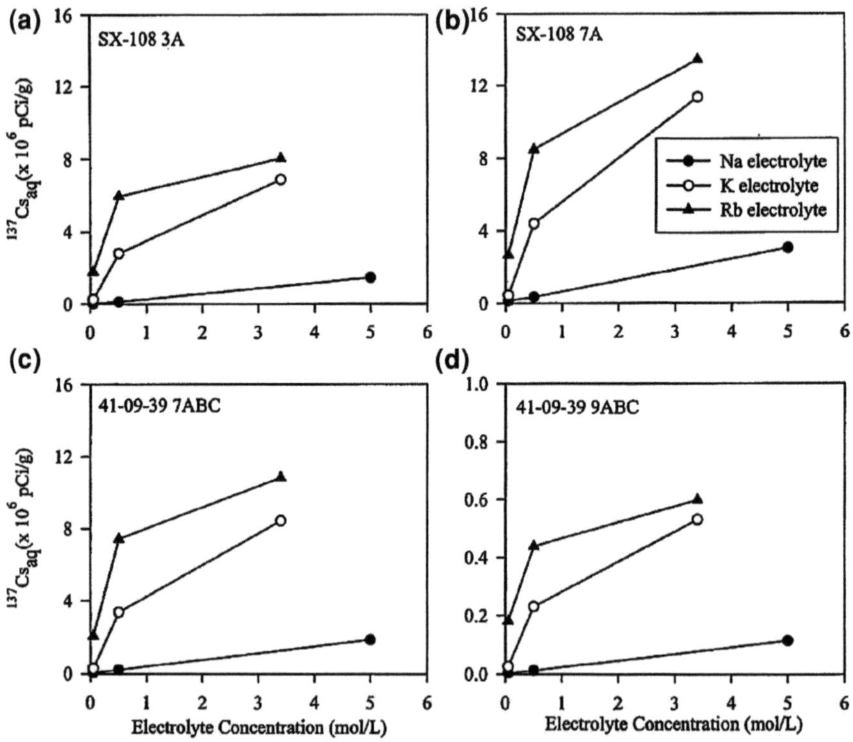

Fig. 8.45 $^{137}Cs^+$ desorption from the contaminated Hanford sediments after 6 days of equilibration, as a function of exchanging cations and their concentrations. Reprinted from Liu et al. (2003). Copyright 2003 with permission of Elsevier

in the concentration of metallic cations in solution, even if there is no modification in the background electrolyte concentration. Release may also result from the addition of protons or complexing molecules to the background electrolyte. This process sometimes is used as an extraction technique for metallic cations, which render mobile some strongly sorbed cations; but their passage into the solution implies partial destruction of the solid phase.

Liu et al. (2003) studied radiocesium desorption from subsurface pristine and contaminated micaceous sediments at the Hanford site, United States. Some of these sediments were, in the past, contaminated accidentally by nuclear wastes containing alkaline $^{137}Cs^+$. The desorption of $^{137}Cs^+$ was measured in solutions of Na^+, K^+, Rb^+, and NH_4 electrolytes of variable concentration and pH and in the presence of a strong Cs^+-specific sorbent. Desorption of $^{137}Cs^+$ from the contaminated sediment exhibits two distinct phases: an initial instantaneous release followed by a slow kinetic process (Fig. 8.45). The extent of $^{137}Cs^+$ desorption increases with increasing electrolyte concentration, following the trend $Rb^+ > K^+ > Na^+$ at a neutral pH. The extent and rate of $^{137}Cs^+$ desorption is

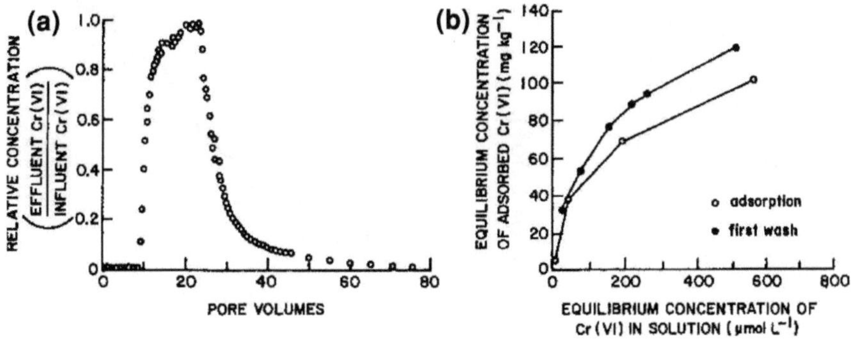

Fig. 8.46 Adsorption and desorption of Cr(VI) by a telluride alluvium in **a** a flow-through column experiment and **b** on alluvium under batch conditions (Stollenwerk and Grove 1985)

influenced by surface configuration, intraparticle diffusion, and the collapse of edge-interlayer sites in solutions of K^+, Rb^+, and NH_4. Moreover, Liu et al. (2003) showed that only 40 % of the $^{137}Cs^+$ adsorbed in a subsurface sediment, contaminated over a 30-year period, was desorbed by exchange with the electrolyte solutions. This value increased up to 80 % after long-term contact with acidified ammonium oxalate. Desorption studies with Cs^+-spiked pristine sediment, equilibrated over short duration, indicated that adsorbed Cs^+ is fully exchangeable with Na^+ solution but becomes less exchangeable when placed in K^+ and Rb^+ electrolyte solutions. This effect was attributed to the collapses of edges and partially expanded interlamellar regions, which result from saturation of the exchange complex with poorly hydrated Rb^+ and K^+ cations.

In many cases, a trace element retained on the subsurface solid phase may undergo chemical reactions that induce a hysteresis phenomenon during the release process. A relevant example of hysteresis due to precipitation of some of the initial contaminants is given by the behavior of Cr(VI), an industrial contaminant, which in the subsurface environment may be subject to reduction reactions. When an available source of electrons is present, such as OM, Cr(VI) is reduced to Cr(III); the rate of this reaction increases with decreases in pH (Ross et al. 1981).

Stollenwerk and Grove (1985) report the adsorption and desorption of Cr(VI) in an alluvial aquifer. From Fig. 8.46a, we see that, over the first ~10 pore volumes, all the Cr(VI) in water contaminant was adsorbed by the alluvium. A rapid increase in the effluent concentration of Cr(VI) then occurred, until the capacity of alluvium for contaminant retention was exhausted (~25 pore volumes). Leaching the alluvium column with 10 pore volumes of Cr-free water caused the release of about 50 % of the adsorbed Cr(VI), and further leaching with 80 pore volumes of groundwater, over 232 days, removed only an additional 34 % of the adsorbed contaminant. Stollenwerk and Grove (1985) attributed the difficulty in removing part of the adsorbed Cr(VI) to the presence of specific adsorption sites and possible reduction to Cr(III) followed by precipitation.

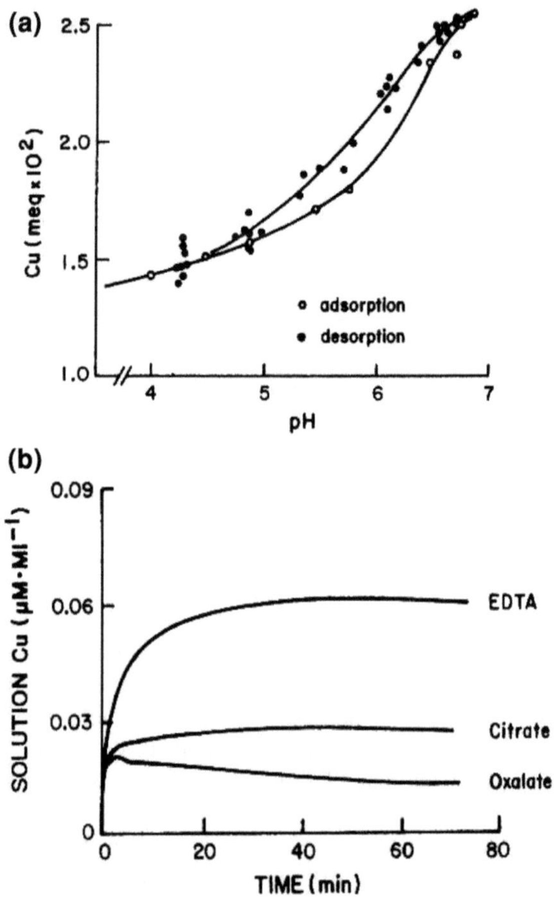

Fig. 8.47 Sorption of Cu^{2+} on montmorillonite as a function of **a** pH and **b** time-dependent release of soil-absorbed Cu by EDTA, citrate, and oxalate (Calvet 1989; Lehman and Harter 1984)

A difference in the rate of adsorption and desorption of Cr(VI) by alluvium was also observed in a batch experiment (Fig. 8.46b). On the basis of these two experiments, Stollenwerk and Grove (1985) concluded that the quantity of Cr(VI) adsorbed by alluvium is a function of its concentration as well as of the type and concentration of other anions in solution. The Cr(VI) adsorbed through nonspecific processes is desorbed readily by a Cr-free solution. Stronger bonds that are formed between Cr(VI) and alluvium during specific adsorption result in very slow release of this fraction. The Cr(VI) desorption from the alluvium material illustrates the hysteresis process that results from chemical transformation of a portion of contaminant retained in the subsurface.

The release rate of "nondesorbable," metallic cation contaminants adsorbed on the solid phase can be examined in terms of three environmental factors—pH, the presence of another metallic cation, and ligand presence—all of which vary substantially in subsurface aqueous solutions. A decrease in pH favors the release of cationic metals adsorbed on surfaces. An example of pH effects on Cu^{2+} release for a montmorillonite clay is given in Fig. 8.47a (Calvet 1989). Lehman and Harter (1984) studied the kinetics of copper desorption from soils when various ligands (Na_2-oxalate, Na_3-citrate, and Na_4-EDTA) were added to copper-contaminated samples, in stoichiometric amounts relevant to the Cu addition (i.e., for a charge ratio of 1:1 Cu/ligand). Figure 8.47b shows the time-dependent release of Cu adsorbed on soil in solutions where the three organic cations are present. It is clear that each ligand in the subsurface environment affects Cu desorption differently; Cu release is enhanced in the presence of the EDTA ligand compared to citrate and oxalate. It also appears that Cu readsorbed onto the surface, in the case of Cu-oxalate. An additional factor to consider is the competition among metallic cations for sorption sites. Metals bound through surface complexation can be displaced by other cations.

Sequential disposal of heavy metals on land surfaces may control the pathway of their retention and release in the soil–subsurface system. An example of Pb and Cd retention on and release from soil material is given by Appel et al. (2008). The authors discuss behavior of these heavy metals when added sequentially or concurrently to three tropical soils: oxisol, ultisol, and mollisol. The retention/release patterns of monometal sequential addition of Pb > Cd or Cd > Pb in the three soils are presented in Fig. 8.48.

The differences in sorbed and exchanged Pb, when Pb was added to the soils before Cd as compared to when it was applied after Cd, were insignificant despite the fact that differences were observed between the soils. In contrast, the trend for Cd was different. In the case of oxisol—characterized by a high surface area—a significant decrease in the adsorbed Cd occurred when Cd was added to the soil after Pb application. Lead, which is much less exchangeable than Cd, inhibited Cd sorption. The authors attribute this behavior mainly to chemical characteristics of the heavy metals. Due to its relatively high electronegativity, low pK_h, and small hydrated radius, Pb was sorbed in preference to Cd. Cadmium was retained on the sorption sites of the soil–subsurface mineral phase to a greater extent than when it was added prior to Pb. Appel et al. (2008) suggested that at a contaminated site receiving successive amounts of Pb and Cd at different times, Cd may be more mobile when introduced into soil after Pb.

Retention of organic contaminants on subsurface solid-phase constituents in general is not completely reversible, so that release isotherms differ from retention isotherms. As a consequence, the extent of sorption depends on the nature of the sorbent. Subsurface constituents as well as the types of bonding mechanisms between contaminants and the solid phase are factors that control the release of adsorbed organic contaminants. Saltzman et al. (1972) demonstrated the influence of soil OM on the extent of hysteresis. Adsorption isotherms of parathion showed hysteresis (or apparent hysteresis) in its adsorption and desorption in a water solution. In contrast, smaller differences between the two processes were observed when the soils were pretreated with hydrogen peroxide (oxidized subsamples) to reduce initial OM content. The parathion content of the natural soils was greater (8 % in Golan soil, 24 % in Meron soil, and 31 % in Bet-Guvrin soil) than the oxidized subsamples. Natural peat retained two to three times more parathion than

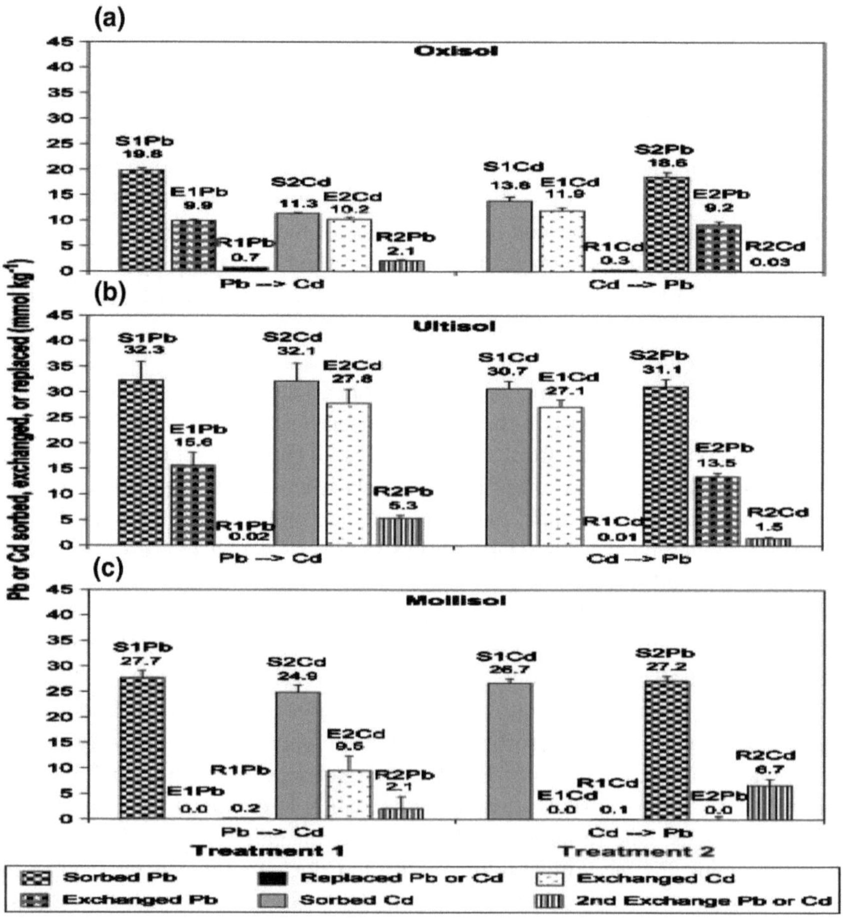

Fig. 8.48 Monometal sequential addition of Pb → Cd or Cd → Pb to the **a** Oxisol, **b** Ultisol, and **c** Mollisol soils. *S*, *E*, and *R* represent the amount of sorbed, exchanged, or replaced metal, respectively. The numbers following the letter designations indicate the order of metal addition to the soils (Appel et al. 2008)

the oxidized one. This behavior shows that parathion–organic complexes are stronger than parathion–mineral ones. Assuming that changes in OM content are solely responsible for changes in contaminant desorption, it is possible to estimate desorption curves for mineral and organic fractions. Figure 8.49 shows the calculated parathion desorption in water solutions from the (a) mineral and (b) organic fractions of Golan and Meron soils. The slopes of the desorption curves of the mineral fraction are rather steep (especially for Golan soil), indicating that adsorption is totally reversible. In the range of the concentration studied, only very small amounts of parathion appear to be released from the OM.

The properties of both OM and clay minerals may affect the release of contaminants from adsorbed surfaces. Zhang et al. (1990) report that desorption (in aqueous

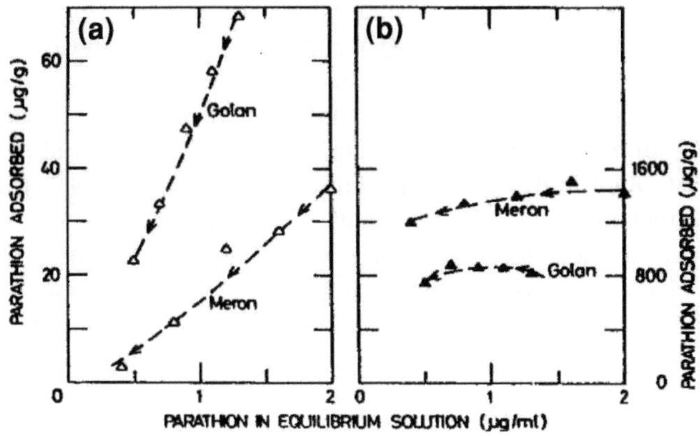

Fig. 8.49 Calculated parathion desorption from the **a** mineral and **b** organic fractions of Golan and Meron soils. Saltzman et al. (1972). Copyright 1972 American Chemical Society

Fig. 8.50 Phenanthrene adsorption–desorption hysteresis observed for Lachine and Chelsa humic acid aggregate; the hysteresis index is given by n. Reprinted from Huang et al. (2003). Copyright 2003 with permission of Elsevier

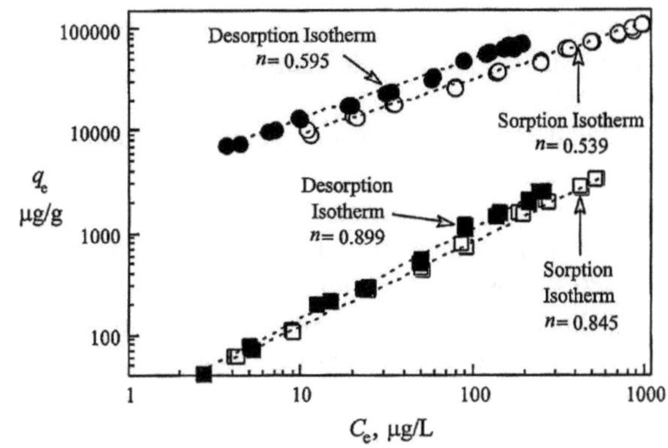

solution) of acetonitrile solvent from homoionic montmorillonite clays is reversible, and hysteresis appears to exist except for K^{+}-montmorillonite. This behavior suggests that desorption may be affected by the fundamental difference in the swelling of the various homoionic montmorillonites, when acetonitrile is present in the water solution. During adsorption, it was observed that the presence of acetonitrile affects the swelling of different homoionic clays. At a concentration of 0.5 M acetonitrile in solution, the layers of K^{+}-montmorillonite do not expand as they would in pure water, while the layers of Ca^{2+}- and Mg^{2+}-montmorillonite expand beyond a partially collapsed state. The behaviors of K^{+}-, Ca^{2+}-, and Mg^{2+}-montmorillonite are different from the behavior of these clays in pure water. Na^{+}-montmorillonite is not affected by acetonitrile presence in an aqueous solution.

The heterogeneity of subsurface OM also may influence the release of adsorbed organic contaminants. Huang and Weber (1998) present adsorption–desorption isotherms of phenanthrene for a kerogen (Lachine) and a humic acid (Chelsea).

We see from Fig. 8.50 that both sorbents exhibit adsorption–desorption hysteresis. Compared to the humic acid, the kerogen exhibits a greater degree of hysteresis, as indicated by a higher hysteresis index.

The quality of desorption water is another factor affecting the release of organic contaminants from adsorbed surfaces. For example, Barriuso et al. (1992) demonstrated that DOM in desorbing aqueous solutions increases the release of atrazine and carbetamide adsorbed on soils. Desorption of diquat and paraquat herbicides was also affected significantly by the salt concentration of the aqueous extract (Kookana and Aylmore 1993). Both Ca^{2+} and Na^{+} cations compete for sorption sites with these herbicides, but Na^{+} is not as influential as Ca^{2+}. Desorption of diquat was higher than that of paraquat for all salt concentrations.

8.6 Bound Residues

Formation of bound residues is related mainly to the fate of crop protection chemicals and other toxic waste organics in the biologically active soil surface, during and after chemical redistribution in the subsurface over long periods of time.

Bound residues are those chemicals retained in the subsurface matrix in the form of the parent organic contaminant or its metabolites; these residues remain after subsequent extractions, during which the nature of the compound or of the matrix is not altered by the extraction procedure. An example of sequential extraction of pesticide residues from soil, to define the nonextractable contaminant or the bound residue, is described in the laboratory experiment protocols presented by Mordaunt et al. (2005). Six pesticides (atrazine, dicamba, isoproturon, lindane, paraquat, and trifluralin) with various properties were added to an agricultural soil from Terrington, United Kingdom (17 % clay, 2–2.5 % OM, pH 8), and kept under controlled environmental conditions for 90 days. The soil was sampled six times, submitted to sequential solvent extraction procedure, and analyzed for pesticide content during the incubation period. The following steps for sequential extraction were performed:

1. Extraction in 0.01 M $CaCl_2$, and shake for 24 h.
2. Acetonitrile: water (9:1) shake extraction for 24 h.
3. Methanol shake extraction for 24 h.
4. Dichloromethane shake extraction for 24 h.
5. Added ^{14}C-activity combusted to $^{14}CO_2$.

Step 1 simulates the readily available soil fraction, steps 2–4 indicate potentially available soil fractions, and step 5 yields the unextracted residue and completes the mass balance. Note that the solvent used becomes increasingly nonpolar during the extraction sequence. Summary data for the six studied compounds are presented in Fig. 8.51.

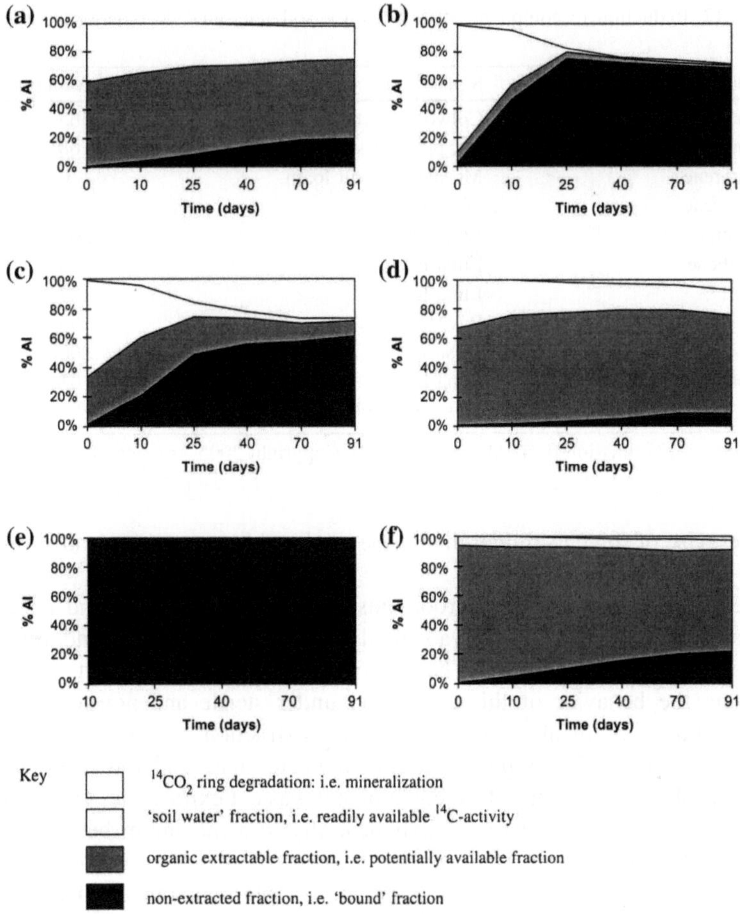

Fig. 8.51 Summary data of pesticide residues in the soil environment, showing the distribution between mineralization and extractability: **a** atrazine, **b** dicamba, **c** isoproturon, **d** lindane, **e** paraquat, and **f** trifluralin. Reprinted from Mordaunt et al. (2005). Copyright 2005 with permission of Elsevier

Mordaunt et al. (2005) categorized three classes of pesticide behavior: type A (atrazine, lindane, and trifluralin), in which ring degradation was limited as well as the formation of nonextractable residues; type B (dicamba, isoproturon), in which 25 % of ^{14}C-activity was mineralized and a large portion became nonextractable after 90 days; and type *C* (paraquat), in which a large portion of contaminant was found nonextractable after 90 days. These results illustrate the effect of extraction sequence by organic solvents on the extractable portion of contaminant from soils. However, because the soil matrix is altered during the extraction procedure, the data cannot be used to define the amount of pesticides remaining in soils as bound

Table 8.12 Reduction in compound availability for soil microbial degradation as a result of aging

Compound	Soil	Aging period (*d*)
Naphthalene	Colwood loam	365
Naphthalene	Mt. pleasant silt loam	68
Phenanthrene	Mt. pleasant silt loam	110
Phenanthrene	16 soils	200
Anthracene	Lima loam	203
Fluoranthene	Lima loam	140
Pyrene	Lima loam	133
Atrazine	Ravenna silt loam	90
Atrazine	16 soils	200
4-Nitrophenol	Lima loam	103
4-Nitrophenol	Edwards muck	103

residues. Use of solvents milder than those used in this experiment lead to different bound residue values.

The effect of native soil microorganisms on formation of bound residues is illustrated by the study of Gevao et al. (2005) on formation and release of "nonextractable" ^{14}C-dicamba. The impact of microorganisms was determined by following the behavior of this compound under sterile and nonsterile regimes during 90 days of incubation, using a mild extraction solvent (0.01 M $CaCl_2$ aqueous solution). The reported results indicate that, one day following the treatment, about 5 % of the added contaminant was not extractable. The fraction of nonextractable dicamba increased exponentially to a maximum between 14 and 21 days after application, followed by a decrease that led, at the end of the experiment (90 days), to about 65 % nonextractable contaminant. This behavior characterizes the nonsterile soil. Different patterns appeared in a sterile incubation: A gradual increase in the amount of nonextractable residue was formed, reaching about 20 % of the initial activity two weeks following treatment. After this period, there were no significant changes in the amount of nonextractable residues formed for the entire incubation period. Additional treatment (e.g., fresh soil added to aged soil) led to the hypothesis that microorganisms play a dual role in the formation and eventual release of nonextractable residues. When a pesticide is added to a catabolically active soil, the degrading microorganisms convert the pesticide to one or more metabolites, which are capable of interacting with subsurface OM. In the case of dicamba, the major metabolite (3, 6 dichlorosalicylic acid) is adsorbed to a much greater extent than the parent compound; this strong adsorption of the metabolite prevents its extraction in $CaCl_2$ solution. As the concentration of the available portion of added pesticide drops, the microbes revert to utilizing subsurface OM for their energy needs. The microbes consume subsurface OM as their primary substrate, so the bound pesticide–metabolite molecules are freed from their entrapment with humic macromolecules.

As they persist or age, organic contaminants in the subsurface become progressively less available. In a review on contaminant bioavailability during aging, Alexander (2000) shows evidence that organic compounds aged in the field are less bioavailable than the same compounds added freshly to samples of the same soil. Table 8.12 compiles the time required for a number of contaminants to become less available for microbial degradation as result of aging. Bioavailability to microorganisms decreases with time, until reaching a value below which further decline is not detectable. The time required to reach this value varies among soils, compounds, and environmental conditions.

Sequestration is considered as the decrease in availability of a compound, for uptake by a living organism and for nonvigorous extraction by an organic solvent. Nam et al. (1998) studied phenanthrene sequestration by fresh and aged soil OM, where the contaminant presence was determined by both bioavailability (e.g., mineralization by an added bacterium) and mild solvent extraction (1-butanol) procedures. Sequestration of phenanthrene as measured by bacteria-induced mineralization was appreciable in samples with >2 % organic C and was not evident in samples with <2 % organic C. Phenanthrene aged 200 days was more slowly degraded than the freshly added compound in soils with >2 % organic C. However, only a small effect was evident when the soil organic C was <2 %. Mild extraction with 1-butanol showed that the quantity removed by the solvent diminished as the compound persisted and the rate of decline in extractability generally diminished with decreasing organic C. Subsequent extraction by Soxhlet led to complete recovery of the compound, showing that the decline in availability comes from sequestration and does not reflect a loss of phenanthrene during aging.

It is questionable whether or not the value obtained by bioassays or the sequestration value can be used to define contaminant-bound residues. Aging–sequestration relationships in the subsurface, as determined through bioavailability, may provide an answer for short-term assessment of the contamination status at a specific site. To predict long-term hazards caused to the subsurface by organic contaminants of anthropogenic origin, where potential changes in environmental parameters may occur, the use of a reliable solvent extraction of bound residues is required

8.7 Contaminants in Urban Geosystems

Land surface deposits in the urban environment comprise a mixture of natural earth materials and disposed residues during many years of human activity. The term "urban geochemistry" was initially used by Thornton in the 1990s and reactivated recently by Jartun and Ottensen (2011) and Lyons and Harmon (2012). Urban geochemistry is characterized by intense toxic chemical disposal on a soil–subsurface landscape already disturbed by previous anthropogenic manipulation. The redistribution of contaminants in an urban environment differs from that occurring in a natural geosystem, due to differential impact of microrelief, water,

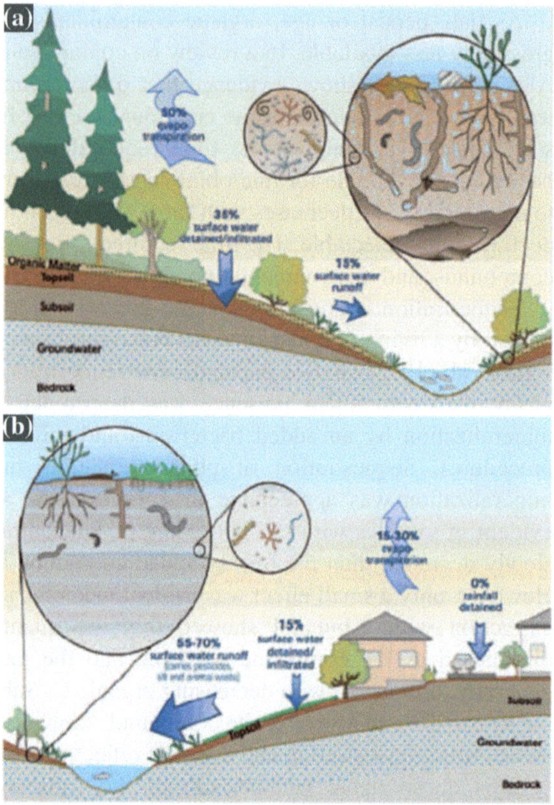

Fig. 8.52 The movement of water into a soil depends on land surface slope, vegetation, degree of surface loading, and soil texture, structure, density, and compaction. More water moves into the soil zone on natural landscapes compared to urban landscapes with disturbed soils. These schematic drawings compare the generalized disposition and movement of incoming water on a natural plant-covered landscape **a** with that in a disturbed urban landscape and **b** with limited vegetation and abundant impervious surfaces (Lyons and Harmon 2012; modified after Scheyer and Hipple 2005)

and vegetation (Fig. 8.52). Urbanization strongly affects the physical and hydrological regime of a natural soil–subsurface and may completely change the pathway of disposed contaminants.

Two specific groups of chemical contaminants that are retained in soil–subsurface systems are heavy metals and metalloids, and petroleum hydrocarbons. Several examples involving these chemicals are discussed here, being representative of urban contamination.

Arsenic contamination of soil–subsurface systems is of great environmental concern. Chirenje et al. (2003) examined arsenic contamination in urban and nonurban areas in Florida (USA). Arsenic presence was determined in 440 urban and 448 uninhabited locations; the distribution of arsenic concentrations is presented in Fig. 8.53. In general, arsenic concentrations in urban areas were higher than those in uninhabited areas. The differences between the two urban sites are due to their specific land uses. The authors noted that soil arsenic concentration in the nonurban areas showed significant correlation with natural soil properties, because they were exposed to a lower disturbance than urban soils. As a

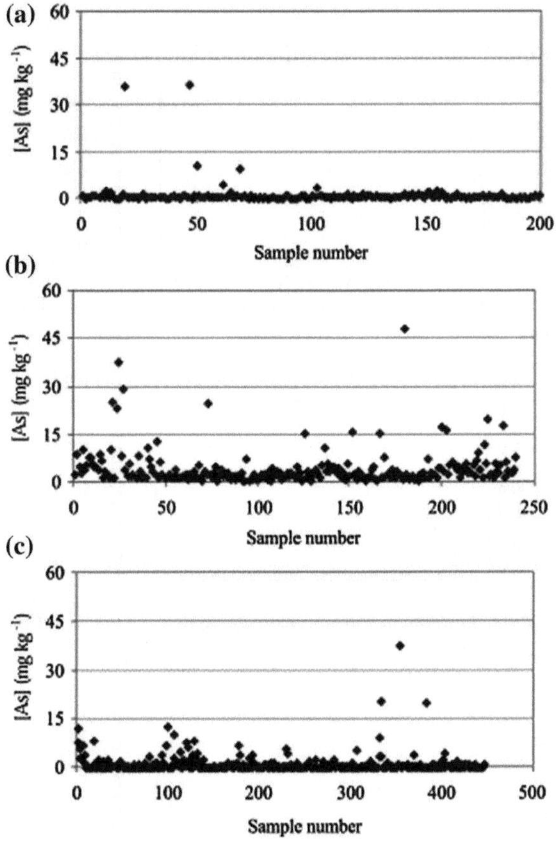

Fig. 8.53 Soil arsenic concentration (raw data) distribution in **a** Gainesville ($n = 200$), **b** Miami ($n = 240$), and **c** nonurban areas ($n = 448$) in Florida (Chirenje et al. 2003)

consequence, arsenic concentrations in urban soils showed greater variation than those from relatively undisturbed nonurban soils.

The potential adsorption of Cd, Cu, Cr, Pb, and Zn heavy metals on a "typical" mixed brownfield deposit from Scotland (UK) is reported by Markiewicz-Patkowska et al. (2005). The urban soil was collected from 15 points within a derelict area formerly used as a mix of residential and workshop/warehouse activities. The soil sample had a sandy loam texture and neutral/slightly acid pH. Fragments of brick, OM, and demolition wastes were clearly visible, suggesting that 10–15 % of the bulk was of anthropogenic origin. The sorption of heavy metals from single and multielement solutions, as a function of their initial concentration and pH, is presented in Fig. 8.54. It is found that adsorption of Cu from a single-element solution was over four times greater than adsorption from a multielement solution, due to competitive effects. In the case of Cr and Zn, release of metal from soil to solution was observed. The adsorption capacity at pH 2 occurred in the order Cr > Cu > Pb > Cd, and at pH 7 Cd > Zn, with precipitation affecting Cu and Pb

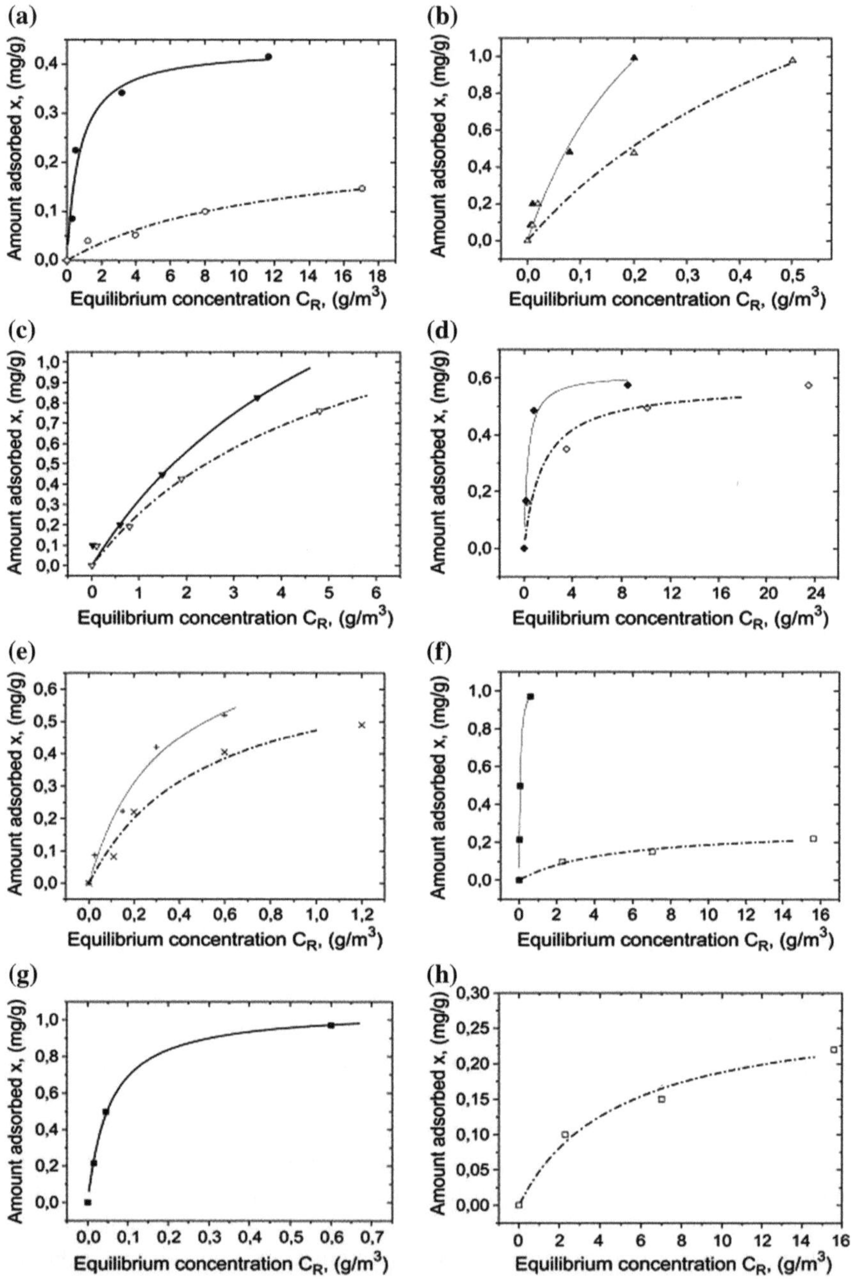

Fig. 8.54 Langmuir adsorption isotherms for the selected metals, at different pH values, adsorption from single- and multielement solutions. Experimental data (*points*), approximation function (*lines*): **a** Cd, pH 2, single (*black circle*), multi (*open circle*), **b** Cd, pH 7, single (*black up-pointing triangle*), multi (*open up-pointing triangle*), **c** Cr, pH 2, single (*black down-pointing triangle*), multi (*open down-pointing triangle*), **d** Pb, pH 2, single (), multi (*open diamond*), **e** Zn, pH 7, single (+), multi (×), and **f** Cu, pH 2, single (*black square*), multi (*open square*) (Markiewicz-Patkowska et al. 2005)

Table 8.13 Range and median of detected PAHs in three urban soils (modified after Morillo et al. 2008)

Compound	Range (and median) (μg/kg, dry weight basis)		
	Glasgow	Ljubljana	Torino
Naphthalene	16.5–940 (68.1)	17.7–38.6 (28.8)	15.3–79.4 (30.0)
Acenaphthene	9.10–848 (25.0)	0.005–19.9 (9.07)	5.69–74.2 (8.82)
Fluorene	13.3–750 (38.1)	6.60–54.3 (10.9)	7.22–241 (11.7)
Phenanthrene	187–9,736 (1379)	21.9–2,776 (84.4)	22.7–2,670 (75.9)
Anthracene	28.6–1,352 (136)	6.02–30.5 (12.0)	5.10–370 (11.6)
Fluoranthene	141–7,249 (1125)	16.1–872 (101)	12.6–3,560 (47.9)
Pyrene	184–7,835 (1124)	15.4–248 (94.9)	9.62–3,460 (71.8)
Benzo[a]anthracene	109–3,928 (680)	11.5–158 (51.7)	8.23–2,040 (40.9)
Chrysene	154–5,002 (856)	12.8–209 (72.3)	8.70–2,430 (41.0)
Benzo[b]fluoranthene	92.1–2,984 (525)	5.21–203 (56.4)	4.68–2,020 (30.7)
Benzo[k]fluoranthene	63.0–1,400 (292)	12.7–90.8 (34.6)	10.4–1,240 (25.8)
Benzo[a]pyrene	132–3,627 (657)	18.4–186 (70.9)	14.1–3,170 (44.5)
Dibenzo[a, h]anthracene	16.7–399 (70.0)	6.23–25.4 (10.9)	5.09–153 (10.8)
Benzo[g, h, i]perylene	117–3,884 (544)	12.8–187 (36.8)	11.9–1,640 (52.8)
Indeno[1, 2, 3-c, d]pyrene	158–2,827 (424)	0.04–185 (66.2)	0.047–364 (63.3)
Σ15 PAHs	1,487–51,822 (8,337)	218–4,488 (791)	148–23,500 (704)

behavior. The urban soil studied displayed some enhanced adsorption as compared to agricultural soils, probably due to the presence of anthropogenic materials.

Contamination of the soil–subsurface environment by PAHs is of major interest due to their potential health hazards. The highest concentration of PAHs occurs in urban areas near point sources and in soils with high OM. A survey of PAH concentrations in European cities was performed by Morillo et al. (2008). Glasgow (UK), Torino (Italy), and Lubljana (Slovenia) were selected for this study, and the same methodology for soil sampling, extraction, and chemical analysis was employed. Median PAH concentrations resulting from analysis of 20 samples in each city are presented in Table 8.13. Despite the wide range of values in PAH concentrations, it is clear that in Glasgow urban soils, the contamination is about 10 times higher than in the other cities studied. The three predominant PAHs were phenanthrene, fluoranthene, and pyrene, representing about 40 % of the total compounds found in the urban soils. Based on these results, the authors consider that PAH contamination of the urban soils is due to pyrogenic origin and is affected by climatic conditions, natural OC of the soil, and proximity to the contamination sources.

Polychlorinated naphthalenes (PCNs) are extremely toxic compounds including congeners with chlorine substitutions in the 2-, 3-, 6-, and 7-positions. These contaminants are found both in urban and rural environments. Krauss and Wilcke (2003) studied the fate of PCNs in soils, investigating their occurrence in urban and rural environments. Their study comprised 49 topsoils from various locations

Table 8.14 Sums of the concentrations of PCNs, PCBs, and PAHs in urban soils of Bayreuth and rural soils under different land use (medians and ranges) (Krauss and Wilcke 2003)

Land use	*n*	Σ35 PCNs (μg/kg)		Σ12 PCBs (μg/kg)		Σ20 PAHs (μg/kg)	
		Median	Range	Median	Range	Median	Range
Urban sites							
Alluvial grassland	10	1.6	0.35–5.7	14.8	3.3–58.1	2.7	0.38–20.7
House garden	10	3.6	0.29–15.4	14.8	2.8–158	10.9	0.81–17.7
Park	9	0.46	0.10–3.1	5.5	0.82–13.7	1.9	0.76–10.5
Roadside	9	0.90	0.28–4.7	14.3	2.2–91.9	2.3	0.63–186
Industrial	7	1.9	0.67–15.0	21.9	2.3–70.2	10.3	2.4–48.9
Agricultural	4	0.10	nd[a]–0.51	1.6	1.1–7.6	0.64	0.28–2.2
Urban area	49	1.3	nd–15.4	13.0	0.82–158	3.4	0.28–186
City center[b]	17	0.74	0.23–7.4	8.7	1.5–20.7	2.7	0.63–20.7
Outskirts[b]	14	0.92	nd–5.6	10.3	0.82–58.1	1.2	0.28–7.2
Rural sites							
Agriculture	6	0.13	nd–0.29	1.6	1.1–2.6	0.30	0.19–0.69
Alluvial grassland	3	0.44	0.07–0.82	1.8	1.3–5.0	0.37	0.16–2.1
Rural area	9	0.13	nd–0.82	1.7	1.1–5.0	0.34	0.16–2.1

[a] Not detected

[b] Without house gardens, industrial sites, and roadside soil

within the city of Bayreuth (Germany) and 9 agricultural soils in the vicinity of the city. It was found that PCN concentrations were five times lower in the rural area than in the central part of the city, characterized by the same natural soil. Table 8.14 shows PCN concentrations in comparison with polychlorinated biphenyls (PCBs) and PAHs found in both urban and rural soils. The human impact on soil contamination is confirmed by the highest concentrations occurring in industrial sites and in house gardens. The increase in PCN ad PCB concentrations in the rural soils may be explained by the more volatile lower molecular weight compounds being easily transported in gaseous phase from urban to rural areas (Krauss and Wilcke 2003).

References

Alexander M (2000) Aging, bioavailability and overestimation of risk from environmental pollutants. Environ Sci Technol 34:4259–4265

Appel C, Ma QL, Rhue RD, Reve W (2008) Sequential sorption of lead and cadmium in three tropical soils. Environ Pollut 155:132–140

Babiarz CL, Hurley JP, Hoffman SR, Andren AW, Shafer MM, Armstrong DE (2001) Partitioning of total mercury and methylmercury to the colloidal phase in freshwaters. Environ Sci Technol 35:4773–4782

Bang JS, Hesterberg D (2004) Dissolution of trace element contaminants from two coastal plain soils as affected by pH. J Environ Qual 33:891–901

Barriuso E, Baer U, Calvet R (1992) Dissolved organic matter and adsorption-desorption of difemuron, atrazine and carbetamide by soils. J Environ Qual 21:359–367

Bar-Yosef B, Afic I, Rosenberg R (1989) Fluoride sorption and mobility in reactive porous media. In: Bar Yosef B, Barrow NJ, Goldschmid J (eds) Inorganic contaminants in the vadose zone. Springer, Heidelberg, pp 75–89

Bennett B, Larter SR (1997) Partition behaviour of alkylphenols in crude/oil brine systems under subsurface condition. Geochim Cosmochim Acta 61:4393–4402

Bergaoui L, Lambert JF, Prost R (2005) Cesium adsorption on soil clay: macroscopic and spectroscopic measurements. App Clay Sci 29:23–29

Buerge-Weirich D, Hari R, Xue H, Behra O, Sigg L (2002) Adsorption of Cu, Cd, and Ni on goethite in the presence of natural ground ligands. Environ Sci Technol 36:328–336

Burns IG, Hayes MHB, Stacey M (1973) Some physico-chemical interactions of paraquat with soil organic materials and model compounds. II. Adsorption and desorption equilibria in aqueous suspensions. Weed Res 13:79–90

Calvet R (1989) Adsorption of organic chemicals in soils. Environ Health Perspect 83:145–177

Chaney RL (1989) Toxic element accumulation in soils and crops protecting soil fertility and agricultural food chains. In: Bar Yosef B, Barrow NJ, Goldschmid J (eds) Inorganic contaminants in the vadose zone. Springer, Heidelberg, pp 140–159

Charlatchka R, Cambier P (2000) Influence of reducing conditions on solubility of trace metals in contaminated soils. Water Air Soil Poll 118:143–167

Chirenje T, Ma LQ, Chen M, Zillioux EJ (2003) Comparison between background concentrations of arsenic in urban and non-urban areas of Florida. Adv Environ Res 8:137–146

Cho HH, Choi J, Goltz MN, Park JW (2002) Combined effect of natural organic matter and surfactants on the apparent solubility of polycyclic aromatic hydrocarbons. J Environ Qual 31:275–280

Coulibaly KM, Borden RC (2004) Impact of edible oil injection on the permeability of aquifer sands. J Contam Hydrol 71:219–237

Donovan WC, Logan TJ (1983) Factors affecting ammonia volatilization from sewage sludge applied to soil in a laboratory study. J Environ Qual 12:584–590

Dror I, Kliger L, Laufer A, Hadas A, Russo D, Yaron B (1999) Behavior of atrazine and terbuthylazine in an irrigated field: I. Soil and management spatial variability. Agrochimica 43:21–29

Dror I, Gerstl Z, Prost R, Yaron B (2000) Behavior of neat and enriched volatile petroleum hydrocarbons mixture in the subsurface during leaching. Land Contam Reclam 8:341–348

Dror I, Gerstl Z, Prost R, Yaron B (2002) Abiotic behavior of entrapped petroleum products in the subsurface during leaching. Chemosphere 49:1375–1388

Dror I, Amitay T, Yaron B, Berkowitz B (2003) A "salt-pump" mechanism for induced intrusion of organic contaminants from marine sources into coastal aquifers. Science 300:950

Engel MH, Macko SA (1993) Organic geochemistry: principles and applications. Plenum, New York

Fine P, Yaron B (1993) Outdoor experiments on enhanced volatilization by venting of kerosene components from soil. J Contam Hydrol 12:335–374

Flühler H, Polomski J, Blaser P (1982) Retention and movement of fluoride in soils. J Environ Qual 11:461–468

Foster KL, Mackay D, Parkerton TF, Webster E, Milford L (2005) Five-stage environmental exposure assessment strategy for mixture: gasoline as a case study. Environ Sci Technol 39:2711–2718

Galin Ts, Gerstl Z, Yaron B (1990) Soil pollution by petroleum products. III. Kerosene stability in soil columns as affected by volatilization. J Contam Hydrol 5:375–385

Gerstl Z, Yaron B (1981) Attapulgite-pesticide interactions. Residue Rev 78:69–99

Gevao B, Jones KC, Semple KT (2005) Formation and release of non-extractable ^{14}Dicamba residues in soil under sterile and non-sterile regimes. Environ Pollut 133:17–24

Giles CH, MacEwan TH, Nakhwa SN, Smith D (1960) Studies in adsorption. Part XI. A system of classification of solution adsorption isotherms and its use in diagnosis and adsorptive mechanism and measurements of specific area of solids. Chem Soc J 3:3973–3993
Gobran G, Miyamoto S (1985) Dissolution rate of gypsum in aqueous salt solutions. Soil Sci 140:89–94
Green CH, Heil DM, Cardon GE, Butters GL, Kelly EF (2003) Solubilization of manganese and trace metals in soils affected by acid mine runoff. J Environ Qual 32:1323–1334
Haitzer M, Aiken GR, Ryan JN (2002) Binding of mercury(II) to dissolved organic matter: the role of the mercury-to-DOM concentration ratio. Environ Sci Technol 36:3564–3570
Hargrove WL (1986) The solubility of aluminum-organic matter and its implication on plant uptake aluminum. Soil Sci 142:179–181
Hayden NJ, Voice TC, Wallace RB (1997) Residual gasoline saturation in unsaturated soil with and without organic matter. J Contam Hydrol 25:271–281
Hayes MHB, Mingelgrin U (1991) Interactions at the soil colloid-solution interface. In: Bolt GH, De Boodt MF, Hayes MF, McBride MB (eds) NATO ASI series—applied science—series F vol 190. Kluwer, Dordrecht, pp 324–401
Huang W, Peng P, Yu Z, Fu J (2003) Effects of organic matter heterogeneity on sorption and desorption of organic contaminants by soils and sediments. Appl Geochem 18:955–972
Huang W, Weber WJ Jr (1998) A distributed reactivity model for sorption by soils and sediments: slow concentration-dependent sorption rate. Environ Sci Technol 32:3549–3555
Jarsjo J, Destouni G, Yaron B (1994) Retention and volatilization of kerosene: laboratory experiments on glacial and postglacial soils. J Contam Hydrol 17:167–185
Jartun M, Ottensen RT (2011) Urban geochemistry. In: Harmon RS, Parker A (eds) Contribution of geochemistry to the study of the earth frontiers in geochemistry. Wiley, Hoboken, pp 221–237. doi:10.1002/9781444329957.ch11
Jayaweera GR, Mikkelsen DS (1991) Assessment of ammonia volatilization from flooded soil systems. Adv Agron 45:303–357
Johnston AE, Goulding KWT, Poulton PR (1986) Soil acidification during more than 100 years under permanent grassland and woodland at Rothamsted. Soil Use Manage 2:3–10
Karlson U, Frankenberger WT (1988) Determination of gaseous Se-75 evolved from soil. Soil Sci Soc Am J 52:678–681
Khachikian C, Harmon TC (2000) Nonaqueous phase liquid dissolution in porous media: current state of knowledge and research needs. Trans Porous Media 38:3–28
Kookana RS, Aylmore LAG (1993) Retention and release of diquat and paraquat herbicides in soils. Austral J Soil Res 31:97–109
Krauss M, Wilcke W (2003) Polychlorinated naphthalenes in urban soils: analysis, concentrations, and relation to other persistent organic pollutants. Environ Pollut 122:75–89
Lai KM, Johnson KL, Scrimshaw MD, Lester JN (2000) Binding of waterborne steroid estrogens to solid phases in river and estuarine systems. Environ Sci Technol 34:3890–3894
Lee LS, Strock TJ, Sarmah AK, Rao PSC (2003) Sorption and dissipation of testosterone, estrogens, and their primary transformation products in soils and sediment. Environ Sci Technol 37:4098–4105
Lehman RG, Harter RD (1984) Assessment of copper-soil bond strength by desorption kinetics. Soil Sci Soc Am J 43:769–772
Lin ZQ, Schemenauer RS, Cervinka V, Zayed A, Lee A, Terry N (2000) Selenium volatilization from a soil—plant system for the remediation of contaminated water and soil in the San Joaquin Valley. J Environ Qual 29:1048–1056
Lindsay WL (1979) Chemical equilibria in soils. Wiley, New York
Liu C, Zachara JM, Smith SC, McKinley JP, Ainsworth CC (2003) Desorption kinetics of radiocesium from subsurface sediments at Hanford Site USA. Geochim Cosmochim Acta 67:2893–2912
Loffredo E, Senesi N (2006) Fate of anthropogenic organic pollutants in soils with emphasis on adsorption/desorption processes of endocrine disruptor compounds. Pure App Chem 78:947–961

Lyons WB, Harmon RS (2012) Why urban geochemistry? Elements 8:439–444
Maqueda C, Undabeytia T, Morillo E (1998) Retention and release of copper on montmorillonite as affected by the presence of pesticide. J Agric Food Chem 46:1200–1204
Markiewicz-Patkowska JA, Hursthouse A, Przybyla-Kij H (2005) The interaction of heavy metals with urban soils: sorption behaviour of Cd, Cu, Cr, Pb and Zn with a typical mixed brownfield deposit. Environ Inter 31:513–521
Master Y, Laughlin RJ, Stevens RJ, Shaviv A (2004) Nitrite formation and nitrous oxide emissions as affected by reclaimed effluent application. J Environ Qual 33:852–860
McBride MB (1989) Reactions controlling heavy metals solubility in soils. Adv Soil Sci 10:1–47
McBride MB (1994) Environmental chemistry of soils. Oxford University Press, Oxford
Melkior T, Yahiaoi S, Motellier S, Thoby D, Tevissen E (2005) Cesium sorption and diffusion in Bure mudrock samples. Appl Clay Sci 29:172–186
Mordaunt CJ, Gevao B, Jones KC, Semple KT (2005) Formation of non-extractable pesticide residues: observations on compound differences, measurement and regulatory issues. Environ Pollut 133:25–34
Morillo E, Romero AS, Madrid L, Villaverde J, Maqueda C (2008) Characterization and sources of PAHs and potentially toxic metals in urban environments of Sevilla (Southern Spain). Water Air Soil Pollut 187:41–51
Mulder J, van Grinsven JJM, Breemen N (1987) Impacts of acid atmospheric deposition on woodland soils. III. Aluminum chemistry. Soil Sci Soc Am J 51:1640–1646
Nam K, Chung N, Alexander M (1998) Relationship between organic matter content of soil and the sequestration of phenanthrene. Environ Sci Technol 32:3785–3788
Ni N, El-Sayed MM, Sanghvi T, Yalkowsky SH (2000) Estimation of the effect of NaCl on the solubility of organic compounds in aqueous solutions. J Pharm Sci 89:1620–1625
Park SK, Bielefeldt A (2003) Equilibrium partitioning of non-ionic surfactant and pentachlorophenol between water and a non-aqueous phase liquid. Water Res 37:3412–3420
Peck AM, Hornbuckle KC (2005) Gas-phase concentrations of current-use pesticides in Iowa. Environ Sci Technol 39:2952–2959
Prost R, Gerstl Z, Yaron B, Chaussidon J (1977) Infrared studies of parathion attapulgite interaction. In: Behavior of pesticides in soils. Israel-France symposium INRA, Versailles, pp 108–115
Qin H, Yakoyama Y, Fan Q, Iwatani H, Tanaka K, Sakaguchi A, Kanai Y, Zhu J, Onda Y, Takahashi Y (2012) Investigation of cesium adsorption on soil and sediment samples from Fukushima Prefecture by sequential extraction and EXAFS technique. Geochem J 46:297–302
Ross DS, Sjogren RE, Bartlett EJ (1981) Behavior of chromium in soils: IV. Toxicity to microorganisms. J Environ Qual 10:145–148
Rytwo G, Tropp T, Serban C (2002) Adsorption of diquat, paraquat and methyl green on sepiolite: experimental results and model calculations. Appl Clay Sci 20:273–282
Saltzman S, Kliger L, Yaron B (1972) Adsorption-desorption of parathion as affected by soil organic matter. J Agric Food Chem 20:1224–1227
Saltzman S, Yariv S (1976) Infrared and X-ray study of parathion montmorillonite sorption complexes. Soil Sci Soc Am J 35:700–705
Sanudo-Wilhelmy SA, Rossi FK, Bokuniewicz H, Paulsen RJ (2002) Trace metal levels in groundwater of a coastal watershed: importance of colloidal forms. Environ Sci Technol 36:1435–1441
Scheyer JM, Hipple KW (2005) Urban soil primer. United States Department of Agriculture, Natural Resources Conservation Service, National Soil Survey Center, Lincoln, Nebraska. http://soils.usda.gov/use
Schmidt TC, Kleinert P, Stengel C, Goss KU, Haderlein SB (2002) Polar fuel constituents—compound identification and equilibrium partitioning between nonaqueous phase liquid and water. Environ Sci Technol 36:4074–4080
Song J, Peng P, Huang W (2002) Black carbon and kerogen in soils and sediments. I. Quantification and characterization. Environ Sci Technol 36:3960–3967

Sorensen H, Pedersen KS, Christensen PL (2002) Modeling gas solubility in brine. Org Geochem 33:635–642

Sparks DL (2005) Sorption–desorption, kinetics. In: Hillel D et al (eds) Encyclopedia of soils in the environment. Elsevier Ltd, Oxford, pp 556–561

Spencer WF, Cliath MM (1969) Vapor densities of dieldrin. Environ Sci Technol 3:670–674

Spencer WF, Cliath MM (1973) Pesticide volatilization as related to water loss from soil. J Environ Qual 2:284–289

Sposito G, Martin Neto L, Yang A (1996) Atrazine complexation by soil humic acids. J Environ Qual 25:1203–1209

Sterling MC Jr, Bonner JS, Page CA, Ernest ANS, Autenrieth RL (2003) Partitioning of crude oil polycyclic aromatic hydrocarbons in aquatic systems. Environ Sci Technol 37:4429–4434

Stollenwerk KC, Grove DB (1985) Adsorption and desorption of hexavalent chromium in an alluvial aquifer near Telluride, Colorado. J Environ Qual 14:150–155

Strawn DG, Sparks DI (2000) Effects of soil organic matter on the kinetics and mechanisms of Pb(II) sorption and desorption in soil. Soil Sci Soc J 64:144–156

Stumm W, Morgan JJ (1996) Aquatic chemistry, 3rd edn. Wiley, New York

Taylor AW, Spencer WF (1990) Volatilization and vapor transport processes. In: Cheng HH (ed) Pesticides in the soil environment (Soil Sci Soc Amer Book Ser 2). Madison, Wisconsin, pp 213–369

Tengen I, Doerr H, Muennich KO (1991) Laboratory experiments to investigate the influence of microbial activity on the migration of cesium on a forest soil. Water Air Soil Poll 57(58):441–449

Thomas GW, Yaron B (1968) Adsorption of sodium from irrigation water by four Texas soils. Soil Sci 106:213–220

Thorpe KL, Cummings RI, Hutchinson TH, Scholze M, Brighty G, Sumpter JP, Tyler CR (2003) Relative potencies and combination effects of steroidal estrogens in fish. Environ Sci Technol 37:1142–1149

Tsai WT, Lai CW, Hsien KJ (2003) Effect of particle size of activated clay on the adsorption of paraquat from aqueous solution. J Coll Interface Sci 263:29–34

Turner A (1996) Trace-metal partitioning in estuaries: importance of salinity and particle concentration. Mar Chem 54:27–39

Turner A, Millward GE, LeRoux SM (2001) Sediment-water partitioning of inorganic mercuries in estuaries. Environ Sci Technol 35:4648–4654

Turner A, Martino M, Le Roux SM (2002) Trace metal distribution coefficients in the Mersey Estuary UK: evidence for salting out of metal complexes. Environ Sci Technol 36:4578–4584

Undabeytia T, Morillo E, Ramos AB, Maqueda C (2002) Mutual influence of Cu and a cationic herbicide on their adsorption–desorption processes on two selected soils. Water Air Soil Poll 137:81–94B

van Grunsven HJM, van Riensdijk WH, Otjes R, van Beemer N (1992) Rates of aluminum dilution in acid sandy soils observed in column experiments. J Environ Qual 21:439–447

Vinten AJ, Mingelgrin U, Yaron B (1983) The effect of suspended solids in waste water on soil hydraulic conductivity I. Suspended solid labeling method II. Vertical distribution of suspended solids. Soil Sci Soc Am J 47:402–412

Waite DT, Cessna AJ, Grover R, Kerr LA, Snihura AD (2004) Environmental concentrations of agricultural herbicides in Saskatchewan, Canada: bromoxynil, dicamba, diclofop, MCPA, triluralin. J Environ Qual 33:1616–1628

Wolters A, Linnemann V, Herbst M, Klein M, Schaffer A, Vereecken H (2003) Pesticide volatilization from soil: lysimeter measurements versus predictions of European registration models. J Environ Qual 32:1183–1193

Xiang HF, Banin A (1996) A solid-phase manganese fractionation changes in saturated arid-zone soils: pathways and kinetics. Soil Sci Soc Am J 60:1072–1080

Xiao B, Yu Z, Peng P, Song J, Huang W (2004) Black carbon and kerogen in soils and sediments. 2. Their role in phenanthrene and naphthalene sorption equilibria. Environ Sci Technol 38:5842–5852

Xu D, Zhou X, Wang X (2008) Adsorption and desorption of Ni^{2+} on Na-montmorillonite: effects of pH, ionic strength, fulvic acid, humic acid addition sequences. Appl Clay Sci 39:133–141

Yaron B (1978) Organophosphorus pesticide-clay interactions. Soil Sci 125:412–417

Yaron B, Dror I, Graber E, Jarsjo J, Fine P, Gerstl Z (1998) Behavior of volatile organic mixtures in the soil environment. In: Rubin H, Narkis N, Carberry J (eds) Soil and aquifer pollution. Springer, Heidelberg, pp 37–58

Yaron B, Saltzman S (1978) Soil-parathion surface interactions. Residue Rev 69:1–34

Yaron-Marcovich D, Dror I (2007) Berkowitz B (2007) Behavior and stability of organic contaminant droplets in aqueous solutions. Chemosphere 69:1593–1601. doi:10.1016/j.chemosphere.2007.05.056

Zhang ZZ, Sparks DL, Pease RA (1990) Sorption and desorption of acetonitrile on montmorillonite from aqueous solutions. Soil Sci Soc Am J 54:351–356

Zhong L, Mayer AS, Pope GA (2003) The effects of surfactant formulation on non-equilibrium NAPL solubilization. J Contam Hydrol 60:55–75

Part IV
Contaminant Transport from Land Surface to Groundwater

In the previous sections of this book, we focused on the nature of contaminants and the geochemical reactions that can occur in the subsurface environment. Chemical compounds introduced into infiltrating water or in contact with soil or rock surfaces are subject to chemically and biologically induced transformations. Other compounds are retained by the soil constituents as sorbed or bound residues. Thus, in terms of geochemical interactions and reactions among dissolved chemical species, interphase transfer occurs in the form of dissolution, precipitation, volatilization, and various forms of physicochemical retention on the solid surfaces.

These phenomena do not occur in a "static" domain: chemical compounds migrate and are redistributed along the soil profile, down to the water table region, and within the fully saturated aquifer zone, by flowing water. The extent of this redistribution and the kinetics of the geochemical interactions are controlled by the very nature of fluid flow in porous media, the water chemistry, and of course the properties of the soil and contaminant(s).

To describe and quantify these complex dynamics, models are used. Modeling of contaminant transport involves formulation of a conceptual framework and corresponding quantitative relationships that lead to determination of contaminant distributions over space and time. Models can also be used to investigate the relative influence of different physical and (geo)chemical mechanisms on contaminant transport and to assist in designing management and remediation strategies.

The next chapters consider the dynamics of flow and transport of water and chemical compounds, as they migrate from land surface to the groundwater regime. In Chap. 9, we focus on water flow in the partially saturated zone and through the capillary fringe into the saturated regime below the water table. Chapter 10 then treats (non-chemically reactive) transport mechanisms that govern migration of chemical species, while Chap. 11 integrates the effects of water flow, chemical transport, and geochemical interactions through the subsurface environment. Finally, Chap. 12 considers selected research findings, providing specific examples of the transport behaviors discussed here.

Chapter 9
Water Flow in the Subsurface Environment

Before considering the processes and means to quantify contaminant transport from land surface to groundwater, we first consider water movement in this region; clearly, water movement plays a key role in contaminant migration. This book focuses on chemical transport from land surface to the water table, and as such, we do not dwell on flow through the saturated zone. Numerous books on groundwater and contaminant hydrology are readily available.

9.1 Water Flow in Soils and the Vadose Zone

Soils, typically, are not fully saturated by water; the soil layer and the region reaching to the water table contain water contents below full saturation. These regions usually are referred to as the *vadose zone* and said to be "unsaturated," but they are more correctly considered "partially saturated." The degree of saturation is the ratio of the volume of water to the pore volume within the porous medium. Saturation levels usually are a few percent at land surface (or even zero in perpetually dry arid zones) and increase slowly with depth until the region of the capillary fringe (water table), where it increases rapidly to 100 %.

The thickness of the vadose zone can vary widely, from just below the soil layer to depths of tens or hundreds of meters. However, regardless of the actual thickness of this zone, essentially the same mechanisms govern distribution of water and patterns of water and chemical movement. Rather than advance downward, water and chemicals generally follow nonuniform, preferential pathways, through macropores, soil cracks and aggregates, fissures, solution channels, root paths, and wormholes. As a consequence, downward migration of contaminants from the land surface usually is highly nonuniform, influenced strongly by movement through preferential pathways. The term *fingering* therefore is often used to describe this movement of water (and dissolved contaminants) and other liquids (NAPLs); fingering is a basic feature during infiltration, as well as in the processes of drainage and imbibition. Air–water and NAPL–water interactions are treated as displacement phenomena between two immiscible phases.

B. Berkowitz et al., *Contaminant Geochemistry*,
DOI: 10.1007/978-3-642-54777-5_9, © Springer-Verlag Berlin Heidelberg 2014

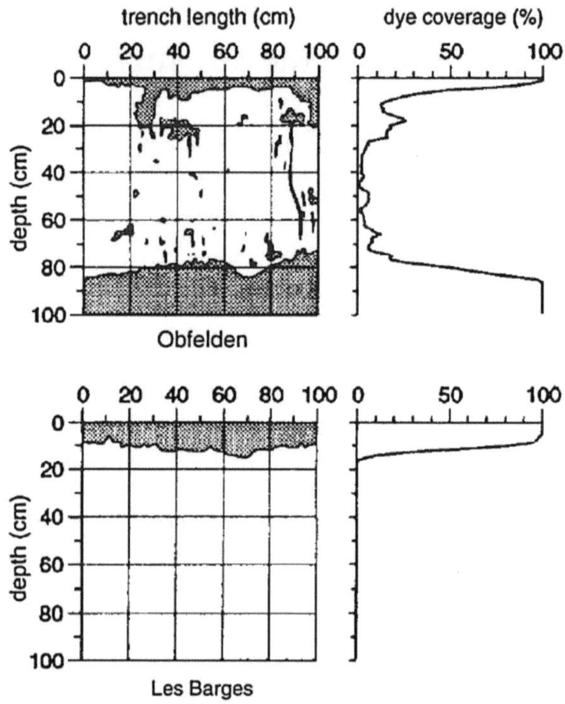

Fig. 9.1 Patterns of vertical water flow in a porous medium with "wet" initial conditions; 40 mm of water containing Brilliant Blue FCF as a dye were applied to the ground surface. The *left figures* show the pattern of infiltrating water, while the *right plots* show one-dimensional (averaged) profiles of dye coverage with depth. Reproduced by permission of American Geophysical Union. Flury et al. (1994), doi:10.1029/94WR00871. Copyright 1994 American Geophysical Union

Immiscible two-phase displacement occurs in two forms: drainage, which occurs when a wetting phase is displaced by a nonwetting phase, and imbibition, which is the opposite process, i.e., a wetting phase displacing a nonwetting phase. The contact angle that each liquid phase forms with the grains of the solid matrix determines which of the two liquid phases is the wetting one. The basic mechanisms of immiscible displacement during drainage and imbibition are different (Lenormand 1990; Blunt and Scher 1995). In drainage, the nonwetting fluid must exceed the pressure in the wetting fluid within the throat by a value equal to the capillary pressure. Subsequently, it spontaneously invades the adjacent pore. This invading behavior is defined as piston-like invasion. In imbibition, the invading behavior is affected by the pore geometry, the number of filled throats attached to a pore, and film flow controlled by the roughness of the solid. Wetting invasion therefore is characterized by pinch-off behaviors (Vizika and Payatakes 1989) in addition to piston-like invasion.

Important studies of field-scale, water infiltration patterns are provided by Flury et al. (1994). Figures 9.1 and 9.2 show typical infiltration patterns of water, in various soils, for different degrees of initial saturation. A wide range of infiltration patterns is exhibited by these soils, from strong preferential flow to relatively stable, uniform flow (e.g., Fig. 9.1, lower plot).

A key control on preferential flow is the heterogeneity of soils in the vadose zone. This heterogeneity is omnipresent, at scales ranging from the pore level to

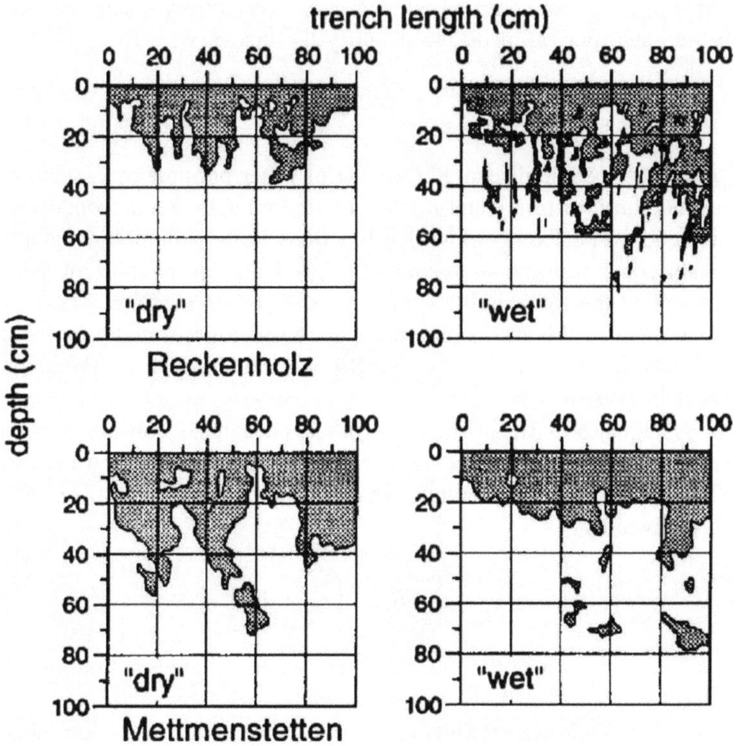

Fig. 9.2 Patterns of vertical water flow in a porous medium, as a function of the initial water content. Shown here are two patterns of water infiltration, in two different "dry" and "wet" soils, following infiltration of 40 mm of water containing Brilliant Blue FCF as a dye applied to the ground surface (Flury et al. 1994). Reproduced by permission of American Geophysical Union. Flury et al. (1994), doi:10.1029/94WR00871. Copyright 1994 American Geophysical Union

the scale of an entire geological layer or aquifer. Heterogeneity of natural porous media is affected by the various features just mentioned (including macropores, fissures, fractures, roots, and wormholes), as well as by variability in mineral composition and mineral distribution at greater depth (recall Chap. 1). Fracture networks dominate preferential flows in clay soils. Also, the soil region (top meters of the subsurface zone) is particularly susceptible to preferential flow caused by "biopores."

Notwithstanding the natural heterogeneity of the subsurface, we can usefully consider "homogeneous" (bulk, effective) descriptions for at least some problems, especially for water flow (but less so for contaminant migration; see Sect. 10.1). Therefore, two basic approaches to modeling generally are used to describe and quantify flow and transport: continuum-based models and pore-network models. We discuss each of these here.

The traditional, continuum-based approach uses Darcy's law, modified for partially saturated porous media, to quantify the flux of water:

$$q = -K(\theta)dh(\theta)/dx \tag{9.1}$$

where q denotes specific discharge (volume of water per unit cross-sectional area of porous medium, per time), and $K(\theta)$ and $h(\theta)$ are the hydraulic conductivity and hydraulic head, respectively, which are functions of the volumetric water content θ. The water content (or moisture content), θ, is defined as the ratio of the volume of pore water to the total (bulk) volume of the porous medium. The functional dependence of K on θ is based on empirical considerations, and a range of functional forms have been suggested (e.g., Brooks and Corey 1966; Mualem 1976; van Genuchten 1980).

Richards' equation (Richards 1931), based on a mass conservation balance, together with Eq. 9.1, can be used to describe the transient flow of water through a partially saturated porous medium. In one dimension (vertically), Richards' equation is given as

$$\frac{\partial \theta}{\partial t} = \frac{\partial}{\partial z}\left[K_z(\theta)\left(\frac{\partial h}{\partial z} + 1\right)\right] \tag{9.2}$$

This equation incorporates bulk parameters and provides a continuum-level, averaged quantification of flow.

Pore-scale network models (including percolation models), on the other hand, are best suited for analyzing fluid and contaminant distribution and movement within pores and clusters of pores. Such models are particularly effective for capturing fingering processes during infiltration, drainage, and imbibition. These models conceptualize a porous material as an arrangement of pore throats (tubes or cones) connected by pores (spheres); extensive discussion of these networks is given in Berkowitz and Ewing (1998). Description of flow patterns is then achieved by solving for the fluid advance within these networks, using local scale Darcy's law (Eq. 9.1), for either fully (e.g., Margolin et al. 1998) or partially saturated systems (e.g., Blunt and Scher 1995). In both cases, flow patterns demonstrating strong fingering and preferential flow arise naturally.

9.2 Flow Through the Capillary Fringe

The capillary fringe (CF), which is the region where saturation increases quickly, near the water table, deserves special consideration. Flow through the vadose zone commonly is characterized by a mean vertical flow, with substantial temporal and spatial variability. Within this zone, water pressures are less than atmospheric, and the focus is on issues of fingering of water fronts and characterization of the relationships among moisture content, matric potential, and relative permeability.

In contrast, flow in the groundwater zone generally is characterized by pores that essentially are saturated with water. The water within the pores typically is at pressures greater than atmospheric. Flow no longer is predominantly vertical but reacts to local or regional recharge and discharge fluxes, including natural fluxes such as recharge from precipitation or discharge to surface water.

The CF constitutes a relatively narrow zone that is the critical connection between the vadose zone and aquifers. Sediments immediately below the water table, for example, are characterized by pore-water pressures greater than atmospheric pressure and nearly full saturation by the water phase in most portions of the medium. Also present are local zones of entrapped air in the form of bubbles or small inclusions of sediments at low moisture content. The region immediately above the water table contains water that essentially saturates the pores, although the pore-water pressures are less than atmospheric. The CF, combined with the region immediately below the water table, may affect, far more significantly than usually is assumed, the natural geochemical and microbial conditions present in the region of transition from the vadose zone to the saturated groundwater zone.

The literature demonstrates the potential impact of flow within the CF on hydraulics in the subsurface. Most studies of the CF have focused on responses to pumping from groundwater wells, hydrologic response of streams, behavior of seepage faces, and exfiltration to excavations (e.g., Abdul and Gillham 1984; Gillham 1984; Mixon 1984; Nwankwor et al. 1992; Zhang et al. 1999). The CF also has been shown to affect groundwater response to water table dynamics, particularly in response to infiltration events (e.g., Rosenberry and Winter 1997; Nielsen and Perrochet 2000). However, to date, little information is available on the impact, at the local scale, of the CF on exchange of water across the water table or the influence of heterogeneity of grain size on behavior at the microscale within the CF.

The dynamics of water flow therefore are a combination of those governing flow in the partially saturated zone (essentially vertical, downward flow) and flow in saturated zones (aquifers), which can be fully three-dimensional. In general, the modeling approaches mentioned in Sect. 9.1 are applicable here—continuum models and pore-scale network models—although detailed quantification of flow and transport in the CF has received only limited attention.

Silliman et al. (2002) use a fully saturated–unsaturated model of flow within the combined vadose zone, CF, and saturated groundwater zone to reproduce patterns of tracer transport in a laboratory flow cell. Extrapolation to the field scale indicates that CF flow, while of local importance, contributes insignificantly to regional volumetric flow. This was demonstrated quantitatively by Silliman et al. (2002), who simulated horizontal flux through a coupled saturated and partially saturated medium, for a series of (homogeneous) sediment types ranging from sand to clay. The total horizontal flux above the water table was estimated and compared to the flux observed below the water table for aquifer thicknesses of 1–100 m. The calculated percentages of total horizontal flux in the CF, relative to the region below the water, were as high as $\sim$6 % for a 1-m-thick loam aquifer and as low as $\sim$0.04 % for a 100-m-thick sand aquifer. However, lateral flow in the CF can be significant at local scales and in the presence of aquifer heterogeneities. Moreover, even a relatively minimal contribution to flow in the CF can translate into significant transport processes within the CF.

An alternative to continuum-scale models is offered by network models. Network-scale models are particularly useful for investigating small-scale (or pore-scale) flow dynamics in porous media, including processes such as drainage and imbibition. Ronen et al. (1997) used a detailed network model by Blunt and Scher (1995) to capture key features of the CF. The model accounted for capillary-controlled water displacement processes in the presence of a gravitational field. Structures generated by the numerical simulations illustrated the irregularity of the upper "surface" of the CF and regions of trapped air within the water phase above the water table. Moreover, analysis of horizontal cuts through the three-dimensional CF structure demonstrated that "islands" of water and trapped air can coexist in the CF; connections among islands of water permit horizontal flow in the CF. In fact, the simulations indicated that island coverage of $\sim$50 % leads to a connected phase, in qualitative agreement with field measurements.

Berkowitz et al. (2004) suggest that physical heterogeneity in the sediments within the vicinity of the water table can lead to (1) increased flow and exchange of chemical constituents between the CF and the underlying saturated zone; (2) preferential transport of chemicals moving into the CF during infiltration events; (3) enhanced horizontal chemical flux above the water table, providing an opportunity for significant horizontal motion of contaminants without the possibility for sampling, e.g., via groundwater piezometers; and (4) increased contact between gas (both trapped and free flowing) and water phases in the region bounding the water table. Such phenomena may drive a number of transport and reaction processes, including (1) nutrient and oxygen availability to microbes in both the CF and the upper portion of the groundwater zone, (2) geochemical alteration of the CF and the shallow regions of the groundwater zone, and (3) delivery of volatiles within the groundwater zone into contact with the air phase in the vadose zone.

References

Abdul AS, Gillham RW (1984) Laboratory studies of the effects of the capillary fringe on streamflow generation. Water Resour Res 20:691–698

Berkowitz B, Ewing RP (1998) Percolation theory and network modeling applications in soil physics. Surv Geophys 19:23–72

Berkowitz B, Silliman SE, Dunn AM (2004) Impact of the capillary fringe on local flow, chemical migration, and microbiology. Vadose Zone J 3:534–548

Blunt MJ, Scher H (1995) Pore-level modeling of wetting. Phys Rev E 52:6387–6403

Brooks RH, Corey AT (1966) Properties of porous media affecting fluid flow. J Irrig Drain Div Am Soc Civil Eng 92:61–88

Flury M, Fluhler H, Jury W, Leuenberger J (1994) Susceptibility of soils to preferential flow of water: a field study. Water Resour Res 30:1945–1954

Gillham RW (1984) The effect of the capillary fringe on water-table response. J Hydrol 67:307–324

Lenormand R (1990) Liquids in porous media. J Phys 2:SA79–SA88

Margolin G, Berkowitz B, Scher H (1998) Structure, flow, and generalized conductivity scaling in fracture networks. Water Resour Res 34:2103–2121

Mixon FO (1984) Saturated and capillary fringe ground-water behavior near an excavation. Ground Water 26:148–155

Mualem Y (1976) Hysteretical models for prediction of the hydraulic conductivity of unsaturated porous media. Water Resour Res 12:1248–1254

Nielsen P, Perrochet P (2000) Watertable dynamics under capillary fringes: experiments and modelling. Adv Water Resour 23:503–515

Nwankwor GI, Gillham RW, van der Kamp G, Akindunni FF (1992) Unsaturated and saturated flow in response to pumping of an unconfined aquifer—field evidence of delayed drainage. Ground Water 30:690–700

Richards LA (1931) Capillary conduction of liquids through porous mediums. Physics 1:318–333

Ronen D, Scher H, Blunt M (1997) On the structure and flow processes in the capillary fringe of phreatic aquifers. Transp Porous Media 28:159–180

Rosenberry DO, Winter TC (1997) Dynamics of water-table fluctuations in an upland between two prairie-pothole wetlands in North Dakota. J Hydrol 191:266–289

Silliman SE, Berkowitz B, Simunek J, van Genuchten MT (2002) Fluid flow and chemical migration within the capillary fringe. Ground Water 40:76–84

van Genuchten MT (1980) A closed-form equation for predicting the hydraulic conductivity of unsaturated soils. Soil Sci Soc Am J 44:892–898

Vizika O, Payatakes AC (1989) Parametric experimental-study of forced imbibition in porous media. Physicochem Hydro 11:187–204

Zhang H, Barry DA, Hocking GC (1999) Analysis of continuous and pulsed pumping of a phreatic aquifer. Adv Water Resour 22:623–632

Chapter 10
Transport of Passive Contaminants

A tremendous amount of research has been devoted to quantifying and modeling transport processes in the vadose zone, with readily available scientific literature (journals and textbooks) extending over the last half century. Modeling is used to quantify the dynamic redistribution of chemicals along the near-surface and deeper-subsurface profile, which often also is subject to reactive chemical processes including sorption, dissolution or precipitation, and volatilization.

There are fundamental differences in the conceptual approaches to quantification and model application between water flow and contaminant transport situations. While bulk and effective parameters can be very useful for quantifying water flow—especially at the field scale, where heterogeneity effects can be extreme and extend to geological features with controlling lengths from meters to kilometers—treating contamination in terms of "averages" is less meaningful. To illustrate this argument, consider that, in terms of water flow, the average (vertical, downward) volumetric flow in a partially saturated soil column can be sufficient to estimate water recharge to the water table. However, in terms of contaminant migration, there usually is a complex interaction of mechanisms as contaminants are transported by advection, diffusion, and dispersion (see Sect. 10.1) and as contaminants interact geochemically with the solid matrix (and possibly with other dissolved constituents). Together with the finger flow phenomena discussed in Chap. 9, it is questionable if one can even define an "average arrival time" of a contaminant to the water table. In terms of water quality, the arrival of only a few percent of the total amount of released contaminant to the water table is sufficient to make the water unpotable; in this context, average contaminant arrival time is of less relevance.

In this chapter, we examine the various mechanisms that influence chemical redistribution in the subsurface and the means to quantify these mechanisms. The same basic principles can be applied to both saturated and partially saturated porous media; in the latter case, the volumetric water content (and, if relevant, volatilization of NAPL constituents into the air phase) must be taken into account. Also, such treatments must assume that the partially saturated zone is subject to an "equilibrium" (steady state) flow pattern; otherwise, for example, under periods of heavy infiltration, the volumetric water content is both highly space and time dependent. When dealing with contaminant transport associated with unstable

B. Berkowitz et al., *Contaminant Geochemistry*,
DOI: 10.1007/978-3-642-54777-5_10,

water infiltration processes, other quantification methods (e.g., using network models) can be applied. Similar alternative approaches are also appropriate for modeling displacement of immiscible contaminant phases (NAPLs) in water. We deal with these problems in Sect. 11.4.

The vast majority of literature on quantifying transport processes has been considered in the framework of laboratory experiments. Field experiments, which often display fundamental differences in transport behavior relative to laboratory experiments, are inevitably subject to serious uncertainties, relating to initial and boundary conditions, medium heterogeneity, and experimental control. A major aspect—and difficulty—lies in integrating laboratory and field measurements and upscaling small-scale laboratory measurements to treatment of field-scale phenomena.

Throughout the following sections, we consider mechanisms relevant to both the (partially saturated) vadose and capillary fringe zones as well as to saturated zones.

10.1 Advection, Dispersion, and Molecular Diffusion

At the scale of soil and rock pores, the two principal transport mechanisms are molecular diffusion and advection. At larger scales, transport is usually quantified in terms of an additional parameter, referred to as *mechanical dispersion.*

Molecular diffusion is a process in which chemical species move from regions of high concentration to regions of low concentration by Brownian motion. The rate of movement is directly proportional to the concentration gradient, normal to the direction of movement. Distribution of chemicals in the near surface and subsurface is not uniform, so that molecular diffusion is an ever-present mechanism.

In bulk water (free solution), the diffusive flux, J_d, given in units of mass per area per time, is related to the concentration by Fick's first law,

$$J_d = -D\frac{dc}{dx}, \tag{10.1}$$

where D is the diffusion coefficient in free solution, c denotes chemical concentration in the water, and dc/dx is the concentration gradient. In a porous medium, the diffusion coefficient is modified and called the *apparent diffusion coefficient.* The apparent diffusion coefficient, D^*, is smaller because the solid matrix forces ions to follow longer paths of diffusion. Usually, we write $D^* = \omega D$, where $\omega < 1$ is an empirical coefficient. Laboratory studies typically indicate that $0.01 < \omega < 0.5$ for nonsorbing ions in soils and rock materials.

In partially saturated media, the diffusion coefficient also is a function of the volumetric water content, θ. Calvet (1984) showed that the variation in soil water content influences the apparent diffusion coefficient for organic contaminants in two ways: by changing the ratio of gas diffusion of volatilizable pollutants to liquid diffusion, because the air-filled porosity is affected, and by modifying pollutant adsorption at low water contents, because of competition between organic and water molecules.

From Eq. (10.1), together with the equation of continuity, we can derive Fick's second law, otherwise known simply as the diffusion equation; in one-dimensional form, this is given by

$$\frac{\partial c}{\partial t} = D * \frac{\partial^2 c}{\partial x^2}. \tag{10.2}$$

This equation relates the temporal concentration of a diffusing chemical to its location in space. In real soil and aquifer materials, the diffusion coefficient can be affected by the temperature and properties of the solid matrix, such as mineral composition (which affects sorption, a process that can be difficult to separate from diffusion), bulk density, and critically, water content.

Advection (or convection) is the process by which chemicals are transported by the average (or bulk) water velocity. Thus, the advective flux, J_{adv}, is described simply by

$$J_{\mathrm{adv}} = qc = -c\left[K(\theta)\frac{dh(\theta)}{dx}\right], \tag{10.3}$$

where q denotes the specific discharge given by Darcy's law (Eq. 9.1). As for Eq. (10.1), the advective flux is given in units of mass per area per time, that is, the mass of chemical passing through a unit cross-sectional area of porous material per unit time. Combining the diffusive and advective fluxes in Eqs. (10.1) and (10.3), using, e.g., a volumetric mass conservation, leads to the (one-dimensional form of the) advection–diffusion equation:

$$\frac{\partial(\theta c)}{\partial t} = -\frac{\partial}{\partial x}(v\theta c) + \frac{\partial}{\partial x}\left[D * \theta\frac{\partial c}{\partial x}\right], \tag{10.4}$$

where v denotes the average water velocity.

Clearly, particularly, because of the complex topology of the three-dimensional network of pores that make up soils and aquifer materials, large-scale (larger than the pore scale) contaminant spreading also is driven by local velocity fluctuations around q, so that mechanical dispersion must be considered as well. Mechanical dispersion accounts for the local variations in water flux about the average, advective flux. It can be considered an artifact of upscaling, that is, of the use of a continuum-level averaged ("homogenized") transport equation. Considering transport processes at the local or pore-scale level, this parameter often is neglected, which is well justified if the advective flux is characterized at a sufficiently high resolution.

The traditional, advection–dispersion equation, a generalization of Eq. (10.4), is then written in one-dimensional form, as

$$\frac{\partial(\theta c)}{\partial t} = -\frac{\partial}{\partial x}(v\theta c) + \frac{\partial}{\partial x}\left[D_k\theta\frac{\partial c}{\partial x}\right], \tag{10.5}$$

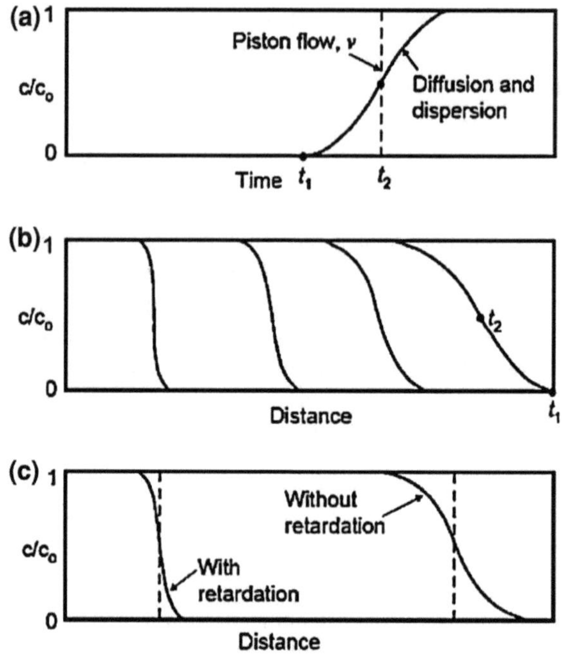

Fig. 10.1 Effect of different mechanisms on behavior of contaminants advancing through a column of porous material; the relative concentration is given by c/c_o: **a** temporal breakthrough curves at the column outlet, showing effects of diffusion and dispersion; **b** spatial concentration profiles along the column, at different times; **c** spatial concentration profiles illustrating effects of retardation caused by contaminant absorption

where D_h is now the hydrodynamic dispersion, and

$$D_h \equiv \alpha_L v + D*, \tag{10.6}$$

with α_L being the longitudinal dispersivity coefficient. This formulation relies on the fundamental assumption that mechanical dispersion is a Fickian process, which can be described similarly to molecular diffusion (i.e., using Eq. 10.1).

Note that Eq. (10.5) is written to allow the velocity to vary as a function of location; typical application of the advection–dispersion equation assumes the velocity and the hydrodynamic coefficients to be constant. Moreover, the time dependence of these parameters arises when flow (infiltration) is unsteady or transient; in these cases, the contact time between contaminants and the solid matrix (and any immobile water within it) is too short to allow an equilibrium to be reached.

Figure 10.1 shows the behavior of contaminant breakthrough curves (concentration as a function of time, at the column outlet) as well as spatial concentration profiles, in a water-saturated column initially at concentration $c = 0$, with a (constant) inlet concentration c_o. Piston flow, shown in Fig. 10.1a, refers to the hypothetical situation of purely advective transport of a contaminant, given by the fluid velocity v, in the absence of diffusion, dispersion, and any reactive mechanisms. In reality, diffusion and dispersion (and often reactive mechanisms) are always present, so that breakthrough curves generally take on an "S" shape with varying degrees of elongation of the early arrival times and late time tails

(Fig. 10.1a). Here, the average contaminant advance, given by the point at which the relative concentration $c/c_o = 0.5$, corresponds to the average water velocity, v.

Figure 10.1b shows spatial concentration profiles within the column, at different "snapshots" in time. Note that the profile spreads with increasing travel distance (and thus with increasing time). The positions (distances) noted by the points t_1 and t_2 correspond to the times t_1 and t_2 shown in Fig. 10.1a. The effect of retardation, caused by the additional mechanism of chemical adsorption, is shown in Fig. 10.1c; both the average velocity of the contaminant (corresponding to the point $c/c_o = 0.5$) and the degree of spreading around this value are reduced. This behavior is discussed further in Sect. 11.1.

While the advection–dispersion equation has been used widely over the last half century, there is now widespread recognition that this equation has serious limitations. As noted previously, laboratory and field-scale application of the advection–dispersion equation is based on the assumption that dispersion behaves macroscopically as a Fickian diffusive process, with the dispersivity being assumed constant in space and time. However, it has been observed consistently through field, laboratory, and Monte Carlo analyses that the dispersivity is not constant but, rather, dependent on the time or length scale of measurement (Gelhar et al. 1992), leading to what has been referred to as the *scale effect.* The spatial dependence of transport coefficients is usually attributed to the existence of hydraulic conductivity fields with coherence (correlation) lengths varying over many scales. We observe that chemical transport at the field scale now is often recognized to be non-Fickian, so that one can argue that the "anomalous," non-Fickian transport actually is the norm rather than the exception.

Therefore, Eq. 10.5 is limited in its applicability, as are variations of this equation such as the mobile–immobile one (see Sect. 10.2). We discuss non-Fickian transport in detail in Sect. 10.3.

10.2 Preferential Transport

As noted in Sect. 10.1, heterogeneities play a dominant role in the migration of contaminants in the subsurface. Nonuniform, preferential patterns of flow and transport are ubiquitous. It is important to recognize that, at the field scale, contaminant movement generally is very difficult to anticipate. In natural soils and aquifer materials, macropores, soil cracks and aggregates, fissures, solution channels, root paths, and wormholes, as well as variable mineral composition (e.g., clay aggregates and lenses) all serve to make nonuniform patterns of flow highly ubiquitous (recall also Chap. 9). Moreover, soils that have relatively high clay contents are subject to shrinkage and cracking under repeated cycles of wetting and drying. During contaminant infiltration, cracks and larger pores control the water distribution and limit reaction surfaces to smaller areas. As a consequence, downward migration of contaminants from the land surface usually is highly nonuniform, influenced strongly by movement through preferential pathways.

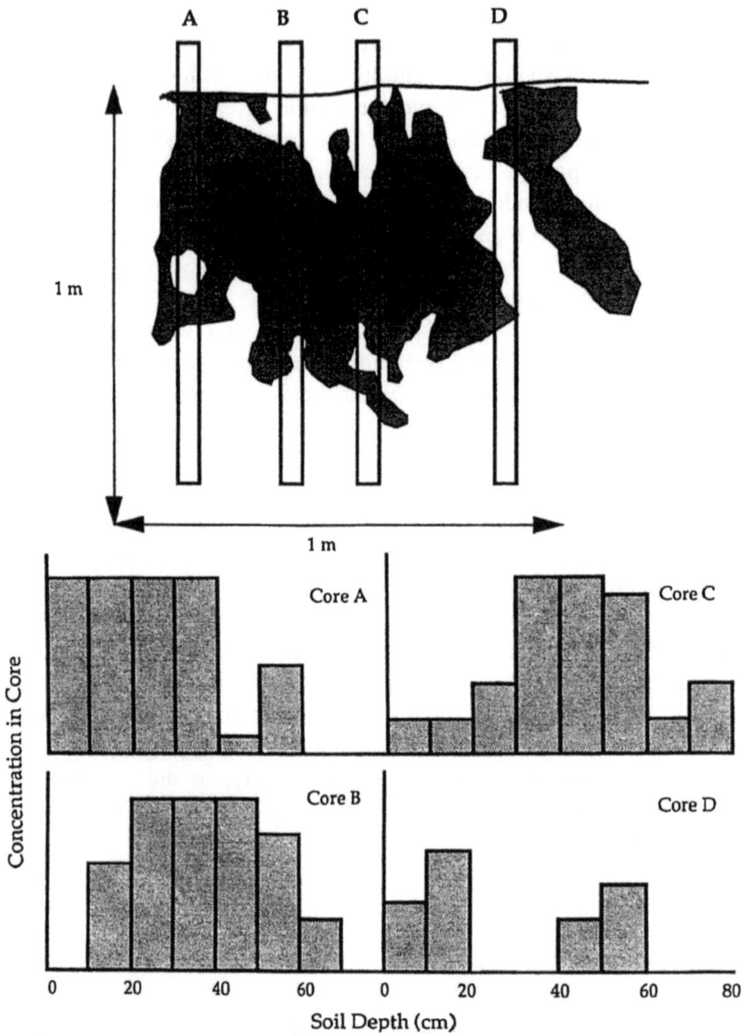

Fig. 10.2 Pattern of contaminant advance (water containing dye) in a partially saturated soil. Note the high degree of spatial variability in the pattern, both laterally and with depth. Reprinted from Ghodrati and Jury (1992). Copyright 1992 with permission of Elsevier

Figure 10.2 shows a typical pattern of contaminant migration through the upper meter of a partially saturated soil.

To quantify such transport, the advection–dispersion equation, which requires a narrow pore-size distribution, often is used in a modified framework. Van Genuchten and Wierenga (1976) discuss a conceptualization of preferential solute transport through "mobile" and "immobile" regions. In this framework, contaminants advance mostly through macropores containing mobile water and diffuse into and out of relatively immobile water resident in micropores. The mobile–immobile model involves two coupled equations (in one-dimensional form):

$$\begin{aligned} \frac{\partial(\theta_{\mathrm{im}}c_{\mathrm{im}})}{\partial t} + \frac{\partial(\theta_m c_m)}{\partial t} &= -\frac{\partial}{\partial x}(v\theta_m c_m) + \frac{\partial}{\partial x}\left[D * \theta_m \frac{\partial c_m}{\partial x}\right], \\ \frac{\partial(\theta_{\mathrm{im}}c_{\mathrm{im}})}{\partial t} &= \varepsilon(c_m - c_{\mathrm{im}}) \end{aligned} \tag{10.7}$$

where c_{m} and c_{im}, and θ_{m} and θ_{im} denote the mobile and immobile zone concentrations and volumetric moisture contents, respectively, and ε is a mass-transfer coefficient. The mobile–immobile approach, while applied frequently, is limited by its reliance on the Fickian-based advection–dispersion equation. We consider alternatives to this approach in Sect. 10.3.

As an alternative to continuum-based treatments, and to capture the pore-scale details of such transport, a wide variety of network modeling approaches can be applied, such as those described at the end of Sect. 9.1. A detailed analysis of dispersion phenomena in porous media using network models and percolation theory is presented by Sahimi (1987) and Sahimi and Imdakm (1988), who derive a set of relationships relevant to the geometrical and hydraulic characteristics of capillary tube networks and use them to establish various expressions for describing dispersion of contaminants under various conditions and assumptions. For example, the relationship between mean square contaminant displacement and time yields different exponents, depending on the flow conditions (e.g., as characterized by the Peclet number, $\mathrm{Pe} \equiv vl/D$, where v is mean water velocity, l is the characteristic length—roughly, the mean pore size—and D is the diffusion coefficient; this dimensionless number gives the ratio of advective to diffusive forces). In particular, such random networks can be used to examine advective transport through the main conducting path, with diffusive transport into dead ends and stagnant regions, similar to the mobile–immobile conceptualization of transport (Koplik et al. 1988).

We discuss network models further in Sect. 11.4, in the context of immiscible displacements.

10.3 Non-Fickian Transport

The previous sections discussed the advection–dispersion equation and variants such as the mobile–immobile conceptualization, which are based on the key assumption that mechanical dispersion is Fickian. In other words, the advection–dispersion equation (Eq. 10.5) is strictly valid only under perfectly homogeneous flow conditions. This requires that the length scale of all heterogeneities be much smaller than the length scale of the domain. In terms of real transport problems, the contaminant migration time must be sufficiently large before the advection–dispersion equation correctly applies. In laboratory-scale columns, homogeneous

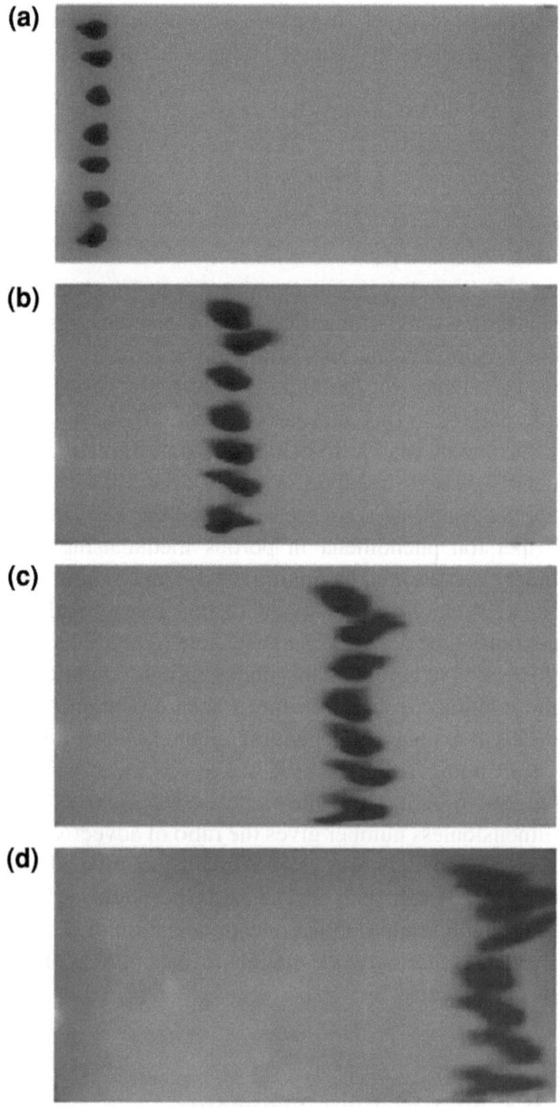

Fig. 10.3 Photographs of a homogeneous, saturated sand pack with seven dye tracer point injections being transported, under a constant flow of 53 ml/min, from *left* to *right*; times at **a** $t = 20$ **b** $t = 105$ **c** $t = 172$ **d** $t = 255$ min after injection. Internal dimensions of the flow cell are 86 cm (length), 45 cm (height), and 10 cm (width). Reprinted from Levy and Berkowitz (2003)

transport conditions generally are reached far more quickly than in large, field-scale applications, because natural heterogeneity appears on virtually all scales Jury and Fluhler (1992). As noted by Berkowitz et al. (2006), small-scale heterogeneities do not simply average out.

Even "perfectly homogeneous" soils and aquifer materials display heterogeneity, and preferential paths for water flow and tracer transport are present. For example, as shown in Fig. 10.3 and contrary to Fickian transport, contaminant plumes are not symmetrical ellipses nor are the different plumes identical to each other. Moreover, measurements of contaminant breakthrough curves in such "homogeneous," 20-cm-to-meter-length flow cells have been shown to display "anomalous" (non-Fickian) early time arrivals and late time tails (Levy and Berkowitz 2003; Cortis et al. 2004). Detailed analysis shows that the motion and spreading of these chemical plumes are characterized by distinct temporal scaling; that is, the time dependence of the spatial moments does not correspond to the normal (or Gaussian) distribution that would otherwise lead to Fickian transport.

To account for the effect of a sufficiently broad, statistical distribution of heterogeneities on the overall transport, we can consider a probabilistic approach that will generate a probability density function in space (s) and time (t), $\psi(s, t)$, describing key features of the transport. The effects of multiscale heterogeneities on contaminant transport patterns are significant, and consideration only of the mean transport behavior, such as the spatial moments of the concentration distribution, is not sufficient. The continuous time random walk (CTRW) approach is a physically based method that has been advanced recently as an effective means to quantify contaminant transport. The interested reader is referred to a detailed review of this approach (Berkowitz et al. 2006).

A variety of specific mathematical formulations of the CTRW approach have been considered to date, and network models have also been applied (Bijeljic and Blunt 2006). A key result in development of the CTRW approach is a transport equation that represents a strong generalization of the advection–dispersion equation. As shown by Berkowitz et al. (2006), an extremely broad range of transport patterns can be described with the (ensemble-averaged) equation

$$u\tilde{c}(s,u) - c_o(s) = -\tilde{M}(u)[v_\psi \cdot \nabla\tilde{c}(s,u) - D_\psi : \nabla\nabla\tilde{c}(s,u)], \tag{10.8}$$

where the Laplace transform of a function $f(t)$ is denoted by $\tilde{f}(u)$, $-c_o(s)$ represents the initial condition, and the "memory function" is defined as

$$\tilde{M}(u) \equiv \bar{t}u\frac{\tilde{\psi}(u)}{1 - \tilde{\psi}(u)}, \tag{10.9}$$

where $\bar{t}$ denotes a characteristic time. Here, it is important to recognize that the "transport velocity" v_ψ is distinct from the "average water velocity" v, whereas in the classical advection–dispersion picture, these velocities are identical. Similarly, the "dispersion," D_ψ, has a different physical interpretation than in the usual advection–dispersion equation definition. More important, a key feature of Eq. (10.8) is that it encompasses various common models, such as multirate and mobile–immobile transport equations (Eq. 10.7) for specific examples of $\tilde{M}(u)$ $(or\, \psi(s,t))$, together with other simplifications. For example, under perfectly homogeneous conditions, $\tilde{M}(u) = 1$, and Eq. (10.8) is formally equivalent to the classical advection–dispersion equation.

The CTRW approach accounts naturally for transport in preferential pathways, with mass transfer to "stagnant" and slow flow regions; the CTRW can account

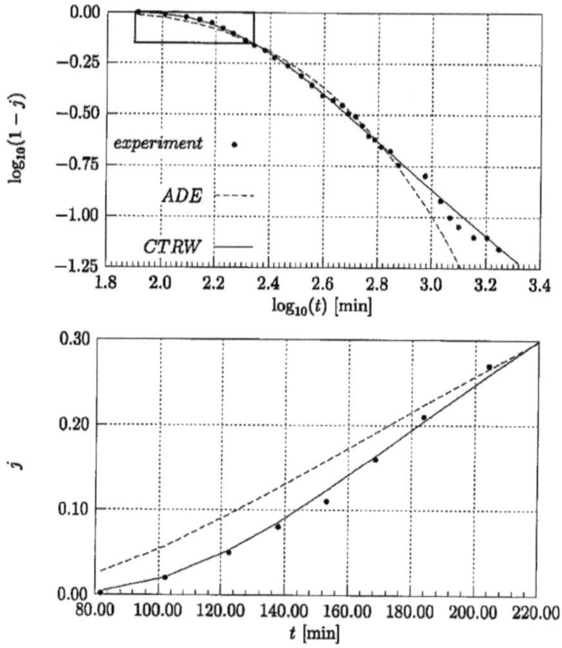

Fig. 10.4 Measured breakthrough curve of bromide with CTRW and advection–dispersion equation (ADE) fits. Here, the quantity j represents the normalized, flux-averaged concentration: (*top*) complete breakthrough curve (*bottom*) region identified by the bold-framed rectangle in the top plot. Note the difference in scale units between the plots. Pressure head $h = -10$ cm; water velocity $v = 2.82$ cm/h. The dashed line is the best advection–dispersion equation solution fit. The solid line is the best CTRW fit. (Cortis and Berkowitz 2004)

for these physical transport mechanisms, as well as other factors that influence transport of reactive contaminants, such as sorption.

Of specific interest here are the analyses by Cortis and Berkowitz (2004) of transport in partially saturated, laboratory columns. Three typical breakthrough curves from a series of miscible displacement experiments in partially saturated soils, presented by Nielsen and Biggar (1962) and Jardine et al. (1993), were reanalyzed using the CTRW approach. Jardine et al. (1993) measured breakthrough curves on undisturbed cylindrical soil columns (8.5 cm diameter, 24 cm length). The soil columns were saturated with 0.05 M $CaCl_2$ from the bottom then allowed to drain. Bromide was used as the nonreactive, passive tracer. Cortis and Berkowitz (2004) reexamined three breakthrough curves obtained from three different degrees of saturation. As shown in Fig. 10.4, the CTRW solutions were found to reproduce the breakthrough behavior far more effectively than the advection–dispersion equation solution. Nielsen and Biggar (1962) reported systematic deviations in the calculated parameter values using the advection–dispersion equation from the experimental data, which displayed non-Fickian transport behavior. Again, the CTRW solutions were found to reproduce the breakthrough behavior far more effectively.

Another application of the CTRW approach is to quantifying transport of contaminants through a catchment, as they interact between surface water (streams) and the underlying porous medium. The distribution of travel times determines the retention of pollutants until they eventually are released to a river or lake, and it controls the transport of chemicals in subsurface hydrological systems. The nature of this distribution determines the expected ecological impact of contaminants. Long-term, time-series measurements of chloride, a natural passive tracer, in rainfall and runoff in catchments, were reported by Kirchner et al. (2000). The empirically derived distribution of tracer travel times was found to follow a power law, indicating low-level contaminant delivery to streams over a very long time. Scher et al. (2002) used a CTRW framework to develop a model that explains such tracer transport in a catchment with surface and subsurface interactions, and in particular, the extremely long chemical retention times in catchments.

References

Berkowitz B, Cortis A, Dentz M, Scher H (2006) Modeling non-Fickian transport in geological formations as a continuous time random walk. Rev Geophys 44, RG2003. doi:10.1029/2005RG000178

Bijeljic B, Blunt MJ (2006) Pore-scale modeling and continuous time random walk analysis of dispersion in porous media. Water Resour Res 42:W01202. doi:10.1029/2005WR004578

Calvet R (1984) Behavior of pesticides in the unsaturated zone: adsorption and transport phenomenon. In: Yaron B, Dagan G, Goldchmid J (eds) Pollutants in porous media. Springer, Berlin, pp 143–151

Cortis A, Berkowitz B (2004) Anomalous transport in "classical" soil and sand columns. Soil Sci Soc Amer J 68:1539–1548 Erratum: 69:285 (2005)

Cortis A, Chen Y, Scher H, Berkowitz B (2004) Quantitative characterization of pore-scale disorder effects on transport in "homogeneous" granular media. Phys Rev E 70:041108. doi:10.1103/PhysRevE.70.041108

Gelhar LW, Welty C, Rehfeldt KR (1992) A critical review of data on field-scale dispersion in aquifers. Water Resour Res 28:1955–1974

Ghodrati M, Jury WA (1990) A field study using dyes to characterize preferential flow of water. Soil Sci Soc Am J 54:1558–1563

Ghodrati M, Jury WA (1992) A field study of the effects of soil structure and irrigation method on preferential flow of pesticides in unsaturated soil. J Contam Hydrol 11:101–125

Jardine PM, Jacobs GK, Wilson GV (1993) Unsaturated transport processes in undisturbed heterogeneous porous media. I. inorganic contaminants. Soil Sci Soc Am J 57:945–953

Jury WA, Fluhler H (1992) Transport of chemicals through soil: mechanisms, models and field applications. Adv Agron 47:142–202

Kirchner JW, Feng X, Neal C (2000) Fractal stream chemistry and its implications for contaminant transport in catchments. Nature 403:524–527

Koplik J, Redner S, Wilkinson D (1988) Transport and dispersion in random networks with percolation disorder. Phys Rev A: Math Gen 37:2619–2636

Levy M, Berkowitz B (2003) Measurement and analysis of non-Fickian dispersion in heterogeneous porous media. J Contam Hydrol 64:203–226

Nielsen DR, Biggar JW (1962) Miscible displacement in soils. III. Theoretical considerations. Soil Sci Soc Am Proc 26:216–221

Sahimi M (1987) Hydrodynamic dispersion near the percolation threshold: scaling and probability densities. J Phys A: Math Gen 20:L1293–L1298

Sahimi M, Imdakm AO (1988) The effect of morphological disorder on hydrodynamic dispersion in flow through porous media. J Phys A: Math Gen 21:3833–3870

Scher H, Margolin G, Metzler R, Klafter J, Berkowitz B (2002) The dynamical foundation of fractal stream chemistry: The origin of extremely long retention times. Geophys Res Lett 29, doi:10.1029/2001GL014,123

van Genuchten M, Th Wierenga PJ (1976) Mass transfer studies in sorbing porous media. I. analytical solutions. Soil Sci Soc Am Proc 40:473–480

Chapter 11
Transport of Reactive Contaminants

The previous chapter focused on the physical mechanisms (advection, diffusion, and dispersion) and the physical characteristics of the subsurface (heterogeneity) that control the dynamics of contaminant transport from the land surface to the water table. In addition, contaminants are subject to a range of chemical interactions with other (dissolved) chemical species present in the subsurface, colloids, and the porous matrix itself. The previous sections of this book dealt with such reactions, including sorption (by various types of bonding), decay, degradation, complexation, precipitation, dissolution, and volatilization as well as interactions (and transport) with migrating colloids. These reactions thus influence—and are influenced by—advective, diffusive, and dispersive transport mechanisms. To include the effects of these reactions in quantifying the dynamics of contaminant transport, additional terms can be included in the transport equations surveyed in Chap. 10. In most cases, the resulting transport equations contain relatively simple terms that account for chemical species loss from or entry to the aqueous solution.

It should be emphasized that, to date, the ability to quantify the complex chemical reaction phenomena that occur in the subsurface and also integrate the variability in flow behavior caused by natural heterogeneity and fluctuating boundary (land surface) conditions remains very limited. As a consequence, developing and improving the predictive capabilities of models are the areas of active research.

11.1 Contaminant Sorption

Sorption of contaminants can be included in the advection–dispersion equation by introducing a retardation factor:

$$\frac{\partial}{\partial t}\left(c + \frac{\rho_b S}{\theta}\right) = \frac{\partial}{\partial x}(vc) + \frac{\partial}{\partial x}\left[D_h \frac{\partial}{\partial x}\right], \tag{11.1}$$

B. Berkowitz et al., *Contaminant Geochemistry*,
DOI: 10.1007/978-3-642-54777-5_11, © Springer-Verlag Berlin Heidelberg 2014

where ρ_b denotes bulk mass density of the porous medium and S is the adsorbed phase concentration. The sorbed phase S subsequently is quantified in terms of an isotherm, as discussed in Chap. 5. Usually, the simple linear Freundlich isotherm is assumed to be valid, so that $S = K_d c$, where K_d is the distribution coefficient (or sorption coefficient). More complicated adsorption–desorption terms can be incorporated, for example, for the case of different forward and reverse sorption rates. Then, Eq. (11.1) can be written in terms of a retardation coefficient, R, in the form

$$R\frac{\partial c}{\partial t} = -\frac{\partial}{\partial x}(vc) + \frac{\partial}{\partial x}\left[D_h\frac{\partial}{\partial x}\right], \tag{11.2}$$

where

$$R \equiv 1 + \frac{\rho_b K_d}{\theta}. \tag{11.3}$$

Thus, the travel time for an adsorbed chemical, t_a, can be related to the travel time for a mobile, conservative (nonsorbing) chemical, t_a, by

$$t_a = (1 + \rho_b K_d/\theta)t_m \equiv Rt_m. \tag{11.4}$$

Equation (11.4) also can be written as $v_a = vR$, where v_a denotes the average velocity of a sorbing contaminant. Thus, the concentration profile of a sorbing contaminant is retarded, relative to a nonsorbing contaminant, as shown schematically in Fig. 10.1c.

Sorption can be included in the more general CTRW transport equation discussed in Sect. 10.3. Margolin et al. (2003) show the relation between macroscopic transport behaviors of passive and sorbing (reactive) tracers in heterogeneous media. In the framework of CTRW, they formulate the sorption process using a "sticking" rate and a "sticking" time distribution and derive a relation between the distributions of the sorbing and nonsorbing tracer in terms of these quantities. Alternatively, sorption can be included directly in the definition of the transition time distribution, $\psi(s, t)$. This approach is valid when chemical sorption acts only as a relatively gentle "modifying" mechanism on the overall transport, in which the long-time tail in a breakthrough curve is further delayed. In other physical situations, however, transport may be governed by two highly distinct rate spectra for transport, such as reactive tracers undergoing (slow) sorption (adsorption–desorption) during migration in a heterogeneous advective flow field (e.g., Starr et al. 1985). Berkowitz et al. (2008) show that this "two-scale" behavior can be quantified by appropriate modification of the governing transport equation.

Regardless of the transport equation considered, the major effect of sorption on contaminant breakthrough curves is to delay the entire curve on the time axis, relative to a passive (nonsorbing) contaminant (Rubin et al. 2012). Because of the longer residence time in the porous medium, advective–diffusive–dispersive interactions also are affected, so that longer (non-Fickian) tailing in the breakthrough curves is often observed.

11.2 Colloids and Sorption on Colloids

In previous chapters, we discussed the role of colloids and colloid-like materials, as both carriers of adsorbed chemicals and contaminants in their own right. It is well recognized that contaminants farther can interact with and adsorb onto migrating colloids, and thus advance much farther or deeper into the subsurface than would be expected if the role of colloids were ignored (e.g., Saiers and Hornberger 1996). For example, an organic colloid particle can act as a sorbent for a neutral organic molecule, thus facilitating advective transport of the neutral molecule.

In this situation, transport equations similar to those discussed previously can be applied. For example, by assuming sorption to be essentially instantaneous, the advective–dispersion equation with a reaction term (Saiers and Hornberger 1996) can be considered. Alternatively, CTRW transport equations with a single $\psi(s, t)$ can be applied or two different time spectra (for the dispersive transport and for the distribution of transfer times between mobile and immobile—diffusion, sorption—states can be treated; Berkowitz et al. 2008; Bijeljic et al. 2011).

The transport of colloids in porous media is usually considered to be controlled by four mechanisms: advection, dispersion, straining, and physicochemical particle–surface interactions (McDowell-Boyer et al. 1986; Ryan and Elimelech 1996); see also Sect. 5.7.3. *Straining* refers to the permanent physical trapping of colloids in narrow pore throats and is a key mechanism, particularly in porous media with grain diameters smaller than 20–50 times the colloid diameter (McDowell-Boyer et al. 1986). Physicochemical interactions with the porous medium occur when colloids approach grain surfaces due to pore-scale diffusion, sedimentation, or inertial forces. These interactions lead to permanent or temporary attachment to the porous medium solid phase and are controlled by electrostatic forces (including London–van der Waals forces). Such electrostatic forces are explained in terms of electric double-layer theory (Israelachvili 1991). Particles deposit in the secondary energy minimum of the electric double layer at the grain surface. Moreover, deeper secondary minima closer to the solid surfaces occur at higher ionic strength (Redman et al. 2004). Thus, changes in solution chemistry promote colloid mobilization (deposition) by altering the double-layer potential energy.

The transport behavior of colloids commonly is modeled by colloid filtration theory (CFT) (Yao et al. 1971), which is based on extension of the common advection–dispersion equation. The one-dimensional advection–dispersion–filtration equation is written

$$\frac{\partial c}{\partial t} = -v\frac{\partial c}{\partial x} + \frac{\partial}{\partial x}\left[D_h \frac{\partial c}{\partial x}\right] - \lambda v c \tag{11.5}$$

where λ is known as the filtration coefficient, which is assumed to be constant in time and space. The CFT allows determination of λ from the physical properties of the colloid and the porous medium:

$$\lambda = \frac{3}{2}(1 - n)\frac{L}{d_c}\alpha_c \eta, \tag{11.6}$$

where n is the porosity, L is the length of the porous medium column, d_c is the median grain size diameter (Rajagopalan and Tien 1976), α_c is the collision efficiency (an empirical, fitting constant), and η is the collector efficiency (estimated by various means, as discussed by Logan et al. (1995) and Tufenkji and Elimelech 2004. The CFT assumes homogeneity of particles and the porous matrix and leads to an expected fast exponential concentration decay along the colloid travel path. The CFT can be modified to account for the occurrence of nonideal behavior, for example, by combining Eq. (11.5) with a model for a first-order, rate-limited adsorption–desorption process (Eq. 11.1) (Toride et al. 1995).

A major problem with these approaches, however, lies in the complexity and nonuniqueness involved with identification of parameterizations for processes of particle straining, deposition, and detachment. An alternative to CFT-based theories is given by Amitay-Rosen et al. (2005), who suggest a simple phenomenological model of particle deposition and porosity reduction that avoids these difficulties.

As noted by Cortis et al. (2006), CFT and modifications of standard filtration models invariably predict an exponential decay of the colloid concentration with distance. However, similar to the discussion of non-Fickian transport in Sect. 10.3, power-law tails in breakthrough curves are observed frequently in experiments (e. g., in the context of bacterial "biocolloids," see Albinger et al. 1994; Martin et al. 1996; Camesano and Logan 1998; Baygents et al. 1998; Bolster et al. 1998; Redman et al. 2001). Cortis et al. (2006) propose a generalized model based on CTRW theory that captures these power-law tails, together with the full evolution of breakthrough curves. The CTRW filtration model is found to be controlled by three parameters, which are related to the overall breakthrough retardation (R), the slope of the power-law tail (β), and the transition to a decay slower than t^{-1}.

11.3 Dissolving and Precipitating Contaminants

Dissolution and precipitation can occur as contaminants travel from the land surface to groundwater aquifers. These processes can affect water chemistry, and they can significantly modify the physical and chemical properties of porous media (Lasaga 1984; Palmer 1996; Dijk and Berkowitz 1998, 2000; Darmody et al. 2000). Under some conditions, large quantities of mass can be transferred between the liquid and solid mineral phases.

A number of experimental and theoretical studies analyzed the influence of dissolution processes on the physical and chemical properties of soluble porous matrices (Fogler and Rege 1986; Daccord and Lenormand 1987; Hoefner and Fogler 1988; Daccord 1987; Daccord et al. 1993). Similarly, several theoretical studies provided some understanding of the evolution of hydraulic conductivity caused by precipitation processes (Novak and Lake 1989; Novak 1993; Bolton et al. 1996, 1997, Dijk and Berkowitz 1998).

Loss or gain of dissolved chemical species from soil water by precipitation or dissolution, respectively, usually is accounted for by adding a simple sink-source term in the advection–dispersion equation:

$$\frac{\partial}{\partial t}(\theta c) = -\frac{\partial}{\partial x}(v\theta c) + \frac{\partial}{\partial x}\left[D_h \theta \frac{\partial c}{\partial x}\right] - \sum_{i=1}^{n} Q_i, \qquad (11.7)$$

where Q_i denotes sink (or source) terms that account for, for example, contaminant degradation, plant uptake, volatilization, or precipitation (dissolution). Alternatively, focusing on precipitation–dissolution at the pore scale, the hydrodynamic dispersion coefficient D_h can be replaced by the coefficient of molecular diffusion, so that Eq. (11.7) can be applied as an advection–diffusion equation.

Dissolution is a relatively simple mechanism, and so application of an advection–diffusion equation modified by a source term, or a similarly modified non-Fickian (CTRW) transport equation (Hornung et al. 2005), can effectively capture chemical transport patterns. On the other hand, the interplay between the chemical and physical aspects of mineral precipitation is more complex. In some instances, during infilling of fissures and pore spaces by mineral precipitation, porosity is reduced uniformly; in other cases, pockets of high porosity or unfilled fissures may remain. In addition to the chemical kinetics of precipitation, the flow dynamics and transport also are important factors in determining the resulting patterns of porosity and mineral deposition (Dijk and Berkowitz 1998). Note also that modeling approaches similar to those for treating precipitating contaminants can be employed to account for degradation and volatilization of contaminants in water.

For a given mineral to precipitate, a solution must be oversaturated with respect to the mineral being deposited (i.e., the ion product must exceed the solubility product), and the precipitation process must be kinetically favorable. The most "conventional" model for describing the deposition of minerals within rock formations maintains that supersaturated fluids pass through a porous or fractured rock, precipitating minerals along the way. However, recent experimental studies demonstrate that a number of difficulties are inherent in such a simplified picture of mineralization. Lee and Morse (1999) and Hilgers and Urai (2002) found that when supersaturated fluids flowed through artificial fractures, most of the mineral deposition occurred within several centimeters of the inlet.

Fluids become supersaturated, and thus, mineral deposition occurs, by three main mechanisms: (1) changes in fluid pressures and temperatures during flow through the porous medium, (2) dissolution of a particular mineral in the matrix that results in the fluid being supersaturated with respect to another mineral (e.g., dissolution of calcite and precipitation of gypsum; Singurindy and Berkowitz 2003), and (3) mixing-induced supersaturation. This third mechanism occurs when two initially saturated or undersaturated fluids of different chemical compositions or temperatures are mixed, resulting in a new solution that is oversaturated. While widely discussed in the geochemical literature, only limited reference to mixing-induced precipitation has been made in hydrological studies.

In addition to considering the mode by which supersaturation is reached, the physical changes in the rock resulting from precipitation also crucially affect the dynamics of precipitation. Porosity can be reduced by crystal growth, and this can have an effect on additional physical parameters in the porous matrix, such as the specific surface area. Both porosity and specific surface area are important parameters in subsurface systems, as they determine to a large extent both the permeability of the rock and the chemical kinetics of mineral precipitation. Therefore, both porosity and specific surface area must be incorporated into mass conservation and transport equations. Changes in porosity can be related to the volumetric changes in the amount of mineral precipitated in a relatively straightforward manner, but changes in specific surface area are more difficult to quantify.

One approach to quantitatively relate changes in porosity to changes in specific surface area is to assume that the porous medium is composed of spherical grains. The resulting specific surface area per unit volume of rock $\bar{s}$, is related to the porosity, ϕ, by the relationship (Lichtner 1988)

$$\bar{s} = \bar{s}_o \left(\frac{1 - \phi}{1 - \phi_o} \right)^{2/3}, \tag{11.8}$$

where $\bar{s}_o$ is the specific surface area of the rock at a porosity of ϕ_o. Alternatively, if it is assumed that the pores can be approximated by spheres, the specific surface area can be expressed by Kieffer et al. (1999)

$$\bar{s} = \bar{s}_o \left(\frac{\phi}{\phi_o} \right)^{2/3}. \tag{11.9}$$

Therefore, in the first model, the specific surface area increases with decreasing porosity, while in the second, the opposite relationship is specified. While some attempts have been made to experimentally verify these models in individual rock types (Kieffer et al. 1999; Jové Colon et al. 2004), the data concerning a wide range of rocks and precipitation–dissolution reactions remain limited.

Emmanuel and Berkowitz (2005) examined the dynamics and patterns of changing porosity during mixing-induced precipitation, using a two-dimensional numerical model that simulates mixing-induced precipitation in both homogeneous and heterogeneous porous media. The role of specific surface area also was explored. The precipitation of a and b in equimolar amounts to form a solid ab was considered, a scenario that ensures that $R_a = R_b$. The sink term can then be assumed to have the generic form often valid for near-equilibrium conditions (Morse and Arvidson 2002):

$$R = -\bar{s}K_r\left(\frac{C_aC_b}{K_{sp}} - 1\right)^n, \tag{11.10}$$

where K_r is a kinetic rate coefficient, K_{sp} is the solubility product of the solid *ab* , and *n* is an empirical reaction order. The rate of change of porosity can be derived from the sink term by dimensional analysis. As *R* is defined as the number of moles precipitated per unit time per unit volume of fluid, multiplying by both porosity and molar volume (M_v) yields the volumetric rate of change of precipitated material per unit volume of bulk matrix:

$$\frac{\partial \phi}{\partial t} = \phi M_v R. \tag{11.11}$$

Equation (11.10) is valid for near-equilibrium systems, so that it can be applied to mass-transfer-limited processes, that is, systems in which the large-scale rate of change of porosity is limited by the solute flux rather than the reaction kinetics.

Application of this model showed that the complete reduction in porosity near the inlet does not lead to clogging of a system: nonsupersaturated fluids are able to flow around closed regions and into more permeable areas where mixing and subsequent deposition can occur. Thus, mixing-induced precipitation is expected to reduce porosity in concentrated regions rather than lead to evenly precipitated material throughout the porous domain.

11.4 Transport of Immiscible Liquids

Petroleum products, synthetic organic solvents, and other toxic organic compounds dissolved in organic solvents, generally referred to as nonaqueous phase liquids (NAPLs), are a key class of contaminants in the subsurface. These compounds are generally insoluble or only slightly soluble in water. As a consequence, NAPLs remain as distinct liquid phases as they are transported downward from land surface to the water table; this migration is governed mostly by the density and viscosity of the NAPL. Specific contaminants often are classified as dense NAPLs (DNAPLs), such as trichloroethylene and carbon tetrachloride, or as light NAPLs (LNAPLs), such as oil and gasoline, according to their density relative to water. DNAPLs sink through the water table region, downward through the fully saturated zone, while LNAPLs remain in the capillary fringe, "floating" on the surface of the free water. A schematic illustration of LNAPL and DNAPL transport from land surface is shown in Fig. 11.1.

Figure 11.1 provides only a coarse-level picture of NAPL migration. Depending on the viscosity and density ratios, between the NAPL and the resident water, as well as the properties (physical and chemical) of the porous medium, a variety of (unstable) fingering and (stable) fronts can develop with depth. More specifically, the degree of fingering in the displacement and the actual rate of downward (and

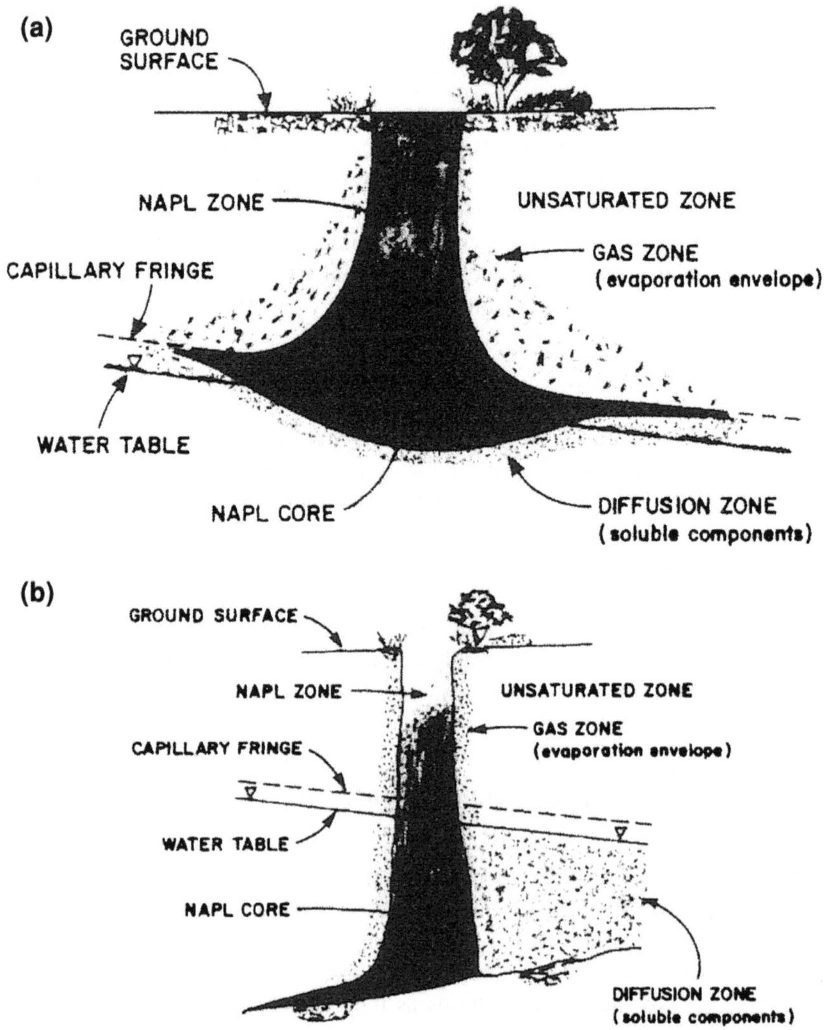

Fig. 11.1 Schematic representation of NAPL movement from land surface to the water table region. **a** LNAPL movement and **b** DNAPL movement (Abriola and Pinder 1985). Reproduced by permission of American Geophysical Union. Abriola and Pinder (1985), Copyright 1985, American Geophysical Union

lateral) migration are governed by (1) the properties (e.g., surface wetting) of the NAPL itself, (2) the NAPL density and viscosity relative to the resident pore water, (3) the actual moisture content, (4) the properties (physical, such as pore size, and chemical, such as wettability) of the porous matrix, and (5) the inlet/infiltration condition and NAPL volume (Schwille 1988). Volatilization of components in the NAPL also affects the rate and pattern of migration, as properties of surface wetting, density, and viscosity change over time. Figures 11.2 and 11.3 show migration

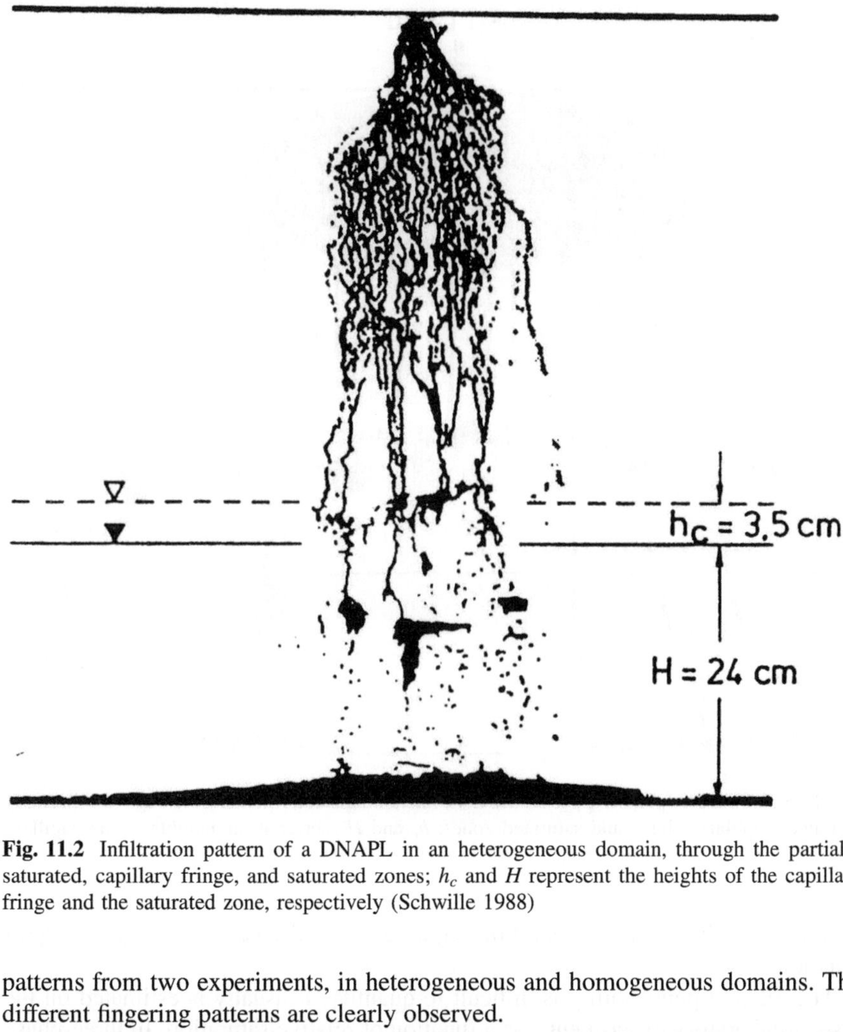

Fig. 11.2 Infiltration pattern of a DNAPL in an heterogeneous domain, through the partially saturated, capillary fringe, and saturated zones; h_c and H represent the heights of the capillary fringe and the saturated zone, respectively (Schwille 1988)

patterns from two experiments, in heterogeneous and homogeneous domains. The different fingering patterns are clearly observed.

In partially saturated media, the moisture content affects NAPL migration significantly. Abriola and Pinder (1985) suggested that the flux of a phase p (NAPL, water, or air), J_p, can be written as

$$J_p = -k_p K_{\text{sp}}^h \left[\frac{\mathrm{d}h_p}{\mathrm{d}x} + \frac{\rho_p}{p^*} \right], \tag{11.12}$$

where k_p is the relative permeability of phase p in the porous matrix, K_{sp}^h is the saturated hydraulic conductivity for phase p in the matrix, h_p is the pressure head of phase p, ρ_p is the density of the phase p, and p^* is a scaled reference phase density. Clearly, the actual pressure head in each phase depends on the fluid configuration within the pores. Flux equations for each of the three phases can be

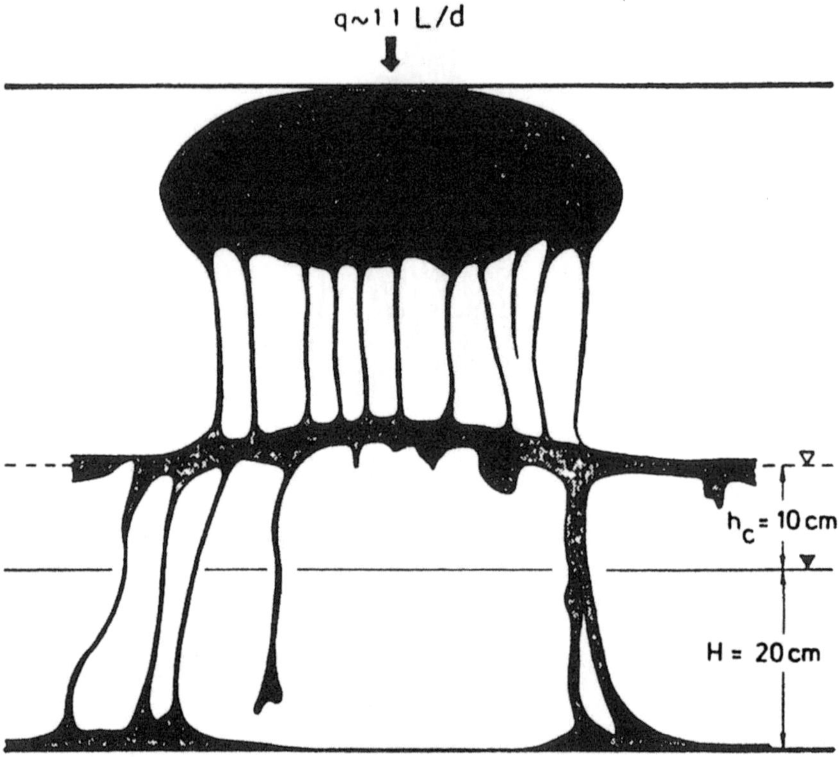

Fig. 11.3 Infiltration pattern of a DNAPL in an homogeneous domain, through the partially saturated, capillary fringe, and saturated zones; h_c and H represent the heights of the capillary fringe and the saturated zone, respectively (Schwille 1988)

combined with mass conservation equations to derive governing transport equations.

The relative permeability is difficult to quantify. It usually is estimated on the basis of laboratory experiments, as a function of relative saturation. In three-phase NAPL–water–air systems, each relative permeability is dependent on the relative saturation of each of the phases. Modeling based on empirical considerations can be employed. For example, Blunt (2000) discusses a model to estimate three-phase relative permeability, based on saturation-weighted interpolation of two-phase relative permeabilities. The model accounts for the trapping of the NAPL and air (gas) phases and predicts three-phase relative permeabilities for different porous medium and NAPL properties. However, experimental determination of two- and three-phase relative permeabilities is a particularly complex issue that remains a subject of basic research.

Another aspect of NAPL migration in the subsurface environment, different from movement of dissolved contaminants, is that, as it advances downward through the partially saturated zone, a fraction of the NAPL phase remains trapped in some pores, as a coating on pore walls and as microdroplets trapped within pore spaces by interfacial surface tension. This NAPL fraction is referred to as the *residual saturation.* The presence of this fraction may contribute volatilized phases and dissolved components during subsequent water infiltration and redistribution. In addition, if the residual fraction is hydrophobic, its presence further affects subsequent water and contaminant movement. The flow and distribution of the liquid phases are controlled by the movements of the interfaces between the nonwetting and wetting phases that result from changes in pressure and saturation (Blunt and Scher 1995; Reeves and Celia 1996; Cheng et al. 2004).

Quantification of NAPL transport usually is considered by using advection–dispersion equations for each of the water and NAPL phases and defining relative permeabilities (as noted previously). Alternatively, if the emphasis is on contamination of water from a dissolving NAPL source (e.g., from residual NAPL in the vadose zone or NAPL pools and localized leaks), then transport equations containing a simple source term, such as Eq. (11.7), can be considered.

The previously mentioned approaches are useful when quantification at the average ("effective") continuum ("macroscale") level is appropriate. However, pore-scale network models are more appropriately used to examine fluid distribution within pores and clusters of pores (recall Sect. 9.1). NAPL migration through the vadose zone undergoes significant fingering (unstable infiltration), and because it is a nonwetting liquid relative to the solid phase (and immiscible in water), NAPL displacement of water is in effect a drainage process, as discussed in Sect. 9.1. Pore-scale information such as NAPL geometry or interfacial configuration (or the interfacial area between the nonwetting and wetting liquids) is important (Miller et al. 1990; Powers et al. 1992; Cheng et al. 2004; Ovdat and Berkowitz 2006). Knowledge of this interfacial area is particularly useful for quantifying processes such as contaminant adsorption, dissolution, volatilization, and enhanced oil recovery (Reeves and Celia 1996; Saripalli et al. 1997; Johns and Gladden 1999; Schaefer et al. 2000; Jain et al. 2003). NAPL migration at the pore scale is illustrated in Fig. 11.4.

At the pore level, fluids are conveniently characterized by three main parameters. The viscosity ratio is given by $M = \mu_d/\mu_r$, where μ_d and μ_r are the viscosities of the displacing and resident fluids, respectively, while the capillary number is defined as $\mathrm{Ca} = q\mu_d/\gamma$, where q is the specific discharge of the displacing fluid and γ is the interfacial tension. The bond number, $\mathrm{Bo} = gr^2\Delta\rho/\gamma$, where g is the gravitational acceleration, r is the characteristic pore radius, and $\Delta\rho$ is the fluid density difference, is then introduced to account for buoyancy-gravity forces. The usual convention is that $\mathrm{Bo} > 0$ (dense fluid displaced by light fluid) stabilizes the interface, whereas $\mathrm{Bo} < 0$ (light fluid displaced by dense fluid) destabilizes it.

Lenormand et al. (1988) pioneered experiments on immiscible displacements in two-dimensional, artificial porous media, using horizontal models with negligible inertial and gravity forces and developed a numerical model to describe fluid displacement in such porous media, accounting for capillary and viscous forces. Comparison between the experiments and numerical simulations led Lenormand

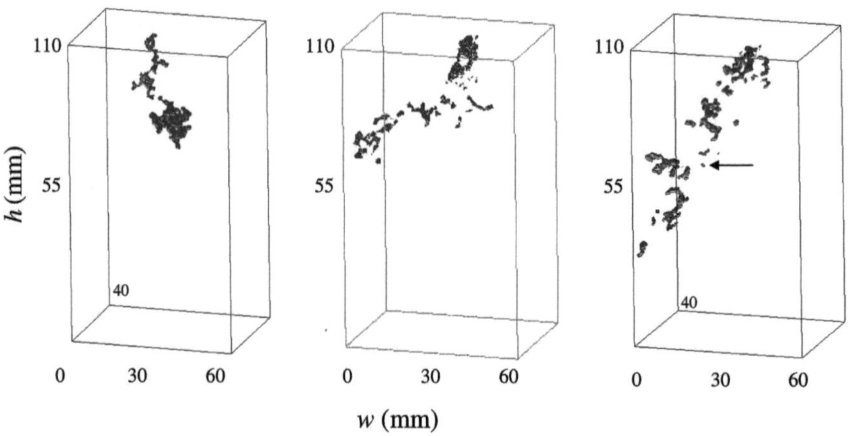

Fig. 11.4 Pore-scale, three-dimensional images of nonwetting fluid invading a wetting fluid, under different conditions of capillary and buoyancy-driven forces and for a similar injected volume. The invasion behavior is characterized by a wide range of displacement patterns and fragmented clusters, one of which is indicated by an *arrow*. (Ovdat and Berkowitz 2006) Reproduced by permission of American Geophysical Union. Ovdat H, Berkowitz B (2006)

et al. (1988) to define a Ca-M phase diagram that specifies regions in which stable and unstable displacements occur. At high M and low Ca, capillary-driven fingering occurs, whereas at low M and high Ca, viscous fingering dominates. Ewing and Berkowitz (1998) extended this phase diagram to three dimensions to account for buoyancy-gravity forces.

We note also that invasion percolation (IP) models (Wilkinson and Willemsen 1983; Furuberg et al. 1988) have been used extensively to simulate the slow displacement of a wetting fluid by a nonwetting fluid in porous media, when capillary forces dominate over viscous forces. In such models, the fluid–fluid displacement process is represented by a cluster growth process on a lattice that at breakthrough generates a self-similar fractal. The IP algorithm generates structures that quantitative analysis reveals to be similar to patterns formed during fluid–fluid displacement processes. While the IP model originally was developed for neutrally buoyant systems, accounting for capillary forces, it has since been modified to include the effects of buoyancy on the drainage process (Wilkinson 1984, 1986). Lenormand and Zarcone (1985) measured directly the structure of injected clusters in transparent resin used as porous media.

Finally, it should be noted that NAPL infiltration is accompanied by dissolution of some components into the aqueous phase, volatilization of other components into the air phase of the partially saturated zone, and possibly organic complexation with resident soil (Jury and Fluhler 1992). Mass flux of volatilized phases in the near surface can be included in transport equations, as discussed in Sect.11.3.

11.5 Transport of Contaminants by Runoff

11.5.1 Overland Runoff

The fraction of rainfall or irrigation water that flows over a land surface from higher to lower elevations, known as *overland runoff*, is an additional pathway for contaminant transport. Runoff occurs when the amount of rain or irrigation water is greater than the soil infiltration capacity. The formation of a crust on the soil surface is a major contributor to runoff formation in arid and semiarid zones, because it decreases the infiltration capacity. The soil crust is a thin layer (0–3 mm) with a high density, fine porosity, and low hydraulic conductivity compared to the underlying soil. This "skin" forms as a result of falling raindrops or sodification of soil clays.

Overland runoff may be expressed as a conservation of mass in a flow domain, where the excess rainfall or irrigation rate, R_e, is equal to the difference between the rainfall and infiltration rates:

$$R_\mathrm{e} = \frac{\partial h}{\partial t} + \nabla \cdot q, \tag{11.13}$$

where h is the flow depth and q is the unit discharge. The excess rainfall may be determined directly, while the infiltration rate may be estimated by empirical or physical models. Equation 11.13 is adjusted to consider spatial variability and land hydraulic properties, to reduce errors in estimating runoff infiltration (Wallach et al. 1997; Michaelides and Wilson 2007).

After sediment has been detached by rainfall impact during an erosion event, the majority of it returns to the soil surface after being carried some distance downslope. Through the continuous addition of sediment to the water layer, by impact of raindrops, fine sediment forms a "suspended load" with a settling velocity less than the typical velocity of whirling eddies in flowing water. This velocity is called the *friction velocity* (Rose 1993). The rate of deposition, d_i, defined as the mass of sediment reaching the soil surface per unit area per time (kg/m^2/s), is given by

$$d_i = v_i c_i, \tag{11.14}$$

where c_i is the sediment concentration and v_i is the settling velocity. The rate of rainfall detachment and redetachment is proportional the rainfall rate, raised to some power that has been found empirically to be close to unity (Hairsine and Rose 1991).

When water flows over a contaminated land surface, pollutants released from higher elevations are transported, as dissolved solute or adsorbed on suspended particles, and accumulate at lower elevations. This behavior is reflected in the spatial variability of contaminant concentration, which affects contaminant redistribution with depth following leaching. If a sorbed contaminant is not of uniform concentration across all soil-size ranges but is higher in the fine sediment fraction, the deposition of this soil fraction controls contaminant redistribution in the subsurface.

The extent of transport of dissolved contaminants in overland runoff is controlled by the topography and morphology of the land (also affected by anthropogenic activity), the depth of chemical incorporation into soil, and the time between rainfall initiation and surface runoff commencement. In addition to these factors, transport of adsorbed contaminants on suspended particles is affected by rainfall intensity, which favors soil erosion.

In general, soluble and nonreactive contaminants are found mainly in dissolved form in runoff water. For example, a large percentage (up to 90 %) of the most soluble herbicides present in the soil layer may be partitioned in overland flowing water. A substantial portion of dissolved nitrogen (8–80 %) and phosphorus (7–30 %) also may be transported in runoff water (Menzel et al. 1978; Hubbard et al. 1982; McDowell et al. 1984). Experiments comparing concentrations of soluble bromide in overland runoff water, as a function of soil type, infiltration rate, and contaminant incorporation depth, under similar "rainfall" conditions, ranged from 0.1 to 14 % (Ahuja 1982; Ahuja and Lehman 1983). Bromide was released to runoff from depths up to 2.0 cm, and concentrations decreased exponentially below the thin crust layer. In treatments where infiltration was impeded due to the presence of a deeper clay layer, the amount of bromide in the runoff water was greater than in treatments with a high infiltration rate. The authors suggested that the mechanism of chemical transfer from below the soil surface to runoff is through a turbulent diffusion process caused by raindrop impact. Based on field experiment results, Walton et al. (2000) stated that the type of soil and its hydraulic and structural properties are the main factors that determine the amount of bromide transported in runoff water. These results are in agreement with Wallach et al. (1997), who quantify overland flow and total runoff volume by considering a spatially variable infiltration rate along the slope. The effect of spatial variability of infiltration on the uncertainty in predicting runoff from sites with distinct topographic characteristics was discussed by Michaelides and Wilson (2007). They found that increasing the range of spatial correlation of infiltration rates leads to increased uncertainty in modeled runoff.

A reactive contaminant may be adsorbed on the soil surface prior to rainfall then, following rainfall that causes erosion, the soil is transported by runoff water in the form of suspended particles redistributed on the land surface. In general, the settling velocity distribution during runoff indicates that the finer particles are resettled initially (Proffitt et al. 1991), although the details of the settling process are affected by different environmental factors, such as soil type and rainfall rate.

11.5.2 Urban Runoff

Urban (municipal) runoff contaminants like surfactants, pesticides, and pharmaceutical products and contaminants associated with highway construction and maintenance are nonpoint sources that contribute significant amounts of pollution to the land surface and thus, indirectly, to the subsurface. Runoff pollution is associated with rainwater or melting snow, which washes impermeable urban surfaces as roads, bridges, parking lots, and rooftops. Suspended solids carried by runoff water represent the major source of urban contamination. For example, Wiesner et al. (1995) estimates that over 40 % of the suspended solids entering Galveston Bay (Texas) are of urban origin. Particle size distributions in runoff, taken under storm and ambient conditions, show that more than 90 % of the particles are between 0.45 and 2 μm. The particulate phase includes 1–10 % organic carbon, with heavy metal functional groups containing detectable levels of Fe, Ba, Cu, Mn, Pb, and Zn. Zinc and barium were distributed bimodally with respect to the size fractions, while lead and iron were associated almost exclusively with the largest size fraction. These results show that suspended particles from runoff include heavy metals. The association of heavy metals with different sizes of suspended particles may affect their spatial distribution on the disposal areas.

A wide range of transport models are available for predicting water contamination and flow under urban runoff. Models based on conventional methods for runoff generation and routing were reviewed critically by Eliott and Trowsdale (2007). These authors suggest that future models on urban runoff should include a broad range of contaminants and improve the representation of contaminant transport.

References

Abriola LM, Pinder GF (1985) A multiphase approach to the modelling of porous media contamination by organic compounds. Water Resour Res 21:11–18

Ahuja LR (1982) Release of soluble chemicals from soil to runoff. Trans Am Soc Agric Eng 25:948–953

Ahuja LR, Lehman OR (1983) The extent and nature of rainfall-soil interaction in the release of soluble chemicals. J Environ Qual 12:34–44

Albinger O, Biesemeyer BK, Arnold RG, Logan BE (1994) Effect of bacterial heterogeneity on adhesion to uniform collectors by monoclonal populations. FEMS Microbiol Lett 124:321–326

Amitay-Rosen T, Cortis A, Berkowitz B (2005) Magnetic resonance imaging and quantitative analysis of particle deposition in porous media. Environ Sci Technol 39:7208–7216

Baygents JC, Glynn JRJ, Albinger O, Biesemeyer BK, Ogden KL, Arnold RG (1998) Variation of surface charge density in monoclonal bacterial populations: implications for transport through porous media. Environ Sci Technol 32:1596–1603

Berkowitz B, Emmanuel S, Scher H (2008) non-Fickian transport and multiple rate mass transfer in porous media. Water Resour Res 44: doi:10.1029/2007WR005906

Bijeljic B, Rubin S, Scher H, Berkowitz B (2011) Non-Fickian transport in porous media with bimodal structural heterogeneity. J Contam Hydrol 120(121):213–221

Blunt MJ (2000) An empirical model for three-phase relative permeability. SPE J 5:435–445

Blunt MJ, Scher H (1995) Pore-level modeling of wetting. Phys Rev E 52:6387–6403

Bolster CH, Hornberger GM, Mills AL, Wilson JL (1998) A method for calculating bacterial deposition coefficients using the fraction of bacteria recovered from laboratory columns. Environ Sci Technol 32:1329–1332

Bolton EW, Lasaga AC, Rye DM (1996) A model for the kinetic control of quartz dissolution and precipitation in porous media flow with spatially variable permeability: formulation and examples of thermal convection. J Geophys Res 101:22157–22187

Bolton EW, Lasaga AC, Rye DM (1997) Dissolution and precipitation via forced-flux injection in the porous medium with spatially variable permeability: kinetic control in two dimensions. J Geophys Res 102:12159–12172

Camesano T, Logan B (1998) Influence of fluid velocity and cell concentration on the transport of motile and non-motile bacteria in porous media. Environ Sci Technol 32:1699–1708

Cheng JT, Pyrak-Nolte LJ, Nolte DD, Giordano NJ (2004) Linking pressure and saturation through interfacial areas in porous media. Geophys Res Lett 31:L08502

Cortis A, Harter T, Hou L, Atwill ER, Packman AI, Green PG (2006) Transport of *Cryptosporidium parvum* in porous media: long-term elution experiments and continuous time random walk filtration modeling. Water Resour Res 42:W12S13

Daccord G (1987) Chemical dissolution of a porous medium by a reactive fluid. Phys Rev Lett 58:479–482

Daccord G, Lenormand R (1987) Fractal patterns from chemical dissolution. Nature 325:41–43

Daccord G, Lietard O, Lenormand R (1993) Chemical dissolution of a porous medium by a reactive fluid, 2, Convection vs. reaction behavior diagram. Chem Eng Sci 48:179–186

Darmody RG, Thorn CE, Harder RL, Schlyter JPL, Dixon JC (2000) Weathering implications of water chemistry in an arctic-alpine environment, North Sweden. Geomorphol 34:89–100

Dijk P, Berkowitz B (1998) Precipitation and dissolution of reactive solutes in fractures. Water Resour Res 34:457–470

Dijk P, Berkowitz B (2000) Buoyancy-driven dissolution enhancement in rock fractures. Geol 28:1051–1054

Eliott AH, Trowsdale SA (2007) A review of models for low impact urban stormwater drainage. Environ Mod Softw 22:394–405

Emmanuel S, Berkowitz B (2005) Mixing-induced precipitation and porosity evolution in porous media. Adv Water Resour 28:337–344

Ewing RP, Berkowitz B (1998) A generalized growth model for simulating initial migration of dense non-aqueous liquids. Water Resour Res 34:611–622

Fogler HS, Rege SD (1986) Porous dissolution reactors. Chem Eng Commun 42:291–313

Furuberg L, Feder J, Aharony A, Jøssang T (1988) Dynamics of invasion percolation. Phys Rev Lett 61:2117–2120

Hairsine PB, Rose CW (1991) Rainfall detachment and deposition: modeling the physical processes. Proc Soil Sci Soc Am 55:320–324

Hilgers C, Urai JL (2002) Experimental study of syntaxial vein growth during lateral fluid flow in transmitted light: first results. J Struct Geol 24:1029–1043

Hoefner ML, Fogler HS (1988) Pore evolution and channel formation during flow and reaction in porous media. AIChE J 34:45–54

Hornung G, Berkowitz B, Barkai N (2005) Morphogen gradient formation in a complex environment: an anomalous diffusion model. Phys Rev E 72:041916

Hubbard RK, Erickson AE, Ellis BG, Wolcot AR (1982) Movement of diffuse source pollutants in small agricultural watersheds of the Great Lakes Basin. J Environ Qual 11:117–123

Israelachvili JN (1991) Intermolecular and surface forces. Elsevier, New York

Jain V, Bryant S, Sharma M (2003) Influence of wettability and saturation on liquid–liquid interfacial area in porous media. Environ Sci Technol 37:584–591

Johns ML, Gladden LF (1999) Magnetic resonance imaging study of the dissolution kinetics of octanol in porous media. J Colloid Interface Sci 210:261–270

Jové Colon CF, Oelkers EH, Schott J (2004) Experimental investigation of the effect of dissolution on sandstone permeability, porosity, and reactive surface area. Geochim Cosmochim Acta 68:805–817

Jury WA, Fluhler H (1992) Transport of chemicals through soil: mechanisms, models and field applications. Adv Agron 47:142–202

Kieffer B, Jové CF, Oelkers EK, Schott J (1999) An experimental study of the reactive surface area of the Fontainebleau sandstone as a function of porosity, permeability, and fluid flow rate. Geochim Cosmochim Acta 63:3525–3534

Lasaga A (1984) Chemical kinetics of water-rock interactions. J Geophys Res 89:4009–4025

Lee YJ, Morse JW (1999) Calcite precipitation in synthetic veins: implications for the time and fluid volume necessary for vein filling. Chem Geol 156:151–170

Lenormand R, Zarcone C (1985) Invasion percolation in an etched network: measurement of fractal dimension. Phys Rev Lett 54:2226–2229

Lenormand R, Touboul E, Zarcone C (1988) Numerical models and experiments on immiscible displacement in porous media. J Fluid Mech 189:165–187

Lichtner PC (1988) The quasistationary state approximation to coupled mass transport and fluidrock interaction in porous media. Geochim Cosmochim Acta 52:143–165

Logan BE, Jewett DG, Arnold RG, Bouwer EJ, Omelia CR (1995) Clarification of clean-bed filtration models. J Environ Eng ASCE 121:869–873

Margolin G, Dentz M, Berkowitz B (2003) Continuous time random walk and multirate mass transfer modeling of sorption. Chem Phys 295:71–80

Martin MJ, Logan BE, Johnson WP, Jewett DG, Arnold RG (1996) Scaling bacterial filtration rates in different sized porous media. J Environ Eng 122:407–415

McDowell LL, Willis GH, Murphree CE (1984) Plant nutrient yields in runoff from a Mississippi delta watershed. Trans ASAE 27:1059–1066

McDowell-Boyer L, Hunt JR, Sitar N (1986) Particle transport through porous media. Water Resour Res 22:1901–1921

Menzel RG, Rhoade ED, Onless AE, Smith SJ (1978) Variability of annual nutrient and sediment discharge in runoff from Okalahoma cropland and rangeland. J Environ Qual 7:401–406

Michaelides K, Wilson MD (2007) Uncertainty in predicted runoff due to patterns of spatially variable infiltration. Water Resour Res 43:W02415

Miller CT, Poirier-McNeill MM, Mayer AS (1990) Dissolution of trapped nonaqueous phase liquids: mass transfer characteristics. Water Resour Res 26:2783–2796

Morse JW, Arvidson RS (2002) The dissolution kinetics of major sedimentary carbonate minerals. Earth Sci Rev 58:51–84

Novak CF (1993) Modelling mineral dissolution and precipitation in dual-porosity fracture-matrix system. J Contam Hydrol 13:91–115

Novak CF, Lake LW (1989) Diffusion and solid dissolution/precipitation in permeable media. AIChE J 35:1057–1072

Ovdat H, Berkowitz B (2006) Pore-scale study of drainage displacement under combined capillary and gravity effects in index-matched porous media. Water Resour Res 42:W06411

Palmer AN (1996) Rates of limestone dissolution and calcite precipitation in cave streams of east-central New York state, northern section. Geol Soc Am 28:89

Powers SE, Abriola LM, Weber WJ Jr (1992) An experimental investigation of nonaqueous phase liquid dissolution in saturated subsurface systems—steady state mass transfer rates. Water Resour Res 28:2691–2705

Proffitt APB, Rose CW, Hairsine PB (1991) Rainfall detachment and deposition. Experiments with low slopes and significant water depth. Soil Sci Soc Am J 55:325–332

Rajagopalan R, Tien C (1976) Trajectory analysis of deep bed filtration with the sphere-in-cell porous media model. AIChE J 3:523–533

Redman JA, Grant SB, Olson TM, Estes MK (2001) Pathogen filtration, heterogeneity, and the potable reuse of wastewater. Environ Sci Technol 35:1798–1805

Redman JA, Walker SL, Elimelech M (2004) Bacterial adhesion and transport in porous media: role of the secondary energy minimum. Environ Sci Technol 38:1777–1785
Reeves CP, Celia MA (1996) A functional relationship between capillary pressure, saturation, and interfacial area as revealed by a pore-scale network model. Water Resour Res 32:2345–2358
Rose CW (1993) The transport of adsorbed chemicals in eroded sediments. In: Russo D, Dagan G (eds) Water flow and solute transport in soils. Springer, Heidelberg, pp 180–199
Rubin S, Dror I, Berkowitz B (2012) Experimental and modeling analysis of coupled non-Fickian transport and sorption in natural soils. J Contam Hydrol 132:28–36
Ryan JN, Elimelech M (1996) Colloid mobilization and transport in groundwater. Colloids Surf A 107:1–56
Saiers JE, Hornberger GM (1996) The role of colloidal kaolinite in the transport of cesium through laboratory sand columns. Water Resour Res 32:33–41
Saripalli PK, Kim H, Suresh P, Rao C, Annable DM (1997) Measurements of specific fluid–fluid interfacial areas of immiscible fluids in porous media. Environ Sci Technol 31:932–936
Schaefer C, DiCarlo DA, Blunt MJ (2000) Determination of water–oil interfacial area during three-phase gravity drainage in porous media. J Colloid Interface Sci 221:308–312
Schwille F (1988) Dense chlorinated solvents in porous and fractured media. Lewis Publishers CRC Press, Boca Raton
Singurindy O, Berkowitz B (2003) Evolution of hydraulic conductivity by precipitation and dissolution in carbonate rock. Water Resour Res 39:1016
Starr RC, Gillham RW, Sudicky EA (1985) Experimental investigation of solute transport in stratified porous media: the reactive case. Water Resour Res 21:1043–1050
Toride N, Leij F, van Genuchten M (1995), The CXTFIT code for estimating transport parameters from laboratory or field tracer experiments, version 2.0. Res. Rep. 137, U.S. Salinity Lab, Riverside
Tufenkji N, Elimelech M (2004) Correlation equation for predicting single-collector efficiency in physicochemical filtration in saturated porous media. Environ Sci Technol 38:529–536
Wallach R, Grogorin G, Rivlin (Byk) J (1997) The errors in surface runoff prediction by neglecting the relationship between infiltration rate and overland flow depth. J Hydrol 200:243–259
Walton RS, Volker RE, Bristow KL, Smettem KRJ (2000) Experimental examination of solute transport by surface runoff from low-angle slopes. J Hydrol 233:19–36
Wiesner MR, Characklis G, Brejchova D (1995) Colloidal contaminants in urban runoff. In: Proceedings of the 21st annual RREL research symposium, Cincinnati, OH, April, United State Environmental Protection Agency, EPA/600/R-95/012
Wilkinson D (1984) Percolation model of immiscible displacement in the presence of buoyancy forces. Phys Rev A 30:520–531
Wilkinson D (1986) Percolation effects in immiscible displacement. Phys Rev A 34:1380–1391
Wilkinson D, Willemsen JF (1983) Invasion percolation: a new form of percolation theory. J Phys A 34:1380–1391
Yao K-M, Habibian MT, O'Melia CR (1971) Water and waste water filtration: concepts and applications. Environ Sci Technol 5:1105–1112

Chapter 12
Selected Research Findings: Contaminant Transport

Many factors affect the transport of contaminants in the subsurface, including the (spatially and temporally variable) hydraulic and physicochemical characteristics of the solid phase and the properties of water and the contaminants themselves. In this chapter, we focus on several specific groups of contaminants, including heavy metals and metalloids, nanomaterials, pesticides, and petroleum hydrocarbons, as representative examples of reactive (nonconservative) contaminant transport.

12.1 Heavy Metals and Metalloids

During their transport in soil and subsurface environments, heavy metals and metalloids (recall Sect. 3.5) may be involved in a series of surface-induced, chemical, and biological reactions. Solid-phase properties, water-phase composition, and microorganisms presence control the retention and mobility of heavy metals and metalloids, and their complexation, during transport through the soil and subsurface. Important factors affecting heavy metal and metalloid mobility in the soil–subsurface system are the pH of the ambient solution, sorbent nature, and presence and concentration of organic and inorganic ligands, including humic and fulvic acids, and root exudates. Chemical and biochemical processes involved in heavy metal and metalloid mobility in soil–subsurface systems are sorption–desorption, oxidation–reduction, complexation–dissociation, and precipitation–dissolution. Preferential flow of water into the soil–subsurface system induces an increase in heavy metal and metalloid transport, either as solute or adsorbed on colloidal materials. For an extended presentation of this topic, the reader is directed to the reviews of Kretzschmar et al. (1999), Violante et al. (2010) and Selim (2013).

B. Berkowitz et al., *Contaminant Geochemistry*,
DOI: 10.1007/978-3-642-54777-5_12, © Springer-Verlag Berlin Heidelberg 2014

Table 12.1 Lead compositions (mg/kg) in cultivated soils (Zhang 2003)

	1913	1923	1933	1944	1955	1967	1974	1985	1995
CC[a]	26	28	32	35	38	34	42	38	42
	2[b]	*2*	*2*	*2*	*2*	*2*	*3*	*2*	*3*
CO	26	28	30	35	36	35	34	37	34
	2	*2*	*2*	*2*	*2*	*2*	*2*	*2*	*2*
COH	20	23	29	41	33	33	34	38	34
	1	*1*	*2*	*3*	*2*	*2*	*2*	*2*	*2*
	19[c]	20	19	19	23	26	–[d]	–	34
	1	*1*	*1*	*1*	*1*	*2*			*2*
	21[e]	20	19	18	18	20	19	–	20
	1	*1*	*1*	*1*	*1*	*1*	*1*	*1*	*1*

[a] Subplots with continuous corn (CC), rotation of corn and oats (CO) and rotation of corn, oats, and hay (COH)
[b] Number in italics is the uncertainty for the value above
[c] Data for soils with depth from 0.18 to 0.33 m
[d] Data not available
[e] Data for soils with depth from 0.33 to 0.51 m

12.1.1 Heavy Metals

An example of heavy metal deposition and transport in soils with time is given by Zhang (2003), who surveyed 100 years of atmospheric lead contamination of Mollisols in the Morrow plots of Champaign, Illinois, USA. Data on Pb content of agricultural soils between the years 1900 and 2000 under different cropping systems are presented in Table 12.1. Based on these data, the authors report that Pb content in surface soils increased with time during the first half of the twentieth century and slowly decreased in the second half.

Lead disposed on the land surface began to be transported downward into the A horizon, in appreciable amounts, in the 1950s; the background value of 20 mg/kg in the upper soil B horizon, at ~60 cm depth, remained relatively constant (Fig. 12.1). These data show that more than 50 % of the atmospherically derived Pb was transported into the soil–subsurface system. However, it is not clear what precise mechanisms affected the Pb mobility.

Often, heavy metals and metalloids reach the land surface and subsurface regions not as isolated compounds but as a mixture of compounds with different properties. For example, while Pb transport occurs mainly in colloidal forms, other heavy metals such as Zn and Cd exhibit downward movement in dissolved forms. The characteristics of the porous medium solid phase, the chemistry of transporting water phase, and biogeochemical redox processes affect heavy metal and metalloid transport in the soil–subsurface system. Here, we present several particular cases of heavy metal and metalloid aqueous transport in the soil–subsurface system, where the driving forces are well defined.

In the soil–subsurface system, colloid-facilitated transport can be a major pathway for strongly adsorbed contaminants. Operationally defined as particles

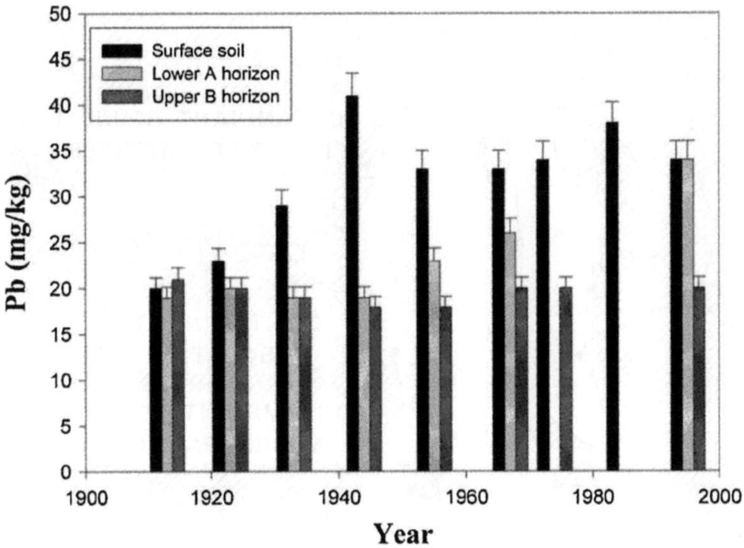

Fig. 12.1 Lead contents (mg/kg) in surface and subsurface soils during 100 year of cultivation (Zhang 2003)

between 1 μm and 1 nm in size, colloids in the subsurface may be of mineral, organic, or biological origin. Colloidal fractions in the soil–subsurface solid phase may interact with heavy metals, inducing an increase in their mobility. A laboratory experiment by Grolimund et al. (1996) examined colloid-facilitated aqueous transport of lead in a silty loam soil column and developed a schematic representation of colloid-facilitated contaminant transport (Fig 12.2). The contaminant is transported either dissolved in the liquid phase or as adsorbed to the surface of the solid phase.

The decrease in ionic strength may favor the condition for enhanced soil colloid release and transport. Figure 12.3 shows the experimental results. The soil columns were first saturated with Na^+ and contaminated with $PbCl_2$. Under these conditions, Pb was strongly adsorbed on the soil. Subsequently, leaching the column with a $CaCl_2$ solution led to a decrease in the ionic strength, and two fronts moving through the porous media were observed. Colloid release occurs between these two fronts, associated with a significant release of a colloid-bound Pb. The transport of colloid-bound Pb increased, confirming the pathway of heavy metal and metalloid colloid-facilitated transport.

An example of colloid-facilitated water transport of Cu and Zn is given by Karathanasis (1999). The transport of heavy metals adsorbed on colloidal fractions of six soils from Kentucky, USA, through undisturbed soil columns was studied, in comparison with the transport of dissolved compounds. Properties of the soils and soil colloids used in the experiment are reproduced in Table 12.2. The sorption

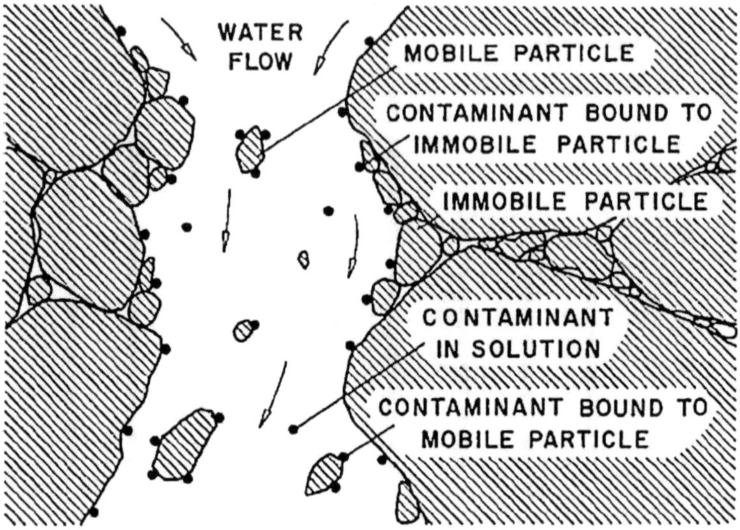

Fig. 12.2 Schematic representation of colloid-facilitated contaminant transport within a pore of a water-saturated porous medium in the subsurface. Contaminants (*filled circle*) are either dissolved in the liquid phase or adsorbed to the surfaces of the solid phase. The entire particular phase is commonly assumed to be at rest, but it is possible that colloidal particles disperse in the liquid phase and provide a rapid transport pathway for the contaminant (Grolimund et al. 1996). Reprinted with permission from Grolimund et al. (1996). Copyright 1996 American Chemical Society

affinity of both heavy metals increases with organic carbon, surface area, and pH of the colloid fraction, and is greater for Cu than for Zn.

Colloid-mediated transport of Cu and Zn was established by eluting the contaminants disposed on two soil columns (Loradale and Maury) with an aqueous solution. Figures 12.4 and 12.5 show breakthrough curves for Cu and Zn eluted in the absence and in the presence of colloids. The presence of colloids enhanced metal transport through the soil columns, with the total eluted heavy metal (colloid-bound + soluble) being greater than that of the control treatment (heavy metal without colloid). The colloid-mediated metal transport was as high as 50 times that of the control. Karathanasis (1999) notes the effect of soil composition and structure, and of the properties of the heavy metals themselves, on their colloid-enhanced transport. The author suggests that a stronger metal affinity for colloid surfaces than for the soil matrix appears to be the dominant mechanism facilitating heavy metal transport. However, physical exclusion, competitive adsorption, and metal solubility are additional factors contributing to the extent of colloid-mediated transport.

Soil colloid release and subsequent aqueous transport are strongly affected by the solution pH; as such, the extent of facilitated transport of an adsorbed heavy metal is pH dependent. A particular case of facilitating Pb release from a contaminated podzolic acid soil by increasing pH is reported by Klitzke et al. (2008),

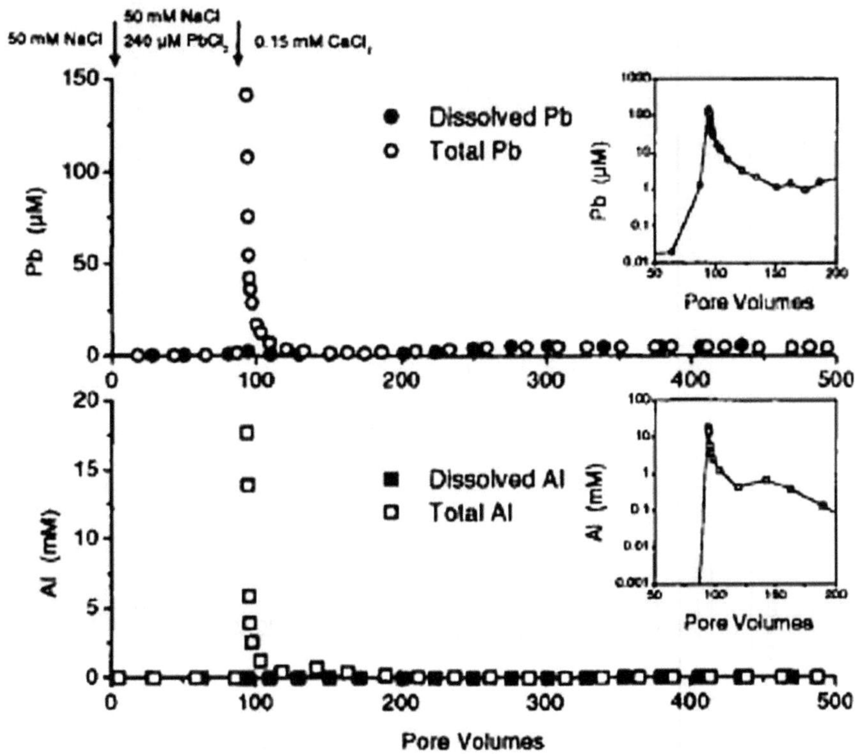

Fig. 12.3 Transport of Pb^{2+} in association with in situ mobilized colloids in a packed, Na^{+}-saturated soil column following a decrease in solution ionic strength. Colloidal Al was used as a key element to quantify the colloid concentration. The *inlets* show that elevated colloidal Al and Pb^{2+} concentrations were observed for at least 50 pore volumes after the change in feed solution (Kretzschmar et al. 1999). Reprinted from Kretzschmar et al. (1999). Copyright 1999 with permission from Elsevier

who examined pH ranging between 3 and 7. It was observed that both dissolved lead and lead adsorbed on colloids increase with an increase in pH (Fig. 12.6a); this is related to the organic content of the adsorbing phase (Fig. 12.6b). Specifically, the colloidal Pb fraction increased continuously from almost zero at pH 3 to 1,200 mg/kg at pH 7. Klitzke et al. (2008) also found that the ratio between colloidal and dissolved Pb in soil suspension is affected by the counter-ion valency: Pb shows a much stronger mobilization in the presence of the monovalent K^{+} than in the presence of the divalent Ca^{2+} (Fig. 12.7).

In the previous examples, we discussed the case of heavy metal-facilitated transport when they are bound on colloidal fractions in the soil–subsurface domain. However, in many instances, heavy metals are disposed on the land surface together with organic residues of urban or agricultural origin. Sewage sludge disposal, for example, encompasses simultaneous disposal on land surface

Table 12.2 Physicochemical and mineralogical properties of soils and colloids (Karathanasis 1999)

Properties	Soils		Colloids					
	Loradale	Maury	Beasley	Loradale	Maury	Shrouts	Smithdale	Waynesboro
Clay (%)	21	35	–	–	–	–	–	–
Hydraulic conductivity (cm/min)	0.3 ± 0.2	2.6 ± 1.7	–	–	–	–	–	–
Bulk density (g/cm^3)	1.4	1.6	–	–	–	–	–	–
Mean colloid diameter (nm)	–	–	220	300	700	270	400	1,050
Organic C (%)	2.1	0.5	0.8	3.4	0.8	0.8	0.5	0.4
pH	6.3	5.8	6.2	6.7	6.3	5.8	5.9	5.2
CEC (cmol/kg)[a]	25.2	21.9	63.4	81.8	47.5	46.4	37.2	29.0
Extractable bases (cmol/kg)	15.0	10.1	26.5	29.2	20.6	17.3	18.4	8.1
Extractable Fe (mg/g)	6.5	8.3	15.9	15.9	16.2	16.4	67.9	75.7
Extractable Al (mg/g)	4.4	2.8	6.1	5.2	10.9	9.2	57.1	61.3
Surface area (m^2/g)	–	–	386	186	190	123	121	114
Electrophoretic mobility (cm/v/s)	–	–	–1.8	–1.9	–1.4	–1.6	–0.7	–0.8
Smectite + vermiculite (%)	–	–	60	–	–	17	–	–
HISM + HIV (%)[b]	–	–	–	44	46	–	29	21
Mica (%)	–	–	20	15	10	60	7	11
Kaolinite (%)	–	–	16	35	36	20	58	56
Quartz (%)	–	–	4	6	8	3	6	12

[a] CEC cation-exchange capacity

[b] HISM hydroxyinterlayered smectite; *HIV* hydroxyinterlayered vermiculite; mean colloid diameter is expressed on a mass basis as measured by a microscan particle-size analyzer

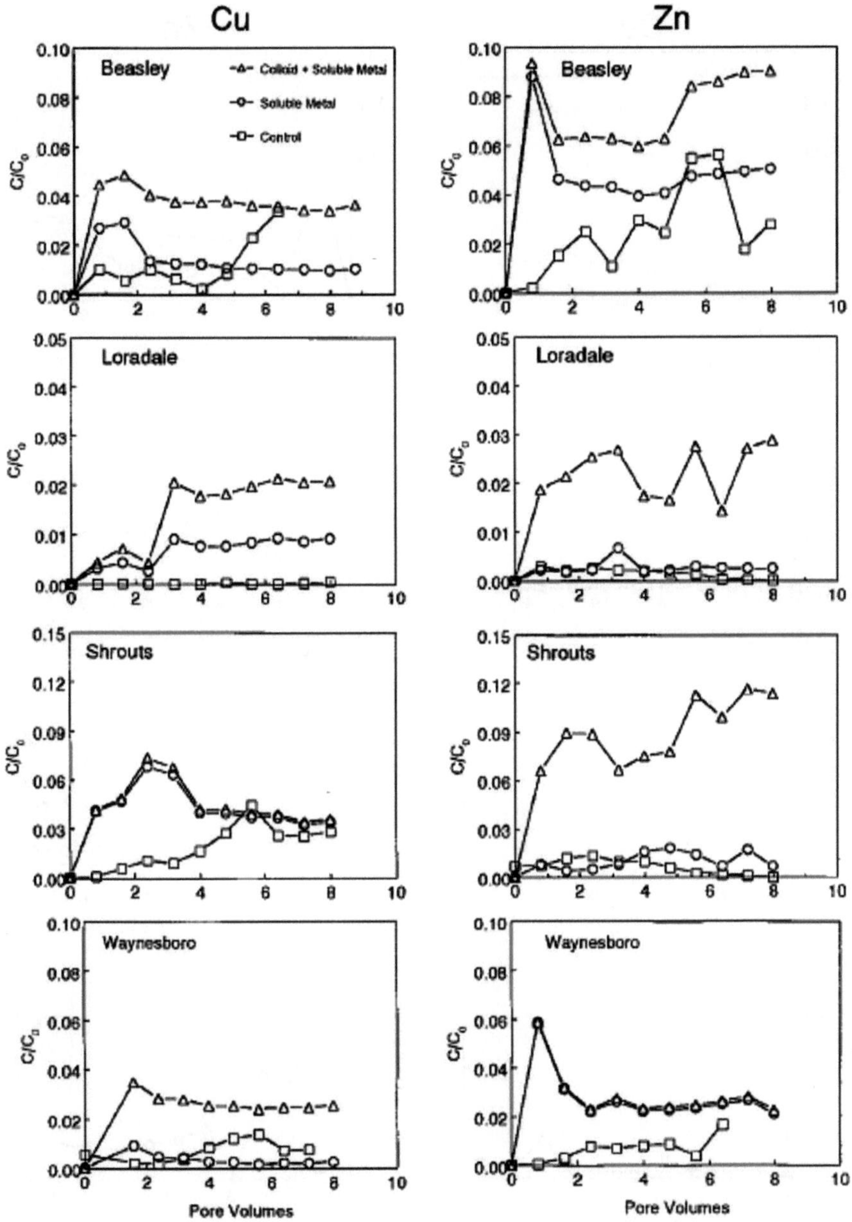

Fig. 12.4 Breakthrough curves for Cu and Zn eluted in the presence or absence of colloids from Loradale soil columns (Karathanasis 1999)

of anthropogenic organic substances and heavy metals, which may be transported into the soil–subsurface system via rainfall or irrigation. Under a typical soil pH, dissolved organic matter is able to form complexes with heavy metals which

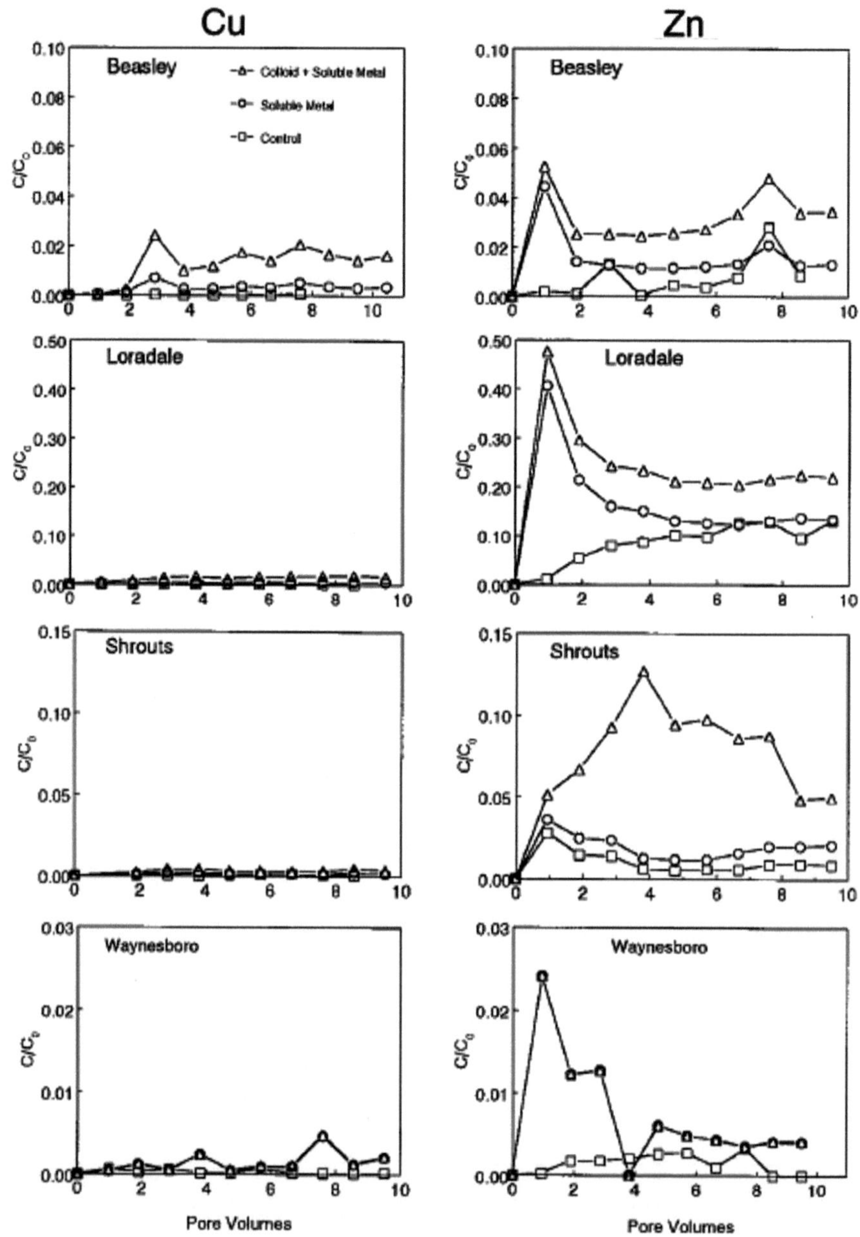

Fig. 12.5 Breakthrough curves for Cu and Zn eluted in the presence or absence of colloids from Maury soil columns (Karathanasis 1999)

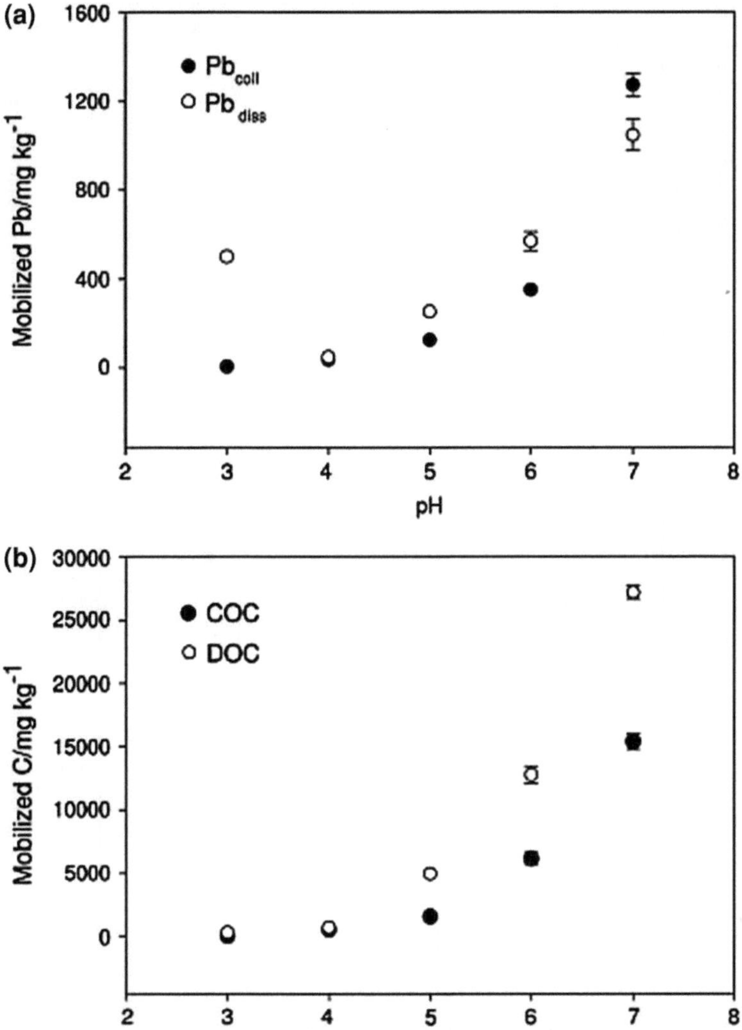

Fig. 12.6 Mobilized **a** Pb and **b** C of soil suspensions extracted from soil 1 at various pHs of the KOH treatment. Pb_{coll} colloidal Pb; Pb_{diss} dissolved Pb; COC colloidal organic carbon; DOC dissolved organic carbon (Klitzke et al. 2008). Reprinted from Klitzke et al. (2008). Copyright 2008 with permission of Wiley

exhibit significant mobility through the soil–subsurface system. The column study of Ashworth and Alloway (2004) on sludge disposal on sandy loam soil illustrates this process. Water-extractable component concentrations of the sewage sludge and characteristic properties of the soil used in the experiment are presented in Table 12.3.

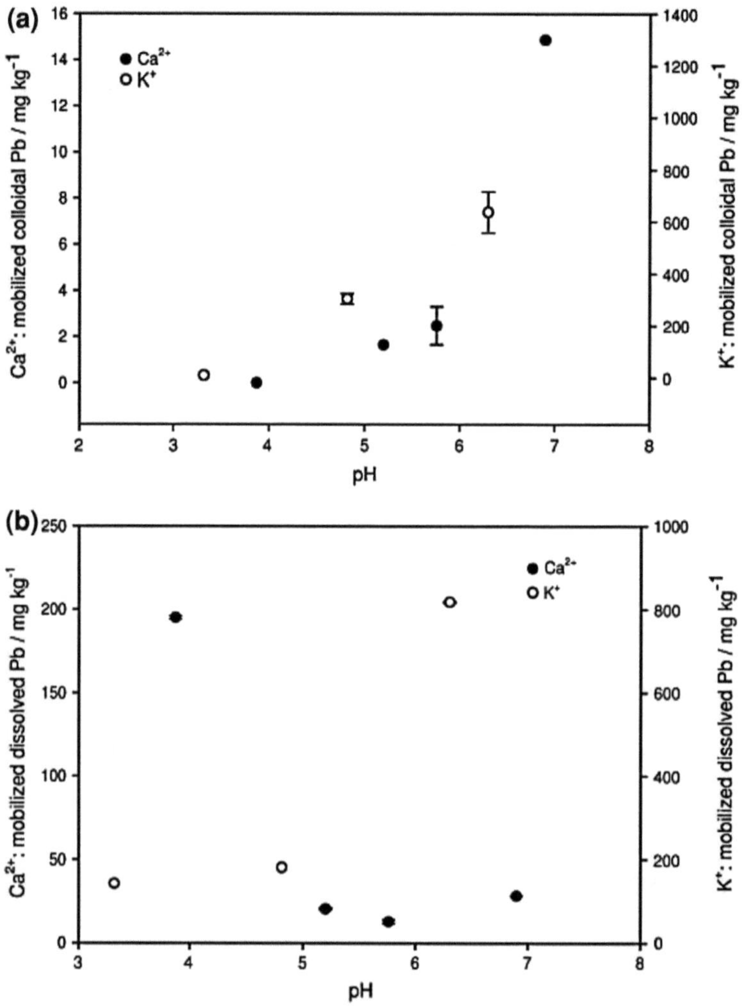

Fig. 12.7 Concentrations of **a** colloidal Pb and **b** dissolved Pb of soil suspensions extracted from soil at various pHs by KOH and $Ca(OH)_2$ (Klitzke et al. 2008). Reprinted from Klitzke et al. (2008). Copyright 2008 with permission of John Wiley and Sons

The redistribution under leaching of nickel was then tested in column experiments, using a sandy loam topsoil of the Sonning Series (Reading, Berkshire, UK). Leachate samples were collected during leaching with 10 pore volumes. The pH values of the control and sludge-amended columns ranged between 7 and 8 and remained consistent through leaching. Column effluent concentrations of dissolved organic carbon (DOC) and Ni through the control and sludge-amended soil columns are presented in Fig. 12.8. Ashworth and Alloway (2004) observed that the Ni-sludge breakthrough curve is similar to the DOC-sludge curve, and that in both

Table 12.3 **A** Water-extractable component concentrations (on a dry solids basis) of the experimental sewage sludge. **B** Selected physical and chemical characteristics of the Sonning soil (Modified after Ashworth and Alloway 2004)

A		B	
Cu^{2+} (mg/kg)	905	CEC (cmolc/kg)	17.8
Ni^{2+} (mg/kg)	2,215	Sand (g/100 g)	69
Zn^{2+} (mg/kg)	2,315	Silt (g/100 g)	19
Al^{3+} (mg/kg)	1.2	Clay (g/100 g)	12
Ca^{2+} (mg/kg)	3,105		
Fe^{3+} (mg/kg)	1.9	pH (deionized water)	6.7
K^{+} (mg/kg)	510	Total organic C (g/100 g)	2.1
Mg^{2+} (mg/kg)	412.5	Total Cu (mg/kg)	12.0
Na^{+} (mg/kg)	1,165	Total Ni (mg/kg)	14.7
Chloride (mg/kg)	865	Total Zn (mg/kg)	56.9
Phosphate (mg/kg)	ND	Dissolved organic C (mg/kg)	101.0
Sulfate (mg/kg)	4,620	Dissolved Cu (mg/kg)	59.8
Nitrate (mg/kg)	ND	Dissolved Ni (mg/kg)	34.4
Inorganic carbon (mg/kg)	376	Dissolved Zn (mg/kg)	814.1
Organic carbon (mg/kg)	4,395		
pH	6.84		

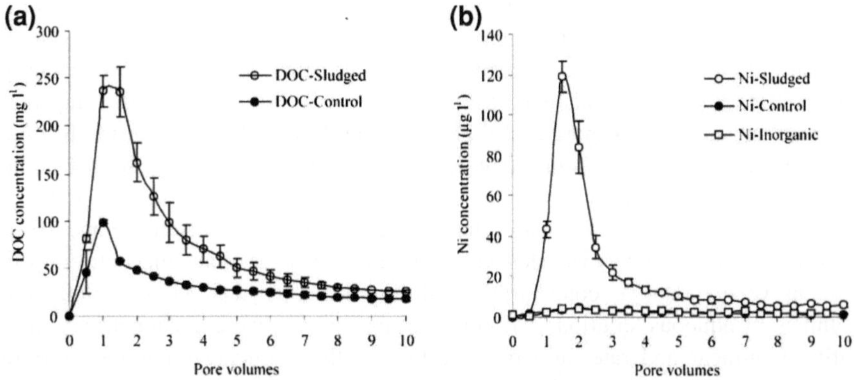

Fig. 12.8 Elution profiles of **a** DOC and **b** Ni concentration through both the control and sludge-amended soil columns. *Error bars* represent the range of values found (modified after Ashworth and Alloway 2004). Reprinted from Ashworth and Alloway (2004). Copyright 2004 with permission from Elsevier

cases, these components are mobile in the soil. To emphasize the role of organics on Ni mobility, the authors added to the control an additional variable, amending the soil with metal nitrate. From Fig. 12.8b, it may be seen that the inorganic amendment did not contribute to the Ni mobility in the soil.

An example of large-scale effects of sludge disposal on heavy metal transport in soils and groundwater is reported by Behbahaninia et al. (2010). Wastewater and sludge collected from a wastewater treatment plant in Teheran (Iran) were disposed of on four experimental plots containing silty loam soils from the Varamin plain area, and heavy metal transport with depth was determined. The metal distribution of Cr, Cu, and Pb under various treatments is shown in Fig. 12.9. It can be seen that under both wastewater and sludge disposal treatments, heavy metals are transported with depth, relative to the control case where the contaminants remain in the soil upper layer.

12.1.2 Metalloids

Arsenic (As) is a highly toxic metalloid with various oxidation states. For example, arsenite [As(III)] may be oxidized to arsenate [As(V)] as a result of oxidation by either mineral or microbial environments (Jones et al. 2012). A conceptual model showing biotic and abiotic processes of arsenic fate in the soil–subsurface system is depicted in Fig. 12.10.

The mobility of arsenic in soil–subsurface systems is affected by its oxidation state, the constituents of the adsorbing phase, and the pH of pore water. An example of metalloid redistribution in such systems is given by Williams et al. (2003), who report results of a laboratory study on coupled adsorption and transport of As(V) in a heterogeneous soil containing iron oxide. In contact with the soil, As(V) was subject to reactive interactions that affected the oxyanion arsenate mobility. Adsorption experiments showed that As(V) adsorbed strongly and nonlinearly to the soil, with natural pH of 4.5 (Fig. 12.11), indicating that arsenate subsurface transport would be significantly retarded. In addition to time and concentration, it was found that As(V) adsorption was affected by pH, ionic strength, and competing anions. For example, Williams et al. (2003) found that higher adsorption at low pH and a dramatic decrease at pH between 9 and 10.

The mobility of As(V) was determined by measuring arsenate concentrations in effluent versus relative concentrations during leaching with up to 4,000 pore volumes of aqueous solution. The breakthrough curves were asymmetrical, indicating nonlinear and rate-limited adsorption, with significant retardation due to chemical interactions with the subsurface materials. Detectable As(V) concentrations were not determined in the column effluent until aqueous leaching of about 200 pore volumes (Fig. 12.12). The experiment also illustrated the presence of an irreversibly adsorbed and/or a slowly desorbing fraction. After the appearance of the As(V) peak, the concentration of the arsenate began to decrease as the influent concentration became an As(V)-free solution. At this point, the amount of As(V) recovered was only 44 % of the total arsenate input. As a consequence, a significant fraction of As(V) was irreversibly adsorbed on the soil, affecting the pathway of this contaminant in the soil–subsurface system.

Arsenate concentration affects its transport in the soil–subsurface region. Yolcubal and Akyol (2008) found that the degree of arsenate retardation in a carbonate-rich soil decreased with increasing As(V) concentration. The

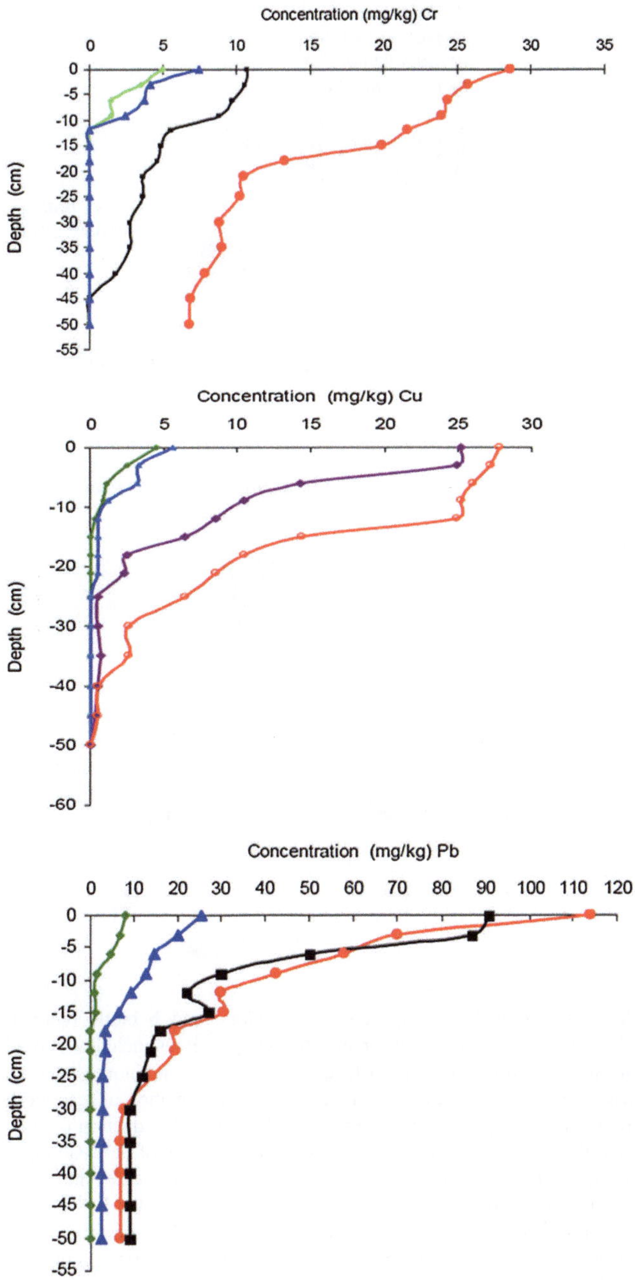

Fig. 12.9 Transport of heavy metals in soil profile *blue triangle*—tap water (control); *red circle*—effluent; *green diamond*—sludge and tap water; *black square*—sludge +effluent (modified after Behbahaninia et al. 2010)

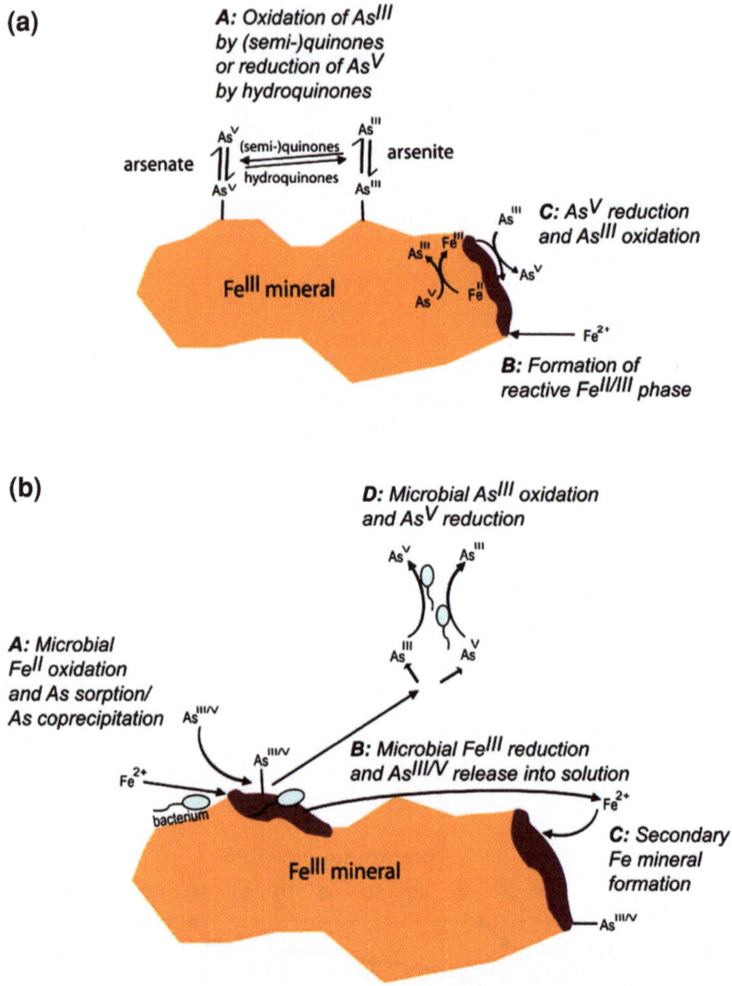

Fig. 12.10 Conceptual model showing selected **a** abiotic and **b** biotic redox processes that influence the fate of arsenic in the environment. Abiotic processes include (*A*) oxidation of As^{III} and reduction of As^{V} (dissolved or surface-bound) by semiquinones and hydroquinones, respectively (as present, e.g., in humic substances) and (*B*) formation of reactive $Fe^{II/III}$ mineral phases (e.g., by microbially produced Fe^{2+}) that can lead to As^{V} reduction or As^{III} oxidation as illustrated in (*C*). Biotic processes include (*A*) microbial oxidation of Fe^{II} leading to Fe^{III} mineral formation and sorption or coprecipitation of As^{III} or As^{V} to/with these Fe^{III} minerals (oxidation of Fe^{II} can also occur abiotically), (*B*) microbial reduction of As-loaded Fe^{III} minerals leading to release of As^{III} or As^{V} into solution, (*C*) formation of secondary Fe minerals by reaction of microbially released Fe^{2+} with remaining Fe^{III} minerals leading to sorption/coprecipitation of As, and finally (*D*) direct microbial oxidation of As^{III} or reduction of As^{V} for either dissimilation or detoxification (Borch et al. 2010). Reprinted with permission from Borch et al. (2010). Copyright 2010 American Chemical Society

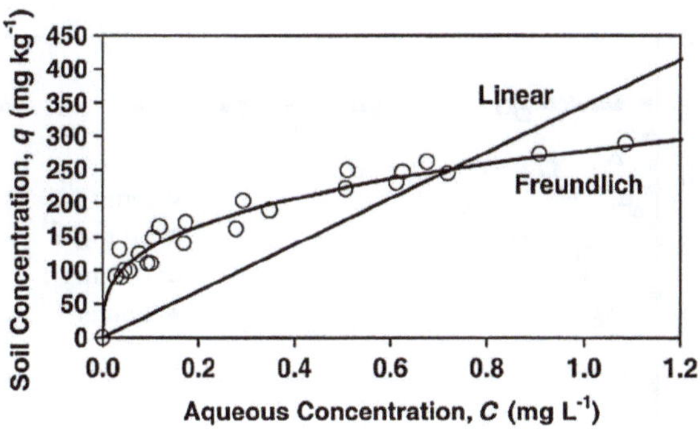

Fig. 12.11 Adsorbed solid versus aqueous As(V) concentration (pH 4.5, 0.01 M $NaNO_3$) (Williams et al. 2003)

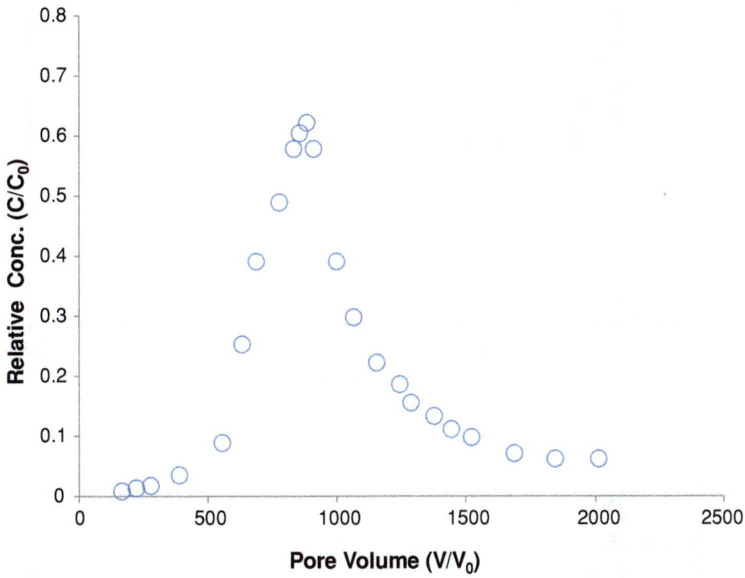

Fig. 12.12 Trend of As(V) concentration in the effluents of a soil (sand 31 %, silt 50 %, clay 19 %) column leached with an aqueous solution (pH = 4.5, 0.01 M $NaNO_3$, C_0 = 1 mg/L, pore water velocity = 0.53 cm/min) (Modified after Williams et al. 2003)

breakthrough curves of As(V) at different initial concentration, in comparison with that of a nonreactive tracer (Cl^-), are shown in Fig. 12.13. Arsenate transport was significantly retarded with respect to the tracer breakthrough curve. At higher concentrations, arsenate eluted first but at lower concentrations; the breakthrough curves exhibit a greater tailing. The nonideal transport of As(V) is explained by the coupled effect of nonlinear and rate-limited sorption. Yolcubal and Akyol (2008)

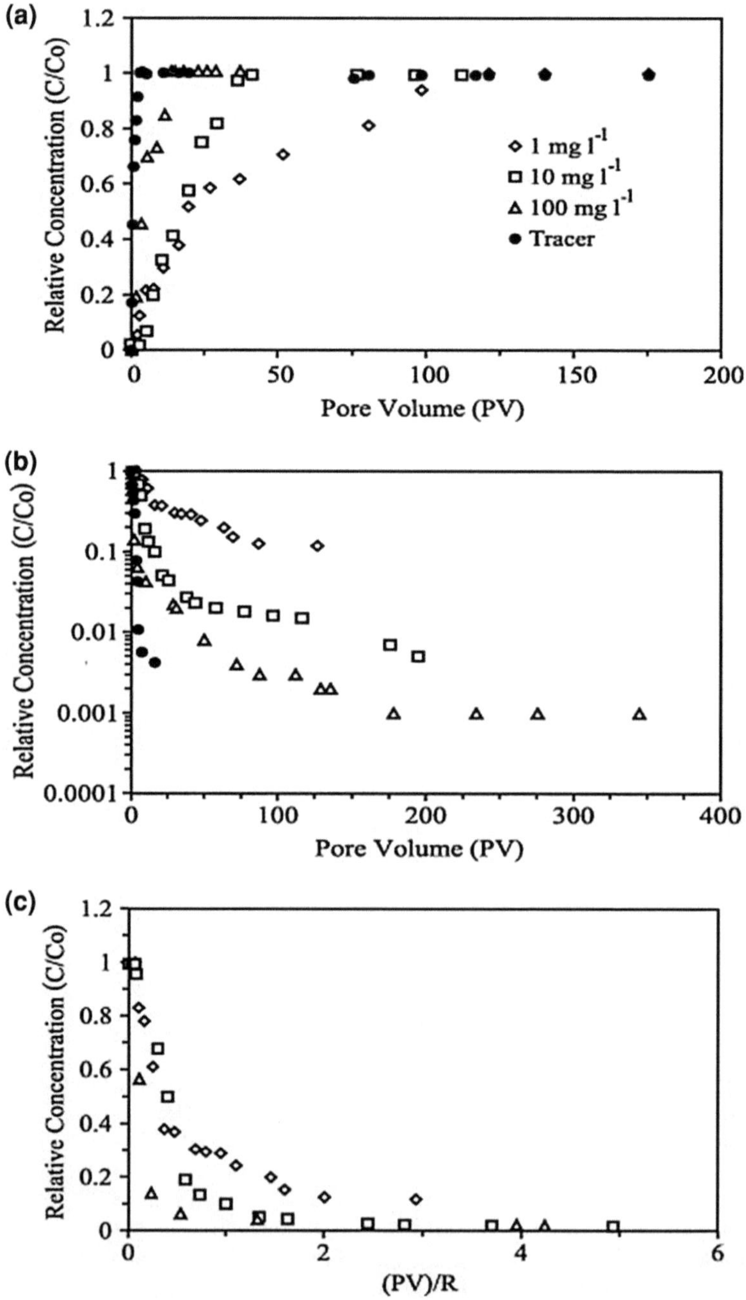

Fig. 12.13 **a** Arrival and **b** elution of the arsenate breakthrough curves at the different arsenate initial concentrations (1 PV $\approx$ 1.16 h), relative to those of the tracer (Cl^-). **c** Comparison of the elution curves of the arsenate breakthrough curves at different C_0. Pore volume was normalized by dividing the corresponding retardation factors (R). Reprinted from Yolcubal and Akyol (2008). Copyright 2008 with permission from Elsevier

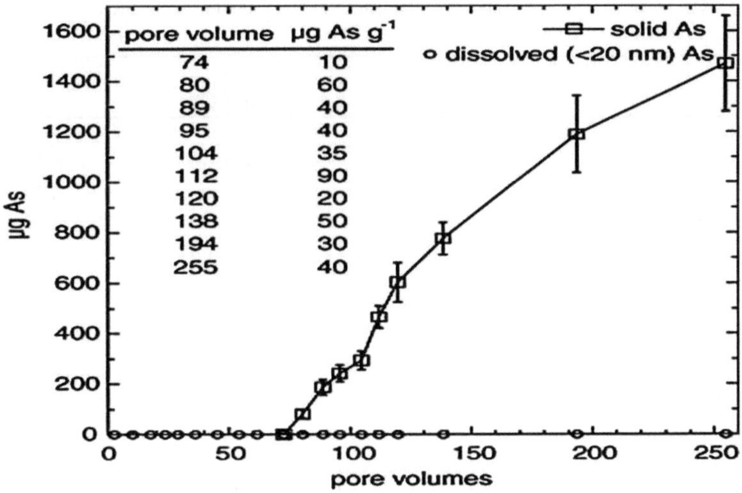

Fig. 12.14 Release of As during leaching of mine tailings with a solution containing 500 µM of both oxalate and citrate. Error bars represent 95 % confidence intervals ($\pm 3.18\sigma$) calculated for all measurements inherent to these data. The inset table summarizes the concentrations of As in the colloid composites and analytical error is less than the significant figures reported (Slowey et al. 2007). Reprinted from Slowey et al. (2007). Copyright 2007 with permission from Elsevier

show that in the case of a calcareous soil and due to a limited concentration of Fe and Al oxyhydroxide, the contribution of nonlinear sorption to the retardation of arsenate is negligible.

Metal presence in geological deposits may sometimes be associated with As presence (Smedley and Kinniburgh 2002). In particular, abandoned mine tailings are often an important source of arsenic contamination in soil–subsurface systems. For example, transport of As(V) adsorbed on suspended particles, from mine tailings, was investigated at a contaminated site at Clear Lake, California, USA Slowey et al. 2007). The mineralogy of the studied site includes the presence of multiple particles of jarosite, a basic hydrous sulfate and iron, which may serve as a potential host of As. It was determined that arsenic concentration increases as particle size decreases in the following dry-sieved fractions: particle sizes of 500–2,000 µm, 250–500 µm, 150–250 µm, 75–150 µm, 45–75 µm, and <45 µm yield arsenic concentrations of 78, 92, 110, 125, 150, and 152 µg/g, respectively.

As(V) transport was examined in columns packed with mine tailings and leached with an electrolyte solution, at a pH of 5.7. Almost all of the measurable As(V) removed from the column was adsorbed on solid state, and only traces of dissolved As(V) were observed (Fig. 12.14). Chemical analysis of the column effluent indicated that all of the mobilized As was particulate, representing 6 % of the total As through 255 pore volumes of leaching. Slowey et al. (2007) conclude that the main mode of arsenic transport from the mine tailings occurs as As-bearing colloidal particles, including As(V) sorbed to Fe(III)-hydroxides and

Table 12.4 Characteristics of ambient water samples used in the experiments (Modified after Keller et al. 2010)

	pH	TOC µM C	Ionic strength eq/L
Santa Barbara seawater	8.05	54	7.07×10^{-1}
Artificial seawater	7.99	55.9	6.39×10^{-1}
Bodega Bay seawater	8.08	131.8	6.79×10^{-1}
UCSB lagoon	8.9	522.6	2.15×10^{-1}
Santa Paula groundwater	7.9	842.4	9.09×10^{-2}
Santa Clara river water	8.33	163.8	1.84×10^{-2}
Estero effluent	7.68	378	3.17×10^{-2}
Mesocosm wastewater	7.07	691.8	3.18×10^{-3}
Storm runoff	7.09	1,564	9.63×10^{-3}
Mesocosm freshwater	8.38	5,283	7.18×10^{-3}

coprecipitated within hydrosulfates (e.g., jarosite). They suggest that As release from soils and sediments contaminated with tailings, and consequently their transport, are controlled by either acid or reductive dissolution of the host phase.

12.2 Engineered Nanoparticles (ENPs)

Engineered nanoparticles (ENPs), or nanomaterials (recall Sect. 3.6), are reactive pollutants that, after reaching the land surface, are subject to environmentally induced transformations that may affect their redistribution in the soil–subsurface system. The mobility of ENPs through porous media is governed by ENP composition and morphology, the solid-phase physicochemical status, the chemistry of the pore water (e.g., pH, ionic strength, and the presence of natural organic matter), and by biochemically induced processes. In contrast to the transport of primary particles, ENPs tend to aggregate into clusters having a size of several microns. As a consequence, changes in ENP stability and aggregation status are crucial factors in defining ENP transport in the vadose zone (Hotze et al. 2010).

12.2.1 Fate of ENPs in Aqueous Solutions

Because the pathway of ENP transport may be affected by changes in ENP aggregation status during their contact with the soil–subsurface environment, we consider it useful to present an example of the relationship between metal oxide nanoparticle stability and aggregation. Keller et al. (2010) measured electrophoretic mobility, state of aggregation, and rate of sedimentation of TiO_2, ZnO, and CeO_2 ENPs as affected by the chemistry of environmental aqueous matrix (mainly by natural organic content, ionic composition, and ionic strength) as described in Table 12.4.

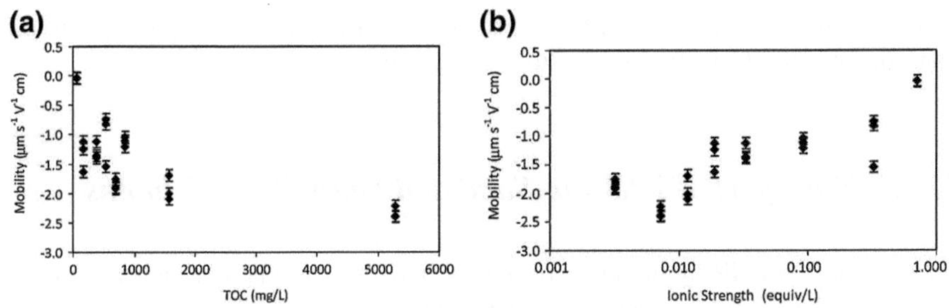

Fig. 12.15 Electrophoretic mobility of TiO_2, ZnO, and CeO_2 in nine different waters as a function of **a** TOC; and **b** ionic strength (Keller et al. 2010). Reprinted with permission from Keller et al. (2010). Copyright 2010 American Chemical Society

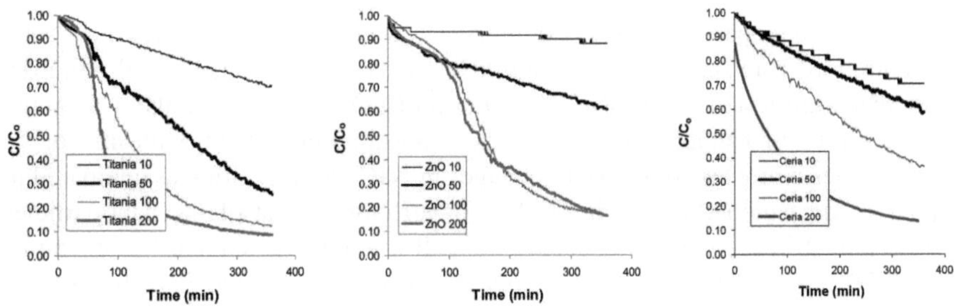

Fig. 12.16 Sedimentation of TiO_2, ZnO, and CeO_2 in seawater at 4 different initial nanoparticle concentrations (10, 50, 100, and 200 mg/L) (Keller et al. 2010). Reprinted with permission from Keller et al. (2010). Copyright 2010 American Chemical Society

The electrophoretic mobilities of these metal oxide nanoparticles, in aqueous solutions of different origin and composition, are presented in Fig. 12.15. The electrophoretic mobility was dependent on the presence of natural organic matter (represented by Total Organic Carbon, TOC) (Fig. 12.15a) and ionic strength (Fig. 12.15b), and not affected by the pH (data not shown). This observation was explained by the following trends: (1) as TOC concentration increases, the charge on the nanoparticles becomes more negative, and (2) as ionic strength increases, the nanoparticle charge is neutralized more effectively.

Keller et al. (2010) report that the aqueous composition affects the sedimentation and aggregation status of TiO_2, ZnO, and CeO_2. The highest rate of sedimentation was observed in an aqueous environment characterized by low TOC and high ionic strength; note that high concentrations of suspended nanoparticles were considered. The rate of sedimentation decreased rapidly as nanoparticle aggregates were removed from solution, lowering the concentration of dispersed nanoparticles (Fig. 12.16). The chemistry of the aqueous solution controls the extent of sedimentation. The availability of organic molecules which may be adsorbed on the surface of metal oxide nanoparticles provides a barrier for their aggregation and consequently the formation of a stable suspension.

12.2.2 Transport of ENPs in Sand and Glass Bead Columns

The transport of ENPs in inert porous media, in particular, glass beads, or sand, has been studied extensively. Only a few examples are presented here, but many more publications discuss this subject; information on additional types of ENPs, solution composition, and flow rate is easily available in the relevant literature. The transport of various metal oxide nanoparticles in saturated sand columns was investigated by Ben-Moshe et al. (2010). In this study, the effects of several factors, including flow rate, NaCl, and humic acid concentrations, and type of metal oxide nanoparticle, were examined. It was shown that increasing NaCl concentration from 0.01 to 0.1 M caused a reduced total elution of all nanoparticles that were tested (TiO_2, ZnO, Fe_3O_4, and CuO). However, it also was noted that each type of nanoparticle was affected to a different extent, with iron oxide nanoparticles being less sensitive to changes in NaCl concentration as compared to copper oxide nanoparticles (Fig. 12.17). In all cases, most of the nanoparticles were retained in the column and only a small fraction was eluted. For example, only 1.5 % of the amount of iron oxide nanoparticles injected into the column eluted for the 0.01 M NaCl experiment, and only 0.2 % of the injected amount eluted for the 0.1 M NaCl experiment.

Further examination of iron oxide nanoparticle transport (Ben-Moshe et al. 2010) demonstrated the strong effect of humic acid presence, as shown in Figs. 12.18 and 12.19. For the case of relatively low ionic strength (0.01 M NaCl), the total amount of iron oxide nanoparticle elution from the column increased from 1.5 % for the case of water without humic acid to elution of 75 % of the total amount for the case of 60 mg/L humic acid. In addition to a large increase in mobility even for addition of 10 mg/L humic acid, it was also shown that once the particles are mobile, they move relatively quickly, with a peak at 1 pore volume and a long tail that continues until 3.5 pore volumes have flowed through the column (Fig. 12.18). The results suggest that humic acids can stabilize the iron oxide nanoparticles in solution, and this stability in turn increases dramatically the mobility of the particles. Because most natural water resources, and particularly soil solutions, contain dissolved organic matter, these results suggest that the mobility of nanoparticles must be tested against real solutions.

When high ionic strength solutions and humic acids are combined, a complex picture of nanoparticle elution can be observed (Ben-Moshe et al. 2010). As demonstrated in Fig 12.19, long multipeak elution patterns were found. Unlike the "simple" one-peak elution for the low ionic strength experiment shown in Fig. 12.18, the nanoparticles for this case eluted over much longer times (or pore volumes) and multipeak patterns were found when the concentration of the humic acid increased above 10 mg/L. These results can be explained by formation and

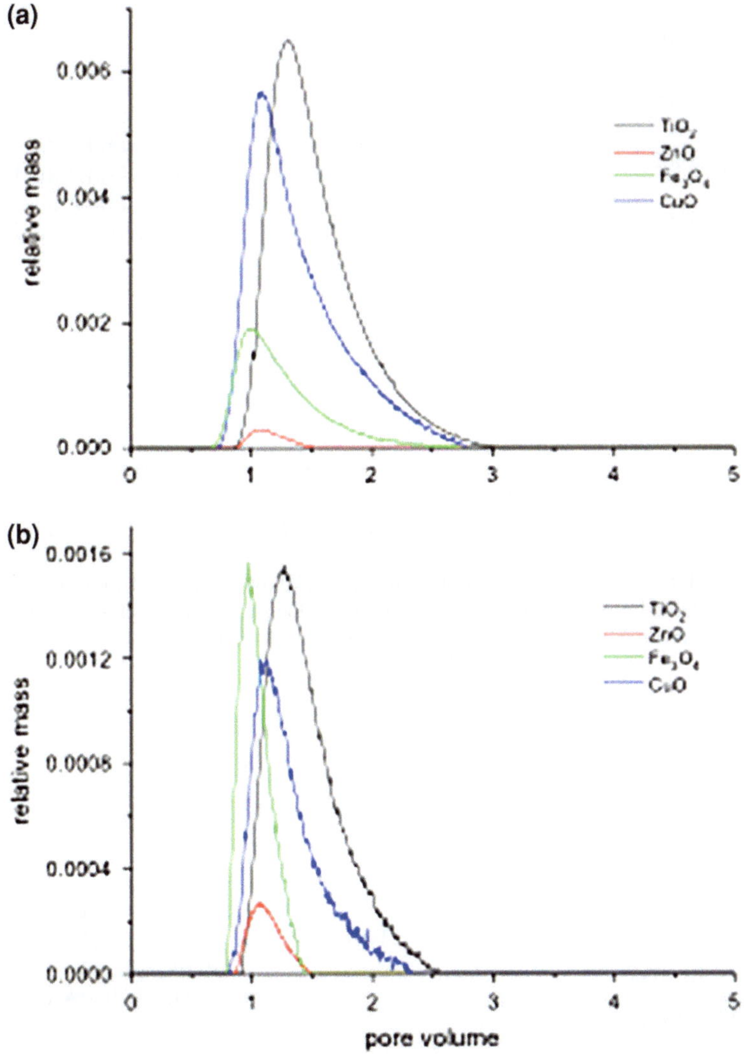

Fig. 12.17 Breakthrough curves for metal oxide nanoparticles at pH 7 with **a** 0.01 M and **b** 0.1 M NaCl (Ben-Moshe et al. 2010). Reprinted from Ben-Moshe et al. (2010). Copyright 2010 with permission from Elsevier

breaking of nanoparticle aggregates during the transport which release to solution small nanoparticles that are more mobile than the aggregates. Another explanation is that the larger nanoparticle aggregates move much more slowly through the column, so that elution patterns are affected by different mobilities associated with a range of aggregate sizes.

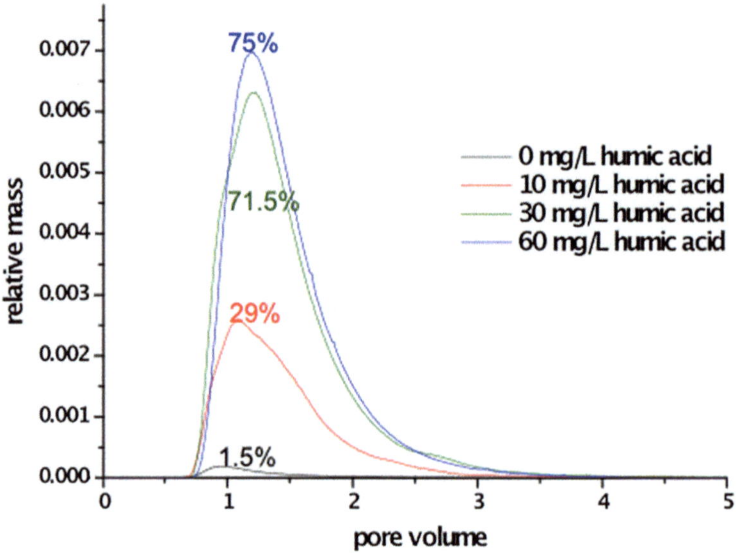

Fig. 12.18 Effect of humic acid concentration on breakthrough curves of iron oxide nanoparticles with 0.01 M NaCl. Numbers above each line are the sum % of particles eluted during the experiment (modified after Ben-Moshe et al. 2010). Reprinted from Ben-Moshe et al. (2010). Copyright 2010 with permission from Elsevier

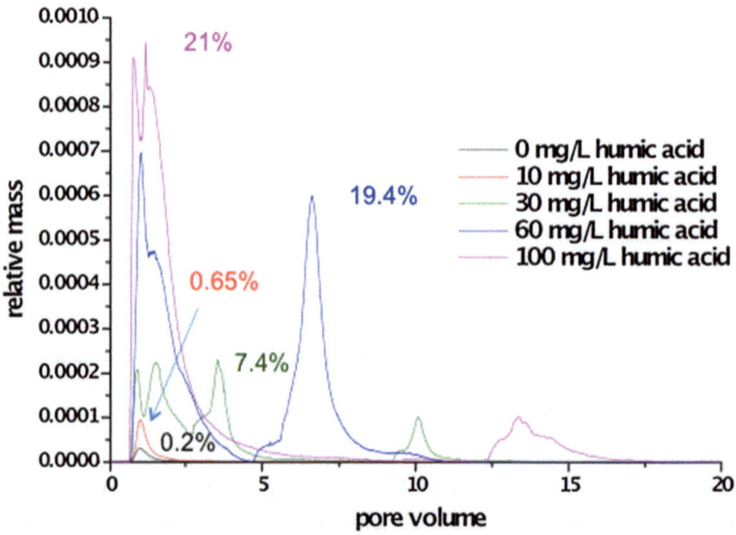

Fig. 12.19 Effect of humic acid concentration on breakthrough curves of iron oxide nanoparticles with 0.1 M NaCl. Numbers above each line are the sum % of particles eluted during the experiment (modified after Ben-Moshe et al. 2010). Reprinted from Ben-Moshe et al. (2010). Copyright 2010 with permission from Elsevier

Overall, Ben-Moshe et al. (2010) showed that the composition of the solution has a strong impact on the mobility of nanoparticles, both in terms of mobility and retention in the column and in terms of elution patterns.

Large commercial use of nanoscale fullerenes (nC_{60})—an hydrophobic molecule composed of 60 carbons atoms arranged in a spherical shape—suggests their potential release to land surface. The inherent solubility of fullerene (nC_{60}) in water is very low, but stable fullerene aqueous suspensions may be formed in the presence of commercial or natural organic compounds. Under these conditions, the transport of fullerenes through the soil–subsurface region may occur. Some results on the influence of environmental conditions on fullerene (nC_{60}) transport in the geochemical system are presented below.

The effects of flow conditions on the transport of fullerene (nC_{60}) nanoparticles in quartz sand were studied in column experiments by Li et al. (2008). Four different fractions of Ottawa sand and two pore water velocities were tested for fullerene breakthrough curves (Fig. 12.20). For the lower flow rate ($\sim$m/d), breakthrough of nC_{60} aggregates was observed only in the columns packed with 20–30- and 40–50-mesh Ottawa sand (Fig. 12.20a, b). The retention of nC_{60} aggregates, expressed as the retained fullerene mass per gram of dry sand, is related in Fig. 12.20c, d to the distance from the column inlet. Columns packed with coarse sands generally exhibited a slower attachment than those packed with finer sands. A coupled mathematical model incorporating nonequilibrium attachment kinetics and a maximum retention capacity is used to provide good agreement between experimental and simulated results.

Additional factors that influence fullerene transport in porous media are the electrolyte species and concentration in aqueous solutions. For example, Wang et al. (2008) studied the influence of electrolyte presence in leaching water on the retention and transport of fullerene nanoparticles in water-saturated quartz sand. Figures 12.21 and 12.22 show fullerene concentrations in effluents, and their retention profiles, when an initial pulse injection of fullerene into a quartz column is leached with $CaCl_2$ (Fig. 12.21) and NaCl (Fig. 12.22) aqueous solutions. As the electrolyte concentration increased from 1 to 100 mM, the change in nC_{60} particle diameter was minimal for the NaCl case; but in the presence of $CaCl_2$, the nC_{60} particle diameter increased by more than sevenfold. At low ionic strength (3.05 mM), fullerene particles were readily transported through 40–50-mesh quartz sand after leaching with 1.5 pore volumes of nC_{60} suspension. At a tenfold higher ionic strength and in a column packed with fine Ottawa sand (100–140 mesh), the transport of fullerene was drastically reduced due to a strong retention in sand (higher than 95 % from the amount initially added). Based on the above results, Wang et al. (2008) underlined the strong dependence of fullerene transport on aqueous solution chemistry.

Carbon nanotubes are nanomaterials formed entirely of carbon; they are expected to have broad industrial applications which may lead to unavoidable releases into the soil–subsurface environment. The transport of single-walled carbon nanotubes (SWNTs) was tested by Jaisi et al. (2008) in saturated laboratory columns packed with quartz sand. SWNTs were added and then subjected to

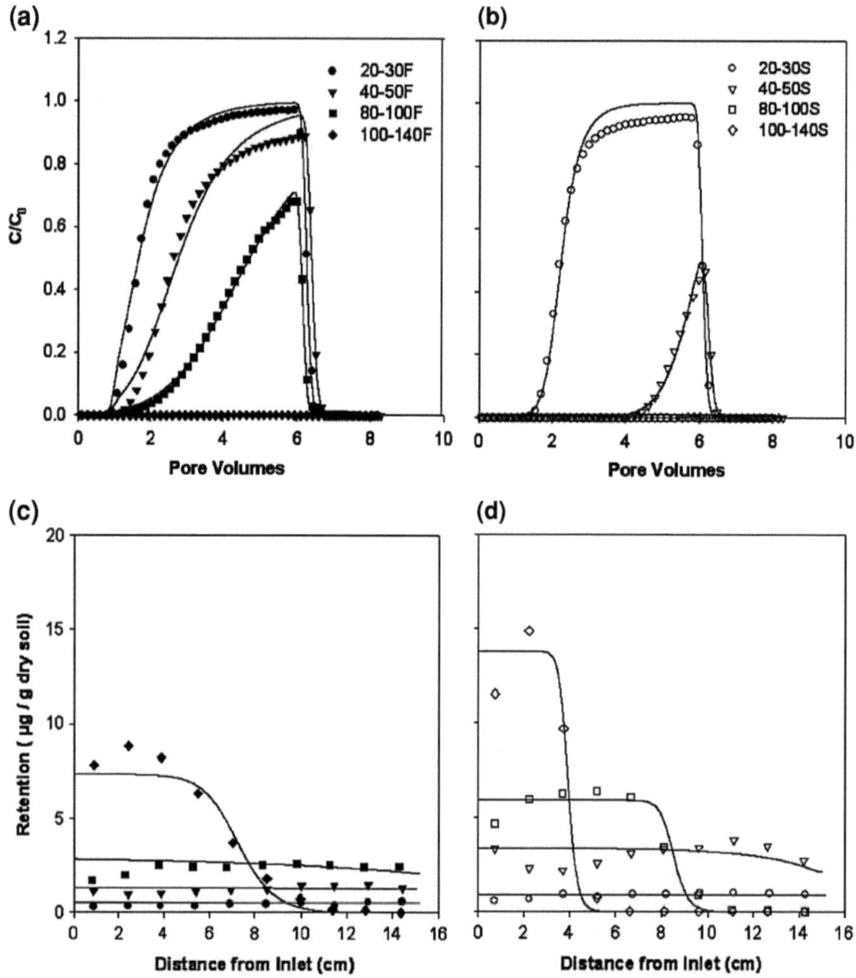

Fig. 12.20 Measured and simulated breakthrough curves (**a** and **b**) and retention profiles (**c** and **d**) for representative nC_{60} transport experiments at pore water velocities of 8 and 1 m/d in water-saturated columns packed with four different size fractions of Ottawa sand. Aqueous nC_{60} suspensions contained 1.0 mM $CaCl_2$ as a background electrolyte, buffered to pH 7 with 0.065 M $NaHCO_3$ (Li et al. 2008). Reprinted with permission from Li et al. (2008). Copyright 2008 American Chemical Society

leaching by solutions of various chemical compositions. Figure 12.23 shows the SWNT breakthrough curves from elution with aqueous solutions of KCl with various concentrations, followed by leaching with deionized water. Similar breakthrough curves are shown in Fig. 12.24 for transport in the presence of Suwannee River humic acid (SRHA). These results indicate that deposited SWNTs are released from quartz sand upon introduction of low ionic strength monovalent

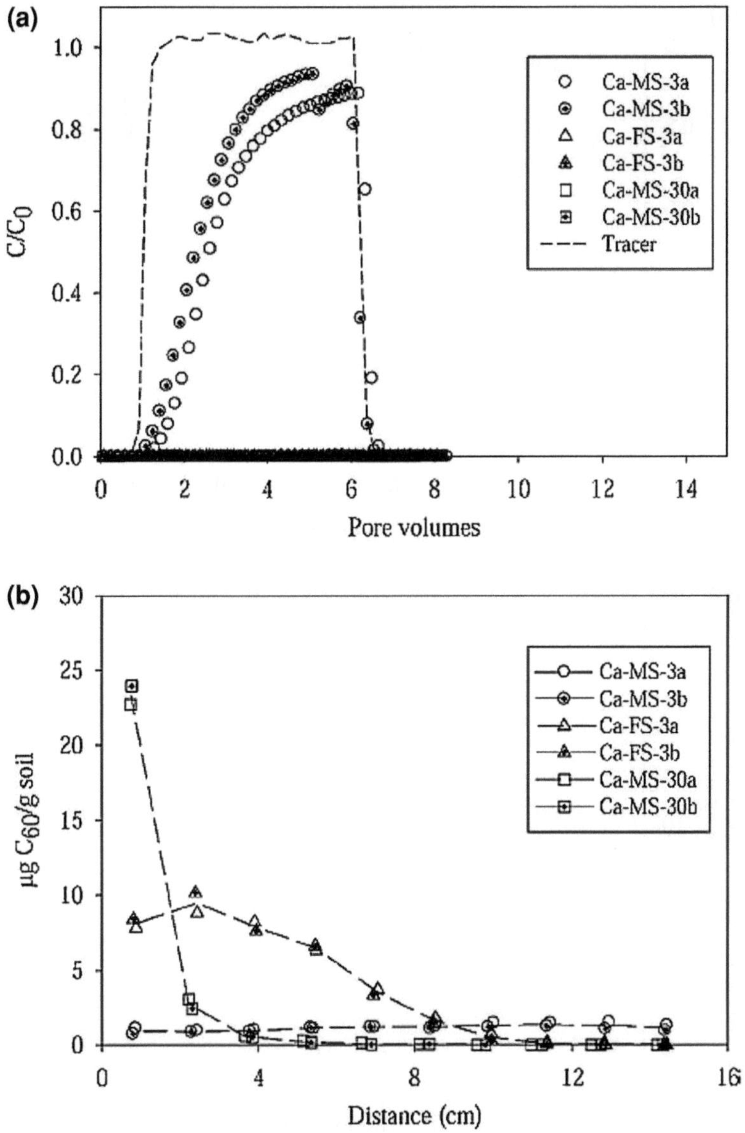

Fig. 12.21 Effluent concentrations (**a**) and retention profiles (**b**) after pulse injections of fullerene nanoparticle (nC_{60}) suspensions in columns packed with either 40–50-mesh or 100–140-mesh quartz sand with $CaCl_2$ as the background electrolyte (Wang et al. 2008). Reprinted with permission from Li et al. (2008). Copyright 2008 American Chemical Society

(KCl) solutions, and consequently favor transport through the inert porous medium. In contrast, SWNTs in the quartz column in the presence of calcium ions are not released even when humic acid is present in the eluting solution.

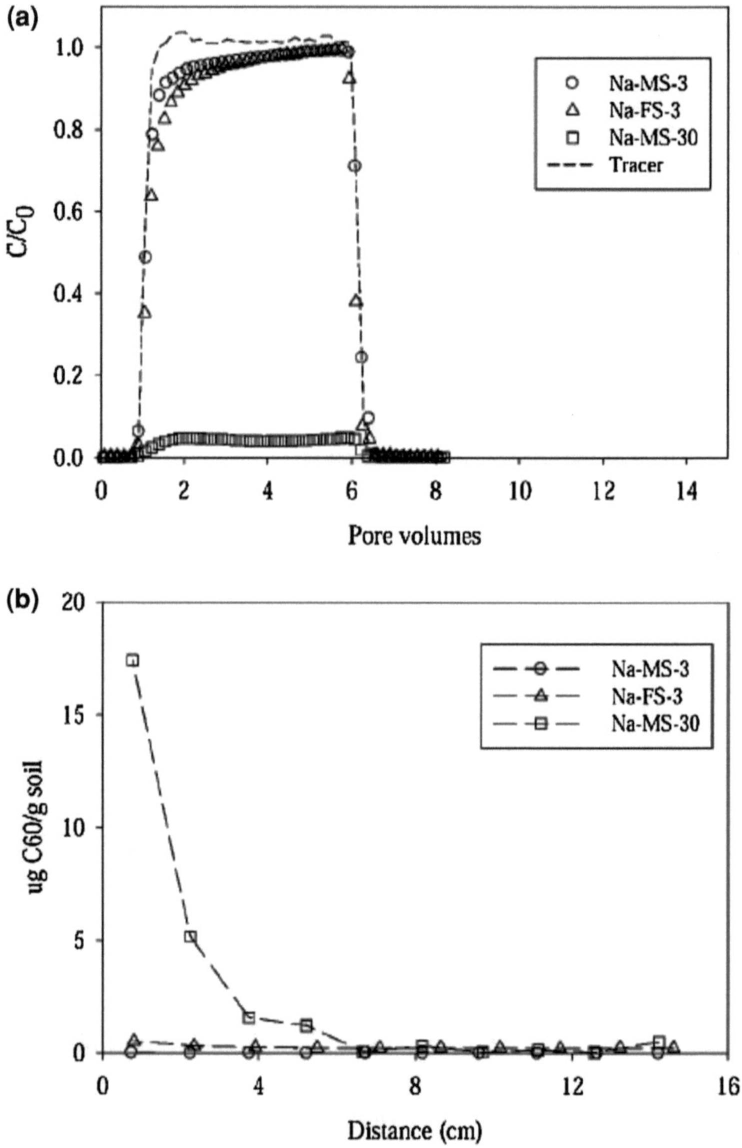

Fig. 12.22 Effluent concentrations (**a**) and retention profiles (**b**) after pulse injections of fullerene nanoparticle (nC_{60}) suspensions in columns packed with either 40–50-mesh or 100–140-mesh quartz sand with NaCl as the background electrolyte (Wang et al. 2008). Reprinted with permission from Li et al. (2008). Copyright 2008 American Chemical Society

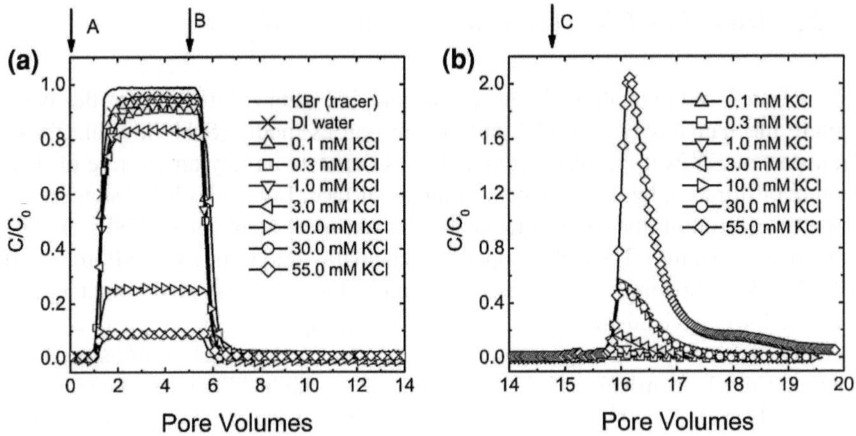

Fig. 12.23 **a** SWNT breakthrough curves at various KCl concentrations as well as a tracer (KBr) breakthrough curve. **b** SWNT release curves following elution with deionized water. *Step A* deposition of SWNTs in the packed quartz sand bed at different ionic strengths (KCl) plus 0.1 mM $KHCO_3$ as a buffer (pH 7.0). *Step B* elution with the same electrolyte solutions (without SWNTs). *Step C* elution with deionized water (pH unadjusted). Experimental conditions are: SWNT concentration = 87 (±4) mg/L, approach velocity = 0.021 (±0.002) cm/s, packed column length = 6.3 cm, porosity = 0.37, mean grain diameter = 264 μm, and pH 7.0, temp. = 23 °C (Jaisi et al. 2008). Reprinted with permission from Jaisi et al. (2008). Copyright 2008 American Chemical Society

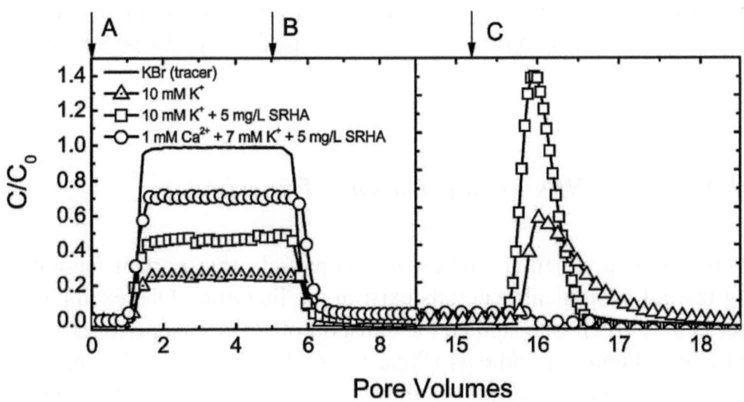

Fig. 12.24 SWNT breakthrough curves in the presence of 5 mg/L Suwannee River humic acid (*SRHA*) presented as normalized SWNT effluent concentration as a function of the number of pore volumes pumped through the packed quartz sand bed. Humic acid concentration is based on SRHA dry mass (corresponding to 2.54 mg/L as TOC). The breakthrough curves for a tracer (KBr) and SWNTs at 10 mM of KCl (with no SRHA) are also included. Column experiments were carried out at pH 7.0 (adjusted by adding 0.1 mM $KHCO_3$ as a buffer) (Jaisi et al. 2008). Reprinted with permission from Jaisi et al. (2008). Copyright 2008 American Chemical Society

12.2.3 Modeling ENP Transport in Porous Media

Gravitational sedimentation, interception, and Brownian diffusion are the basic capture mechanisms of colloidal particles in porous media (Schrick et al. 2004). Transport of ENPs in porous media and soils depends largely on the rate of their capture or filtration by the stationary granular surface. Lin et al. (2010) suggest use of particle filtration theory to understand the transport behavior of ENPs in soil–subsurface systems. They developed a schematic diagram for ENP transport (Fig. 12.25) showing that sedimentation (Fig. 12.25a) and interception (Fig. 12.25b) mechanisms mainly control the transport of larger pristine nanoparticles and nanoparticle aggregates. In the case of nanoparticles, diffusion is the dominant mechanism (Fig. 12.25c), because of the high diffusivity of nanoparticles which leads to a higher incidence of collisions with the porous media solid phase. When ENPs aggregate, they become microscale particles, which are retained by the porous medium according to a physical screening (Fig. 12.25d).

Particle filtration processes are determined largely by colloid–porous medium surface interactions and may be affected by particle surface chemistry, water pH, ionic strength, and organic matter content. In addition, physical parameters such as particle size and morphology, fluid velocity, and soil–subsurface water temperature should be considered in defining ENP breakthrough capability.

Characteristics of the soil–subsurface system influence ENP filtration. Both physical (e.g., porosity and aggregate size) and surface properties (e.g., charge) of the porous media solid phase affect ENP transport through soil–subsurface systems. For example, when soil particles are negatively charged, ENPs with high negative charge are more mobile and may be transported through the soil–subsurface down to the groundwater zone, while positively charged ENPs will be retained in the soil upper layer.

12.2.4 Fate of ENPs in Soil Column Experiments

Relatively few experimental studies of nanoparticle transport in laboratory columns of natural geological materials exist in the literature. One example is given by Sagee et al. (2012), who studied the transport of silver nanoparticles (AgNPs) in a set of soil column experiments (Figs. 12.26, 12.27, 12.28). The effect of soil aggregate size on AgNP transport is shown in Fig. 12.26a as can be seen from the similar breakthrough of the AgNPs in all three cases, the nanoparticle transport rate was not affected by the size of the aggregate. Moreover, the amount of AgNPs moving through the soil column is affected by the aggregate size and that the larger the aggregates, the more nanoparticles are eluted. For the coarse (largest aggregate) soil, about 70 % of the AgNPs that enter the soil column elute, while for the fine (smallest aggregates) soil, only about 30 % of the nanoparticles entering the soil column eluted. Sagee et al. (2012) noted that the AgNPs moved through

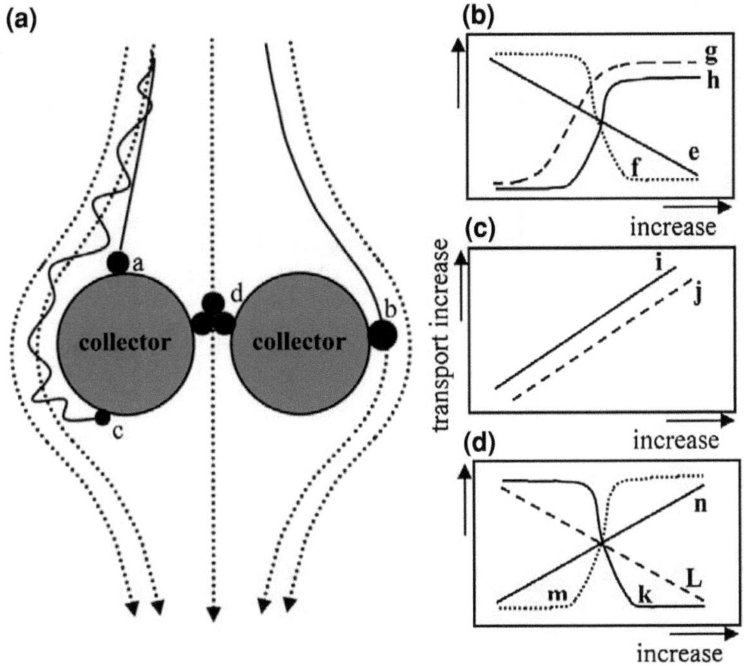

Fig. 12.25 Schematic diagrams for transport of engineered nanoparticles (*NPs*) in porous media. The filtration mechanisms are illustrated in (**a**), modified from Elimelech et al. (1995), whereas the influences of properties of NPs, collectors, and solution condition on the transport are presented in (**b**, **c**), and (**d**), respectively. The *black spheres*, *dotted lines*, and *thick lines* in (**a**) stand for NPs, the fluid streamlines, and the particle path (trajectory), respectively. The NPs can attach to the collector through (*a*) gravitational sedimentation, (*b*) interception, and (*c*) Brownian diffusion. (*d*) Large aggregates may be physically retained by small pores. Transport of NPs through the porous media (normally negatively charged) decreases with increasing (*e*) size, and (*f*) zeta potential of the NPs. The transport increases with increasing surface hydrophilicity of (*g*) the NPs and (*h*) coating of dispersants. The transport increases with increasing (*i*) size of the collectors, and (*j*) porosity of the media. The transport decreases with increasing (*k*) ionic strength, and (*l*) temperature of the fluid solution and increases with increasing (*m*) natural organic matter content of the fluid, and (*n*) fluid velocity (Lin et al. 2010)

the soil column relatively quickly, and after one pore volume reached the maximum eluting concentration. This "faster than conservative tracer" transport was explained by preferential flow of the AgNPs due to size exclusion and electrostatic repulsion of the negatively charged AgNPs from the soil aggregates. This explanation was further supported by comparison with the transport of AgNPs and tracer (bromide) in the same soil, as well as AgNP transport in a sand column, as shown in Fig. 12.26b. In both cases, the transport exhibited classical behavior, with no retention of the bromide in soil or the AgNPs in sand; the transport rate is indicated by $C/C_0 = 0.5$ for 1 pore volume.

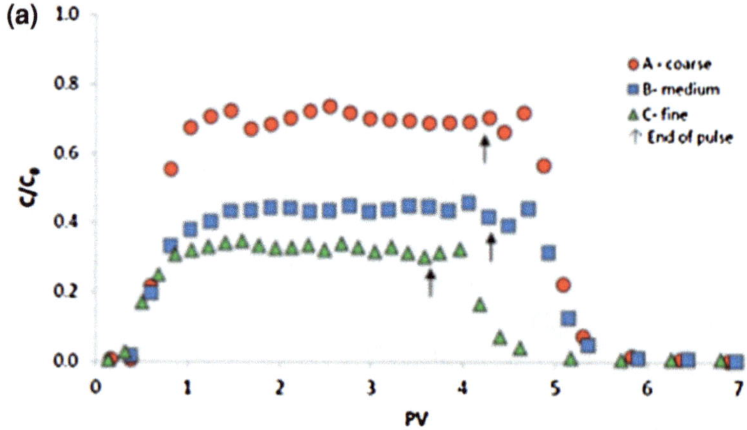

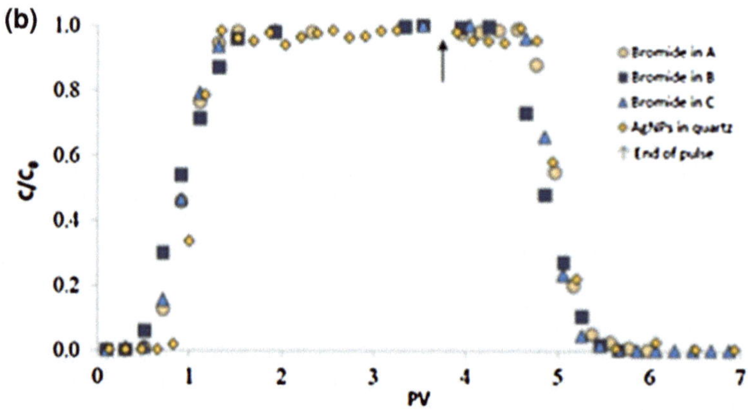

Fig. 12.26 Breakthrough curves for AgNPs in different soil mixtures **a** coarse: soil fraction of 250–1,000 μm, medium: 70 wt. % of the 250–1,000 μm and 30 % of the <250 μm diameter fraction, and fine: containing equal weights of the two fractions; Darcy velocity 0.66 cm/min. Average aggregate size is in the order coarse > medium > fine. **b** Breakthrough curves for bromide in the same soil mixtures coarse, medium, and fine, and for AgNPs in pure quartz (Darcy velocity 0.66 cm/min) (Sagee et al. 2012). Reprinted from Sagee et al. (2012). Copyright 2012 with permission from Elsevier

These results indicate that the AgNPs interact with the soil, influencing both retention and rate of transport. Another indication of AgNP-soil interactions is seen in Fig. 12.27, where AgNP transport in solutions with and without humic acid is compared. The presence of humic acids increases the concentration of eluting AgNPs from less than 50 to ~75 % of the initial concentration. This increase suggests that humic acids act as surfactants that stabilize nanoparticles in solution

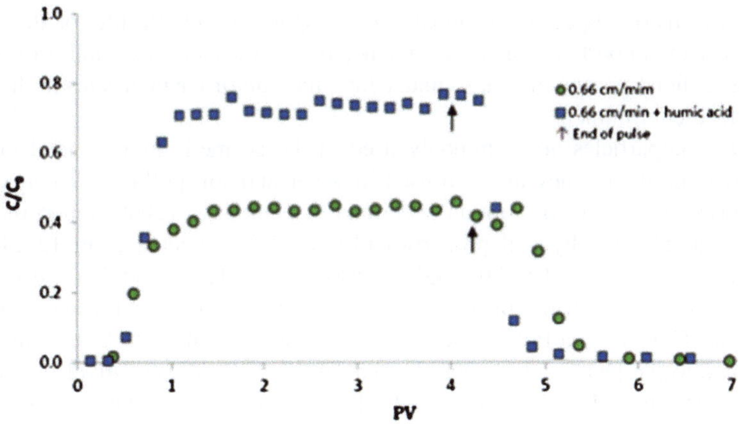

Fig. 12.27 Breakthrough curves for AgNPs in soil (Darcy velocity 0.66 cm/min) with and without the presence of humic acid in solution (Sagee et al. 2012). Reprinted from Sagee et al. (2012). Copyright 2012 with permission from Elsevier

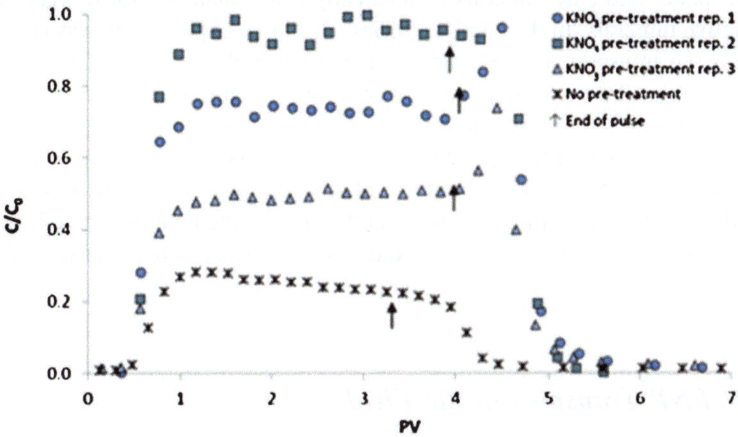

Fig. 12.28 Breakthrough curves for AgNPs in soil (Darcy velocity 0.17 cm/min) in columns pre-treated with potassium nitrate to remove soil residual chlorine, and a control column (Sagee et al. 2012). Reprinted from Sagee et al. (2012). Copyright 2012 with permission from Elsevier

and reduce their interactions (and thus retention) in the soil column. On the assumption that residual chlorides in the soil are the major components involved in the interaction with the nanoparticles, Sagee et al. (2012) tested the effect of washing the soil with nitrate solution (Fig. 12.28); nitrates were added in excess to the soil to replace all chloride residuals. The concentration of eluted AgNP increased dramatically in all cases, indicating that the chemical composition of the soil has a major impact on nanoparticle transport in soil. Results of this study emphasize the importance of studying nanoparticle transport in natural soils, rather than in artificial porous media, to understand the fate of nanoparticles in the soil–subsurface environment.

TiO_2 nanoparticles are commonly used in the cosmetic industry and in photocatalytic products; they are also used in water and soil pollutant decontamination. Based on a series of column experiments, Fang et al. (2009) report on TiO_2 mobility as affected by soil properties (Table 12.5) collected from 12 Chinese provinces. From the breakthrough curves of a TiO_2 aqueous suspension (Fig. 12.29), the effect of soil properties on ENP retention and transport may be seen. The TiO_2 nanoparticle content in soil suspensions after 24 h was positively correlated to the DOC and the clay content of the soils, and negatively correlated to ionic strength, pH, and zeta potential. The authors found that TiO_2 nanoparticles are generally stable in soil suspensions, and that higher stability leads to higher mobility. In a number of soils, the travel distances exceeded 30 cm, suggesting a potential risk of ENP movement down to the groundwater region.

In contrast to the results of fullerene transport through quartz sand, the studies of Wang et al. (2010) showed that transport of nC_{60} in natural soils containing organic matter and clay minerals is drastically lower than previously reported for nonreactive minerals. In this study, a series of column experiments was performed to assess the transport and retention of nanoscale fullerene aggregates in water-saturated soils. Two types of soils were used—Appling and Webster—which contain 0.75 and 3.33 % organic carbon, respectively. Complete retention of nC_{60} was observed even after the applied aqueous nC_{60} suspension pulse to Appling soil increased from 5 to 65 pore volumes. It also was found that the increase in pulse size led only to a minor increase in travel distance in the column, from 3 to 8 cm. Addition of 20 mg/L humic acid to the influent suspension increased fullerene travel transport in the Appling soil, but did not result in breakthrough.

12.2.5 ENP Transport in the Field

We consider two experiments related to zero valent zinc and copper oxide nanoparticle transport in field soils, as examples of potential nanomaterial transport pathways in soil–subsurface systems. Zinc oxide (ZnO) is one of the ENPs used frequently in personal care products, pharmaceuticals, catalysis, and environmental remediation products. Similarly, copper oxide (CuO) is used in electronics, ceramics, films, polymers, inks, metallics, lubricant oil, coatings, and health care products. Both types of ENPs may reach the land surface via sewage water, and waste disposals will be retained on the upper layers of the soil or transported with depth in the soil–subsurface system. Their prevalence virtually ensures that these ENPs may be considered chemical environmental contaminants.

The experiment of Collins et al. (2012) was performed on containers packed with an agricultural soil rich in total organic matter (13,100 mg/kg) with a pH of

Table 12.5 Selected properties of soils and soil solutions (Fang et al. 2009)

Soil										Soil suspension				
Soil	pH	CEC (cmol/kg)	SOM (%)	Zeta potential (mv)	Texture % (v/v)			Taxonomy[a]	Main mineralogy	pH	DOC (mg/L)	Ionic strength (mM)	Mg^{2+} (mg/L)	Ca^{2+} (mg/L)
					Clay (<2 μm)	Silt (2–20 μm)	Sand (20–2,000 μm)							
BJ	7.26	12.5	2.96	−18.2	19.4	47.8	32.8	Silt-sandy-loam	Feldspar, quartz, vermiculite	7.12	105.3	0.58	3.99	16.53
HLJ	6.39	20.2	6.86	−22.8	13.1	30	56.9	Loam	2:1 Montmorillonite and illite	6.15	149.4	0.44	7.28	0.21
JX	6.32	13.3	1.82	−25.9	32.3	47.4	20.3	Silt-clay-loam	Kaolinite, hematite	6.30	80.8	0.15	1.18	2.29
AH	6.67	14.2	2.51	−22.3	32.2	59.1	8.7	Silt-clay-loam	Kaolinite	6.58	84.7	0.32	1.95	5.59
SX	6.98	16.9	1.14	−20.5	45.6	23.5	30.9	Clay-loam	Quartz, mica, feldspar, chlorite	6.94	158.3	0.35	4.61	6.88
CD	7.4	18.9	4.46	−17.6	19.8	48	32.2	Silt-loam	Quartz, montmorillonite, illite	7.35	75.6	1.40	2.53	19.45
NMG	7.96	24.9	3.81	−17.2	13.3	23.1	63.6	Sandy-loam	Quartz, mica	8.32	62.8	1.23	2.93	26.92
JS	8.14	7.5	1.1	−17.4	7.8	34.5	57.7	Sandy-loam	Mica, vermiculite, kaolinite	8.50	87.6	1.01	1.38	19.52
CQ	7.99	27.3	1.1	−17.8	27.7	41.5	30.8	Clay-loam	Quartz, plagioclase, illite	8.22	86.4	1.09	2.11	24.64
HN	8.05	18.5	1.28	−16.9	10.7	72.3	17	Silt-loam	Mica, kaolinite, vermiculite	8.36	60.3	1.36	5.24	22.65
SD	7.5	9.5	1.97	−16.8	22.4	55.6	22	Silt-loam	Vermiculite, montmorillonite	7.85	57.7	3.08	8.16	45.41

(continued)

Table 12.5 (continued)

Soil										Soil suspension				
Soil	pH	CEC (cmol/kg)	SOM (%)	Zeta potential (mv)	Texture % (v/v)			Taxonomy[a]	Main mineralogy	pH	DOC (mg/L)	Ionic strength (mM)	Mg^{2+} (mg/L)	Ca^{2+} (mg/L)
					Clay (<2 μm)	Silt (2–20 μm)	Sand (20–2,000 μm)							
LZ	7.97	18.1	0.96	−16.5	20.5	56.7	22.8	Silt-sandy-loam	Quartz, illite, calcite, kaolinite	8.58	51.1	4.95	8.59	72.97

Reprinted from Fang et al. (2009). Copyright 2009 with permission from Elsevier

[a] Taxonomy was classified based on particle size distributions

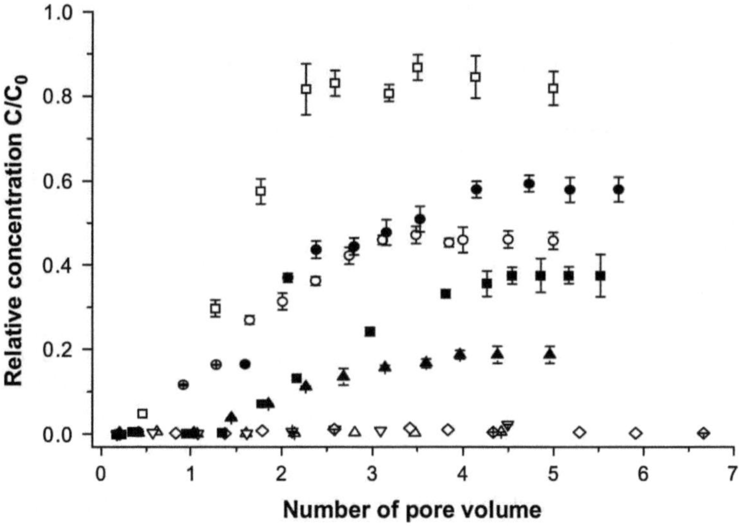

Fig. 12.29 Breakthrough curves of TiO_2 suspensions in various soil columns. *Small open square* BJ soil, *big open circle* HLJ soil, *up-pointing triangle* JX soil, *down-pointing triangle* AH soil, *white diamond* SX soil, *black square* CD soil, *black up-pointing triangle* NMG soil, *black circle* JS soil. For further soil details, see Table 12.5 (Fang et al. 2009). Reprinted from Fang et al. (2009). Copyright 2009 with permission from Elsevier

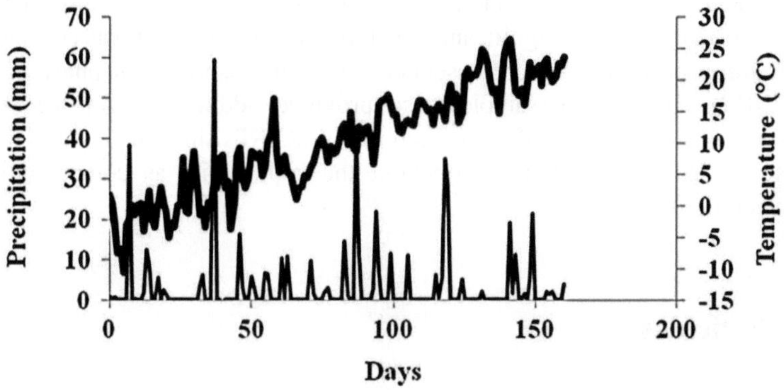

Fig. 12.30 Temperature and accumulative precipitation at the experimental site during field study on the transport of copper oxide (CuO) and zinc oxide (ZnO) (Collins et al. 2012)

7.5. The CuO and ZnO ENPs were disposed on the soil surface and allowed to freely migrate with depth while exposed to environmental field conditions over 160 days. Temperature and cumulative precipitation during the experiment at the field study site are presented in Fig. 12.30.

Background levels of CuO and ZnO in the soils were determined prior to ENP disposal, and CuO and ZnO redistribution with depth was determined at 1, 7, 30,

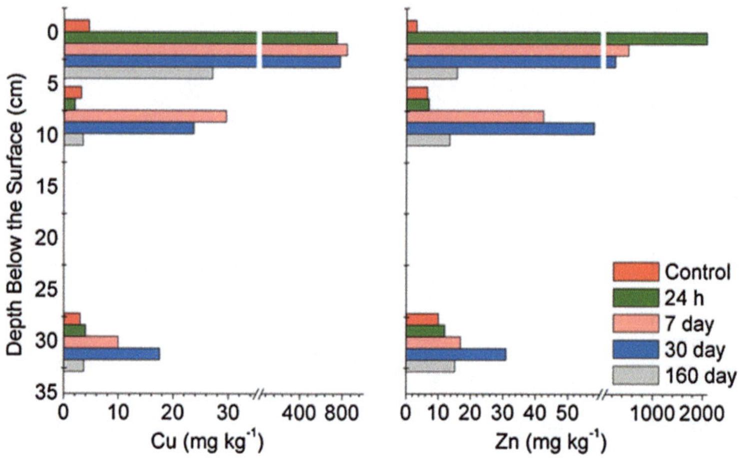

Fig. 12.31 Transport of Cu and Zn in soil cores as a function of depth (Collins et al. 2012)

and 160 days following addition of contaminants. The results showed no significant difference in the concentration of Cu and Zn in the surface layer and only modest variations in the second horizon at the third sampling period, compared with the 7-day samples (Fig. 12.31). In the third horizon, however, the concentration of Zn and Cu doubled. The significant increase in Cu ENPs is clearly demonstrated, suggesting continuous vertical transport of ENPs. Zinc speciation in the third horizon indicates significant dissolution of the ZnO and was corroborated by the strong post-edge feature associated with an adsorbed Zn phase (Collins et al. 2012). The 30-day samples also provided additional evidence for the increased mobility of ZnO NPs compared with CuO NPs since there was a marked increase in Zn concentrations throughout the soil profile as compared to Cu (Collins et al. 2012).

12.3 Pesticides

We discuss a series of experiments on redistribution of organic contaminants (pesticides) in irrigated fields between the years 1973 and 1999, reported in a number of research studies (Yaron et al. 1974; Gerstl and Yaron 1983; Yaron and Gerstl 1983; Indelman et al. 1998; Russo et al. 1998; Toiber-Yasur et al. 1999; Dror et al. 1999). The experimental fields were located at the Gilat and Bet-Dagan ARO-experimental stations, Israel, on loessial sandy loam soil (organic matter of 0.6 %; pH 8.4) and on sandy loam Mediterranean red soil (organic matter ranging from 1.2 to 0.7 %, from soil surface to 130 cm depth; pH ranging from 7.7 to 8.2). Under sprinkler or trickle irrigation and rainfall, the redistribution with depth of a number of nonconservative pesticides (Table 12.6), applied together with passive

Table 12.6 Pesticides used in field experiments

Pesticides	Formula	Solubility in water (mg/L)	Chemical name
Atrazine	$C_8H_{14}ClN_5$	30	2-Chloro-4-ethylamino-6-isopropylamino-1,3,5-triazine
Azinphosmethyl	$C_{10}H_{12}N_3O_3PS_2$	30	O,O-Dimethyl S-[(4-oxo-1,2,3-benzotriazin-3 (4H)-yl) methyl] phosphorodithioate
Bromacil	$C_9H_{13}BrN_2O_2$	813–815	5-Bromo-3-sec-butyl-6-methyluracil
Napropamide	$C_{17}H_{21}NO_2$	73	(R,S)-N,N-diethyl-2-(1-naphthyloxy) propionamide
Parathion	$C_{10}H_{14}NO_5PS$	24	O,O-Diethyl O-4-nitrophenyl phosphorothioate
Prometryn	$C_{10}H_{19}N_5S$	33–48	N^2,N^4-Diisopropyl-6-methylthio-1,3,5-triazine-2,4-diamine
Tetradifon	$C_{12}H_6Cl_4O_2S$	<0.1	4-Chlorophenyl 2,4,5-trichlorophenyl sulfone
Terbuthylazine	$C_9H_{16}ClN_5$		N^2-Tert-butyl-6-chloro-N^4-ethyl-1,3,5-triazine-2,4-diamine

(conservative) soluble tracers (chloride or bromide), was determined periodically. Transport of contaminants was determined for both nonpoint (areal) and point application on the field surface.

Field transport of a persistent organochlorine (tetradifon) and a degradable organophosphorous (azinphosmethyl) insecticide was examined under dry and wet irrigation treatments followed by winter precipitation (Yaron et al. 1974). In the dry irrigation treatment, the moisture deficit was replenished periodically in the 0–60-cm soil layer; in the wet treatment, an additional 30 % volume of water was applied in the same irrigation cycles. Redistribution profiles of the two insecticides in the 0–30 cm (a and b) and 0–110 cm (c) soil during the irrigation season are presented in Fig. 12.32. We see no significant differences in azinphosmethyl concentration between the two irrigation treatments. Azinphosmethyl concentration decreased as a result of volatilization or biologically induced degradation. Tetradifon, however, exhibited a concentration decrease under the wet irrigation regime, on account of its downward movement. Tetradifon is persistent and relatively soluble in water, and it was found in the soil profile up to a depth of 110 cm after the rainfall period (when the amount of leaching water, irrigation and rainfall, was 618 mm for the dry treatment and 770 mm for the wet treatment; see Fig. 12.32c). While in the dry irrigation treatment, the concentration front reaches a depth of 60 cm, the wet treatment caused the front to reach a depth of 90 cm. Significantly, the winter rainfall was sufficient to eliminate concentration differences created by the irrigation regimes.

In a separate study, the effect of spatial variability on the transport of nonconservative and conservative contaminants, under leaching by irrigation and rainwater, was studied in an experimental 175-m^2 plot located on a sandy loam Mediterranean red soil. The redistribution of two nonconservative herbicides

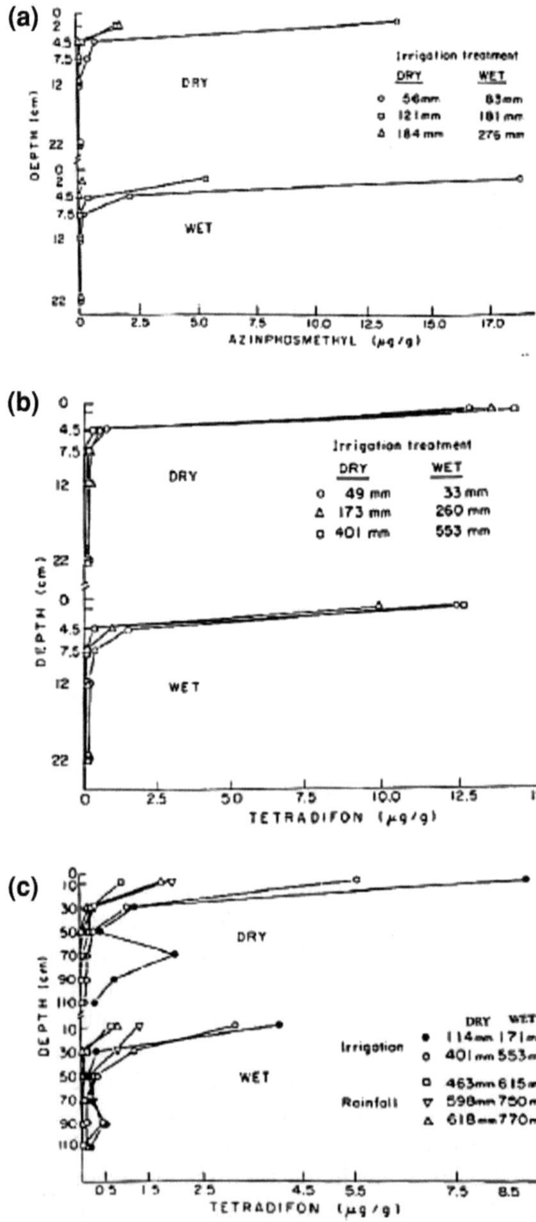

Fig. 12.32 Transport of **a** azinphosmethyl, and of **b** tetradifon in the loessial loamy soil *upper layer* (0–30 cm) under two irrigation regimes (*dry* and *wet*) and **c** tetradifon to a depth of 110 cm under irrigation and rainfall (Yaron et al. 1974)

(terbuthylazine and bromacil) and of the conservative $CaBr_2$ was measured to a depth of 120 cm, following two sprinkler irrigations of 73 and 78 mm, respectively. In addition, the field was sampled to a depth of 400 cm during the rainy season that followed (with a total of 470 mm rainfall). Cumulative irrigation and

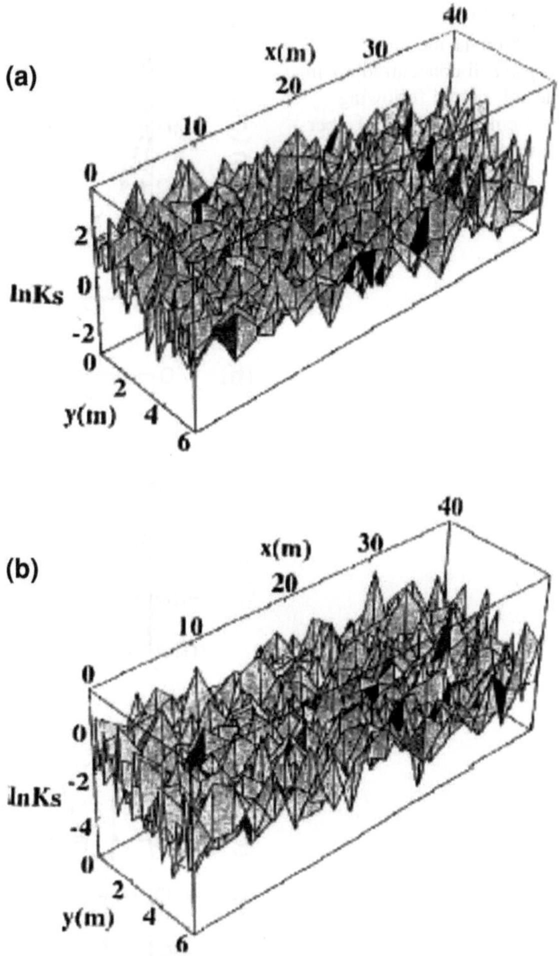

Fig. 12.33 Distribution of ln Ks (saturated hydraulic conductivity) at two different soil depths, **a** 0.3 m, and **b** 1 m, obtained from a single realization of conditional simulations of ln Ks (Toiber-Yasur et al. 1999)

rainfall leaching occurred over 250 days. The experimental results are summarized in Tables 12.7 and 12.8 and Figs. 12.33 and 12.34.

The spatial distributions of bromacil and terbuthylazine adsorption–desorption coefficients, both vertically and laterally, in the experimental field are given in Tables 12.7 and 12.8. The effect of vertical variability is shown in Table 12.7, while the lateral spatial variability is shown in Table 12.8. The vertical and lateral spatial variabilities were defined on the basis of either the measured adsorption coefficient (K_d), as generated from adsorption isotherms on soil profiles, or on adsorption coefficients on soil organic matter (K_{oc}), calculated as adsorption on organic carbon per unit weight of soil. We see that both vertical (Table 12.7) and lateral (Table 12.8) variability of soil affect the adsorption coefficients. A comparison between the bromide (conservative) and the two nonconservative herbicides distributions with depth after about 900 mm of leaching is shown in Fig. 12.34. We see

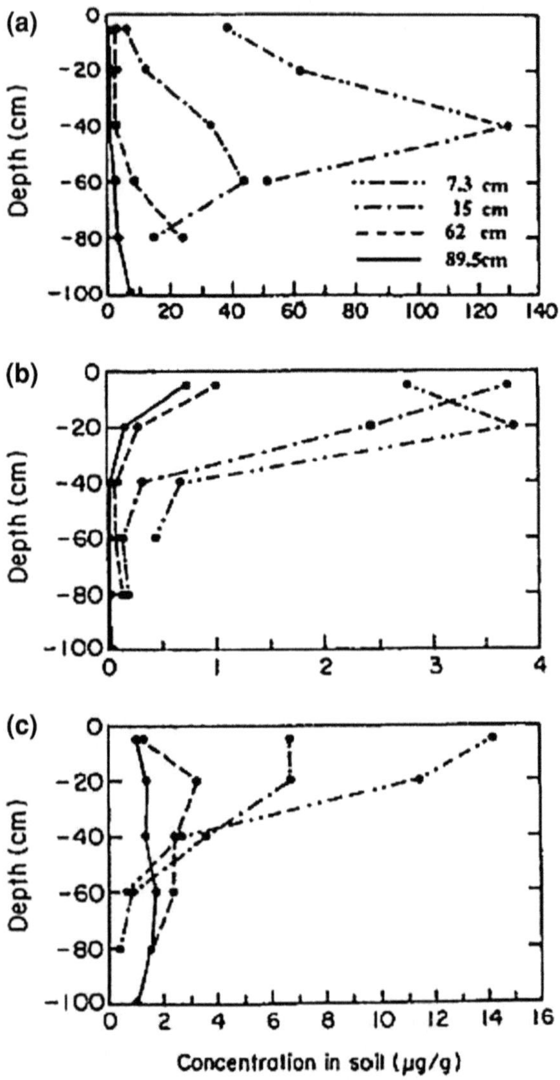

Fig. 12.34 Average **a** $CaBr_2$, **b** terbuthylazine, and **c** bromacil concentrations in the soil profile following 89.5 cm of leaching (Toiber-Yasur et al. 1999)

that, in the case of bromide, there is a continuous displacement of the center of mass with cumulative infiltration. In contrast, the bulk of the herbicide contaminant mass remains in the upper soil layer, with very little displacement.

Figure 12.35 shows the spatial distribution of terbuthylazine and bromacil concentrations in a vertical cross section, after leaching by successive irrigations. Differences between the concentrations of the two herbicides may be explained in terms of the chemical properties and the soil spatial variability. The bulk of the terbuthylazine remained close to the surface, being strongly affected by adsorption.

Table 12.7 Adsorption ($K_{d(\mathrm{ads})}$, K_{oc})–desorption $K_{d(\mathrm{des})}$ coefficients for bromacil and terbuthylazine, as affected by the vertical variability of the soil profile (Toiber-Yasur et al. 1999)

Soil depth (cm)	Bromacil			Terbuthylazine		
	K_{oc} (mL/g)	$K_{d(\mathrm{des})}$ (mL/g)	$K_{d(\mathrm{ads})}$ (mL/g)	K_{oc} (mL/g)	$K_{d(\mathrm{des})}$ (mL/g)	$K_{d(\mathrm{ads})}$ (mL/g)
0–10	56	1.7	0.7	164	4.6	2.0
10–30	65	1.5	0.8	249	3.9	2.9
30–50	61	1.2	0.5	255	2.4	1.9
50–70	96	2.6	0.6	356	2.4	2.1
70–90	77	1.1	0.5	304	2.5	2.1
90–110	111	1.9	0.7	358	3.4	2.3
110–130	108	1.7	0.7	365	3.1	2.4

Notation K_d adsorption coefficient; K_{oc} adsorption on organic carbon per unit weight of soil

Table 12.8 Effect of field site spatial variability on adsorption coefficients of terbuthylazine in soil upper layer (Toiber-Yasur et al. 1999)

Sample location	Soil organic matter (%)	K_d	K_{oc}
2	1.19	1.59	133.27
7	1.01	1.39	137.35
9	1.04	1.14	109.19
14	1.12	1.54	136.53
17	1.18	1.54	131.06
18	1.05	1.32	125.15
19	0.89	1.06	117.92
10	1.22	2.00	163.10

Sample location was chosen randomly. *Notation*: K_d adsorption coefficient; K_{oc} adsorption on organic carbon per unit weight of soil

In contrast, bromacil has a higher solubility in water and is leached downward under the influence of the spatial variability in hydraulic conductivity.

Concentrations of terbuthylazine and bromacil in the experimental field at a depth of 400 cm, after leaching (irrigation and rainfall) of 110 cm water, are shown in Fig. 12.36. The total amount of bromacil and terbuthylazine recovered from the soil profile after this leaching period was much smaller than the amount of chemicals applied initially. This result suggests that the reduction in concentration of herbicide due to degradation is greater than that caused by leaching. Similar results were obtained when an additional herbicide, atrazine, was applied together with terbuthylazine on a neighboring experimental plot and leached by irrigation and rainwater during a two-year period (Dror et al. 1999).

The field experiments from the Bet-Dagan site were used to test different theoretical models for field-scale chemical transport. One study expanded a simple column model for flow and transport in partially saturated soils (Bresler and Dagan 1983), by using a stochastic analysis (Indelman et al. 1998). The approach ignored

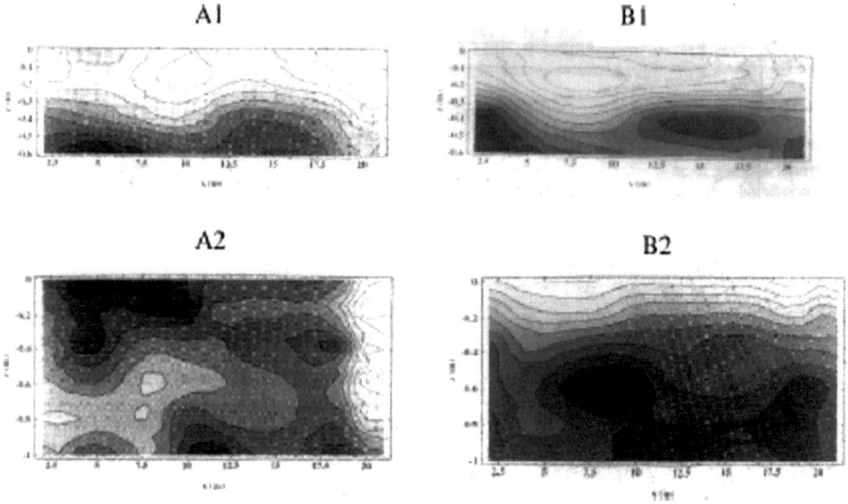

Fig. 12.35 Distribution of the resident concentration of **A** terbuthylazine and **B** bromacil in a vertical cross section after leaching with (**A**1, **B**1) 7.3 cm and (**A**2, **B**2) 89.5 cm of applied water. Contour spacing is 0.1 times the mean concentration (Toiber-Yasur et al. 1999)

the effects of contaminant dispersion and assumed that transport is controlled by three mechanisms: advection by water flow, equilibrium sorption, and exponential mass loss. Transport of conservative contaminants (advection only) was modeled by averaging the solution over probability density functions for the various model parameters (including, saturated hydraulic conductivity, water content, and rate of water application); experimental data for bromide concentrations were used to constrain parameter value estimates. The transport of two reactive chemicals, bromacil and terbuthylazine, then was modeled. The calculated and measured concentrations of bromide, bromacil, and terbuthylazine, 18 and 32 days after chemical application in the irrigated field, are shown in Fig. 12.37.

The analysis was limited in part by the scarcity of measurements, and clear discrepancies between measured and calculated values may be observed. As discussed in Chap. 10, tailing effects often are due to non-Fickian transport behavior, which was not accounted for in this model. Interestingly, the field-scale retardation coefficient values of the reactive contaminants were smaller by an order of magnitude than their laboratory values, obtained in an accompanying experiment.

Another modeling analysis is presented by Russo et al. (1998), who examined field transport of bromacil by application of the classical one-region, advection–dispersion equation (ADE) model and the two-region, mobile–immobile model (MIM); recall Sects. 10.1 and 10.2. The analysis involved detailed, three-dimensional numerical simulations of flow and transport, using in situ measurements of hydraulic properties and adsorption–desorption and degradation rate coefficients obtained from laboratory measurements.

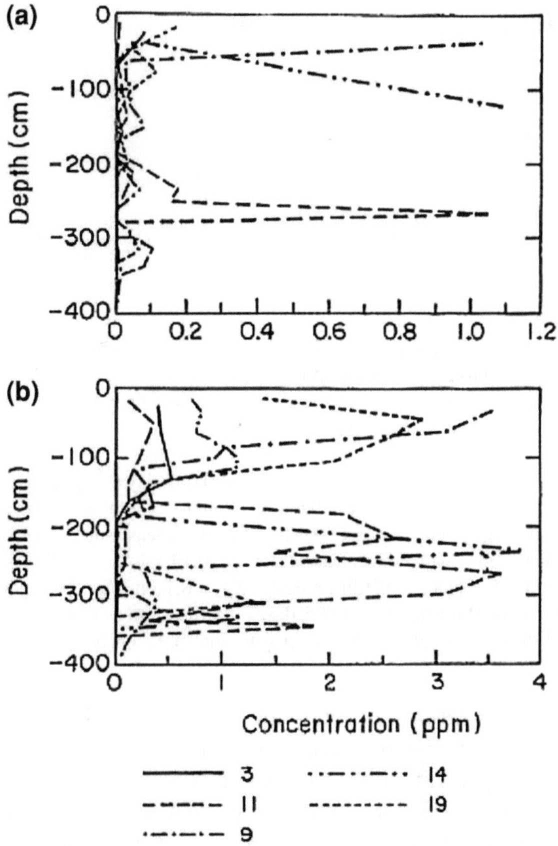

Fig. 12.36 Concentrations of **a** terbuthylazine and **b** bromacil in the experimental fields (5 core samples, 400 cm depth) one year after application of chemicals and leaching with 110 cm of irrigation and rainwater (Toiber-Yasur et al. 1999)

Comparisons between the results obtained from the field-scale experiment and the calculated bromacil concentrations using the ADE and MIM approaches (0–200 cm depth) are presented in Fig. 12.38. The analysis is restricted to characterizing the measured bromacil concentration profiles expressed (in terms of mass of soil) in terms of horizontal spatial averages at a given time. Both measured and simulated concentration profiles demonstrate that the downward movement of bromacil is limited by its adsorption to the soil. Russo et al. (1998) note that, for a given flow regime, mass exchange between the mobile and immobile regions retards bromacil degradation and affects the spatial distribution of bromacil. At relatively large travel times, mass exchange also affects the spatial moments of the distribution and increases the skewing of the bromacil breakthrough curves and the prediction uncertainty, as compared to soils that contain only a single (mobile) region. At relatively short travel times, both the ADE and MIM approaches yield similar results. In the case of the Bet-Dagan field experiment, the transport of bromacil under leaching is somewhat better described by the MIM than by the ADE model. However, neither approach properly captures the complete profile.

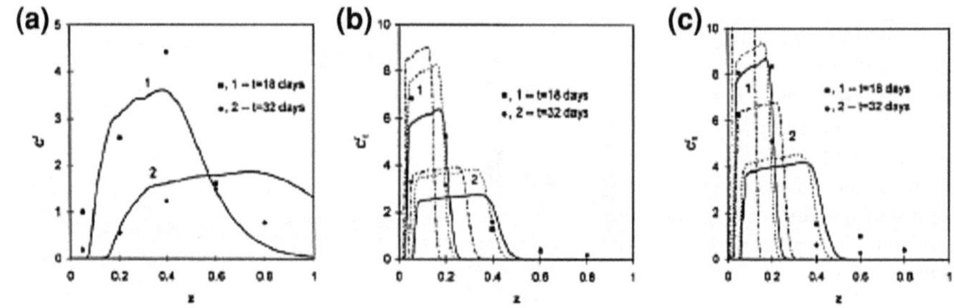

Fig. 12.37 **a** Calculated (*solid curves*) and measured (*points*) mean bromide concentrations, normalized by the mass of applied solute per unit soil surface, at times $t = 18$ and 32 days; **b** Mean bromacil concentrations, normalized by the mass of applied solute per unit soil surface, at times $t = 18$ and 32 days (*solid curves* best-fit based on least squares of differences between computed and measured mean concentrations, $K_d = 0.14$ mL/g, $\lambda = 0.022$ days^{-1}; *dashed curves* based on sum of squares of differences, $K_d = 0.16$ mL/g, $\lambda = 0.012$ days^{-1}; *dotted-dashed curves* $K_d = 0.24$ mL/g, $\lambda = 0.022$ days^{-1}; *points* measurements); **c** Mean terbuthylazine concentration, normalized by the mass of applied solute per unit soil surface at times $t = 18$ and 32 days (*solid curves* best-fit based on least squares of differences between computed and measured mean concentrations, $K_d = 0.14$ mL/g, $\lambda = 0.005$ days^{-1}; *dashed curves* based on sum of squares of differences, $K_d = 0.16$ mL/g, $\lambda = 0.005$ days^{-1}; *dotted-dashed curves* $K_d = 0.29$ mL/g, $\lambda = 0.005$ days^{-1}; *points* measurements). Reprinted from Indelman et al. (1998). Copyright 1998 with permission of Elsevier

The limitations of both the ADE and MIM formulations were discussed in Chap. 10; models that describe non-Fickian transport behavior remain to be further developed and applied to such field studies.

Two particular aspects of the transport of degradable contaminants were considered in laboratory experiments that used soil originating from the field experiments described in the previous sections. Studies on diffusion of degradable insecticides were performed in diffusion cells, while the spatial redistribution of pesticides from a point source was measured in specially designed pans (60 cm high, 40 cm diameter). Periodic sampling and contaminant analysis enabled visualization of the contaminant transport pathway.

Diffusion of degradable pesticides in a natural soil may be tested against diffusion in the same, biologically inactivated, soil. Gerstl et al. (1979a, b) investigated diffusion of parathion in a loessial sandy loam soil (Gilat) before and after "sterilization." In the sterilized soil, it was found that the apparent diffusion coefficient (D) of parathion over a wide range of moisture contents varies between 0.66×10^{-7} and 3.39×10^{-7} cm^2/s. The main factors affecting parathion diffusion in a sterile soil are the adsorption coefficient of the compound on the solid phase, the moisture content, and the impedance factor (which is determined by the tortuosity of the porous medium). In the presence of microorganisms, parathion is degraded as it diffuses in soil.

Figure 12.39 shows an example of parathion distribution in sterilized and natural, biologically active Gilat soil columns. We see that, at relatively early times, when the effect of decomposition is minimal, the parathion distribution is

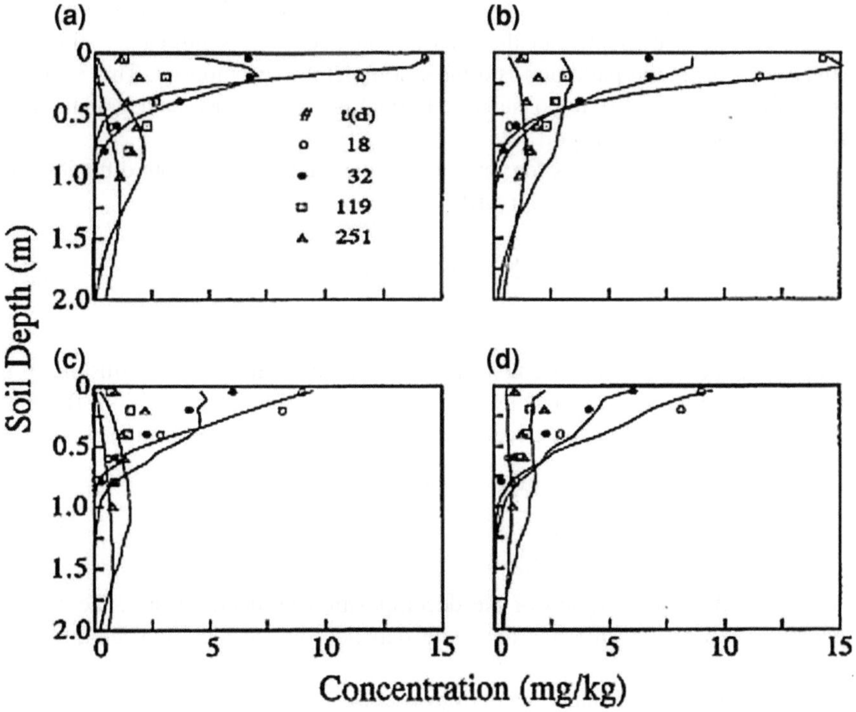

Fig. 12.38 Profiles of means (**a**, **b**) and standard deviations (**c**, **d**) of the bromacil concentrations at four different time points. *Solid curves* denote simulated profiles obtained from the advection–dispersion equation (**a**, **c**) and the mobile–immobile model (**b**, **d**). The different symbols denote measured profiles at different times. Reprinted from Russo et al. (1998). Copyright 1998 with permission of Elsevier

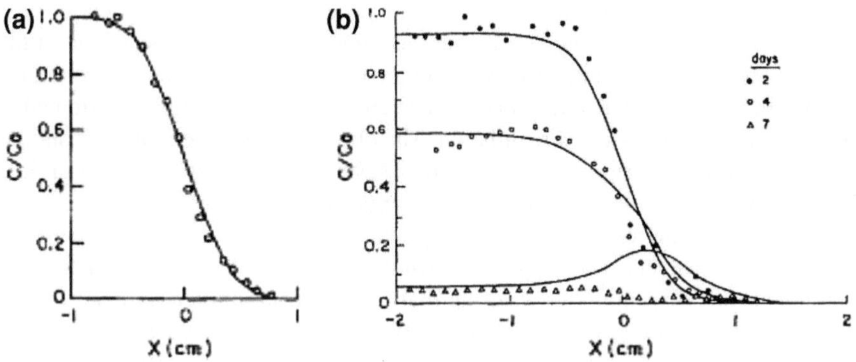

Fig. 12.39 Distributions of parathion in **a** biologically inert (sterile) Gilat soil (20 % moisture content, 1.4 g/cm^3 bulk density) after 3.03 days, and **b** biologically active Gilat soil (34 % soil moisture content, 1.4 g/cm^3 bulk density) after 2, 4, and 7 days. The solid curves were calculated using $D = 1.67 \times 10^{-7}$ cm^2/s; the points represent experimental measurements (Gerstl et al. 1979a)

similar to that in the sterile soil. After four days, the effect of microbial activity on decomposition is evident, and the distribution pattern is significantly different. After seven days, the parathion is almost completely decomposed. This example emphasizes the necessity to consider additional processes, such as degradation, in analyses of pollutant transport.

An equation that accounts for parathion transport, subject to both diffusion and microbial decomposition, can be written as

$$\frac{\partial c}{\partial t} = D\frac{\partial^2 c}{\partial x^2} - R_{xt}, \quad (12.1)$$

where c is the concentration, D is the coefficient of molecular diffusion, x is distance, t is time, and R_{xt} denotes the rate of microbial decomposition. Because parathion is converted quantitatively, mole for mole, its decomposition product is

$$\frac{\partial c'}{\partial t} = D'\frac{\partial^2 c'}{\partial x^2} - R_{xt}, \quad (12.2)$$

where c' is the concentration of the decomposition product. Clearly, the rate of decomposition at any distance and time depends on the local concentration of parathion and on microbial activity.

The transport of a contaminant from a *point source* into the subsurface was examined for two herbicides (napropamide and bromacil), by introducing them through a single-drip irrigation source. Lateral and vertical sampling was performed to determine the spatial distribution, over time, of water and contaminants; results for one-time and one-application rate are shown in Fig. 12.40. In general, napropamide was moderately adsorbed by the soils and concentrated around the emitter. At a high application rate, the compound moved laterally more than vertically, and at a lower application rate, it penetrated to a greater depth. After several cycles of wetting and drying, napropamide was partially leached out of the emitter zone. The amount leached is related to the soil hydraulic properties: Bromacil, which is weakly adsorbed by the soil, was more uniformly distributed in the soil but did exhibit slight retardation. The effect of application rate was similar to that observed for napropamide, and after several cycles of wetting and drying, bromacil was completely leached from vicinity of the emitter.

Preferential flow of bromacil and napropamide through undisturbed cores of structured Evsham clay soil (Aquic Eutrochrept) from Oxfordshire, England, under continuous and discontinuous watering regimes and with different initial moisture contents, was reported by White et al. (1986). A conservative solute, chloride, was used as to trace water movement. Adsorption coefficients range between 1.7 and 2.0 L/kg for bromacil and between 17.8 and 22.7 L/kg for napropamide. In a dry soil, bromacil attains equilibrium with the solid phase almost instantaneously, while napropamide shows a delay of 2–3 h. In an initially wet soil, napropamide does not reach equilibrium with the solid phase even after 48 h.

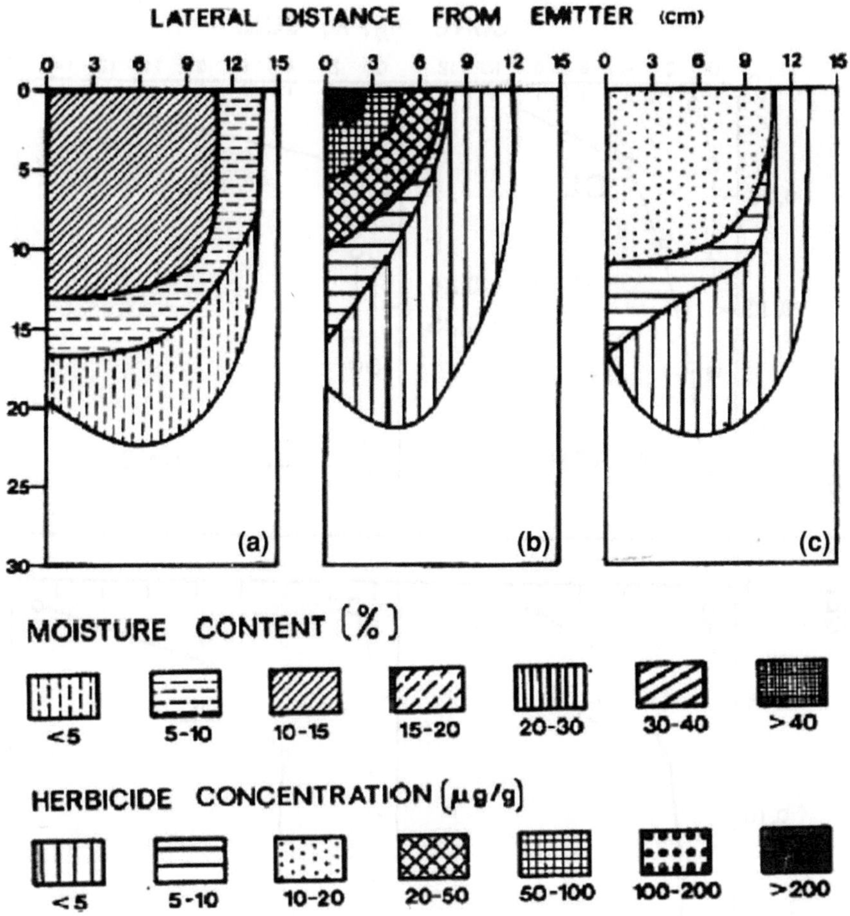

Fig. 12.40 Distribution of **a** moisture, **b** napropamide, and **c** bromacil 24 h after application of the herbicide solution to Bet-Dagan soil, at 4 L/h from a point source (Gerstl et al. 1979a)

Differences in the leaching of bromacil and napropamide through the undisturbed soil columns were observed; in particular, there is a striking effect of initial soil water content on herbicide transport (Fig. 12.41). Pre-wetted soil was found to retain more herbicide than initially dry soil, and the highest concentration of herbicide was consistently found in the uppermost soil layer. Napropamide has a high-adsorption coefficient, and this, in general, should be the dominant process. However, because adsorption equilibrium is not attained instantaneously and water flow is rapid especially in the dry soil, leaching also occurs. This is a typical case of preferential flow. Discontinuous leaching of the pre-wetted soil permits more of both herbicides to be retained in the surface layer than continuous leaching, presumably because diffusion of the herbicides into aggregates occurs during the interval between leachings. The results of White et al. (1986) clearly show that, for

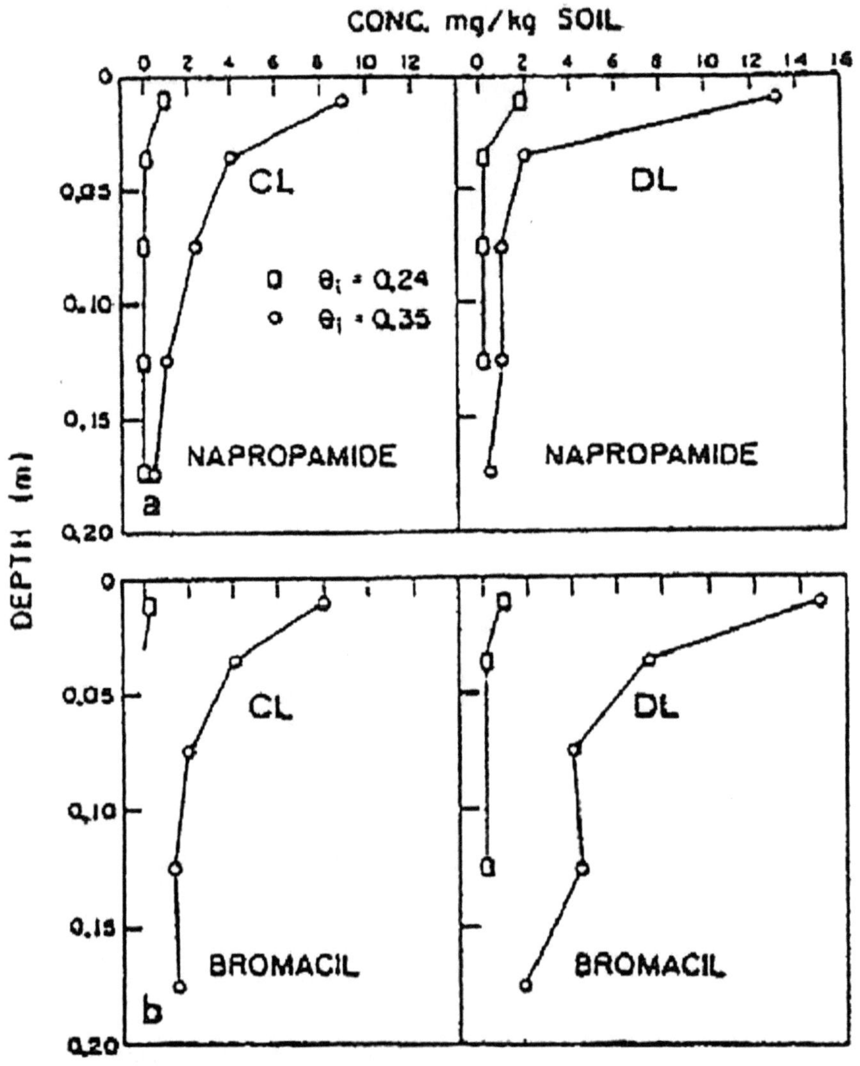

Fig. 12.41 Distribution of napropamide and bromacil in undisturbed soil columns, under continuous (*CL*) and discontinuous (*DL*) leaching (White et al. 1986). *Square* and *round symbols* denote two different initial soil moisture contents

an undisturbed structured soil, even strongly sorbing chemicals are vulnerable to leaching through macropores, cracks, and other channels. Similar behavior in napropamide transport was observed in a field experiment in California conducted by Jury et al. (1986).

Another field experiment investigating preferential flow of atrazine, napropamide, and prometryn, under various agricultural irrigation practices, is reported by Ghodrati and Jury (1990). The field site, on a loamy sand soil, included the

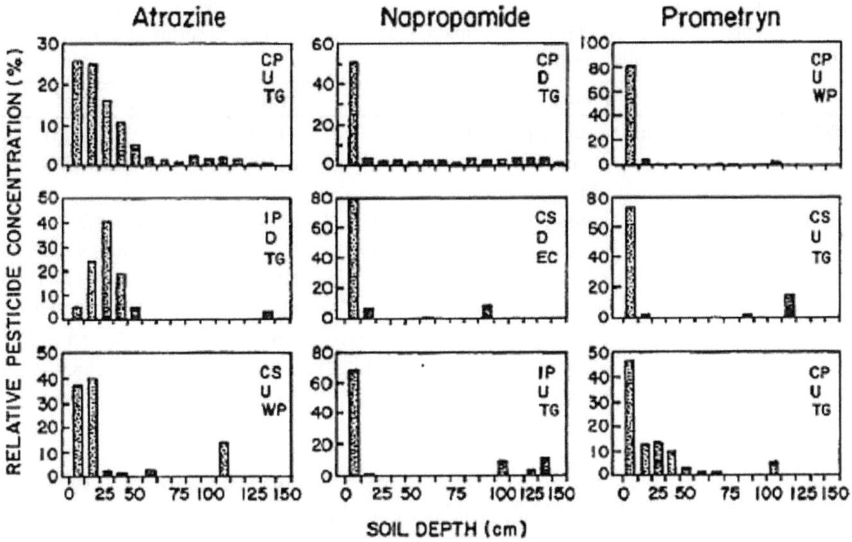

Fig. 12.42 Examples of pesticide concentration versus depth, indicating preferential flow of each pesticide in individual field plots. *Symbols*: *C* continuous; *I* intermittent; *P* ponding; *S* sprinkling; *TG* technical grade; *EC* emulsifiable concentrate; *WP* wettable powder; *U* undisturbed; *D* disturbed (Ghodrati and Jury 1990)

following treatments: two formulations of each herbicide, two soil-surface preparations, and four flow conditions involving continuous or intermittent sprinkler and flood irrigation. Immediately after herbicide application, the soil was irrigated and sampled to a depth of up to 150 cm after six days. From the results presented in Fig. 12.42, we see that the chemicals were transported to a depth of more than 100 cm; according to their retardation factor values, these chemicals would be expected to be far less mobile. Differences in the redistribution with depth were due to the various treatments, but the general trend of downward transport was attributed to preferential flow.

12.4 Colloid-Mediated Transport of Organic Contaminants

Organic contaminants retained on colloid surfaces may be transported by flowing water through the vadose zone and reach the groundwater or be relocated on the land surface by runoff processes. A series of field and laboratory experiments identified the spatial redistribution of contaminants by surface runoff and contamination of the subsurface by vertical transport of colloids. These findings demonstrate that colloid-mediated transport of contaminants is a major environmental hazard.

Table 12.9 Colloidal (1 μm–1 nm) and dissolved (<1 nm) molybdate-reactive phosphorus (MRP) and major cation concentrations in runoff, generated by simulated sprinkler irrigation onto three semiarid, arable soils (Turner et al. 2004)

	Filter size	Portneuf	Warden	Greenleaf
MRP, % of MRP <1 μm	1 μm–1 nm	13–22 (18)	11–39 (24)	37–56 (45)
	<1 nm	78–87 (82)	61–89 (76)	44–63 (55)
	μM			
MRP	1 μm–1 nm	0.5–3.1 (1.4)	0.2–1.5 (0.5)	0.3–0.7 (0.5)
	<1 nm	2.1–16.6 (6.8)	0.5–5.8 (1.8)	0.4–0.8 (0.6)
Al	1 μm–1 nm	2.2–22.2 (9.3)	ND[a]–4.5 (1.5)	1.1–7.8 (4.1)
	<1 nm	ND–0.7 (ND)	ND–3.0 (0.7)	ND–3.0 (1.1)
Ca	1 μm–1 nm	3.5–61.1 (31.7)	0.5–12.7 (7.2)	2.5–10.5 (8.2)
	<1 nm	62.1–156.2 (127.8)	2.3–15.0 (9.0)	53.2–68.9 (62.1)
Fe	1 μm–1 nm	0.7–5.4 (2.2)	ND–2.2 (0.7)	0.5–13.1 (4.5)
	<1 nm	ND–0.2 (ND)	ND–1.6 (0.4)	ND–0.2 (ND)
K	1 μm–1 nm	2.6–14.1 (6.1)	1.0–17.1 (7.4)	<0.3–1.5 (0.8)
	<1 nm	19.7–63.7 (35.6)	4.9–28.9 (16.9)	29.9–30.1 (38.1)
Mg	1 μm–1 nm	2.1–7.8 (5.4)	ND–3.7 (2.1)	3.3–7.8 (5.4)
	<1 nm	13.2–40.3 (27.6)	0.8–3.7 (2.5)	12.8–18.5 (15.6)
Si	1 μm–1 nm	5.0–59.5 (27.1)	0.7–14.2 (5.3)	4.6–22.1 (13.5)
	<1 nm	13.9–51.3 (35.6)	4.6–16.0 (8.2)	19.6–29.2 (24.6)

Values are the range of concentrations from runoff generated on dry and wet soils, with the mean value in parentheses. Values for the Portneuf soil are flow-weighted means of six samples taken at 5-min intervals during 30 min of continuous runoff. Values for the Warden and Greenleaf soils are means of samples taken from each of two subplots after 30 min of continuous runoff
[a] Not detectable (less than the limit of detection)

12.4.1 Surface Runoff

Transport of contaminants by surface runoff is illustrated in the experimental results of Turner et al. (2004), which deal with the colloid-mediated transfer of phosphorus (P) from a calcareous agricultural land to watercourses. Colloidal molybdate-reactive phosphorus (MRP) was identified by ultrafiltration associated with particles between 1 μm and 1 nm in diameter. Colloidal P compounds can constitute a substantial component of the filterable MRP in soil solution and include primary and secondary P minerals, P occluded or adsorbed on or within mineral or organic particles, and biocolloids (Kretzschmar et al. 1999).

The experiments were conducted on three plots of semiarid soils from the western United States, characterized by various concentrations of $CaCO_3$ and organic carbon. Surface runoff, generated by simulated sprinkler irrigation, was applied twice on each plot: first to yield a moisture content of 2–5 % then again on the same plot at a moisture content of 17–29 %. Colloidal (1 μm–1 nm) and dissolved (<1 nm) MRP and major cation concentrations in the generated runoff are presented in Table 12.9. In the runoff colloids, MRP concentrations range between 0.16 and 3.07 μM, constituting 11–56 % of the MRP in the <1 μm fraction. Turner et al. (2004) suggested that the soils may have two size fractions of colloidal MRP: a large fraction (0.2–1.0 μm) associated with Al and Fe, which contains phosphate adsorbed to fine clays, and a smaller fraction (0.3–3 nm) probably associated with Ca and Mg (suggesting the presence of calcium phosphate and magnesium phosphate minerals). Large concentrations of colloidal MRP in runoff suggest significant phosphate mobility, which may lead to phosphate enrichment of surface water and negative effects on water quality, or greater vertical phosphate mobility into the subsurface.

12.4.2 Transport in the Subsurface

Colloid-facilitated transport of organic compounds occurs frequently in the subsurface. Organic compounds with very high K_d values (and low solubility) are virtually immobile in the subsurface. As such, their presence in the groundwater suggests that they migrate through the subsurface together with colloidal materials rather than as dissolved solutes. Karathanasis et al. (2005), in an experiment carried out on undisturbed soil monoliths, investigated the ability colloidal fraction of organic biosolid applied to the soil as agricultural amendments to be transported through the soil–subsurface cross section. Water-dispersible colloids fractionated from lime-stabilized biosolids (LSB), aerobically digested biosolids (ADB), and poultry manure biosolids (PMB) were applied to the soil monoliths, and effluents were monitored for heavy metal and colloid concentration over 16–24 pore volumes of leaching. Breakthrough curves for Cl^- and colloids eluted from three soils experiments are presented in Fig. 12.43. Colloid breakthrough curves generally were irregular but in all cases showed a slower breakthrough than the conservative Cl^- tracer, indicating significant interaction with soil during leaching. Irregularity of the breakthrough curves may be also affected by different preferential flows in each soil.

Early experimental tests on vertical transport of pesticides through soils demonstrated this phenomenon (Vinten et al. 1983). Figure 12.44 demonstrates vertical transport of an herbicide (paraquat) and a very persistent insecticide (DDT) adsorbed on suspended materials through various soil columns leached with aqueous solution. Figure 12.44a shows the results of an experiment using ^{14}C-labeled paraquat adsorbed on a clay mineral (Li-montmorillonite) suspension through a soil column. When the suspension medium was distilled water, 50 % of the pesticides penetrated beyond 12 cm. Under these conditions, clay remains dispersed and pesticide is readily transported through the soil. However, for a suspension medium with an electrolyte concentration of 1 mM $CaCl_2$, paraquat remains in the upper 1-cm layer. The high calcium concentration results in rapid immobilization of the clay in the soil through flocculation, and consequently, little pesticide transport occurs.

The redistribution of DDT with depth also was tested, in the presence of organic suspended solids from sewage effluents. Figure 12.44b shows a range of behaviors

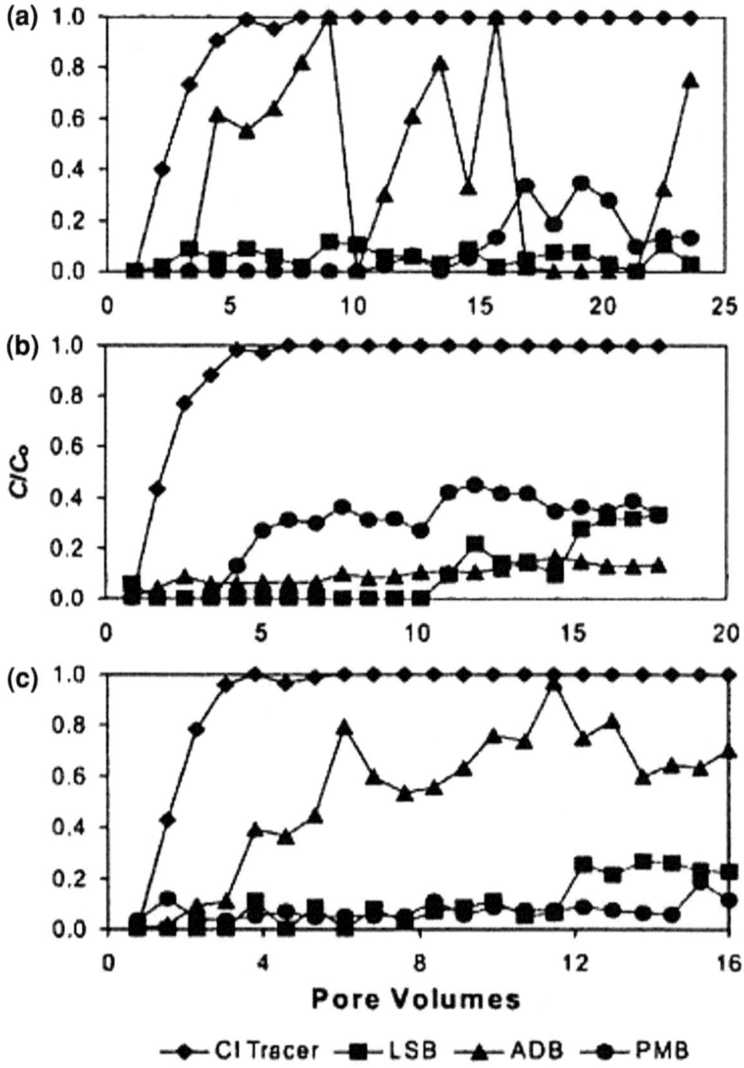

Fig. 12.43 Breakthrough curves for a conservative tracer (Cl^-) and lime-stabilized biosolid (*LSB*), aerobically digested biosolid (*ADB*), and poultry manure biosolid (*PMB*) colloids eluted from intact **a** Maurry, **b** Woolper, and **c** Bruno soil monoliths [*C* effluent concentration, C_o influent concentration (Karathanasis et al. 2005)]

of ^{14}C-labeled DDT observed in three soils with different properties. In the low-porosity Gilat soil (silt loam), the flow rate was slow and little DDT transport occurred; only 3 % of the amount applied reached a depth of 5.4 cm. In the sandy loam Bet-Dagan soil, considerable DDT transport was observed. In this case, the flow rate was higher and the organic colloids were more mobile: 18 % of the

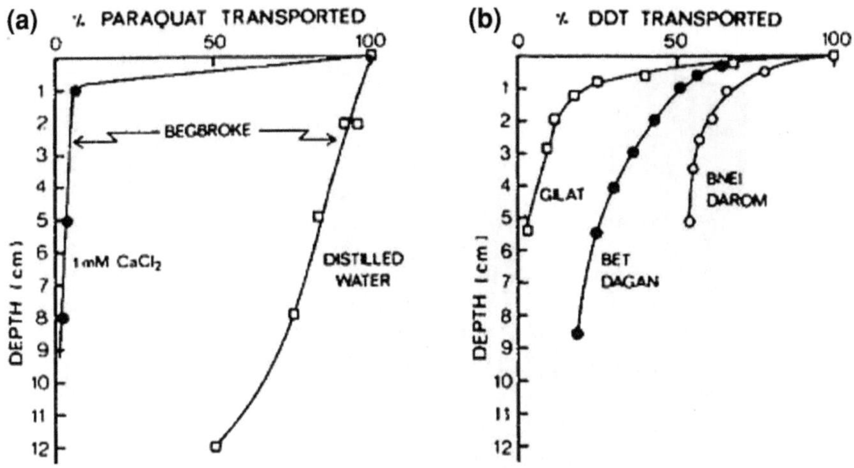

Fig. 12.44 Vertical transport of paraquat and DDT adsorbed on suspended particles (in percentage of total applied). **a** ^{14}C-labeled paraquat adsorbed on Li-montmorillonite, **b** ^{14}C-labeled DDT adsorbed on suspended solids in sewage effluent. Reprinted with permission from Vinten et al. (1983). Copyright 1983 American Chemical Society

amount applied reached a depth of 9 cm. In the coarse sands from Bnei Darom, 54 % of the applied DDT was transported deeper than 5 cm. These differences in transport of DDT adsorbed on organic colloids were due to differences in mobility of the solids, which are related directly to the soil hydraulic properties and the flow rate.

In another experiment, colloid-facilitated vertical transport of the organic contaminant (herbicide) atrazine, which is frequently found in groundwater, was tested under a corn field irrigated with secondary effluents containing organic colloidal suspensions (Graber et al. 1995). Figure 12.45 shows the distribution of atrazine in the field irrigated with secondary effluent and the high-quality water. In the effluent-irrigated field, atrazine was found to a depth of nearly 400 cm. Peaks in atrazine concentration often correspond to peaks of soil organic carbon, suggesting that enhanced transport is affected at least in part by complexation with effluent-borne organic matter. In a control field where high-quality water was used for irrigation, atrazine was concentrated only in the upper 100 cm, as soil organic carbon decreased with increasing depth.

12.5 Transport of Nonaqueous-Phase Liquids

The behavior of nonaqueous-phase liquids (NAPLs) as they enter the partially saturated subsurface from a land surface source follows two well-defined scenarios: in one case, the physical properties of the NAPL remain unchanged, while

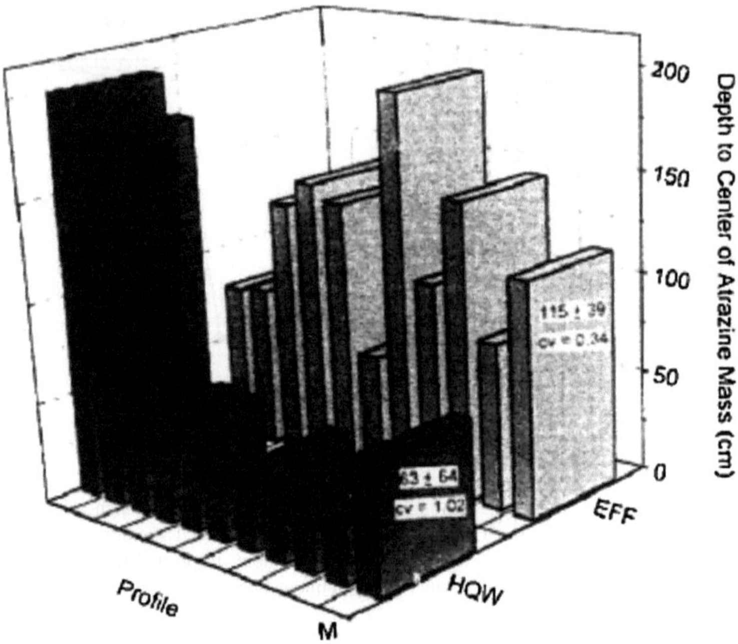

Fig. 12.45 Mean (*M*) center of atrazine mass and individual centers of mass for ten high-quality water (*HQW*) and ten effluent (*EFF*) profiles (Graber et al. 1995)

in the second case, NAPL properties are altered during transport. In the case of dense NAPLs, the contaminant plume reaches the aquifer and is subject to long-term, continuous, slow "local" redistribution due to groundwater flushing-dissolution processes. These plumes become contamination source zones that evolve over time, often with major negative impacts on groundwater quality.

Experiments carried out in Canada, Israel, Sweden, and the United States, are used here to illustrate aspects of NAPL transport and redistribution in the subsurface.

12.5.1 Infiltration into the Subsurface

Viscosity Effect The infiltration and redistribution of two hydrocarbons in moist silt loam and loamy sand soils, with viscosities of 4.7 (soltrol) and 77 (mineral oil) times greater than that of water, was reported by Cary et al. (1989). The spatial distribution of the two hydrocarbons and water is presented in Fig. 12.46: the infiltration rate of hydrocarbons is inversely related to their viscosity. The mineral oil remains in the upper layer of the soil column, while the soltrol redistribution is similar to that of water.

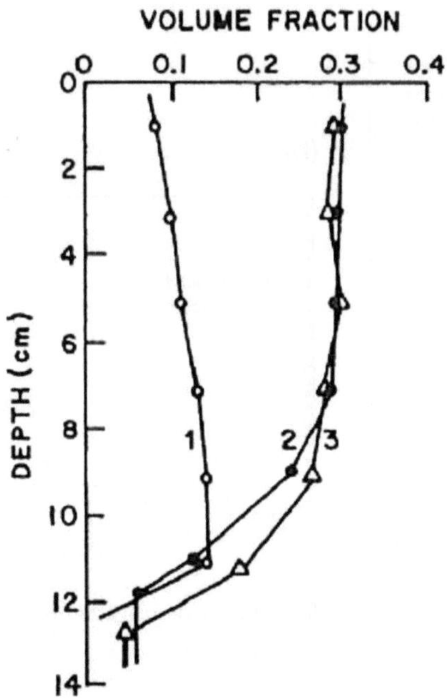

Fig. 12.46 Distribution of (*1*) mineral oil, (*2*) soltrol-220, and (*3*) water in a silt loam soil 8 h after adding water to dry soil and 4 h after adding the hydrocarbons (Cary et al. 1989)

When the NAPL consists of a mixture of hydrocarbons with different vapor pressure, the liquid composition is affected by changes in temperature occurring during infiltration into the subsurface. Increased temperature enhances the partial volatilization of constituents in the liquid mixture, leading to viscosity changes, and thus affecting transport of the remaining liquid phase. Figure 12.47a shows the effect of viscosity changes on the conductivity of kerosene, a petroleum product containing more than 100 hydrocarbons, in three types of soils with different physicochemical properties. Changes in kerosene composition, due to volatilization of the light fractions, result in increased viscosity and decreased mass flow through porous media of the residual NAPL.

Soil Moisture Effect The effects of soil moisture content and soil texture and pore size distribution on kerosene conductivity were studied by Gerstl et al. (1994). Figure 12.47b shows kerosene conductivity in sand, loam, and clay soil columns, as the initial moisture content varies from 0 % (oven-dried) to field capacity.

The kerosene conductivity of the soil was affected strongly by the soil texture and initial moisture content. In sand, the kerosene conductivity was affected slightly by the initial moisture content, as high as 70 % field capacity, but decreased thereafter. The kerosene conductivity of the loam soil was similar in oven-dried and air-dried soils and increased significantly in soils at 70 % and full

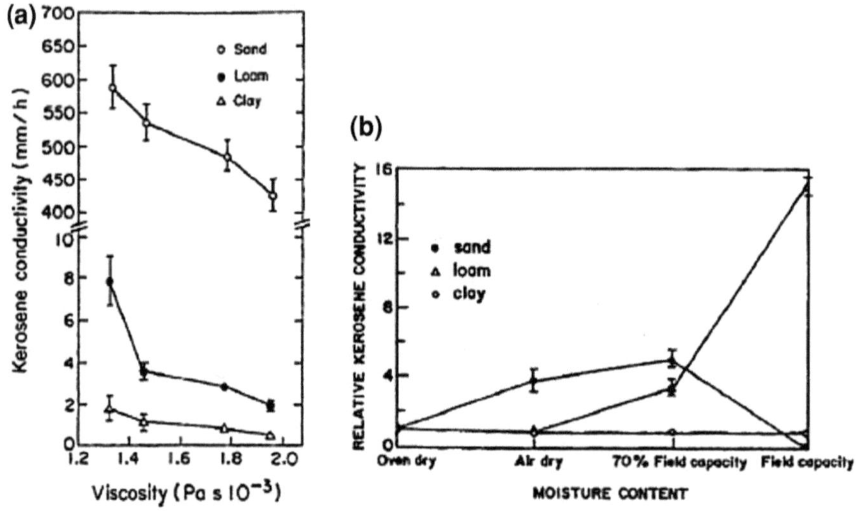

Fig. 12.47 Kerosene conductivity as affected by **a** viscosity and **b** soil moisture content (Gerstl et al. 1994)

field capacity. No kerosene flow was observed in the clay soils from oven-dried to field capacity moisture contents.

These behaviors can be explained as follows. First, the decrease in kerosene conductivity in sand is caused by a mechanically induced change in the original porosity, which occurs during leaching to reach a field capacity moisture content. In the loamy soils, the porosity of the system is hardly affected, as the moisture content increases toward field capacity, and instead, there is a reduced resistance to the flow of kerosene. This phenomenon flow can be attributed to the Yuster effect (Rose 1993), which is a lubrication effect occurring when a thin layer of water coats the surface, thus reducing the resistance to flow and preventing interaction between the solid phase and kerosene molecules. When the loamy soil is oven-dried, kerosene infiltrating into the soil is adsorbed on the pore surfaces, creating a wall effect that retards movement by formation of a viscous layer. The kerosene conductivity is not changed in clay soils, regardless of the moisture content. Initially, when the soil humidity ranges between oven-dried and air-dried, kerosene adsorption on the surface of the solid material controls its flow, and subsequent pores plugging by water impede the kerosene flow.

Retention in Porous Media As NAPL migrates through the subsurface, some of it becomes entrapped according to its retention capacity. In addition, NAPL constituents may become redistributed in the gaseous phase.

The extent of kerosene trapping was determined quantitatively in a series of laboratory and outdoor experiments with Swedish soils (Jarsjo et al. 1994), yielding an empirical equation for the kerosene residual content as a function of soil composition:

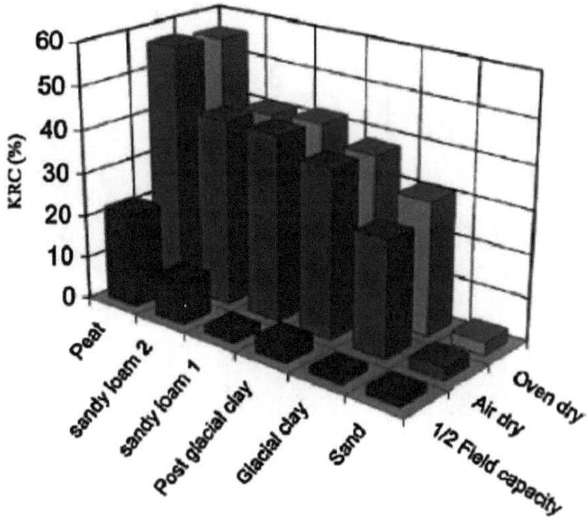

Fig. 12.48 Kerosene retention capacity (KRC) as affected by soil type and soil moisture conditions, expressed as volume of kerosene per bulk volume of air-dry soil. Reprinted from Jarsjo et al. (1994). Copyright 1994 with permission of Elsevier

$$\text{KRC} = 0.13 \times (\%\text{clay}) + 1.46 \times (\%\text{OM}) - 0.32 \times (\%\text{moisture}) + 4.31, \quad (12.3)$$

where OM denotes organic matter. The KRC is negatively correlated with the soil moisture content and positively correlated with clay and organic matter contents. An illustration of the effects of moisture content and soil type on kerosene retention capacity is shown in Fig. 12.48. As soil moisture increases from oven-dried to 50 % field capacity, the KRC retention capacity decreases markedly in all soils studied. It may be assumed that water retained in small soil pores reduces the effective soil pore volume, thus decreasing capacity of the soil to entrap NAPLs. At relative humidities of about 50 % field capacity (most relevant to a soil environment in a temperate climate), a monolayer (or more) of water evidently is presented n external surfaces of macro- and mesopores, while the micropores and some mesopores are filled with water. These results imply that, for characteristically high soil water contents in humid zones, differences between soils on NAPL transport are reduced. In arid and semiarid areas, these differences are significantly greater because the small water content in some soils allows access of NAPL to a larger pore volume.

12.5.2 Transport of Soluble NAPL Fractions in Aquifers

As discussed previously, NAPLs are transported through the partially saturated zone to the groundwater (Feenstra et al. 1991). Natural groundwater flow through a NAPL source zone forms plumes of aqueous-phase contamination that can extend over aquifer volumes much larger than the initial source zone. Simultaneously, the

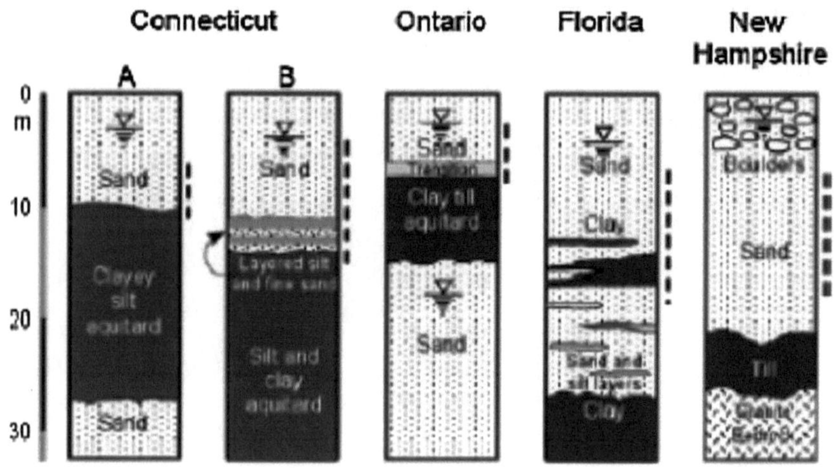

Fig. 12.49 *Columns* representing the geology of five study sites. The depth zone of DNAPL contamination investigated at each site is indicated by the *dashed lines* (Parker et al. 2003)

NAPL source ages, its geometry changes, and the amount of soluble NAPL fraction in water may decrease.

Contamination of sandy aquifers by DNAPL chlorinated solvents was investigated at five field study sites in Canada and the United States (Parker et al. 2003). Figure 12.49 shows the general geological setting of the studied sites. In all the sites, the sand deposits forming the aquifer extend up to the ground surface, so that DNAPLs disposed on the surface infiltrate only a few meters to reach the water table. Contamination of the study sites started in the 1950s and 1960s, from the discharge of individual DNAPLs (TCE or PCE) used as routine solvents in metal industries. Because the DNAPLs entered the subsurface at these sites decades ago, the DNAPL zones aged due to groundwater dissolution.

It was assumed that initially most DNAPL was found in the coarser grained zones and more rapid groundwater flow in these zones caused preferential removal of DNAPL. Groundwater flow is horizontal through the DNAPL source zone, which creates the down-gradient contaminant plume. Illustration of the evolution of a layered DNAPL source zone, showing complete dissolution of the residual trails, is given in Fig. 12.50. The major finding of the analysis is that DNAPL originally located in coarser grained layers was removed by groundwater dissolution. The present-day DNAPL in the source zone resides only in isolated, thin horizontal layers and represents only the less soluble or leachable remnants of the original contaminant zone. DNAPL located in the less permeable layers or in layered transition zones at the bottom of the aquifer is subject to a slower dissolution rate.

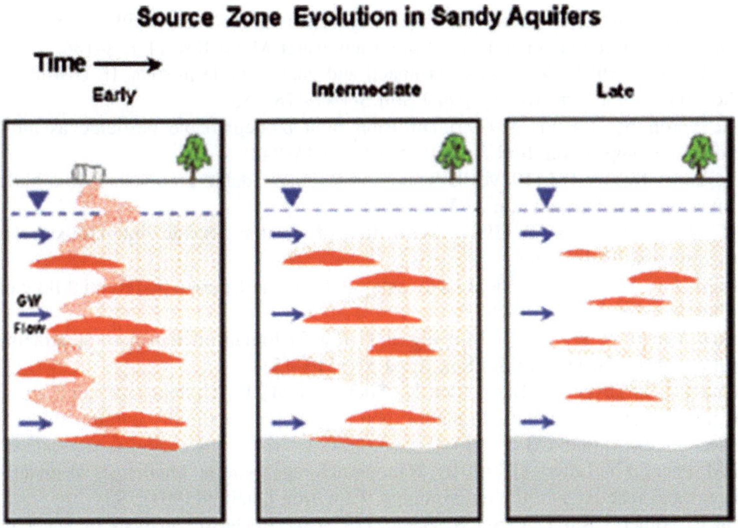

Fig. 12.50 The evolution of layered DNAPL source zones, showing complete dissolution of residual trails, shrinkage of some layers, and complete removal of others due to a decrease in groundwater flushing (Parker et al. 2003)

References

Ashworth DJ, Alloway BJ (2004) Soil mobility of sewage sludge-derived dissolved organic matter, copper, nickel and zinc. Environ Pollut 127:137–144

Behbahaninia A, Mirbagheri SA, Nouri J (2010) Effects of sludge from wastewater treatment plants on heavy metals transport to soil and groundwater. Iran J Environ Health Sci Eng 7:401–406

Ben-Moshe T, Dror I, Berkowitz B (2010) Transport of metal oxide nanoparticles in saturated porous media. Chemosphere 81:387–393

Borch T, Kretzschmar R, Kappler A, van Cappellen P, Ginder-Vogel M, Voegelin A, Campbell K (2010) Biogeochemical redox processes and their impact on contaminant dynamics. Environ Sci Technol 44:15–23

Bresler E, Dagan G (1983) Unsaturated flow in spatially variable fields. Water Resour Res 19:421–435

Cary JW, Simmons CS, McBride JF (1989) Oil infiltration and redistribution in unsaturated soils. Soil Sci Soc Am J 53:335–342

Collins D, Luxton T, Kumar N, Shah S, Walker VK, Shah V (2012) Assessing the impact of copper and zinc oxide nanoparticles on soil: a field study. PLoS ONE 7(8):42663

Dror I, Kliger L, Laufer A, Hadas A, Russo D, Yaron B (1999) Behavior of atrazine and terbuthylazine in an irrigated field. I. Soil and management spatial variability. Agrochimica 43:21–29

Elimelech M, Gregory J, Jia X, Williams RA (1995) Particle deposition and aggregation: measurement, modelling, and simulation, Butterworth-Heinemann. England, Oxford. ISBN 075067024X

Fang J, Shan XQ, Wen B, Lin JM, Owens G (2009) Stability of Titania nanoparticles in soil suspensions and transport in saturated homogeneous soil columns. Environ Pollut 157:1101–1109

Feenstra S, MacKay CM, Cherry JA (1991) Presence of residual NAPL based on organic chemical concentrations in soil samples. Groundwater Monit Rev 11:128–136
Gerstl Z, Yaron B (1983) Behavior of bromacil and napropamide in soils. II. Distribution after application from a point source. Am J Soil Sci 47:478–483
Gerstl Z, Yaron B, Nye PH (1979a) Diffusion of a biodegradable pesticide as affected by microbial decomposition. Soil Sci Soc Am J 43:843–848
Gerstl Z, Yaron B, Nye PH (1979b) Diffusion of a biodegradable pesticide: I. In a biologically inactive soil. Soil Sci Soc Am J 43:839–842
Gerstl Z, Galin TS, Yaron B (1994) Mass flow of volatile organic liquid mixture in soils. J Environ Qual 23:487–493
Ghodrati M, Jury WA (1990) A field study using dyes to characterize preferential flow of water. Soil Sci Soc Am J 54:1558–1563
Graber ER, Gerstl Z, Fischer E, Mingelgrin U (1995) Enhanced transport of atrazine under irrigation with effluent. Soil Sci Soc Am J 59:1513–1519
Grolimund D, Borkover M, Barmettler K, Sticher H (1996) Colloid-facilitated transport of strongly sorbing contaminants in natural porous media: a laboratory column study. Environ Sci Technol 30:3118–3123
Hotze EM, Phenrat T, Lowry CV (2010) Nanoparticle aggregation: challenges to understanding transport and reactivity in the environment. J Environ Qual 39(1909):1924
Indelman P, Toiber-Yasur I, Yaron B, Dagan G (1998) Stochastic analysis of water flow and pesticides transport in a field experiment. J Contam Hydrol 32:77–97
Jarsjo J, Destouni G, Yaron B (1994) Retention and volatilization of kerosene: Laboratory experiments on glacial and post glacial soils. J Contam Hydrol 235:625–631
Jaisi DP, Saleh NB, Blake RE, Elimelech M (2008) Transport of single-walled carbon nanotubes in porous media: filtration mechanism and reversibility. Environ Sci Technol 42:8317–8323
Jones JL, Lafferty BL, Sparks DL (2012) Additive and competitive effects of bacteria and Mn oxides on arsenite oxidation kinetics. Environ Sci Technol 46:6548–6555
Jury WA, Elabad H, Reskelo M (1986) Field study of napropamide movement through unsaturated soils. Water Resour Res 22:749–755
Karathanasis AD (1999) Subsurface migration of copper and zinc mediated by soil colloids. Soi Sci Am Soc J 63:830–838
Karathanasis AD, Johnson DMC, Matocha CJ (2005) Biosolid colloid mediated transport of copper zinc and lead in waste-amended soils. J Environ Qual 34:1153–1164
Keller AA, Wang H, Zhou D, Lenihan SH, Cherr G, Cardinale BJ, Miller R, Ji Z (2010) Stability and aggregation of metal oxide nanoparticles in natural aqueous matrices. Environ Sci Technol 44:1962–1967
Klitzke S, Lang F, Kaupenjohan M (2008) Increasing pH releases colloidal lead in a highly contaminated forest soil. Eur J Soil Sci 59:265–273
Kretzschmar R, Borkovec M, Grolimund D, Elimelech M (1999) Mobile subsurface colloids and their role in contaminant transport. Adv Agron 66:121–192
Li Y, Wang Y, Pennell KD, Abriola LM (2008) Investigation of the transport and deposition of fullerene (C60) nanoparticles in quartz sands under varying flow conditions. Environ Sci Technol 42:7174–7180
Lin D, Tian X, Wu F, Xing B (2010) Fate and transport of engineered nanomaterials in the environment. J Environ Qual 39:1896–1908
Parker BL, Cherry JA, Chapman SW, Guilbeault MA (2003) Review and analysis of chlorinated solvent dense nonaqueous phase liquid distribution in five sandy aquifers. Vadose Zone J 2:116–137
Rose CW (1993) The transport of adsorbed chemicals in eroded sediments. In: Russo D, Dagan G (eds) Water flow and solute transport in soils. Springer, Heidelberg, pp 180–199
Russo D, Toiber-Yasur I, Laufer A, Yaron B (1998) Numerical analysis of field scale transport of bromacil. Adv Water Resour 21:637–647
Sagee O, Dror I, Berkowitz B (2012) Transport of silver nanoparticles (AgNPs) in soil. Chemosphere 88:670–675

Schrick B, Hydutsky BW, Blough JL, Mallouk TE (2004) Delivery vehicles for zerovalent metal nanoparticles in soil and groundwater. Chem Mater 16:2187–2193

Selim HM (2013) Transport and retention of heavy metal in soils: competitive sorption. Adv Agron 119:275–308

Slowey AJ, Johnson SB, Newville M, Brown GE Jr (2007) Speciation and colloid transport of arsenic from mine tailings. Appl Geochem 22:1884–1898

Smedley PL, Kinniburgh DG (2002) A review of the source, behavior and distribution of arsenic in natural waters. Appl Geochem 17:517–568

Toiber-Yasur I, Hadas A, Russo D, Yaron B (1999) Leaching of terbuthylazine and bromacil through field soils. Water Air Soil Poll 113:319–335

Turner BL, Kay MA, Westermann DT (2004) Colloid phosphorus in surface runoff and water extracts from semiarid soils of the western United States. J Environ Qual 33:1464–1472

Vinten AJ, Yaron B, Nye PH (1983) Vertical transport of pesticides into soil when adsorbed on suspended particles. J Agric Food Chem 31:661–664

Violante A, Cozzolino V, Perelomov L, Caporale AG, Pgna M (2010) Mobility and bioavailability of heavy metals and metalloids in soil environment. J Soil Sci Plant Nutr 10(3):268–292

Wang Y, Li Y, Kim H, Walker SL, Abriola ML, Pennell KD (2010) Transport and retention of fullerene nanoparticles in natural soils. J Environ Qual 39:1925–1933

Wang Y, Li Y, Pennell KD (2008) Influence of electrolyte species and concentration on the aggregation and transport of fullerene nanoparticles in quartz sand. Environ Tox Chem 27:1860–1867

White RE, Dyson GS, Gerstl Z, Yaron B (1986) Leaching of herbicides through undisturbed cores of a structured clay soil. Soil Sci Soc Am J 50:277–282

Williams LE, Barnett MO, Kramer AT, Melville JO (2003) Adsorption and transport of Arsenic(V) in experimental subsurface system. J Environ Qual 32:841–850

Yaron B, Gerstl Z (1983) Herbicide residues in soils following point source application. Pesticide chemistry, human welfare and the environment. Pergamon Press, Oxford, pp 207–212

Yaron B, Bielorai H, Kliger L (1974) Fate of insecticides in an irrigated field—azinphosmethyl and tetradifon cases. J Environ Qual 3:413–417

Yolcubal I, Akyol NH (2008) Adsorption and transport of arsenate in carbonate-rich soils: Coupled effects of nonlinear and rate-limited sorption. Chemosphere 73:1300–1307

Zhang Y (2003) 100 years of Pb deposition and transport in soils in Champaign, Illinois, USA. Water Air Soil Pollut 146:197–210

Part V
Transformations and Reactions of Contaminants in the Subsurface

After reaching the subsurface, contaminants are partitioned among the solid, liquid, and gaseous phases. A fraction of the contaminated gaseous phase is transported into the atmosphere, while the remaining part may be adsorbed on the subsurface solid phase or dissolved into the subsurface water. Contaminants dissolved in the subsurface aqueous phase or retained on the subsurface solid phase are subjected, over the course of time, to chemical, biochemical, and surface-induced degradation, which also lead to formation of metabolites.

In many cases, the subsurface is contaminated by a mixture of toxic chemicals with components having different physical and chemical properties. Therefore, a contaminant undergoes transformations controlled by the properties of each substance, the characteristics of the subsurface, and the ambient conditions. Physicochemically mediated and biologically mediated degradation are the main processes involved in such transformations.

From an environmental point of view, the transformation of contaminants is not restricted to molecular changes. Rather, we also must consider all deviations from the original properties of a contaminant that may be relevant to its behavior in the subsurface. For example, in addition to degradation-induced transformation, contaminants may change their physicochemical characteristics as a result of specific reactions with other natural or anthropogenic chemicals found in the subsurface environment. To be more specific, complexation of a contaminant with existing natural ligands may lead to a change in its solubility, thus affecting its transport and redistribution in the vadose zone. The volatilization of high vapor pressure compounds from a mixture of immiscible (NAPL) contaminants, such as petroleum products, leads to changes in the retention or transport behavior of remaining constituents.

As a first approximation, we consider the main subsurface transformation processes to comprise reactions leading to chemical transformation or degradation and metabolite formation in the liquid phase or the solid–liquid interface and reactions resulting in complexation of chemicals, which in turn lead to a change in their physicochemical properties.

The upper part of the subsurface is characterized by enhanced biological activity. Therefore, contaminant transformation in this zone proceeds mainly by microbial processes, which are often faster than chemical ones. In the deeper subsurface, biological activity often is reduced and, therefore, degradation proceeds abiotically, at a much slower rate. It should be emphasized, however, that contaminant transformation is related mainly to the formation of metabolites with properties different from those of the parent material (sometimes more polar, more soluble in water, and even more toxic), which may reach the groundwater. Although abiotic and biologically mediated contaminant transformation may occur simultaneously in the subsurface aqueous solution and in the solid–liquid interface, they are discussed separately for didactic purposes.

Environmental conditions cause changes in the initial properties of organic and inorganic contaminants and affect their persistence in the subsurface. Relevant external factors include temperature and solar radiation, while the principal subsurface properties include water chemistry, the surface properties of the solid phase, and bioactivity. In Part V, we focus on research findings that illustrate abiotic and biologically mediated degradation of individual contaminants occurring in the liquid phase and at the solid–liquid interface, as well as examples of multiple-component contaminant transformation.

Chapter 13
Abiotic Contaminant Transformations in Subsurface Water

13.1 Abiotic Contaminant Transformations in Natural Subsurface Water

Natural subsurface water is considered here as the (bulk) body of water that completely or partially fills the porous matrix, from the soil surface to the groundwater zone. This water may be flowing or relatively static, having accumulated above an impermeable subsurface.

Abiotic transformation of contaminants in subsurface natural waters results mainly from hydrolysis or redox reactions and, to lesser extent, from photolysis reactions. Complexation with natural or anthropogenic ligands, as well as differential volatilization of organic compounds from multicomponent liquids or mixing with toxic electrolyte aqueous solutions, may also lead to changes in contaminant properties and their environmental effects. Before presenting an overview of the reactions involved in contaminant transformations, we discuss the main chemical and environmental factors that control these processes.

13.1.1 Factors Affecting Contaminant Transformations

The chemistry of subsurface water is a main factor to be considered when dealing with contaminant transformation processes. The acidity (pH) of subsurface aqueous solutions can vary over a wide range of values. For example, the pH of swamp water can be as low as 3.5, because of a high concentration of humic and fulvic acids in a eutrophic lake, but may reach a pH as high as 9.5 (Hutchinson 1957). In general, climatic conditions and properties of the surrounding porous media greatly affect the pH of subsurface water. For contaminants containing acid or basic functional groups, the pH of water controls species dissociation-association behavior and thus affects contaminant half-life (Bender and Silver 1963).

Variations in pH induce acid–base mediated hydrolysis. For base-catalyzed hydrolysis, for example, the rate equation for an organic contaminant is

B. Berkowitz et al., *Contaminant Geochemistry*,
DOI: 10.1007/978-3-642-54777-5_13,

$$(-\mathrm{d}P/\mathrm{d}t) = k[\mathrm{OC}][\mathrm{OH}^-] \tag{13.1}$$

where P is the product, k denotes a second-order rate constant, [OC] is the organic contaminant concentration and $[OH^-]$ is the hydroxide ion concentration (see also Sect. 13.1.2). Subsurface waters contain organic and inorganic species (Table 13.1), which, as a function of their character, affect hydrolysis reactions in different ways. The presence of carbonate, fulvic acid, and silicate favor acid–base reactions in aerobic waters. In anaerobic interstitial waters, these reactions are induced by the presence of ammonia, carbonate, fulvic acid, phosphate, and sulfides. Both dissolved and suspended organic material can enhance or impede hydrolysis of selected organic contaminants. For example, dissolved fulvic acids accelerate atrazine hydrolysis, while humic acids retard alkaline hydrolysis of the hydrophobic n-octyl ester of 2, 4-D herbicide (Perdue and Wolfe 1983).

Metal ions in aerobic, natural waters, such as Cu^{2+}, Fe^{3+}, Mn^{2+}, Mg^{2+}, and Ca^{2+}, may catalyze hydrolysis of organic contaminants. Blanchet and St. George (1982), for example, showed that interaction of organophosphate esters with Cu^{2+} and Mn^{2+} led to the hydrolysis of pesticides. However, similar studies with Mg^{2+} and Ca^{2+} did not induce any transformation process.

The main environmental factors that control transformation processes are temperature and redox status. In the subsurface, water temperature may range from 0 °C to about 50 °C, as a function of climatic conditions and water depth. Generally speaking, contaminant transformations increase with increases in temperature. Wolfe et al. (1990) examined temperature dependence for pesticide transformation in water, for reactions with activation energy as low as 10 kcal/mol, in a temperature range of 0–50 °C. The results corresponded to a 12-fold difference in the half-life. For reactions with an activation energy of 30 kcal/mol, a similar temperature increase corresponded to a ~2,500-fold difference in the half-life. The Arrhenius equation can be used to describe the temperature effect on the rate of contaminant transformation, k:

$$k = A\mathrm{e}^{-Ea/\mathrm{RT}}, \tag{13.2}$$

where Ea is the activation energy for the reaction, R is the gas constant, T is the absolute temperature, and A is a constant.

The redox potential in subsurface water varies with alterations from aerobic to anaerobic conditions. In and around anaerobic environments, conditions for reduction exist and contaminants are transformed accordingly. Under aerobic conditions, O_2 is the predominant oxidation agent (mainly through biological processes), because the transformation of contaminants is mainly through oxidative pathways. Aerobic and anaerobic states may occur both in surface waters and in deeper subsurface water.

Table 13.1 Concentration of selected organic and inorganic species in interstitial water, compared to river water and seawater (Wolfe et al. 1990)

Species	World average (mol/L)		
	Interstitial water	River	Seawater
Ammonia	2.00×10^{-4}		
Carbonate	9.09×10^{-3}	9.78×10^{-4}	2.33×10^{-3}
Fulvic acid	2.00×10^{-4}	1.00×10^{-4}	5.00×10^{-6}
Phosphate	1.80×10^{-5}		2.84×10^{-6}
Sulfide	3.96×10^{-7}		
Silicate		2.18×10^{-4}	1.03×10^{-4}
Borate			4.10×10^{-4}

13.1.2 Reactions in Natural Waters

Hydrolysis is a bond-making-bond-breaking process, which may be described by the equation

$$RX + H_2O \rightarrow ROH + H^+ + X \tag{13.3}$$

where R denotes an organic moiety. The general expression for the observed global rate constant K_{obs} is given by

$$K_{obs} = K_H[H^+] + K_{OH}[OH^-] + K_w + \sum_i (K_{HA}[HA] + K_A[A^-])_i, \tag{13.4}$$

where K_H and K_{OH} are specific acid-base catalyzed second-order rate constants; K_w is the neutral hydrolysis rate constant; K_{HA} and K_A are, in general, acid-base catalyzed rate constants; and the summation is over all components i. In Eq. (13.4), $[H^+]$ and $[OH^-]$ are the hydrogen and hydroxyl ion concentrations (activities), respectively, while [HA] and $[A^-]$ are, respectively, the concentrations (activities) of acid and bases in the reaction mixture. The term K_{obs} can include contributions from acid–or base-catalyzed hydrolysis, nucleophilic attack by water, or catalysis by buffers in the reaction medium.

Abiotic hydrolysis of pollutants in subsurface waters is pH dependent. The predominant pathways are acid-catalyzed, base-mediated, and neutral (pH-independent) hydrolysis. The acid-catalyzed hydrolysis reaction rate is dependent on proton concentration increases with a decrease in pH. This behavior occurs because the proton is not consumed in the reaction.

Increases in pH as a direct proportional augmentation of the hydroxyl ion activity leads to a base-mediated hydrolysis process. In this case, the hydroxyl behaves as a nucleophile and is consumed in the reaction. Neutral and alkaline hydrolysis are the most frequent reactions over the common environmental pH ranges. The relation between first-order hydrolysis rate constants and the pH often is presented as a pH rate profile (Wolfe et al. 1990).

Both inorganic (e.g., metals) and organic substances may be subject to a hydrolysis reaction in waters. Examples of several hydrolyzable functional groups are given in Table 13.2. Water is a weak acid and the acidity of the water molecules in the hydration shell of a metal ion usually is greater than that of the water. The acidity of aqueous metal ions is expected to increase with a decrease in the radius and an increase in the charge of the central ion. In the case of Fe(III), for example, hydrolysis can extend beyond the uncharged species $Fe(OH)_3(H_2O)_{3(s)}$, to form anions such as ferrate (FeO_4^{2-}). All hydrated ions, in principle, can donate a larger number of protons than that corresponding to their charge and can form anionic hydroxo-metal complexes (Stumm and Morgan 1996).

The rate of hydrolysis of an organic contaminant also may be affected by the pH, due to specific acid–base effects or changes in compound speciation. A change in the pH can shift the equilibrium in favor of the charged or uncharged species, which often have different hydrolysis rate constants. Under drastic reaction conditions (i.e., extremely low or high pH, high temperature), many of the major organic contaminants, such as pesticides, undergo hydrolysis. Functional groups of organic substances susceptible to hydrolysis include carboxylic acid esters, organophosphates, amides, anilides, carbamates, triazines, oximes, and nitriles. The role of hydrolysis in the overall transformation process, however, depends on the rate of other degradation processes that may occur simultaneously, such as photolysis, biolysis, or redox reactions. If, in the liquid phase, additional organic or inorganic chemicals of natural or anthropogenic origin are present, contaminant hydrolysis could be affected by the presence of other solutes. This is the case of dissolved metals or humic acids acting on the hydrolysis of organic toxic elements present in the same water phase. This effect, however, has only minor significance. Perdue and Wolfe (1983) considered the maximum predicted buffer-catalysis contribution to be ≤ 10 % of the uncatalyzed process.

Redox reactions in natural waters involve the transfer of electrons between chemical species or changes in the oxidation state of species involved in the reaction. Specifically, oxidation describes the loss of electrons by a molecule, atom, or ion, while reduction describes the gain of electrons by a molecule, atom, or ion. Therefore, *oxidation* is defined as an increase in oxidation number, while *reduction* is defined as a decrease in oxidation number. Differences in the oxidation states of natural waters generally exist between surface and ground waters, between locally inter aggregate stagnant and flowing waters, or in stagnant waters obstructed by biological or vegetative cover.

Reductants and *oxidants* are defined as electron donors and proton acceptors (Sect. 2.2.2). Because there are no free electrons, every oxidation is accompanied by a reduction and vice versa. In aqueous solutions, proton activities are defined by the pH:

$$pH = -\log[H^+]. \tag{13.5}$$

Similarly, we can define a convenient parameter for the redox intensity:

Table 13.2 Examples of hydrolyzable functional groups (Larson and Weber 1994)

1. Halogenated aliphatics

Nucleophilic substitution:

$$RCH_2X \xrightarrow{H_2O,\ OH^-} RCH_2OH + HX$$

Elimination:

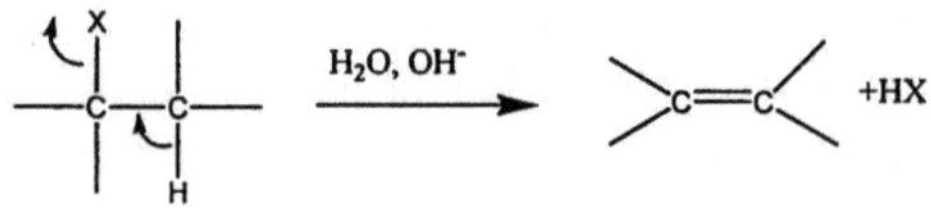

2. Epoxides:

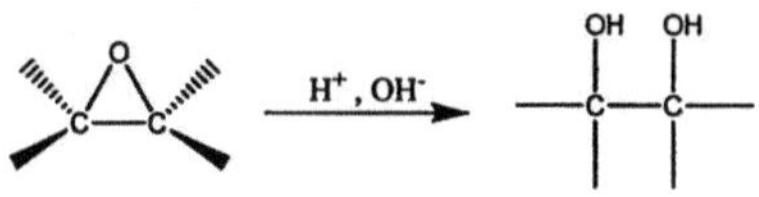

3. Organophosphorus esters:

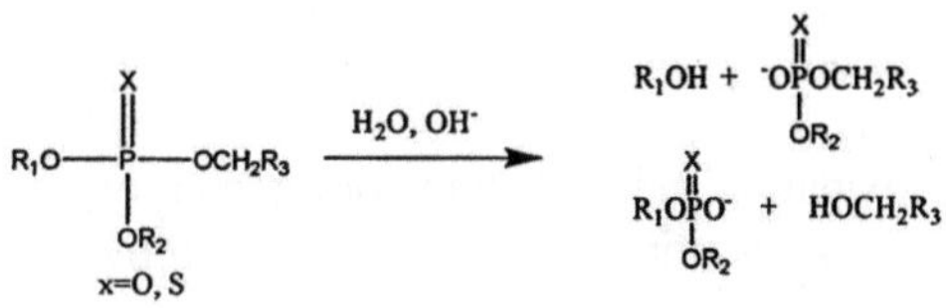

4. Carboxylic acid esters:

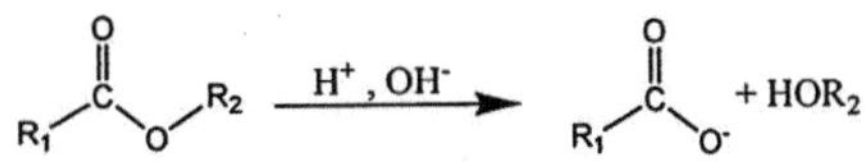

5. Anhydrides:

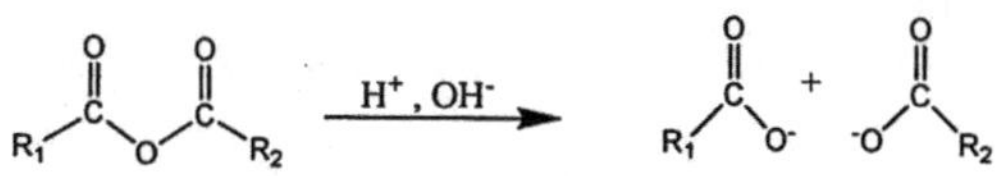

6. Amides:

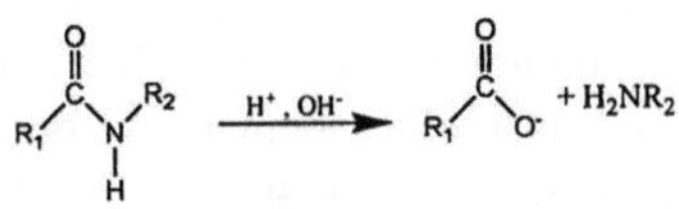

7. Carbamates:

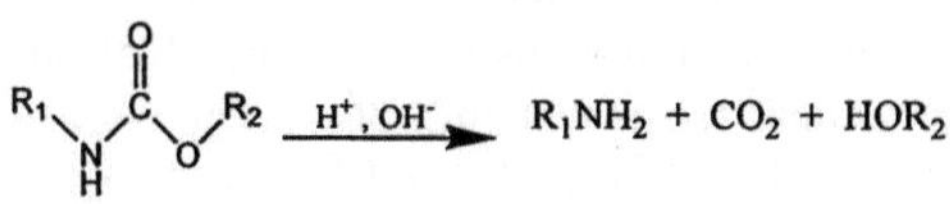

8. Ureas:

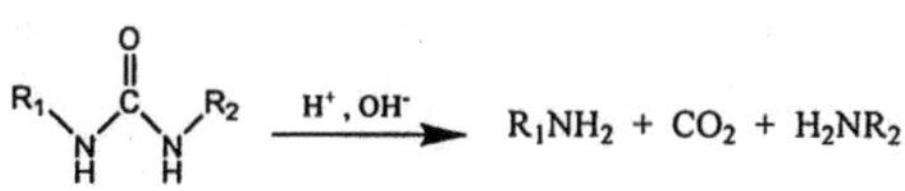

$$pE = -\log[e^-], \tag{13.6}$$

where pE gives the (hypothetical) electron (e^-) activity at equilibrium and measures the relative tendency of a solution to accept or transfer electrons.

Stumm and Morgan (1996) showed that C, N, O, S, Fe, and Mn are the main elements participating in aquatic redox processes. Table 13.3 presents equilibrium constants for several redox processes relevant to natural waters. The symbol pE°(W) expresses the redox situation in natural waters; that is, the values for pE°(W) apply to the electron activities for unit activities of oxidants and reductants in natural water at standard conditions (pH = 7.0 and 25 °C). Values for pE°(W) at 25 °C and pH = 7 can be determined according to

$$pE^\circ(W) = pE^\circ + (n_H/2)\log\ K_w, \tag{13.7}$$

where n_H is the number of moles of protons exchanged per mole of electrons and K_w denotes the equilibrium constant for the redox reaction in natural water. A list of pE°(W) values is given in Table 13.3, showing the oxidizing intensity at standard conditions.

Reduction-or oxidation-induced transformations of contaminants occur in subsurface water as a function of environmental aerobic or anaerobic conditions. The presence or lack of O_2 is the determining factor in defining the transformation pathways. Redox reactions are driven by microbial activity and through abiotic processes. In both cases, though, the overall reactions may be very similar. Hence, most of the general redox reactions discussed here also are relevant for the biologically mediated processes considered later in this chapter and vice versa. We discuss reduction processes in a more detailed manner here; oxidation mechanisms are discussed later in Chap. 15, when considering biologically mediated transformations.

Reduction occurs when there is a transfer of electrons from an electron donor, or a reducing agent, to an electron acceptor, or oxidizing agent. Reducing environments are very common in the subsurface, being present in groundwater, bottom sediments, and anaerobic stagnant waters. Naturally occurring reducing agents constitute a complex array of species, ranging from chemical or "abiotic" reagents like sulfide minerals, reduced metals such as Fe and Mn, and natural organic matter, to biological systems such as microbial populations. In addition, extracellular biochemical substances may act as catalysts for reduction reactions, such as metalloporphyrins, corrinoids, and bacterial transition-metal coenzymes that abound in the subsurface. The relationship among these various reducing agents in the subsurface environment is quite complex. For example, chemical species, such as reduced metals and sulfide ions, may result directly from microbial metabolism (Larson and Weber 1994). A list of common reduction reactions is given in Table 13.4. Reduction of organic pesticides, for example, includes reactions such as dehalogenation of alkanes, nitroreduction to the corresponding amine, azo reduction to an hydrazo or amino group, quinine reduction

Table 13.3 Examples of equilibrium constants of redox processes pertinent in aquatic conditions (25 °C) (Stumm and Morgan 1996)

Reaction	pE° (≡log K)	pE° (W)
$\frac{1}{4}O_2(g) + H^4 + e = \frac{1}{2}H_2O$	+20.75	+13.75
$\frac{1}{5}NO_3^- + \frac{6}{5}H^+ + e = \frac{1}{10}N_2(g) + \frac{3}{5}H_2O$	+21.05	+12.65
$\frac{1}{8}NO_3^- + \frac{5}{4}H^+ + e = \frac{1}{8}NH_4^+ + \frac{3}{8}H_2O$	+14.90	+6.15
$\frac{1}{6}NO_2^- + \frac{4}{3}H^+ + e = \frac{1}{6}NH_4^+ + \frac{1}{3}H_2O$	+15.14	+5.82
$\frac{1}{2}CH_3OH + H^+ + e = \frac{1}{2}CH_4(g) + \frac{1}{2}H_2O$	+9.88	+2.88
$\frac{1}{4}CH_2O + H^+ + e = \frac{1}{4}CH_4(g) + \frac{1}{4}H_2O$	+6.94	−0.06
$\frac{1}{2}CH_2O + H^+ + e = \frac{1}{4}CH_3OH$	+3.99	−3.01
$\frac{1}{8}SO_4^{2-} + \frac{5}{4}H^+ + e = \frac{1}{8}H_2S(g) + \frac{1}{2}H_2O$	+5.25	−3.50
$\frac{1}{8}SO_4^{2-} + \frac{9}{8}H^+ + e = \frac{1}{8}HS^- + \frac{1}{2}H_2O$	+4.25	−3.75
$\frac{1}{8}S(s) + H^+ + e = \frac{1}{2}H_2S(g)$	+2.89	−4.11
$\frac{1}{8}CO_2(g) + H^+ + e = \frac{1}{8}CH_4(g) + \frac{1}{4}H_2O$	+2.87	−4.13
$\frac{1}{6}N_2(g) + \frac{4}{3}H^* + e = \frac{1}{3}NH_4^+$	+4.68	−4.68
$\frac{1}{2}(NADP^+) + \frac{1}{2}H^+ + e = \frac{1}{2}(NADPH)$	−2.0	−5.5
$H^+ + e = \frac{1}{2}H_2(g)$	0.0	−7.00
Oxidized ferredoxin + e = reduced ferredoxin	−7.1	−7.1
$\frac{1}{4}CO_2(g) + H^+ + e = \frac{1}{24}(glucose) + \frac{1}{4}H_2O$	−0.20	−7.20
$\frac{1}{2}HCOO^- + \frac{3}{2}H^+ + e = \frac{1}{2}CH_2O + \frac{1}{2}H_2O$	+2.82	−7.68
$\frac{1}{2}CO_2(g) + \frac{1}{2}H^+ + e = \frac{1}{24}HCOO^-$	−4.83	−8.33

to semiquinones or hydroquinones, and sulfone reduction to sulfoxide or sulfide (Macalady et al. 1986).

Reductive Dehalogention Reductive dehalogenation is a general phenomenon in many subsurface environments (e.g., anaerobic sediments, soils, groundwater, aquifers). For many substances containing one or more halogenated functional groups, reduction is the pathway of choice for degradation. Note that some metabolites of halogenated compounds may be as toxic (or even more toxic) than the parent compound. Reductive dehalogenation of halo-aliphatic substances generally is initiated by formation of carbon radicals through electron transfer, which can then (1) abstract an H atom from a suitable donor (hydrogenolysis) (2) form a C=C double bond (dehydrogenation) (3) induce radical coupling (dimerization), or (4) form a C=C double bond through elimination of vicinal halides (vicinal dehalogenation). The reduction of halo-aromatic compounds occurs mainly through hydrogenolysis. It usually is a slower process than aliphatic dehalogenation, and in many cases, occurs through microbially mediated reactions.

Table 13.4 Reductive transformation known to occur in natural reducing environments (Larson and Weber 1994)

1. Reductive dehalogenation:

Hydrogenolysis: $R-X + 2e^- + H^+ \longrightarrow R-H + X^-$

Vicinal dehalogenation:

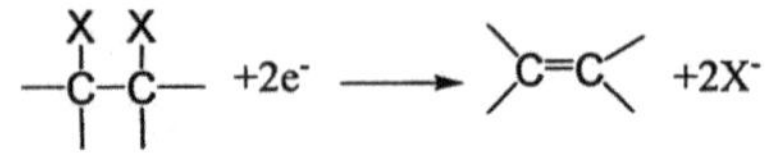

2. Nitroaromatic reduction:

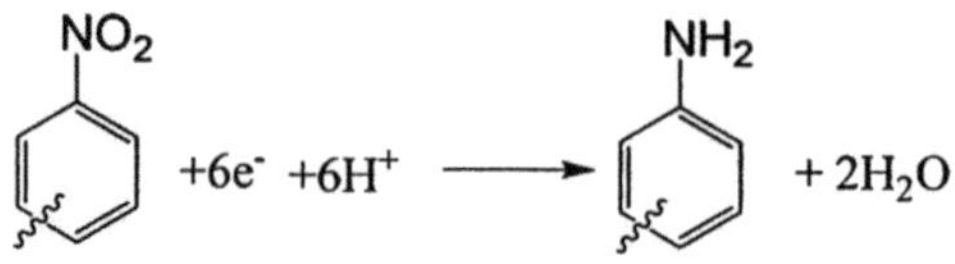

3. Aromatic azo reduction:

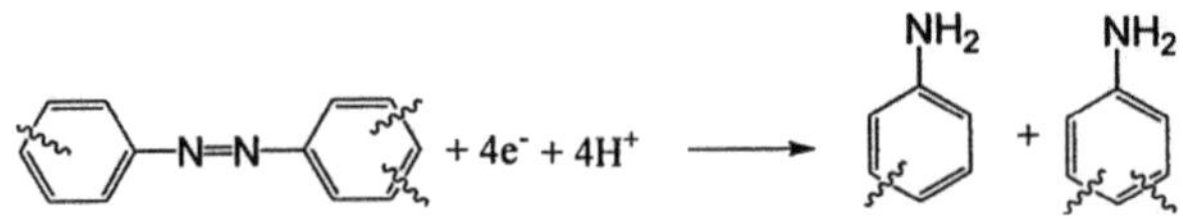

4. Sulfoxide reduction: $R_1-S(=O)-R_2 + 2e^- + 2H^+ \rightleftharpoons R_1-S-R_2 + H_2O$

5. *N*-nitrosoamine reduction:

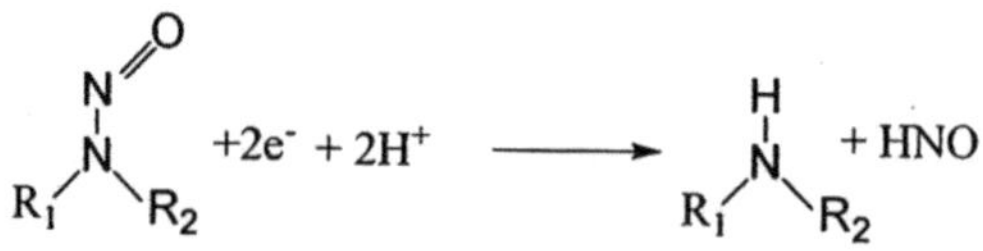

6. Quinone reduction:

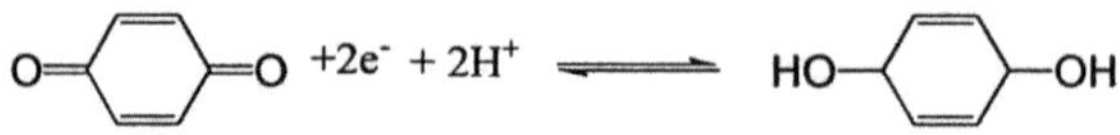

7. Reductive dealkylation: $R_1-X-R_2 + 2e^- + 2H^+ \longrightarrow R_1-XH + R_2H$

X= NH, O, or S

Nitroaromatic Reduction Nitroaromatics constitute an important class of potential environmental contaminants, because of their wide use in agrochemicals, textile dyes, munitions, and other classes of industrial chemicals. Reduction of nitroaromatics produces amines, through a series of electron transfer reactions with nitroso and hydroxylamines as intermediates (Fig. 13.1). Compared to the parent nitroaromatic compound, all intermediates typically reduce readily (Larson and Weber 1994).

Aromatic Azo Compounds Reduction of aromatic azo compounds involves a four-electron process that proceeds through a short-lived intermediate, hydrazobenzene, and ends with complete reductive cleavage of the azo linkage and formation of aromatic amines.

Sulfoxide Reduction Sulfoxide reduction is a two-electron-transfer reversible reaction resulting in thioethers. Organic sulfoxides are used mainly as agrochemicals, and their reduction (abiotic and microbially mediated) has been found in anaerobic soils, sediments, and groundwater (Larson and Weber 1994).

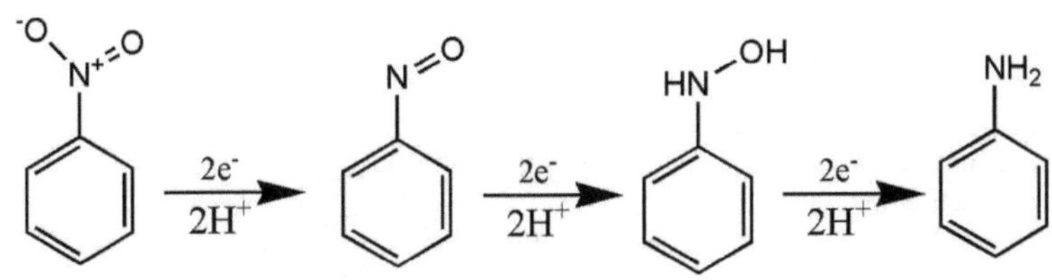

Fig. 13.1 Reduction of nitrobenzene to aniline

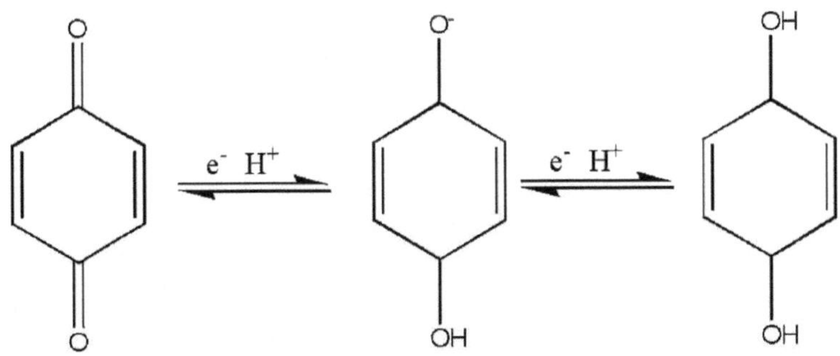

Fig. 13.2 Reduction of quinone to semiquinone and hydroquinone through two one-electron transfer steps (Larson and Weber 1994)

***N*-nitosoamine Reduction** The reduction of *N*-nitrosoamines can occur across either the N–N bond or the N–O bond. Both reactions ultimately result in the formation of the parent amine and ammonia.

Quinone Reduction This is a reversible, one-electron transfer reaction to the semiquinone radical, followed by a second, reversible electron transfer that results in the formation of hydroquinone, as shown in Fig. 13.2.

Reductive Dealkylation Reductive dealkylation involves replacement of an alkyl group on a heteroatom by hydrogen. The reaction is considered to be mainly biologically mediated and usually is important in the subsurface for transformation of agrochemicals.

Photolysis Abiotic oxidation occurring in surface water is often light mediated. Both direct oxidative photolysis and indirect light-induced oxidation via a photolytic mechanism may introduce reactive species able to enhance the redox process in the system. These species include singlet molecular O_2, hydroxyl-free radicals, super oxide radical anions, and hydrogen peroxide. In addition to the photolytic pathway, induced oxidation may include direct oxidation by ozone (Spencer et al. 1980) autooxidation enhanced by metals (Stone and Morgan 1987) and peroxides (Mill et al. 1980).

Photolysis is an environmental process in which a substrate in a natural aqueous solution is subjected to ultraviolet (UV) or visible light, causing its transformation. Sunlight at the surface of the earth consists of direct and scattered light entering a

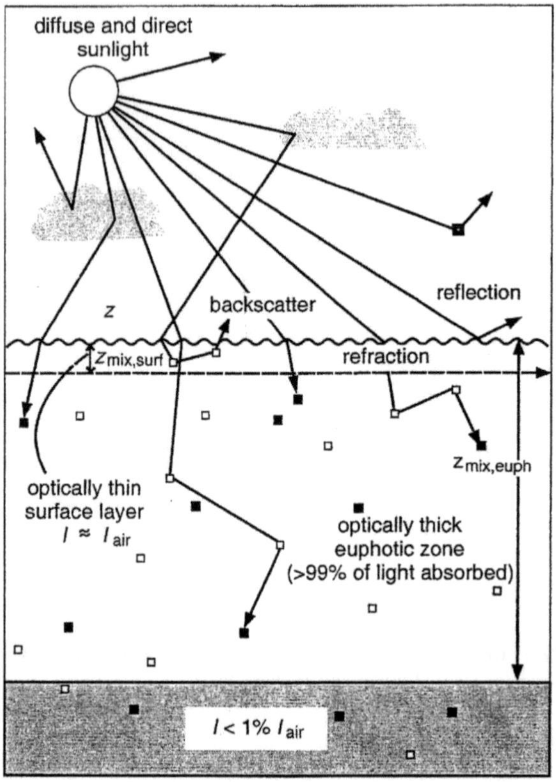

Fig. 13.3 Fate of photons in a natural water body. *Open Square* reflective particles; *Filled Square* absorptive particles or molecules (Schwarzenbach et al. 2003)

water body at various angles. From Fig. 13.3, the distribution of photons in a natural water body may be seen. The rate of transformation may be enhanced in the presence of additional reactants of natural or anthropogenic origin. Contaminant photolysis in water has been studied extensively over the last three decades (e.g., Zepp 1980; Miller and Crosby 1983; Zafiriou 1984; Leifer 1988; Acher and Saltzman 1989; and Schwarzenbach et al. 2003).

The solar spectrum at a particular location depends on a large range of factors, including season of the year, time of day, weather conditions, and atmospheric pollution. The intensity of solar irradiance depends on the optical thickness of the atmosphere, which together with ozone absorption and molecular scattering, are strongly wavelength dependent. Direct and indirect photoreactions can be distinguished.

Direct photolysis processes involve the direct adsorption of light by the substrate

$$P + \text{light} \rightarrow P^* \rightarrow \text{products} \qquad (13.8)$$

where P is a photoreactive substance and P* is the excited state that reacts to form products. The photolysis rate is related directly to the adsorption spectrum of contaminants having the spectral distribution of sunlight. Because radiation at wavelengths below 290 nm is adsorbed by atmospheric ozone, organo or organometallic contaminants that do not absorb radiation above this wavelength do not undergo direct photolysis.

The rate of any type of photoreaction depends also on the rate of light absorption by the photoactive species. The light transmission characteristics of an aqueous system must be determined to account for the degree to which system components other than the contaminant itself attenuate radiation in the water column. Summing over all wavelengths absorbed by all contaminants and natural organic compounds in the water body gives the total rate constant of light absorption.

Indirect photolysis occurs when a species other than the organic or organometallic toxic compound absorbs sunlight and initiates reactions leading to contaminant transformation. Natural products present in natural waters, such as humic substances, clay colloids, and transition metals, absorb the radiation required for subsequent reactions and act as natural sensitizers. Electrically excited molecules are capable of greatly accelerating the process and determine a number of light-induced transformations of toxic organic chemicals.

Singlet oxygen is a selective oxidant that reacts at appreciable rates with electron-rich groups. Singlet oxygen is formed when an excited sensitizer transfers its energy to ground-state (triplet) oxygen. In natural waters, the sensitizers are generally related to dissolved humic substances. In addition, any substance dissolved in water, on adsorbing sunlight, crosses to a triplet state and potentially produces singlet oxygen; such substances include hydrogen peroxide, superoxide, alkyl peroxy radicals, and hydroxyl radicals.

In natural water, singlet oxygen originating from humic substances has been shown, for example, to oxidize thioether pesticide contaminants such as disulfoton (Zepp et al. 1981). Irradiation of dilute hydrogen peroxide in the presence of various non-sunlight-absorbing herbicides results in enhanced oxidation of these substances (Draper and Crosby 1981).

Coordinative interactions in natural waters change as a result of a variation in coordinative species or coordination number, which in turn leads to a transformation of contaminant properties. Any combination of cations with molecules or anions containing free pairs of electrons (bases) is called *coordination* (or *complex formation*). The coordination can be electrostatic, covalent, or a mixture of both. The metal cation is called the *central atom,* and the anion or molecule with which it forms a coordinative compound is referred to as a *ligand.*

The actual form in which a contaminant molecule or ion is present in natural water, as result of a change in the coordinative relationship, emphasizes a specific *chemical speciation.* A *chemical species* is defined by IUPAC "as the isotopic composition, electronic or oxidation state, and/or complex or molecular structure," and the *speciation of an element* as "the distribution of an element amongst defined chemical species in a system" (Templeton et al. 2000).

Speciation is important for understanding the behavior of toxic heavy metals, because different chemical species may behave differently in the subsurface environment (including subsurface water). Subsurface water contains a broad spectrum of dissolved and colloidally dispersed natural products, so it is possible that the same molecule exists in one or more species. For example, in evaluating the presence of heavy metals in subsurface water, one should recognize that the subsurface environment has a direct influence on heavy metal speciation and, thus, on the environmental impact. By complexation with the various components of water, the initial properties of a contaminant can be changed. A free metal ion in an aqueous solution is more biologically active than an ion adsorbed on suspended particles or complexed to other species.

Natural inorganic ligands of heavy metals in subsurface water, which are present in a concentration of about 1 mM, include nitrite, sulfate, chloride, carbonate, and bicarbonate. These potential ligands generally are efficient only under special conditions. For example, in an alkaline environment, carbonate and bicarbonate can be significant complexors of transition metals like Cu^{2+} or the uranyl ion, UO_2^{2+}, and cadmium may be complexed with Cl^- or SO_4^{2-} to form $CdCl^+$ or $CdSO_4$. Additional inorganic ligands, at a micromolecular level, include phosphate and fluoride, which have high affinities for Fe^{3+} and Al^{3+} (Sauve and Parker 2005).

Dissolved organic chemicals acting as potential ligands may be found in the subsurface (including the root zone). This group is very diverse and comprises many compounds like sugars, organic acids, phenols, and lipids. Not all natural organic acids found in subsurface water are able to function as ligands for heavy metals. The common amino acids, such as a root exhudate, are too weak to enhance metal complexation. Low-molecular-weight organic acids (e.g., citric acid), however, exhibit affinity for metals due to their carboxylic functionality. While the aliphatic organic acids are very active in the speciation of metals, in most environmental situations, the ability of phenolic acids to form ligands is of minor significance. On the other hand, the presence of high-molecular-weight compounds in natural waters, such as fulvic or humic acids, leads to significant heavy metal speciation; this is an effect explained by the affinity of carboxyl groups.

By complexation with various components of the natural subsurface solution, the initial properties of a toxic molecule can be changed. These transformations involve adsorption on the subsurface solid phase, transport into the saturated or partially saturated zone, and contaminant half-life in the subsurface environment.

It is operationally difficult to distinguish between dissolved and colloidally dispersed substances. For example, colloidal metal-ion precipitates occasionally have particle sizes smaller than 100 A, sufficiently small to pass through a membrane filter, and organic substances can exist as a stable colloidal suspension. Information on the types of species encountered under different chemical conditions (type of complexes, their stabilities, rate of formation) is a prerequisite to better understanding of the transformation in properties of toxic chemicals in a water body.

Liquid Mixture Transformation Changes in the composition and concentration of volatile fractions of a *volatile organic liquid mixture* (VOLM) contaminating a subsurface water occur when the (immiscible with water) liquid is volatilized into the gaseous phase of the porous medium and subsequently transported into the atmosphere. Fine and Yaron (1993) showed that the volatilization of each hydrocarbon from kerosene (containing more than 100 hydrocarbons), added to a subsurface, occurred at different rates. The mass transfer of spilled VOLM from natural water bodies into the subsurface and near surface atmosphere and the effect of this process on their distribution between the gaseous, liquid and solid phases have been the subject of a series of studies (e.g., Mackay and Yeun 1983; Burris and Macintyre 1986; Nye et al. 1994; Yaron et al. 1998; Dror 2005). The rate of volatilization is controlled by the vapor pressure of the components and environmental conditions. The less-volatile components of a VOLM are transported as liquid into the porous medium. During their transport, differential dissolution of the component mixture into the subsurface water or adsorption on the subsurface solid phase occurs simultaneously with the transport of the volatile fraction in the gaseous phase. Adsorption leads to retardation of volatilization but not to cessation of the process. A change in the VOLM composition occurs as result of this process.

Adapting the approach of Nye et al. (1994) to subsurface conditions, the following stages in contaminant transformations of VOLMs can be identified:

1. A volatilization-induced VOLM depletion stage of the components, characterized by a high vapor pressure.
2. A VOLM depletion stage caused by dissolution in the liquid and adsorption on solid phases of the components, characterized by aqueous solubility.
3. A VOLM enrichment stage caused by desorption of the previously adsorbed fraction on the solid phase.
4. A liquid depletion stage of the gas molecules sorbed on solids or dissolved in water phases.

In stages 2 and 3, the components accumulate or deplete according to their aqueous solubility, volatility, diffusivity, and cosolvent presence. The transformation rate of VOLMs occurs accordingly. In this system, two distinct liquid phases can be recognized: the first is a VOLM, insoluble in water and having chemical properties different from the original VOLM; the second is a mixed aqueous solution of partially miscible organic compounds dissolved in water from the original VOLM.

The *chemodynamic properties* of a multiphase liquid are transformed with respect to the initial state of their components. In the case of VOLMs, both chemical and physical properties of the initial liquid are changed. A second type of liquid transformation is that of water with a given quality, when it mixes with water of different chemical composition. For example, consider the case of a mixture of two aqueous solutions having a similar total salt content and different electrolyte composition, with one having a high Na content and low Ca concentration (SAR = 2.5) and the other having a low Na and high Ca content

(SAR = 2.5). The result is an aqueous solution with a sodium adsorption ratio (SAR) between the volumes of water mixed together. In both cases, with organic and inorganic contaminants, the chemodynamic properties and the quality of the resulting water change.

13.2 Selected Contaminant Transformations in Sediments and Groundwater

Special consideration should be given to the transformation of contaminants in sediments and groundwater. Under saturated conditions, the solid phase may function as a sink, reservoir, and reactor for contaminants. Contaminant presence, persistence, and transformation in the water phase is controlled by the chemistry of the water body, the surface properties of the materials forming the solid phase (sediments or suspended particles), and environmental conditions (temperature and aerobic or anaerobic status).

Pore waters of confining beds in an aquifer (e.g., Black Creek) may contain relatively high concentrations ($\sim 100\ \mu M$) of dissolved organic acid anions, such as acetate and formate, whereas aquifer water contains relatively low concentrations of organic acid anions ($\sim 1\ \mu M$). In addition, confining bed pore waters also may contain sulfate in higher concentrations, that is, $\sim 100\ \mu M$, while sulfate concentrations in aquifers are only about $\sim 5\ \mu M$ (McMahon and Chapelle 1991; McMahon et al. 1992). Discussing these findings for the Black Creek aquifer, Chapelle (2005) assumed that diffusion of DIC (dissolved inorganic carbon), together with DOC (dissolved organic carbon), sulfate, and cations (Mg and Ca) from confining bed pore waters, provide an electron donor (organic carbon) and an electron acceptor (sulfate) for microbial metabolism. Additional inorganic carbon enhances magnesium-calcite dissolution-precipitation, as driven by microbially produced carbon dioxide. Under these conditions, microbial processes occurring in the aquifer (sulfate reduction), as well as microbial processes in confining beds (organic matter fermentation), have an important impact on contaminant transformation in sediments and groundwater.

13.2.1 pH and Hydrolysis Reactions

The pH in ponds and rivers generally is within one pH unit of that of the underlying water. Wolfe et al. (1990) explain these differences by noting that pH values measured in sediments are a composite of the pH of the interstitial water and the pH in the vicinity of the charged surface. Because most solids in sediment systems exhibit a negative charge, the pH near the surface becomes lower than in the bulk water phase. Partially saturated conditions in sediments favor biological activity, leading to carbonate formation and a rise in pH.

In addition to the environmental pH and type of hydrolysis, contaminant hydrolysis in sediments is controlled by the properties of the contaminant molecule and the sediment constituents. In natural sediment, it is difficult to determine if hydrolysis is a biologically mediated or an abiotic reaction. A simple test consists of measuring the contaminant disappearance rate in the sediment and in distilled water, with a pH adjusted to that of the sediment. If the two disappearance rates are similar, the process can be assumed to be abiotic, while if the rate is greater in the sediment system, the hydrolysis is biologically mediated.

In a sediment system, the hydrolysis rate constant of an organic contaminant is affected by its retention and release with the solid phase. Wolfe (1989) proposed the hydrolysis mechanism shown in Fig. 13.4, where P is the organic compound, S is the sediment, P:S is the compound in the sorbed phase, k^1 and k^{-1} are the sorption and desorption rate constants, respectively, and k_w and k_s are the hydrolysis rate constants. In this proposed model, sorption of the compound to the sediment organic carbon is by a hydrophobic mechanism, described by a partition coefficient. The organic matrix can be a reactive or nonreactive sink, as a function of the hydrolytic process. Laboratory studies of kinetics (e.g., Macalady and Wolfe 1983, 1985; Burkhard and Guth 1981), using different organic compounds, show that hydrolysis is retarded in the solid-associated phase, while alkaline and neutral hydrolysis is unaffected and acid hydrolysis is accelerated.

Based on these results and on additional publications (e.g., Konrad and Chester 1969; Armstrong and Konrad 1974), Wolfe et al. (1990) suggested that the tendency of an organic contaminant to hydrolyze in sediments is influenced by the proximity of the molecule to the solid surface in the interstitial water and the direct effect of solid particles on the susceptibility of adsorbed toxic organic molecules to hydrolyze. Because abiotic hydrolysis of organic contaminants in the sediment liquid phase, in many cases, exhibits the same degradation products as those in clear water, it might be suggested that the solid phase in a sediment system alters only the rate of hydrolysis, not the reaction pattern. Some constituents of the sediment solid phase, however, may enhance hydrolysis of organic contaminants adsorbed on their surfaces. For example, in heterogeneous hydrolysis of nonionic pesticides, significant participation by acid groups belonging to solid phase organic matter was observed.

13.2.2 Redox State and Reactions

Redox activity in sediments may be defined by an equilibrium approach, considering that the Eh of an aqueous system is bound in the upper boundary by the oxidation of water and in the lower boundary by reduction of water. The term Eh is defined as the redox potential relative to the standard hydrogen half-cell (see Sect. 2.2.2). Measurements of Eh in sediments of ponds and rivers show that the Eh stabilizes at a depth of about 1 cm and remains stable to a depth of 5–6 cm (Wolfe et al. 1986). Sediments are strongly electron buffered, which is shown by titration

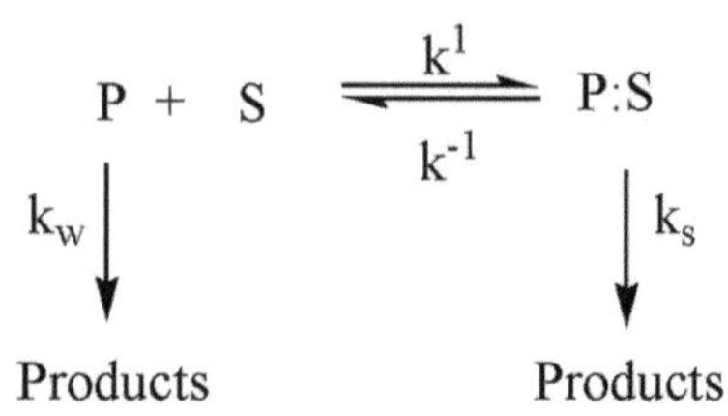

Fig. 13.4 Proposed two-step hydrolysis mechanism (Wolfe 1989)

with chemical oxidants. Although molecular O_2 oxidizes sediments, which affects their reduction capacity, this reaction occurs slowly (Macalady et al. 1986).

The redox state also is defined by a kinetic approach, when the electron acceptors are used predominantly by microorganisms found in the system. It is difficult to define the terminal electron-accepting processes. Chapelle (2005) considers that microbially mediated redox processes tend to become segregated into discrete zones. At the sediment-water interface, oxic metabolism predominates. This oxic zone may comprise zones dominated by nitrate, manganese, or ferric iron reduction. The redox zonation is a result of the ecology of aquatic sediments. The reduction rate of sediment oxidation is explained by intrinsically slow reaction rates of O_2 with reducing moieties in the sediment or by slow diffusion of O_2 into the sediment. In anoxic sediments, oxidation of organics is carried out in the food chain; fermentative microorganisms partially oxidize organic matter with the production of fermentation products, such as acetate and hydrogen. These fermentation products then are consumed by terminal electron-accepting microorganisms such as Fe(III) or sulfate reducers.

Wolfe (1989) suggested a model to describe abiotic reduction in sediments, where a nonreactive sorptive site and an independent reactive sorptive site are considered. The nonreactive sorptive sink is consistent with partitioning of the contaminant to the organic carbon matrix of the solids. The model is described by Fig. 13.5 where P:S′ is the compound at the reactive sorbed site; P is the compound in the aqueous phase; S and S′ are the sediments, P:S is the compound in the non-reactive sink; k^2, k^{-2}, k^1, and k^{-1} are the sorption–desorption rate constants, and k_c, k_w, and k_s are the respective reaction rate constants. If the reaction constants k_w and k_s are neglected, two rate-limiting situations are observed: transport to the reactive site and reduction at the reactive site. The available kinetic data, however, do not allow one to distinguish between the two mechanisms.

An example of a redox transformation in natural water (oxidation) and sediments (reduction) of an organ chlorinated contaminant (aldicarb insecticide) is given in Fig. 13.6.

Because hydrocarbon mixtures originating from petrogenic or pyrogenic residues are a major group of groundwater contaminants, which originate mainly from human activity, these compounds are considered to illustrate transformation of organic molecules in the subsurface environment. These hydrocarbon mixtures generally contain a diverse group of compounds, whose behaviors, persistence, and transformation in the subsurface and in groundwater are dissimilar. In the partially saturated subsurface, a range of selective processes, such as solid phase

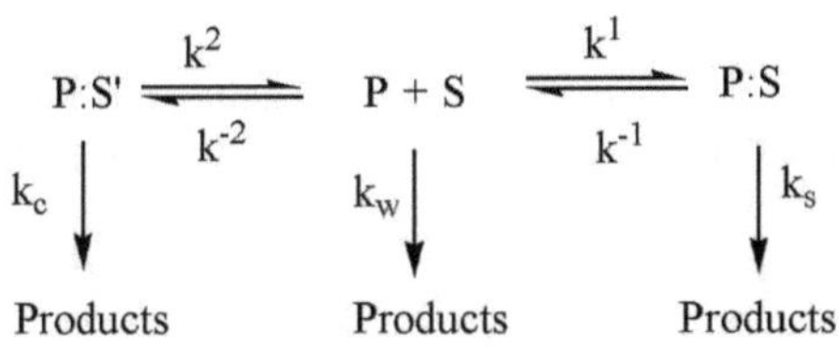

Fig. 13.5 Proposed model for abiotic reduction in a sorbing sediment system (Wolfe 1989)

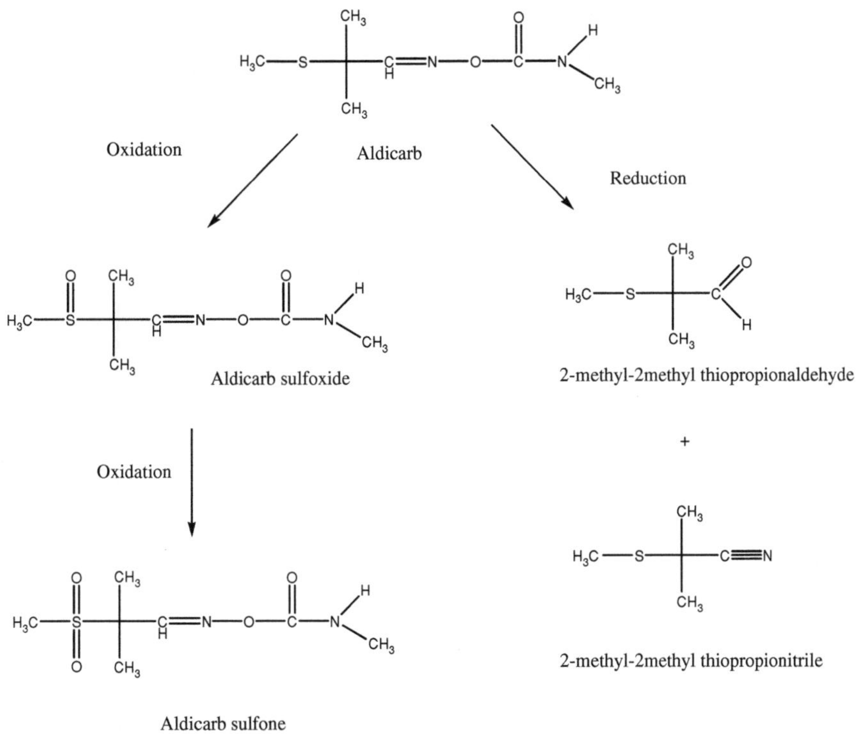

Fig. 13.6 Redox transformation pathways of aldicarb (Macalady et al. 1986; Wolfe et al. 1986)

retention, volatilization, dissolution, photolysis and surface-, chemically, and biologically induced degradation can affect hydrocarbons in groundwater. Biodegradation is a major transformation process of petroleum hydrocarbons.

Both aerobic and anaerobic biological degradation control the persistence of hydrocarbons in groundwater, such as the n-alkane group. A generalized biodegradation pathway is presented in Fig. 13.7. The aerobic pathway shows conversion of an alkane chain to fatty acids, fatty alcohols, and aldehyde and carboxylic acids, which are then channeled into the central metabolism for subsequent β-oxidation.

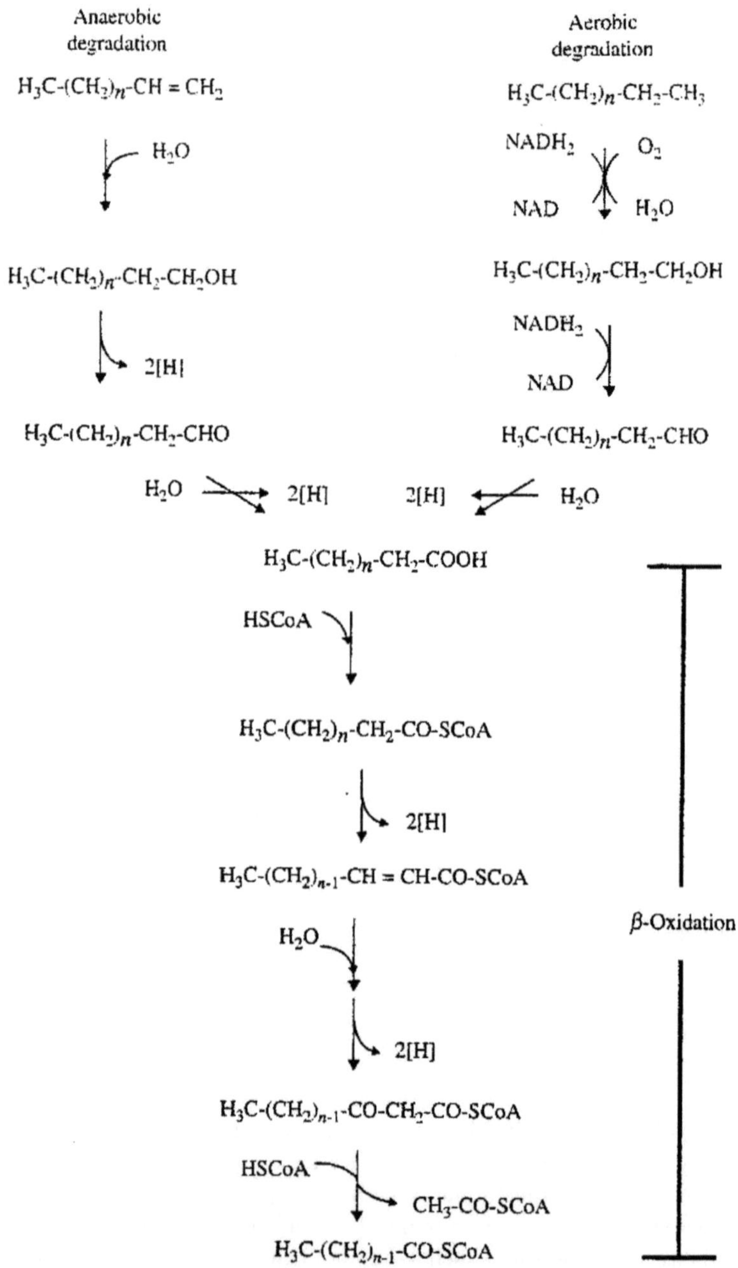

Fig. 13.7 Generalized aerobic and anaerobic biodegradation pathways for *n*-alkanes. Reprinted from Abrajano et al. (2005). Copyright 2005 with permission of Elsevier

The degradation pathway involves independent oxidation to fatty acids, followed by *β*-oxidation.

Anaerobic degradation proceeds with nitrate, Fe^{3+}, or sulfate as the terminal electron acceptor, without any intermediate, such as alcohols. Sulfate reducers apparently show specificity toward utilization of short chain alkanes. Susceptibility to *n*-alkane degradation is an inverse function of chain length. Branched alkanes are less susceptible than straight-chain *n*-alkanes, and the most resilient saturated components are the pristine and phytane isoprenoids (Wang et al. 1998).

As with aliphatic hydrocarbons, oxidative biodegradation of aromatic compounds requires insertion of oxygen into the molecule to form catechol. The susceptibility to biodegradation increases with decreasing molecular weight and degree of alkylation. The most easily degradable polyaromatic hydrocarbons mentioned are the alkyl homologues of dibenzothiophene, fluorine, phenanthrene, and chrysene (Wang et al. 1998).

Microbially induced degradation is recognized as the major mechanism in the transformation of PAHs in the aquatic environment (NRC 2000). In general, PAH degradation rates are a factor of 2–5 slower than degradation rates of monoaromatic hydrocarbons and of similar magnitude as for high-molecular-weight *n*-alkanes (C_{15}–C_{36}). Under similar aerobic conditions, the most rapid biodegradation of PAHs occurs at the water-sediment interface. Prokaryotic microorganisms metabolize PAHs primarily by an initial dioxygenase attack to yield cis-dihydrodiols and finally catechol. General aerobic biodegradation pathways for aromatic hydrocarbons are shown in Fig. 13.8.

Higher-molecular-weight PAHs, such as pyrene, benzo(a)pyrene, and benzo(e)pyrene, exhibit a high resistance to biodegradation. PAHs with three or more condensed rings tend not to act as a sole substrate for microbial growth and require cometabolic transformations. Neilson and Allard (1998) report a cometabolic reaction of pyrene, 1,2-benzanthracene, 3,4-benzopyrene, and phenanthrene in the presence of either naphthalene or phenanthrene. However, the co metabolic reactions are very slow in natural ecosystems.

Natural attenuation processes occurring in groundwater, which include dilution, sorption, volatilization, and biodegradation, may affect the downward migration of hydrocarbon plumes over time. Cozzarelli et al. (1999) report such behavior in the configuration of a petroleum hydrocarbon plume in groundwater, due to rupture of an oil pipeline that occurred in Minnesota in 1979. The oil spill induced formation of an oil lens floating on the water table. The behavior of this lens was studied between 1980 and 1990. In 1980, the dissolved hydrocarbon plume was composed mainly of soluble benzene, toluene, ethylbenzene, and xylene (BTEX), and developed a downward gradient. From groundwater samples collected in 1985, it was observed that the BTEX plume stopped spreading. The dynamic steady state of the plume reflected a balance between the rate at which soluble hydrocarbons spread into the groundwater and the rate at which the hydrocarbons were

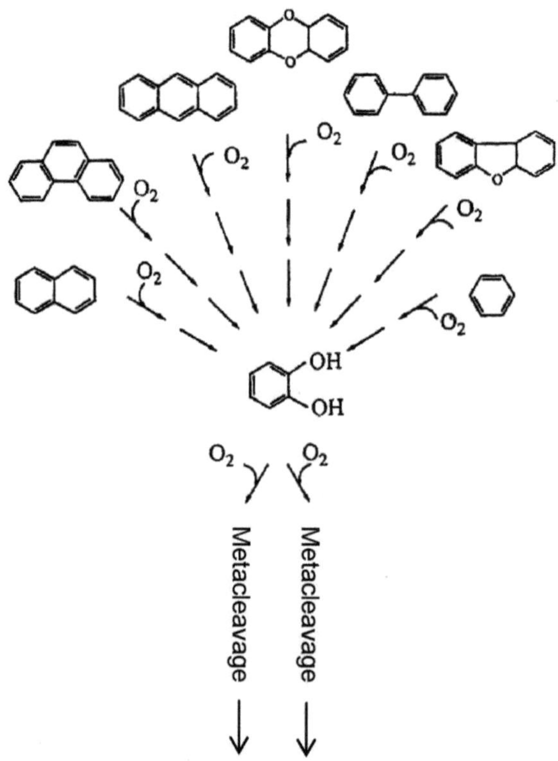

Fig. 13.8 Generalized aerobic biodegradation pathways for aromatic hydrocarbons. Reprinted from Abrajano et al. (2005). Copyright 2005 with permission of Elsevier

consumed by microorganisms or affected by additional processes causing natural attenuation (Baedecker et al. 1993).

Persistence and degradation of chlorinated solvent contaminants in groundwater systems have been the object of a large number of investigations over the last 20 years. In flowing groundwater systems, chlorinated solvents, such as trichloroethylene (TCE) and perchloroethylene (PCE), act as persistent contaminants. The degradation of TCE and PCE, and their transformation into more lightly chlorinated ethenes such as cis-dichloro ethene (DCE) and vinyl chloride (VC), are depicted in Fig. 13.9. Chlorinated ethenes may be subject to microbial degradation processes, including reductive dechlorination and aerobic and anaerobic oxidation. In many natural environments, the initial reductive dechlorination drives the transformation of PCE and TCE to DCE and VC, respectively. The combined effects of methane-oxidizing cometabolism and anaerobic oxidation lead to the transformation of DCE and VC into carbon dioxide, chloride, and water. Because these biodegradation processes are all redox processes, the efficiency of biodegradation is very sensitive to reduction-oxidation conditions (Chapelle 1996).

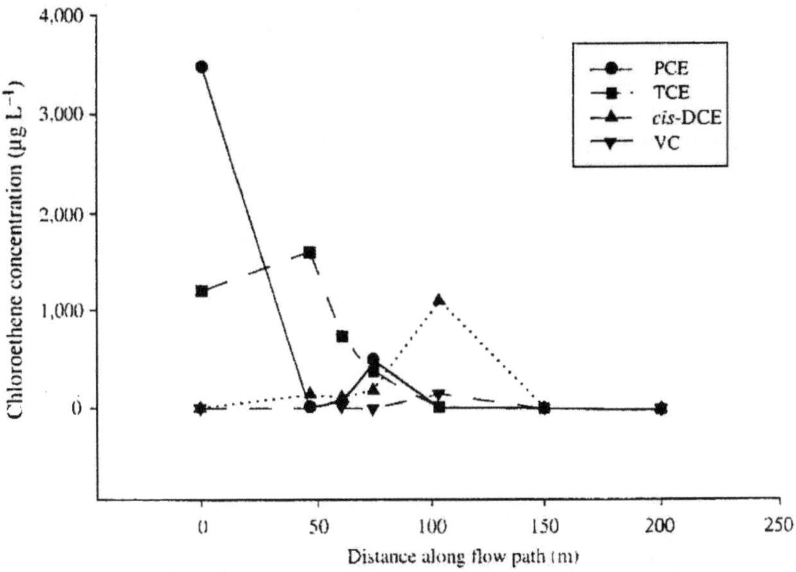

Fig. 13.9 Observed transformations of chlorinated ethenes in groundwater system. (Chapelle and Bradley 1998)

References

Abrajano TA, Yan B, O'Malley V (2005) High molecular weight petrogenic and pyrogenic hydrocarbons in aquatic environments In: Drever JI (ed) Surface and ground water, weathering and soils. Treatise on Geochemistry, vol 5. pp 475–509

Acher A, Saltzman S (1989) Photochemical inactivation of organic pollutants from water. In: Gerstl Z, Chen Y, Mingelgrin U, Yaron B (eds) Toxic organic chemicals in porous media. Ecological Studies 73. Springer, Berlin, Heidelberg, pp 302–320

Armstrong DE, Konrad JG (1974) Nonbiological degradation of pesticides. In: Guenzi WD (ed) Pesticides in soil and water. Soil Science Society America, Wisconsin

Baedecker MJ, Cozzarelli IM, Eganhouse RP, Siegel DI, Bennett PC (1993) Crude-oil in a shallow sand and gravel aquifer. 3. Biogeochemical reactions and mass-balance modeling in anoxic groundwater. Appl Geochem 8:569–586

Bender ML, Silver MS (1963) The hydrolysis of substituted 2-phenyl-1,3-dioxane. J Am Chem Soc 85:3006–3010

Blanchet PF, St. George A (1982) Hydrolysyis of chloropyrifos and chloropyrifos-methyl in the presence of copper. Pestic Sci 13:85–91

Burkhard N, Guth JA (1981) Chemical hydrolysis of 2-chloro-4,6 bis(alkylamino)-1,3,5- triazine herbicide and their breakdown in soil under the influence of adsorption. J Pestic Sci 12:45–52

Burris DR, McIntyre G (1986) Solution of hydrocarbons in a hydrocarbon-water system with changing phase composition due to evaporation. Environ Sci Technol 20:296–299

Chapelle FH (1996) Identifying redox conditions that favor the natural attenuation of chlorinated ethenes in contaminated groundwater systems. In: Symposium on natural attenuation of chlorinated organics in ground water. EPA/540/R-96/509, pp 17–20

Chapelle FH (2005) Surface and ground water, weathering, and soils. In: Drever JI. (ed) Geochemistry of ground water. Treatise of Geochemistry, vol 5. pp 25–449

Chapelle FH, Bradley PM (1998) Selecting remediation goals by assessing the natural attenuation capacity of groundwater systems. Bioremediat J 2(3–4):227–238

Cozzarelli IM, Herman JS, Baedecker MJ, Fischer JM (1999) Geochemical heterogeneity of a gasoline contaminated aquifer. J Contam Hydrol 40:261–284

Draper WM, Crosby DG (1981) Hydrogen peroxide and hydroxyl radicals: intermediates in photolysis reactions in water. J Agric Food Chem 29:699–702

Dror I (2005) Fate of petroleum hydrocarbons in the subsurface environment: abiotic processes. In: Livingston JV (ed) Trends in agriculture and soil pollution research. Nova Science, New York, pp 1–42

Fine P, Yaron B (1993) Outdoor experiments on enhanced volatilization by venting on kerosene components from soils. J Contam Hydrol 12:355–374

Hutchinson GE (1957) The thermal properties of lakes. A treatise on limnology, vol 1. Wiley, New York, pp 426–540

Konrad JG, Chester G (1969) Soil degradation of diazinon, an organophosphate insecticide. J Agri Food Chem 17:226–230

Larson AR, Weber EJ (1994) Reaction mechanisms in environmental chemistry. CRC Press, Boca Raton

Leifer A (1988) The kinetics of environmental aquatic photochemistry. Amer Chem Soc, Washington

Macalady DL, Wolfe NL (1983) New perspectives on the hydrolytic degradation of the organophosphorothionate insecticide chloropyrifos. J Agric Food Chem 31:1139–1147

Macalady DL, Wolfe NL (1985) Effects of sediment sorption and abiotic hydrolysis. J Agric Food Chem 33:167–173

Macalady DL, Tratnyek PG, Grundi TJ (1986) Abiotic reduction reactions of anthropogenic organic chemicals in anaerobic systems. A Critical Rev J Contam Hydrol 1:1–28

Mackay D, Yeun TK (1983) Mass transfer coefficient correlation for volatilization of organic solutes from water. Environ Sci Technol 13:211–212

McMahon PB, Chapelle FH (1991) Geochemistry of dissolved inorganic carbon in a coastal plain aquifer. J Hydrol 125:109–135

McMahon PB, Chapelle FH, Falls WF, Bradley PM (1992) Role of microbial processes in linking sand stone diagenesis with organic-rich clay. J Sedimen Petrol 62:1–10

Mill T, Hendry DG, Richardson H (1980) Free radical oxidants in natural waters. Science 207:886–887

Miller GC, Crosby DG (1983) Pesticide photoproducts: generation and significance. J Toxicol Clin Chem 19:707–727

Neilson AH, Allard A (1998) Microbial metabolism of PAHs and heteroarenes. In: Neilson AH (ed) The handbook of environmental chemistry, vol 3. Springer, Heidelberg, pp 224–273

NRC (2000) Natural attenuation for groundwater remediation, Committee on Intrinsic Remediation, Water Science and Technology Board, National Research Council

Nye PH, Yaron B, Galin Ts, Gerstl Z (1994) Volatilization of a multi component liquid through dry soils: testing a model. Soil Sci Soc Am J 58:269–277

Perdue EM, Wolfe NL (1983) Prediction of buffer catalysis in field and laboratory studies of pollutant hydrolysis reactions. Environ Sci Technol 17:635–642

Sauve S, Parker DR (2005) Chemical speciation of trace elements in soil solution. In: Tabatabai MA, Sparks DL (eds) Chemical processes in soils. Soil Sci Soc Am Book Series no 8, Madison, Wisconsin

Schwarzenbach RP, Gschwend PM, Imboden DM (2003) Environmental organic chemistry, 2nd edn. Wiley-Interscience, Hoboken

Spencer WF, Shoup TD, Spear RC (1980) Conversion of parathion to paraoxon on soil dusts and clay-minerals as affected by ozone and UV-light. J Agric Food Chem 28:366–371

Stone AT, Morgan JJ (1987) Reductive dissolution of metal oxides. In: Stumm W (ed) Aquatic surface chemistry: chemical processes at the particle water interface. Wiley, New York, pp 221–254

Stumm WS, Morgan JJ (1996) Aquatic chemistry, 3rd edn. Wiley, New York

Templeton DF, Ariese R, Cornelis LG, Danielson H, Muntau H, Van Leeuwen H, Lobinski R (2000) Guidelines for terms related to chemical speciation and fractionation. Pure Appl Chem 72:1453–1470

Wang Z, Fingas M, Blenkinsopp S, Landriault M, Sigouin L, Fought J, Semple K, Westlake SWS (1998) Comparison of oil composition changes due to biodegradation and physical weathering in different oils. J Chromatogr A809:89–107

Wolfe NL (1989) Abiotic transformation of toxic organic chemicals in liquid phase and sediments. In: Gerstl Z, Chen Y, Mingelgrin U, Yaron B (eds) Toxic organic chemicals in porous media. Springer, Heidelberg, pp 136–148

Wolfe NL, Kitchens BE, Macalady DL, Grundi TJ (1986) Physical and chemical factors that influence the anaerobic degradation of methyl parathion in sediment system. Environ Sci Technol 5:1019–1026

Wolfe NL, Mingelgrin U, Miller GC (1990) Abiotic transformation in water, sediments and soils. In: Cheng HH (ed) Pesticides in soil environment. Soil Sci Soc Am Book Series no 2, Madison, Wisconsin, pp 104–169

Yaron B, Dror I, Graber E, Jarsjo J, Fine P, Gerstl Z (1998) Behavior of volatile organic liquid mixtures in the soil environment. In: Rubin H, Narkis N, Carberry J (eds) Soil and aquifer pollution. Springer, Berlin

Zafiriou OC (1984) A bibliography of references in natural waters photochemistry. Tech. Mem. WHOI-2 1984 Woods Hole Oceanographic Inst., Woods Hole, Massachusetts

Zepp RG (1980) Assessing the photochemistry of organic pollutants in aquatic environments. In: Haque R (ed) Dynamics exposure and hazard assessment of toxic chemicals. Ann Arbor Sci, Ann Arbor, pp 69–110

Zepp RG, Baughman GL, Schlotzhauer PF (1981) Comparison of photochemical behavior of various humus substances in water: sunlight induced reactions of aquatic pollutants photosensitized by humic substances. Chemosphere 10:109–117

Chapter 14
Abiotic Transformation at the Solid–Liquid Interface

14.1 Catalysis

The aqueous environment in the vicinity of the subsurface solid matrix is different from that of the bulk water. The electric field emanating from the charged solid may strongly affect polarizable species of contaminants, and therefore, their potential for abiotic transformation is much greater at the solid–liquid interface than in natural bulk waters. Molecules in direct contact with reactive solid constituents often are subject to catalytic properties of the surface or interact with available adsorbing sites and can undergo a variety of transformations. Abiotic catalysis reactions of contaminants at the solid–liquid interface have been discussed at length in the literature, including in a number of reviews (Chaussidon and Calvet 1974; Mingelgrin and Prost 1989; Wolfe et al. 1990; Huang 2000).

We refer to a catalyst as an earth material that enables and/or enhances a chemical reaction without undergoing any permanent chemical change. Catalysts provide alternative reactive pathways by which a reaction reaches a local equilibrium, although it does not alter the position of equilibrium (Daintith 1990).

If the catalyst is in the same phase as the reactant (e.g., dissolved metals catalyzing transformation of dissolved organic substances), the catalysis is called *homogeneous.* When the catalytic process is determined by a catalyst in a different phase than the reactant (e.g., solid metal oxides catalyzing transformation of dissolved organic or inorganic substances), the catalysis is called *heterogeneous.* In this case, the catalyzed reaction steps occur very close to the solid surface; the reactions may be between the molecules adsorbed on the catalyst surface or may involve the top-most atomic layer of the catalyst.

A heterogeneous natural system such as the subsurface contains a variety of solid surfaces and dissolved constituents that can catalyze transformation reactions of contaminants. In addition to catalytically induced oxidation of synthetic organic pollutants, which are enhanced mainly by the presence of clay minerals, transformation of metals and metalloids occurs with the presence of catalysts such as Mn-oxides and Fe-containing minerals. These species can alter transformation pathways and rates through phase partitioning and acid–base and metal catalysis.

B. Berkowitz et al., *Contaminant Geochemistry*,
DOI: 10.1007/978-3-642-54777-5_14, © Springer-Verlag Berlin Heidelberg 2014

Catalysis by proton transfer is significant in the subsurface and associated environment and is common in homogeneous reactions. The strength of the acid or base is determined by the ionization constant, while its efficiency as a catalyst is controlled by the reaction rate. This relation, known as the *Brönsted catalysis law,* is expressed as

$$k_a = C_A K_a^{\alpha} k_b = C_B K_b^{\beta}, \tag{14.1}$$

where k_a and k_b are the rate constants (catalytic constants) for acid and base catalytic reactions, k_a and k_b are the acid and base ionization constants, and C_A, C_B, α, and β are constants representing characteristics of the reactions, the solvent, and the temperature, respectively.

Normally α and β are positive with values between 0 and 1. Low values reflect low sensitivity of the catalytic constant to the strength of the catalyzing acid or base, while high α and β values indicate an inverse catalytic pathway. In acid catalysis of organic molecules, the proton located on negatively charged molecules reduces the negative charge, so that the transfer of electrons is facilitated. It can be assumed that, under such conditions, a metal ion that generally acts as an acid will form a metal–organic complex, which reduces the negative charge and enhances the electron transfer. Unlike a proton, the metal ion can be stabilized by other ligands. Some metal ions, especially of the transition series, have several stable oxidation states that enable them to act as catalysts in redox reactions; these ions can catalyze a wide variety of transformation reactions of organic and inorganic contaminants (Huang 2000).

14.2 Surface-Induced Transformation of Organic Contaminants

The spatial distribution of ions and their charge are affected strongly by the electric field emanating from charged surfaces. As a result, some organic contaminants in direct contact with these surfaces can undergo transformation by catalytic processes.

Clay minerals behave like Bronsted acids, donating protons, or as Lewis acids (Sect. 6.3), accepting electron pairs. Catalytic reactions on clay surfaces involve surface Bronsted and Lewis acidity and the hydrolysis of organic molecules, which is affected by the type of clay and the clay-saturating cation involved in the reaction. Dissociation of water molecules coordinated to surface, clay-bound cations contributes to the formation active protons, which is expressed as a Bronsted acidity. This process is affected by the clay hydration status, the polarizing power of the surface bond, and structural cations on mineral colloids (Mortland 1970, 1986). On the other hand, ions such as Al and Fe, which are exposed at the edge of mineral clay colloids, induce the formation of Lewis acidity (McBride 1994).

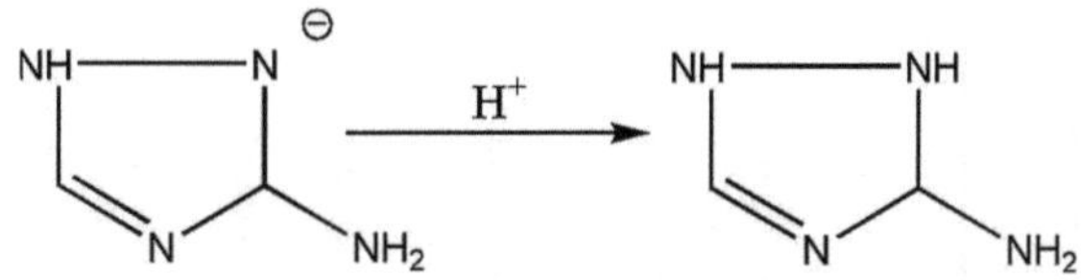

Fig. 14.1 Conversion of 3-aminotriazole to the imino form (Russel et al. 1968)

Many nonionic organic contaminants require extreme acid conditions to accept H^+ ions. In clays, the extent of protonation is related to the electronegativity and polarizing power of structural metal cations, in the order: $H^+ > Al^{3+} > Fe^{3+} > Mg^{2+} > Ca^{2+} > Na^+ > K^+$ (McBride 1994). Mineral surface acidity also catalyzes hydrolysis of organic contaminants in the subsurface. This transformation pathway depends both on the type of clay mineral forming the solid phase and on the clay-saturating cation. For example, Mg^{2+}-montmorillonite exhibits much weaker catalytic capacity than Ca^{2+}-montmorillonite; and Ca^{2+}-beidellite and Ca^{2+}-nontronite exhibit a lower catalytic capacity than Ca^{2+}-montmorillonite (Mortland and Raman 1967). As clay surfaces become drier, the protons become concentrated in a smaller volume of water, and the surface acidity increases to an extreme value. Under these conditions, even a very weak base can be protonated. Chaussidon and Calvet (1965) showed that amines adsorbed on montmorillonite undergo transformation on dehydration of the clay. Catalytic hydroxylation of an organic molecule (e.g., atrazine) on H^+-montmorillonite involves the substitution of a chlorine atom by a hydroxyl ion; the degradation product apparently remains adsorbed on the clay as the keto form of the protonated hydroxy analog (Russel et al. 1968). These authors also showed that Mg^{2+}-montmorillonite converts the amino form of 3-aminotriazole to the imino form, as shown in Fig. 14.1.

Surface-catalyzed degradation of pesticides has been examined in the context of research on contaminant–clay interactions. Such interactions were observed initially when clay minerals were used as carriers and diluents in the crop protection industry (Fowker et al. 1960). Later specific studies on the persistence of potential organic contaminants in the subsurface defined the mechanism of clay-induced transformation of organophosphate insecticides (Saltzman et al. 1974; Mingelgrin and Saltzman 1977) and s-triazine herbicides (Brown and White 1969). In both cases, contaminant degradation was attributed to the surface acidity of clay minerals, controlled by the hydration status of the system.

Rearrangement reactions catalyzed by the clay surface were observed for parathion (an organophosphate pesticide) when it was adsorbed on montmorillonite or kaolinite in the absence of a liquid phase. The rate of rearrangement reactions increased with the polarization of the hydration water of the exchangeable cation (Mingelgrin and Saltzman 1977). Table 14.1 summarizes a series of reactions catalyzed by clay surfaces, as reported in the literature.

Dissolved metals and metal-containing surfaces play an important role in the transformation of organic contaminants in the subsurface environment. Metal ions can catalyze hydrolysis in a way similar to acid catalysis. Organic hydrolyzable compounds susceptible to metal ion catalysis include carboxylic acids, esters,

Table 14.1 Selected examples of reactions catalyzed by clay surfaces (Wolfe et al. 1990)

Substrate	Reaction	Type of clay	Remarks
Ethyl acetate	Hydrolysis	Acid clays	–
Sucrose	Inversion	Acid clays	–
Alcohols, alkene	Ester formation	Al-montmorillonite	–
Organophosphate esters	Hydrolysis	Cu- and Mg-montmorillonite, Cu-beidellite Cu-nontronite	Cu-mont. better catalyst than other Cu-clays or Mg-mont.
Organophosphate esters	Hydrolysis and rearrangement	Na-, Ca-, and Al-kaolinites and bentonites	Room and other temperatures. No liquid phase
Phosmet	Hydrolysis	Homoinic montmorillonite	In suspension
Ronnel	Hydrolysis and rearrangement	Acidified bentonite; dominant exchangeable cations: H, Ca, Mg, Al, Fe(III)	Al-catalyzed suggested reaction
s-Triazines	Hydrolysis	Montmorillonitic clay	Cl-analog degrades faster than methoxy or methoxythio compounds
Atrazine	Hydroxylation	H-montmorillonite	–
Atrazine	Hydrolysis	Al- and H-montmorillonite; montmorillonitic soil clay	Ca- and Cu-clays much weaker catalysts
1-(4-methoxy phenyl)-2, 3-epoxy propane	Hydrolysis	Homoionic montmorillonites; Na-kaolinite	–
DDT	Transformation to DDE	Homoionic clays	Na-bentonite is a better catalyst than H-bentonite
Urea	Ammonification	Cu-montmorillonite	Air-dried clay at 20 °C
Aromatic amines	Redox	Clays (review)	In the presence of metal ions that are good redox agents
3, 3′, 5, 5′ Tetra methyl benzidine	Redox	Hectorite	O_2 oxidizing agent
Pyridine derivatives, olefines, dienes	Oligomerization	Clays (review)	–
Glycine	Oligomerization	Na-kaolinite, Na-bentonite	–
Styrene	Polymerization	Palygorskite, kaolinite, and montmorillonite	Weakest catalyst is montmorillonite
Fenarimol	Dialdehyde formation	Homoionic montmorillonite	–

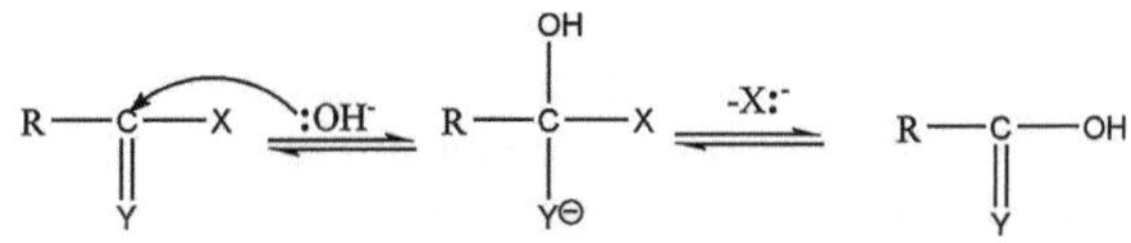

Fig. 14.2 Illustration of the S_N2 mechanism (nucleophilic substitution) for hydrolysis reaction (Huang 2000)

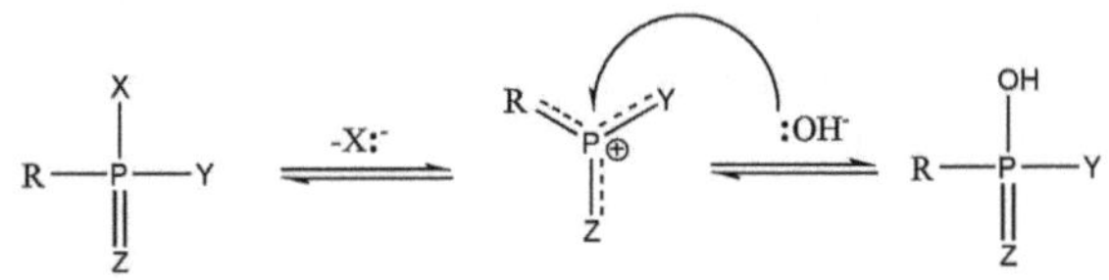

Fig. 14.3 Illustration of the S_N1 mechanism (nucleophilic substitution) for hydrolysis reaction (Huang 2000)

amides, anilides, and phosphate-containing esters. Metal ions and protons coordinate to the organic contaminant, so that electron density is shifted away from the site of nucleophilic attack to facilitate the reaction.

Metal ion-induced catalysis generally occurs via complexation of the reactant molecule. Stone et al. (1993) formulate a general pathway of the process as follows: (1) Complex formation constants increase as the charge-to-radius ratio of the metal ion increases, (2) polarizable metals and ligands exhibit complex stability through covalent bond formation, and (3) competition for the metal among available ligands becomes greater as complex formation constants increase. The properties of metal ions, in general, and transition metal ions, in particular, make them good catalysts for a broad range of organic and inorganic reactions in the subsurface environment. These reactions are described at length in the literature (e.g., McBride 1994; Smolen and Stone 1998). Catalysis by surface-bound metals is observed when all participating reactants are significantly adsorbed, and when rate constants for the reaction at the solid–liquid interface exceed those in the surrounding liquid phase. Lewis acid properties of metals, for example, are significant in mineral-catalyzed hydrolysis reactions. Mineral-catalyzed degradation occurs mainly for the hydrolysable organics that have a structure suitable for complexation with the surface metal cations.

In an extensive review on abiotic catalysis, Huang (2000) noted that the reactivity of hydrolyzable organic contaminants arises from the presence of electron-deficient (electrophilic) sites within the molecules. Figures 14.2 and 14.3 show the patterns of reactivity in two cases of nucleophilic substitution and monomolecular nucleophilic substitution. The S_N2 mechanism (nucleophilic substitution) involves attack of the electrophilic sites by OH^- or H_2O, generation of a higher coordination number intermediate, subsequent elimination of the leaving group, and the formation of an hydrolysis product (Fig. 14.2).

In the case of monomolecular nucleophilic substitution (S_N1), the reaction proceeds with the loss of the leaving group to generate a lower coordination number intermediate, followed by generation of the hydrolysis product by nucleophilic addition, as shown in Fig. 14.3.

Humic substances are also able to enhance abiotic transformation of organic substances, such as anthropogenic pollutants, at the solid–liquid interface. The transformation of organic pollutants adsorbed on organic matter surfaces occurs because the natural organic fraction contains many reactive groups that are known to enhance chemical changes in several families of organic substances, and humic substances provide a strong reducing capacity (Stevenson 1982). The presence of relatively stable free radicals in the fulvic and humic acid fractions of subsurface organic matter further supports enhanced abiotic transformations of many organic contaminants.

Transformation of toxic organic chemicals by humic substances at the solid–liquid interface may occur mainly through hydrolysis, as discussed in the reviews of Senesi and Chen (1989) and Wolfe et al. (1990).

Hydrolysis of triazine herbicides, for example, describes humic-induced transformation of toxic organic molecules. The catalytic effects of the chloro-s-triazine herbicides on dechloro-hydroxylation were determined by Armstrong et al. (1967).

They found that the formation of H-bonding between the ring or side-chain nitrogen of s-triazine and the humic acid surface causes electron withdrawal from the electron-deficient carbon atom that already is surrounded by electronegative nitrogen and chlorine atoms. As a consequence, the weak nucleophile, water, replaces the chlorine atom and increases the rate of hydrolysis. The formation of H-bonding between humic acids and atrazine was suggested as being responsible for the observed decrease in the activation energy barrier of the reaction. The catalytic effect of humic acids depends not only on the number of effective acid groups but also on their arrangement in the humic acid molecule (Li and Felbeck 1972). Based on knowledge of the type, number, and pK_a values of the acidic functional groups in a quantitatively characterized fulvic acid, Gamble and Khan (1985) confirmed that hydrogen ions and undissociated carboxyl groups are the only catalytic agents for atrazine hydrolysis.

Purdue and Wolfe (1983) proposed a general mechanism for the effects of subsurface organic matter on the hydrolysis of hydrophobic organic contaminants. The suggested mechanism is derived from a combination of processes that describe, separately, partitioning equilibrium, acid–base catalysis, and micellar catalysis. The resulting model indicates that the overall reaction rates of toxic organic chemicals can be attributed almost totally to partitioning equilibrium and micellar catalysis.

Nonspecific enhanced surface transformation is observed on charged surfaces found in the vicinity of the surface that are not specific adsorption sites. Probably the most significant phenomenon in the interfacial region, near the surface of charged solids, is the strong dependence of the concentration of charged solutes on the distance from the surface. The concentration in the interfacial region of charged inductors, catalysts, reactants, and products therefore can be different from their concentration in the bulk solution. Another factor that can affect organic transformations in the interfacial region near charged particles is the aforementioned influence of the electric field on the polarization and dissociation of the solute and solvent.

Table 14.2 Estimated trace element concentrations in soil solutions (based on Senesi and Loffredo 2005; Sauve and Parker 2000)

Element	Soil solution μg/L
As	0.5–60,000
Cd	0.01–5,000
Cr	2–500
Cu	5–10,000
Ni	0.5–5,000
Pb	0.5–500
Se	1–100,000
Zn	1–100,000

14.3 Surface-Induced Interactions of Inorganic Contaminants

Metal ion complexation to natural organic components in the solid phase is a major example of abiotic interactions of inorganic contaminants in the subsurface. Through these interactions, initial metal ion relationships of the original compounds are changed and contaminant retention, persistence, and transport in the environment exhibit different behaviors. As mentioned previously (see Part I), subsurface organic compounds that form complexes with toxic metal ions include natural humic substances (about 80 %), additional organic substances of biological origin (such as aliphatic and amino acids, polysaccharides, and polyphenols), and xenobiotic organic chemicals released on the land surface by accident or intentionally.

Potentially toxic compounds in the subsurface, such as Cd^{2+}, Pb^{2+}, or Hg^{2+}, which are generally found in very low concentrations, are considered "soft cations" (Buffle 1988). These ions have strong affinity to intermediate and soft ligands and therefore bond to them covalently. "Borderline" cations, which embrace transition metals such as Cu^{2+} and Zn^{2+}, exhibit affinity for the soft cations as well as for alkaline earth compounds. The order of donor atom affinity for soft metals is $O < N < S$. Functional groups present in subsurface organic matter that show affinity for "soft" and "borderline" metals are shown in Table 14.2.

References

Armstrong DE, Chester G, Harris RF (1967) Atrazine hydrolysis in soil. Soil Sci Soc Am Proc 31:61–66

Brown CB, White JL (1969) Reactions of 12-triazines with soil clays. Soil Sci Soc Am Proc 33:863–867

Buffle J (1988) Complexation reactions in aquatic systems: an analytical approach. Ellis Horwood, Chichester, England

Chaussidon J, Calvet R (1965) Evolution of amine cations adsorbed on montmorillonite. J Phys Chem 69:265–268

Chaussidon J, Calvet R (1974) Catalytic reactions on clay surfaces. Third international congress of pesticides chemistry (IUPAC), Helsinki. In: Coulton F, Albaky NY, Konle F (eds) Environmental quality and safety, vol 3. Gerg Thiem, Stuttgart

Daintith J (1990) A concise dictionary of chemistry. Oxford University Press, Oxford

Fowker FM, Benesi HA, Ryland RB, Sawyer WM, Detling KD, Folkemer FB, Johnson MR, Sun YP (1960) Clay catalyzed decomposition of insecticides. J Agric Food Chem 8:203–210

Gamble DS, Khan SU (1985) Atrazine hydrolysis in soils—catalysis by the acidic functional-groups of fulvic-acid. Can J Soil Sci 65:435–443

Huang OM (2000) Abiotic catalysis. In: Sumner ME (ed) Handbook of soil science. CRC Press Boca Raton, Florida, pp 303–327

Li GC, Felbeck GT Jr (1972) A study of the mechanism of atrazine adsorption by humic acid from muck soil. Soil Sci 113:140–148

McBride MB (1994) Environmental chemistry of soils. Oxford University Press, Oxford

Mingelgrin U, Prost R (1989) Surface interaction of toxic organic chemicals with minerals. In: Gerstl Z, Chen Y, Mingelgrin U, Yaron B (eds) Toxic organic chemicals in porous media. Springer, Berlin, pp 91–136

Mingelgrin U, Saltzman S (1977) Surface reactions of parathion on clays. Clays Clay Miner 27:72–78

Mortland MM, Raman KV (1967) Catalytic hydrolysis of some organophosphate pesticides by copper. J Agric Food Chem 15:163–167

Mortland MM (1970) Clay-organic complexes and interactions. Adv Agron 22:75–117

Mortland MM (1986) Mechanism of adsorption of nonhumic organic species by clays. In: Huang PM, Schnitzer M (eds) Interaction of soil minerals with natural organics and microbes. Soil Sci Soc Am, Madison, Wisconsin, pp 59–76

Purdue EM, Wolfe NL (1983) Prediction of buffer catalysis in field and laboratory studies. Environ Sci Technol 17:635–642

Russel JD, Cruz M, White JL (1968) The adsorption of 3-aminitriazole by montmorillonite. J Agric Food Chem 16:21–24

Saltzman S, Mingelgrin U, Yaron B (1974) The surface catalyzed hydrolysis of parathion on kaolinite. Soil Sci Soc Am Proc 38:231–234

Sauve S, Parker DR (2000) Chemical speciation of trace elements in soil solution. In: Tabatabai MA, Sparks DL (eds) Chemical processes in soils, vol 8., Soil Sci Soc Am Book Series, Madison, Wisconsin

Senesi M, Chen Y (1989) Interaction of toxic organic chemicals with humic substances. In: Gerstl Z, Chen Y, Mingelgrin U, Yaron B (eds) Toxic organic chemicals in porous media. Springer, Berlin, pp 37–91

Senesi N, Loffredo E (2005) Metal ion complexation by soil humic substances. In: Tabatabai MA, Sparks DL (eds) Chemical processes in soils, vol 8., Soil Sci Soc Am Book Series, Madison, Wisconsin, pp 563–618

Smolen JM, Stone AT (1998) Organophosphorous ester hydrolysis catalyzed by dissolved metals and metal containing surfaces. In: Huang PM et al (eds) Soil chemistry and ecosystem health, vol 52., Soil Sci Soc Am Spec Publ, Madison, Wisconsin, pp 157–171

Stevenson FJ (1982) Humus chemistry. Wiley, New York

Stone AT, Torrents A, Smolen J, Vasudevan D, Hadley J (1993) Adsorption of organic-compounds processing ligand donor groups at the oxide-water interface. Environ Sci Technol 27:895–909

Wolfe NL, Mingelgrin U, Miller GC (1990) Abiotic transformation in water, sediments and soils. In: Cheng HH (ed) Pesticides in soil environment, vol 2., Soil Sci Soc Am Book Series, Madison, Wisconsin, pp 104–169

Chapter 15
Biologically Mediated Transformations

15.1 Subsurface Microbial Populations

The subsurface microbial population includes a fascinating array of organisms, with diverse capabilities for deriving energy from the metabolism of organic and inorganic compounds (Alexander 1980). This population is active mainly in soils forming the upper layer of the subsurface and decreases with depth. Contaminants reaching the land surface, such as industrial solvents and agrochemicals, may be incorporated into the biological body or redistributed in subsurface phases.

Microbial life is adapted to the available energy and nutrient supply under varying environmental conditions. A contaminant reaching the subsurface may become a source of energy for the biomass. When a source of energy is offered to a specific microbial population, the population rapidly increases, enhancing the activity for utilizing the available energy source (Keeney 1983). Energy sources and environmental conditions interact to determine the micropopulation ecology for microbially mediated transformation of subsurface contaminants. Table 15.1 shows the main microbial reactions in relation to energy.

15.2 Processes Governing Contaminant Attenuation

The relationship between microbial physiology and biologically mediated transformation of contaminants is summarized by Azadpour-Keely et al. (1999), who reviewed microbial processes affecting natural attenuation of toxic organic chemicals in the subsurface. Information encoded in the DNA (deoxyribonucleic acid) of unicellular bacteria is transferred through RNA (ribonucleic acid) to the ribosome to produce proteins or enzymes. These enzymes induce organocarbon degradation, which is used by bacteria as a source of energy. The electrons or reducing equivalents (hydrogen or electron-transferring molecules) produced are transferred to a terminal electron acceptor (TEA), and during the transfer process, energy is produced and utilized by the cell.

B. Berkowitz et al., *Contaminant Geochemistry*,
DOI: 10.1007/978-3-642-54777-5_15, © Springer-Verlag Berlin Heidelberg 2014

Table 15.1 Classification of the main microbial reactions in relation to (*top*) different classes of energy sources, and (*bottom*) reactions that influence the fate of elements and compounds in the subsurface (Delwich 1967; Alexander 1977)

Class	Electron		
	Donor	Acceptor	Products
Photoautotrophic	H_2O H_2S, H_2R	CO_2	$(HCHO)_n$ and other reduced compounds
Respiration aerobic	Organic compounds	O_2	$CO_2 + H_2O$
Anaerobic	Organic compounds	Many ranging from the same molecule, another molecule, CO_2 or inorganic compound $CO_2 + H_2O$	Many ranging from reduced compounds, oxidized compounds, CO_2 and H_2O
Chemoautotrophic	Inorganic compound	O_2 or another inorganic compound	Oxidized inorganic compound

Reaction
1. Mineralization of inorganic ions during organic matter decomposition when amount of element present is in excess of microbial demand
2. Immobilization of inorganic ions to organic forms to satisfy needs of microbial growth
3. Oxidation of inorganic ions, particularly as energy sources by autotrophs, although numerous heterotrophs, also oxidize compounds but do not obtain energy from the process
4. Reduction of oxidized elements, particularly when an alternate electron acceptor to O_2 is required. However, some reduction reactions occur in which the oxidized species are not needed as an electron acceptor
5. Indirect transformations leading to changes in the microenvironment (pH, depletion of O_2 to lower Eh, addition of CO_2)
6. Production or degradation of organic ligands
7. Methylation to produce more mobile and/or volatile compounds

Bacteria generally are grouped into three categories, according to TEA utilization: *aerobes,* which utilize molecular oxygen as a TEA (and without molecular oxygen these bacteria are not capable of degradation); *aerobes/anaerobes,* which can utilize molecular oxygen or in its absence may switch to nitrate, manganese oxides, or iron oxides as electron acceptors; and *anaerobes,* for which oxygen is toxic, and which utilize nitrate, sulfate, or carbon dioxides or other electron acceptors as TEA.

Aerobic respiration is one of the major processes mediating chemical transformation, mainly through the activity of heterotrophic microorganisms. Carbon turnover in the subsurface, for example, involves chemicals such as N, S, and P. The large polymeric molecules, which constitute the bulk of biological residues, often are used as a source of energy; they ultimately become humic substances. These substances may further affect the fate of contaminants that originate from anthropogenic activity, by the interactions discussed previously. Aerobic respiration also causes changes in the molecular structure of organic agrochemicals. Autotrophic and chemoautrophic reactions are among the most important aerobic processes in nitrification. Autotrophic transformation controls the oxidation of S in

SO_4^{2-} or $S_2O_3^{2-}$, organic Fe compounds in Fe precipitates, and arsenite (mobile and toxic) in arsenate (nontoxic). Aerobic respiration also has indirect effects on the transformation of chemicals. The release of CO_2, which affects redox potential and pH, leads to the transformation of metal ions by chelation.

Anaerobic metabolism occurs under conditions in which the O_2 diffusion rate is insufficient to meet the microbial demand, and alternative electron acceptors are needed. The type of anaerobic microbial reaction controls the redox potential (Eh), the denitrification process, reduction of Mn^{4+} and SO_4^{2-}, and the transformation of selenium and arsenate. Keeney (1983) emphasized that denitrification is the most significant anaerobic reaction occurring in the subsurface. Denitrification may be defined as the process in which N-oxides serve as terminal electron acceptors for respiratory electron transport (Firestone 1982), because nitrification and NO_3^- reduction to NH_4^+ produce gaseous N-oxides. In this case, a reduced electron-donating substrate enhances the formation of more N-oxides through numerous electrocarriers. Anaerobic conditions also lead to the transformation of organic toxic compounds (e.g., DDT); in many cases, these transformations are more rapid than under aerobic conditions.

Microbial methylation is a reaction that affects mainly properties of toxic, inorganic trace elements, which involves the addition of a methyl group to the contaminant molecule. It occurs under aerobic or anaerobic conditions. Mercury methylation, for example, occurs under both conditions and leads to the release of mercury into the atmosphere.

15.3 Biotransformation of Organic Contaminants

The microbial metabolic process is the major mechanism for the transformation of toxic organic chemicals in the subsurface environment. The transformation process may be the result of a primary metabolic reaction, when the organic molecule is degraded by a direct microbial metabolism. Alternatively, the transformation process may be an indirect, secondary effect of the microbial population on the chemical and physical properties of the subsurface constituents. Bollag and Liu (1990), considering behavior of pesticides, defined five basic processes involved in microbially mediated transformation of toxic organic molecules in the soil upper layer environment. These processes are described next.

Biodegradation is a process in which toxic organic molecules serve as substrate for microbial growth. In this case, organic molecules are used by one or more interacting microorganisms and metabolized into CO_2 and inorganic components. In this way, microorganisms obtain their requirements for growth and toxic organic molecules are completely decomposed, without producing metabolites (which in some cases could be more toxic than the parent material). From an environmental point of view, this process is highly effective and desirable; the presence of a biodegradable compound in a subsurface site may enhance the proliferation of active microbial populations and consequently increase the rate of decomposition of additional contaminants that enter the system.

Cometabolic transformations include the degradation of toxic organic molecules by microorganisms that grow at the expense of a substrate other than the toxic organic one. This process, in which enzymes involved in catalyzing the initial reaction are lacking in substrate specificity, may lead to the formation of intermediate products that, in some cases, are more toxic than the parent material. These products cause an adverse environmental impact and may inhibit microbial growth and metabolism. Environmental factors and the contaminant concentration may affect the nature of the metabolism; different microorganisms can metabolize different toxic organic molecules. Wang et al. (1984) found that isopropyl N-phenylcarbamate (IPC) is mineralized at low concentrations, while at higher concentrations, it is converted to organic products by cometabolism. In general, cometabolism does not result in extensive degradation of a specific organic contaminant. However, different microorganisms can transform a molecule by sequential cometabolic attacks, and cometabolic products of one organism can be used as a growth substrate for another organism.

Polymerization, or *conjugation,* is the process in which toxic organic molecules undergo microbially mediated transformation by oxidative coupling reactions. In this case, a contaminant or its intermediate product(s) combines with itself or other organic molecules (e.g., xenobiotic residues, naturally occurring compounds) to form larger molecular polymers that can be incorporated in subsurface humic substances.

Cellular accumulation is an additional microorganism-mediated transformation pathway of organic contaminants in the subsurface environment. The rate of accumulation differs for various organisms and depends on the type and concentration of toxic organic chemical in the surrounding medium. Microbial uptake is a passive absorption process and not active metabolism. Studies by, for example, Johnson and Kennedy (1973) and Paris and Lewis (1976) proved that dead, autoclaved cells accumulate similar amounts of toxic organic molecules to living organisms. These results suggest that cell accumulation is not an induced metabolic process but an absorptive one. For some specific cases, such as the pesticides fenitrothion and DDT, accumulation is greater in dead cells than living organisms (Kikuchi et al. 1984). Once accumulated in microbial cells, toxic organic chemicals may be degraded. The organophosphate pesticide fensulfothion, for example, is metabolized by the bacterium *Klebsiella pneumoniae,* the transformation product fensulfothion sulfide being also bound by both living and dead cells (Timms and MacRae 1982).

Nonenzymatic transformation of toxic organics is an indirect process that occurs in the subsurface as a result of microbially induced changes in environmental parameters such as pH and redox potential. The activity of microorganisms leads to changes in pH, for example, due to biochemical processes such as degradation of proteins or oxidation of organic N to nitrite and nitrate, sulfide to elemental S, or ferrous sulfate to ferric ion. Changes in pH may induce the transformation of exogenous organic contaminants in the aqueous and solid

Table 15.2 Oxidation reactions in microbial pesticide metabolism (Bollag and Liu 1990)

Hydroxylation	
	$RCH \rightarrow RCOH$
	$ArH \rightarrow ArOH$
N-dealkylation	
	$RNCH_2CH_3 \rightarrow RNH + CH_3CHO$
	$ArNRR' \rightarrow ArNH_2$
β-Oxidation	
	$ArO(CH_2)_nCH_2CH_2COOH \rightarrow ArO(CH_2)_nCOOH$
Decarboxylation	
	$RCOOH \rightarrow RH + CO_2$
	$ArCOOH \rightarrow ArH + CO_2$
	$Ar_2CH_2COOH \rightarrow Ar_2CH_2 + CO_2$
Ether cleavage	
	$ROCH_2R' \rightarrow ROH + R'CHO$
	$ArOCH_2R \rightarrow ArOH + R'CHO$
Epoxidation	

Oxidative coupling	
	$2ArOH \rightarrow (Ar)_2(OH)_2$
Sulfoxidation	
	$RSR' \rightarrow RS(O)R'$ or $RS(O_2)R'$

phases. Reductive reactions occurring in a saturated subsurface, due to microbial activity, can lower the redox potential to a range of 0 to −100 mV (Parr and Smith 1976) and lead to the transformation of organic chemicals.

Oxidation of organic contaminants by microorganisms is one of the basic metabolic reactions in the subsurface and involves the presence of a group of oxidative enzymes such as peroxidases, lactases, and mixed-function oxidases. Major oxidative reactions that may occur in the subsurface are presented and explained in Table 15.2.

Hydroxylation can occur on the aromatic ring, on aliphatic groups, and on alkyl side chains. It makes these compounds more polar, so that their solubility increases. Hydroxylation (Fig. 15.1) is one of the most common first steps in contaminant transformation, and it begins by the addition of an hydroxyl group.

N-dealkylation results from an alkyl substitution on an aromatic molecule, which is one of the first places where microorganisms initiate catabolic transformation of atrazine, a xenobiotic molecule (Fig. 15.2). It is a typical example of a reaction leading to transformation of pesticides such as phenyl ureas, acylanilides, carbamates, s-triazines, and dinitranilines. The enzyme mediating the reaction is a mixed-function oxidase, requiring a reduced nicotinamide nucleotide as an H donor.

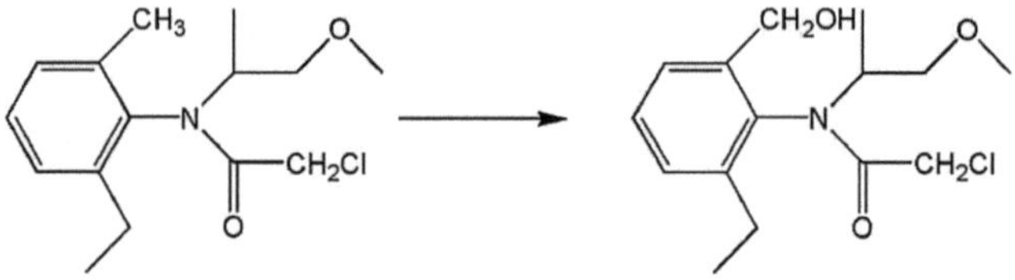

Fig. 15.1 Hydroxylation of metolachlor-soil actinomycete (Bollag and Liu 1990)

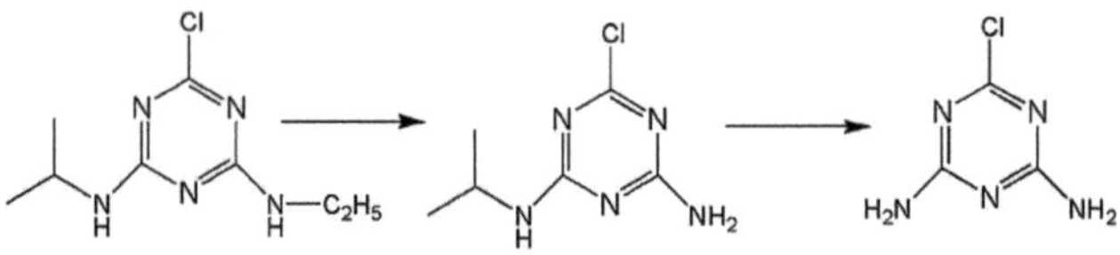

Fig. 15.2 N-dealkylation of atrazine. Reproduced with permission Bode et al. (2003). Copyright Society of Chemical Industry. Permission granted by John Wiley & Sons Ltd on behalf of the SCI

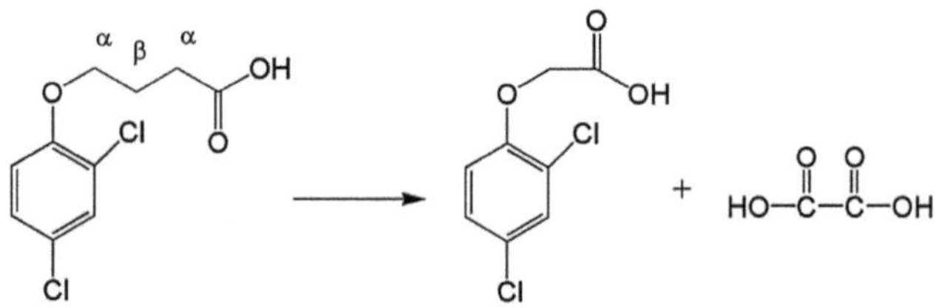

Fig. 15.3 The β-oxidation of 2,4-DB, a phenoxy herbicide (Bollag and Liu 1990)

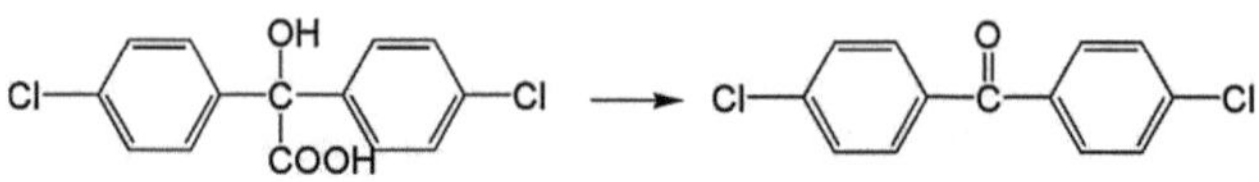

Fig. 15.4 Decarboxylation of 4,4′-dichlorobenzilic acid (Bollag and Liu 1990)

The *β-oxidation* reaction occurs, for example, when contaminants contain a fatty acid chain. In this case, the reaction proceeds by the stepwise cleavage of two carbon fragments of the fatty acid and ceases when the chain length is two or four carbons (Fig. 15.3). The implementation of β-oxidation requires two protons on both α- and β-carbons and is barred when these carbons are substituted. This reaction was found initially with ω-phenoxyalkanoic acids and is similar in bacteria, actinomycetes, and fungi; the general pathway was described by Loos (1975).

Decarboxylation designates the loss of a carboxyl group, as a result of enzymatic microbial activity. For example, the carboxyl group of aliphatic carboxylic acids differentially degrades the molecule (Fig. 15.4). In a subsurface environment characterized by extensive microbial activity, catalytic decarboxylation for both naturally occurring and exogenous organic compounds may occur.

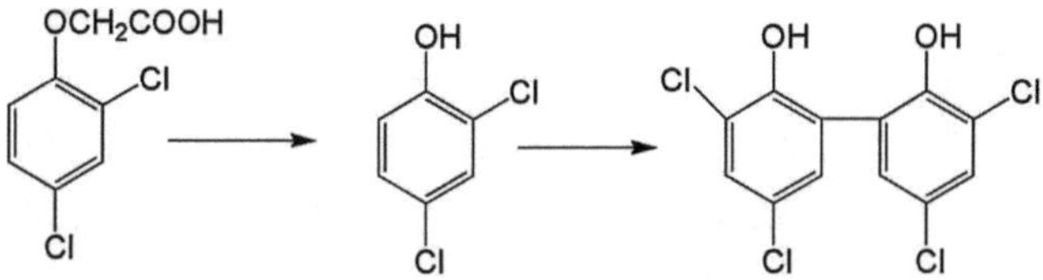

Fig. 15.5 Oxidative coupling of 2,4 dichlorophenol (Bollag and Liu 1990)

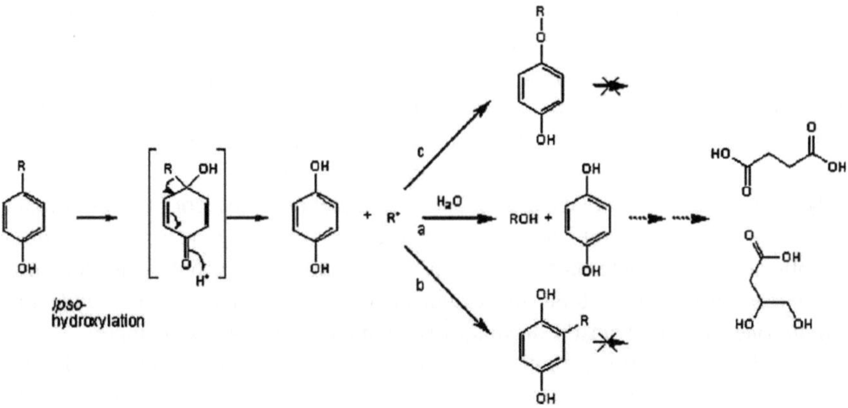

Fig. 15.6 Aromatic ring cleavage degradation pathways of endocrine disruptor nonylphenol isomers by *Sphingomonas* sp. strain TTNP3 (Corvini et al. 2006)

Epoxidation reactions define the insertion of an oxygen atom into a carbon–carbon double bond, leading to the formation of a product with toxicity greater than the parent chemical. Highly toxic organochlorinated pesticides are subject to epoxidation in the presence of various microorganisms, such as *Aspergillus* and *P. niger, A. flavus, Penicillium chrysogenum* and *P. notatum,* which catalyze the reaction (Milles et al. 1969).

Oxidative coupling involves condensation reactions catalyzed by phenol oxidases. In oxidative coupling of phenol, for example, arloxy or phenolate radicals are formed by the removal of an electron and a proton from an hydroxyl group. The herbicide 2,4-D is degraded (Fig. 15.5) to 2,4 dichlorophenol, which can be oxidatively coupled by phenol oxidases (Bollag and Liu 1990).

Aromatic and *heterocyclic cleavage* involves a hydrocarbon being separated from an oxygen atom, which functions to link it to another moiety of the molecule; this process may lead to a decrease in the initial toxicity. The metabolic effects of microorganisms differ with the molecular configuration of the product, which affect aromatic or heterocyclic ring cleavage differently.

Aromatic ring cleavage is a microorganism-mediated catabolic process (Fig. 15.6). The type of linkage, the specific substituents, their position, and their number determine the susceptibility of an aromatic ring to fission. Usually, the substituents must be modified or removed, and an hydroxyl group inserted in an

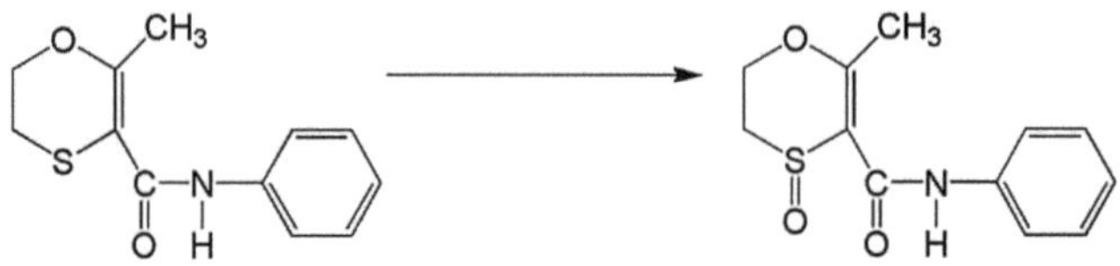

Fig. 15.7 Sulfoxidation of carboxine by the fungus *Ustilage maydis* (Bollag and Liu 1990)

appropriate position before oxygenase enzymes can cause ring cleavage. Dehydroxylation usually is essential for enzymatic cleavage of the benzene ring under aerobic conditions. The hydroxyl groups must be placed either ortho or para to each other, probably to facilitate the shifts of electrons involved in the ring fission. Dioxygenases are the enzymes responsible for ring cleavage, and they can cause ortho-(intradiol) or meta-(extradiol) fission of a catechol, forming *cis, cis*-muconic acid or 2-hydroxymuconic, and semialdehyde, respectively.

Heterocyclic ring cleavage also occurs as a metabolic, microorganism-mediated process. For contaminants having a heterocyclic ring, the degradation path is complicated by the heteroatoms, usually N, O, and S, contributing to decomposition reactions through their individual characteristics. These compounds may contain one or more (mostly aromatic) rings having five or six members.

Sulfoxidation reactions are characterized by enzymatic conversion of a divalent compound to sulfoxide (Fig. 15.7) or, in some cases, to sulfone ($S \rightarrow SO \rightarrow SO_2$). The degradation also may be catalyzed by minerals, converting organic sulfides (thioesters) and sulfites to the corresponding sulfoxides and sulfates. Because it is difficult to determine if the reaction is chemically or biologically induced, microbially mediated sulfoxidation in the subsurface environment can be established only when a biocatalyst is found.

Reduction reactions mediated by microorganisms may include the reduction of nitro bonds, sulfoxide reduction, and reductive dehalogenation. Reduction of the nitro group to amine involves the intermediate formation of nitrase and hydroxyamino groups. Selected reductive reactions may involve the saturation of double bonds, reduction of aldehydes to alcohols or ketones to secondary alcohols, or of certain metals. The main reductive processes in the subsurface environment have been discussed earlier in this chapter.

Hydrolytic reactions involve organic toxic molecules that have ether, ester, or amide linkages. In the case of hydrolytic dehalogenation, a halogen is exchanged with an hydroxyl group. This reaction is mediated by hydrolytic enzymes, excreted outside the cells by microorganisms. In general, enzymes involved in hydrolytic reactions include esterase, acrylamidase, phosphatase, hydrolase, and lyase. Bollag and Liu (1990) emphasized that it is often difficult to determine the original catalyst of the reaction, because specific environmental conditions or secondary effects of microbial metabolism create conditions conducive to hydrolysis. Table 15.3 summarizes microbially mediated hydrolytic and reductive reactions of synthetic pesticides that reach the subsurface via land application.

Table 15.3 Reductive and hydrolytic microbially mediated reactions of synthetic pesticides

Reduction of nitro group	
	$RNO_2 \rightarrow ROH$
	$RNO_2 \rightarrow ROH_2$
Reduction of double bond or triple bond	
	$Ar_2C{=}CH_2 \rightarrow Ar_2CHCH_3$
	$RC{\equiv}CH \rightarrow RCH{=}CH_2$
Sulfoxide reduction	
	$RS(O)R' \rightarrow RSR'$
Reductive dehalogention	$Ar_2CHCCl_3 \rightarrow Ar_2CHCHCl_2$
Ether hydrolysis	$ROR' + H_2O \rightarrow ROH + R'OH$
Ester hydrolysis	
	$RC(O)OR' + H_2O \rightarrow RC(O)OH + R'OH$
Phosphor-ester hydrolysis	$(RO)_2P(O)OR' + H_2O \rightarrow (RO)_2P(O)OH + R'OH$
Amide hydrolysis	
	$RC(O)NR'R'' + H_2O \rightarrow RC(O)OH + HNR'R''$
Hydrolytic dehalogenation	
	$RCl + H_2O \rightarrow ROH + HCl$

R organic moiety, Ar aromatic moiety (Bollag and Liu 1990)

Synthetic reactions lead to contaminant transformation when organic chemicals in the subsurface merge together or are linked to natural organic compounds. Synthetic reactions may be divided into conjugation reactions, which involve the merger of two substances, and condensation reactions, which yield oligomeric or polymeric compounds (Bollag and Liu 1990). Conjugation reactions, such as methylation and acetylation, commonly occur during microbial metabolism of xenobiotics. Microbial phenoloxidases and peroxidases catalyze the transformation of phenolic compounds to polymerized products.

15.4 Biotransformation of Inorganic Contaminants

The microbially mediated transformation of inorganic contaminants encompasses a broad spectrum of compounds. Here, we survey only a few contaminants that have a major impact on the subsurface, such as nitrates, phosphates, and toxic metals. In the subsurface, the microbial population distribution generally decreases with depth, and the moisture content and aerobic–anaerobic states fluctuate with time, affected by climatic conditions, depth of groundwater, and human intervention.

Nitrification–denitrification involves the conversion of NH_4^+ to NO_2^-, the oxidation of NO_2^- to NO_3^-, and the reduction of NO_3^- to NO_2^-. The gases N_2O and N_2 are used in the microbially mediated processes involved in the nitrification–denitrification phenomenon.

Nitrification is associated with chemoautotrophic bacteria, which under aerobic conditions derive their energy from the oxidation of NH_4^+ to NO_2^-. *Nitrobacter* is

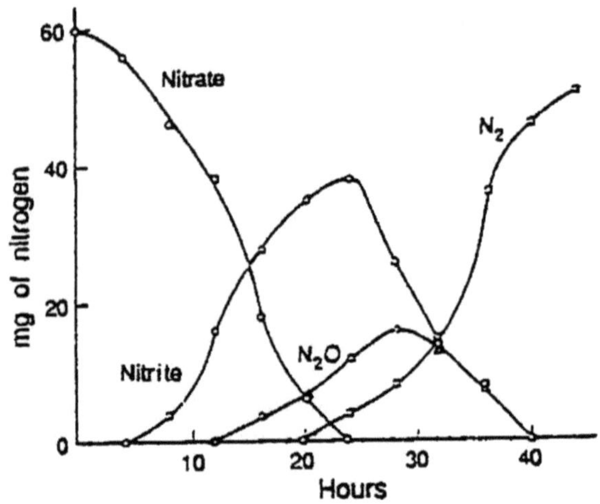

Fig. 15.8 Products formed during denitrification in Melville loam, pH 7.8 (Cooper and Smith 1963)

the soil bacterium that oxidizes nitrite NO_2^- to NO_3^-. In most habitats, this bacterium is found together with *Nitrosomonas, Nitrospira,* or *Nitrosovibrio,* which oxidize ammonia (NH_3) to the nitrite required for NO_3^- formation. Nitrification is affected by the subsurface pH (with an optimum value varying between 6.6 and 8.0) and the subsurface water–air ratio. Once NO_3^- is formed, it becomes subject to transformation by microorganism-mediated denitrification to gaseous oxides of nitrogen and to N_2. The NO_3^- may be taken up by organisms and used in synthesis of amino acids (assimilatory reduction), or in the absence of O_2, it may be used by microorganisms as an electron acceptor by reduction to NH_4^+ (Paul and Clark 1989). Enzymatic denitrification is the result of assimilatory reduction of NO_3^- by microorganisms and dissimilatory reduction of nitrate to ammonium. This is accomplished by specific organisms in the absence of O_2. Figure 15.8 shows the following sequence of identifiable products formed during denitrification: NO_3^-; NO_2^-; N_2O; N_2.

Under field conditions, not all intermediate products are converted to N_2. Nitrate reductase, for example, causes a decrease in the enzymatic activity. Denitrificaton in the absence of oxygen is caused by a large number of bacteria; Table 15.4 lists the main microorganisms capable of denitrification.

Denitrification occurs only in the presence of oxidized nitrogen and in an environment with limited O_2 (which prevails in the subsurface). Because denitrification is an enzyme-mediated reaction, the substrate concentration functions as a rate-determining factor. The dominant denitrifying bacteria are heterotrophic. The favored environmental conditions for the growth of denitrifying bacteria include a neutral pH (6–8), a favorable water–air (oxygen) ratio, and a subsurface temperature between 20 and 30 °C.

Phosphorus in the subsurface originates from a natural parent material or anthropogenic application on land surface (e.g., fertilizers, pesticides, surfactant

Table 15.4 Bacteria capable of denitrification (Firestone 1982)

Genus	Interesting characteristics of some species
Alcaligenes	Commonly isolated from soils
Agrobacterium	Some species plant pathogens
Azospirillum	Capable of N_2 fixation, commonly associated with grasses
Bacillus	Thermophilic denitrifiers reported
Flavobacterium	Denitrifying species isolated
Halobacterium	Requires high salt concentrations for growth
Hyphomicrobium	Grows on one-carbon substrates
Paracoccus	Capable of both lithotrophic and heterotrophic growth
Propionibacterium	Fermentors capable of denitrification
Pseudomonas	Commonly isolated from soils
Rhizobium	Capable of N_2 fixation in symbiosis with legumes
Rhodopseudomonas	Photosynthetic
Thiobacillus	Generally grow as chemoautotrophs

products, sludge, and effluents). This element may be found in inorganic or organic forms, which are in a dynamic equilibrium with dissolved P in the subsurface liquid phase.

Phosphorus transformations in the subsurface environment, known as the P cycle, depend on their original state. Subsurface transformation of phosphorus is due to fundamental processes such as mineral equilibrium (dissolution–precipitation) and interactions between P in solution and the solid phase (adsorption–desorption). Alternatively, biologically mediated conversion of P between inorganic and organic forms (mineralization-immovability) and bonding of inorganic P by organic ligands (complexation) may affect P transformation in the subsurface. Figure 15.9 shows the phosphorus forms found in the upper layer of the subsurface, as result of various transformations occurring during the P cycle.

In subsurface aqueous solutions, P may be found primarily as PO_4^{3-} and to lesser extent as HPO_4^{2-} and $H_2PO_4^{-}$. In general, orthophosphate species in solution vary in relative concentration as a function of pH. An increase in pH causes larger concentrations in the secondary species and changes the ratio of primary to secondary orthophosphates. For example, at a pH of 4.0–5.5, the predominant P form is $H_2PO_4^{-}$, while at pH > 8.0, HPO_4^{2-} species predominate. Phosphorus species concentrations in subsurface aqueous solutions are affected by anthropogenic factors, either directly by the P species disposed of on a particular site or indirectly via pH changes resulting from municipal, agricultural, or industrial composition of the disposed materials. Ambient temperature and the water–air ratio in the subsurface also control rate factors of biologically mediated P species transformation.

Here, we do not discuss P transformation resulting from chemical equilibrium and adsorption–desorption processes; the reader is directed to the comprehensive review of Sims and Pierzjinski (2005). Instead, we focus on biologically mediated transformation of organic P in the subsurface. The ultimate and "natural" sources of organic P in the subsurface environment are animal and plant residues. These

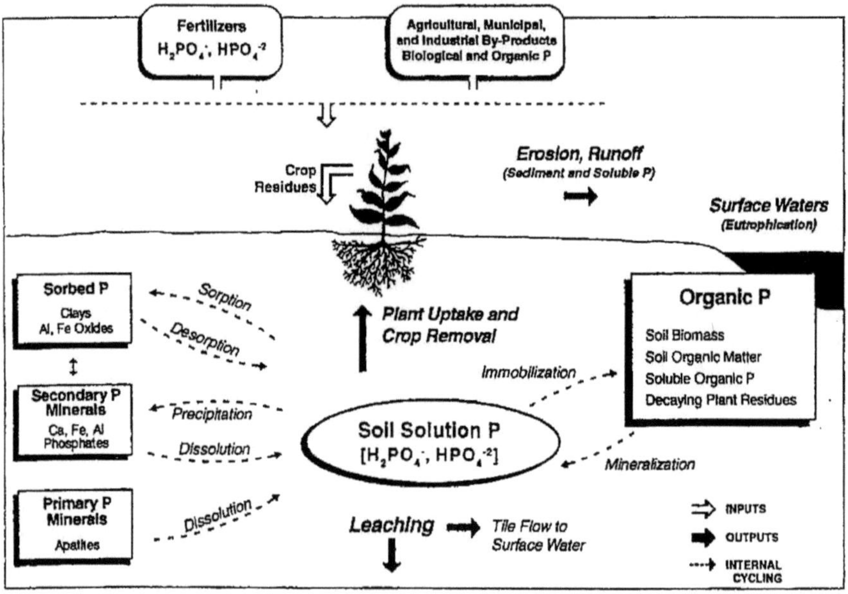

Fig. 15.9 The soil P cycle (Pierzjinski et al. 2000)

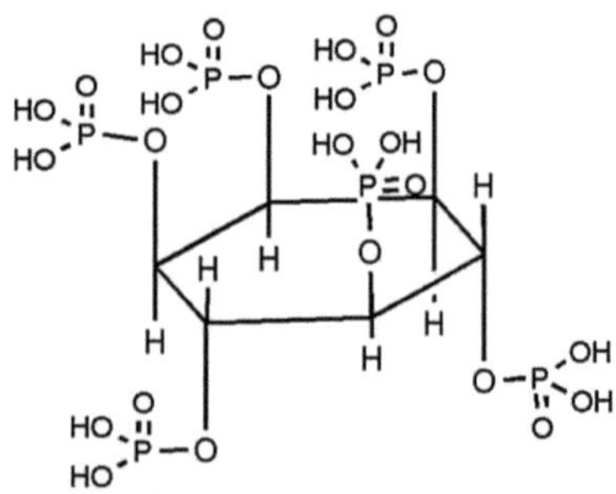

Fig. 15.10 Myo-inositol hexaphosphate structure (Sims and Pierzjinski 2005)

are accompanied by additional phosphorus organic chemicals of anthropogenic origin, disposed of on the land surface or spilled in surface waters. These residues are decomposed by microorganisms, forming inositol phosphate metabolites that generally are resistant to further, biologically induced, decomposition. The basic inositol group is the six C ring structure, hexahydrobenzene. Inositol phosphates are monoesters with the hexaphosphate ester and phytic acid, which are found commonly in soils (Stevenson 1986). An example of myo-inositol hexaphosphate is shown in Fig. 15.10. Mono-, di-, tri-, tetra-, and pentaphosphates, which are degradation products of the hexaphosphate form of myo-inositol, may also be found.

Phospholipids containing phosphatidyl, inositol, lecithin, serine, and ethanolamine (Stevenson 1986) are the second most abundant identifiable form of organic

P in the upper layer of the subsurface. These groups contain glycerol, fatty acids, and phosphate (Sims and Pierzjinski 2005). The P in the structure is a diester, which is more susceptible to degradation in soils than monoesters.

Additional transformable organic P compounds found in the subsurface environment are nucleic acids and phosphonates. Nucleic acids constitute a minor portion of identifiable organic P in the subsurface. They are readily degraded, and their presence in the subsurface environment is due to continuous microorganism production rather than persistence. Phosphonates are organic P compounds with direct C–P bonding, as opposed to the C–O–P bonding typical of most other organo-P molecules commonly found in soils; the main compound reported to date is 2-aminoethyl phosphonic (Newman and Tate 1980).

Sims and Pierzjinski (2005) note that the balance between mineralization (due to a biological process where plants produce enzymes and microorganisms hydrolyze organic compounds, releasing inorganic P into solution) and immobilization (conversion of inorganic P to organic P in biomass) ultimately controls P concentration in the aqueous solution. This concentration, however, is affected by the solution pH as well as by the properties of the solution surrounding the solid phase. The P products produced in the various stages of microbially mediated transformation have different properties, and their transport in the subsurface being affected by the type of speciation involved.

Metal transformation includes two main processes: oxidation–reduction of inorganic forms and conversion of metals to organic complex species (and the reverse conversion of organic to inorganic forms). Microbially mediated oxidations and reductions are the most typical pathway for metal transformation. Under acidic conditions, metallic iron (Fe^0) readily oxidizes to the ferrous state (Fe^{2+}), but at a pH greater than 5, it is oxidized to Fe^{3+}. Under acidic conditions, Fe^{3+} is readily reduced. *Thibacillus ferroxidant* mediates this reaction in an acid environment and derives both energy and reducing power from the reaction.

Paul and Clark (1989) showed that before Fe^{3+} is microbiologically reduced, it is chelated by organic compounds. During oxidation, electrons are moved through an electron transport chain, with cytochrome c being the point of entrance into the transport chain. Oxidation can be caused by direct involvement of enzymes or by microorganisms that raise the redox potential or the pH. Iron reduction occurs when ferric iron (Fe^{3+}) serves as a respiratory electron acceptor or by reaction with microbial end products such as formaldehyde or H_2S. Microbiologically mediated transformation, as affected by pH, also is observed in the oxidation of Mn^{2+}, when both bacteria and fungi can oxidize manganese ions only in neutral and acid environments. Dissimilatory metal reduction bacteria can couple organic matter oxidation to metal contaminant reduction (Lovley 1993). Rates of these reactions depend on the reduction potential of the solid or solution phase metal, the surface area, and the presence of competing terminal electron acceptors.

An additional environmental factor that may affect metal contaminant transformation in the subsurface is the air–water ratio. A toxic metal like mercury does not remain in a metallic form in an anaerobic environment. Microorganisms transform metallic mercury to methylmercury (CH_3–Hg^+) and dimethylmercury

Table 15.5 Range of soil solution concentrations for certain trace elements; estimates based on data (Sauve and Parker 2005)

Element	Soil solution μg/L
As	0.5–60,000
Cd	0.01–5,000
Cr	2–500
Cu	5–10,000
Ni	0.5–5,000
Pb	0.5–500
Se	1–100,000
Zn	1–100,000

(CH_3–Hg–CH_3), which are volatile and absorbable by the organic fraction of the subsurface solid phase or subsurface microorganisms.

The transformation of metal contaminants by complexation with inorganic or organic ligands occurs mainly in the subsurface aqueous solution. The potential inorganic ligands are nitrates, sulfate, chloride, and (bi)carbonate. Even though their concentration is relatively high (i.e., millimolar), they are weak ligands with minimal effect on metal contaminant complexation. Organic compounds, of biological origin present in the subsurface aqueous solution, comprise both weak and strong metal-complexing ligands. Their composition is discussed in Chap. 1. The concentration of trace metals able to be complexed by organic ligands existing in the subsurface aqueous solution may vary between uncontaminated and contaminated sites. Data indicating their range of concentration in soil solutions are presented in Table 15.5.

Speciation is a dynamic process that depends not only on the ligand–metal concentration but also on the properties of the aqueous solution in chemical equilibrium with the surrounding solid phase. As a consequence, the estimation of aqueous speciation of contaminant metals should take into account the ion association, pH, redox status, formation–dissolution of the solid phase, adsorption, and ion-exchange reactions. From the environmental point of view, a complexed metal in the subsurface behaves differently than the original compound, in terms of its solubility, retention, persistence, and transport. In general, a complexed metal is more soluble in a water solution, less retained on the solid phase, and more easily transported through the porous medium.

References

Alexander M (1977) Introduction to soil microbiology, 2nd edn. Wiley, New York

Alexander M (1980) Biodegradation of chemicals of environmental concern. Science 211:132–138

Azadpour-Keeley A, Russell HH, Sewell GW (1999) Microbial processes affecting monitored natural attenuation of contaminants in the subsurface. EPE/54o/S-99/001

Bode M, Stobe P, Thiede B, Schuphan I, Schmidt B (2003) Biotransformation of atrazine in transgenic tobacco cell culture expressing human P450. Pest Manag Sci 60:49–58

Bollag JM, Liu SY (1990) Biological transformation processes of pesticides. In: Chen HH (ed) Pesticide in soil environment Book Series no 2. Soil Sci Soc Am, Madison, Wisconsin
Cooper GS, Smith RL (1963) Sequence of products formed during denitrification in some diverse western soils. Soil Sci Am Proc 27:659–662
Corvini PFX, Schäffer A, Schlosser D (2006) Microbial degradation of nonylphenol and other alkylphenols—our evolving view. Appl Microbiol Biotechnol 72:223–243
Delwich CC (1967) Energy relationships in soil biochemistry. In: McLaren AD, Skujins JJ (eds) soil biochemistry, vol 2. Marcel Dekker, New York, pp 173–193
Firestone MK (1982) Biological denitrification. In: Stevenson FJ (ed) Nitrogen in agricultural soils. Agronomy 22:289–326
Johnson BT, Kennedy JO (1973) Biomagnification of p, p-DDT and methoxychlor by bacteria. Appl Microbiol 26:66–71
Keeney DR (1983) Principles of microbial processes of chemical degradation, assimilation and accumulation. In: Nielsen (ed) Chemical mobility and activity in soil system. ASA Soil Sci Soc Am, Madison, pp 153–164
Kikuchi R, Yasutaniya T, Takimoto Y, Yamada H, Miyamoto J (1984) Accumulation and metabolism of fenitrothion in three species of algae. J Pestic Sci 9:331–337
Loos MA (1975) Phenoxyalkanoic acids. In: Kerney PC, Kaufman DD (eds) Herbicides: chemistry, degradation and mode of action, vol 1. Marcel Dekker, New York, pp 1–128
Lovley DR (1993) Dissimilatory metal reduction. Annu Rev Microbiol 47:263–290
Milles JRW, Tu CM, Harris CR (1969) Metabolism of heptachlor and its degradation products by soil microorganisms. J Econ Entomol 62:1334–1339
Newman RH, Tate KR (1980) Soil-phosphorus characterization by p-31 nuclear magnetic-resonance. Comm Soil Sci Plant Anal 11:835–842
Paris DF, Lewis DL (1976) Accumulation of metoxychlor by microorganisms isolated from aqueous systems. Bull Environ Contam Toxicol 13:443–450
Parr JF, Smith S (1976) Degradation of toxaphene in selected anaerobic soil environments. Soil Sci 121:52–57
Paul EA, Clark FE (1989) Soil microbiology and biochemistry. Academic Press, New York
Pierzjinski GM, Sims JT, Vance GF (2000) Soils and environmental quality, 2nd edn. Lewis Publications, Chelsea, Michigan
Sauve S, Parker DR (2005) Chemical speciation of trace elements in soil solution. In: Tabatabai MA, Sparks DL (eds) Chemical processes in soils Book Series no 8. Soil Sci Soc Am, Madison, Wisconsin
Sims JY, Pierzjinski GM (2005) Chemistry of phosphorous in soils. In: Tabatabai MA, Sparks DL (eds) Chemical processes in soils, vol 8. Soil Sci Soc Am Publ, Madison, pp 151–192
Stevenson FJ (1986) Cycles of soil. Wiley Interscience, New York
Timms P, MacRae IC (1982) Conversion of fensulfothion by *Klebsiella pneumoniae* to fensulfothion sulfide and its accumulation. Aust J Biol Sci 35:661–667
Wang YS, Subba-Rao RV, Alexander M (1984) Effect of substrate concentration and organic and inorganic compound on the occurrence and rate of mineralization and cometabolism. Appl Environ Microbiol 47:1195–1200

Chapter 16
Selected Research Findings: Transformations and Reactions

16.1 Abiotic Alteration of Contaminants

16.1.1 Transformation in Subsurface Water

Both inorganic and organic contaminants may change their specific properties in the subsurface solution (e.g., toxicity, transport, adsorption, persistence) according to the properties of the aqueous medium. The pH of the subsurface solution, for example, may control basic or acid hydrolysis of contaminants, affect their dissolution-precipitation behavior, or influence their transformation under redox-induced processes. Inorganic and organic chemicals present in the subsurface solution may serve as catalysts, and at specific pH values, they can enhance contaminant transformation. Natural and synthetic organic compounds that serve as ligands for inorganic compounds lead to changes in contaminant dissolution in water and, in some cases, serve as sensitizers in photochemically induced degradation of organic pollutants in contaminated surface waters.

The effect of pH on transformation in the subsurface aqueous environment is considered here for the case of two highly toxic biocides (methyl parathion and acrolein) and for an industrial semivolatile organic pollutant (tribromoneopentyl alcohol, or TBNPA). The examples are based on the work of Guo and Jans (2006), Oh et al. (2006) and Ezra et al. (2005).

Acrolein (CH_2=CHCHO, also known as 2-propenal) is a α,β-unsaturated aldehyde that can be transformed reductively to saturated or unsaturated alcohols by reduction of the C=O or C=C double bonds (Claus 1998). In addition, α,β-unsaturated aldehydes may undergo hydration reactions in aqueous solutions. It was observed that, under acidic ($pH < 1$) or basic ($pH > 12$) conditions, acrolein is hydrated to 3-hydroxypropanal (Jensen and Hashtroudi 1976). In a natural subsurface environment, where pH may range from 6.5 to 8.5, the hydration rate of acrolein increases with the pH and its half-life decreases. Based on an experiment to analyze effects of iron on acrolein transformation, Oh et al. (2006) note that, under acidic conditions (e.g., $pH = 4.4$), acrolein disappears rapidly from solution in the presence of elemental iron (Fig. 16.1). Moreover, the formation of

B. Berkowitz et al., *Contaminant Geochemistry*, 397
DOI: 10.1007/978-3-642-54777-5_16,

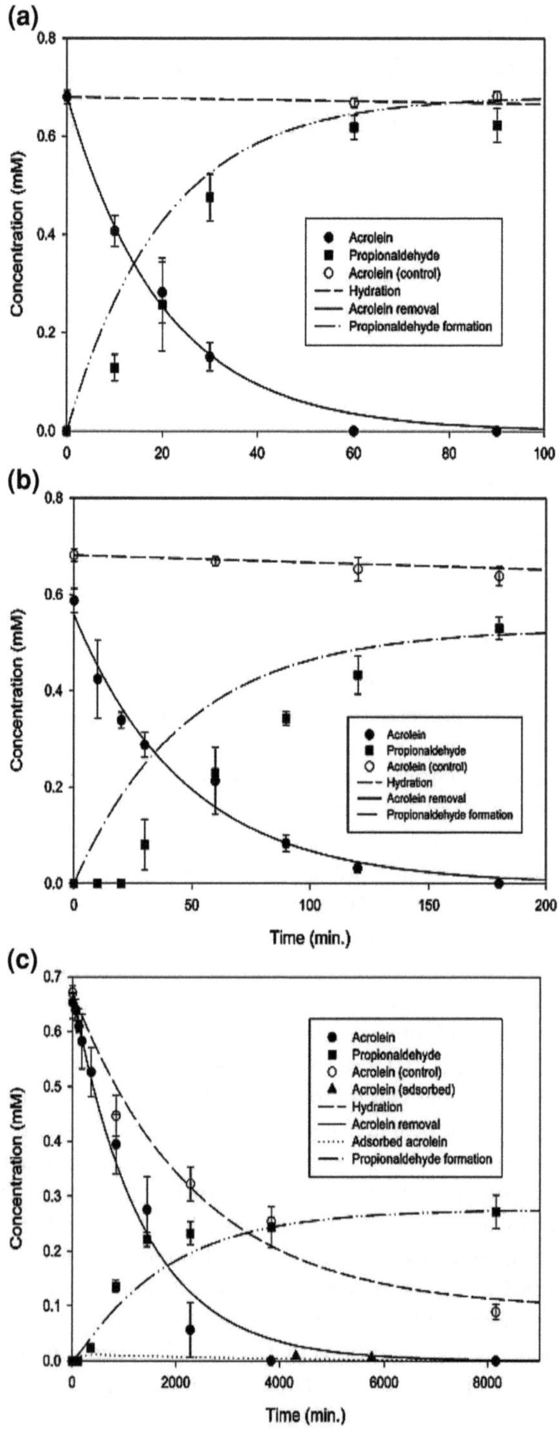

Fig. 16.1 Reduction of acrolein with granular iron at **a** pH = 4.4 ± 0.5, **b** pH = 6.3 ± 0.3, **c** pH = 7.4 ± 0.1. Reprinted with permission from Oh et al. (2006). Copyright 2006 American Chemical Society

propionaldehyde indicates that removal is due to reduction of the C=C double bond and that the hydration of acrolein at this pH is relatively low.

At pH 6.3, acrolein reduction was slower than at a lower pH (4.4). The incomplete mass balance suggests that either an intermediate accumulated to a greater extent or adsorption to iron was greater at pH 6.3 (Fig. 16.1b). Oh et al. (2006) suggest it possible for acrolein to adsorb on iron surfaces through chemical interaction, due to the π acidity of the C=C double bond, which allows the compound to bind to transition metals having a filled orbital of π symmetry. Acrolein reduction by elemental iron is conceptualized as occurring in four sequential steps: transport from solution to iron surface, adsorption to iron, reduction of adsorbed acrolein to propionaldehyde, and release of the degradation product to solution. At pH 7.4, hydration of acrolein is no longer negligible compared to reduction (Fig. 16.1c), and the initial lag before propionaldehyde formation increased to 120 min. Based on these experiments, Oh et al. (2006) suggest that acrolein reduction under acidic or neutral pH and iron presence involves chemisorption to the iron surface, followed by reduction of the adsorbed phase.

The degradation of methyl parathion, an organophosphorus pesticide, as affected by pH was studied in simulated subsurface aqueous solutions containing hydrogen sulfide and natural organic matter (Guo and Jans 2006). Methyl parathion can reach the subsurface aqueous environment via drainage water, runoff, or spray drift and may be reduced under the presence of natural organic matter (which can act as a reducing agent). In a subsurface anoxic environment, favorable conditions are formed for the development of micropopulations able to reduce sulfur species such as hydrogen sulfide. Guo and Jans (2006) show that the degradation rate of methyl parathion in an aqueous solution, in the presence of hydrogen sulfide, is slow but increases significantly with the addition of natural organic matter, such as humic or fulvic acids.

Guo and Jans (2006) also examine the effect of pH on the degradation rate constant, k'_{obs}. Figure 16.2 illustrates that k'_{obs} increases with increasing pH in the range of pH 5.5–8.3 and drops abruptly at pH between 8.3 and 9.5. The observed pH dependence indicates that, at higher pH values, the reactive intermediate is not formed to the same extent. Guo and Jans (2006) consider that two reaction products are produced in a reaction of methyl parathion with hydrogen sulfide, in an aqueous system containing natural organic matter (NOM). Comparing the same reactions, but without the NOM, it is observed that the degradation rate constants are in the same range at pH higher than 8.3. The two postulated reaction mechanisms for the degradation are nucleophilic substitution at the methoxy-carbon and nitro group reduction. The overall degradation rate constant, k'_{obs}, is expressed as the sum of the nucleophilic substitution and nitro group reduction rate constants. Desmethyl methyl parathion is the predominant product detected during such methyl parathion transformation. The presence of 4-nitrophenol, detected in minute concentrations in this system, may have originated from impurities in the original pesticide used in the experiment.

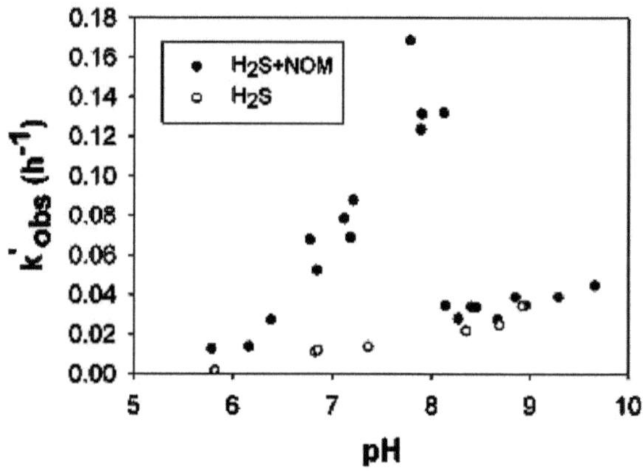

Fig. 16.2 Degradation rate constant (k'_{obs}) for methyl parathion as a function of pH, in aqueous 5.0 mM hydrogen sulfide with and without natural organic matter (NOM), at 25 °C. Reprinted with permission from Guo and Jans (2006). Copyright 2006 American Chemical Society

Tribromoneopentyl alcohol (TBNPA) is an organic halide used by the bromine industry as a reactive intermediate for high molecular weight flame retardants. This potential contaminant decomposes in an aqueous solution under basic conditions (pH from 7.0 to 9.5), by a sequence of reactions that release one bromide ion at each stage. TBNPA has an estimated half-life of about 100 years, it is highly soluble in water (2 g/L at 25 °C), and with its degradation products, it forms an environmental hazard. Ezra et al. (2005) investigated the chemical transformation of 3-bromo-2, 2-bis(bromomethyl)-propanol (TBNPA) in a laboratory experiment simulating conditions below an industrial conglomerate located on a fractured chalk aquitard (Negev, Israel). The natural groundwater under the contaminated location is saline (magnesium chloride type) with a high content of calcium and sulfate; it has a pH ranging between 7 and 8 and contains an enormous variety of organic and inorganic pollutants of industrial origin.

Experimental results of Ezra et al. (2005) show that TBNPA slowly decomposes in an aqueous solution under basic conditions (pH from 7.0 to 9.5) by a sequence of hydrolysis reactions that release one bromide ion at each stage. The sequence of three daughter products is BBMO, BMHMO, and DOH. The first product, BBMO (3,3(bromomethyl) oxetane) contains two equivalent $-CH_2Br$ groups and two equivalent CH_2O fragments bonded to the same carbon atom. The second hydrolysis intermediate is BMHMO (3-bromomethyl-3-hydroxymethyloexane). Because one Br^- ion is released to solution for each molecule that is decomposed, the last daughter product, DOH, is a bromine-free compound identified as 2,6-dioxaspiro 3,3 heptane. The pseudo-first-order rate constant of the decomposition of TBNPA increases linearly with the pH. Analyzing groundwater samples from monitoring wells by GC–MS, the TBNPA decomposition product BBMO was found, although it was not among the chemicals produced or used in the industrial complex covering the area studied. Based on this finding, the authors suggest that the BBMO found in the studied aquitard is a by-product of spontaneous transformation of TBNPA which is the most abundant semivolatile organic pollutant at that site.

Reduction and oxidation reactions in the subsurface environment lead to transformation of organic and inorganic contaminants. We consider chromium (Cr) as an example of an inorganic toxic chemical for which both oxidation and reduction processes may transform the valence of this element, in subsurface aqueous solutions, as a function of the local chemistry.

The most stable oxidation states of chromium in the subsurface environment are Cr(III) and Cr(VI), the latter being more toxic and more mobile. The oxidation of Cr(III) in subsurface aqueous solutions is possible in a medium characterized by the presence of Mn(IV) oxides. Eary and Rai (1987), however, state that the extent of Cr(III) oxidation may be limited by the adsorption of anionic Cr(VI) in acidic solutions and the adsorption and precipitation of various forms of $Cr(OH)_x$. These authors also report a rapid quantitative stoichiometric reduction of aqueous Cr(VI) by aqueous Fe(II), in a pH range covering the acidity variability in the subsurface even in oxygenated solutions.

The effect of the subsurface environment on the kinetics of aqueous Cr(VI) reduction and oxidation was reported by Kozuh et al. (2000), based on a laboratory experiment where four soils (Table 16.1) with different chemistry where used as Cr(VI) incubation media. The concentrations of added Cr(VI) were 1, 10, 25, and 50 μg/g of dry soil. The kinetics of soluble Cr(VI) reduction when added to the four soils, at a concentration of 1 μg/g of dry soil, are shown in Fig. 16.3. The reduction reactions are expressed by nonlinear losses (Fig. 16.3a), with the highest loss in the peat and lowest loss in the cambisols. Therefore, the degree of reduction depends mostly on the organic matter content. The decrease in Cr(VI) concentration is rapid during the first three days of incubation and much slower thereafter. From the $\ln(c/c_o)$ versus time plots of (Fig. 16.3b), Kozuh et al. (2000) suggest that Cr(VI) reduction follows a first-order reaction during the first three days of the experiment but offer no explanation for the subsequent behavior. We see that reduction takes place rapidly in soils with high organic matter. Additional similar experiments show, however, that there is no reduction in concentration of soluble Cr(VI) in the presence of soluble organic matter extracted from peat and clay soil, even after 10 days of incubation. Based on these findings, the authors suggest that the reduction of soluble Cr(VI) probably requires the involvement of available solid organic matter.

Oxidation of Cr(III) occurs when the subsurface environment is characterized by high Mn(IV, III) oxide content and low organic matter content. Kinetics of soluble Cr(III) oxidation in four subsurface materials, with different exchangeable Mn(IV) contents, are reported by Kozuh et al. (2000). Results are presented in Fig. 16.4 for the addition of 100 μg (per g of dry material) of soluble Cr(III), at a constant moisture and temperature; for this case, the oxidation was measured over a 10-day period. The results indicate that oxidation of soluble Cr(III) occurs in the three soils (clay, sand, and cambisols) with low organic matter (<3 %) and high exchangeable Mn(IV) oxides but does not occur in the "organic" peat with low Mn(IV). The concentrations of oxidized Cr(III) achieved maximum values within two days and then decreased slightly over time. The decreases in concentration are a consequence of competition between oxidation and reduction processes of

Table 16.1 Characteristics of the investigated soils and parameters affecting redox of chromium

Soil type	Moisture (%)	pH of soil	TOC (%)	SOM (%)	Exchangeable Cr (μg/g)	Total Cr (μg/g)	Exchangeable manganese (IV) oxides (μg/g)	Oxidative capacity of soil for Cr(III) (μg/g)
Clay	22.0	7.3	1.52	6.8E−5	<5.0	57.7	399	8.96
Peat	50.2	7.0	41.1	4.3E−4	<5.0	57.7	60	<0.05
Sand	13.0	7.4	3.0	1.7E−4	<2.5	48.8	150	3.48
Cambisols	24.4	5.4	0.22	4.6E−5	<5.0	144	234	7.44

TOC Total Organic Carbon, SOM Soluble Organic Matter. Reprinted with permission from Kozuh et al. (2000). Copyright 2000 American Chemical Society

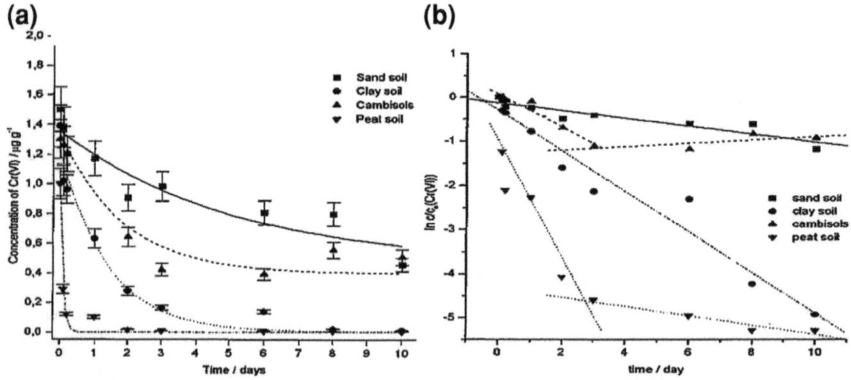

Fig. 16.3 **a** Kinetics of reduction of soluble Cr(VI) added concentrations of 1 μg/g Cr(VI) in various soils and **b** first-order plots of ln[relative Cr(VI) concentration] versus time for the reduction of Cr(VI) in various soils. The slope of the lines is equal to the first-order rate constants (k_{exp}) for the case of 1 μg/g Cr(VI) added. Reprinted with permission from Kozuh et al. (2000). Copyright 2000 American Chemical Society

chromium in the soils. Therefore, the results of Kozuh et al. (2000) confirm that reduction and oxidation of soluble chromium can occur in natural soils, but that reduction of Cr(VI) dominates over oxidation of Cr(III). The composition (organic matter and presence of Mn(IV) oxides) of the soils and environmental conditions affect the extent of these processes.

Arsenic contaminants may be found in the aquatic and terrestrial environments as a result of anthropogenic inputs and weathering of primary materials. It is known (e.g., Oscarson et al. 1983; Tournassat et al. 2002) that in such environments, manganese oxides like birnessite (δ-MnO_2) directly and rapidly oxidize As(III) to As(V). However, As(III) oxidation can be inhibited in sediments when additional natural materials lead to coating of MnO_2 by $CaCO_3$ (Oscarson et al. 1983).

Power et al. (2005) show the effect of pH and initial As(III) concentration on the kinetics of arsenite oxidation at birnessite–water interfaces, when a competitive metal (e.g., Zn) is present in an adsorbed or nonadsorbed state (Fig. 16.5).

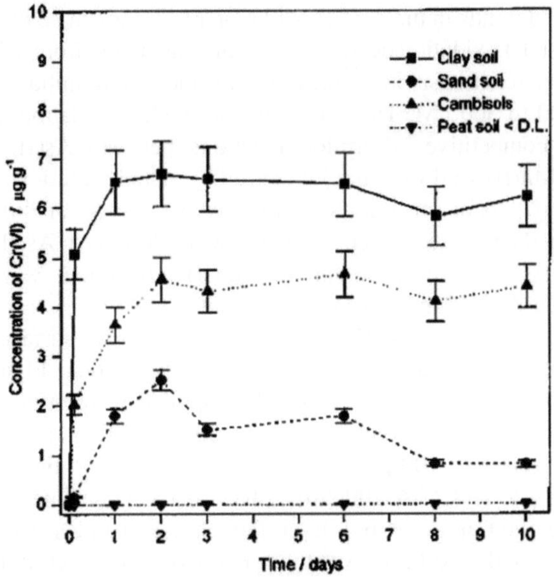

Fig. 16.4 Kinetics of oxidation of soluble Cr(III) in various soils, at initial Cr(III) concentrations of 100 μg/g. Reprinted with permission from Kozuh et al. (2000). Copyright 2000 American Chemical Society

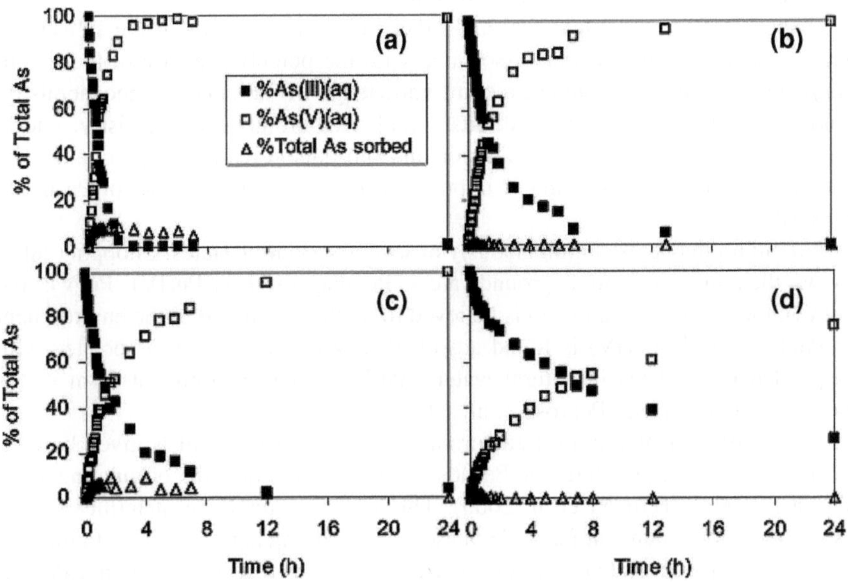

Fig. 16.5 Percent of dissolved $As(III)_{(aq)}$, $As(V)_{(aq)}$, and adsorbed As during As(III) oxidation kinetics on birnessite (suspension density 0.1 g/L, in 0.01 M NaCl, and N_2 atmosphere) as a function of pH and initial As(III) concentration, $[As(III)]_i$, **a** pH 4.5, $[As(III)]_i^2 = 100\,\mu M$; **b** pH 4.5, $[As(III)]_i = 300\ \mu M$; **c** pH 6.0, $[As(III)]_i = 100\ \mu M$; **d** pH 6.0, $[As(III)] = 300\ \mu M$. Reprinted with permission from Power et al. (2005). Copyright 2005 American Chemical Society

Two well-defined trends in the As(III) oxidation reactions can be distinguished: (1) the extent of As(III) oxidation decreases with increasing pH from 4.5 to 6.0 and (2) oxidation on a percent basis is suppressed with increasing initial As(III) concentration from 100 to 300 μM. The pH effects on As(III) oxidation may have been influenced by competitive adsorption reactions between As(III) and reaction products (e.g., Mn(II)) and were not influenced by arsenic solution speciation. The suppressed As(III) oxidation rate constant may be a result of differences in the amount of Mn(II) release, which compete with dissolved As(III) species for unreacted Mn(IV) surface sites, and of Mn(II) adsorption, which inhibit the reaction between As(III) and Mn(IV) surface sites.

The extent and rate of As(III) oxidation on birnessite surfaces are affected strongly by sorbed or competitive metal ligands in solution. Figure 16.6 shows As(III) oxidation when Zn is pre-adsorbed or applied in solution. The abbreviations shown in the figure denote specific reaction conditions used. For example, "PAs100ph45" refers to 100 μM of Zn pre-sorbed prior to the 100 μM As(III) addition at pH 4.5, and "SAs100ph6" refers to the simultaneous 100 μM Zn/100 μM As(III) addition at pH 6.0. Even though adsorbed Zn was present in the system, As(III) readily oxidized over time. However, Power et al. (2005) suggest that Zn is likely to form inner-sphere complexes on birnessite surfaces and chemisorbed Zn ions inhibit electron transfer reactions. When Zn was present, As(III) oxidation was further suppressed by nonadsorbed and pre-adsorbed Zn, compared with the control system, but the pre-adsorbed system was more effective in interfering with electron transfer reactions.

Disposal of spent nuclear fuel and other radioactive wastes in the subsurface and assessment of the hazards associated with the potential release of these contaminants into the environment require knowledge of radionuclide geochemistry. Plutonium (Pu), for example, exhibits complex environmental chemistry; understanding the mechanism of Pu oxidation and subsequent reduction, particularly by Mn-bearing minerals, is of major importance for predicting the fate of Pu in the subsurface.

Plutonium may exist simultaneously in several oxidation states. Choppin (2003) shows that, in oxic natural groundwaters, Pu may exist as Pu(IV), Pu(V), and Pu(VI); the most common form is believed to be Pu(IV), found in the environment as $PuO_{2(s)}$. While Pu(IV) is found adsorbed to solid particles or associated with suspended particulates in natural waters, Pu(V) is the predominant form in the natural aqueous phase (Penrose et al. 1987).

An example of plutonium transformation in the environment is given by considering the oxidation states of Pu adsorbed on a natural tuff, originating from Yucca Mountain (Powell et al. 2006). This tuff contains trace quantities of Mn oxides and more abundant Fe oxide phases. After adding aqueous Pu(V) to the tuff, elemental maps generated with micro-XRF (X-ray fluorescence) imaging, Pu was observed to preferentially associate with Mn oxides. Figure 16.7 shows L_3-edge X-ray absorption near-edge structure (XANES) spectra versus the relative XANES edge energy for sorbed Pu on Yucca Mountain tuff, for measurements made six months and two years after Pu(V) application. Six months after application, Pu(V)

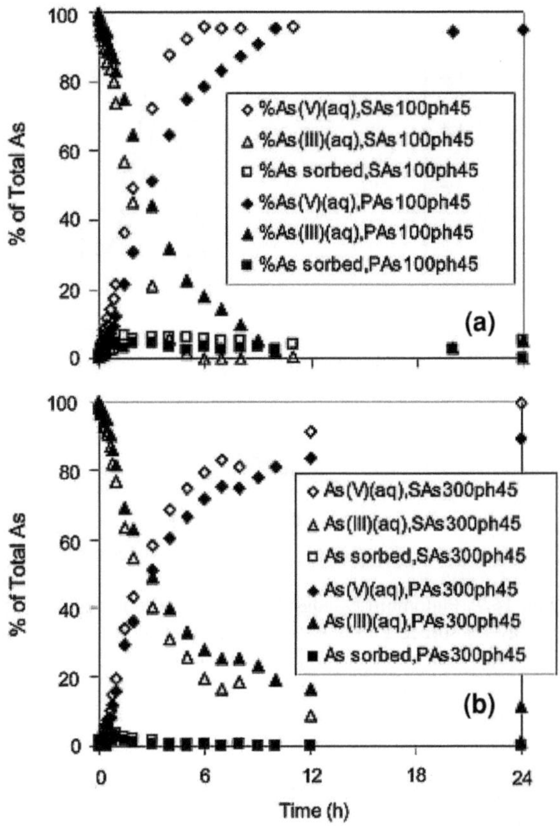

Fig. 16.6 Effects of pre-adsorbed Zn(II) versus Zn(II)/As(III) simultaneous treatment on As(III) oxidation kinetics on birnessite surfaces (pH 4.5, suspension density = 0.1 g/L, in 0.01 M NaCl, total Zn(II) concentration of 100 μM, and N_2 atmosphere). Percent As(III) depletion, As(V) release, and total As adsorption are shown as a function of time (h). **a** Initial As(III) concentrations: $[As(III)]_i$ = 100 μM; **b** $[As(III)]_i$ = 300 μM. Reprinted with permission from Power et al. (2005). Copyright 2005 American Chemical Society

was transformed into Pu(VI) by oxidation; but two years later, due to a reduction process, Pu(VI) was transformed into Pu(IV). Plutonium reduction on the mineral surface is dependent on both Pu speciation/hydrolysis at the mineral surface as well as the redox capacity of iron minerals on the surface (Powell et al. 2006).

Subsurface environments under anoxic conditions may contain high levels of Fe(II) on the solid phase or dissolved within immobile pore water or groundwater. The role of Fe(II) species in reductive transformation reactions of organic and inorganic contaminants in the subsurface was reviewed by Haderlein and Pecher (1988). A major finding of current studies is that Fe(II) associated with solid phases is much more reactive than Fe(II) present in dissolved forms (e.g., Erbs et al. 1999; Hwang and Batchelor 2000).

Klupinski et al. (2004) report a laboratory experiment on the degradation of a fungicide, pentachloronitrobenzene ($C_6Cl_5NO_2$), in the presence of goethite and iron oxide nanoparticles; this study was intended to illustrate the fate of organic agrochemical contaminants in an iron-rich subsurface. To compare the effects of iron with and without a mineral presence, experiments were performed using Fe(II) goethite and Fe(II) with no mineral phase added. The degradation kinetics of

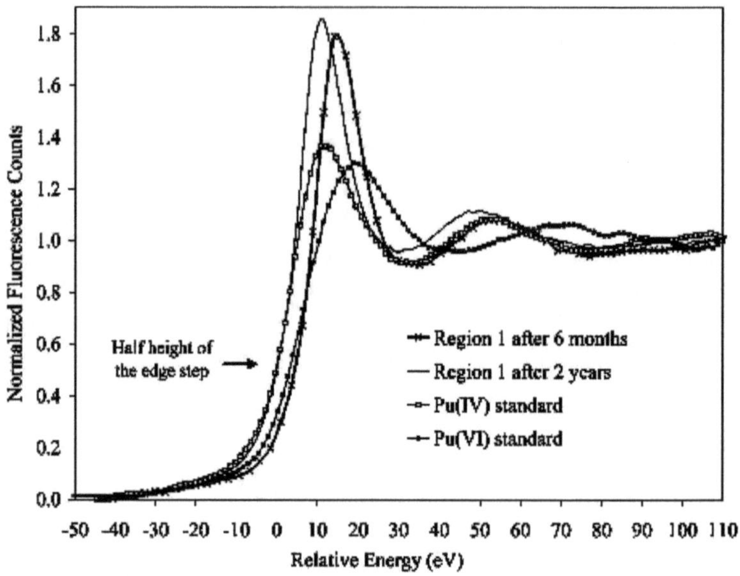

Fig. 16.7 Plutonium L_3-edge XANES spectra plotted with respect to the relative XANES edge energy for sorbed Pu on Yucca Mountain Tuff at 6 months and 2 years. All spectra taken after two years indicated an average oxidation state of Pu(IV); those taken after 6 months had average oxidation states predominantly of Pu(V) and Pu(VI). Reprinted with permission from Powell et al. (2006). Copyright 2006 American Chemical Society

$C_6Cl_5NO_2$ in reactions and the change in sorbed Fe(II), $[Fe(II)]_s$ as a function of pH are shown in Fig. 16.8. The primary degradation route for $C_6Cl_5NO_2$ occurs through a surface-mediated reaction with Fe(II); the final product is $C_6Cl_5NH_2$, with an intermediate product believed to be phenylhydroxylamine (C_6Cl_5NHOH).

Comparing degradation kinetics for reactions with different goethite contents (Fig. 16.9), Klupinski et al. (2004) found that $[Fe(II)]_s$ changes linearly with goethite content. The rate constant (k_A) decreased by more than a factor of 10 as the goethite content was adjusted from 100 to 10 mg/L, and it decreased by an even larger factor as the goethite content was further lowered to 1 mg/L. In an additional experiment, it was observed that the rate constant is affected by decreasing $[Fe(II)]_s$ while increasing the amount of goethite, which reveals that the change in the first-order rate constant for $C_6Cl_5NO_2$ reduction, k_A, is proportionally greater than the change in $[Fe(II)]_s$.

Klupinski et al. (2004) conclude that the reduction of nitroaromatic compounds is a surface-mediated process and suggest that, with lack of an iron mineral, reductive transformation induced only by Fe(II) does not occur. However, when $C_6Cl_5NO_2$ degradation was investigated in reaction media containing Fe(II) with no mineral phase added, a slow reductive transformation of the contaminant was observed. Because the loss of $C_6Cl_5NO_2$ in this case was not described by a first-order kinetic model, as in the case of high concentration of Fe(II), but better by a

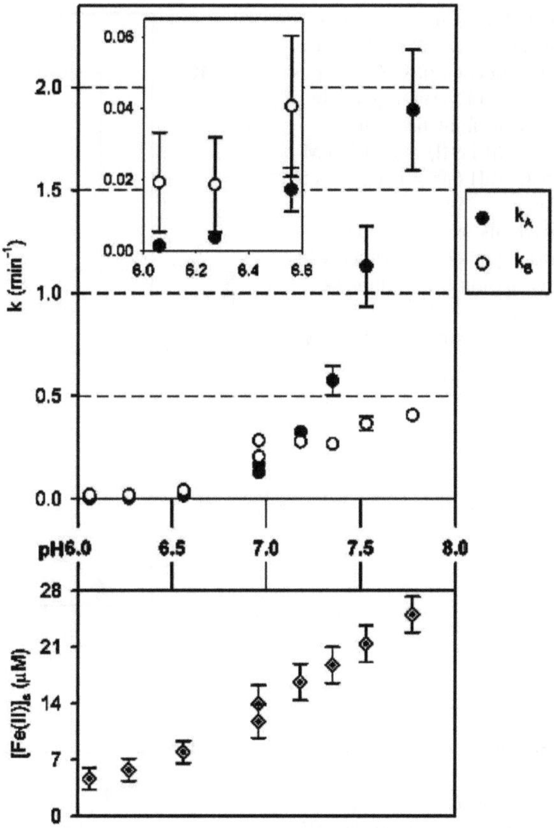

Fig. 16.8 Change in the first-order rate constants k_A and k_B for reduction of $C_6Cl_5NO_2$ and C_6Cl_5OH, respectively, and change in sorbed Fe(II), $[Fe(II)]_s$, as a result of varying pH in media with 473 μM Fe(II), 100 ± 11 mg/L goethite and 200 mM NaCl. Error bars indicate 95 % confidence intervals. When not shown, error bars are smaller than symbols. Data at the lower pH values are also plotted in the inset graph, showing that all rate constants are statistically greater than zero. Reprinted with permission from Klupinski et al. (2004). Copyright 2004 American Chemical Society

zero-order kinetic description, Klupinski et al. (2004) suggest that degradation in these systems in fact is a surface-mediated reaction. They note that, in the reaction system, trace amounts of O_2 oxidize Fe(II), which form in situ suspended iron oxide nanoparticles that serve as surface catalysts for $C_6Cl_5NO_2$. This behavior is supported by time profile results (Fig. 16.10) that show the reaction to be autocatalytic.

Reduction by Fe(II) results in an increase in the amount of iron oxides, which favor further reaction. Such autocatalytic behavior characterizes the oxidation of Fe(II) by O_2 and explains $C_6Cl_5NO_2$ reduction by Fe(II) in the absence of an iron mineral phase. Generalizing this behavior, it can be assumed that Fe(III) colloids derived from Fe(II) oxidation in subsurface anoxic systems, together with other colloids, affect the environmental persistence of nitroaromatic contaminants. Colon et al. (2006), for example, elucidate factors controlling the transformation of nitrosobenzenes and *N*-hydroxylanilines, which are the two intermediate products formed during the reduction of nitroaromatics in ferric oxide systems. Nitrosobenzenes are reduced by both Fe(II) solution and Fe(II)-treated goethite suspensions at pH 6, while *N*-hydroxyl anilines are reduced only by Fe(II)-treated

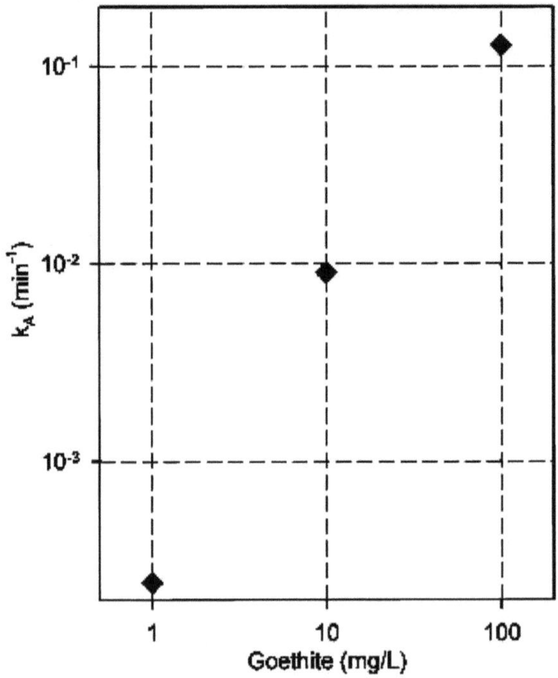

Fig. 16.9 Change in first-order rate constant (k_A) for the reduction of $C_6Cl_5NO_2$ as a result of varying goethite content in media with 473 µM Fe(II) and 200 mM NaCl (pH 6.96). Error bars to indicate 95 % confidence intervals would be smaller than symbols. Reprinted with permission from Klupinski et al. (2004). Copyright 2004 American Chemical Society

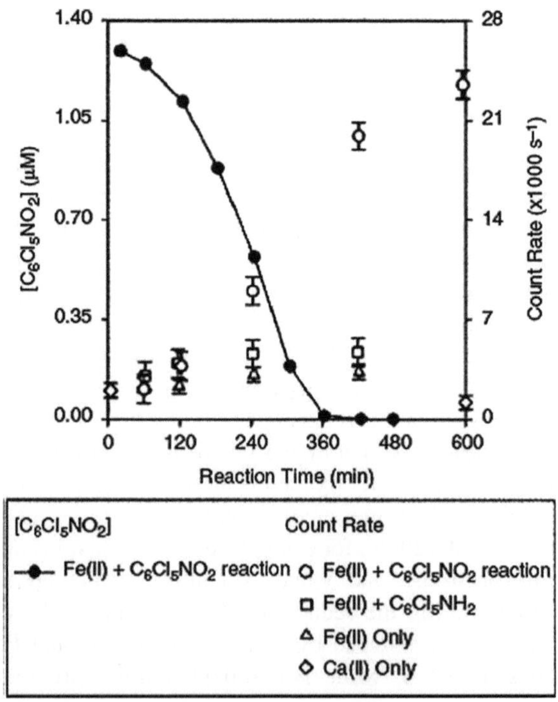

Fig. 16.10 Plot showing kinetics of $C_6Cl_5NO_2$ reduction (*filled circles*) occurring in conjunction with increasing photon correlation spectrometry (*PCS*) count rates (*open circles*), which are indicative of particle formation, in reaction with 0.80 mM Fe(II) (pH 7.0) (for clarity, the symbols showing measured values of [$C_6Cl_5NO_2$] are connected point to point). The other open symbols show PCS count rates in nonreaction mixtures (i.e., without $C_6Cl_5NO_2$) containing either 0.80 mM Fe(II) (pH 7.0) or 0.80 mM Ca(II) (pH 7.0). Reprinted with permission from Klupinski et al. (2004). Copyright 2004 American Chemical Society

Table 16.2 Characteristics of common iron(hydr)oxide minerals

Name (formula)	p^{H}_{ZPC}	BET surface (m^2/g)	$[\equiv FeOH]_t$ (nm^{-2})	XRD	SEM
Goethite (α-FeOOH)	7.8	17.5	5.5	√	Acicular; needles with ~0.3–1.5 μm length and ~0.1 μm diameter
Hematite (α-Fe_2O_3)	8.3–9.3	13.7	0.7	√	Hexagonal platy; single crystals form larger aggregates
Lepidocrocite (γ-FeOOH)	7.3	17.6	1.67	√	Irregular plates
Magnetite (Fe_3O_4)	6.4–6.9	8.1	9.4	√	Partly rounded octahedral crystals
Hematite (α-Fe_2O_3)	8.48	109	2.07	√	Spherical particles, 10–12 nm

goethite. The nitrosobenzene reactivity trend indicates that electron-withdrawing groups in the para position increase their rate of reduction.

Pecher et al. (2002) show how the uptake of ferrous iron from aqueous solution, by iron oxides, leads to the formation of a variety of reactive surface species that are capable of reducing polyhalogenated methanes (PHMs). The iron oxides used in the experiments and their characteristics are shown in Table 16.2. The PHMs studied include bromodichloromethane ($CHBrCl_2$), chlorodibromomethane ($CHBr_2Cl$), bromoform ($CHBr_2$), tetrachloromethane (CCl_4), hexachloroethane (HCE), fluorotribromomethane ($CFBr_3$), bromotrichloromethane ($CBrCl_3$), and dibromodichloromethane (CBr_2Cl_2).

Reduction of these PHMs took place in experiments containing both Fe(II) and iron oxide minerals, under anoxic conditions. The transformation of PHMs by surface-bound Fe(II) generally follows a pseudo-first-order kinetic rate law, expressed by

$$\ln(c/co) = -k_{\text{obs}}t \tag{16.1}$$

where k_{obs} is the pseudo-first-order rate constant. Figure 16.11 shows an evaluation of the rate law for the reduction of CBr_2Cl_2 by Fe(II), in the presence and the absence of goethite. This iron mineral was considered as the "master system," and assays with other mineral oxides were performed to obtain complementary data.

The values of the reaction rate, k_{obs}, for polyhalogenated alkanes in Fe(II)/goethite suspensions are noted in Table 16.3 together with their pseudo-rate constants and half-lives. The reaction rates are affected by contact time, sorption density, and solution pH. Pecher et al. (2002) note that a contact time of 20 h is necessary to achieve equilibrium for the uptake of Fe(II) by iron oxides, the uptake being formed by a fast initial step followed by a slower second phase. An increase in the density of Fe(II) at the surface of the mineral, especially after saturation of available surface sites, is the determining factor in quantifying transformation rates

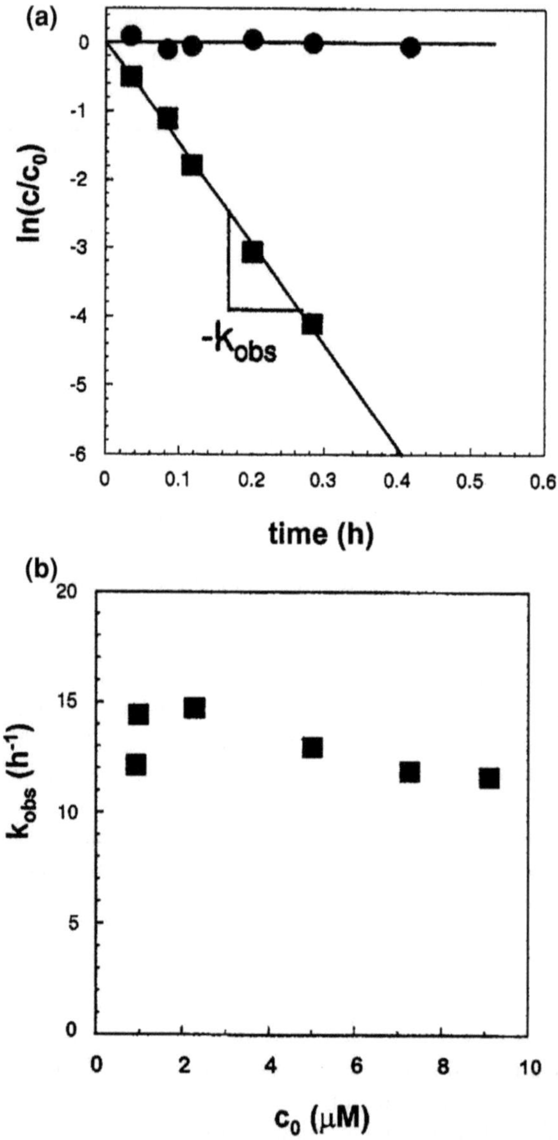

Fig. 16.11 Evaluation of the rate law of the reduction of CBr_2Cl_2 by Fe(II) in suspensions of goethite (25 m^2/L; $Fe(II)_{tot}$ of 1 mM; pH 7.2; ionic strength 20 mM; at 25 °C). The contact time of Fe(II) with iron oxide before addition of PHM was >24 h. **a** semilogarithmic plot of relative concentration (CBr_2Cl_2, $c_0 = 2.3$ μM) versus time in the presence (*filled square box*) and absence (*filled circle*) of goethite; **b** plot of pseudo-first-order rate constants for CBr_2Cl_2, k_{obs}, versus initial concentrations of CBr_2Cl_2 in suspension of Fe(II) and goethite. Reprinted with permission from Pecher et al. (2002). Copyright 2002 American Chemical Society

Table 16.3 Names, abbreviations, pseudo-first-order rate constants, and half-lives of polyhalogenated alkanes in Fe(II)/goethite suspension

Compound	Abbreviation	K_{obs} (h^{-1})	s^a	n^b	$t_{½}$ (h)
Bromodichloromethane	$CHBrCl_2$	0.0013	0.0002	4	533
Chlorodibromomethane	$CHBr_2Cl$	0.0029	0.001	4	239
Bromoform	$CHBr_3$	0.0048	0.002	2	144
Tetrachloromethane	CCl_4	0.016	0.005	7	43
Hexachloroethane	HCE[c]	0.0501		1	13.8
Fluorotribromomethane	$CFBr_3$	0.506	0.06	2	1.37
Bromotrichloromethane	$CBrCl_3$	3.577	1.52	2	0.19
Dibromodichloromethane	CBr_2Cl_2	11.296	3.28	8	0.06

Reprinted with permission from Pecher et al. (2002). Copyright 2002 American Chemical Society
Experimental conditions: 25 m^2/L goethite, pH 7.2, $t_{eq} > 24$ h. $Fe(II)_{tot} = 1$ mM
[a] Standard deviation
[b] number of replicates
[c] $t_{eq} = 5$ h

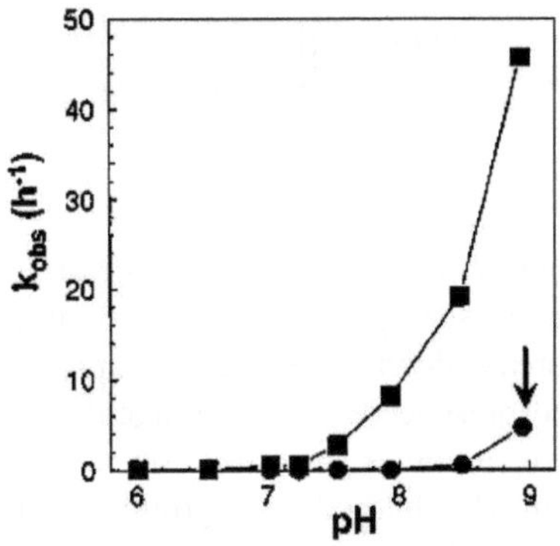

Fig. 16.12 Pseudo-first-order rate constants, k_{obs}, for the transformation of $CFBr_3$ ($c_0 = 6.5$ μM) in suspensions of goethite (25 m^2/L; $Fe(II)_{tot}$ of 1 mM; at 25 °C; ionic strength = 20 mM) as a function of solution pH (*filled square box*). The contact time of Fe(II) with iron oxide before addition of $CFBr_3$ was >24 h. Also shown are k_{obs} values for the control experiments in the absence of goethite (*filled circle*). The precipitate formed in the pH 8.9 control (*arrow*) was identified as a form of green rust. Reprinted with permission from Pecher et al. (2002). Copyright 2002 American Chemical Society

of PHMs by Fe(II) in iron oxide systems. The effect of pH on k_{obs} is illustrated in Fig. 16.12 for the reduction of $CFBr_3$ in an iron(II)-goethite suspension. We see that reaction rates increase almost exponentially with pH. Also, surface-bound Fe(II) species formed at high pH, under conditions favoring the surface precipitates, are more reactive than isolated surface complexes at lower pH or low surface coverage. Pecher et al. (2002) suggest that a variety of ferrous iron species coexist as potential reductants for PHMs at iron oxide surfaces. The reactivity of these species with respect to transformation of PHMs depends primarily on the sorption density of Fe(II) at the mineral surfaces, the concentration of Fe(II) in solution, the solution pH, and the contact time between dissolved Fe(II) and the oxide surfaces.

16.1.2 Surface-Induced Transformation on Clays

Transformation of organic contaminants adsorbed on clay materials is a surface-mediated process controlled by the molecular structure, the type of clay, and the clay-saturating cation, with the rate of contaminant conversion affected by the hydration status of the system. Studies on surface interactions of clays with organophosphorous pesticides (Yaron 1978; Yaron and Saltzman 1978) are used here to illustrate this mechanism. Of the many organophosphorus pesticides studied, parathion–clay compatibility is presented as an example.

Parathion, a heavily used pesticide with strongly adverse environmental effects, is a member of the phosphoric acid esters group with a general formula of the type

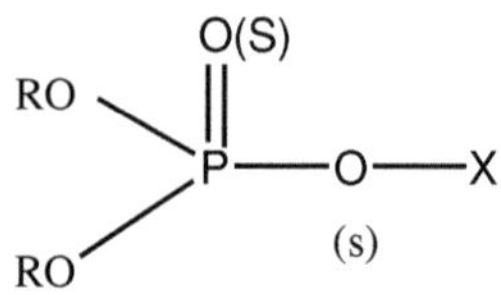

(16.2)

where R is an alkyl group and X is an organic radical. These esters are stable at neutral or acidic pH but are susceptible to hydrolysis in the presence of alkalines when the P–O–X ester breaks down. The rate of the process is related to the nature of the constituent X, the presence of catalytic agents, pH, and temperature. The mechanism of hydrolysis involves an attack on a relatively positive site, the phosphorus, by the negatively charged OH^- group. In the case of parathion, X in Eq. 16.2 is *p*-nitrophenol. This group increases the positive character of the phosphorus and, thus, its susceptibility to a nucleophilic attack. In contact with clay surfaces, nonionic organic molecules such as parathion are retained at the surface by different mechanisms, mainly by the ion-dipole or coordination interaction. The size of the adsorptive molecule, the nature of the clay-saturating cation, and the amount of water associated with the cation are determining factors in the adsorption and subsequent conversion of organic molecules.

Saltzman et al. (1974) compare the persistence of parathion on a glass surface and adsorbed on an oven-dried Ca^{2+}-kaolinite clay (Fig. 16.13). Parathion is relatively stable on a glass surface, but it breaks down partially in an aqueous solution with pH 8.5 and degrades much more when adsorbed on dry Ca^{2+}-kaolinite. The differences in degradation of parathion in water and on the clay surface suggest a strong catalytic activity of the Ca^{2+}-kaolinite.

Parathion molecules react with dissociated water, and hydrolysis may occur at the clay surface according to

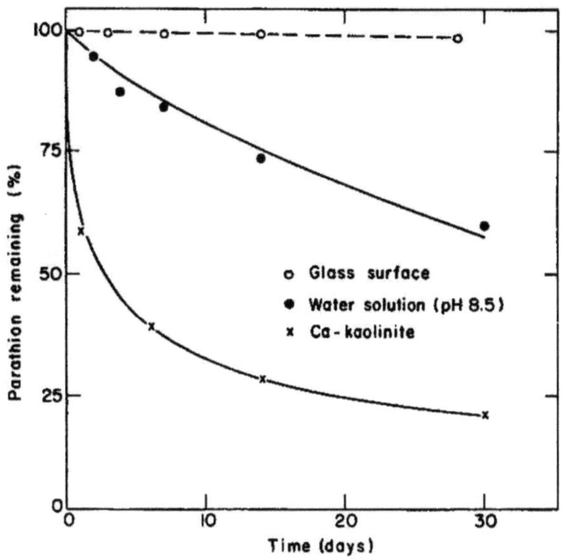

Fig. 16.13 Parathion losses from a glass surface, a water solution, and a dry Ca^{2+}-kaolinite at 40 °C (Saltzman et al. 1974)

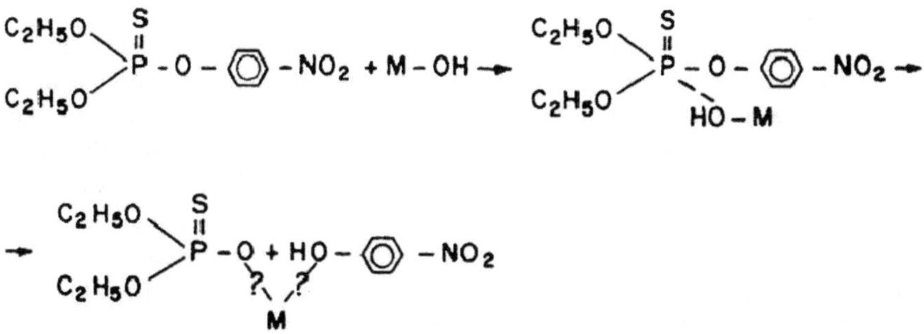

(16.3)

where M represents an exchangeable cation. In other experiments, Saltzman et al. (1974) found that, after 60 days, the quantities of parathion converted to diethyl phosphate were 72 % for Ca^{2+}-kaolinite, 34 % for Na^{+}-kaolinite, and 19 % for Al^{3+}-kaolinite. The amount of *p*-nitrophenol recovered in the case of Na^{+}- and Ca^{2+}-kaolinite was equal to that of the ^{14}C-labeled fraction of the water-soluble product, indicating that hydrolysis was the only path of degradation and extraction was apparently complete. In addition to the availability of OH^{-}, the cation charge and radius affect the rate of hydrolysis.

Surface catalysis affects the kinetics of the process as well. Saltzman et al. (1974) note that in the case of Ca^{2+}-kaolinite, parathion decomposition proceeds in two stages with different first-order rates (Fig. 16.14). In the first stage, parathion molecules specifically adsorbed on the saturating cation are quickly hydrolyzed by

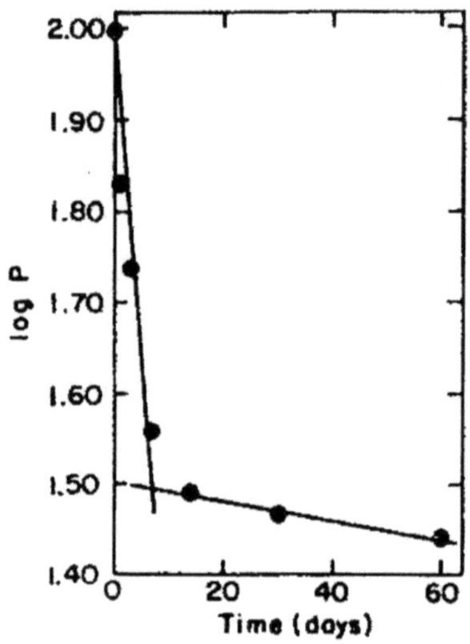

Fig. 16.14 Kinetics of parathion degradation on kaolinite (Saltzman et al. 1974)

contact with the dissociated hydration water molecules. In the second stage, parathion molecules that might have been initially bound to the clay surface by different mechanisms are very slowly hydrolyzed, as they reach active sites with a proper orientation.

Parathion hydrolysis on clay surfaces also is affected by environmental factors, such as temperature and water content. A rise in temperature generally enhances parathion hydrolysis on kaolinite, but the effect is greater in Na^{+}-kaolinite than Ca^{2+}-kaolinite. These differences are due to the different hydrolysis pathways in the presence of Na^{+}- and Ca^{2+}-saturating cations. In the limit of sorbed water, the addition of water to kaolinite affects degradation kinetics and degradation rates. A slight increase in the moisture content, above that corresponding to sorbed water, results in a steep decrease in the degradation rate. To emphasize the relationship between degradation rate and moisture content, Saltzman et al. (1974) plotted (Fig. 16.15) the parathion remaining in Na^{+}-, Ca^{2+}-, and Al^{3+}-kaolinites after a 15-day incubation period against water content. The functional dependence of degradation on moisture content on the various homoionic clays is similar. The presence of free water, at about 11 % moisture content, almost completely hinders the catalytic effect of the clay surface. Over a wide range of moisture contents, the rate of parathion degradation is similar for Na^{+}- and Ca^{2+}-kaolinites, showing that, in some cases, the effect of water may be stronger than that of the cation. For Al^{3+}-kaolinite, the opposite is true: The cation seems to be the determining factor of the degradation rate.

In addition, the type of clay can affect the catalysis of organophosphate contaminants. As shown in the case of parathion transformation on kaolinite, the main feature of the degradation process is the direct hydrolysis of the phosphate ester bond, the catalysis being strongly moisture and cation dependent. The rearrangement of organophosphorus on kaolinite is a secondary process with a much

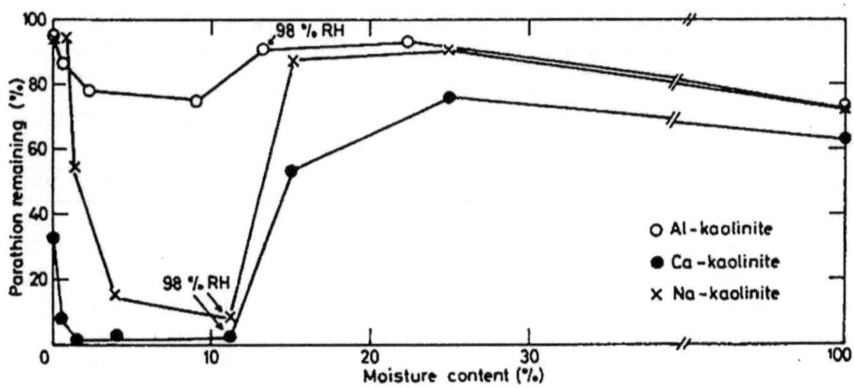

Fig. 16.15 Parathion degradation on kaolinite as affected by clay water content; incubation period 15 days (Saltzman et al. 1974)

slower rate on the direct hydrolysis. Rearrangement and hydrolysis of organophosphates also occurs on smectites and attapulgite clays, their extent being a function of contaminant structure and of the hydration status of the clays (Yaron 1978; Yaron and Saltzman 1978). In the case of pirimiphos ethyl degradation on Na^+-smectite, the dominant mode of loss is the direct hydrolysis of the phosphate ester bond with the pyrimidine ring of the compound. The high rate of hydrolysis does not enable observation of the rearrangement process. The studies of Saltzman and Mingelgrin (1978) on parathion–smectite and Gerstl and Yaron (1978) on parathion–attapulgite interactions show that these clays act mainly as catalysts for the rearrangement of parathion molecules. In this case, parathion is rearranged as an S-ethyl isomer and appears together with hydrolysis. The hydration status of the clay surface, different for each type of clay and its saturating cation, defines the type of hydrolysis: directly or through a rearrangement process.

When organic contaminants contact clay layers coated by natural organic matter, surface interactions between the contaminant and the porous medium exhibit different patterns. Yaron (1975, 1979) reported abiotic transformation of organophosphates, in general, and parathion, in particular, in a large number of soils with various types of clay, amounts of clay, and amounts of organic matter. Table 16.4 includes data on soil properties and water-soluble parathion degradation products, formed in sterilized soils after 130 days of incubation at 22 °C. The soils were initially air-dried and subsequently reached wet conditions by adding distilled water (50 % w/w).

Despite the differences in soil properties, *p*-nitrophenol was recovered in all the soils studied, together with water-soluble diethyl-thiphosphate. This result proves that hydrolysis is the main degradation path, independent of the nature of the soil. However, both the type of clay and the presence of organic matter affect the amount of degraded parathion. This behavior is illustrated in Fig. 16.16.

Table 16.4 Percent of remaining parathion and of its degradation products after 130 days of incubation, in 14 sterile soils with different chemical and mineralogical properties (Yaron 1975)

Soil type	Predominant clay	Clay (%)	Organic matter (%)	pH	Parathion remaining (%)		Diethyl thiophosphoric	
					Dry	Wet	Dry	Wet
Sandy regosol	Mont.	2.9	0.45	8.2	96.8	79.0	330	2,200
Loessial light brown	Kaol.	16.9	0.66	8.2	88.0	84.0	1,300	1,700
Grumusolic brown	Mont.	40.0	0.08	7.9	83.0	90.8	1,700	960
Red terra rossa	Kaol.	75.5	1.07	6.8	77.0	87.0	2,400	1,400
Reddish brown grumusol	Mont.	57.6	1.50	7.6	82.8	90.8	1,800	960
Red terra rossa	Kaol.	76.6	1.89	6.9	81.1	99.0	1,900	100
Basaltic brown Mediterranean	Kaol.	22.4	2.47	6.4	92.0	97.5	840	10
Hamra	Kaol.	9.7	2.75	6.8	92.0	97.5	840	260
Basaltic brown Mediterranean	Kaol.	32.0	3.61	6.5	84.0	97.4	1,700	270
Calcareous brown Mediterranean	Mont.	42.9	3.94	7.5	91.7	98.8	870	130
Reddish brown terra rossa	Mont.	65.3	4.10	7.7	96.8	98.0	400	200
Reddish brown terra rossa	Mont.	71.1	4.94	7.5	96.0	99.7	420	30
Red terra rossa	Kaol.	68.5	4.98	7.5	80.7	97.8	2,000	220
Brown rendzina	Mont.	46.5	12.10	7.3	95.7	99.8	450	20

The conversion of parathion is affected by soil constituents in the order kaolinite > smectite > organic matter and is related inversely to the adsorption affinity of these materials for this contaminant (Mingelgrin and Saltzman 1977). Although the extent of hydrolysis differs among soils, it is apparent that soil-surface degradation of parathion is caused by hydrolysis of the phosphate ester bond. The presence of water in the soil–parathion system led to a decrease in the surface-induced degradation of the organic molecule and diminished the effect of natural soil properties on its persistence.

16.1.3 Photochemical Reactions

Photochemical transformation of contaminants occurs in surface waters or when adsorbed on atmospheric particulates, which originate from the subsurface solid phase and are redeposited on land surfaces. We consider three studies: the first on photochemical reactions of drugs in natural waters, the second on the photodegradation of a polycyclic aromatic hydrocarbon adsorbed on a particulate, and the third on the effect of sunlight on photodegradation of a flame retardant (BDE-209) adsorbed onto clay minerals, metal oxides, and sediments.

The photolysis of a class of sulfa drugs containing six-membered heterocyclic substituents (sulfamethazine, sulfamerazine, sulfadiazine, sulfachloropyridazine, and sulfadimethoxine) was investigated by Boreen et al. (2005) in natural lake

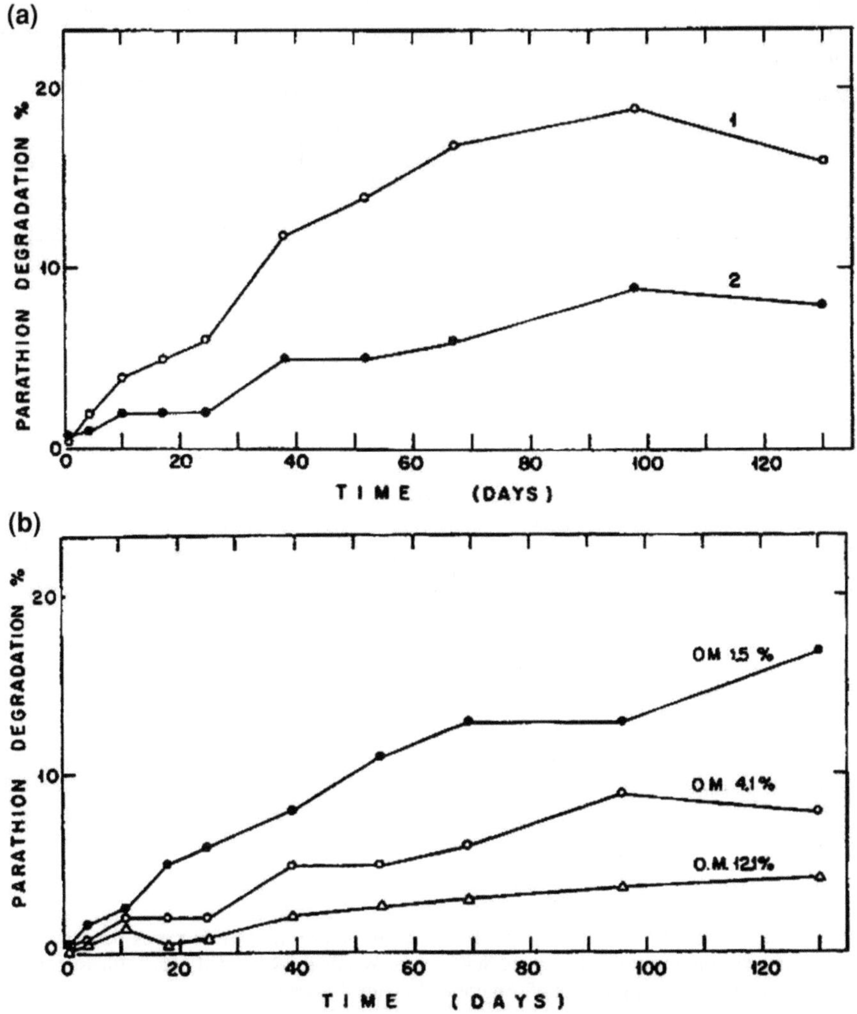

Fig. 16.16 Percentage of water-phase parathion degradation product recovered from montmorillonitic (1) and kaolinitic (2) soils, with **a** the same organic matter (OM) content and **b** different organic matter content (Yaron 1975)

water and in deionized water. The general structures of the sulfa drugs are presented in Table 16.5. The natural water (Lake Josephna, St. Paul, Minnesota) is characterized by pH 8 and 0.2 μm filtered DOC of 5.9 mg/L.

Photolysis of sulfa drugs in the natural lake water was enhanced significantly, with the exception of sulfadimethoxine. Figure 16.17, for example, shows the photodegradation of sulfamethazine in deionized water and natural lake water; photolysis gave rise to identical photoproducts. The primary product formed in

Table 16.5 General structure of a sulfa drug, structure of five sulfa drugs containing six-membered heterocyclic R substituents, and measured pK_a values; pK_a values were calculated using a spectrophotometry titration

Compound	Structure	$pk_{a,1}^{a}$	$pk_{a,2}^{a}$
	General structure 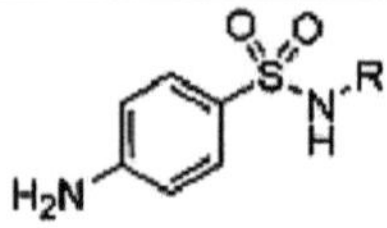		
1 Sulfamethazine	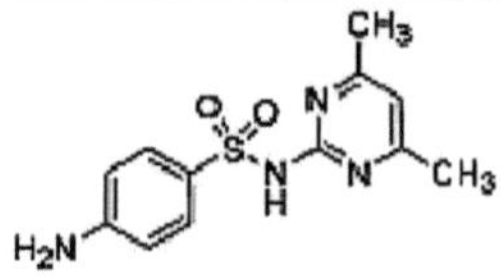	2.6 ± 0.2	8 ± 1
2 Sulfamerazine	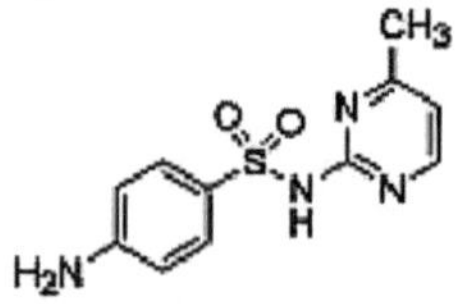	2.5 ± 0.7	7 ± 1
3 Sulfadiazine	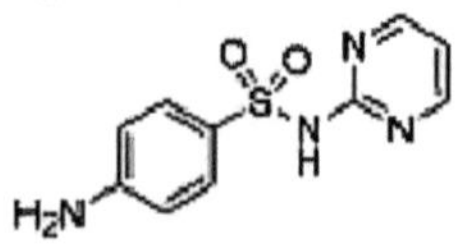	2 ± 1	6.4 ± 0.6
4 Sulfachloropyridazine	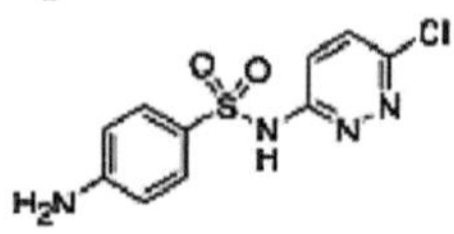	2 ± 3	5.9 ± 0.3
5 Sulfadimethoxine	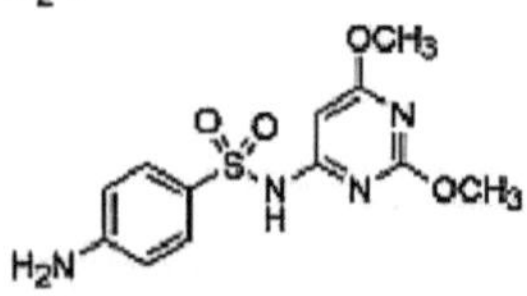	2.9 ± 0.5	6.1 ± 0.2

Errors represent the 95 % confidence levels, and both the pK_a values and the associated errors were obtained from fits of the data using Scientist for Windows. Reprinted with permission from Boreen et al. (2005). Copyright 2005 American Chemical Society

both direct and indirect photodegradation, for all five studied compounds, was identified as a sulfur dioxide extrusion product. Based on accurate mass balance, Boreen et al. (2005) suggest several structures for the photoproducts, such as 4-(2-imino-4,6-dimethilpyridimin-1-(2H)-yl)aniline derived from sulfamethazine.

Photochemical transformation of benzole (e)pyrene (BeP), a polycyclic aromatic hydrocarbon, adsorbed on silica gel and alumina surfaces, is reported by Fioressi and Arce (2005). This study was designed to define the atmospheric degradation of the PAH adsorbed on particulates that originated from wind erosion of the land surface. It can be assumed that similar behavior occurs at the land surface–atmosphere interface.

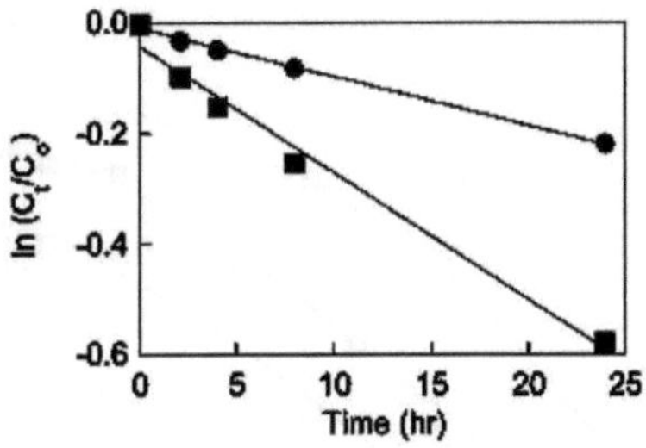

Fig. 16.17 Representative plot of the behavior of sulfa drugs obtained from photolysis on a turntable apparatus, under four Pyrex-filtered 175-W medium-pressure Hg-vapor lamps. Photodegradation of sulfamethazine in deionized H_2O (*filled circle*) and enhanced photodegradation in natural lake water (*filled square box*). Reprinted with permission from Boreen et al. (2005). Copyright 2005 American Chemical Society

In general, photodegradation of adsorbed PAHs occurs mainly through single, oxygen-mediated pathways or a radical intermediate mechanism. In the experiment of Fioressi and Arce (2005), samples of BeP in solution or adsorbed on silica gel or alumina surfaces were irradiated with a 1,000 W Xe (Hg) lamp; a glass filter was used to isolate the wavelength range between 250 and 359 nm. Irradiation of BeP adsorbed on silica gel and alumina surfaces, exposed to air, led to a decrease in intensity of absorption and of emission bands, thus demonstrating its photodegradation. The process followed a first-order kinetic, with an apparent rate constant of 0.173 min^{-1} for BeP adsorbed on silica, and of 0.089 min^{-1} for BeP adsorbed on alumina.

The effects of surface nature and loading on the photodegradation rate of BeP adsorbed on silica and alumina are shown in Fig. 16.18. The type of adsorbent affects the rate of BeP photodegradation. On silica gel, the rate is approximately two times greater than that on an alumina surface after 12 h of irradiation. Differences between photodegradation of BeP in solution and adsorbed on particulates are clearly reflected in the degradation product. Diones, diols, and hydroxyl derivates were identified as the major photoproducts of irradiated BeP adsorbed on particulates. The distribution of photoproducts for samples of BeP adsorbed on alumina was, however, different from that for silica gel under similar surface loading. The photodegradation rate also was affected significantly by an increase in subsurface loading and a decrease in BeP loading (Fioressi and Arce 2005).

Irradiation of BeP in hexane solution did not lead to the formation of the major derivates produced on adsorbed phases. This result suggests that BeP in nonpolar solvents photodegrades through a different reaction pathway than when adsorbed on surfaces. The type of solvent may affect, on the other hand, BeP photodegradation. In acetonitrile, for example, the photodegradation rate is faster than in hexane, and dione has been detected as one of the products.

Decabromodiphenyl ether (BDE-209) is a major industrial product from the polybrominated diphenyl ethers used as flame retardants; derivatives of this product

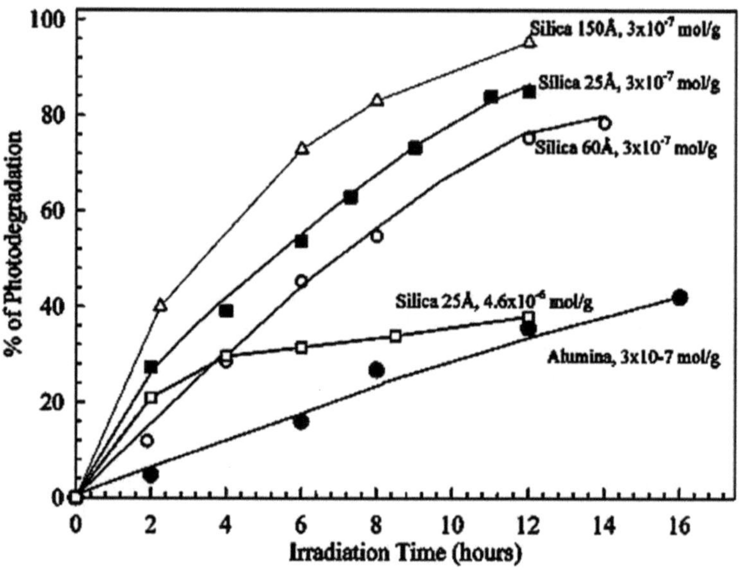

Fig. 16.18 Effect of surface nature and loading on the photodegradation rate of adsorbed BeP. The photolysis experiments were carried out in a rotary cell, and the BeP remaining in the sample was determined by HPLC. Uncertainties in measurements are in the range of 4–10 % of the absolute values. Reprinted with permission from Fioressi and Arce (2005). Copyright 2005 American Chemical Society

have been detected in the environment. After exposure to the land surface, these contaminants adsorb on soil materials and may reach the atmosphere as particulate matter; these particulates are subsequently subject to photolytic reactions. In this context, Ahn et al. (2006) studied photolysis of BDE-209 adsorbed on clay minerals, metal oxides, and sediments, under sunlight and UV dark irradiation. Dark and light control treatments during UV and sunlight irradiation showed no disappearance of BDE-209 during the experiments. Data on half-lives and rate constants of BDE-209 adsorbed on subsurface minerals and sediments, as determined by Ahn et al. (2006) and extracted from the literature, are shown in Table 16.6.

As a general pattern, we see that photodegradation is greater under UV light than sunlight irradiation. After 100 days of sunlight irradiation, the half-lives of BDE-209 are 216 days adsorbed on montmorillonite, 408 days adsorbed on kaolinite, and 990 days adsorbed on sediments. There are no significant losses of BDE-209 adsorbed on aluminum hydroxide, ferrihydrite, or birnessite. The longer half-life of BDE-209 adsorbed on sediments is explained by the presence of organic matter. Organic matter may inhibit photodegradation of organic chemicals either by shielding the contaminant from available light or by quenching the excited states of the organic molecules before they react to form products (Bachman and Patterson 1999). Fourteen days of UV irradiation cause the

Table 16.6 Half-lives *($t_{1/2}$)* and rate constants *(k)* of BDE-209 adsorbed on minerals and sediments, with the assumption of first-order reaction

Sorbents/solvents	UV		Sunlight	
	$t_{1/2}$	k (d^{-1})	$t_{1/2}$	k (d^{-1})
Montmorillonite	36 days	0.0192 ± 0.0067	261 days	0.0032 ± 0.0009
Kaolinite	44 days	0.0158 ± 0.0012	408 days	0.0017 ± 0.0003
Sediment	150 days	0.0046 ± 0.0003	990 days	0.0007 ± 0.0003
	40–60 h		80 h	
	53 h		81 h	
Al hydroxide	178 days	0.0039 ± 0.0007		–[a]
Mn dioxide (birnessite)	1423 days	0.0005 ± 0.0002		–[a]
Fe oxide (ferrihydrite)		–[a]		–[a]
Sand	12 h		37 h	
	12 h		37 h	
			533 h	
Soil	150–200 h			
	185 days			
Silica gel	<0.25 h			
Toluene	<0.25 h			
Hexane			<0.2 h	0.0011 (s^{-1})
Methanol/water	0.5 h			
Methanol	0.85 h			
Tetrahydrofuran	1–1.5 h			

Reprinted with permission from Ahn et al. (2006). Copyright 2006 American Chemical Society
[a] $k \approx 0$

following order of BDE-209 loss from the adsorbents: montmorillonite > kaolinite > sediment > aluminum hydroxide > birnessite. Ahn et al. (2006) consider that reductive debromination is the result of the electron-donating ability of clay minerals during the irradiation process. The fact that montmorillonite and kaolinite have different bonding affinities for BDE-209 but degrade it at relatively similar rates suggests that bonding forces do not affect the photodegradation rate.

BDE-209 debromination during photolysis leads to the formation of less brominated congeners. The appearance of these photoinduced products is faster under UV irradiation than under sunlight. Figure 16.19 shows gas chromatograms over time of BDE-209 adsorbed on montmorillonite, during 101 days of irradiation by sunlight, and compared with a control held in darkness. Over time, the areas of peaks for nona-, octa-, and heptabromodiphenyl ethers increased in the irradiated samples and remained at "trace" values in the dark control. Ahn et al. (2006) note a comparable but slower time trend for BDE-209 adsorbed on kaolinite.

These results indicate that the first step in BDE-209 photodegradation is the loss of one bromine atom to form nonabromodiphenyl ether congeners, followed by the subsequent formation of octabromodiphenyl ethers.

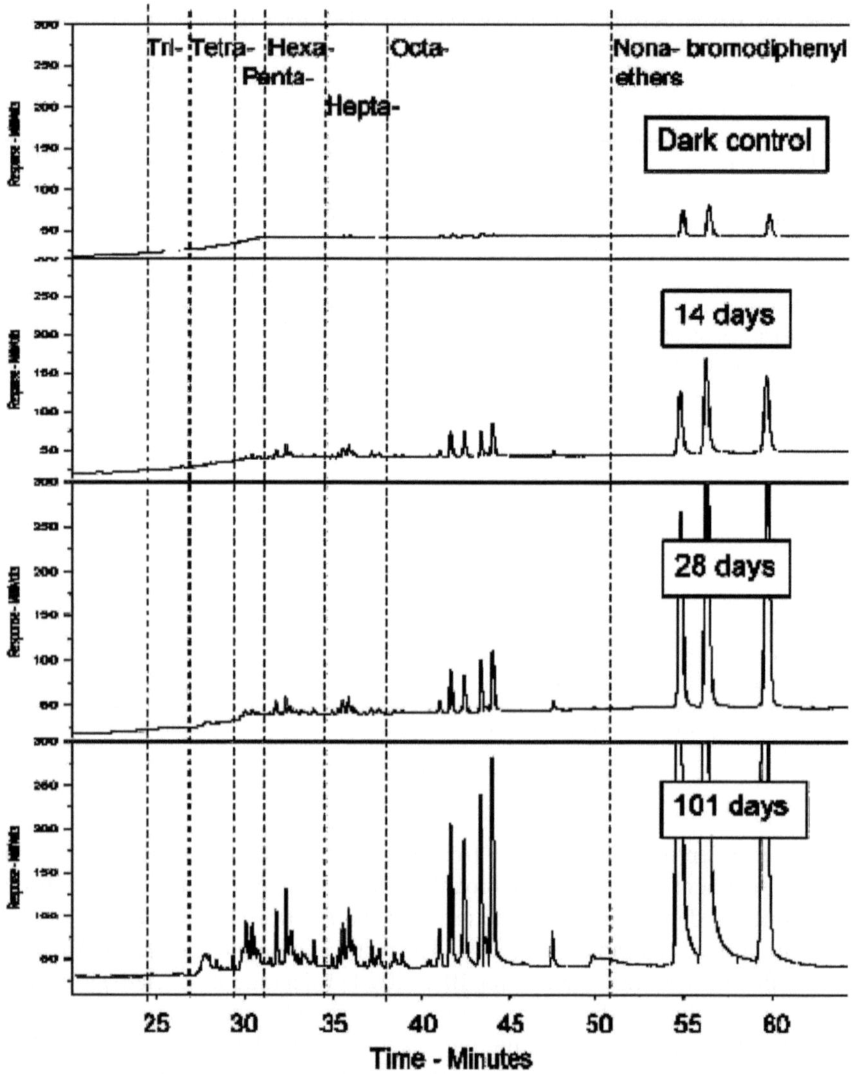

Fig. 16.19 GC-ECD chromatograms showing appearance of polybrominated diphenyl ether (*PBDE*) congeners, after sunlight irradiation of BDE-209 (retention time = 88.7 min) adsorbed on montmorillonite, at different times. Reprinted with permission from Ahn et al. (2006). Copyright 2006 American Chemical Society

16.1.4 Ligand Effects

Speciation induces transformation of contaminants, affecting mainly their retention and redistribution in the subsurface. As an illustration, we consider three studies that deal with speciation of trace metals and organic contaminants in the subsurface.

Subsurface contamination by uranium wastes and contaminant speciation during transport from a wastewater pond (originating from a plutonium production plant) to groundwater were studied by Catalano et al. (2006). Land disposal of basic sodium aluminates and acidic U(VI)–Cu(II) and their redistribution in the vadose zone resulted in development of a groundwater uranium plume. The solid-phase speciation of uranium from the base of the pond, through the subsurface, to the groundwater region was investigated by analyzing a depth sequence of sediments. The mineralogy of the sediments was relatively uniform (quartz, plagioclase feldspar with minor muscovite, chlorite, hornblende, and smectite) except for the calcite content, which was greater in the upper sediment below the pond and decreased with depth. The uranium concentration is positively correlated with the calcite content. Near-surface sediments contain uranium coprecipitated with calcite and formed due to overneutralization of the waste pond with NaOH. At intermediate depths in the vadose zone, meta-torbernite $Cu(UO_2PO_4)_2 8H_2O$ was detected, which presumably precipitated during pond operation. In the deeper zones and the groundwater, uranium was predominantly sorbed onto phyllosilicates.

Based on EXAFS (extended X-ray adsorption fine structure) spectra, Catalano et al. (2006) suggest that the chemical speciation of uranium changed systematically with increasing depth, even though XANES (X-ray adsorption near-edge structure) spectroscopy indicates uranium presence throughout the subsurface, primarily as U(VI). The phases present in the sample, as evaluated by PCA (principle component analysis) of the EXAFS spectra, include two primary components. In small concentrations, additional phases also may be found. Because adsorbed U(VI) is the dominant uranium species in the contaminated sediment, Catalano et al. (2006) suggest that adsorption–desorption processes dominate the future fate and transport of uranium in the site. Ca^{2+} and CO_3^{2-} in the local groundwater, at concentrations close to saturation with respect to calcite, complex with U(VI), thus inhibiting its reaction with the solid phase and favoring contaminant transport to the groundwater.

Natural organic matter, in general, and humic substances, in particular, may affect the fate and transport of heavy metals in the subsurface by speciation of toxic chemicals. Binding constants of divalent mercury (Hg^{2+}) by subsurface organic matter were studied by Khwaja et al. (2006). The formation constants of mercury in humic substances extracted from peat (PHA) and from a soil with high soil organic matter (SOM) content were reported. The authors used a competitive ligand exchange method with DL-penicillamine, a synthetic thiol amino acid that has Hg^{2+} binding abilities, to determine distribution coefficients (K_{oc}) for Hg^{2+}. The formation constants for PHA and SOM were calculated assuming that Hg^{2+} bonds to two thiol groups of bidentate sites. Khwaja et al. (2006) show a linear increase in log K_{oc} in the pH range 1.9–5.8 and the slope of pH versus log K_{oc} was 2.68. This indicates that two or more protons are released when each Hg^{2+} is bound to two thiol groups.

The effect of solution chemistry on the speciation of the organic contaminant 1-naphthol (1-hydroxynaphthalene) and its complexatiom with humic acid is reported by Karthikeyan and Chorover (2000). The complexation of 1-naphthol

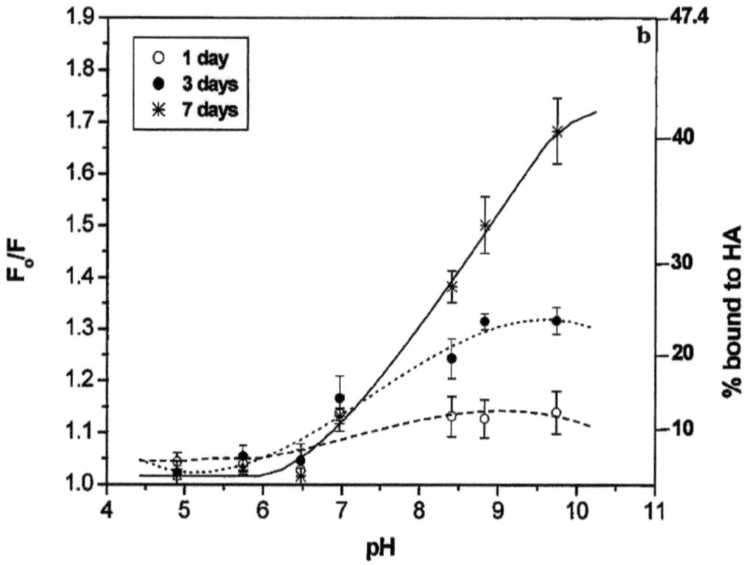

Fig. 16.20 Fluorescence quenching (FQ) of 1-naphthol in the presence of HA as a function of pH and reaction time (1-naphthol = 8 μmol/L; HA = 11 ppm C; ionic strength of 0.1 M LiCl); F_0 and F denote fluorescence intensities in the absence and in the presence of the quencher (HA), respectively. Reprinted with permission from Karthikeyan and Chorover (2000). Copyright 2000 American Chemical Society

with humic acid (HA) was studied during seven days of contact, as a function of pH (4–11), ionic strength (0.001 and 0.1 M LiCl), and dissolved O_2 concentration (DO of 0 and 8 mg/L) using fluorescence, UV absorbance, and equilibrium dialysis techniques. In a LiCl solution, even in the absence of HA, oxidative transformation of 1-naphthol mediated by O_2 was observed. In addition, the presence of humic acid in solution, in the absence of DO, was found to promote 1-naphthol oxidation. These reactions are affected by the solution chemistry (pH, ionic strength, and cation composition).

Because the mechanisms of 1-naphthol complexation with HA obtained by using these three techniques exhibit similar pathways, we present the results only from fluorescence spectroscopy. The ratio of fluorescence intensity in the absence (F_o) and in the presence (F) of the quencher (HA) over time, as affected by pH and ionic strength, is illustrated in Fig. 16.20. The fluorescence intensity of a fluorophore in the absence of a quencher is directly proportional to its concentration in solution, and therefore time-dependent changes in F_o can be used to assess the stability of 1-naphthol under different pH and ionic strength. Quenching (FQ) of 1-naphthol fluorescence by humic acid increased with equilibration time from one to seven days. This time-dependent relationship was found to result from weak complexation of 1-naphthol by humic acid and by its oxidative transformation (slow reaction), resulting in the formation of secondary products that are more reactive with the HA. From Fig. 16.20, we see that the time-dependent increase in quenching of 1-naphthol fluorescence by HA occurs. This time-dependent increase provides indirect evidence of the operating mechanisms, which may be summarized as follows: direct sorption–complexation of 1-naphthol to HA is a relatively fast reaction that is completed within 24 h, followed by sorption of oxidative products characterized by a longer contact time.

16.1.5 Multiple-Component Contaminant Transformation

Subsurface transformations of contaminants comprising multiple components are reflected in the composition of the residual contamination products, which may have different physical and chemical properties than the original pollutant. Differential partitioning dissolution and volatilization of component mixtures are the main abiotic processes leading to alteration of the original pollutant.

One of the most ubiquitous multiple-component contaminants that reaches the soil and deeper subsurface layers is crude oil and its refined products. In the subsurface, these contaminants are transformed differently by various mechanisms (Cozzarelli and Baher 2003). Crude oil contains a multitude of chemical components, each with different physical and chemical properties. As discussed in Chap. 4, the main groups of compounds in crude oils are saturated hydrocarbons (such as normal and branched alkanes and cycloalkanes without double bonds), aromatic hydrocarbons, resins, and asphaltenes, which are high molecular weight polycyclic compounds containing nitrogen, sulfur, and oxygen.

Bennett and Larter (1997) discuss the effect of partitioning dissolution in an aqueous phase of alkylphenol. Specifically, they show that the depletion of this crude oil component affects the chemical composition of the original pollutant. Partitioning at equilibrium can be considered the maximum dissolution value of a compound under optimal solvation conditions. Partitioning dissolution is obtained by "washing" the crude oil with saline water at variable temperature and pressure conditions, similar to those in the subsurface. The data reported were obtained using a partition device able to simulate the natural environmental conditions of a crude oil reservoir. The alkylphenol partition coefficients between crude oil and saline subsurface water were measured as a function of variation in pressure, temperature, and water salinity. Preliminary trials proved that the experimental device did not allow alkylphenol losses due to volatilization.

The crude oil used by Bennett and Larter (1997) was a typical North Sea oil generated from Upper Jurassic, Kimmeridge clay formation source rocks. The alkylphenol distribution in a sample of (Miller) crude oil, determined using solid-phase extraction (SPE), is shown in Fig. 16.21. The crude oil is dominated by phenol and cresol and contains appreciable quantities of dimethylphenols. The concentrations of 2,3-, 3,4-, and 3,5-dimethylphenol also include a contribution from 2-, 3-, and 4-ethylphenols, because they coelute under the conditions employed (Bennett et al., 1996).

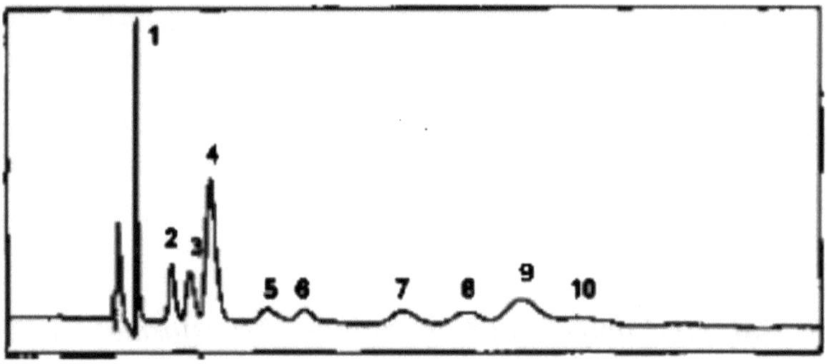

Fig. 16.21 Reversed-phase HPLC-ED chromatogram of alkylphenols from SPE extract of Miller oil. See Table 16.7 for identification of peaks. Reprinted from Bennett and Larter (1997). Copyright 1997 with permission of Elsevier

Table 16.7 shows the depletion of alkylphenols from crude oil, by dissolution in water, after equilibration of the crude oil with deionized water at a pressure of 25 bar and at 25 and 125 °C. Based on steric considerations, Bennett and Larter (1997) showed that the aqueous affinity of *o*-cresol (hindered hydroxyl group) is expected to be lower than *m*- and *p*-cresol (unhindered), because the close proximity of the methyl group to the active hydroxyl functionality may decrease H-bonding interactions with water. The calculated ratio of hindered to unhindered cresols (*o*-cresol/(*m*-cresol + *p*-cresol), abbreviated to *o*-/(*m*- + *p*-)) in the equilibrated oil and water, with respect to the ratio determined in the original crude oil (Table 16.7), reveals a slightly higher value for the oil. This indicates that *o*-cresol is preferentially retained in the oil; it has a lower relative value in the water sample, suggesting more favored *m*- and *p*-cresol partitioning into water.

Overall, dissolution in the electrolyte solutions of all studied compounds led to substantial decreases in their concentrations in the original crude oil. However, the rate of decrease was different for each compound, in accordance with the molecular properties. Bennett and Larter (1997) note that, in comparison with the original crude oil, the concentration of phenol decreased relative to cresol. The change in phenol content relative to cresol is highlighted by measuring the ratio (phenol/*o*- + *m*- + *p*-cresol), which shows that the reduction of more hydrophilic phenol relative to cresol is in accord with their relative dissolution in water.

Bennett and Larter (1997) also studied the solvation of alkylphenols in crude oil–water systems at equilibrium to obtain partitioning coefficients under variable temperature, pressure, and water salinity concentration. Alkylphenol depletion from crude oil, expressed by phenol, cresols, and 3,5 dimethyl phenol, versus temperature in a range of 25–125 °C, is given in terms of partition coefficient (*P*) values (Fig. 16.22). Partition coefficient values increase with addition of alkyl groups to the phenol nucleus. Note that the alkylphenol partition coefficient curves for different isomers tend to converge at higher temperatures and, as a

Table 16.7 Comparisons of alkylphenol concentrations (mg/L), determined by RP-HPLC-ED after isolation by SPE of Miller oil, equilibrated oils at 25 °C and 125 °C, and corresponding water samples (Bennett and Larter 1997)

	Compounds	Original Miller oil	Equilibrated Oil	25 °C equilibrated Water	Sum[1]	Percentage Recovery	Equilibrated Oil	125 °C equilibrated Water	Sum[1]	Percentage Recovery
1	Phenol	2.92	1.72	1.117	2.79	96	1.51	1.37	2.88	99
2	m-cresol	1.98	1.42	11.35	1.77	89	1.35	0.55	1.88	95
3	p-cresol	2.34	1.59	0.38	1.97	84	1.51	0.64	2.15	92
4	o-cresol	8.82	6.94	1.16	7.10	80	6.81	1.81	8.62	98
5	3,5-DMP[a]	1.90	2.04	0.15	2.19	115	2.01	0.30	2.31	122
6	3,4-DMP[a]	1.55	1.34	0.08	1.42	92	1.08	0.18	1.26	81
7	2,5-DMP	2.37	2.49	0.16	2.65	112	2.86	0.21	3.07	130
8	2,3-DMP[a]	3.75	3.00	0.19	3.19	85	2.21	0.36	2.57	69
9	2,4-DMP	6.94	7.63	0.36	7.99	115	6.64	0.63	7.27	105
10	2,6-DMP	6.77	6.69	0.20	6.89	102	5.09	0.29	5.38	79
	Ratios									
	P/(*o*- + *m*- + *p*-)	0.22	0.17	0.57	0.26		0.16	0.46	0.23	
	o-/(*m*- + *p*-)	2.04	2.31	1.59	1.90		2.38	1.52	2.14	

[a] Coelution of compounds determined by cochromatography with authentic standards: 3,5-DMP + 4-ethylphenol; 3,4-DMP + 3-ethylphenol; 2,3-DMP + 2-ethylphenol

[1] Sum of concentrations in equilibrated oil and corresponding water. Reprinted from Bennett and Larter (1997). Copyright 1997 with permission of Elsevier

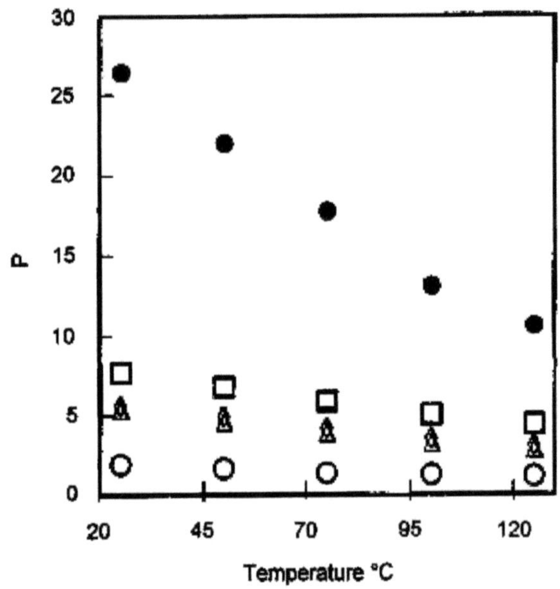

Fig. 16.22 Plot of partition coefficient (*P*) versus temperature for alkylphenol (*open circle*), *m*- (*open triangle*). *p*- (*open diamond*), *o*- (*open square*) cresols, and 3,5-DMP (*filled circle*) after 1:1 equilibration of Miller oil and (deionised) water at 25 bars. Reprinted from Bennett and Larter (1997). Copyright 1997 with permission of Elsevier

consequence, differences between phenol and *p*-cresol decrease with increases in temperature. Similar results for oil–deionized water and oil–brine experiments show that increasing temperature leads to a decrease in partition coefficient values.

It is known that pressure may affect the solubility of a compound. Analysis of alkylphenol partitioning in crude oil reservoir–water systems, as affected by pressure, was performed by Bennett and Larter (1997). In these experiments, the presence of gases was neglected; and a pressure of 340 bars, similar to that encountered in North Sea crude oil reservoirs, was chosen as the highest value. Alkylphenol partition coefficients measured at pressures of 25, 200, and 340 bars, as a function of temperature, are shown in Fig. 16.23. Over the pressure range studied, only slight variations in the partition coefficients were observed at a temperature of 25 °C; the phenol partition coefficient decreased from 1.95 at a pressure of 25 bars to 1.77 at 340 bars. Similar low differences in the effect of pressure on partitioning values were measured at a temperature of 125 °C. Bennett and Larter (1997) conclude that a pressure lower than 340 bars and a temperature lower than 125 °C do not significantly affect the partitioning of alkylphenols in a crude oil–water system. On the other hand, *o*-cresol shows sensitivity to temperatures over the range 20–125 °C but is insensitive to pressure.

The effect of brine salinity on the dissolution of alkylphenols, from crude oils into water with variable electrolyte concentration (0–100 mg/L NaCl), leading to depletion of the original pollutant, is presented in Fig. 16.24. By increasing the salt concentration in the "washing" water, the partition coefficient of alkylphenols increases, indicating preference for the petroleum phase at higher water salinity.

Fig. 16.23 Plot of partition coefficients (*P*), for alkylphenol (*shaded symbols*) and *o*-cresol (*open symbols*) at 25 (*open circle*), 200 (*open circle*), and 340 (*open square*) bar pressure versus temperature after 1:1 Miller oil/brine (20 g/L NaCl) equilibration. Reprinted from Bennett and Larter (1997). Copyright 1997 with permission of Elsevier

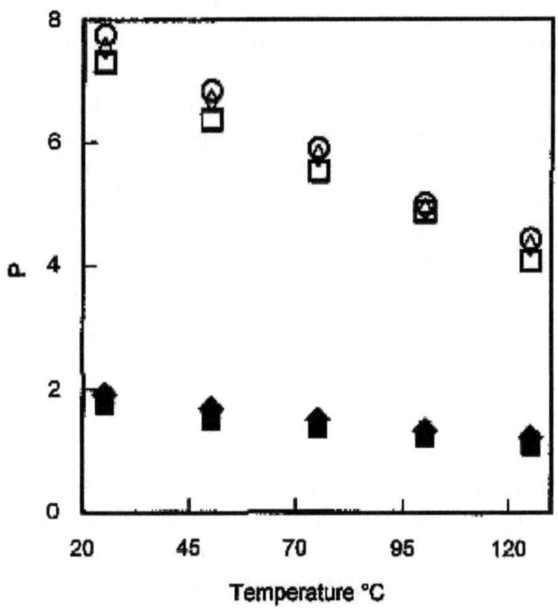

Fig. 16.24 Plot of variation of partition coefficients (l:1, Miller oil/brine) for phenol (*open circle*) *m*- (*open triangle*), *p*- (*open diamond*), and *o*- (*open square*) cresols versus brine salinity (g/L NaCl) at 25°C and 25 bar. Reprinted from Bennett and Larter (1997). Copyright 1997 with permission of Elsevier

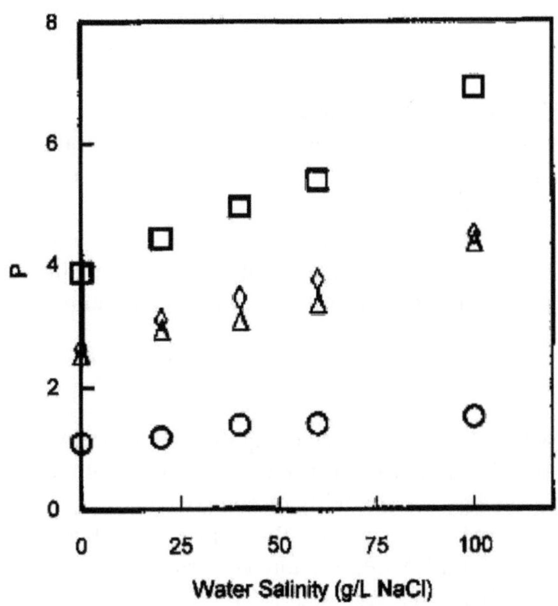

This trend is opposite to the effect of temperature; in the deep subsurface characterized by the presence of saline water, where the temperature also may be high, opposing effects of increasing temperature and salinity may compensate one

another (Bennett and Larter 1997). The decreasing concentration of alkylphenols with increases in water salinity is explained by the well-known salting-out effect (Sect. 6.5).

Another example of transformation in the composition of an industrial petroleum product, due to dissolution-partitioning processes, is reported by Yaron et al. (1998) and Dror et al. (2000). Figure 16.25 shows the gas chromatograms of a neat kerosene and the kerosene extracted from an aqueous electrolyte solution (0.01 M NaCl) after 100 h of contact at an ambient temperature of 22 °C. From analysis of the chromatograms, it is evident that aliphatic and branched aliphatic hydrocarbons in the range of C_9–C_{16} constitute the major group of components in neat kerosene; there is only a minor set of aromatic compounds. In the aqueous electrolyte solution, on the other hand, aromatic compounds, especially the branched benzenes and naphthalene, make up the majority of compounds in the aqueous phase. The C_9 and C_{10} aliphatic and branched aliphatic components do not appear in the chromatogram of aqueous kerosene. Aromatic components are several orders of magnitude more soluble than aliphatic components. As such, aromatic compounds that are minor constituents of neat kerosene dissolve in water much more readily than in the major group of aliphatic compounds.

Broholm et al. (2005) discuss a field experiment at Borden, Ontario, where the transformation of a mixture of solvents forming a contamination source is reflected in the presence of solvents in an aquifer. A chlorinated solvent mixture containing 2.0 L of trichloroethylene (TCE), 0.5 L chloroform (TCM), and 2.5 L of tetrachloroethylene (PCE) was released into a sandy aquifer, and the development of a dissolved-phase plume was studied over one year. Using a multiple-component dissolution model, the authors estimated the mass of multiple-component dense nonaqueous-phase liquid (DNAPL) source in the groundwater and the changes in the source composition over time. When a multiple-component DNAPL is dissolved in an aqueous phase, the most soluble compounds are depleted fastest and the DNAPL composition changes. Estimation of the mass of residual solvent is based on a theory of dissolution from a multiple-component mixture of organic compounds, for which the dissolved concentrations are described by Raoult's law:

$$C_i = x_i S_i \tag{16.4}$$

where C_i is the aqueous equilibrium concentration of compound i (called the *effective solubility*), x_i is the mole fraction of compound i in organic phase, and S_i is the aqueous solubility of compound i.

Based on this theory, Broholm et al. (2005) present the theoretical dissolution curve for a DNAPL mixture consisting of 10 % TCM, 40 % TCE, and 50 % PCE by volume (see Fig. 16.26a). TCM, which is the most soluble compound, depletes quickly, while PCE, the less soluble fraction, is depleted slowly. Based on these curves, concentration ratios between TCE and TCM, PCE and TCM, and PCE and TCE were determined as a function of the source depletion (Fig. 16.26b).

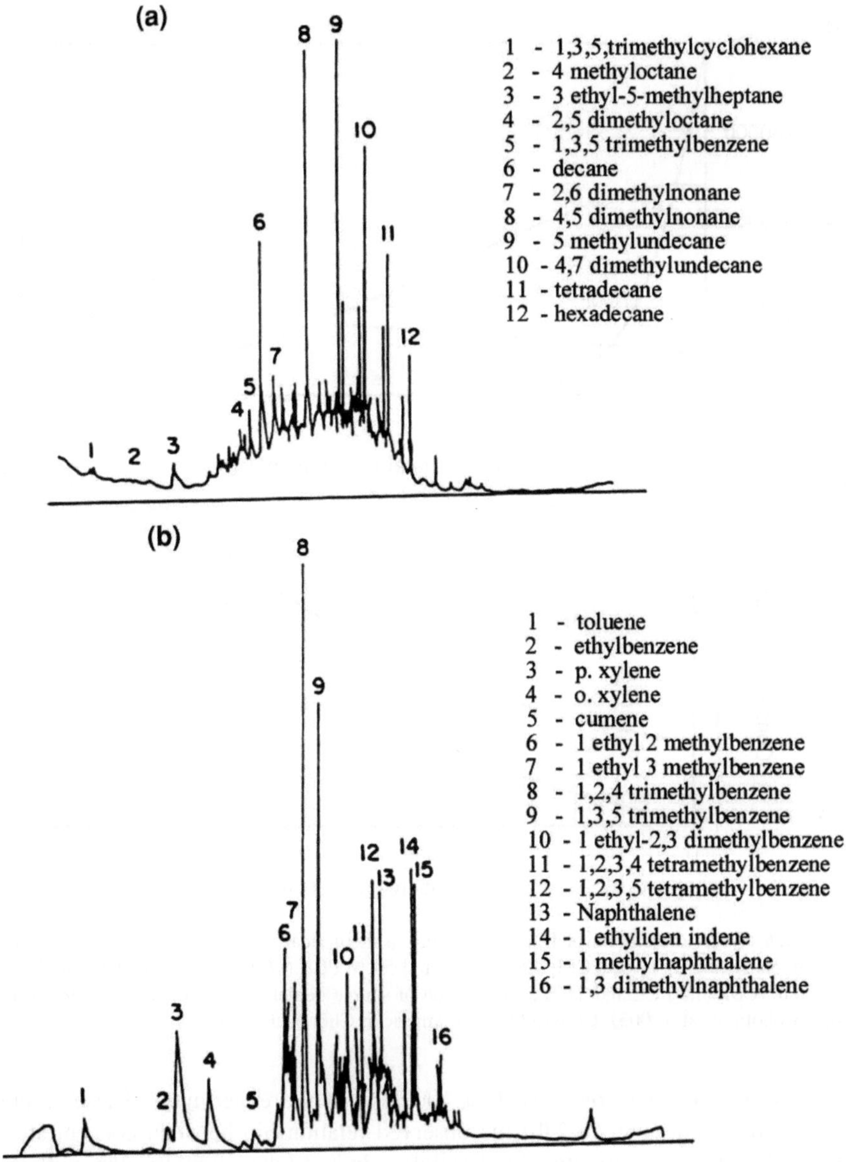

Fig. 16.25 Gas chromatograms of **a** neat kerosene and **b** kerosene dissolved in 0.01 M NaCl aqueous solution (Yaron et al. 1998)

Results of the field experiment are shown in Fig. 16.27a, which is based on combined discharge from five extraction wells. After about 50 days, TCM concentrations decreased. In contrast, concentrations of TCE fluctuated but remain relatively high. PCE concentrations continued to increase over time, exhibiting a

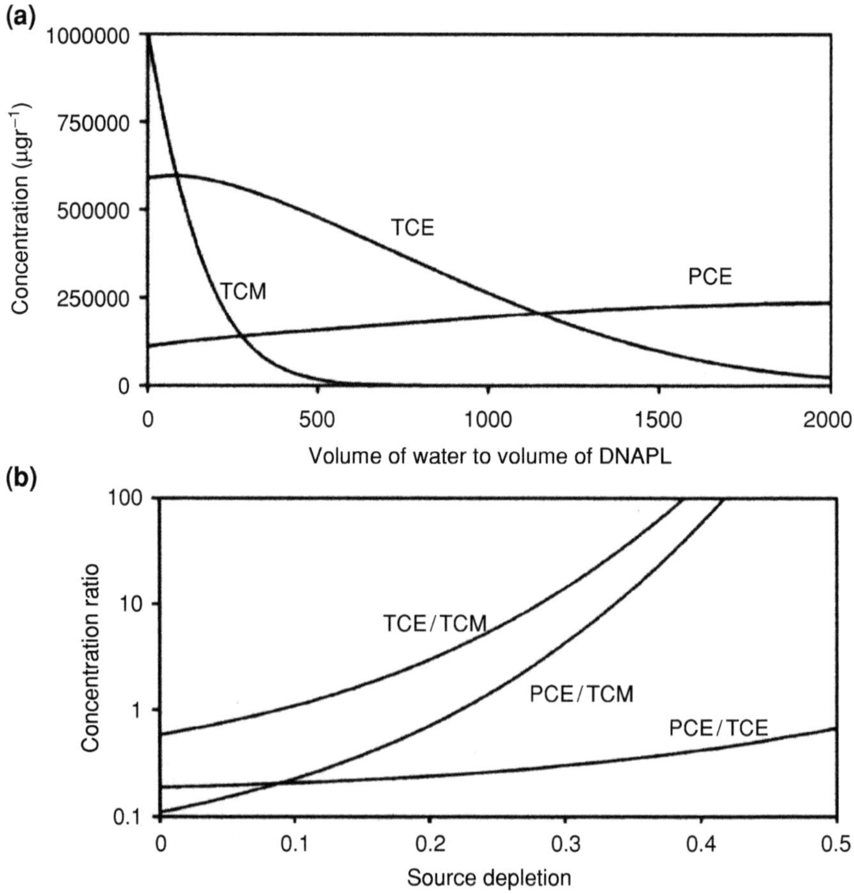

Fig. 16.26 Water dissolution of DNAPL mixture, **a** theoretical dissolution curve for a DNAPL mixture consisting of 10 % TCM, 40 % TCE, and 50 % PCE (v/v); **b** ratios of TCE and TCM, PCE and TCM, and PCE and TCE as a function of source depletion. Reprinted with permission from Broholm et al. (2005). Copyright 2005 American Chemical Society

higher dissolution rate over the first 100 days of the experiment. These results were used to plot (Fig. 16.27b) the observed relationship between concentration ratio and source transformation by dissolution-induced depletion, together with the equivalent theoretical relationships. Source depletion was calculated from the cumulative mass removed, as determined from monitoring of effluent at specific times, divided by the initial source mass.

The results of a field-scale test at the Borden site are reported by Rivett and Feenstra (2005). This experiment used a DNAPL multicomponent system similar to that of Broholm et al. (2005), with the difference being that the contamination source was emplaced below the water table and exposed to natural groundwater flow during a period of three years. The contamination source contained 23 kg of

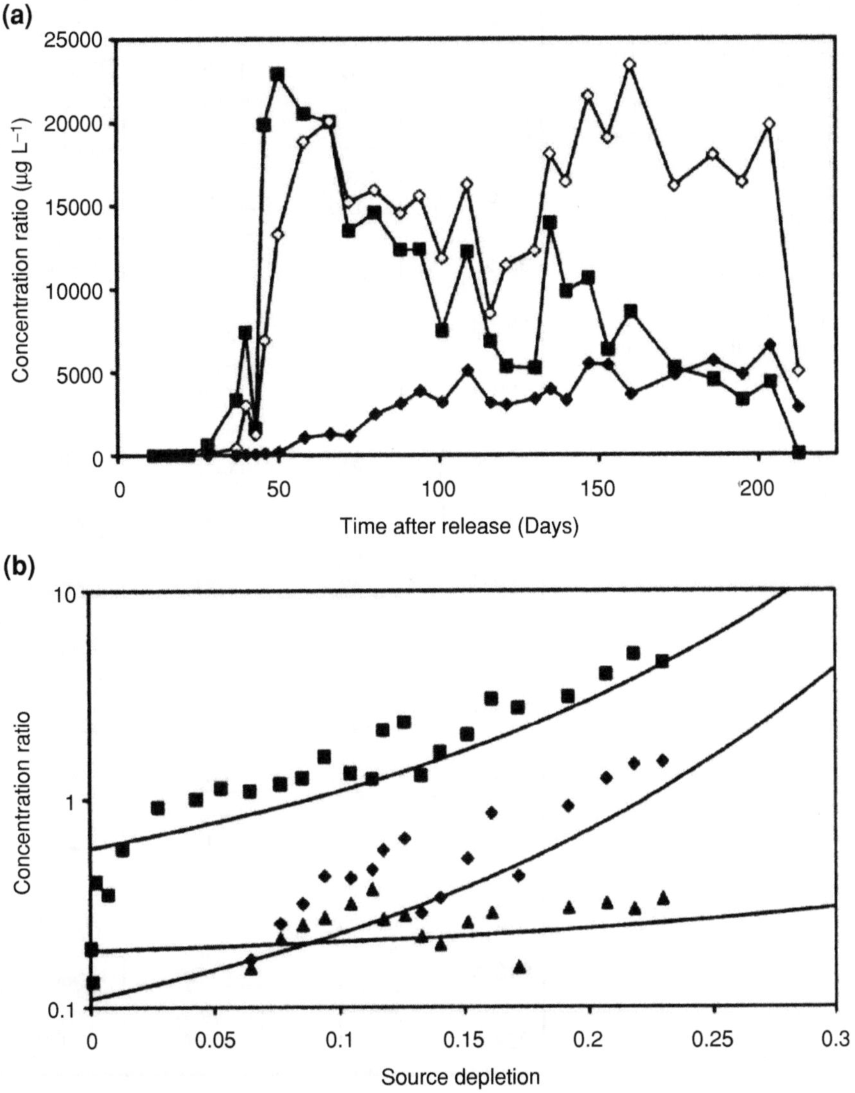

Fig. 16.27 **a** Effluent concentration of TCM (*filled square box*), TCE (*open triangle*), and PCE (*filled diamond*) as a function of time. **b** Ratios of TCE and TCM (*filled square box*), PCE and TCM (*filled diamond*), and PCE and TCE (*filled triangle*) as a function of source depletion. The solid lines show the theoretical curves for the three ratios. Reprinted with permission from Broholm et al. (2005). Copyright 2005 American Chemical Society

trichloromethane (TCM), trichloroethylene (TCE), and perchloroethylene (PCE) mixture. The dynamics of the dissolved-phase concentration plume, over more than 900 days, are presented in Fig. 16.28. We see that irregularly shaped plumes exhibit significant lateral movement and are correlated with changes in

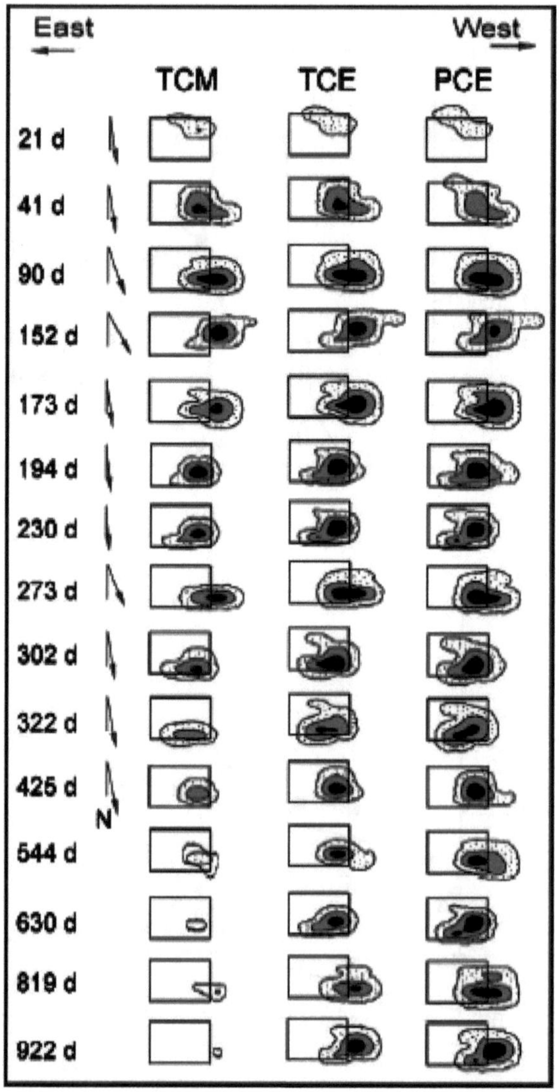

Fig. 16.28 Dissolved-phase concentration plumes showing 0.5, 0.15, and 0.015 (and 0.0015 for last four TCM dates) contours. The inner, darkest shade contour is the maximum concentration. The source is 1 m upgradient of the 1 × 1.5 m areal faces shown here, and groundwater flow directions relative to north (normal to source face) are also shown. Reprinted with permission from Rivett and Feenstra (2005). Copyright 2005 American Chemical Society

groundwater flow direction. The dissolution patterns of the various components differ spatially and temporally, according to their chemical properties, despite the relative homogeneity of the source and aquifer.

Volatilization of the light fraction of a liquid mixture in the subsurface brings about changes in both the physical and chemical properties of the residual liquid. This process is illustrated in a series of experiments investigating kerosene behavior (e.g., Dror et al. 2000, 2001, 2002; Fine and Yaron, 1993; Galin et al. 1990; Gerstl et al. 1994; Jarsjo et al. 1994, 1997; Yaron et a1. 1998). The differential composition of kerosene during volatilization by up to 50 % (w/w),

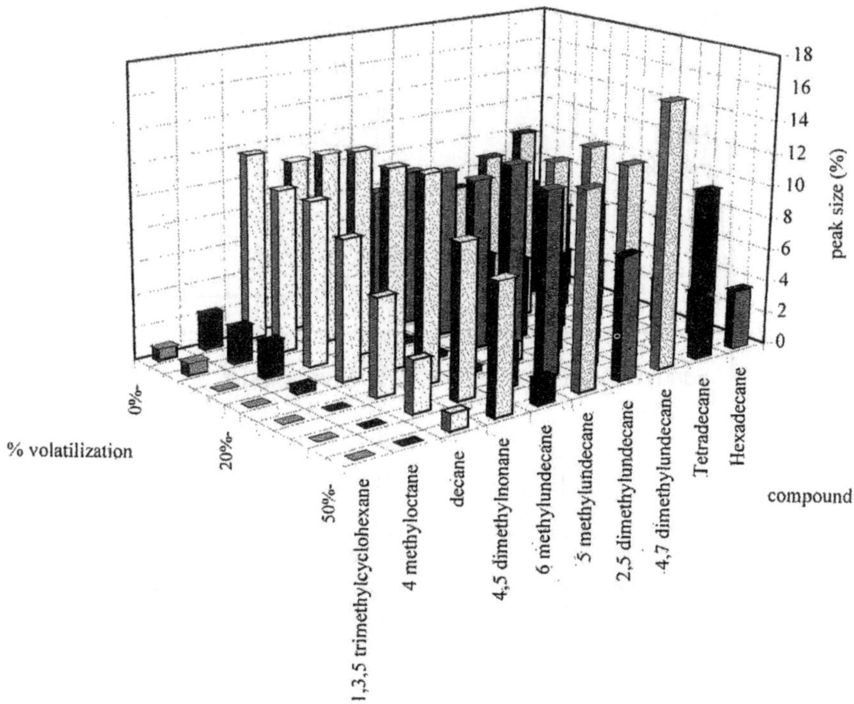

Fig. 16.29 Major components remaining from kerosene during the volatilization process (Yaron et al. 1998)

Table 16.8 Effect of volatilization on physical properties of residual kerosene

Amount volatilized (%)	Density (g/cm)	Surface tension (N/m)	Viscosity (Pa s $\times$ 10^{-3})
0	0.805	2.75	1.32
20	0.810	2.78	1.48
40	0.818	2.80	1.78
60	0.819	2.78	1.96

Reprinted from Galin et al. (1990). Copyright 1990 with permission of Elsevier

expressed by the concentration of selected components, is illustrated in Fig. 16.29. It is seen that the lighter fractions evaporate at the beginning of the volatilization process, with additional components volatilizing as the total amount of evaporation increases. This behavior leads to a relative increase in the heavier fractions of kerosene in the remaining liquid.

Changes in the chemical composition of the kerosene during volatilization also affect the physical properties of this petroleum product. Table 16.8 summarizes the effect of volatilization on kerosene viscosity, surface tension, and density when 20, 40, and 60 of the initial amount has been removed by the partial transfer of light hydrocarbon fractions to the atmosphere. Only the liquid viscosity is affected, with

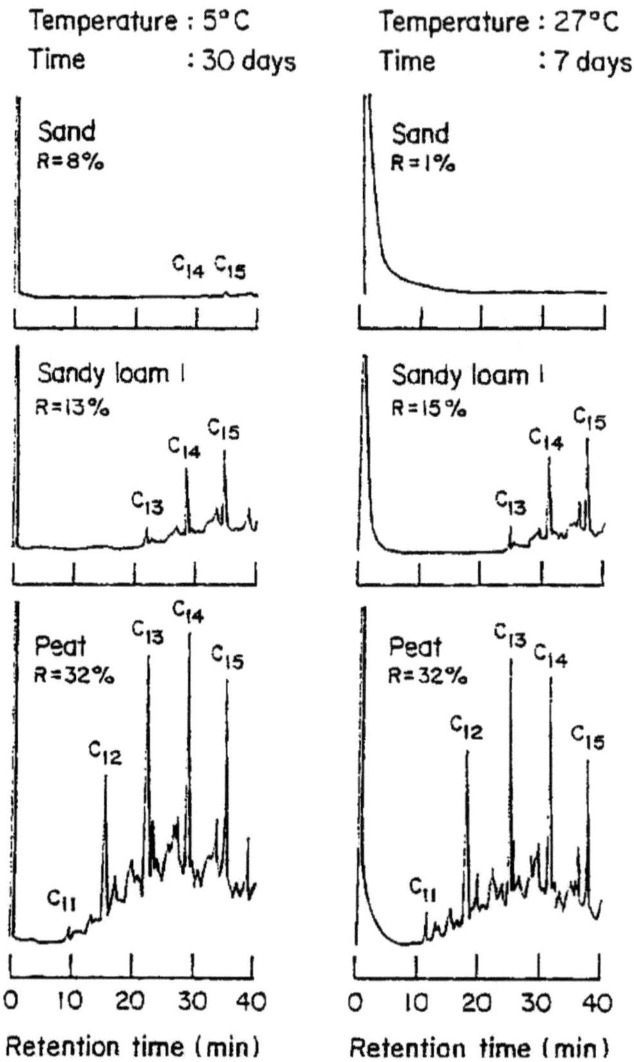

Fig. 16.30 Gas chromatograms of kerosene recovered from sand, sandy loam, and peat after volatilization at 5 and 27 °C for 30 and 7 days, respectively: *R* denotes remaining kerosene (% of initial amount). Reprinted from Jarsjo et al. (1994). Copyright 1994 with permission of Elsevier

volatilization having a negligible effect on the density and surface tension of the kerosene.

Once reaching the subsurface, kerosene is subject to chemical and physical changes by volatilization, and the rate and the extent of this process controlled by the properties of the porous medium and the ambient temperature. Different volatilization through the subsurface leads to changes in the residual kerosene concentration. Figure 16.30 shows gas chromatograms in sand, sandy loam, and peat after volatilization at 5 and 27 °C for 30 and 7 days, respectively. The lighter components disappeared from the sand and the sandy loam, and the residual kerosene composition is similar regardless of temperature. The relative concentration of the light fractions (C_{10}–C_{12}) diminished with time in all cases, while the concentrations of the heavy fractions (C_{14} and C_{15}) increased. Galin et al. (1990) determined the relationship between viscosity of the liquid and percent of kerosene remaining in coarse, medium, and fine sand following volatilization. The highest increase in viscosity occurs for coarse sand with an initial kerosene concentration equivalent to the residual retention capacity, while the lowest increase occurs in fine sand. Viscosity increases with the time of volatilization, in the order coarse > medium > fine sand, for each initial concentration.

These findings deal mainly with the volatilization of various components of petroleum products that may be subject to nonideal equilibrium partitioning and limitations of diffusion. An example of typical volatilization behavior of binary NAPL mixtures entrapped at residual saturation, within partially saturated porous media, is given by Abriola et al. (2004). They examine volatilization of binary mixtures of styrene and toluene, and styrene and PCE, from columns packed uniformly with aquifer grains of various mineral composition. The average sand porosity, residual water, and residual NAPL saturation in sand columns were 0.382, 0.146, and 0.111, respectively. Various gas flow rates were used in a steady state as well as in transient volatilization experiments, and component concentrations and mass transfer coefficients were determined. Abriola et al. (2004) concluded that, under steady state conditions, NAPL saturation and composition do not change appreciably. Under a low pore gas velocity, as in the transient case, the initial component concentrations are proportional to their mole fractions. When the more volatile components deplete, a sharp rise in the concentration and mole fraction of the remaining component is observed, which may reach a near-saturation level.

16.2 Biomediated Transformation of Contaminants

We examine two groups of contaminants that exhibit biologically mediated transformation in the subsurface: petroleum hydrocarbons and pesticides. Both aerobic and anaerobic pathways for their biodegradation are considered.

16.2.1 Petroleum Products

A hydrocarbon mixture may be biodegraded in both aerobic and anaerobic environments. The literature on this subject is vast and deals with many specific aspects; here, biologically mediated transformation of petroleum hydrocarbons is discussed on the basis of the review by Cozzarelli and Baher (2003). Hydrocarbons

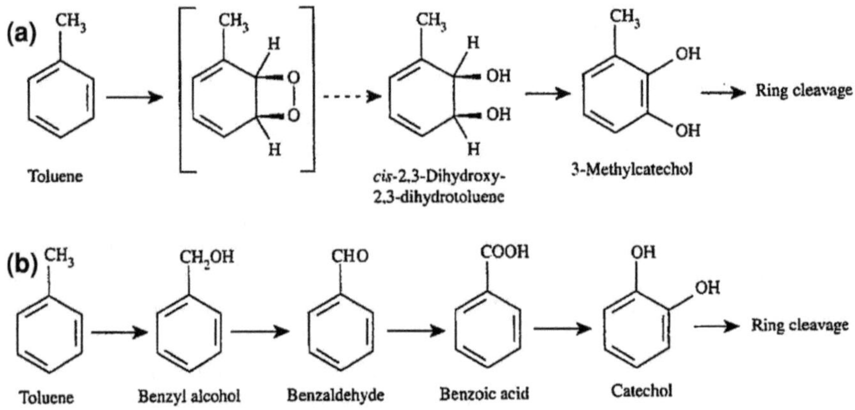

Fig. 16.31 Aerobic toluene degradation pathway by **a** aromatic ring attack by dioxygenation, and **b** side-chain attack by stepwise oxidation (Smith 1990)

are degraded in aerobic environments by a wide variety of bacteria and fungi that contain the genetic capability to incorporate molecular oxygen into the hydrocarbon system. Both prokaryotic and eukaryotic microorganisms have the enzymatic potential to oxidize aromatic hydrocarbons. While bacteria utilize the compounds as a sole source of carbon and energy, fungi cometabolize aromatic hydrocarbons to hydroxylated products. The first step in fungal metabolism is the formation of an epoxide; bacteria initiate the oxidation of unsubstituted aromatic compounds by incorporating both atoms of molecular oxygen into the aromatic ring, to form a cis-dihydrodiol and by further oxidation that leads to catechol formation (Gibson and Subramanian 1984; Reineke 2001).

Aerobic conditions for biologically mediated toluene degradation favor an aromatic ring attack by dioxygenation, and a side-chain attack by stepwise oxidation (Smith 1990). These degradation pathways are illustrated in Fig. 16.31. In these cases, the presence of an alkyl substituent group on the benzene ring allows microorganisms an additional site of attack; oxidation of this chain results in the formation of benzyl alcohols, benzaldehydes, and alkylbenzoic acids. The oxidation of straight-chain alkanes proceeds by oxidation of the terminal ethyl group, a process mediated by a class of enzymes called *oxygenases*. The most degradable *n*-alkanes are the C_{10}–C_{18} compounds (Fig. 16.32). The intermediate products include alcohols, aldehydes, and carboxylic acids (Bouwer and Zehnder 1993). Aerobic degradation reactions require available moisture, nitrogen, and phosphorus in addition to molecular oxygen; nutrient availability is the limiting factor for biodegradation when oxygen and moisture are plentiful.

Anaerobic conditions often develop in hydrocarbon-contaminated subsurface sites due to rapid aerobic biodegradation rates and limited supply of oxygen. In the absence of O_2, oxidized forms or natural organic materials, such as humic substances, are used by microorganisms as electron acceptors. Because many sites polluted by petroleum hydrocarbons are depleted of oxygen, alternative

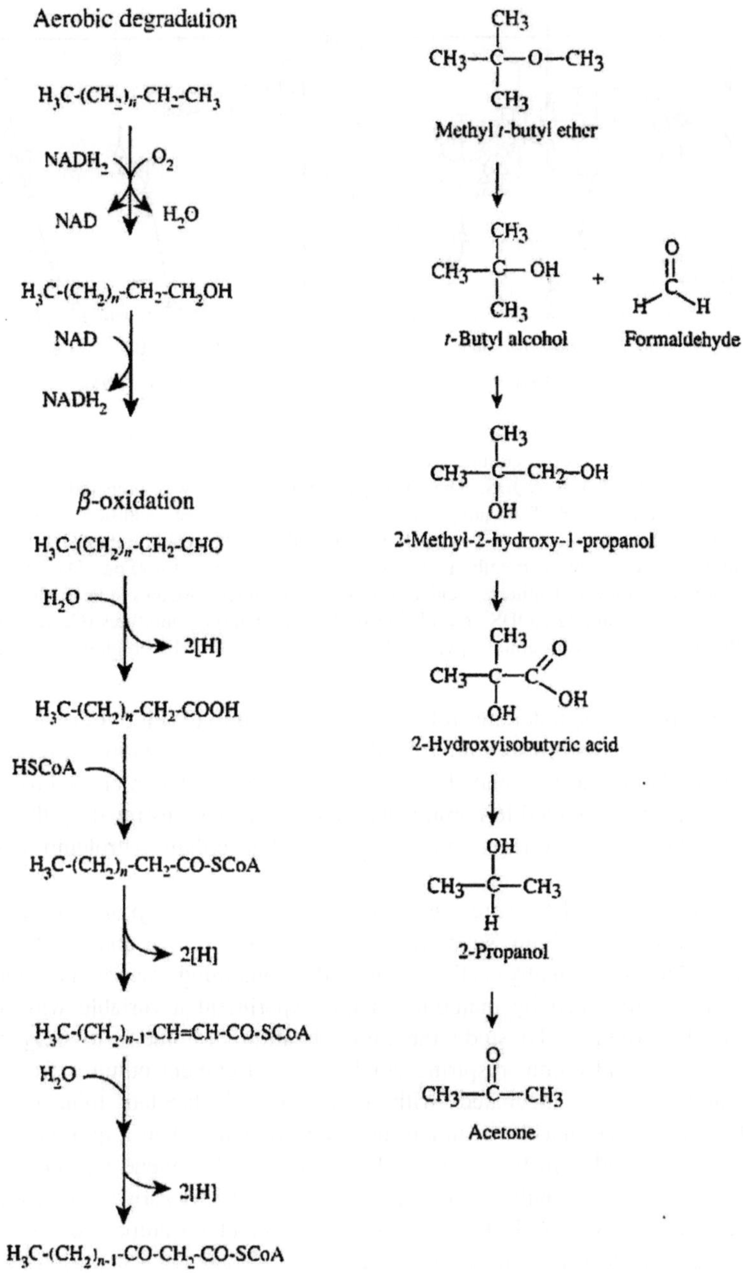

Fig. 16.32 Alkane degradation under aerobic conditions, showing incorporation of oxygen from molecular oxygen into the aliphatic compound, producing a fatty acid. Fatty acids are oxidized further by β-oxidation: [H] indicates reducing equivalents that are either required or formed in the reaction step (Bouwer and Zender 1993)

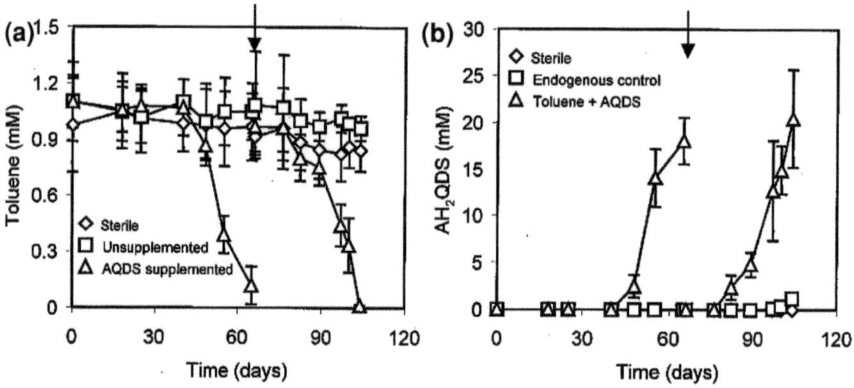

Fig. 16.33 Simultaneous **a** toluene conversion and **b** AQDS reduction by Amsterdam petroleum harbor sediment in anaerobic culture bottles containing bicarbonate-buffered basal medium, supplemented with 25 mM AQDS. The unsupplemented control was prepared in the same manner but without AQDS. The endogenous control (without toluene addition) contained the same amount of hexadecane (0.2 % [vol/vol]) as that used for toluene addition. AQDS reduction was quantified spectrophotometrically as the increase in absorbance at 450 nm. Data are means and standard deviations for triplicate incubations in each treatment. Arrows indicate the addition of fresh medium containing AQDS and toluene in depleted bioassay mixtures (Cervantes et al. 2001). Reprinted with permission. Copyright American Society for Microbiology

degradation pathways under anaerobic conditions tend to develop. Cervantes et al. (2001) tested the possibility of microbially mediated mineralization of toluene by quinones and humus as terminal electron acceptors. Anaerobic microbial oxidation of toluene to CO_2, coupled to humus respiration, was demonstrated by the use of enriched anaerobic sediments (e.g., from the Amsterdam petroleum harbor). Natural humic acids and humic quinone moiety models (anthraquinone-2,-6-disulfonate, denoted AQDS) were used as terminal electron acceptors. Toluene was added to the basal medium, supplemented by HA or AQDS, and incubated for 120 days. The biologically active sediments containing bicarbonate-buffered degradation were tested by including in the experiment a variable with sterile (autoclaved) sediment. To study the mineralization of toluene to CO_2, under anoxic quinone and humus-respiring conditions, enrichment cultures from sediment samples were incubated with uniformly ^{13}C-labeled toluene. From Fig. 16.33A, we see that the sediment degraded toluene when a quinone moiety (AQDS) was included in the medium. In the sediment, toluene was completely eliminated after two months of incubation. Figure 16.33B shows a concomitant reduction of AQDS to AH_2QDS. The same treatment on autoclaved sediment, however, did not lead to toluene reduction.

To validate mineralization of toluene to CO_2 under anoxic quinone and humus-respiring conditions, Cervantes et al. (2001) performed additional experiments using enriched phosphate-buffered basal sediments from Amsterdam petroleum harbor. After two weeks of incubation, 85 % of added ^{13}C-labeled toluene was observed as $^{13}CO_2$. Enriched sediment converted ^{13}C-labeled toluene to $^{13}CO_2$ in

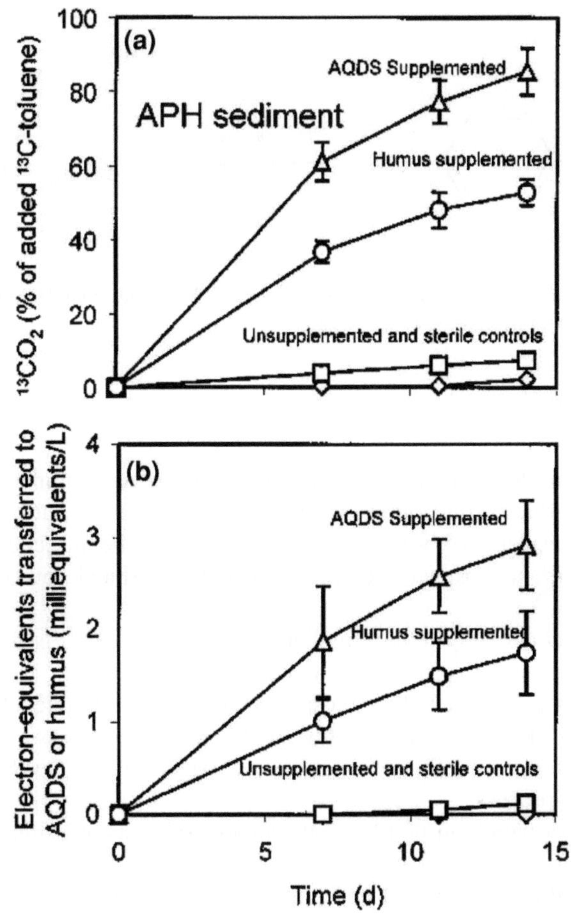

Fig. 16.34 Mineralization of [^{13}C] toluene to $^{13}CO_2$ (**a**) coupled to the reduction of AQDS or humus (**b**) by enriched Amsterdam petroleum harbor (APH) (**a, b**) sediments in anaerobic culture bottles containing phosphate-buffered basal medium supplemented with AQDS (5 mM) or with highly purified soil humic acids (12 g/L). Uniformly ^{13}C-labeled toluene was added at an initial concentration of 100 μM relative to the liquid volume. Unsupplemented controls were prepared in the same manner but without AQDS and humus. All data were corrected relative to the endogenous control (without ^{13}C-labeled toluene addition). Data are means and standard deviations for triplicate incubations in each treatment: *d* denotes days (Cervantes et al. 2001). Reprinted with permission. Copyright American Society for Microbiology

media supplemented with AQDS or with humic acid (Fig. 16.34a). There was negligible recovery of $^{13}CO_2$ in the endogenous and sterile controls. The conversion of ^{13}C-labeled toluene to $^{13}CO_2$ was coupled to an increase in electrons recovered as AH_2QDS or as reduced humus (Fig. 16.34b). However, there was no toluene reduction in autoclaved sediments. These results indicate that humic substances may contribute significantly to the transformation of contaminants under anoxic conditions, by serving as electron acceptors.

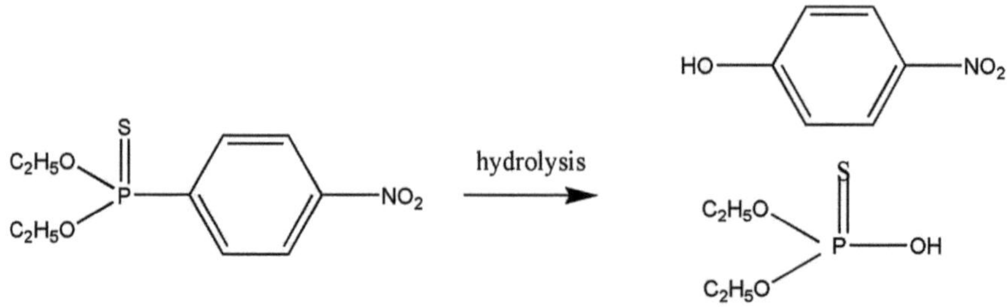

Fig. 16.35 Pathway of parathion degradation in soil. Reprinted from Nelson et al. (1982). Copyright 1982 with permission of Elsevier

16.2.2 Pesticides

Crop protection chemicals are an important group of contaminants that exhibit biologically mediated transformation in aerobic or anaerobic subsurface environments. We consider two well-known contaminants: the insecticide parathion, which is an organophosphate compound, and the herbicide atrazine, from the triazine group.

Parathion (*O*,*O*-diethyl *O*-*p*-nitrophenyl phosphorothioate) is degraded in the near-subsurface aerobic environment via hydrolysis, where two degradation products are observed, diethylthiophosphoric acid and *p*-nitrophenol, according to the schematic pathway described in Fig. 16.35. Abiotic hydrolysis of parathion in the subsurface is a result of a surface-mediated transformation (see Sect. 16.1) or a biodegradation process.

Hydrolysis of parathion in a loessial semiarid soil was investigated by Nelson et al. (1982). They found that *Arthrobacter sp.* hydrolyzed parathion rapidly in sterilized, parathion-treated soil under aerobic conditions (20 % w/w water content). This bacterium was isolated from a silty loam, sierozem soil of loessial semiarid origin (Gilat). It uses parathion or its hydrolysis product, *p*-nitrophenol, as the sole carbon source. However, when parathion hydrolysis causes the amount of *p*-nitrophenol to reach a concentration greater than 1 mM or if the concentration is greater than 1 mM in the case of a single application of *p*-nitrophenol, the hydrolysis product becomes noxious to the bacteria and their growth is inhibited.

In an accompanying laboratory study on the kinetics of biologically induced hydrolysis of parathion, Nelson et al. (1982) found that bacteria populations increased to a maximum four to five days after parathion application, with the increase proportional to the concentration of parathion, followed by a decline. Figure 16.36 shows this behavior in remoistened Gilat soil after application of parathion in amounts of 10–160 μg/g dry soil.

The rate of parathion hydrolysis is independent of the parathion concentration; the rate of formation of the hydrolysis product, *P,* is described by

$$\mathrm{d}P/\mathrm{d}t = 1/Y\,\mathrm{d}m/\mathrm{d}t + A(m - m_o) \tag{16.5}$$

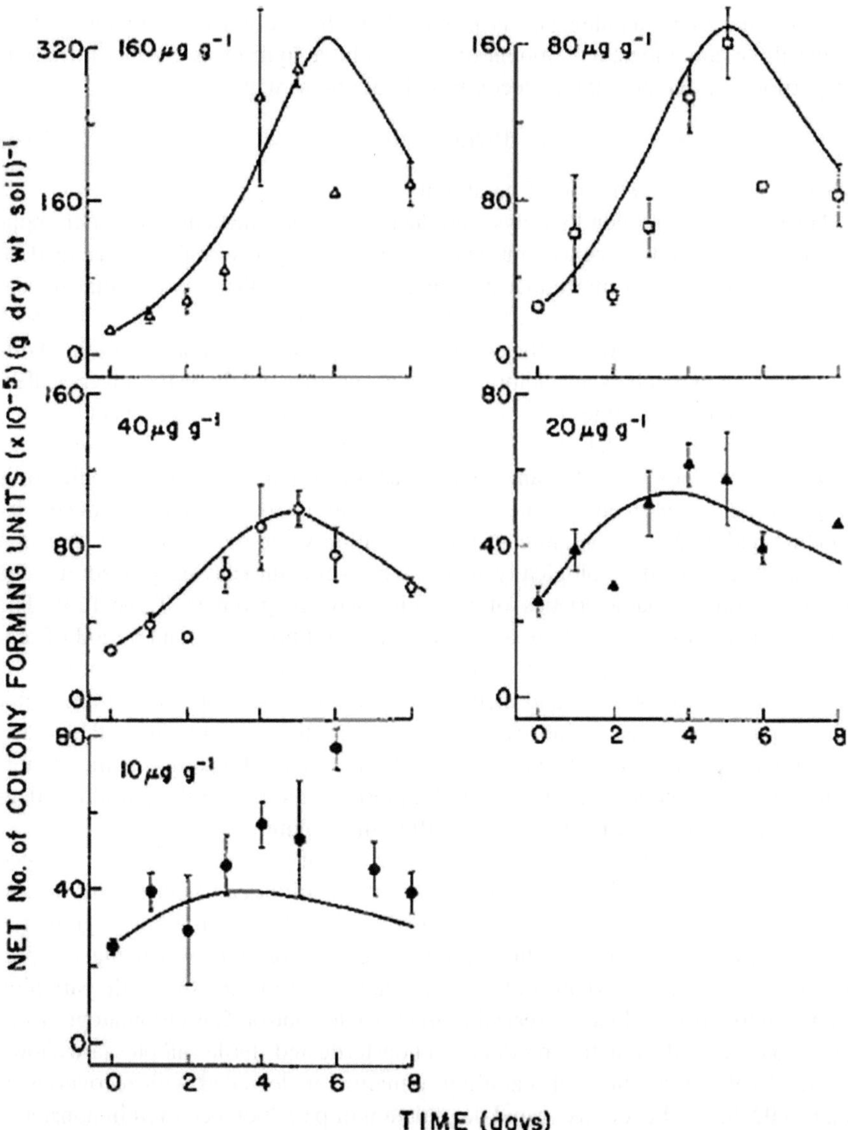

Fig. 16.36 Change in bacteria populations in remoistened Gilat soil after application of 10–160 μg parathion per g dry soil. Plotted points are means of three replicates ± standard error. Continuous curves represent model simulations. Values obtained in control soils to which hexane alone was added have been subtracted. Reprinted from Nelson et al. (1982). Copyright 1982 with permission of Elsevier

where t is time in days, P is concentration of diethylthiophosphate (per μg dry weight soil), m is the number of bacteria (per 10^{-5} g dry weight soil), m_o is the initial value of m, and Y is the yield of bacteria (μg hydrolysis product formed). Some hydrolyzing bacteria are present in the soil initially and the equation assumes that their maintenance is provided for by the existing organic matter. Under these conditions, the amount of water-soluble hydrolysis product formed is proportional to the amount of decomposed parathion, with

$$\mathrm{d}P/\mathrm{d}t = -\mathrm{d}c/\mathrm{d}t \qquad (16.6)$$

where c is the concentration of parathion.

Formation of parathion hydrolysis products in remoistened Gilat soil after application of parathion at concentrations of 10–160 μg/g dry soil is illustrated in Fig. 16.37. These data imply that parathion provides a substance essential for the growth of a portion of the near-surface microbial population, but that these increased numbers can be sustained for only a short period. A possible explanation for the short-lived increase in numbers, in addition to lack of substrate, might be the adverse effect of the parathion metabolite *p*-nitrophenol when it reaches a specific concentration.

When parathion reaches a flooded near-surface site, anaerobic conditions develop. In a study on instantaneous degradation of parathion in anaerobic soils, Wahid et al. (1980) equilibrated parathion with soils previously reduced by flooding and analyzed contaminant transformation when the soils were kept in a natural state or made biologically inactive by a pre-autoclaving procedure. The physicochemical characteristics of the soil used are given in Table 16.9. The extent of parathion degradation, as a function of soil properties and period of pre-flooding, is shown in Fig. 16.38. The soils were first flooded for different periods ranging from 0 to 190 days and then equilibrated with parathion for 30 min, after which parathion degradation was measured. Following flooding, the redox potential dropped (up to −130 mV) at a faster rate in the Pokkali, organic matter-rich, soil. Aminoparathion and desethyl aminoparathion were the parathion degradation products determined in the equilibrating solution.

To ascertain whether instantaneous degradation of parathion in the pre-reduced soils was chemical or biological, parathion was equilibrated with nonsterile and sterile soils pre-reduced by flooding. Biologically mediated parathion degradation was determined by analyzing the equilibrating solution after sterilizing the pre-flooded laterite and Pokkali soils by autoclaving. Soils treated with parathion immediately prior to flooding served as the aerobic control. Gas chromatograms of parathion and its degradation products, in nonsterile and sterile samples, are shown in Fig. 16.39. In aerobic soil, parathion remained undegraded with recoveries as high as 92–95 %. In contrast, rapid degradation of parathion occurred in nonsterile samples of pre-reduced soils. Parathion concentrations declined to 13.8 and 52.1 % in Pokkali and Laterite soils, respectively. Wahid et al. (1980) determined that aminoparathion and desethyl aminoparathion were the parathion degradation products in the equilibrating solution. Both degradation products are found in pre-reduced Pokkali soils, but only aminoparathion appears in the pre-reduced Lateritic soil. Parathion was not degraded in autoclaved samples of pre-reduced samples, proving that anaerobic parathion degradation in a pre-flooded near-surface layer is a biologically mediated process.

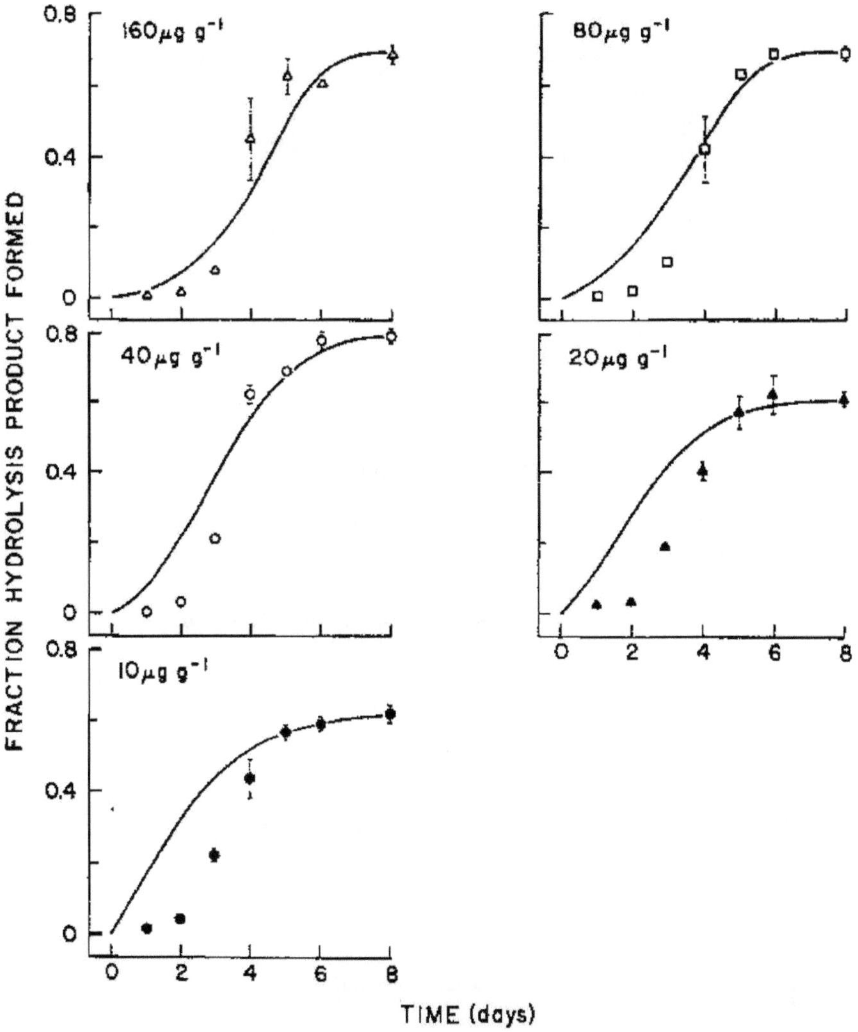

Fig. 16.37 Formation of parathion hydrolysis product (fraction of initial ^{14}C label appearing in the aqueous extract) in remoistened Gilat soil after application of 10–160 μg parathion per g dry soil. Plotted points are means of three replicates ± standard error. Continuous curves represent model simulations. Reprinted from Nelson et al. (1982). Copyright 1982 with permission of Elsevier

Table 16.9 Physicochemical characteristics of soils used in experiments (Wahid et al. 1980)

Soil	pH	Organic matter	Clay	Free Fe
Alluvial (no. 10)	6.2	0.75	15.6	0.83
Laterite (no. 11)	6.3	2.88	23.6	2.40
Pokkali (no. 13)	5.2	5.52	45.6	1.38

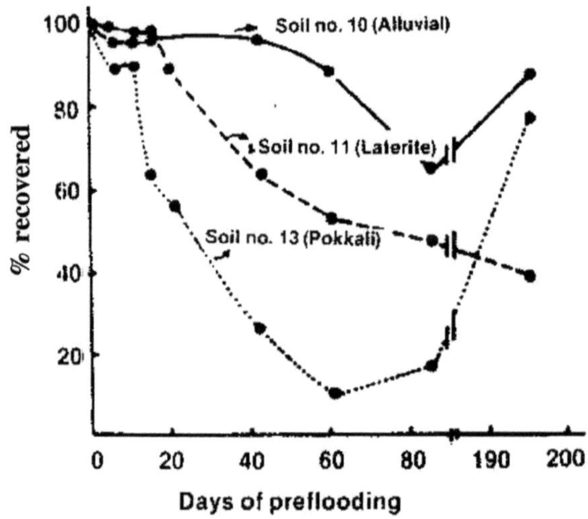

Fig. 16.38 Parathion recovered after 30-min equilibration with soils pre-flooded for different periods (Wahid et al. 1980)

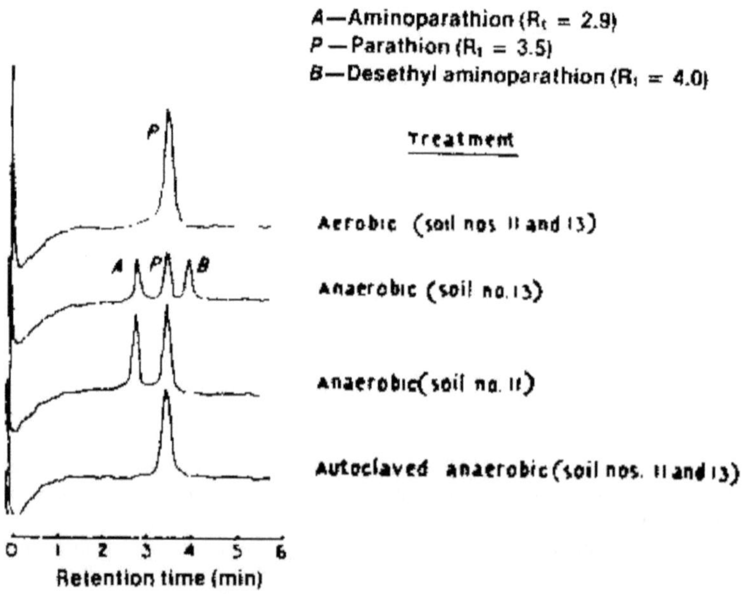

Fig. 16.39 Gas chromatograms of parathion and its products formed in nonsterile aerobic soil and in nonsterile and sterile (autoclaved) samples of pre-reduced (anaerobic) soil after 30-min equilibration (Wahid et al. 1980)

Atrazine (6-chloro-N^2-ethyl-N^4 isopropyl-1,3,5-triazine-2,4-diamine) is a herbicide from the family of *s*-triazines, used frequently for weed control and applied areally or as a point source on the land surface. The abiotic transformation of

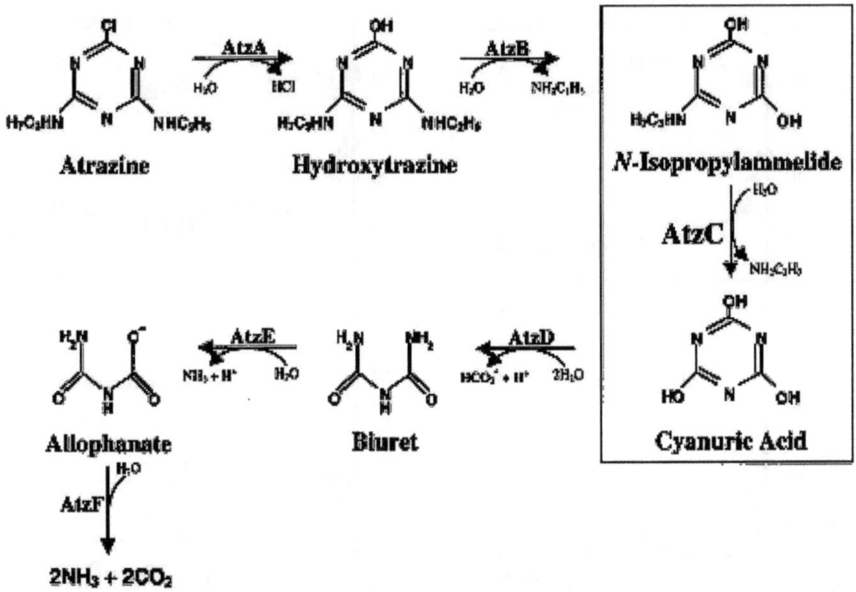

Fig. 16.40 Atrazine catabolic pathway in *Pseudomonas* sp. strain ADP (Shapir et al. 2002). Reprinted with permission. Copyright American Society for Microbiology

atrazine in near-surface environments may result from hydrolytic processes due to high or low pH or from surface-catalyzed hydrolysis as a result of hydrogen bonding between carboxyl groups in organic matter and atrazine ring nitrogen atoms. Changes induced in atrazine metabolism by the *Pseudomonas sp.* strain ADP (Mandelbaum et al. 1995) are given by Shapir et al. (2002) in Fig. 16.40. The primary microbial degradation pathway is *N*-dealkylation of the side chain, to produce deethylatrazine and deisopropylatrazine (chlorohydrolase (AtzA)), followed by a second reaction, where AtzB catalyzes the hydrolysis of hydroxyatrazine to yield izopropylamelide. The third metabolic step utilizes *N*-isopropylammelide isopropylaminohydrolase (AtzC) to hydrolytically remove *N*-isopropylamine and produce cyanuric acid. It is believed that the advent of bacterial atrazine catabolism requires the presence of AtzA, AtzB, and AtzC enzymes (Seffernick and Wackett 2001).

The persistence (half-life) of atrazine in the subsurface is governed by chemically and biologically mediated transformations. Because the solubility of atrazine is relatively high ($\sim$30 mg/L) compared with its toxicity level in water (5 μg/L), atrazine has become a hazard to groundwater quality. Atrazine has been detected in groundwater more than any other crop protection chemical; two examples of atrazine persistence–transformation in aquifer environments are discussed next.

Pang et al. (2005) examined atrazine transformation in groundwater and aquifer materials, considering the following treatments: atrazine with natural and sterilized groundwaters and atrazine with natural and sterilized aquifer sands and groundwaters. The studied systems closely mimicked natural groundwater and aquifer

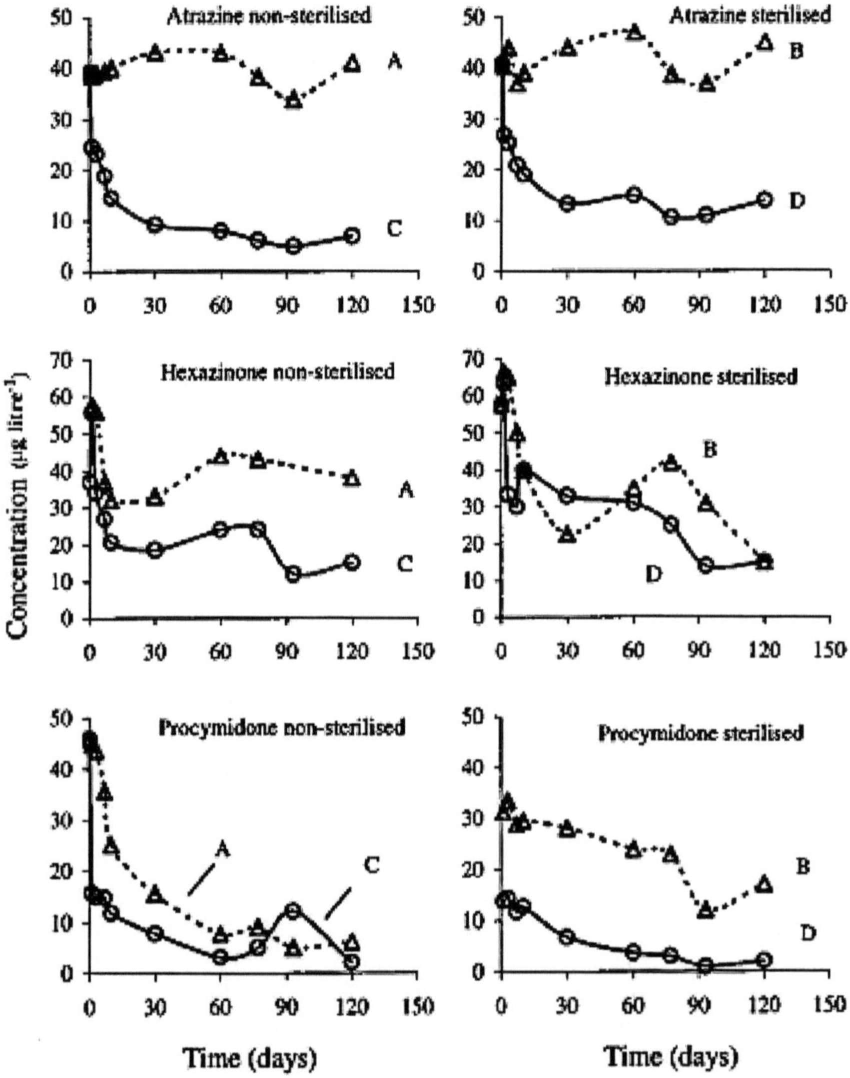

Fig. 16.41 Reduction of pesticide concentrations during degradation experiments: (*open triangle*) groundwater; (*open circle*) groundwater with aquifer sand: *A* = unsterilized groundwater, *B* = sterilized groundwater, *C* = unsterilized groundwater with unsterilized aquifer sand, *D* = sterilized groundwater with sterilized aquifer sand. Reproduced with permission (Pang et al. 2005). Copyright Society of Chemical Industry. Permission granted by John Wiley & Sons Ltd. on behalf of the SCI

environments, the experiments conducted at a flow rate, temperature, and pH level typical in the field. Only chemical degradation of atrazine occurs in sterilized aquifer sands and groundwater, so that differences in behavior between natural and sterilized media represent biologically mediated degradation. Figure 16.41 illustrates the dynamics of atrazine concentration in natural (A, C) and sterilized (B, D) groundwater (A, B) and in groundwater with aquifer sand (C, D), over 120 days. Atrazine concentration does not show any changes during the 120-day period, in either natural or sterilized waters; there is no chemical or biological degradation. The degradation rate of atrazine in the natural groundwater–aquifer sand mixture is greater than that obtained for the sterilized matrix, proving the occurrence of a biologically mediated process.

Pang et al. (2005) assumed that atrazine degradation follows a first-order decay law,

$$c = c_o \exp(-\lambda t) \tag{16.7}$$

where c is the atrazine concentration at time t (mg/L), c_o is the initial atrazine concentration (mg/L), λ is the degradation rate coefficient (d^{-1}), and t is time (d). The calculated chemical and biological degradation rate, λ, and the half-life, $t_{1/2}$ (d), of atrazine and other pesticides in the groundwater and sand mixtures are given in Table 16.10. In this particular case, biologically mediated degradation of atrazine is about 8 % greater than chemical degradation, and this result is reflected in the half-life of the compound.

The degradation of atrazine in saturated aquifer materials was studied under static and recirculating laboratory conditions by Schwab et al. (2006). Aquifer and soil samples were collected from south central Kansas (Hutchinson) and from the floodplain of the Kansas River (Topeka and Ashland). In general, groundwater used in the experiments was collected from the same location and depth as the aquifer solid material. The majority of the aerobic microorganisms from the aquifers were bacteria and actinomycetes, although the population of actinomycetes was 100-fold less than in soils. No information on anaerobic microorganism activities in the aquifers is presented. The contribution of biodegradation to atrazine transformation was determined by comparing natural and sterilized (autoclaved) variables. The experiments were performed temperatures of 15 and 25 °C. The degradation patterns of atrazine incubated in natural and sterilized aquifers are presented in Fig. 16.42.

Atrazine degradation was different for the sterilized and natural aquifer treatments after the first 80–100 days, at which point atrazine disappearance continued in natural aquifers but almost ceased in the sterilized system. This behavior may be explained by a joint chemical and biological effect in the first 100-day period. After 100 days of incubation, chemical processes cease and atrazine is transformed only through a minor biologically mediated process. Circulation of the water through aquifer material resulted in the highest disappearance of atrazine. Under recirculating conditions, no difference between atrazine degradation in sterilized and nonsterilized aquifers was observed, indicating that chemical degradation is the predominant degradation pathway. It is likely that the recirculating pathway extends contact with active sites of the aquifer solid phase, favoring continued

Table 16.10 Degradation rates and half-lives determined from batch incubation experiments at pH 5.33–5.95

Description	Data[a]	Degradation rate, λ (day^{-1})[b]			Half-life, $t_{1/2}$ (day)[b]		
		Alrazine	Hexazinone	Procymidone	Atrazine	Hexazinone	Procymidone
(1) Degradation in the liquid phase							
Total degradation in groundwater	A	0.00E+00 (0.00E+00–1.09E−03)	5.18E−03 (0.00E+00–1.22E−02)	3.32E−02 (2.61E−02–4.37E−02)	Infinite (636-infinite)	134 (56.8-infinite)	20.8 (15.9–27.6)
Chemical degradation in groundwater	B	0.00E+00 (0.00E+00–8.23E−04)	8.95E−03 (3.23E−03 02)–1.60E−02)	5.72E−03 (3.07E−03–8.38E−03)	Infinite (842-infinite)	77 (43.3–215)	121 (82.7–226)
Biodegradation in groundwater	A–B	0.00E+00 (0.00E+00–2.67E−04)	– –	2.75E−02 (2.20E−02–3.53E−02)	Infinite (2596–infinite)	– –	25.2 (19.6-31.5)
(2) Reduction in the mixture of liquid and solid phases							
Total degradation in groundwater and sand	C	1.95E−02 (1.72E−02–2.21E−02)	8.30E−03 (2.67E−03–1.50E−02)	1.28E−02 (8.99E−03–1.72E−02)	36.5 (31.4–40.3)	83.5 (46.2–260)	54.1 (40.3–77.1)
Chemical degradation in groundwater and sand	D	9.26E−03 (7.75E−03–1.18E−02)	5.09E−03 (5.15E−04–9.68E−03)	2.20E−02 (1.84E−02–2.56E−02)	74.8 (64.2–89.4)	196 (71.6–1346)	31.5 (27.1–37.7)
Biodegradation in groundwater and sand	C–D	1.02E−02 (9.45E−03–1.13E−02)	3.21E−03 (2.16E−03–5.32E−03)	– –	67.7 (61.3–73.3)	216 (130–322)	– –

[a] A = nonsterilized groundwater. B = sterilized groundwater. C = nonsterilized groundwater with nonsterilized aquifer sand. D = sterilized groundwater with sterilized aquifer sand. Estimations for C and D were based on date of days 3–120. Sorption onto the glass bottle was not considered so the results of isothermal tests show that sorption of the pesticides onto the glass bottles is negligible

[b] Values in parentheses give the possible range based on analysis errors

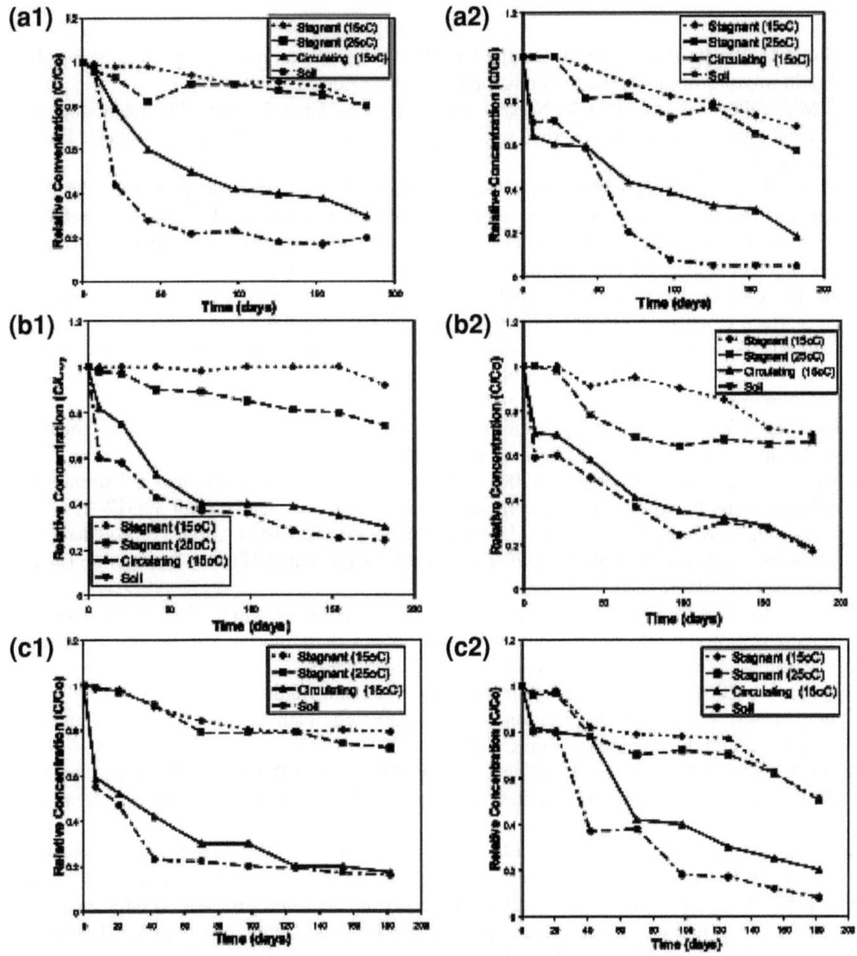

Fig. 16.42 Atrazine degradation in sterilized aquifer material (**a1**, **b1**, **c1**) and in natural (unamended) aquifer material (**a2**, **b2**, **c2**) for reactors under stagnant (15 °C), stagnant (25 °C), and recirculating (15 °C) conditions. Data points are the mean of three replications. The least significant difference between these treatments at the $p < 0.05$ level is 0.075. Figures are identified as **a** Topeka, **b** Ashland, and **c** Hutchinson aquifers (Schwab et al. 2006)

chemically induced degradation. Comparing atrazine degradation behavior in aquifers with those in the near surface, Schwab et al. (2006) show that atrazine degrades faster in soil than in aquifer material; they attribute this behavior to a higher biological population and a higher carbon content in the soil.

References

Abriola LM, Bradford SA, Lang J, Gaither CL (2004) Volatilization of binary nonaqueous phase liquid mixture in unsaturated porous media. Vadose Zone J 3:645–655

Ahn MY, Filley TR, Jafvert CT, Nies L, Hua I, Bezares-Cruz J (2006) Photodegradation of decabromodiphenyl ether adsorbed onto clay minerals, metal oxides, and sediment. Environ Sci Technol 40:215–220

Bachman J, Patterson H (1999) Photodecomposition of the carbamate pesticide caebyfiran: kinetics and the influence of dissolved organic matter. Environ Sci Technol 33:874–881

Bennett B, Bowler BFJ, Larter SR (1996) Rapid methods for the determination of C_0–C_3 alkylophenols in crude oils and water. Annal Chem 68:3697–3702

Bennett B, Larter SR (1997) Partition behaviour of alkylphenols in crude oil brine systems under subsurface conditions. Geochim Cosmochim Acta 61:4393–4402

Boreen AL, Arnold WA, McNeil K (2005) Triplet-sensitized photodegradation of sulfa drugs containing six-membered heterocyclic groups: identification of an SO_2 extrusion photoproduct. Environ Sci Technol 39:3630–3638

Bouwer EJ, Zehnder A (1993) Bioremediation of organic compounds—putting microbial metabolism to work. Bioremediation 11:360–367

Broholm K, Feenstra S, Cherry JA (2005) solvent release into sandy aquifer. 2. Estimation of DNAPL mass based on multiple-component theory. Environ Sci Technol 39:317–324

Catalano JG, McKinley JP, Zachara JM, Heald SM, Smith SC, Brown GE Jr (2006) Changes in uranium speciation through a depth sequence of contaminated Hanford sediments. Environ Sci Technol 40:2517–2524

Cervantes FJ, Dijksma W, Duong-Dac T, Ivanova A, Lettinga G, Field JA (2001) Anaerobic mineralization of toluene by enriched sediments with quinones and humus as terminal electron acceptors. Appl Environ Microbiol 67:4471–4478

Choppin GR (2003) Actinide speciation in the environment. Radiochim Acta 91:645–649

Claus P (1998) Selective hydrogenation of α,β-unsaturated aldehydes and other C=O and C=C bonds containing compounds. Top Catal 5:51–62

Colon D, Weber EJ, Anderson JL, Winget P, Suarez LA (2006) Reduction of nitrobenzenes and N-hydroxylanilines by Fe(II) species: Elucidation of the reaction mechanism. Environ Sci Technol 40:4449–4454

Cozzarelli IM, Baher LM (2003) Volatile fuel hydrocarbons and MTBE in the environment. In: Sherwood Lollar B (ed) Treatise on geochemistry, vol 9. Elsevier, Amsterdam, pp 433–474

Dror I, Gerstl Z, Prost R, Yaron B (2000) Behavior of neat and enriched volatile petroleum hydrocarbons mixture in subsurface during leaching. Land Contam Reclam 8:341–348

Dror I, Gerstl Z, Yaron B (2001) A field experiment on redistribution and dissipation of residual kerosene in soil subsurface. J Contam Hydrol 48:305–323

Dror I, Prost R, Yaron B (2002) Abiotic behavior of entrapped petroleum products in the subsurface during leaching. Chemosphere 49:1375–1388

Eary LE, Rai D (1987) Kinetics of chromium(III) oxidation to chromium(vi) by reaction with manganese-dioxide. Environ Sci Technol 21:1187–1193

Erbs M, Hansen HCB, Olsen CE (1999) Reductive dechlorination of carbon tetrachloride using iron(II) iron(III) hydroxide sulfate (green rust). Environ Sci Technol 33:307–311

Ezra S, Feinstein S, Bilkis I, Adar E, Ganor J (2005) Chemical transformation of 3-bromo-2, 2-bis(bromomethyl)propanol under basic conditions. Environ Sci Technol 39:505–512

Fine P, Yaron B (1993) Outdoor experiments on enhanced volatilization by venting on kerosene components from soils. J Contam Hydrol 12:355–374

Fioressi S, Arce R (2005) Photochemical transformation of benzo(e)pyrene in solution and adsorbed on silica gel and alumina surfaces. Environ Sci Technol 39:3646–3655

Galin T, Gerstl Z, Yaron B (1990) Soil pollution by petroleum products: Kerosene stability in soil column as affected by volatilization. J Contam Hydrol 5:375–385

Gerstl Z, Galin T, Yaron B (1994) Mass flow of volatile organic liquid mixture in soils. J Environ Qual 23:487–493
Gerstl Z, Yaron B (1978) Behavior of parathion on attapulgite. In: Banin A (ed) Agrochemicals in soils. Springer, Heidelberg
Gibson DT, Subramanian V (1984) Microbial degradation of aromatic hydrocarbons. In: Gibson DT (ed) Microbial degradation of organic compounds, vol 13. Marcel Dekker, New York, pp 181–252
Guo XF, Jans U (2006) Kinetics and mechanism of the degradation of methyl parathion in aqueous hydrogen sulfide solution: investigation of natural organic matter effects. Environ Sci Technol 40:900–906
Haderlein SB, Pecher K (1988) Pollutant reduction in heterogeneous Fe(II)/Fe(III) systems. In: Sparks DL, GrundlT (eds) Kinetics and mechanisms of reactions at the mineral/water interface, vol 715. ACS symposium series. Washington, DC, pp 342–357
Hwang I, Batchelor B (2000) Reductive dechlorination of tetrachloroethylene by Fe(II) in cement slurries. Environ Sci Technol 34:5017–5022
Jarsjo J, Destouni G, Yaron B (1994) Retention and volatilization of kerosene: laboratory experiments on glacial and post glacial soils. J Contam Hydrol 17:167–185
Jarsjo J, Destouni G, Yaron B (1997) On the relation between viscosity and hydraulic conductivity values for volatile organic liquid mixtures in soils. J Contam Hydrol 25:113–127
Jensen JL, Hashtroudi H (1976) Base-catalyzed hydration of α,β-unsaturated ketones. J Organic Chem 41:3299–3302
Karthikeyan KG, Chorover J (2000) Effects of solution chemistry on the oxidative transformation of 1-naphtol and its complexation with humic acid. Environ Sci Technol 34:2939–2946
Khwaja AR, Bloom PR, Brezonik PL (2006) Binding constants of divalent mercury in soil humic acids and soil organic matter. Environ Sci Technol 40:844–849
Klupinski TP, Chin YP, Traina SJ (2004) Abiotic degradation of pentachloronitrobenzene by Fe(II): reactions on goethite and iron oxide nanoparticles. Environ Sci Technol 38:4353–4360
Kozuh N, Stupar J, Gorenc B (2000) Reduction and oxidation processes of chromium in soils. Environ Sci Technol 34:112–119
Mandelbaum RT, Allan DI, Wackett LP (1995) Isolation and characterization of a *Pseudomonas sp.* that mineralizes the s-triazine herbicide atrazine. Appl Environ Microbiol 61:1451–1457
Mingelgrin U, Saltzman S (1977) Surface reactions of parathion on clays. Clays Clay Miner 27:72–78
Nelson LM, Yaron B, Nye PH (1982) Biologically-induced hydrolysis of parathion in soil. Soil Biol Biochem 14:223–227
Oh SY, Lee J, Cha DK, Chiu PC (2006) Reduction of acrolein by elemental iron: kinetics, pH effect, and detoxification. Environ Sci Technol 40:2765–2770
Oscarson DW, Huang PM, Liaw WK, Hammer UT (1983) Kinetics of oxidation of arsenite by various manganese dioxides. Soil Sci Soc Am J 47:644–648
Pang LP, Close M, Flintoft M (2005) Degradation and sorption of atrazine, hexazinone and procymidone in coastal sand aquifer media. Pest Man Sci 61:133–143
Pecher K, Haderline SB, Schwarzenbach RP (2002) Reduction of polyhalogenated methanes by surface-bound Fe(II) in aqueous suspensions of iron oxides. Environ Sci Technol 36:1734–1741
Penrose WR, Metta DN, Hylko JM, Rinkel LA (1987) The reduction of plutonium (V) by aquatic sediments. J Environ Radioact 5:169–184
Powell BA, Duff MC, Kaplan DI, Field RA, Newville M, Hunter BD, Bertsch PM, Coates JT, Serkiz SM, Sutton RS, Triay IR, Vaniman DT (2006) Plutonium oxidation and subsequent reduction by Mn(IV) minerals in Yucca Mountain tuff. Environ Sci Technol 40:3508–3514
Power LE, Arai Y, Sparks DL (2005) Zinc adsorption effect on arsenite oxidation kinetics at the birnessite water interface. Environ Sci Technol 39:181–187
Reineke W (2001) Aerobic and anaerobic biodegradation potentials of microorganisms. In: Beek B (ed) The handbook of environmental chemistry, vol 2. Springer, Berlin

Rivett MO, Feenstra S (2005) Dissolution of an emplaced source of DNAPL in a natural aquifer setting. Environ Sci Technol 39:447–455
Saltzman S, Mingelgrin U (1978) Montmorillonite parathion interactions in aqueous suspensions as affected by mode of preparation. In: Banin A (ed) Agrochemicals in soils. Springer, Berlin
Saltzman S, Mingelgrin U, Yaron B (1974) The surface catalyzed hydrolysis of parathion on kaolinite. Soil Sci Soc Am Proc 38:231–234
Schwab AP, Splichal PA, Banks MK (2006) Persistence of atrazine and alachlor in ground water aquifers and soil. Water Air Soil Poll 171:203–235
Seffernick JL, Wackett LP (2001) Rapid evaluation of bacterial catabolic: a case study with atrazine chlorohydrolase. Biochemistry 40:12747–12753
Shapir N, Osborne JP, Johnson G, Sadowsky MJ, Wackett LP (2002) Purification, substrate range, and metal center of AtzC: the n-isopropylammelide aminohydrolase involved in bacterial atrazine metabolism. J Bacteriol 184:5376–5384
Smith MR (1990) The biodegradation of aromatic hydrocarbons by bacteria. Biodegradation 1:191–206
Tournassat C, Charlet L, Bosbach D, Manceau A (2002) Arsenic (III) oxidation by birnessite and precipitation of manganese (II) arsenate. Environ Sci Technol 36:493–500
Wahid PA, Ramakrishna C, Sethunathan N (1980) Instantaneous degradation of parathion in anaerobic soils. J Environ Qual 9:127–130
Yaron B (1975) Chemical conversion of parathion on soil surface. Soil Sci Soc Am Proc 39:639–643
Yaron B (1978) Some aspects of surface interactions of clays with organophosphorus pesticides. Soil Sci 125:210–216
Yaron B (1979) Chemical conversion of pesticides in the soil medium: the organophosphates. In: Geissbuhler H (ed) Advances in pesticide science. Pergamon Press, Oxford, pp 577–584
Yaron B, Dror I, Graber E, Jarsjo J, Fine P, Gerstl Z (1998) Behavior of volatile organic liquid mixtures in the soil environment. In: Rubin H, Narkis N, Carberry J (eds) Soil and aquifer pollution. Springer, Berlin
Yaron B, Saltzman S (1978) Soil–parathion surface interactions. Residue Rev 69:1–34

Part VI
Contaminant Induced Irreversible Changes in the Soil–Subsurface Regime

Current studies on contaminant interactions in the soil–subsurface region focus mainly on contaminant retention, persistence and transport, and on potential remediation of polluted geological media. Soil–subsurface region properties altered by human intervention are often considered as deviations from a normal geochemical environment, which will disappear via natural processes or following specific remediation procedures. When writing the first edition of this book (2008), during a comprehensive literature review, we observed that in addition to their environmental effects, contaminants may lead—under specific conditions—to irreversible changes in soil–subsurface properties. We subsequently developed this recognition in a number of publications (Yaron et al. 2008, 2009, 2010, 2012). In this second edition of the book, we include Part VI, which discusses irreversible changes that occur in groundwater and in the matrix properties of the vadose zone, due to anthropogenic chemical contamination. Here, we use the term "contaminants of anthropogenic origin" to refer to inorganic and organic compounds that reach the subsurface region during agricultural and soil remediation management, and to industrial and municipal residues that are accidentally released or deliberately disposed of on the land surface (Chaps. 3 and 4).

The literature contains a broad spectrum of research results on contaminant-subsurface interactions. The main emphasis is on contaminant fate as affected by the solid and water phases of the vadose zone. There are, however, specific anthropogenic contamination impacts which may lead to long-term changes in the natural properties of the subsurface geochemical system. These changes are often resistant to remediation procedures and natural attenuation, and are thus irreversible at least on a human lifetime scale. While natural processes that affect the genesis of subsurface formation factors act slowly, over a prolonged "geological" timescale, chemical contaminants may rapidly change the natural soil–subsurface regime. For this reason, chemical contamination may be considered as an additional factor in contemporary soil formation.

In support of this conceptual approach, we select from the literature research results that illustrate various aspects of contaminant-induced, irreversible changes in soil–subsurface solid and aqueous phases.

Chapter 17
Contaminant-Induced Irreversible Changes in Groundwater Chemistry

Chemical contaminants disposed of on the land surface may be transported to the groundwater, interacting with the solid phase or largely bypassing the vadose zone by migrating rapidly through preferential flow paths. The rate and magnitude of transport in preferential flow paths are governed by the properties of the solid phase and the contaminants themselves.

The background chemistry of groundwater is defined by the properties of the aquifer solid phase and by natural processes that occur in the groundwater environment. Groundwater chemistry may, however, be altered by chemicals originating from anthropogenic activities, via areal source (diffusive) or point source contamination. Because many groundwater aquifers are open systems, the extent and persistence of groundwater contamination are governed largely by the low rate of groundwater flow; the slow movement of contaminants results in relatively low mixing rates and dilution. Slow groundwater flow and contaminant mixing rates allow contamination to remain localized over long periods of time (Pye and Patrick 1983), while the effect of biological processes on natural attenuation of contaminants decreases (Hemme et al. 2010).

A review of groundwater remediation efforts in the United States, using various techniques, shows that of 1,700 polluted sites on the US EPA's national priority list, only 363 have reached drinking water quality (Ehlers and Kavanaugh 2013). These authors also quote a US National Research Council Committee's report on contaminated groundwater sites, which notes that persistent, significant limitations will prevent achievement of drinking water standards in many contaminated aquifers over a time scale of at least 50–100 years. As a consequence, the composition and concentration of groundwater contaminants often remain close to their initial state. Under such conditions, we can consider groundwater contamination as irreversible at least on a human lifetime scale.

In this chapter, we present examples of several specific irreversible changes in the chemistry of groundwater which result from anthropogenically induced contamination.

B. Berkowitz et al., *Contaminant Geochemistry*,
DOI: 10.1007/978-3-642-54777-5_17,

17.1 Groundwater Salinization

In their overview of global saline groundwater, van Weert et al. (2009) consider terrestrial anthropogenic origins to be one of the main factors of occurrence and genesis, identifying irrigation, coastal seawater intrusion, and waste effluent disposal as the main anthropogenic drivers affecting groundwater salinization.

Groundwater salinization occurs by diffuse contamination when saline or sewage water is used for irrigation and by anthropogenically induced intrusion of seawater into coastal aquifers. Additionally, local (point) source groundwater salinization may result from waste disposal on land surface, municipal disinfection of drinking and swimming pools water, and in cold areas by large amounts of salt spread on roads to melt ice covering. Groundwater salinization may also occur from nonirrigated surfaces, when an increase in recharge due to agricultural practices enhances leaching of salts from the unsaturated zone into groundwater (Suarez 1989). In general, due to the magnitude of the process, saline contamination of an aquifer is irreversible at least on a human lifetime scale.

17.1.1 Salinization Under Downward Flow Conditions

Changes in aquifer salinity occur when the electrolyte concentration and composition of water transported from the land surface into an aquifer are different from those of the groundwater. The salinity is controlled both by the quality of the water reaching the land surface and by the amount of salt dissolved from the vadose zone solid phase. Under downward flow conditions, gravitational and viscous forces determine mixing processes and control the pattern of groundwater salinization. For concentrated salt solutions, high density and viscosity are also key parameters. As a consequence, downward displacement of a concentrated salt solution during irrigation by freshwater and freshwater displacement by brines are controlled by a dispersive process defined by the gravity number and mobility ratio (Flowers and Hunt 2007).

A typical example of irreversible change in groundwater salinity, following irrigation with water having a salt concentration higher than that of groundwater, occurred in the Coastal Plain aquifer of Israel. During the years 1940–1950, before the development of intensive irrigated agriculture, the chloride concentration of the Coastal Plain aquifer in Israel was less than 100 mg/L; after 30 years of irrigation, a chloride concentration of 180 mg/L was reached. Mercado et al. (1975) considered the period from 1940–1950 as a hydrologically undisturbed period, where the major causes of groundwater salinity were chloride in rainwater, dry fallout, and leakage from more saline aquifers in the vicinity of the Coastal Plain. In this period, the Coastal Plain aquifer drained to the sea and a natural equilibrium of chloride concentration was achieved. A first cause of groundwater salinization was the increase in aquifer exploitation by intensive irrigated

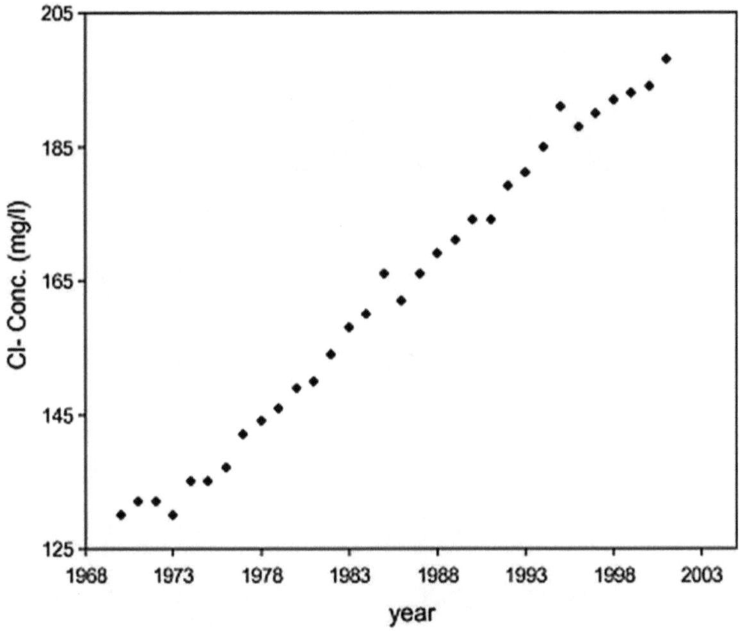

Fig. 17.1 Changes in chloride concentrations in the Israeli coastal aquifer (Gvirtzman 2002)

agriculture; natural seaward flow was reversed, creating conditions of a closed basin with no outlet to the Mediterranean Sea. In the 1960s, an additional source of irrigation water was from the Sea of Galilee, with a chloride concentration of 210–250 mg/L. While the average increase in groundwater chloride concentration between the years 1950–1960 was about 1.5 mg/L/year, the annual chloride input after 1970, using the new source of irrigation water, reached 2.0 mg/L/year. Since the early 1990s, treated sewage water with a chloride content ranging between 190 and 330 mg/L partially replaced the Sea of Galilee irrigation water. Over a 30-year period, and despite the fact that the aquifer salt concentration was diluted by an average rainfall precipitation in the area ranging from 350 to 500 mm/year, the chloride concentration of the coastal aquifer increased from 120 to 180 mg/L (Fig. 17.1).

Many other locations around the world are experiencing irreversible salinization of groundwater following irrigation. For example, in a comprehensive review on the impact of agricultural practices on groundwater salinity, Suarez (1989) notes that irrigation practices have led to groundwater salinization in the San Joaquin Valley, upper Rio Grande, and Maricopa County (USA).

Groundwater salinization under downward flow of rainwater may also occur from nonirrigated agricultural practices that cause natural leaching. Ferguson and Bateridge (1982) determined the balance of salts in cultivated and native glacial till soils from north-central Montana (USA), estimating that after 50 years of

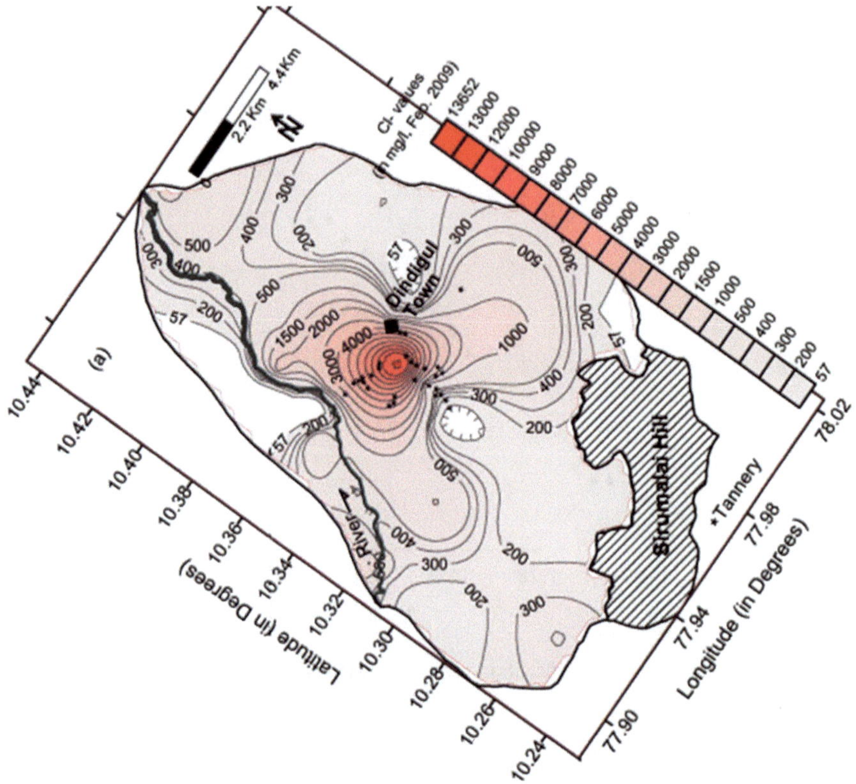

Fig. 17.2 Variation and the distribution of chloride in groundwater under a tannery belt (Mondal and Singh 2011). Reprinted from Mondal and Singh (2011). Copyright (2011) with permission from Elsevier

crop-fallow farming, about 90 t/ha salts had moved under natural precipitation toward the groundwater. A similar effect of farming on salt accumulation in groundwater was observed in Australia, where the chloride concentration beneath a forest catchment was 70–400 mg/L, while that under a cleared farmland was 160–3,000 mg/L (Peck and Hurle 1973).

Downward transport of salts from land surface to groundwater may be observed following waste effluent disposal. An example is the case of groundwater salinization under the disposal of untreated effluents from 80 functioning tannery plants forming a tannery belt in southern India. Mondal and Singh (2011) determined the chloride distribution in groundwater in 29 locations below the tannery belt (Fig. 17.2). Abnormal concentrations of chloride ranging from 57 to 13,652 mg/L were reported.

Flooding of a nonsaline area by saline waters is another aspect of groundwater salinization under downward flow. An example of this salinization pathway is given by McCarthy (2006), who reported increased groundwater salinity following flooding of the Okavango River in the Okavango Delta (Botswana, South Africa).

17.1.2 Salinization Under Lateral Flow Conditions

Groundwater and seawater are primary components of the hydrological system of a coastal aquifer. Seawater intrusion, induced by over-exploitation (over-pumping) of coastal groundwater resources as a result of increasing demand for freshwater, occurs with changes in the pressure balance between groundwater and seawater (Werner et al. 2012 and references therein). Seawater intrusion is expressed by lateral flow into a freshwater aquifer and involves the simultaneous transport of water and dissolved salts; spatiotemporal changes in salinity cause density differences that influence the flow pattern.

Seawater intrusion is a key process that can lead to irreversible changes in coastal aquifer chemistry. The total salt concentration and the ionic composition in the aquifer change as a result of seawater intrusion and may degrade freshwater quality to a point that renders it unportable or environmentally harmful. Moreover, once salts enter an aquifer, they can advance further inland by dispersive and diffusive mixing. We discuss below several examples of lateral saltwater intrusion into a coastal aquifer, selected from different environmental settings.

An example of groundwater salinization by seawater intrusion is given by El Moujabber et al. (2006), who examined a site situated south of Mount Lebanon (Choueifat area) on the Mediterranean coast of Lebanon. The analysis was based on a 3-year, monthly monitoring of six wells, in which groundwater samples were collected and analyzed for chemical content (salt content and composition), and compared to the seawater chemistry. Mediterranean seawater has a salinity of about 40 g/L and exhibits a predominance of Cl^- and Na^+. Chlorides (Cl^-) are in excess of alkali ions (Na^+ and K^+), and Mg^{2+} is greatly in excess of Ca^{2+}, as well as of SO_4^{2-} and HCO_3^-. In contrast, continental groundwater is rich in HCO_3^-, SO_4^{2-}, and Cl^-, with Ca^{2+} the predominant cation accompanied by Mg^{2+}. In general, the alkali ions Na^+ and K^+ are present in smaller amounts.

The measured variation in cation and anion ratios is illustrated in Fig. 17.3. Overall, the curves indicate seasonal variation with highest ratios during the summer months, as a result of intensive groundwater exploitation, and a relative steady state observed during the rainy period, characterized by natural aquifer recharge and lack of need for pumping of irrigation water. However, this steady state period during the winter did not minimize groundwater salinization. More specifically, disregarding the summer months, the Na^+/Ca^{2+} curve reveals a slowly increasing trend of seawater salt penetration into the aquifer over the years. The Mg^{2+}/Ca^{2+} curve shows an overall increase in Mg^{2+}, the SO_4^{2-}/Cl^- curve indicates a SO_4^{2-} decrease coupled with an increase in Cl^-, and the HCO_3^-/Cl^- curve indicates a drop in the concentration of HCO_3^- (contrary to the Cl^- concentration).

Seawater intrusion due to freshwater over-pumping in a different climatic environment is reported by Barlow and Reichard (2010). In a review of saltwater intrusion in coastal regions of North America, these authors examine the case of Cape May County (New Jersey, USA) peninsula. The freshwater aquifer in this peninsula is bounded laterally by saline groundwater, affected by the Atlantic

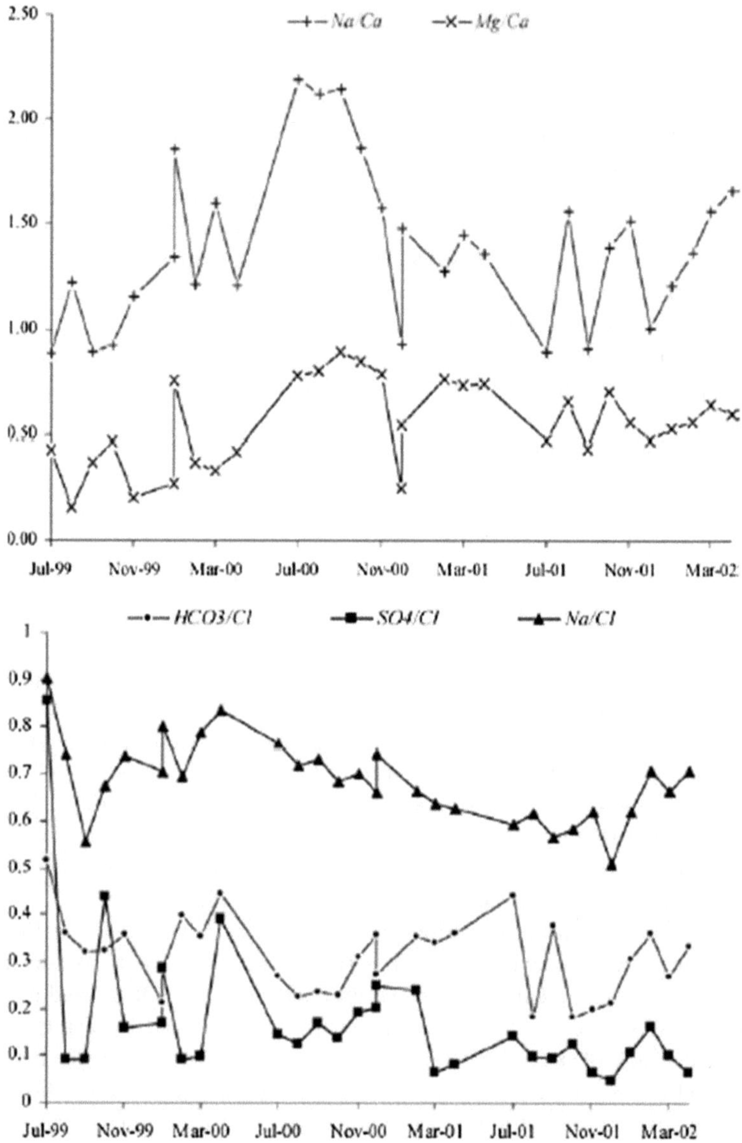

Fig. 17.3 Variation in groundwater ionic ratios from the coastal Choueifat region, south of Beirut, during the period 1999–2002 (El Moujabber et al. 2006)

Ocean and Delaware Bay. Lateral encroachment of ocean saltwater into the confined aquifer occurred as a result of over-exploitation, which began more than 100 years ago. Prior to excessive exploitation, aquifer water levels were above sea level and groundwater flowed to inland, low-lying surface waters. Saltwater intrusion began in 1890 after the start of deep well pumping and the decline of

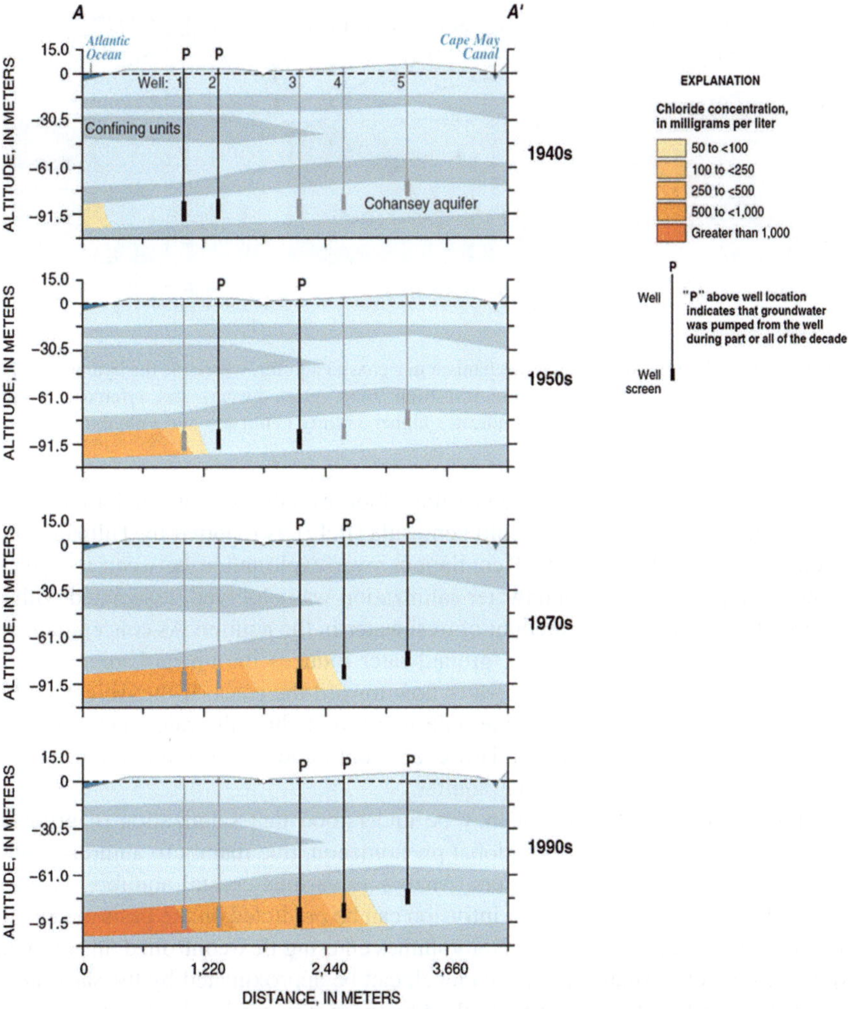

Fig. 17.4 Hydrogeological cross sections showing lateral encroachment of saltwater and salinization of supply wells in the Cohansey aquifer at the Cape May County (New Jersey) well field, 1940s to 1990s (Barlow and Reichard 2010; modified from Lacombe and Carleton 2002)

groundwater levels below sea level (Lacombe and Carleton 2002). After years of continuous exploitation, groundwater levels dropped by up to 30 m, favoring lateral saltwater intrusion into the aquifer. Hydrogeological cross sections showing lateral intrusion of saltwater in the Cohansey aquifer, during 50 years of exploitation, are shown in Fig. 17.4. It may be observed that pumping enhanced saline water intrusion and groundwater salinization in four of the five wells surveyed; aquifer chloride concentrations increased over time, with the deeper aquifer exhibiting the highest chloride concentration.

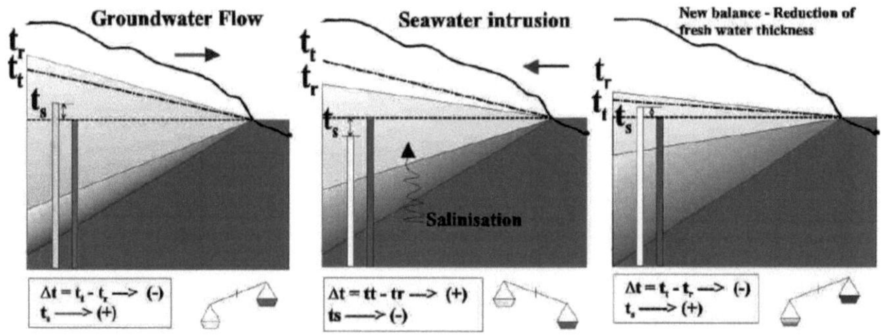

Fig. 17.5 Sketch of the salinization mechanism in a coastal aquifer, where t is the hydraulic head and the subscripts t, s, and r denote transition zone, groundwater, and reference depth, respectively; color code: *darker gray* indicates higher salinity (Tulipano and Fidelibus 2002)

Another example of groundwater salinization by lateral seawater intrusion, in the coastal karstic aquifer of Salento peninsula (Italy), is reported by Tulipano and Fidelibus (2002). As a result of continuous over-exploitation between the years 1974–1996, progressive groundwater salinization was observed, associated with a discernible reduction in the amount of freshwater in the aquifer. As conceptualized in Fig. 17.5, over-exploitation of groundwater induces the upward migration of groundwater and prevents groundwater flow toward the coast. If the weight of the freshwater column reaches a value able to prevent the salt water intrusion, the system may reach equilibrium. However, continuous over-exploitation favors extensive salinization of a coastal aquifer.

From the above examples, it may be understood that salinization of groundwater by seawater intrusion is a global phenomenon, due mainly to anthropogenic over-exploitation of coastal aquifers. Analyzing various coastal aquifers, Werner et al. (2012) suggest that seawater intrusion can be predicted on the basis of several vulnerability indicators: (1) Seawater volumes entering flux-controlled, unconfined aquifers, subject to small rises in sea level, can be approximated by the saltwater–freshwater interface toe length multiplied by the rise in sea level; (2) sea level rise in head-controlled, unconfined aquifers produces higher rates of seawater intrusion (toe movements and seawater volume changes) compared to flux-controlled unconfined aquifers; (3) in unconfined aquifers, flux-controlled boundary conditions produce a larger toe response to changes in recharge than head-controlled boundary conditions; (4) sea level rise is inconsequential in confined aquifers that are flux-controlled. The authors note that in addition to the anthropogenic factor of coastal aquifer over-exploitation, coastal aquifer vulnerability is subject to the threat of sea level rise from climate change.

Under the conditions described in Sects. 17.1.1 and 17.1.2, is groundwater salinization an irreversible process? Theoretically, the level of a freshwater aquifer may increase following a rainy season or by infiltrating the land surface with additional excessive amounts of freshwater. As a result, in coastal aquifers, the higher water table could be expected to drive flow seaward and lead to reversibility of the saltwater intrusion process. However, in reality, excessive pumping occurs mostly in arid or semi-arid areas characterized by little or no rain, so that aquifer restoration is unlikely. In these cases, therefore, we can characterize seawater intrusion and other salinization processes, with associated changes in the groundwater chemistry, as irreversible.

17.2 Groundwater Acidification

Acidification of groundwater is the result of diffuse anthropogenic sources, such as atmospheric industrial emissions or agricultural practices, and of point anthropogenic sources, such as leakage from municipal or industrial wastes. Knutsson (1994 and references therein) provides a short historical review of human-induced atmospheric emissions. As early as the 1730s, Carl von Linne reported damage to vegetation around a copper mine and smelter at Falun, Sweden. The expression "acid rain" (due to emissions of, e.g., sulfur dioxide and nitrogen oxides) was used for the first time by R. A. Smith in 1872 with regard to the acidification of industrial areas in England. Systematic monitoring of airborne substances deposition following acid rains began in Sweden in the middle of the last century; this culminated with Oden (1968), who showed that acid rain deposition over Scandinavia originated mainly from emissions in central and western Europe and the British Isles. Subsequent studies on acid precipitation risk on soil and water environment carried out in North America and Europe showed that the SO_4^{2-} emissions were reduced (e.g., Ulrich 1985; Likens and Bormann 1995). However, accumulated storage of acidity in soil, as found for example in Germany (Alewell et al. 2001), indicates that groundwater acidification remains a long-term threat.

As a result of acid rain, irreversible changes in groundwater chemistry may occur. Acid rain depletes the acid neutralizing capacity of the subsurface environment and may result in groundwater with low pH and enhanced metal concentrations. A few examples follow below. Discussion of acid rain in the context of changes in soil properties appears in Sect. 18.3.1.

Jacks et al. (1984) reported a tendency to lowering of pH in water from shallow wells in western and southwestern Sweden, where the bedrock consists of weathering-resistant minerals and does not act as a buffer on the groundwater acid/alkali ratio. Acid deposition—via acid rain—is highest in this part of Sweden, originating mostly from western and central Europe, as well as from local emissions of sulfur oxides. Because hardness—due to the presence of calcium and magnesium, expressed by equivalent quantity of calcium carbonate—is a routine parameter for characterization of well waters, a correlation between hardness and groundwater acidification was examined. Hardness increase indicates elution of calcium and magnesium ions, which leads to alkalinity decrease and groundwater acidification. Jacks and Knutsson (1982) noted that hardness relative to alkalinity increased markedly in southern Sweden since acid rains began to be recorded (Fig. 17.6).

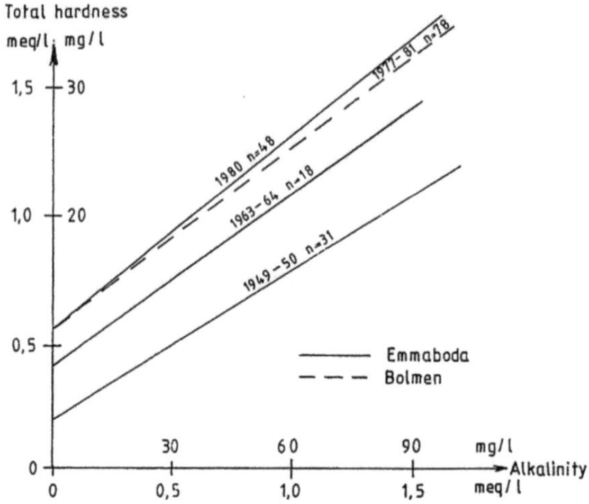

Fig. 17.6 Relationship between total hardness and alkalinity in water at three different times, from sites in southern Sweden (Jacks and Knutsson 1982)

The pathway of groundwater chemistry change due to anthropogenic acidification was depicted by Knutsson (1994) in four distinct stages: (a) Seasonal decreases in groundwater pH, alkalinity, cation concentration, and increases in sulfate concentration, are caused by storms and snowmelt events, expressed by "acid surges." This pattern is shown in Fig. 17.7 where seasonal depressions in pH values are observed; (b) groundwater chemistry is characterized by stable pH and long-term increases in calcium, magnesium, sulfate and nitrate concentrations; (c) groundwater acidification is expressed by long-term reduction in alkalinity and/or pH. This process may occur in noncalcareous sandy aquifers under local (point source) high acid deposition from industrial or municipal origins; and (d) groundwater acidification occurs when alkalinity decreases to zero, pH drops to ~5, and the concentrations of aluminum, nitrate, and sulfate are relatively high (Fig. 17.8).

In a field study of a noncalcareous sandy aquifer in western Denmark, Kjoller et al. (2004) found that acid rain lowered groundwater pH to 4.4, with an Al^{3+} concentration of about 0.2 mM in the upper 3–4 m of the saturated zone. The downward migration rate of the acidification front was 3.5–5.0 cm/year. As a result, the pH in deeper aquifer water rose to 5.2–6.5, following increases in desorbed Ca^{2+} and Mg^{2+} that occurred by exchange with Al^{3+} originating from the acid rain. This pathway of pseudo-irreversible groundwater acidification may lead to complementary changes in vadose zone chemistry, by affecting the distribution of trace metals that are desorbed into groundwater in the acidification zone, and subsequently adsorbed on the solid phase of the aquifer.

Ward et al. (2010) show that the upper layer of groundwater in the Gnagara Mound, Western Australia, is acidifying as a result of acid deposition from air pollution onto the land surface. Profiles of pH indicate that shallow groundwater is acidified to a depth of between 4 and 10 m. In this region, marine-derived salts control the major ion water chemistry, as reflected by high Na^+ and Cl^- levels as

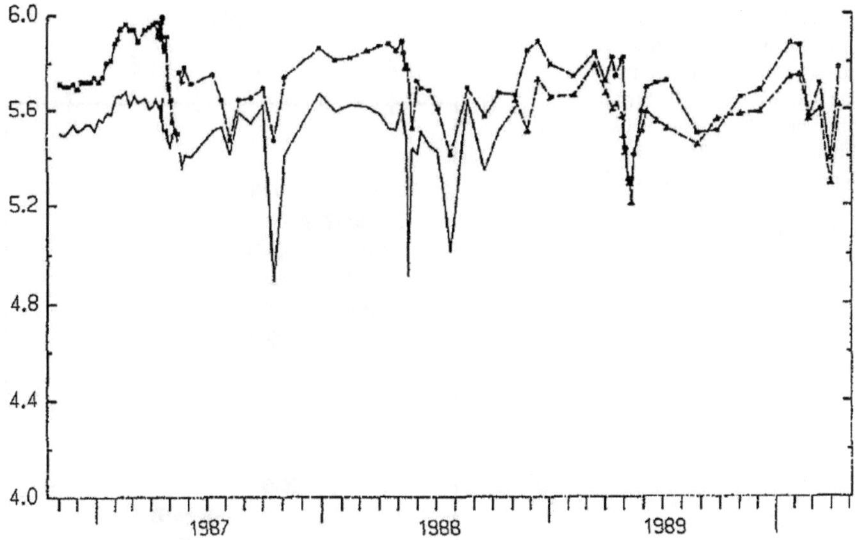

Fig. 17.7 Seasonal fluctuations of pH in three springs in stony sandy till, Lofsdalen, Sweden (Knutsson 1994)

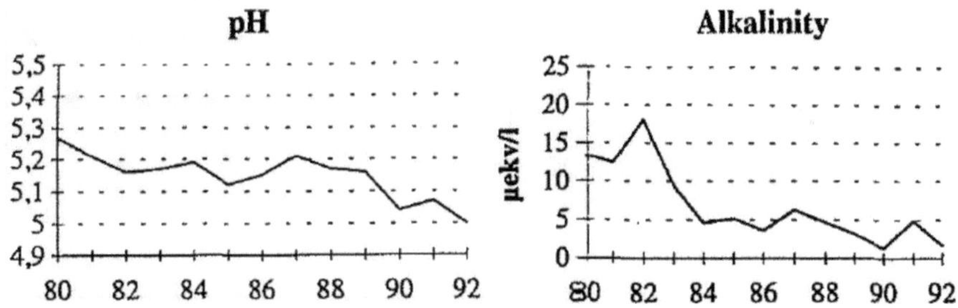

Fig. 17.8 Decreasing pH and alkalinity in groundwater from the research catchment of Birkenes, Norway during the years 1980–1992 (Knutsson 1994)

well as by relatively high values of sulfate. The high SO_4^{2-}/Cl^- ratio suggests a source of sulfur from air pollution in addition to that of marine salts. The acidified groundwater profile in the upper region is followed by an abrupt increase in pH at greater depth. This increase may be explained by the buffering effect of the geochemical environment (Appelo and Postma 2005). Another example of groundwater acidification induced by acid rain is reported by Franken et al. (2009), in a Quaternary sandy aquifer (Hannover, Germany). The average bulk deposition at the study area, from atmospheric acid loads, is about 0.4 kmol/ha/year; this value is comparable to other sites in northern Germany. The pH measured in various locations of the studied area is shown in Fig. 17.9. In the upper 5 m, the pH has a value between 3.5 and 5, while the zone below 6 m is characterized by a pH ranging from 5 to 7.

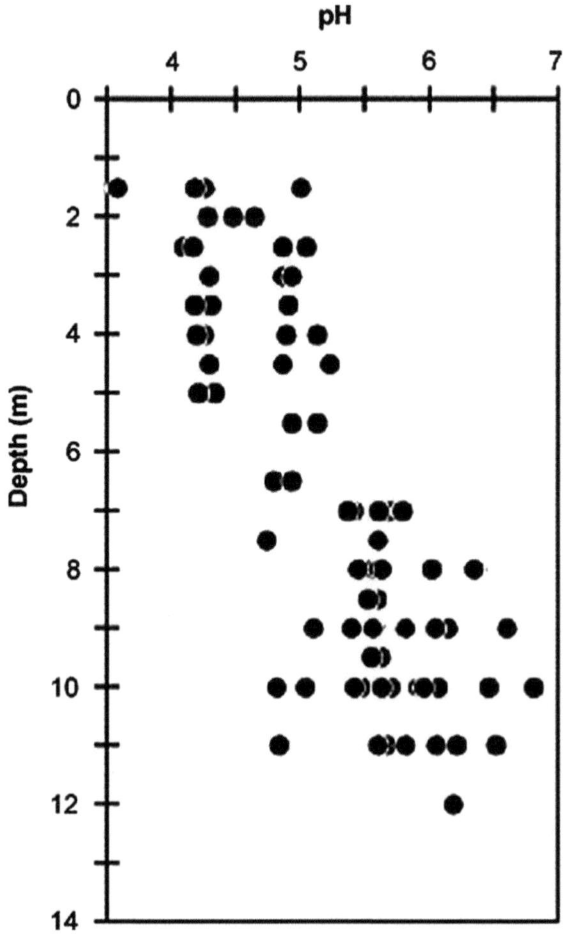

Fig. 17.9 Groundwater pH downstream from a forest area as a function of depth. The data were compiled from different multisampler wells (Franken et al. 2009). Reprinted from Franken et al. (2009). Copyright (2009) with permission from Elsevier

Franken et al. (2009) suggest that the low pH may be associated with the mobilization of Al^{3+} by dissolution of gibbsite and amorphous $Al(OH)_3$:

$$Al(OH)_3 + 3H^+ \leftrightarrow Al^{3+} + 3H_2O$$

or, alternatively, that the pH is in equilibrium with the mineral jurbanite according to the reaction

$$Al(OH)SO_4 \leftrightarrow Al^{3+} + OH^- + SO_4^{2-}.$$

Calculating the saturation index (SI) of groundwater for the above minerals, it is seen (Fig. 17.10) that for jurbanite, the SI of the groundwater remains close to zero over the entire pH range. This indicates that equilibrium with jurbanite controls the pH and Al^{3+} concentration in the groundwater and that an accumulation of SO_4^{2-}

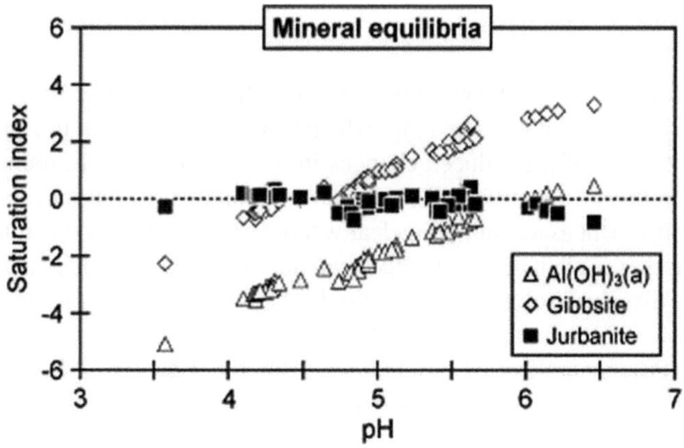

Fig. 17.10 Saturation state of the groundwater downstream from the forest for $Al(OH)_3(a)$, gibbsite, and jurbanite. Saturation index (SI) = log (IAP/K) = 0 indicates equilibrium between the groundwater and the specified mineral, SI > 0 indicates supersaturation, and SI < 0 indicates subsaturation (Franken et al. 2009). Reprinted from Franken et al. (2009). Copyright (2009) with permission from Elsevier

in the aquifer sediment derived from high atmospheric sulfur concentrations and acid rain.

In northern Europe, high rates of acidic deposition between 1950 and 1980 declined drastically as a result of ecological awareness. However, the rate of SO_4^{2-} decline in surface waters is much slower than that of SO_4^{2-} in atmospheric deposition (Alewell et al. 2001). During years of high sulfur deposition, significant amounts of sulfur accumulated in soil remained as a source of groundwater acidification. Despite the fact that the relationship between acid input into aquifers and SO_4^{2-} concentrations in groundwater has changed over time, there is continuous transport by leaching of sulfur from soils into groundwater. Thus, groundwater acidification from anthropogenic sources remains an irreversible process at least on a human lifetime scale.

17.3 Groundwater Contamination by Inorganic Trace Elements

Disposal of inorganic trace contaminants onto the land surface, with subsequent downward transport as dissolved solutes and/or adsorbed on colloids, may lead to irreversible changes in aquifer chemistry. We illustrate such irreversible changes in groundwater chemistry by considering the behavior of radionuclides, heavy metals, and metalloids.

17.3.1 Radionuclides

Radionuclides released to the environment through anthropogenic processes induce groundwater contamination. The long half-life of some species in this group affects the nature of some of the induced changes in the composition of groundwater. For example, significant quantities of elemental plutonium (Pu) have been introduced into the environment as a result of nuclear weapons testing and production, as well as from nuclear power plant accidents. Radionuclides may reach groundwater mainly by migration as dissolved solutes and by colloidal-facilitated transport when insoluble or poorly soluble contaminants are sorbed on colloidal surfaces. A specific example is plutonium (Pu) contamination originating from the Benham, USA underground nuclear test site, which was detected in groundwater collected in wells 1.3 km downgradient (Kersting et al. 1999). More than 90 % of Th, Pu, and other radionuclides found in the aquifer were associated with the transport of naturally occurring colloidal particles (e.g., clays or zeolites, <1 μm in size) in the groundwater.

Additional data on plutonium concentration, solute/colloid distribution, and isotopic composition in groundwater at the Hanford Site (Columbia River, U.S. DOE 100 K-area) are found in Dai et al. (2005). Concentration patterns of ^{239}Pu measured in unfiltered samples of water in a well transect through the site are shown in Fig. 17.11. It was found that ^{239}Pu is associated with the colloidal fraction of the groundwater in 7–29 % of the wells sampled. The authors consider that a high colloidal abundance at the Hanford Site ^{239}Pu may be associated with the oxidation state defined by the geochemical environment.

The oxidation state of plutonium largely controls its migration through sediments and aquifers; plutonium can exist simultaneously in groundwater in four oxidation states: Pu^{III}, Pu^{IV}, Pu^{V}, and Pu^{VI}. In a long-term field lysimeter experiment, Kaplan et al. (2006) observed that Pu moved more rapidly (overall rate of 12 cm/yr) when it was applied initially as Pu^{VI} than as Pu^{III} or Pu^{IV}. The tendency to form complexes, hydrolyze, polymerize, and/or sorb to solids follows the effective charge of the ion: $Pu^{4+} > Pu^{VI}O_2 > Pu^{3+} > Pu^{V}O_2{}^{+}$ (Choppin 1983). Moreover, the plutonium isotopic state may provide more complete knowledge on the contamination source. Fuel-grade Pu has a $^{240}Pu/^{239}Pu$ ratio between 0.07 and 0.18, while reactor-grade Pu has a $^{240}Pu/^{239}Pu$ ratio of 0.18:0.30 (Taylor et al. 2001; Pellaud 2002). Measuring Pu isotope ratios at the Hanford Site 100 K-Area groundwater sample, Dai et al. (2005) evaluated the reactor waste source by comparing to ^{241}Pu (Fig. 17.12). While the overall concentrations of Pu were low, giving general information on Pu contamination of groundwater, the Pu isotopic compositions confirmed the local sources of Pu contamination.

A rapid change in groundwater chemistry was observed under the Hanford reservation nuclear facility (USA) when, in 1984, liquid waste from a uranium extraction operation was routed to a newly constructed crib. Within a few months, uranium concentrations in nearby groundwater monitoring wells increased by 170 times their previous level and reached more than 5000 times the current EPA drinking water standard. Despite intensive reclamation of the polluted site, the

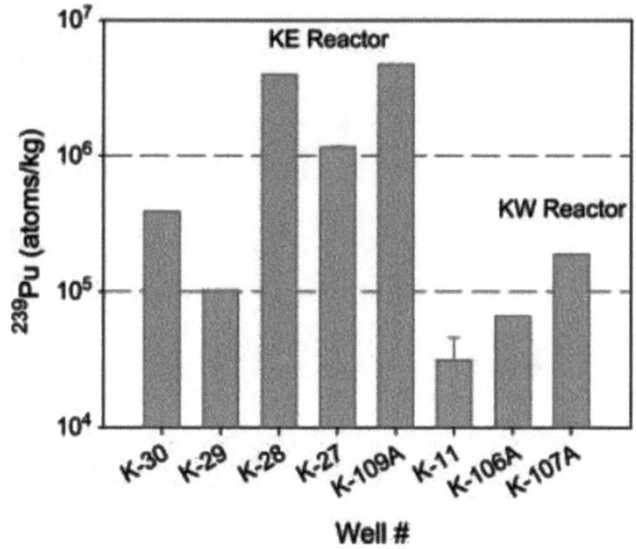

Fig. 17.11 ^{239}Pu concentrations in 100 K-Area wells of the Hanford Site. Transect following the groundwater flow toward the Columbia River. Data include two collections (1997 and 1999) (Dai et al. 2005). Reprinted from Dai et al. (2005). Copyright (2005) with permission from Elsevier

uranium concentration in the groundwater has remained 1,300 times greater than the EPA drinking water standard, indicating a pseudo-irreversible change in the groundwater chemistry (Pedroni 2000).

Natural organic compounds in groundwater such as humic substances facilitate the transport of radionuclides from the land surface into groundwater. As an example, Fig. 17.13 shows the distribution of two actinides, Pu and Am, dissolved in saline groundwater in the presence or absence of fulvic or humic acids. It may be observed that even the percentage of Pu and Am dissolved in the saline groundwater increases in the presence of humic substances. Their distribution varies as a result of different complexation mechanisms.

17.3.2 Heavy Metals

Heavy metals detected in groundwater under heavily urbanized areas are often of anthropogenic origin. Concentrations of heavy metals in groundwater are influenced by the amount of contaminant released to the soil–subsurface zone, by the chemical nature of the groundwater, by the recharge rate, and by the properties of the solid matrix. Heavy metal mobility through porous media may also be linked to interactions between metals and colloids, which themselves are transported through the subsurface.

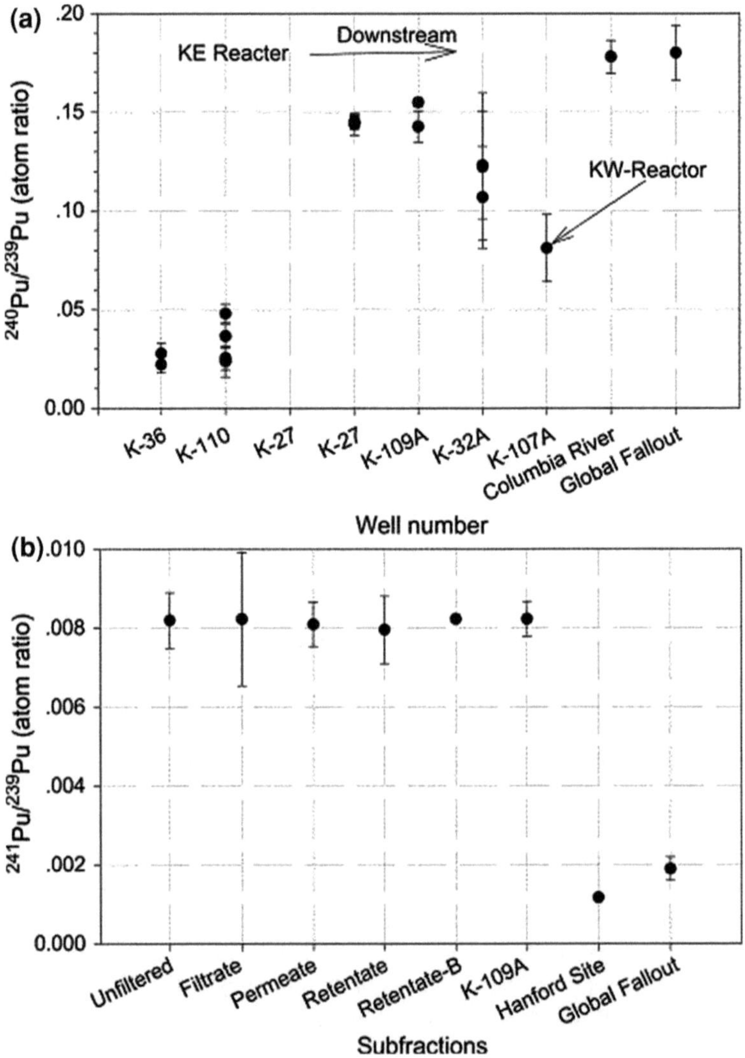

Fig. 17.12 Pu isotope ratios of Hanford Site 100 K-Area groundwater samples. Global fallout ratio from Kelley et al. (1999). (A) $^{240}Pu/^{239}Pu$. (B) $^{241}Pu/^{239}Pu$. Subfraction data are from Well K-27 including the ratio in the redox fraction (referred to as "retentate-B"). The sample from K-109A is unfiltered. Hanford Site refers to typical site operation discharge which contains 0.54 % ^{241}Pu as of June 1968, equivalent to $^{241}Pu/^{239}Pu=0.00117$ (January 2000) (Dai et al. 2005). Reprinted from Dai et al. (2005). Copyright (2005) with permission from Elsevier

Transport into groundwater may also be facilitated by association with microbial populations. In a column experiment, Pang et al. (2005) showed that cadmium (Cd) adsorbed on *Bacillus subtilis* spores or *Escherichia coli* was transported through saturated coarse alluvial gravels, while a complementary

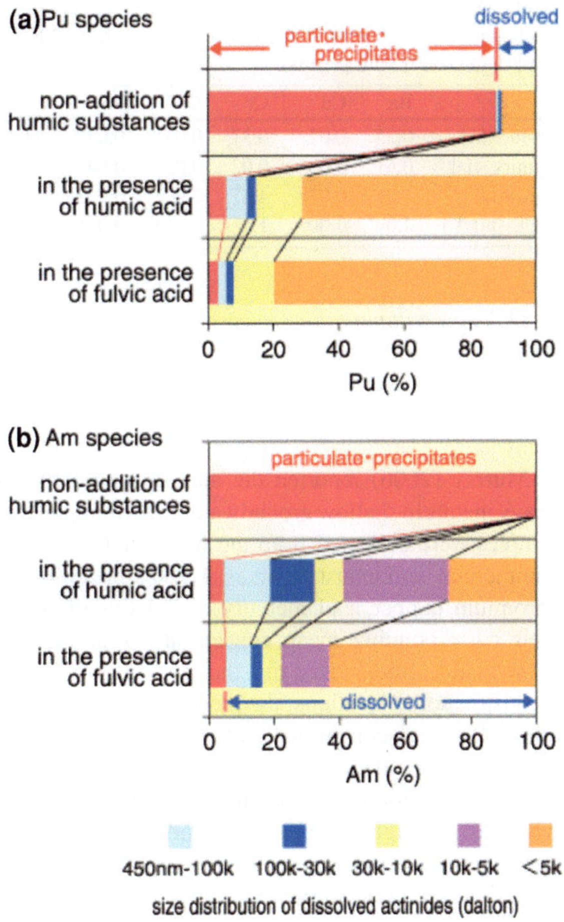

Fig. 17.13 Distribution of Pu and Am in saline groundwater, as affected by the presence of humic and fulvic acids (Japan Atomic Energy Research Institute 2000)

adsorption–desorption study suggested that the rates of Cd retention and release differ not only between bacterial species but also between unattached and deposited bacteria (Pang and Simunek 2006).

When heavy metals reach groundwater, they may become bioavailable and thus toxic to some organisms and/or for groundwater used as drinking water. It should be stressed that once released to the soil–subsurface environment, heavy metals cannot be degraded. The only processes that may affect them are transport, immobilization, and changes in their oxidation state. Heavy metals adsorb onto the mineral fraction of the solid phase and suspended particles in the liquid phase, and become a continuous source of groundwater contamination due to their slow rate of desorption. Therefore, metal contaminants become, essentially, permanent microcomponents of the porous medium and groundwater, which consequently leads to irreversible changes in groundwater composition.

Table 17.1 Mean concentrations of dissolved heavy metals in shallow groundwater in a Michigan urban watershed (modified after Murray 2006). Modified from Murray (2006). Copyright (2006) with permission

	Ba	Cd	Cr	Cu	Pb	Hg	Ni	Se	Ag	Zn
Residential	0.25	0.02	0.37	0.03	0.042	0.0004	NA	0.005	0.009	0.10
Commercial	0.61	0.005	0.02	0.05	0.045	0.015	0.12	0.078	0.008	0.31
Industrial	1.21	0.076	4.87	0.04	0.053	0.10	0.27	0.015	0.02	0.34
MCL (mg/L)	2.0	0.005	1.3	0.015	0.015	0.002	NA	0.05	0.1	5.0

Diffuse contamination: Heavy metals reach groundwater via diffuse sources, when they are distributed over a broad urban area, and via point sources, when leaching from waste disposal sites and landfills. A few examples of irreversible, diffuse and point source groundwater contamination by heavy metals are discussed below.

Murray (2006) reported the anthropogenically induced presence of dissolved heavy metals in shallow groundwater in a Michigan urban watershed. A total of 1,140 samples were collected from the watershed of an urban area divide in residential, commercial, and industrial areas. The samples were analyzed for barium, cadmium, chromium, copper, lead, mercury, nickel, selenium, silver, and zinc, and compared to their corresponding mean concentrations at the near surface (Table 17.1). The comparison demonstrated that the groundwater is significantly polluted with heavy metals; in some locations, the mean concentration of Cr, Cd, and Pb exceeded the allowable mean contamination level (MCL) by 20, 12, and 10 times, respectively. The solubility and mobility of Cr species are specific for each oxidation state. Thus, a high concentration is associated with the presence of toxic, anthropogenic Cr^{6+} which has a high solubility and mobility. The high Cr^{6+} concentration can be attributed to significant releases from industrial sites. Because natural attenuation of Cr, Cd, and Pb contaminants in groundwater is unlikely to occur, such anthropogenically induced contamination can be considered as irreversible.

Diffuse anthropogenic contamination of groundwater by heavy metals in Erasinos Basin (E. Attica, Greece), due to intense urbanization and intense agriculture, was reported by Champidi et al. (2011). The study area occupies 212 km^2 and was sampled at 3 spring outlets, 40 wells, and 12 deep boreholes. Heavy metals (Sr, Fe, Mn, Cu, Cr, Ni, Pb, Cd, and Zn) were detected in the sampled groundwater. In general, the trace element concentrations exceeded the maximum acceptable level according to European Union standard, although significant variation was found among the collected samples: Cr 0.000–0702 ppm; Pb 0.000–0.434 ppm; Fe 0.000–0.271 ppm; Cu 0.003–0.157 ppm; Mn 0.000–0.392 ppm; Zn 0.010–2.343 ppm; Ni 0.000–0.297 ppm; and Sr 0.270–7.240 ppm. By application of R-mode factor analysis, used to characterize relationships among different data sets and variables, it was found that 30 % of the variance in the data matrix may be considered as a "heavy metal factor," and the spatial distribution is depicted in Fig. 17.14. The high heavy metal concentration observed in the western part of the basin is associated with an industrial facility and intense agriculture.

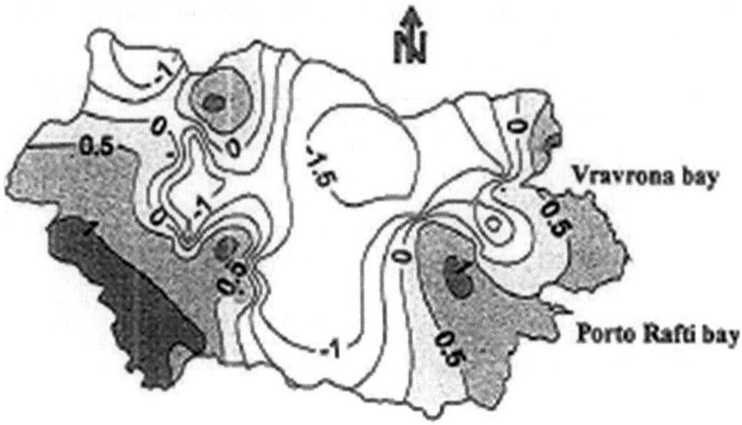

Fig. 17.14 Distribution of "heavy metal factor", based on 30 % of the variance in the data matrix (Champidi et al. 2011)

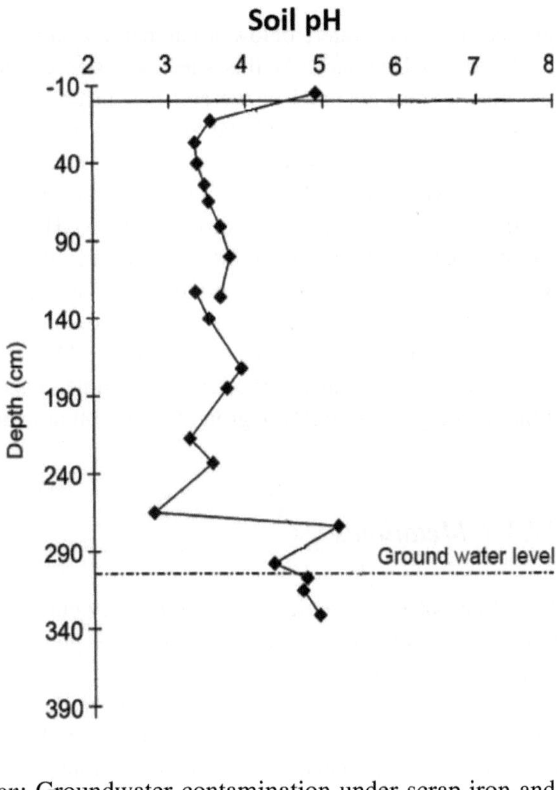

Fig. 17.15 Soil and subsurface pH profiles at a car battery salvage location (Jensen et al. 2000). Reprinted from Jensen et al. (2000). Copyright (2000) with permission from SAGE Publications

Point source contamination: Groundwater contamination under scrap iron and metal recycling facilities, following more than 25 years of activity, was reported by Jensen et al. (2000). Figure 17.15 shows the distribution of pH, from land

Table 17.2 Toxic[a] metal concentrations (μg/L) in groundwater under an abandoned zinc mine, South Korea (Lee et al. 2005)

Species	Mine area	Downgradient area	Comparison area
Al	50–4,680[b] (1,204)[c]	nd[d]–1,240 (217)	nd–30 (30)[e]
Cr	nd	nd	nd
Mn	63–8,286 (1,662)	nd–97 (29)	nd–48 (27)
Fe	160–2,280 (661)	nd–950 (192)	nd–220 (111)
Cu	nd	nd–84 (32)	nd–30 (23)
Zn	50–2,395(1,130)	4–11,280 (490)	7–2,533 (271)
Cd	nd–17 (11)	nd	nd
Pb	nd–9 (8)	nd–90 (90)	nd

[a] Hard, soft, and borderline metals regulated by Korean drinking water standards are all termed as toxic metals in this study
[b] Range of values
[c] Arithmetic mean (censored values are replaced by 0.55 times the lower detection limits for calculation)
[d] Not detected (below detection limit, less than 1 μg/L)
[e] Single value

surface to groundwater, below a car battery salvage unit located on an alluvial sandy soil in Denmark. At this site, increased concentrations of cadmium, zinc, and nickel were detected in the soil and in groundwater at a depth of 3 m. The migration of these metals can be associated with the low soil pH values, ranging between 2.8 and 3.9, likely caused by acid release from the car batteries.

Groundwater contamination by toxic metals around an abandoned zinc mine in South Korea was reported by Lee et al. (2005). High concentrations of Al, Mn, Fe, and Zn were observed in groundwater, especially in the mine area (Table 17.2). Relative metal abundance in groundwater at the mine area shows the trend: Mn > Al > Zn > Fe > Cd > Pb. In the downgradient area, the relative abundance sequence was Zn > Al > Fe > Cu > Mn. It was found that toxic metal content decreases with distance from the mine. This behavior confirmed the zinc mine as the point source for groundwater contamination.

17.3.3 Metalloids

Metalloids such as arsenic and selenium may enrich groundwater, emanating from industrial pollution and from the natural geological environment altered by anthropogenic activities.

In groundwater, arsenic is found mostly in the inorganic forms arsenite (As^{+3}) and arsenate (As^{+5}). Under oxic conditions and at thermodynamic equilibrium, aqueous As is dominated by arsenate oxyanions, while under reducing conditions, the uncharged arsenite species is predominant. The main form of arsenic in groundwater is controlled by the pH. In a reducing environment, dissolved As-S species may precipitate as As sulfide, while in the presence of high amounts of sulfide, the concentration of dissolved As is decreased. Speciation of As is affected by its sorption on surrounding minerals. The presence of amorphous iron oxides, for example, favors strong As adsorption, but this adsorption capacity is controlled by pH and redox conditions (Smedley 2008).

The anthropogenically induced pattern of groundwater arsenic contamination in Bangladesh and West Bengal (India) has been the object of numerous studies (e.g., Nickson et al. 2000; Islam et al. 2004; Harvey et al. 2002, 2006; Polizotto et al. 2006; Neumann et al. 2010). The arsenic originates from sedimentary units within the aquifers that contain small amounts of sedimentary pyrite, which is known to associate arsenic (Nickson et al. 2000). The most accepted explanation for the arsenic contamination of groundwater beneath the Ganges Plain is that it derives from reductive dissolution of arsenic-rich Fe(III) oxyhydroxides in the sediment. The reductive mobilization of arsine is catalyzed by biological activity and presence of electron donors in the sediments, which in turn are derived from delivery of surface organic carbon into subsurface communities (Islam et al. 2004). Additionally, mobilization of sorbed arsenic may also impact arsine release and availability. The fluctuation of the water table, as a result of rainfall and pumping sequences for irrigation, favors cycles of oxidized and anoxic periods. Under anoxic conditions, iron oxides dissolve and arsenic is released into the water phase. The concentration of arsenic in the aquifer, up to 100 m depth, ranges between 0.01 and 10 μM over vertical and horizontal distances (Harvey et al. 2006 and references therein).

Release of natural solid phase arsenic into groundwater has been attributed to various mechanisms, including (a) oxidative or reductive degradation of arsenic-bearing solids, in competition with ligand displacement by phosphate, (b) reductive dissolution of Fe(III) oxides with concomitant arsenic release, (c) arsenic release via redox cycling in surface soils/sediments and transported into the sandy aquifer, and (d) microbially mediated oxidation of organic carbon, inducing release of arsenic from sediments. Arsenic occurrence in groundwater is strongly influenced by anthropogenically induced processes where groundwater irrigated rice fields and constructed ponds (both rich in fertilizers, dissolved organic carbon, and other agrochemicals) contribute most of the recharge to the aquifer.

Neumann et al. (2010) studied the origin of dissolved arsenic under rice cultivation in the Ganges Delta (Bangladesh) and observed that recharge from ponds carries degradable organic carbon into the shallow aquifer and that groundwater pumping and irrigation of pond water increase the depth of the drawn water where dissolved arsenic concentration is greatest. An example of arsenic concentration in seven well clusters within a 0.03 km^2 area of rice cultivation is depicted in Fig. 17.16a, in comparison with the depth profiles of two conservative signatures, chloride and methane (Figs. 17.16b,c). Chemical signatures measured at the depth of the peak arsenic concentration are matched by those measured in the pond surface and sediment pore water. The groundwater residence time in the aquifer—which under current practices is decades or centuries (Harvey et al. 2002)—may be altered under different management schemes. However, the change in groundwater chemistry by added arsenic remains irreversible on a human lifetime scale.

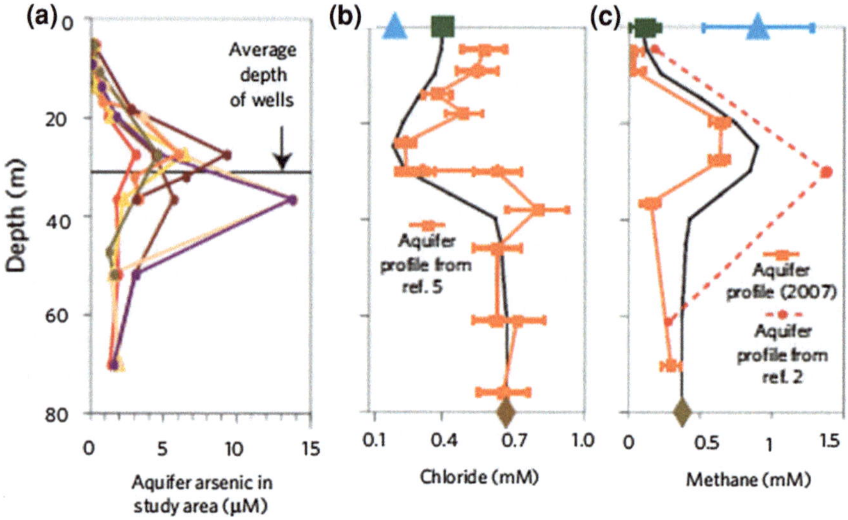

Fig. 17.16 Concentration and recharge profiles of **a** arsenic, **b** chlorides, and **c** methane in groundwater from Munshiganj, Bangladesh in seven well clusters within a 0.03 km^2 (Neumann et al. 2010)

Contamination of groundwater by arsenic is discussed by Appleyard et al. (2006), who examined As enrichment in groundwater in an urban area in Perth (Australia) as a result of exposure of pyritic sediments caused by reduced rainfall and increased groundwater exploitation. SEM analysis of framboidal pyrite within the peat and sand sediments indicates the mineral to be a source of As in shallow groundwater. The oxidation of arsenian pyrite, as described by Smith et al. (2006), is:

$$(\mathrm{FeS}-\mathrm{As})_{(s)}+7/2\mathrm{O}_2+\ \mathrm{H_2O}\ \leftrightarrow\ \mathrm{Fe}^{3+}+\ \mathrm{SO}_4^{2-}+\ \mathrm{AsO}_4^{3-}+2\mathrm{H}^+.$$

Products of this reaction are Fe^{3+}, arsenate as As(V) species, $SO_4{}^{2-}$ and H^+ ions. While no As was detected in groundwater in 1976, excessive dewatering and peat excavation by 2004 led to peak arsenic concentrations in the groundwater, ranging from 1,000 μg/L to 7,000 μg/L. Arsenic concentrations in shallow groundwater fluctuated with time and location (Fig. 17.17), but traces of As always remained. The authors suggest that falling Eh values triggered the release of As from the reduction of Fe(III) oxyhydroxide minerals near the base of the unconfined aquifer, inducing continuous and essentially irreversible (at least on a human lifetime scale) contamination of groundwater.

Groundwater contamination by inorganic speciation forms of arsenic and selenium, following their leaching from a coal ash dumping ground (Poland), is reported by Siepak et al. (2004). Groundwater concentrations of arsenic and selenium, measured over 4 years, and at eight different sampling sites, are shown in Table 17.3. Maximum concentrations of arsenic and selenium were 128.8 and

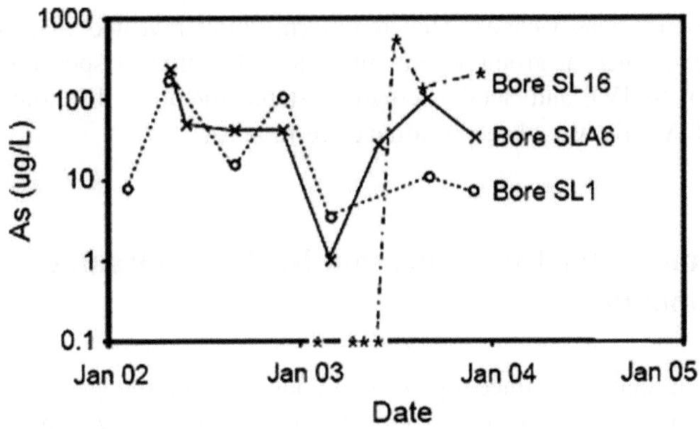

Fig. 17.17 Variation of As concentrations with time in shallow monitoring wells (Appleyard et al. 2006). Reprinted from Appleyard et al. (2006). Copyright (2006) with permission from Elsevier

Table 17.3 Total concentrations of As and Se in water from ash dump ground area (ng/mL) and concentration ratios of different inorganic forms of As and Se at various oxidation levels (Siepak et al. 2004)

	As				Se				As(V)/As(III)	Se(IV)/Se(VI)
Sample	year				year					
	1	2	3	4	1	2	3	4		
	ng/mL									
1	3.10	4.25	2.05	1.50	<0.15	0.20	0.25	<0.15	3.3	-*
2	128.8	86.7	127.1	94.9	18.5	39.1	53.7	52.9	51.7	7.7
A	0.95	0.85	1.05	0.75	<0.15	0.25	0.45	0.40	2.0	1.7
B	0.85	1.60	0.80	0.50	<0.15	0.15	0.25	0.20	2.7	-*
C	1.65	4.60	1.00	0.60	<0.15	0.15	0.25	0.20	1.0	-*
D	1.10	2.15	1.15	1.00	0.20	0.45	0.85	0.65	1.0	2.2
E	32.0	52.5	49.1	39.3	1.75	3.80	6.25	6.00	64.5	1.5
F	0.85	0.80	1.05	1.00	<0.15	<0.15	0.35	0.25	2.3	-*

*Concentration of one of the speciation forms below detection limit

52.9 ng/mL, respectively. In general, the level of contamination varies from site to site and from year to year in all the sampling points analyzed. Arsenic and selenium species were also determined in year 4; the concentration ratios As(V)/As(III) and Se(IV)/Se(VI) are shown in Table 17.3.

Long-term stability of inorganic arsenic and selenium in contaminated groundwater samples from Kelheim, Germany was reported by Rassler et al. (1998). Groundwater analysis showed very high concentrations of iron, manganese, and sulfur, which resulted from the leaching process of pyrites and their oxidation products used as filling materials in road and railway construction. As a result of speciation studies in acidified and nonacidified conditions, it was established that dominating species were arsenate and selenate, while arsenite and selenite presence was weaker. This long-term study revealed that arsenic and selenium persistence in groundwater differs as a function of speciation. Specifically, As(V), Se(IV), and Se(VI) remained stable for up to 12 months whereas oxidation of As(III) was observed after three months.

17.4 Groundwater Contamination by Trace Organic Compounds

Aquifer contamination by trace organic compounds of anthropogenic origin, which can interact with inorganic trace elements, may lead to irreversible changes in groundwater chemistry. In particular, metal ions can catalyze hydrolysis in a way similar to acid catalysis. Organic hydrolyzable compounds susceptible to metal ion catalysis include carboxylic acids, esters, amides, anilides, and phosphate-containing esters. Metal ions and protons coordinate to the organic contaminant so that electron density is shifted away from the site of nucleophilic attack, to facilitate the reaction. Metal ion-induced catalysis generally occurs via complexation of the reactant molecule.

Properties of metal ions, in general, and transition metal ions, in particular, make them good catalysts for a broad range of organic and inorganic reactions in the subsurface environment. These reactions are described in the literature (e.g., Huang 1999; McBride 1994; Smolen and Stone 1998). Catalysis by surface-bound metals on suspended particles is observed when the participating reactants are adsorbed significantly and when rate constants for the reaction at the solid–liquid interface exceed those in the surrounding liquid phase. Lewis acid properties of metals, for example, are significant in mineral-catalyzed hydrolysis reactions. Mineral-catalyzed degradation occurs mainly for the hydrolyzable organics having a structure suitable for complexing with surface metal cations.

Organic contaminants can reach the soil–subsurface zone from a variety of agricultural and urban-industrial practices. We selected from the literature-specific examples of irreversible groundwater contamination following sewage effluent disposal, pesticide application, and pharmaceutical discharge in the urban environment. The extent of the problem is demonstrated by results of the first pan-European reconnaissance survey of the occurrence of polar organic persistent pollutants in European groundwater (Loos et al. 2010). In total, 164 individual groundwater samples from 23 European countries were analyzed for 59 selected organic compounds and their degradates. The most relevant compounds in terms of frequency of detection and maximum concentrations detected were DEET (84 %; 454 ng/L), caffeine (83 %; 189 ng/L), PFOA (66 %; 39 ng/L), atrazine (56 %; 253 ng/L), desethylatrazine (55 %; 487 ng/L), 1H-benzotriazole (53 %; 1032 ng/L), methylbenzotriazole (52 %; 516 ng/L), desethylterbutylazine (49 %; 266 ng/L), PFOS (48 %, 135 ng/L), simazine (43 %; 127 ng/L), carbamazepine (42 %; 390 ng/L), nonylphenoxy acetic acid-NPE1C (42 %; 11 μg/L), bisphenol

A (40 %; 2.3 μg/L), PFHxS (35 %; 19 ng/L), terbutylazine (34 %; 716 ng/L), bentazone (32 %; 11 μg/L), propazine (32 %; 25 ng/L), PFHpA (30 %; 21 ng/L), 2,4-dinitrophenol (29 %; 122 ng/L), diuron (29 %; 279 ng/L), and sulfamethoxazole (24 %; 38 ng/L).

In many cases, groundwater chemistry changes may occur by contamination with mixtures of organic micropollutants, for example, consider the chemical composition of a groundwater contaminant plume in Boston (USA) following disposal of secondary sewage effluents (Barber et al. 1988). About 50 organic compounds were indentified in the contaminant plume, with concentrations ranging between 10 ng/L and 500 μg/L (Table 17.4). The prevailing organic micropollutants found in the contaminated groundwater included di-, tri-tetrachloroethene, *o*- and *p*-dichlorobenzene, C_1 to C_6 alkylbenzenes, and isomers of *p*-nonylphenol.

Barber et al. (1988) found that many trace compounds from chlorinated benzenes, alkylbenzenes, and aliphatic hydrocarbons have persisted for more than 30 years, despite the biodegradation of labile organic compounds. Examination of DCB (dichlorobenzene isomers), NP (nonyphenol isomers), and DTBB (2, 6-di-tert-butylbenzoquinone) concentrations in the contaminated groundwater near the disposal site was similar or even greater than their concentrations in the effluent, indicating little degradation during infiltration through the vadose zone. The long-term persistence of DCB and other potentially degradable compounds (e.g., aromatic and aliphatic hydrocarbons) in the downgradient zone of the aquifer can be attributed to the dynamics of the groundwater system. In this case, carbon sources were limited and the decrease in bacteria with distance from the disposal site led to almost no biologically induced degradation of organic contaminants.

Pesticide contamination of groundwater has been the subject of extensive study and monitoring since the second part of the last century. For example, pesticides that were detected most frequently in European groundwater at concentrations greater than 0.1 μg/L included chloridazon-desphenyl, NPE1C, bisphenol A, benzotriazole N,N′-dimethylsulfamid, desethylatrazine, nonylphenol, chloridazon-methyldesphenyl, methylbenzotriazole, carbamazepine, and bentazone (Loos et al. 2010). In groundwater collected from the wells of seven agricultural areas from 1991 to 1998 in Portugal, detected herbicides included alachlor, atrazine, metolachlor, metribuzine, and simazine, reaching maximum concentration values of 13, 30, 56, 1.4, and 0.4 μg/L, respectively, (Cerejeira et al. 2003). The most frequently detected herbicides were atrazine (64 %), simazine (45 %), and alachlor (25 %), with compounds found in more than 50 % of wells that sampled shallow groundwater beneath agricultural and urban areas.

Cavaller et al. (1991) studied the persistence of chloroacetamide herbicides (alachlor, metolschlor, propanil) and phenoxy herbicides (2, 4-D, dichlorprop) in groundwater samples from three Arkansas (USA) locations. A trend of low degradation rates for low concentrations (1 ppb) was consistent for all five compounds and at all locations over an 18-month period. In contrast, pesticide degradation was found to occur at various rates when the contaminants were applied at higher concentration and kept at a higher temperature (Fig. 17.18). The calculated half-lives ($t_{1/2}$) are given in Table 17.5. Extrapolating over much longer periods of

Table 17.4 Volatile organic compounds identified by close loop stripping and purge and trap gas chromatograph–mass spectrometer (GC-MS) of contaminated groundwater from Boston, USA (Barber et al. 1988). Superscript [a] denotes mixture of several isomers. Reprinted with permission from Barber et al. (1988). Copyright 1988 American Chemical Society

Compound	ng/L
Aromatic Hydrocarbons	
Alkyl compounds	
Benzene	100
C_1 to C_6 benzene[a]	1160
Naphthalene	10
Tetrahydronaphthalene[a]	10
Decahydronaphthalene[a]	20
Methylnaphthalene[a]	10
Dihydro-l*H*-indene	20
C_1 to C_2 dihydro-*H*-indene[a]	110
C_2 phenylmethylbenzene	100
C_1 to C_2 Tetrahydronaphthalene[a]	20
Halogenated compounds	
Chlorobenzene	20
Chloromethylbenzene	50
m-dichlorobenzene	30
p-dichlorobenzene	600
o-dichlorobenzene	670
Trichlorobenzene	10
Iodobenzene	80
Bromobenzene	10
Aliphatic hydrocarbons	
Alkyl compounds	
Trimethylpentane[a]	530
Trimethylpentene[a]	160
C_1 to C_4 cyclohexane[a]	360
C_{12} alkene[a]	100
Limonene	70
Adamantane	30
Halogenated compounds	
Dichloroethene	2.8×10^5
Trichloroethene	9.5×10^4
Tetrachloroethene	1.0×10^6
Bromoadamantane	30
Oxygen-containing compounds	
Dimethylethylphenol	20
Octylphenol	360
p-nonylphenol[a]	790
Nonanal	20
Decanal	20
Undecanal	20
Dodecanal	10
2,6-di-tert-butyl-*p*-benzoquinone	390

(continued)

Table 17.4 (continued)

Compound	ng/L
Dimethylbutanol	70
Methylhexanol	30
Alkyl ester butanoic acid	150
Dibutyl phthalate	450
Alkyl ester propanoic acid	90
Benzeneacetaldehyde	10
Nitrogen-containing compounds	
Methylquinoline	40
Quinolinol	20
C_4 hexanamide	80
C_2 methanamine	10
C_4 quinolinone	40
methyltriazole	30
Sulfur-containing compounds	
Dimethyltrisulfide	10
C_5 tetrahydrothiophene	20

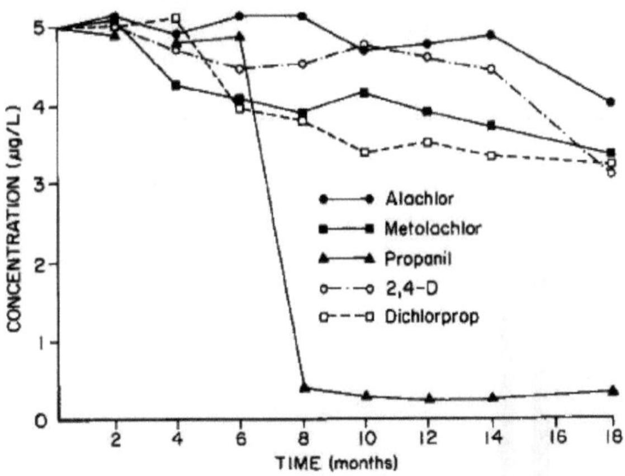

Fig. 17.18 Degradation of herbicides incubated at 22 °C in groundwater from Dumas, Arkansas (Cavaller et al. 1991). Reprinted from Cavaller et al. (1991). Copyright (1991) with permission of John Wiley and Sons

time, it may be seen that in all the cases studied, traces of pesticide will remain in the groundwater. As a consequence, we can consider that initial groundwater chemistry is changed irreversibly on human lifetime scales.

Intensive data collection for the U.S. Geological Survey National Water-Quality Assessment (NAWQA) was completed during 1993–1995 in 20 major hydrologic basins of the United States. These studies indicated that pesticides were commonly detected in shallow groundwater in 54.4 % of the 1,034 sampled sites in agricultural (56.4 %) and urban (34.2 %) locations (Kolpin et al. 1998). The main pesticides detected were atrazine, deethylatrazine, simazine, metolachlor, and prometon. The detection frequencies of pesticides having at least one concentration above 0.01 µg/L are presented in Fig. 17.19.

Table 17.5 Half-lives of selected herbicides in groundwater (Cavaller et al. 1991). Reprinted from Cavaller et al. (1991). Copyright 1991 with permission of John Wiley and Sons

Herbicide	Storage temp °C	Concentration (μg/L)	$t_{½}$ (est.) (*d*)	R^2
Alachlor	15	1.0	1518	0.644
	15	5.0	1446	0.851
	22	1.0	1409	0.693
	22	5.0	808	0.815
Metolachlor	15	1.0	749	0.675
	15	5.0	1074	0.797
	22	1.0	857	0.724
	22	5.0	548	0.894
Propanil	15	1.0	275	0.837
	15	5.0	215	0.871
	22	1.0	259	0.881
	22	5.0	196	0.868
Dichlorprop	15	1.0	1235	0.892
	15	5.0	824	0.857
	22	1.0	1286	0.763
	22	5.0	196	0.898
2,4-D	15	1.0	831	0.757
	15	5.0	1907	0.857
	22	1.0	982	0.763
	22	5.0	969	0.896

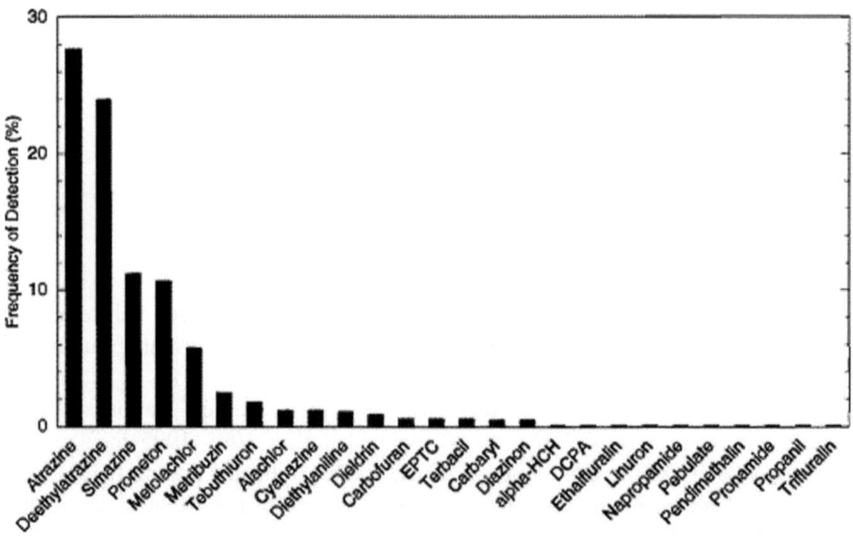

Fig. 17.19 Frequencies of detection of pesticide compounds in shallow groundwater reported in 41 land-use studies, adjusted to a common detection threshold of 0.01 μg/L (Kolpin et al. 1998). Reprinted with permission Kolpin et al. (1998). Copyright (1998) American Chemical Society

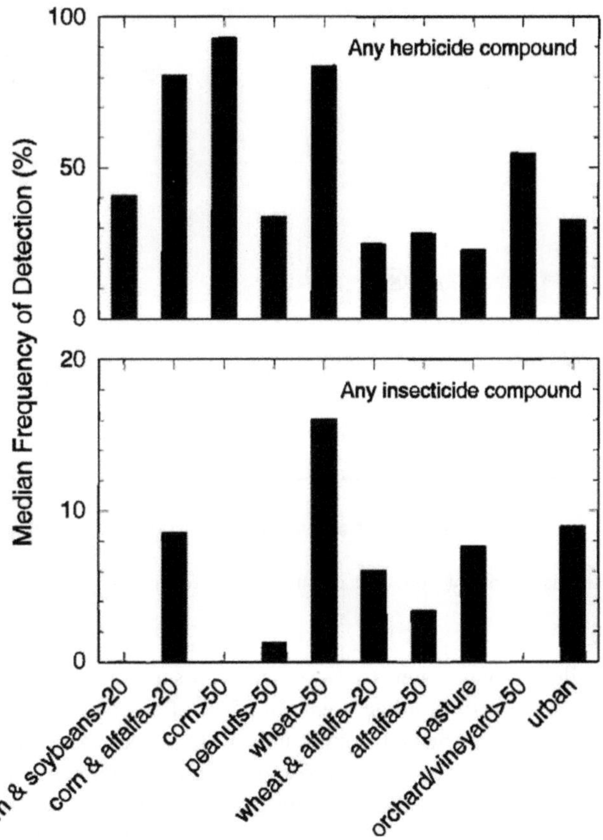

Fig. 17.20 Frequencies of herbicide and insecticide detection in shallow groundwater beneath 10 different land-use settings (Kolpin et al. 1998). Reprinted with permission from Kolpin et al. (1998). Copyright 1998 American Chemical Society

Differences in transport and persistence among the pesticides included in Fig. 17.19 may be explained by variations in the physical and chemical properties of the compounds. Kolpin et al. (1998) related the frequencies of herbicide and insecticide detection in shallow groundwater to land surface use. Although pesticides were found in groundwater beneath different land uses, variations were observed in the frequencies of herbicides and insecticides. For example, urban category ranked seventh in herbicide frequency, but second in frequency of insecticide detection (Fig. 17.20).

Later, Gilliom (2007) reported another 10-year study by the US Geological Survey (NAWQA) which provides a national-scale view of pesticide occurrence in streams and groundwater. Water samples were analyzed for 75 pesticides and 8 degradates currently in use, and bed sediments and fish tissue were analyzed for 32 organochlorine pesticides whose use has been discontinued. In one-third of deeper wells from aquifers used for water supply, at least one pesticide compound was

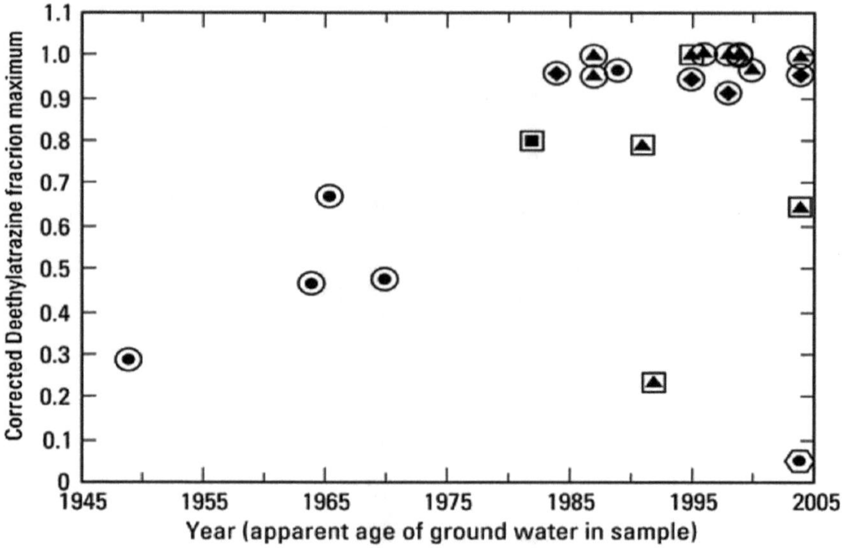

Fig. 17.21 Bivariate plots of corrected deethylatrazine relative to age of groundwater (Steel et al. 2008)

found. In shallow groundwater, 61 % of analyzed samples were from agricultural areas, and 55 % were from urban areas. In 57 % of samples from bed sediments, discontinued organochlorinated compounds were detected, in comparison with 80 % under urban areas. These findings suggest essentially irreversible retention of organochlorine pesticides in the soil–subsurface environment.

The occurrence and fate of 45 pesticides and their degradates in groundwater flow systems within four contrasting agricultural settings from Maryland, Nebraska, California, and Washington were determined by Steel et al. (2008). Of this group, only five parent pesticides and their degradates were detected frequently, comprising two classes of herbicides: triazines and chloroacetanilides. For example, the atrazine degradation product deethylatrazine was detected in about half of the 59 wells across all four study areas (Fig. 17.21). These data show that atrazine degradate is still present in the groundwater composition even >60 years after contamination (groundwater age 1949).

Pharmaceuticals and personal care products (PPCPs), which contaminate soil and groundwater mostly via land disposal of municipal wastewaters, have been recognized during the last two decades as "emerging contaminants" in environmental chemistry. Heberer (2002) reviewed the sources and fate of pharmaceutically active compounds (PhACs) in Europe, South America, and United States, finding that more than 80 pharmaceutical compounds and drug metabolites were detected in aquatic environments in concentrations up to the μg/L level. For example, PhACs such as clofibric acid, carbamazepine, primidone, and iodinated contrast agents were detected in many groundwater samples. A particular case of occurrence and distribution of PhACs in groundwater below a landfill site

Table 17.6 Distribution of different pharmaceutically active compounds identified in groundwater as a function of distance from the Grindsted landfill (Holm et al. 1995). Reprinted with permission from Holm et al. (1995). Copyright 1995 American Chemical Society

		Distance from landfill (m)						
Compound	Depth (m)	0	15	50	115	150	237	260
5.5-diallylbarbituric acid	5.5	–	[b]	[b]	[b]	[b]	[b]	[b]
	7	135	–	10	[c]	[c]	–	–
	8.5	–	205	–	<10	–	–	<10
	10	–	[c]	25	–	–	[c]	–
Propyphenazone	5.5	300	[b]	[b]	[b]	[b]	[b]	[b]
	7	4,000	–	–	[c]	[c]	–	–
	8.5	1,000	–	30	–	<10	<10	–
	10	2,900	[c]	300	70	–	[c]	<10
Aniline	5.5	<10	[b]	[b]	[b]	[b]	[b]	[b]
	7	720	–	–	[c]	[c]	–	–
	8.5	<10	–	70	–	<10	<10	–
	10	1,100	[c]	100	<10	–	[c]	<10
o-chloroaniline	5.5	<10	[b]	[b]	[b]	[b]	[b]	[b]
	7	110	–	–	[c]	[c]	–	–
	8.5	30	–	<10	–	<10	<10	–
	10	15	[c]	10	<10	–	[c]	<10
p-chloroanilina	5.5	<10	[b]	[b]	[b]	[b]	[b]	[b]
	7	20	–	–	[c]	[c]	–	–
	8.5	50	–	<10	–	<10	<10	–
	10	35	[c]	<10	<10	–	[c]	<10
tris(2-methylpropyl) phosphate	5.5	<1	[b]	[b]	[b]	[b]	[b]	[b]
	7	80	–	–	[c]	[c]	–	–
	8.5	<1	–	<1	–	<1	<1	–
	10	70	[c]	8	<1	–	[c]	<1
2-methy1-2-*n*-propyl-1,3-propanediol	5.5	60	[b]	[b]	[b]	[b]	[b]	[b]
	7	18,000	–	–	[c]	[c]	–	–
	8.5	440	–	230	–	<10	<10	–
	10	720	–	80	<10	–	–	<10

[a] All concentrations in μg/L
(–) not analyzed
[b] No sampling point
[c] Not enough water for analysis could be sampled at this depth

(Grinsted, Denmark), where pharmaceutical industry wastes were disposed during 1962–1975, is reported by Holm et al. (1995). The identified pharmaceutical compounds accounted for about 5 % of the nonvolatile organic carbon measured in the groundwater within the first 50 m downgradient of the site. The vertical distribution of the compounds shows a significant variation. At two sampling points (1.5 m depth) at the landfill border, concentrations of 2-methyl-2-*n*-propyl-1,3-propanediol varied from 60 to 18,000 μg/L. The concentration of PhACs decreases with increasing distance from the landfill site (Table 17.6). At a distance of 150 m downgradient of the landfill, no PhACs were detected in the groundwater. However, it should be noted that pharmaceutical residues were found in the groundwater under the landfill 13 years after disposal of PhACs ceased.

The fate of PPCPs during groundwater recharge was investigated by Drewes et al. (2003) at two facilities in the southwestern United States. These facilities employed surface spreading basins characterized by different soil–subsurface conditions and treatment management schemes. The bulk organic and PPCP concentrations in the treated effluent and in groundwater monitoring well samples are presented in Table 17.7. Only organic iodine, carbamazepine, and primidone were found in groundwater after tertiary treatment of the effluent, despite the fact that the effluent comprises a larger number of PPCPs. After subsequent travel of up to eight years, organic iodine, carbamazepine, and primidone were still present in the groundwater, with their concentration decreasing by about 30 %.

Studies of the presence of organic micropollutants in groundwater provide further evidence of irreversibility of the contamination process. Samples of water were collected in 1985 and 2005 from the same monitoring wells located along a plume of groundwater contaminated by 60 years of continuous disposal of a wastewater treatment plant (Barber et al. 1988, 2009). Fifteen percent of more than 200 analyzed organic wastewater contaminants were detected, including the antibiotic sulfamethoxazole (SX), the nonionic surfactant degradation product 4-nonylohenol (NP), the solvent tetrachloroethene (PCE), and the disinfectant 1,4-dichlorobenzene (DCB). The 2005 groundwater sampling confirmed the presence of the contaminants PCE, DCB, and NP that were reported in the original 1985 sampling. For example, the concentrations of NP determined in 2005 were relatively unchanged from the results determined 20 years earlier. Changes in the distribution of PCE, DCB, and NP after 20 years are due to the transport and redistribution of these contaminants in the groundwater profile. Using boron as a transport tracer, Barber et al. (2009) defined two oxic zones in the aquifer: (1) a downgradient transition zone at the interface between the contamination plume and the uncontaminated groundwater, and (2) a contaminated zone located beneath the infiltration beds which were not loaded for 10 years. The pattern of SX transport was nearly coincident with that of boron, while the NP transport pattern demonstrated retardation.

Petroleum hydrocarbons reach the vadose zone and groundwater mainly as a result of leaching from gas stations, by accidental release on land surface, and by seawater migration to coastal shorelines. Groundwater chemistry may be altered by the partial dissolution of petroleum products into the water phase; subsequent groundwater flow may affect the quality of an entire aquifer. The more water-soluble components of petroleum hydrocarbons, such as benzene and the low molecular weight alkylbenzene, are often found in groundwater downgradient from a pollutant source. The fraction of petroleum hydrocarbons that are insoluble with water, called non-aqueous phase liquids (NAPLs), may be also affect groundwater chemistry, by slow volatilization and dissolution, when adsorbed on soluble organics or polyelectrolytes, or as microdroplets incorporated in the water phase.

Table 17.7 Bulk organic and pharmaceuticals and personal care products (PPCPs) concentrations in tertiary-treated effluent and groundwater monitoring well samples (Drewes et al. 2003). Reprinted from Drewes et al. (2003). Copyright 2003 with permission of John Wiley and Sons

		Tertiary-treated effluent	OW2-90′ (6–18 months)	NW-4 (6–18 months)	NW-2 (~16 months)	2U (~years)	6U (>8 years)	Groundwater background
	Mean DOC [mg/L]	5.62 (±0.87)	1.43 (±0.35)	1.49 (±0.26)	1.16 (±0.18)	1.12 (±0.11)	0.9 (±0.12)	0.5
	SUVA [L/mg m]	1.54 (±0.36)	1.87 (±0.38)	1.79 (±0.22)	1.84 (±0.37)	2.46 (±0.61)	2.18 (±0.51)	N/A
Psychomotor stimulants [ng/L]	Caffeine	n.d.	n.d.	n.d.	n.d.	n.d.	n.d.	n.d.
X-ray contrast agents [µg/L]	Organic Iodine [µg I/L]	19.3(Sun) 30.5 (Wed)	15.4	8	8.9	6.7	12.5 (±0.9)	0.6
Lipid regulator [ng/L]	Gemifbrozil	n.d.	n.d.	n.d.	n.d.	n.d.	n.d.	n.d.
	Clofibric acid	n.d.	n.d.	n.d.	n.d.	n.d.	n.d.	n.d.
	Fenofibrate	n.d.	n.d.	n.d.	n.d.	n.d.	n.d.	n.d.
Antiepileptic [ng/L]	Carbamazepine	175 (±28.3)	115	235	125	145	85	n.d.
	Primidone	202 (±24.75)	140	120	160	90	100	n.d.
Analgeise/anti-inflammatory [ng/L]	Diclofenac	n.d.	n.d.	n.d.	n.d.	n.d.	n.d.	n.d.
	Ibuprofen	16 (±0)	n.d.	n.d.	16	n.d.	n.d.	n.d.
	Ketoprofen	n.d.	n.d.	n.d.	n.d.	n.d.	n.d.	n.d.
	Naproxen	8 (±0)	n.d.	n.d.	n.d.	<10	<10	n.d.
	Fenoprofen	n.d.	n.d.	n.d.	n.d.	n.d.	n.d.	n.d.
	Propyphenazone	n.d.	n.d.	n.d.	n.d.	n.d.	n.d.	n.d.
	Meclofenamic acid	n.d.	n.d.	n.d.	n.d.	n.d.	n.d.	n.d.
	Tolfenamic acid	n.d.	n.d.	n.d.	n.d.	n.d.	n.d.	n.d.
Blood viscosity agent [ng/L]	Pentoxifylline	n.d.	n.d.	n.d.	n.d.	n.d.	n.d.	n.d.

n.d.—not detectable; N/A—not available

Table 17.8 Concentrations of petroleum hydrocarbons contaminants from groundwater samples data set collected under 47 service station sites throughout California, USA (Zemo 2006). Reprinted from (Zemo 2006). Copyright 2006 with permission of John Wiley and Sons

	Maximum (mg/L)	Minimum (mg/L)	Average (mg/L)	Vs. Average from Laboratory Studies(%)
TPHg	320	1	86	−14
Benzene	15	0.012	4.9	−83
Toluene	47	0.0003	8.6	−76
Ethylbenzene	6.2	0.022	2.1	−21
Xylenes	33	0.022	9.6	−35
MTBE	310	0.180	48	−78

Table 17.9 Frequency of undissolved petroleum hydrocarbons collected from the service station sites throughout California, USA (Zemo 2006). Reprinted from Zemo (2006). Copyright 2006 with permission of John Wiley and Sons

	Frequency of exceedance of concentration criteria					
	Minimum concentration (%)		Average concentration (%)		Maximum concentration (%)	
Constituent	Lab[1]	Sheen[2]	Lab	Sheen	Lab	Sheen
Ethylbenzene	38	81	12	20	1	0.8
Xylenes	35	86	9	19	0.4	0.4
TPHg	NA	85	9	13	NA	0.6

Note NA = not applicable

[1] Lab = minimum, average, and maximum concentration from laboratory partitioning studies using fresh gasoline (see Table 17.3) used as criteria

[2] Sheen = minimum, average, and maximum concentration from groundwater samples with entrained sheen (see Table 17.4) used as criteria

Petroleum hydrocarbon contamination under downward flow: An example of a large monitoring survey of groundwater chemistry changes due to petroleum product leaks below service stations is given by Zemo (2006). Petroleum hydrocarbon contamination was evaluated on the basis of groundwater monitoring data from 47 service stations throughout California, during the years 1988 to 2002, which were compared to laboratory solubility data. The study considered total petroleum hydrocarbons (TPH) and associated benzene, toluene, ethylbenzene, xylenes (BTEX), and methyl-*tert*-butyl-ether (MTBE) in groundwater samples collected from an aquifer zone where both dissolved and undissolved compounds were present. The minimum, maximum, and average petroleum hydrocarbon concentration values at the sites are presented in Table 17.8. Petroleum hydrocarbon concentrations determined in the laboratory studies are generally higher than those obtained directly from the monitored wells; this can be attributed to in situ contaminant attenuation.

The frequency of undissolved petroleum hydrocarbons at the 47 sites was compared to data from laboratory partitioning studies (Table 17.9). The frequency of entrained undissolved gasoline and compounds in groundwater samples ranged

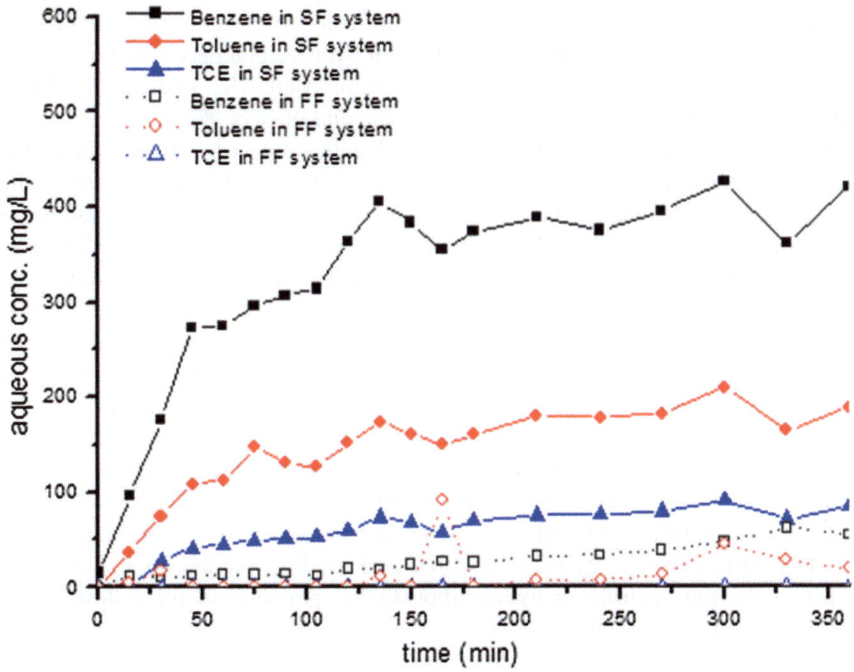

Fig. 17.22 Rate of intrusion of organic compounds from freshwater (FF) and seawater (SF) source cells into freshwater collector cells, expressed as concentrations of selected compounds (in mg/L) (Dror et al. 2003)

from 0.4 to 86 %. These data confirm potential contamination of groundwater by petroleum hydrocarbons below service stations, but also underline differences in concentration values as a function of the distance from the solid phase pollutant within the aquifer. The impact of NAPLs on groundwater chemistry is more evident in the "smear zone" of the aquifer, where the presence of undissolved compounds was higher than that predicted from theoretical solubilities.

Petroleum hydrocarbon contamination under lateral flow: Contamination of seawater by petroleum hydrocarbons raises the possibility of intrusion of organic contaminants into coastal aquifers. Such behavior was observed in a study of intrusion into a freshwater cell of low density (benzene and toluene) and high density (trichloroethylene) petroleum hydrocarbons dissolved in freshwater (FF) and in simulated seawater (SF). From Fig. 17.22, it may be seen that a much higher rate of intrusion and much large amounts of petroleum hydrocarbon were found in the SF system than in FF system (Dror et al. 2003).

The intrusion of petroleum hydrocarbons from seawater into fresh groundwater may occur in response to chemical gradients that are enhanced by the "salting-out effect." This process, which reduces the aqueous solubility of petroleum hydrocarbon in seawater, acts to force petroleum hydrocarbons across the interface between seawater and fresh coastal groundwater. The total petroleum hydrocarbon concentrations in groundwater following the intrusion of contaminated "seawater" are given not only by the amount of petroleum hydrocarbon dissolved but also by the presence of dispersed organic droplets (Yaron-Marcovich et al. 2007). More details on this phenomenon are provided at the end of Sect. 8.3.4.

Petroleum plume breakthrough associated with seawater intrusion along the coast of southwestern Nigeria was reported by Ojuri and Ola (2010). Also, the effect of seawater intrusion on petroleum hydrocarbon plumes in the Mediterranean coastal area of Southern Italy was studied by Mastrocicco et al. (2012), who examined the influence of the density contrast between fresh groundwater and seawater on the transport of petroleum contaminants within an unconfined coastal aquifer. The monitoring site was a petrochemical plant, located on a polluted unconfined sandy and sandy loam aquifer of 10–30 m thickness, which becomes semiconfined in silty clay sediment. Local confined aquifers separated from the polluted unconfined aquifers were also defined. Seawater reaches the polluted site through a canal used permanently for plant cooling purposes. An unknown volume of light nonaqueous phase liquid (LNAPL) contaminating the aquifer in the past decades came into contact with a plume of seawater originating from the cooling canal. Mastrocicco et al. (2012) show that as a result of the density contrast, the effect of the total petroleum hydrocarbon (TPH) plume is to force flow in the bottom part of the aquifer.

Partial attenuation: Natural attenuation of the petroleum hydrocarbon components may be the result of volatilization, adsorption on the aquifer solid phase, or microbial activity. Volatilization leads to removal of a fraction of the compounds, while adsorption of some compounds may be irreversible on specific mineral or organic solid phases; microbial activity may enhance TPH degradation and formation of various metabolites. However, despite the complete eradication of some petroleum product components, TPH residues will remain and lead to irreversible changes in groundwater chemistry at least on the human lifetime scale.

Kerosene is a heavily used petroleum product formed by a mixture of about 100 components. The pathway of kerosene dissolution and volatilization in and from a water solution confirms potential residual kerosene presence in groundwater. Excess aqueous solubility of aromatic constituents is explained by cosolvency due to the presence of polar compounds in TPH mixtures. This cosolvent effect potentially increases aqueous concentration in hydrocarbons (Lee et al. 1992a, b). An extended discussion of kerosene behavior in the soil–subsurface environment is presented in Sect. 16.1.5.

Petroleum hydrocarbon concentrations in groundwater may be also altered by irreversible adsorption on the aquifer solid phase, which contains clays and sands as major constituents. Cosultchi et al. (2004) found that most organic species of petroleum organic compounds remain adsorbed on layered clays after aging. This retention is conditioned by the petroleum paraffinic content and by molecules containing alkane chains, which easy intercalate within the clay interlayers. Petroleum hydrocarbon adsorption may lead to irreversible modification of the clay structure. The partial dissolution of clay matrix is indicated by X-ray diffraction (XRD) as an irreversible modification of the clay structure. Figure 17.23

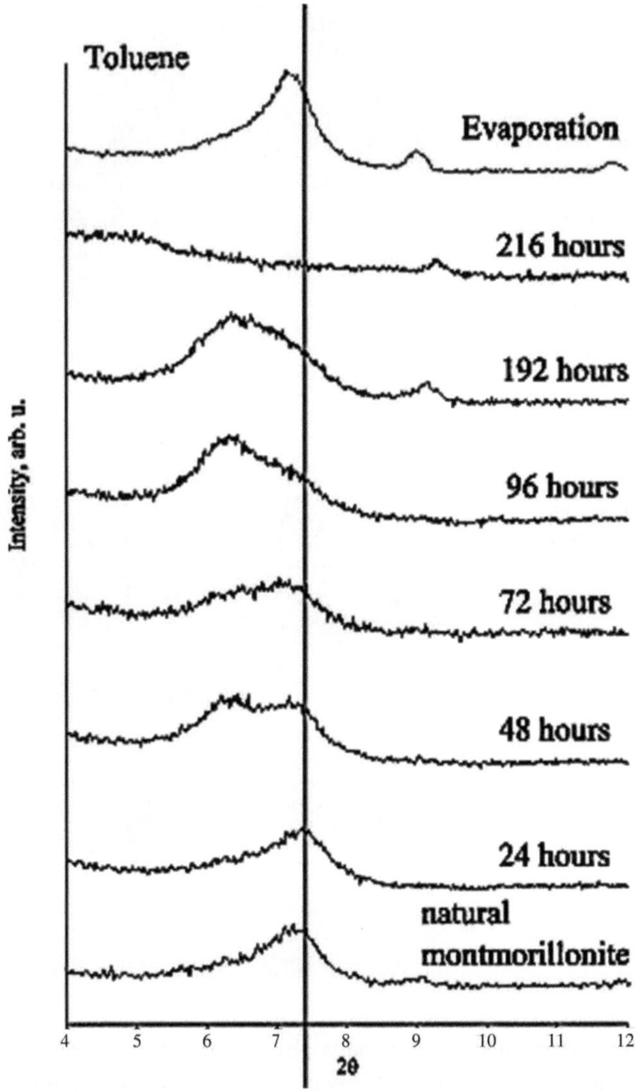

Fig. 17.23 Time sequence X-ray diffraction profiles for toluene uptake on natural montmorillonite clay suspended in groundwater (Cosultchi et al. 2004). Reprinted from Cosultchi et al. (2004). Copyright (2004) with permission from Elsevier

shows the XRD profiles observed for crude oil retention on montmorillonite clay during increasing time sequences. An increase in d-spacing is observed at the beginning of crude oil–clay contact, followed by a decrease due to evaporation, but the d-spacing value (12.11 Å) after 216 h of contact is smaller than that of the original clay. The clay changes are observed in the morphology of crude oil–clay materials (Fig. 17.24).

Fig. 17.24 SEM micrographs of **a** montmorillonite clay showing a poorly layered particle of 40 × 20 μm with irregular surface, and **b** the same sample after impregnation with crude oil, showing a well-layered particle of about 10 μm × 10 μm and surrounded by small particles of about 1 μm of diameter, not observed in the original clay sample. This image shows the dissolution of the clay crystal edges (Cosultchi et al. 2004). Reprinted from Cosultchi et al. (2004). Copyright (2004) with permission from Elsevier

Adsorption of petroleum compounds also occurs in sandy aquifer materials, which are characterized by a much lower surface area. An example is the irreversible sorption of benzene in a sandy aquifer, from an unconfined alluvial plain near the Han River in South Korea (Baek et al. 2003). In a laboratory column experiment, it was observed that peak arrivals following pulse and step injection of KCl and benzene coincide well, although the relative peak concentration of benzene was much lower than that of KCl (Fig. 17.25).

Aerobic and anaerobic biodegradation of hydrocarbons comprising various terminal electron acceptors is considered an important pathway for natural attenuation of petroleum hydrocarbons in groundwater. Most dissolved petroleum hydrocarbons are biodegradable under aerobic conditions (Kao et al. 2010). However, the prevalence of electron acceptors such as nitrate and sulfate, as well

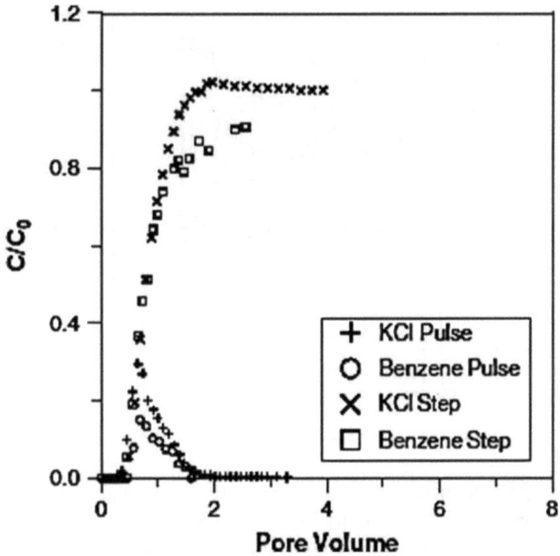

Fig. 17.25 Breakthrough curves of KCl and benzene tracers injected with square pulse and step methods into sandy materials (Baek et al. 2003). Reprinted from Baek et al. (2003). Copyright 2003 with permission of John Wiley and Sons

as ferrous iron and methane, indicates the potential extent of anaerobic biodegradation. However, this type of attenuation may lead only to a partial decrease in petroleum hydrocarbon presence in contaminated groundwater. Suarez and Rifai (2002) evaluated the natural attenuation of a polluted groundwater situated under an industrial facility in a coastal area of the United States. Between 1950 and 1985, the facility produced ethylene, diethylene, and ethylene derivates using raw materials such as crude and refined BTEX and naphthalene. In addition to daily contamination by facility wastes, an accidental pipe rupture in 1977 caused a spill of about 300,000 gallons of BTEX, containing 70 % benzene. This resulted in a petroleum hydrocarbon plume in the groundwater, with maximum benzene and xylene concentrations of 1,780 and 200 mg/L, respectively, in 1979. Due to natural attenuation, the maximum benzene concentration declined to 580 mg/L by 1996 (Fig. 17.26), while that of xylene reached 57 mg/L (Fig. 17.27). In spite of uncertainties associated with the site hydrology, analyses of mass flux proved to be a good tool for assessing the contribution of biodegradation to natural attenuation. The study showed that the total loss of BTEX attributed to biodegradation was 100,000 pounds; from source well data (half-life of 15 years), an attenuation rate of 0.0001 d^{-1} was derived. At this rate of natural attenuation, complete decontamination in the petroleum hydrocarbon polluted area will occur only after about 250 years (Suarez and Rifai 2002). Under such conditions, the groundwater contamination may be considered as irreversible.

Another example of microbiological influence on natural attenuation of petroleum hydrocarbons is found in a monitoring study in a shallow, unconfined aquifer located in Williamtown, Australia (Naidu et al. 2012). The contamination source was a fuel facility, which caused a contamination plume about 30 years ago. Groundwater sampling and analysis revealed that naphthalene and xylene are the

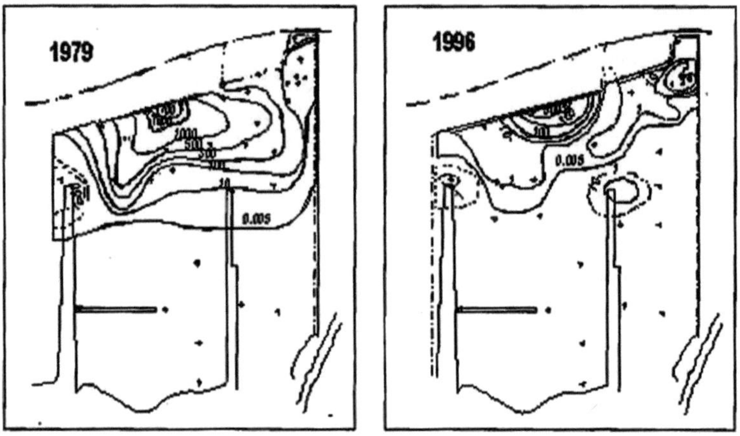

Fig. 17.26 Benzene concentrations from 1979 and 1996 at a polluted groundwater site situated under an industrial facility in a coastal area of United States (Suarez and Rifai 2002). Reprinted from Suarez and Rifai (2002). Copyright 2002 with permission of John Wiley and Sons

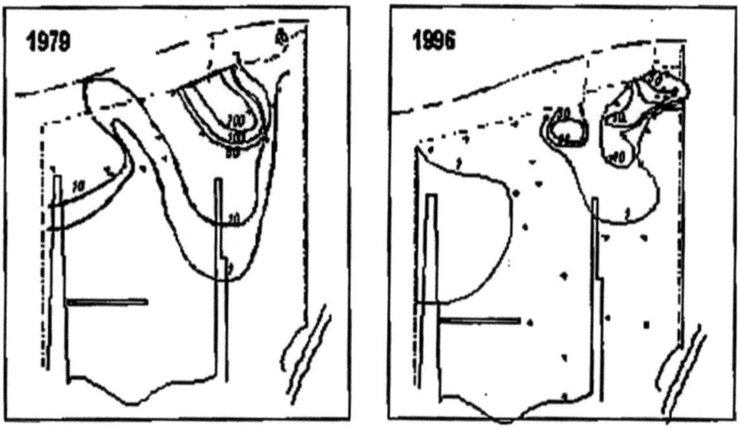

Fig. 17.27 Xylene concentrations from 1979 and 1996 at a polluted groundwater site situated under an industrial facility in a coastal area of the United States (Suarez and Rifai 2002). Reprinted from Suarez and Rifai (2002). Copyright 2002 with permission of John Wiley and Sons

main contaminants. Sulfate concentrations that ranged from 25 to 40 mg/L, and methane presence concurrent with depletion of dissolved oxygen, indicate anaerobic degradation. Dissolved oxygen is low in the source zone and increases to the edge of the plume, showing aerobic degradation. Under no attenuation and based only on hydrological parameters, the calculated maximum plume distance from the spill after 30 years could reach about 11,000 m. However, the plume in the present form has extended only about 170 m from its source. Naidu et al. (2012) attribute this difference to biodegradation, which was confirmed by the presence of microorganisms able to degrade hydrocarbons. Monitoring of dissolved TPH in groundwater during the tears 2003–2010 demonstrated a relatively stable plume that neither shrank nor expanded. The low rate of natural attenuation and the fact that the contamination in the centerline of the plume does not decrease with time suggest that such changes in groundwater chemistry are irreversible on a human lifetime scale.

The above examples show that despite the decrease in the initial contaminant mass due to a natural attenuation, a fraction of the petroleum hydrocarbon presence remains "irreversible," affecting groundwater chemistry.

References

Alewell C, Ambruster M, Bittersohl J, Evans CD, Messenburg H, Moritz K, Prechtel A (2001) Are there signs of acidification reversal in freshwaters of the low mountain ranges in Germany? Hydrol Earth Syst Sci 5:367–378

Appelo CAJ, Postma D (2005) Geochemistry, groundwater and pollution (2nd ed). AA Balkema, Rotterdam, 649 pp

Appleyard SJ, Angeloni J, Watkins R (2006) Arsenic-rich groundwater in an urban area experiencing drought and increasing population density, Perth, Australia. Appl Geochem 21:83–97

Baek DS, Kim SB, Kim DJ (2003) Irreversible sorption of benzene in sandy aquifer materials. Hydrol Process 17:1239–1251

Barber L, Keefe S, LeBlanc D, Bradley P, Chapelle F, Meyer M, Lofton K, Kolpin D, Rubio F (2009) Fate of sulfamethoxazole, 4-nonylphenol and 17β-estradiol in groundwater contaminated by wastewater treatment plant effluent. Environ Sci Technol 43:4843–4850

Barber LB, Thurman EM, Schroeder MP (1988) Long-term fate of organic micropollutants in sewage-contaminated groundwater. Environ Sci Technol 22:205–211

Barlow PM, Reichard EG (2010) Saltwater intrusion in coastal regions of North America. Hydrogeol J 18:247–260

Cavaller TC, Lavy TL, Mattice JD (1991) Persistence of selected pesticides in groundwater samples. Groundwater 29:225–231

Cerejeira MJ, Viana P, Batista S, Pereiera T, Silva E, Valerio MJ, Silva A, Ferreira M, Silva-Fernandez AM (2003) Pesticides in Portuguese surface and ground waters. Water Res 37:1055–1063

Champidi P, Stamatis G, Zagana E (2011) Groundwater quality assessment and geogenic and anthropogenic effect estimation in Erasinos Basin (E. Attica). Eur Water 33:11–27

Choppin GR (1983) Solution chemistry of actinides. Radiochim Acta 32:43–53

Cosultchi A, Bosch P, Lara VH (2004) Adsorption of petroleum organic compounds on natural Wyoming montmorillonite. Colloid Surf A: Physicochem Eng Aspects 243:53–61

Dai M, Buesseler KO, Pike SM (2005) Plutonium in groundwater at the 100 K-Area of the US DOE Hanford Site. J Contam Hydrol 76:167–189

Drewes JE, Heberer T, Rauch T, Reddersen K (2003) Fate of pharmaceuticals during ground water recharge. Groundwater Monitor Remed 23:64–72

Dror I, Amitay T, Yaron B, Berkowitz B (2003) Salt-pump mechanism for contaminant intrusion into coastal aquifers. Science 300:950

Ehlers LJ, Kavanaugh CM (2013) Redefining the end for groundwater remediation. Groundwater 51:170–174

El Moujabber M, Bou Samra B, Darwish T, Atallah T (2006) Comparison of different indicators for groundwater contamination by seawater intrusion on the Lebanese coast. Water Resour Man 20:161–180

Ferguson H, Bateridge T (1982) Salt status of glacial till soils of north central Montana as affected by the crop- fallow system of dryland farming. Soil Sci Soc Am J 46:807–810
Flowers TC, Hunt JR (2007) Viscous and gravitational contributions to mixing during vertical brine transport in water-saturated porous media. Water Resour Res 43:W01407
Franken G, Postma D, Duijnisveld WHM, Bottcher J, Molson J (2009) Acid groundwater in an anoxic aquifer: reactive transport modelling of buffering processes. Appl Geochem 24:890–899
Gilliom RJ (2007) Pesticides in US streams and groundwater. Environ Sci Technol 41:3409–3414
Gvirtzman H (2002) Israel water resources, Chapters in Hydrology and environmental sciences, Yad Ben-Zvi Press, Jerusalem, 301 pp (in Hebrew)
Harvey CF, Swartz CH, Badruzzaman ABM, Keon-Blute N, You W, Ali MA, Jay J, Beckie R, Niedan V, Brabander D, Oates PM, Ashfaque KN, Islam S, Hemond HF, Ahmed MF (2002) Arsenic mobility and ground water extraction in Bangladesh. Science 298:1602–1606
Harvey CF, Ashfaque KN, Yu W, Badruzzaman ABM, Ashraf Ali M, Oates PM, Michael HA, Neumann RB, Beckie R, Islam S, Ahmed MF (2006) Groundwater dynamics and arsenic contamination in Bangladesh. Chem Geol 228:112–136
Heberer T (2002) Occurrence, fate and removal of pharmaceutical residues in the aquatic environment: a review of recent research data. Toxicol Lett 131:5–17
Hemme CL, Deng Y, Gentry TJ, Fields MW, Wu L, Barua S, Barry K, Tringe SG, Watson DB, He Z, Hazen TC, Tiedje JM, Rubin EM, Zhou (2010) Metagenomic insights into evolution of a heavy metal-contaminated groundwater microbial community. ISME J 5:660–672
Holm JV, Rügge K, Bjerg PL, Christensen TH (1995) Occurrence and distribution of pharmaceutical organic compounds in the groundwater downgradient of a landfill (Grindsted, Denmark). Environ Sci Technol 29:1415–1420
Huang OM (1999) Abiotic catalysis. In: Sumner ME (ed) Handbook of Soil Science. CRC Press, Boca Raton, pp 303–327
Islam FS, Gault AG, Boothman C, Polya DA, Charnock JM, Chatterjee D, Lloyd JR (2004) Role of metal-reducing bacteria in arsenic release from Bengal delta sediments. Nature 430:68–71
Jacks G, Knutsson G (1982) Kansligheten for groundvattenforsuring I olika delar av landet (huvudrapport). With English summary. KHM Tek Rap 49, Statens Vattenfallsverk, Vallingby
Jacks G, Knutsson G, Maxe L, Flykner A (1984) Effect of acid rain on soil and groundwater in Sweden. In: Yaron B, Dagan G, Goldschmid J (eds) Pollutants in porous media. Springer, Heidelberg, pp 94–114
Japan Atomic Energy Research Institute—JAERI (2000) Prediction on migration behavior of radionuclides on the basis of their dissolved forms in groundwater. http://jolisfukyu.tokai-sc.jaea.go.jp/fukyu/tayu/ACT00E/11/1105.htm
Jensen DL, Hom PE, Christensen TH (2000) Soil and ground water contamination with heavy metals at two scrap iron and metal recycling facilities. Waste Manage Res 18:52–63
Kao C, Chien H, Surampalli R, Chien C, Chen C (2010) Assessing of natural attenuation and intrinsic bioremediation rates at a petroleum-hydrocarbon spill site: laboratory and field studies. J Environ Eng 136:54–67
Kaplan DI, Demirkanli DI, Gumpas L, Powell BA, Field AR, Molz FJ, Srrkiz SM (2006) Eleven-year field study of Pu migration from Pu III, IV and VI sources. Environ Sci Technol 40:443–449
Kelley JM, Bond LA, Beasley TM (1999) Global distribution of Pu isotopes and ^{237}Np. Sci Total Environ 237/238:483–500
Kersting AB, Efurd DW, Finnegan DL, Rokop DJ, Smith DK, Thompson JL (1999) Migration of plutonium in groundwater at the Nevada test site. Nature 397:56–59
Kjoller C, Postma D, Larsen F (2004) Groundwater acidification and the mobilization of trace metals in a sandy aquifer. Environ Sci Technol 38:2829–2835
Knutsson G (1994) Acidification effects on groundwater-prognosis of the risk for the future. Groundwater Resources at Risk (Proceedings of the Helsinki Conference, June 1994) IAHS Publication no. 222, pp 17
Kolpin DW, Barbash JE, Gilliom RJ (1998) Occurrence of pesticides in shallow groundwater of the United States: initial results from the national water-quality assessment program. Environ Sci Technol 32:558–566

Lacombe PJ, Carleton GB (2002) Hydrologic framework, availability of water supplies and salt intrusion. Cape May County, New Jersey. US Geological Survey science Investigations Report 2004–5067

Lee JY, Choi JC, Yi MJ, Kim JW, Cheon JY, Choi YK, Choi MJ, Lee KK (2005) Potential groundwater contamination with toxic metals in and around an abandoned Zn mine, Korea. Water Air Soil Pollut 165:167–185

Lee LS, Hagwall M, Delfino JJ, Rao PSC (1992a) Partitioning of polycyclic aromatic hydrocarbons from diesel fuel into water. Environ Sci Technol 26:2104–2110

Lee LS, Rao PSC, Okuda I (1992b) Equilibrium partitioning of polycyclic aromatic hydrocarbons from coal tar into water. Environ Sci Technol 26:2110–2115

Likens GE, Bormann FH (1995) Biogeochemistry of a forested ecosystem, 2nd edn. Springer, New York

Loos R, Locoro G, Comero S, Contini S, Schwesig D, Werres F, Balsaa P, Gans O, Weiss S, Blaha L, Bolchi M, Gawlik BM (2010) Pan-European survey on the occurrence of selected polar organic persistent pollutants in groundwater. Water Res 44:4115–4126

Mastrocicco M, Colombani N, Sbarbati C, Petitta M (2012) Assessing the effect of saltwater intrusion on petroleum hydrocarbons plumes via numerical modeling. Water Air Soil Pollut 223:4417–4427

McBride MB (1994) Environmental chemistry of soils. Oxford University Press, Oxford

McCarthy TS (2006) Groundwater in the wetlands of Okavango delta, Botswana, and its contribution to the structure and function of the ecosystem. J Hydrol 320:264–282

Mercado A, Avron M, Kahanowich Y (1975) Groundwater salinity in the coastal plain, Israel. Tahal, Tel Aviv Prog Rep No. 1, 68 pp (in Hebrew)

Mondal NC, Singh VP (2011) Hydrochemical analysis of salinization for a tannery belt in Southern India. J Hydrol 405:235–247

Murray KS (2006) Dissolved heavy metals in shallow ground water in a southeastern Michigan urban watershed. J Am Water Resour Assoc 42:777–792

Naidu R, Nandy S, Megharaj M, Kumar RP, Chadalavada S, Chen Z, Bowman M (2012) Monitored natural attenuation of a long-term petroleum hydrocarbon contaminated site: a case study. Biodegradation 23:881–895

Neumann R, Ashfaque K, Badruzzaman A, Ali M, Shoemaker J, Harvey C (2010) Anthropogenic influences on groundwater arsenic concentrations in Bangladesh. Nature Geosci 3:46–52

Nickson RT, McArthur JM, Ravenscroft P, Burgess WG, Ahmed KM (2000) Mechanism of arsenic release to groundwater, Bangladesh and West Bengal. Appl Geochem 15:403–413

Oden S (1968) The acidification of air and precipitation and its consequences in the natural environment. Bulletin of the ecological research communications NFR. Translation Consultants Ltd, Arlington

Ojuri OO, Ola SA (2010) Modeling LNAPL plume breakthrough and saltwater intrusion for a coastal site in the south-western Nigeria. AU J T 14:119–130

Pang L, Close ME, Noonan MJ, Flinto MJ, van den Brink P (2005) A laboratory study of bacteria-facilitated cadmium transport in alluvial gravel aquifer media. J Environ Qual 34:237–247

Pang L, Simunek J (2006) Evaluation of bacteria-facilitated cadmium transport in gravel columns using the Hydrus colloid-facilitated solute transport model. Water Resour Res 42:W12S10

Peck AJ, Hurle DH (1973) Chloride balance of some farmed and forested catchments in southwestern Australia. Water Resour Res 9:648–5657

Pedroni T (2000) Environmental science in action: Hanford Federal Nuclear Facility. http://www.environmentaleducationohio.org/Biosphere/Case%20Studies/hanford.ht

Pellaud B (2002) Proliferation aspects of plutonium recycling. J Nucl Mater Manage 31:30–38

Polizotto ML, Harvey CF, Li G, Badruzzman B, Ali A, Newville M, Sutton S, Fendorf S (2006) Solid-phase and desorption processes of arsenic within Bangladesh sediments. Chem Geol 228:97–111

Pye VI, Patrick R (1983) Ground water contamination in the United States. Science 221:713–718

Rassler M, Michalke B, Schramel P, Schulte-Hostede S, Kettrup A (1998) Speciation of inorganic arsenic and selenium in contaminated ground water samples—distribution and long-term stability of species. Inter J Environ Anal Chem 72:195–203

Siepak M, Niedzielski P, Bierla K (2004) Determination of inorganic speciation forms of arsenic, antimony and selenium in water from a Grate Ashes dumping ground as an element of hydrogeochemical monitoring of pollution spread. Polish J Environ Studies 13:709–713

Smedley PL (2008) Sources and distribution of arsenic in groundwater and aquifers. In Appelo CAJ (ed) Arsenic in groundwater: a world problem (Chap. 4). Proceedings of the IAH Seminar, Utrecht, Netherlands, Nov 2006

Smith JVS, Jankowski J, Sammut J (2006) Vertical distribution of As(III) and As(V) in a coastal sandy aquifer: Factors controlling the concentration and speciation of arsenic in the Stuarts Point groundwater system, northern New South Wales, Australia. App Geochem 19(9):1479–1496

Smolen JM, Stone AT (1998) Organophosphorous ester hydrolysis catalyzed by dissolved metals and metal containing surfaces. In: PM Huang et al. (eds.) Soil chemistry and ecosystem health, SSSA Spec Publ 52, Madison. pp 157–171

Steel GV, Johnson HM, Sandstrom MW, Capel PD, Barbash JE (2008) Occurrence and fate of pesticides in four contrasting agricultural settings in the United States. J Env Qual 37:1115–1132

Suarez DL (1989) Impact of agricultural practices on groundwater salinity. Agri Ecosys Environ 26:215–227

Suarez MP, Rifai HS (2002) Evaluation of BTEX remediation by natural attenuation at a coastal facility. Groundwater Monitor Remed 22:62–77

Taylor RN, Warneke T, Milton JA, Croudace IW, Warwick PE, Nesbitt RW (2001) Plutonium isotope ratio analysis at femtogram to nanogram levels by multicollector ICP-MS. J Anal Atom Spec 16:279–284

Tulipano L, Fidelibus DM (2002) Mechanism of groundwater salinization in a coastal Karstic subject to overexploitation. In: Proceedings of 17th salt water intrusion meeting, The Netherlands, 6–10 May 2000

Ulrich B (1985) Interaction of indirect and direct effects of air pollutants in forests. In: Troyanowsky C (ed) Air pollution and plants. VCH Verlagsgesellschaft mbH, Weinhaim, pp 149–181

van Weert F, Gun van der J, Reckman J (2009) Global overview of saline groundwater occurrence and genesis. International Groundwater Resources Assessment Centre report GP2009–1, 94 pp

Ward J, Andersen MS, Appleyard S, Clohessy S (2010) Acidification and trace metal mobility in soil and shallow groundwater on the Gnangara mound, Western Australia. World Congress of soil science, soil solutions for a changing world, Brisbane, Australia

Werner DA, Ward JD, Morgan LK, Simmons CT, Robinson NI, Teubner MD (2012) Vulnerability indicators of sea water intrusion. Groundwater 50:48–58

Yaron B, Dror I, Berkowitz B (2008) Contaminant induced irreversible changes in properties of the soil-vadose-aquifer zone: an overview. Chemosphere 71:1409–1421

Yaron B, Dror I, Berkowitz B (2009) Chemical contaminants as factor of soil–subsurface metagenesis. IUSS Bull 115:11–12

Yaron B, Dror I, Berkowitz B (2010) Contaminant geochemistry—a new perspective. Naturwiessenschaften 97:1–17

Yaron B, Dror I, Berkowitz B (2012) Soil subsurface change-chemical pollutant impact. Springer, Heidelberg

Yaron-Marcovich D, Dror I, Berkowitz B (2007) Behavior and stability of organic contaminant droplets in aqueous solutions. Chemosphere 69:1593–1601

Zemo DA (2006) Sampling in the smear zone: evaluation of nondissolved bias and associated BTEX, MTBE, and TPH concentration in ground water samples. Groundwater Monitor Remed 26:125–133

Chapter 18
Contaminant Impacts on the Soil–Subsurface Solid Phase

In this chapter, we consider irreversible changes to the soil–subsurface solid phase matrix and properties that, under specific conditions, may be caused by chemical contamination. We present examples from the literature that show irreversible changes in the soil–subsurface system caused by contaminants originating from agrochemical practices, irrigation with saline water, and disposal of municipal and industrial products. We suggest that chemical contamination of the natural soil–subsurface solid phase may lead to the formation of a "new" soil–subsurface porous medium with matrix and properties different than those of the natural one.

18.1 Impact of Contaminants Originating from Agriculture

A portion of chemicals periodically applied onto the land surface as fertilizers, plant protection chemicals, and reclamation amendments may be retained irreversibly on the soil–subsurface solid phase, leading to changes in natural soil properties. For example, continuing use of ammonia-based fertilizers can induce soil acidity, while long-term application of phosphorus may lead to an increase in bicarbonate-extractable phosphorus relative to untreated plots. Trace elements (micronutrients) may be affected by the soil pH and, consequently, by the long-term addition of nitrogen- and phosphorus-containing compounds (Schwab et al. 1990, and references therein). Soil–subsurface alkaline and acid properties may be changed by chemical amendments such as gypsum and lime, affecting also the fate of other agrochemicals, which together may lead directly or indirectly to changes in soil–subsurface physical properties. We consider several specific examples below.

B. Berkowitz et al., *Contaminant Geochemistry*,
DOI: 10.1007/978-3-642-54777-5_18,

18.1.1 Fertilizers

Nitrogen forms detected in soils include NO_3^-, NO_2^-, exchangeable and mineral-fixed NH_4^+, and nitrous oxide (N_2O_2). Mineral-fixed NH_4^+ is a function of soil type and environmental conditions, comprising 10–50 % of the total N. When large amounts of NH_4^+ are applied, it can, theoretically, become a dominant exchangeable cation and, like Na^+, may favor dispersion on soil colloids. However, in most situations, NH_4^+ is rapidly nitrified to NO_3^- and may cause dispersion of soil colloids only under unusual conditions of low pH and low moisture content (Haynes 1984). In early agricultural management schemes, when this process was ignored, NH_4^+-induced dispersion of soil colloids caused reduced infiltration rates and crust formation. For example, Pilsbury (1947) reported a significant decrease in the soil infiltration capacity of an irrigated Californian soil after application of $(NH_4)_2SO_4$. This resulted in a pronounced decrease of soil pH, a drastic decrease in nitrification, and a large accumulation of NH_4 in the soil.

A decreased nitrification process under low temperature conditions caused soil colloid dispersion, crust formation, and erosion following broad application of NH_4NO_3 on soil in Nebraska (Fox et al. 1952). Soil dispersion and crusting after continual application of $(NH_4)_2SO_4$ on an irrigated soil were also reported by Aldrich et al. (1945). They concluded that under irrigation, dispersed soil colloids were transported at depth, clogging the subsurface pores, and irreversibly changing the infiltration rate of the NH_4^+ applied to the soil. Moreover, soil dispersion and crust formation following NH_4^+ accumulation can stimulate soil erosion, which is also an irreversible process.

Analysis of long-term fertility trials has revealed changes occurring in soil following continuous application of nitrogen fertilizers. For example, Barak et al. (1997) examined long-term soil acidification due to nitrogen fertilizer in fertility trial plots at Arlington Agricultural Research Station, Wisconsin. The trial plots, located on Plano silt loam soil (Argiuoll), were initiated by L. A. Paterson in 1962 and subsequently cultivated under various crops; over a 30-year period, the plots received annual amounts of nitrogen (as ammonium nitrate or urea) of 0, 56, 112, and 168 kg N/ha.

Exchangeable acidity of soils was strongly dependent upon the rate of N fertilization applied during the course of the 30-year fertility trials (Fig. 18.1). Exchangeable acidity of the 112 and 168 kg N/ha/yr treatments averaged 0.48 and 1.58 $cmol_c$/kg greater than the control plot, which did not receive N fertilizers. The increase in exchangeable acidity following N fertilization led to a decline in the soil exchangeable base cations and is related directly to the amount of N applied.

The relationship among exchangeable Ca^{2+}, Mg^{2+} and $K^+ + Na^+ + NH_4^+$, and N fertilizer rate in the long-term fertility plots is presented in Fig. 18.2. Under the higher N treatment, the declines in the soil contents of Ca^{2+} and Mg^{2+} were 31 and 36 %, respectively. No statistically significant response of summed monovalent cations to N fertilizer levels was observed. The cation exchange capacity of the Arlington long-term fertility plots, determined as the sum of exchangeable base

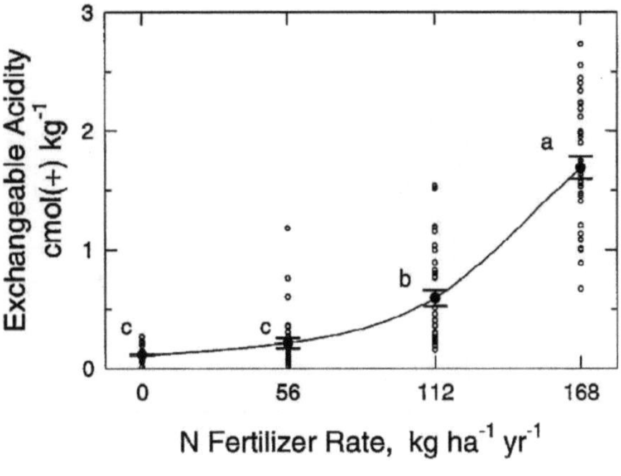

Fig. 18.1 Relationship between exchangeable acidity and N fertilizer rates in long-term fertility plots. Error bars indicate standard deviation of the mean. Means indicated by the same letter are not significantly different at the 5 % level by ANOVA. Cubic spline drawn through treatment means (Barak et al. 1997)

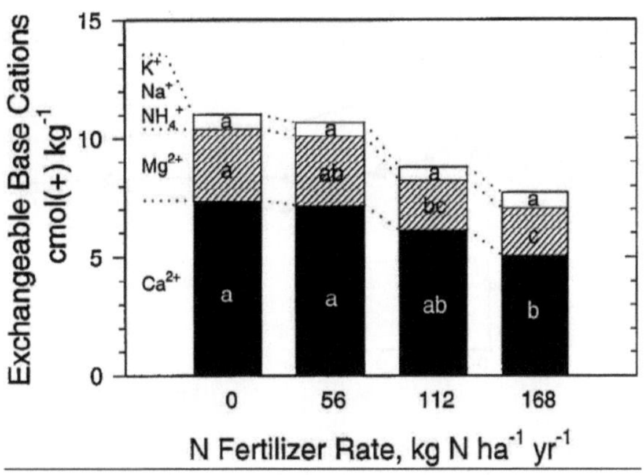

Fig. 18.2 Relationship among exchangeable Ca^{2+}, Mg^{2+}, and K^{+}+Na^{+}+NH_4^{+}, and N fertilizer rate in long-term fertility plots. Mean comparisons are valid for the same measure across treatments. Means indicated by the same letter are not significantly different at the 5 % level by ANOVA (Barak et al. 1997)

cations and exchangeable acidity, showed strong correlation to the N fertilizer rate. Long-term soil acidification due to nitrogen fertilizer input has also been confirmed through similar results obtained in other field trials in North America, reported for Kansas silt loam (Schwab et al. 1989), Saskatchewan loamy soil (Bouman et al. 1995), and Kentucky silt loam soil (Blevins et al. 1977).

Phosphorus forms include organic and inorganic fractions, consisting of compounds characterized by different solubility and availability. Calcium phosphates, which are the predominant stable compounds in soils, are characterized by various solubilities as follows: $H_2PO_4 > H_2PO_4^{2-} > PO_4^{3-}$. Stable phosphate compounds such as tricalcium phosphate and apatite are favored by high pH, while Al- and Fe-phosphate are formed at low pH. In a field experiment testing phosphate application on a red-yellow podzolic soil from North Carolina, Lutz et al. (1966) observed that P-treated plots were moist, loose, and easy to plow, or had a lower bulk density and a higher moisture content, compared with untreated control plots. Later studies (Lutz and Pinto 1965; Lutz et al. 1966) on the effect of phosphorus on soil water retention and soil hardness provide more specific information on this phenomenon.

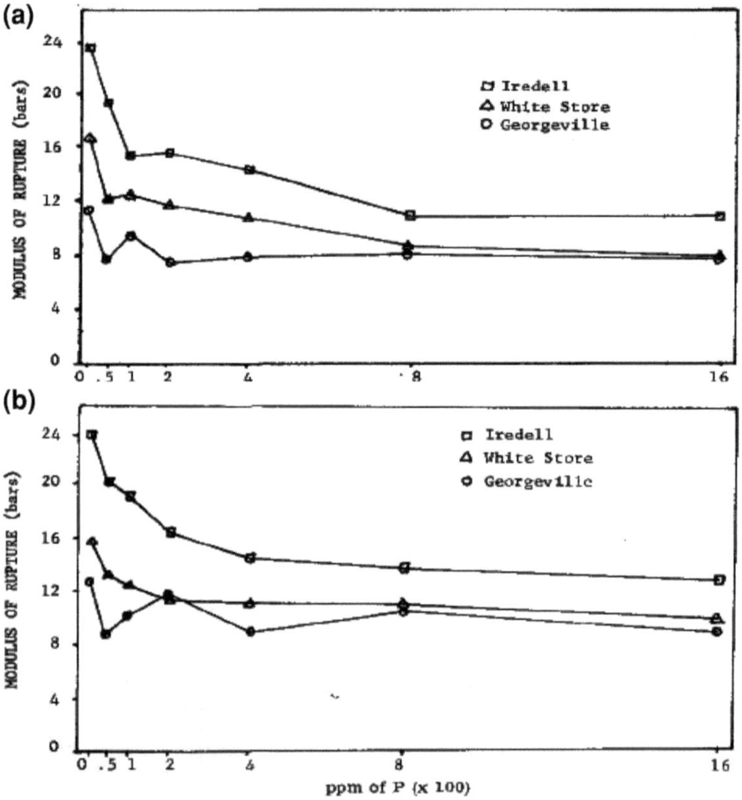

Fig. 18.3 Effect of P on the modulus of rupture of three soils using **a** H_3PO_4, and **b** $Ca(H_2PO_4)_2$ as sources of P (modified after Lutz and Pinto 1965). Copyright 1965 Soil Science Society of America. Reprinted with permission

The effect of phosphorus on soil hardness was determined in laboratory experiments by measuring the modulus of rupture of three soils with different clay mineralogy. The effects of phosphoric acid, H_3PO_4, and monocalcium phosphate, $Ca(H_3PO_4)_2$, on the modulus of rupture of these soils are shown in Fig. 18.3. The modulus of rupture decreased with increasing rates of P; both phosphoric acid and monocalcium phosphate led to greater reductions in the modulus of rupture in soils containing higher amounts of 2:1 type clay. In all soils studied, phosphoric acid had a greater effect than monocalcium phosphate.

The effect of P on water retention properties of soils was determined in laboratory experiment by using the same soils and type of phosphate fertilizer described above. Water retention was measured by determining the percent water held at increasing pressure, from 1/10 to 15 bars. In all soils studied, P increased the water-holding properties appreciably (Lutz et al. 1966). Associated particle charge determination established that P increased the negative charge of the particles and that surface charge is the dominant factor in water retention.

The increase in water retention capacity is associated with the increase in negative charge of the soil particles, and the charge is related closely to the Al-phosphate/Fe-phosphate ratio. Haynes (1984) showed that addition of P to acidic soils results in the precipitation of Al as insoluble Al-phosphates, which act as flocculating and cementing agents in a similar way to hydroxy-Al compounds. Because dispersion and flocculation are, in the case of acid soils, part of the pathway of phosphate-induced changes in soil hardness and water retention capacity, we consider such changes to be irreversible.

18.1.2 Micronutrients

We consider here two representative microelements, Cu and Zn, which may be strongly sorbed onto the surfaces of soil–subsurface layer silicates. Sorption studies in pure systems have shown that the clay fraction of soils binds Cu and Zn by ion exchange, fixation, and hydrolysis reactions, all of which are pH dependent. In a study performed on acidic clay, Cavallaro and McBride (1984) found that the oxide fraction of the soil solid phase most strongly controls Zn and Cu fixation and solubility. In the tested range of pH (4–7), up to 95 % of the sorbed metal was retained irreversibly, exhibiting a hysteretic process. The results of López-Periago et al. (2008) on copper retention kinetics showed that independent of metal loading, the release rate coefficients were more than ten times lower than those of retention, confirming the irreversibility of the process. In the examples that follow, a direct change in soil rheological properties after Zn disposal is noted, while in the case of Cu application, the soil bacterial community is initially affected, leading to a change in the soil aggregation status.

Zinc (Zn^{2+}) effects on soil rheological properties, under different pH and concentration values, were studied by Ouhadi et al. (2006). More specifically, Ouhadi et al. (2006) examined the influence of zinc on the osmotic compressibility and consolidation of bentonite, under variable pH regimes. Changes in the creep behavior or volume strain of earth material, upon exposure to a high concentration of inorganic electrolytes, is explained in terms of the osmotic compressibility (consolidation) phenomenon and the formation of macropores among aggregates (Di Maio 1996; Suzuki et al. 2005). Ouhadi et al. (2006) also showed that the addition to pore water of a trace element such as Zn^{2+}, in increasing amounts, decreases both the soil-clay Atterberg liquid limits (LL) and the plasticity index (PI) of soil bentonite (Table 18.1).

The Atterberg limits are a basic measure of the nature of fine-grained soil clays at various solid/liquid ratios. The liquid limit is the water content at which a soil clay changes from plastic to liquid behavior; the plasticity index is a measure of soil-clay plasticity. The consolidation performance of soil bentonite with different Zn^{2+} concentrations in pore water is presented in Fig. 18.4. It can be observed that the consolidation performance increases with the increase in Zn^{2+} concentration in pore water. Based on this series of rheological tests, Ouhadi et al. (2006) showed that the

Table 18.1 Atterberg liquid limits (LL) and plasticity index (PI) as affected by increasing Zn^{2+} concentrations in bentonite-heavy metal (nitrate forms) mixtures (Ouhadi et al. 2006). Reprinted from Ouhadi et al. (2006). Copyright 2006 with permission from Elsevier

Cation type	Atterberg limits	Contaminant concentration (mol/L)				
		0.005	0.01	0.1	0.5	1.0
Zn^{2+}	LL	310.9	247.4	121.3	101.9	93.2
	PI	274.6	209.9	82.1	57.8	47.4

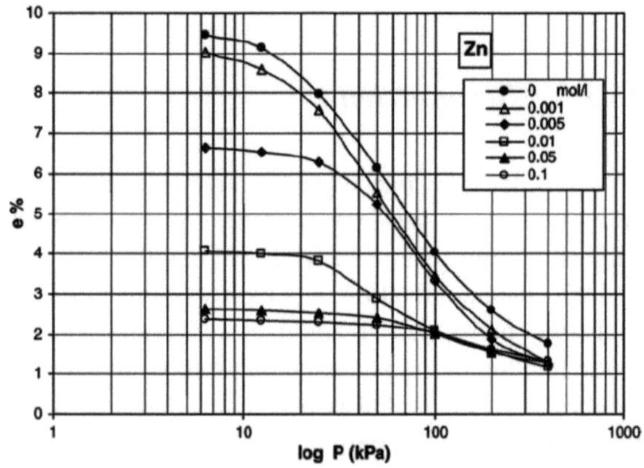

Fig. 18.4 Consolidation performance of bentonite samples, at different concentrations of Zn2 + in pore fluid; e denotes void ratio, and P denotes osmotic pressure (Ouhadi et al. 2006). Reprinted from Ouhadi et al. (2006). Copyright 2006 with permission from Elsevier

presence of Zn^{2+} in pore water causes the formation of a kind of osmotic pressure. This pressure in turn changes the osmotic equilibrium of the natural system and thus affects the natural rheological properties of the soil–subsurface system.

Copper (Cu^{2+}) application as a micronutrient to agricultural lands, in quantities greatly in excess of those required by crops, occurs with sewage sludge application, and on orchard and vineyard growths when Cu^{2+} salts are used as pesticides. Although there are several potential sources of Cu^{2+} applied to soil, Cu^{2+} compounds used as pesticides form the largest input. For example, according to Epstein and Bassein (2001), about 3.8×10^3 ton of Cu^{2+} was used as pesticide in 1997 in California. Copper-induced phytotoxicity has been reported in vineyards, orchards, and sewage sludge farms after long-term application, and copper presence was even considered as a potential crop toxicity indicator (McBride 2001). Copper salts accumulated in soils after long-term application also have a toxic effect on soil–subsurface bacterial communities, affecting their enzymatic activities. In an experiment on tolerance of bacterial communities to long-term copper application in vineyard soils in Spain, Díaz-Raviña et al. (2007) found that increased soil bacterial toxicity to Cu^{2+} was observed at concentrations greater than 100 mg Cu^{2+}/kg soil; copper inhibits enzymatic activities of bacterial communities. Wyszkowska et al. (2005) experimented on heavy loamy and sandy silt loam soils, finding that soil contamination with $CuSO_4 \cdot 5H_2O$ at doses ranging from 150 to 600 mg Cu^{2+}/kg inhibited the activity of dehydrogenase, ureases, acid phosphatases, and alkali phosphatases. Dehydrogenase and urease, which are

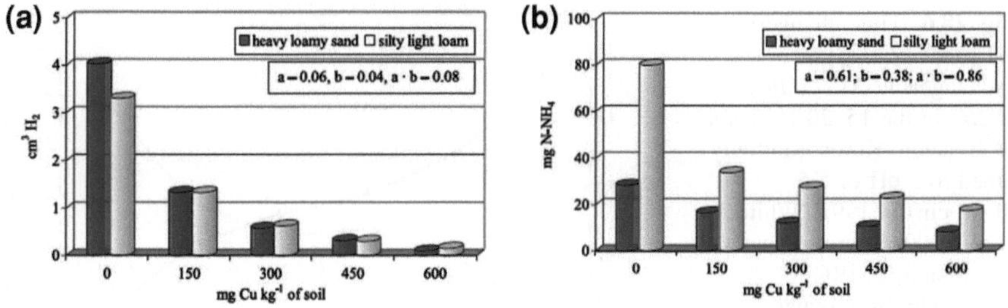

Fig. 18.5 The activity of **a** dehydrogenases (cm^3 H_2/d/kg soil) and **b** urease (mg N–NH_4/h/kg soil), in copper-contaminated soils (modified after Wyszkowska et al. 2005)

indicators of the activity of microorganisms in soil (Kiss 1999), are found to be good sensors for Cu^{2+} contamination of soils. It can be seen from Fig. 18.5 that the inhibition of the activity of these two enzymes, reflecting the decrease in the activity of soil microbial communities, is closely associated with the amount of copper added to the soil; the degree of inhibition varies among soils.

It is known, however, that soil microbial populations are responsible primarily for the formation and stabilization of soil aggregates. Potential binding agents among soil mineral particles include filamentous microbes (fungus hyphae), metabolites produced by the decomposer, and organic substances anthropogenically introduced into the soil–subsurface medium (Guidi 1981; Morel and Guckert 1983). Experimental results confirmed that for soil-sludge mixtures in which microbial activity was suppressed, no aggregative processes were recorded (Metzger et al. 1986; Metzger and Yaron 1987). As a consequence, copper contaminants that impact negatively on soil microbial communities may indirectly alter the process of aggregate formation and aggregate stability in soils. The degree of Cu^{2+} content controls the extent of aggregate formation and stability in contaminated soil, and defines the irreversibility of this process.

18.1.3 Soil–Subsurface Amendments

Properties of agricultural soils exhibiting extreme pH values are improved by adding various forms of lime and gypsum. Successive application of these compounds over many years has led to changes in soil properties which, on a lifetime scale, become irreversible.

Lime Under natural conditions, acidic soils exhibiting a pH $< \sim 4.8$ are formed on weathered acidic rocks. This process occurs in a climatic regime where the amount of rainwater exceeds the amount of water lost by evapotranspiration under vegetation canopies. Soil acidification is a continuous process supported by the natural input of very dilute carbonic acid and the leaching of weathering products. However, soil acidification has increased from anthropogenic activities. On a global scale, the acidification process has affected 10 million ha of cultivated land

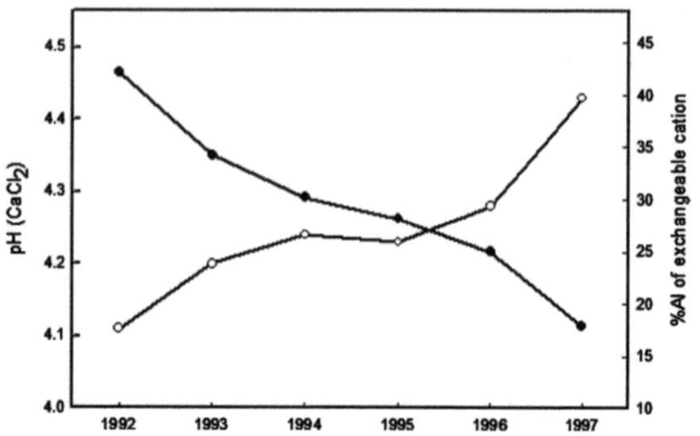

Fig. 18.6 Time change in pH (*open circle*) and exchangeable Al^{3+} (*filled circle*) in the 15–20 cm layer of an acidic Kurosol initially limed to a pH of 5.5 (0–10 cm) in 1992 (White et al. 2000) Reprinted with permission, CSIRO Publishing, http://www.publish.csiro.au/nid/72/paper/EA98013.htm

(Greenland et al. 1997), without considering forest or urban surfaces. In agricultural lands, the process of soil acidification is accelerated by the large amount of N cycling through the system, when substantial amounts of organic products and residues are removed. Permanent acidity may occur when H^+ and OH^- generation during mineral N uptake and transformation is decoupled, enriching the soil with an excess of H^+ (White 2008). Urban and industrial inputs to the soil acidification process are due generally to the release of forms of mineral sulfur and nitrogen, which are converted to sulfuric or nitric acids, leading to acid rain and/or atmospheric dry acid deposition.

Raising the pH of acidic soils through addition of lime materials is a very old procedure. In 1882, Ruffen achieved this goal by adding oyster shells to acidic soil in the coastal plain of Virginia, USA. With time, liming became a routine procedure for altering natural or anthropogenically induced soil acidity. Examples of liming materials used in Australia and New Zealand are reviewed by White et al. (2000). Under field conditions, the positive effect of lime addition consists of a combination of cementing and flocculation actions. Hoyt (1981) found that application of lime to two Canadian soils increased soil $pH_{(water)}$ to the range of 5.5–6.5 causing a significant change in distribution of dry aggregates toward coarser fractions, and increasing resistance of soil to compression. Liming is a slow release process; the time required to change the pH of an acidic soil (Kurosol) by surface application of lime materials is shown in Fig. 18.6, where the change of pH_{Ca} from 4.1 to 5.5 occurred only 5 years after application.

The downward movement of liming products such as OH^- and HCO_3 released from $CaCO_3$, through their diffusion or leaching, is overwhelmed by the buffering capacity of soils. As a consequence, changing soil pH by liming into the subsurface regime can be achieved only by direct subsoil application of liming materials. The slow release properties of the liming material, associated with continuous land application of lime within sustainable agriculture management, can cause irreversible changes to soil acidity.

Increases in pH as a result of liming may cause a dispersion/flocculation process that leads to irreversible changes in soil physical properties. At low pH, acidic soils are normally flocculated because Al^{3+} and H^+ activity in soil solution causes

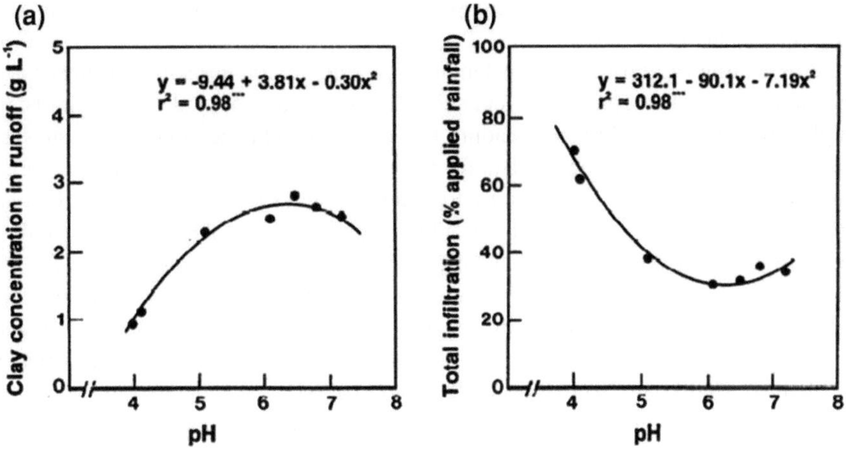

Fig. 18.7 Relationship between pH and clay dispersion expressed as **a** the amount of clay concentrated in runoff and **b** the total infiltration as a percentage of applied rainfall (Haynes and Naidu 1998)

compression of the double layer and flocculation of clays. Liming-induced increases in pH lead to an increase in the negative charge on soil surfaces, and thus to the ratio of negative to positive charges. This in turn causes a decrease in Al^{3+} activity because of aluminum precipitation as hydroxy-Al polymers. As a result, the dominant repulsive forces among particles lead to their dispersion. Under leaching, the dispersed particles formed under liming may result in a surface crust, clog the pore space, and/or be transported and redeposited at depth; these processes irreversibly affect water and solute transport in the soil–subsurface system (Roth and Pavan 1991; Haynes and Naidu 1998; Bednarek and Reszka 2007).

Irreversible effects of liming on the soil structure can also be attributed to the flocculation and cementing actions of applied lime and of newly precipitated iron and aluminum oxides and hydroxides (Haynes and Naidu 1998). The relationship between pH_{KCl} and clay dispersion for a Brazilian Oxisol, expressed as the amount of clay concentrated in runoff and as the total infiltration (as a percentage of applied rainfall), is shown in Fig. 18.7.

Gypsum, a hydrated calcium sulfate, is a naturally occurring mineral that may also be available as a byproduct material (e.g., phosphogypsum, from the phosphorus fertilizer industry). Gypsum has been used on agricultural soils to increase the Ca^{2+} content of acidic soils and to exchange Na^{+} with Ca^{2+} in calcareous saline-sodic soils. Gypsum solubility depends on both the composition of soil solution and the exchange phase, and is lower when applied as mined gypsum than as a byproduct, in the widely used form phosphogypsum. Gypsum dissolution follows first-order reaction kinetics. The electrolyte concentration of water flowing through a layer of gypsum can be described adequately by a transport equation which includes terms to account for advection, diffusion, and dissolution kinetics. For a given rate of water application, the concentration of gypsum in the soil–

subsurface electrolyte solution increases with the amount of gypsum that is applied (Kemper et al. 1975; Oster 1982).

Application of gypsum as a reclamation product to counter soil and subsurface domain acidity may, however, enhance the dispersion of soil-clay colloids in highly weathered acidic soils, leading to irreversible changes in the soil–subsurface system. A possible mechanism for the enhancement of clay dispersion caused by gypsum application was given by Roth and Pavan (1991). They considered that gypsum-induced clay dispersion in acidic soils is due to Al monomers and/or polymers, which act as binding agents, being displaced in the presence of SO_4^{2-} from exchange sites by Ca^{2+}. An alternative mechanism to Ca^{2+}-induced dispersion, following application of gypsum, is the specific adsorption or ligand exchange of SO_4^{2-} which can change the clay particle charge characteristics (Uehara and Gillman 1981; Rao and Sridharan 1984).

The favorable effects of gypsum for reclamation of saline-sodic soils were determined in field experiments carried out as early as the beginning of the last century (e.g., Hilgard 1906; Kelley and Arany 1928), and discussed over the years in comprehensive reviews and books (e.g., US Salinity Laboratory Staff 1954; Bresler et al. 1982; Oster 1982; Shainberg and Letey 1984). Water transmission properties expressed in terms of both hydraulic conductivity (HC) and infiltration rate (IR) may be altered by irrigation with sodic water that originates from a saline aquifer, drainage, or sewage effluents. Both HC and IR are controlled by soil–subsurface clay content and mineralogy, as well by the Na^+ concentration of the incoming water, due to Na^+-induced changes to the physical configuration of the solid phase. Gypsum amendments applied directly on the land surface or added to the incoming water may change the water transmission properties by exchanging Na^+ with Ca^{2+} in sodic soils, or by preventing Na^+ adsorption on the soil solid phase caused by irrigation with sodic water.

Hydraulic conductivity of sodic soils is improved following gypsum application. In a laboratory column experiment, Shainberg et al. (1982) measured the hydraulic conductivity of two sodic soils having an exchangeable sodium percentage (ESP) of 20, by leaching soils with distilled water to which equivalent amounts of $CaCl_2$ and gypsum were added. The effect of varying application of $CaCl_2$ and $CaSO_4$ on the relative hydraulic conductivities of noncalcareous (Golan) and calcareous (Nahal Oz) soils, as a function of cumulative effluent volume, is shown in Fig. 18.8. It can be observed that addition of gypsum yielded a higher hydraulic conductivity than addition of $CaCl_2$.

When applied in the field, gypsum dissolution in incoming water controls the efficiency of this amendment on the HC. Due to the relatively low dissolution rate of gypsum, long-term dissolution in water occurs so that HC changes are irreversible on a human lifetime scale.

Infiltration rate, defined as the volume of water penetrating the soil–subsurface cross section per unit surface area, is controlled by the aggregation status of the upper soil layer and the potential formation of a crust (seal). Irreversible changes in soil structure may occur when clay particles become dislodged due to an increase in the Na^+ fraction in soil and a decrease in the electrolyte level of the soil

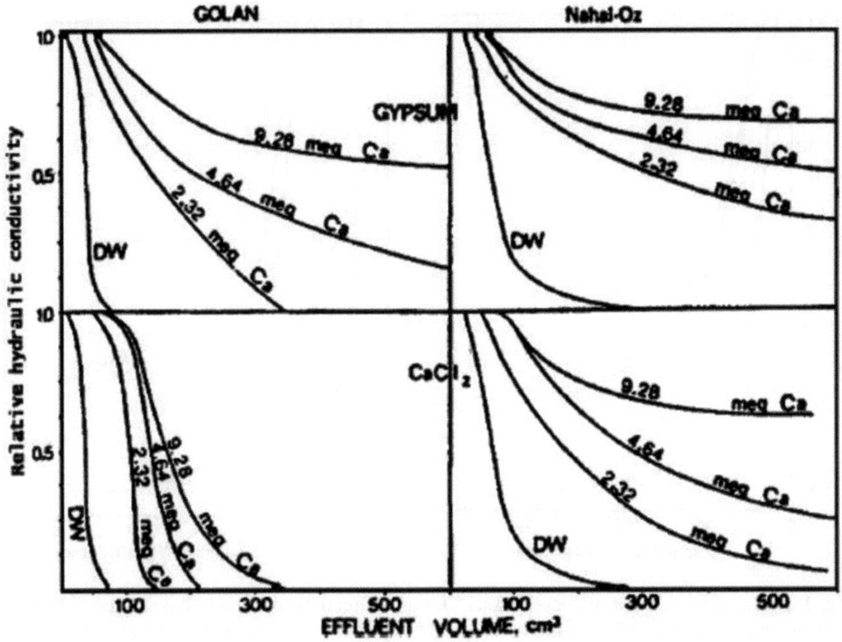

Fig. 18.8 Effect of varying application of $CaSO_4$ and $CaCl_2$ on the relative hydraulic conductivities of Golan (noncalcareous) and Nahal Oz (calcareous) soils, as a function of cumulative effluent volume (Shainberg et al. 1982). Copyright 1982 Soil Science Society of America. Reprinted with permission

solution, or due to the impact of raindrops. Dispersion of soil clays may lead to the clogging of pores beneath the surface, forming a layer of low permeability. Seals (<2 mm) that form have a greater density, higher shear strength, finer pores, and lower permeability than the bulk soil. Addition of gypsum to the soil surface increases electrolyte composition of the incoming water, preventing the breakdown of soil aggregates and decreases in pore size.

The effect of addition of mined gypsum (G) and the industrial byproduct phosphogypsum (PG) on infiltration rates was studied by Levy and Sumner (1998). It was found that PG is more effective than G in enhancing water transmission, and that infiltration rates depend on the amount and method of application of PG, and on cumulative rainfall (Fig. 18.9).

Aggregation status of the soil–subsurface system is a main factor in controlling water transmission properties of the soil–subsurface system. In general, the active pores conducting water are of micrometer scale and are enclosed in aggregates formed by heterogeneous conglomerates in which submicron clay particles are associated (Ahuja et al. 1989). A laboratory study by Lebron et al. (2002) examined changes in aggregate size and geometry of sodic soils amended with gypsum, as reflected in soil water transmission properties. An increase in ESP due to gypsum application was associated with a decrease in larger size aggregates and

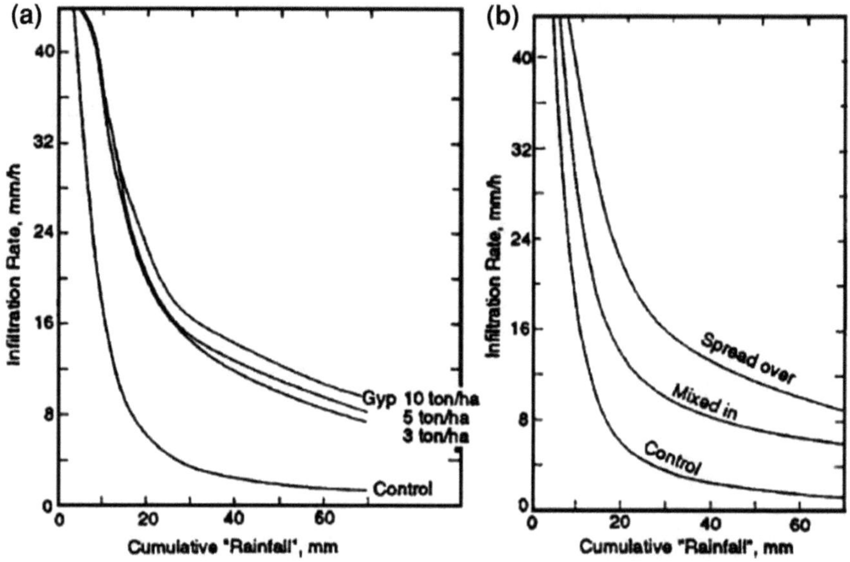

Fig. 18.9 Infiltration rates of loess (Nahal Oz) as a function of cumulative-simulated rainfall: **a** the effect of quantities of applied gypsum, and **b** the effect of method of application, at a rate of 5 t/ha (Agassi and Ben-Hur 1992). Reprinted from Agassi and Ben-Hur (1992). Copyright 1992 with permission from Elsevier

saturated hydraulic conductivity. At the soil water content of saturation, the disruption of aggregates in the soil matrix was to a certain extent irreversible (Lebron et al. 2002). Once an aggregate is broken, the individual particles migrate; and under saturated conditions, formation of new aggregates is unlikely. Thus, addition of gypsum causes irreversible changes to the water transmission properties of sodic soils, by reducing breakdown of larger soil aggregates.

Water erosion and runoff are natural processes occurring on geomorphologic slopes, when the impact of raindrops induces formation of a thin surface layers (seals) on the soil. Under these processes, soil morphology changes due to dispersion of soil particles in the upper section of a slope and deposition of sediments in the lower section. Application of gypsum (both mined and of byproduct origin) on soil surfaces may cause irreversible changes to the patterns of this process.

We consider rain simulator studies on a stable sandy loamy soil, from the coastal plain of Israel, to illustrate the effects of gypsum amendment on water erosion and runoff (e.g., Kazman et al. 1983; Warrington et al. 1989; Agassi et al. 1990; Agassi and Ben-Hur 1991). Figure 18.10 shows the effects of phosphogypsum addition on the infiltration rate and loss of soil caused by rainfall-induced erosion. Seal formation at the soil surface exposed to rainfall is caused by the breakdown of soil aggregates by raindrops, and by physicochemical dispersion of clay soil that migrates and clogs pores beneath the soil surface. The data shown in Fig. 18.10 demonstrate that phosphogypsum spread over the soil surface prevented clay dispersion, thus increasing infiltration and decreasing the depth of runoff. The decrease

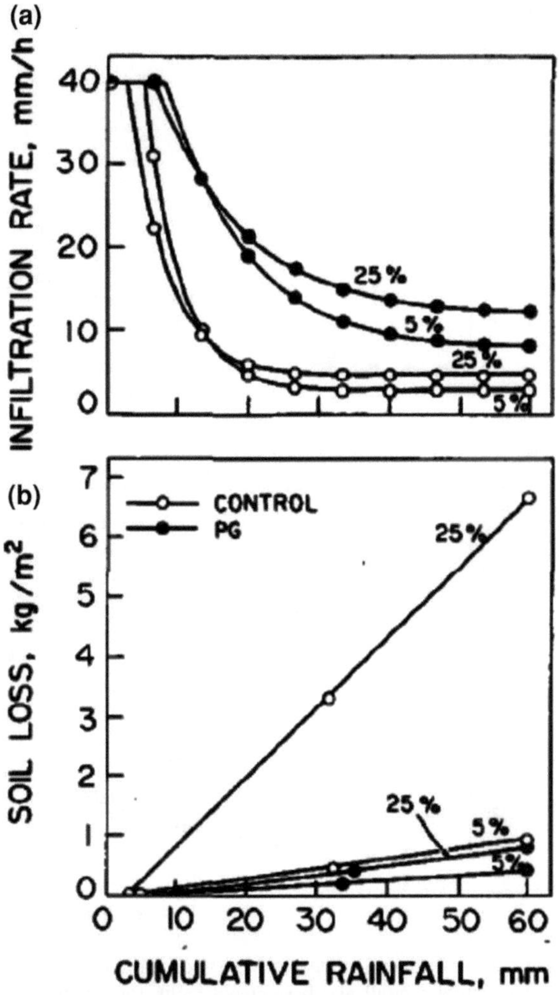

Fig. 18.10 Effect of rainfall, phosphogypsum treatment, and slope (5 and 25 %) on **a** infiltration rate, and **b** soil loss from a sandy loam (Typic Rhodoxeralf) (Warrington et al. 1989). Copyright 1989 Soil Science Society of America. Reprinted with permission

in soil erosion and runoff intensity following phosphogypsum application is explained by the increase of electrolyte concentration in both runoff and percolating water, preventing aggregate dispersion, increasing infiltration rate, maintaining or even improving the roughness of the surface, and avoiding sediment deposit.

Organic polymers In general, organic polymers are applied to the soil surface to prevent soil sealing and clay particle dispersion and runoff, caused mechanically by rain or irrigation water. Addition of organic polymers increases aggregate stability and improves water infiltration. The effects of organic polymers are, however, temporary, and beneficial changes in soil physical properties become irreversible only with continued application of such compounds.

Nonionic polyacrylamides (PAMs) are used as a soil conditioner to stabilize soil aggregates and to flocculate suspended particles. Flanagan et al. (2002) present an example of a polymer-induced effect on soil morphology, showing that PAM

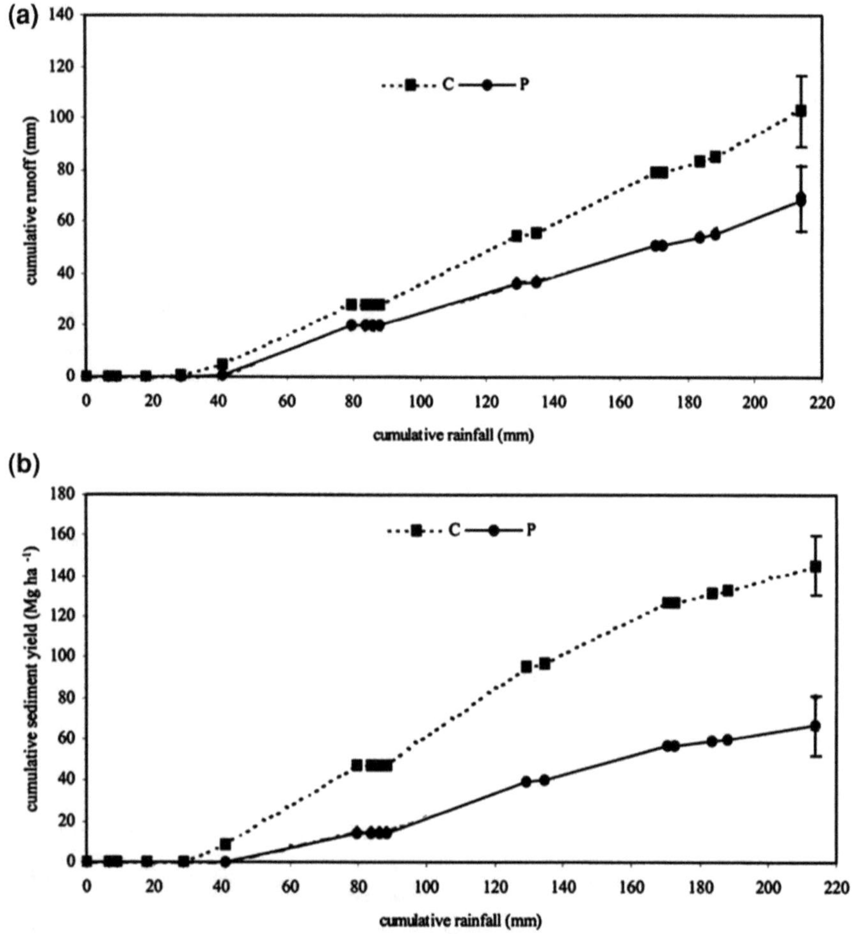

Fig. 18.11 Polyacrylamide (PAM) effects on runoff and soil loss cumulative runoff versus **a** cumulative rainfall and **b** sediment yield, for a natural rainfall study. Bars for final points indicate $P < 0.05$ confidence; C and P are untreated and treated sites, respectively (modified after Flanagan et al. 2002) Originally published as Fig. 1 in Transactions of the ASAE 45 (5):1339–1351. Reprinted with permission of the American Society of Agricultural and Biological Engineers, St. Joseph Michigan

treatment significantly reduced cumulative runoff and soil loss (Fig. 18.11), persisting over about 60 days.

General trends suggest that medium- and high-charge density anionic and neutral PAMs produce the greatest increases in infiltration, compared with controls. Lentz et al. (2000) noted that compared to control and low molecular weight treatments, the medium- and high-molecular weight PAM seals contained a greater number of fluid-conducting pores. Despite the fact that PAM impacts on furrow soils were temporary, the effect on soil morphology due to changes in sediment erosion patterns was irreversible.

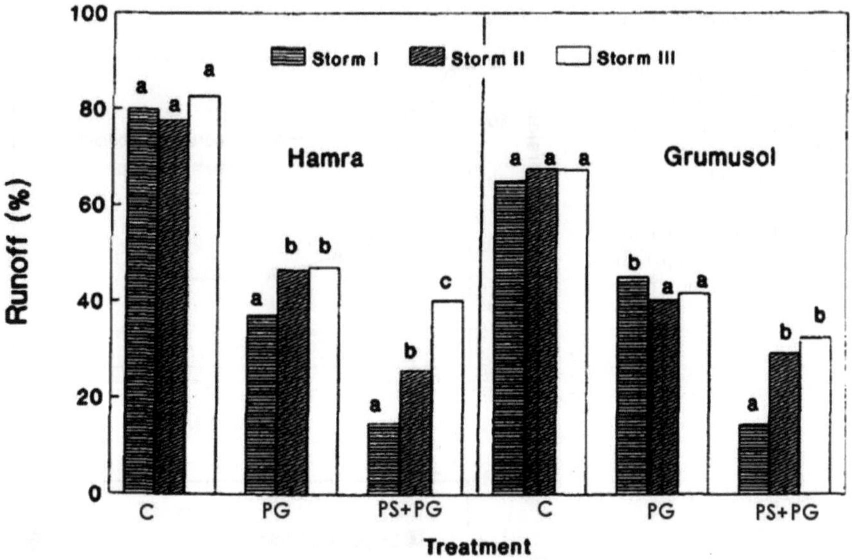

Fig. 18.12 Application of polysaccharide (*PS*), phosphogypsum (*PG*), and polysaccharide with phosphogypsum (*PS* + *PG*) on soil versus runoff, in comparison with untreated site (*C*). Percent runoff for three consecutive 60 mm rainstorms (modified after Warrington et al. 1991) Reprinted from Warrington et al. (1991). Copyright 1991 with permission from Elsevier

Polysaccharides (PSDs) of anthropogenic origin form another group of organic polymers used to prevent or diminish clay particle dispersion and erosion of soils caused by aggregate stabilization and increasing infiltration rate. Ben-Hur and Letey (1989) showed that the type of polysaccharide has a significant effect on the change in infiltration rate (IR); of the polymers tested, only the anionic compound had no effect. The final IR values of the efficient amendments were 19.0, 14.2, and 5.2 mm/h, compared with 4 mm/h for the untreated soil. These results suggest that the electrostatic adsorption of the PSDs on the negative soil-clay surface was the dominant factor in providing stability to the aggregates and a change in the infiltration rate. Furthermore, the soil IR change induced by low charge cationic polymer addition to water during the first application was partially maintained during succeeding untreated water applications, demonstrating stability of soil aggregates against the energy impact of rain/irrigation drops (Ben-Hur et al. 1989).

Another essentially irreversible effect of PSD application is the prevention of soil runoff that occurs as a result of seal formation at the soil surface during rainstorms. Warrington et al. (1991) reported changes in runoff patterns of soils from arid and semiarid areas, as a result of application of an anionic acidic polysaccharide (F-AC) of algae extracellular origin, with a molecular weight of ~70,000 g/mol. The effects on runoff reduction of PSD applied individually or associated with electrolytes released by phosphogypsum dissolution, in three consecutive simulated rains, are shown in Fig. 18.12. A lasting and marked effect

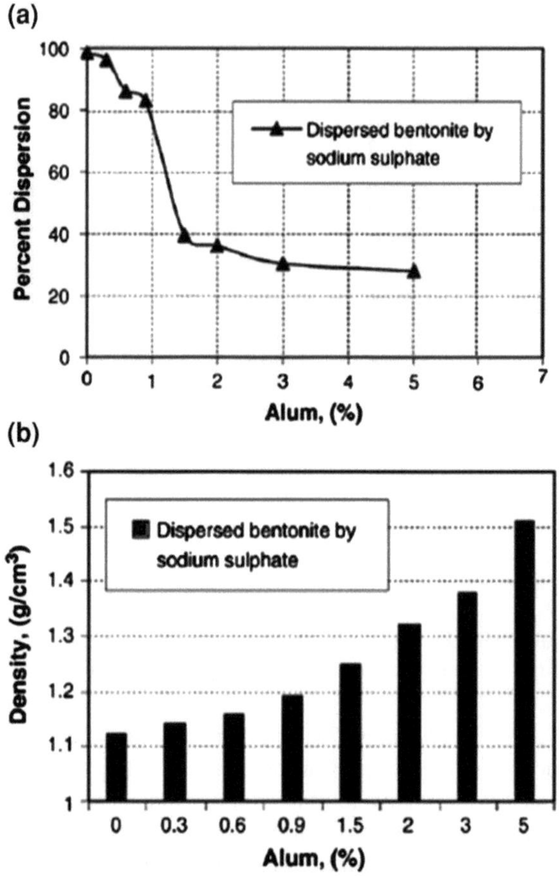

Fig. 18.13 Effect of alum (aluminum sulfate) concentration on percentage on soil bentonite dispersion: **a** percentage of dispersion and **b** density variation in dispersed-treated bentonite (Ouhadi and Goodarzi 2006) Reprinted from Ouhadi and Goodarzi (2006). Copyright 2006 with permission from Elsevier

on runoff reduction in the PSD-phosphogypsum-treated soil, compared with untreated or only phosphogypsum-treated soils, is clearly observed.

PSDs are applied to soils mainly in arid and semiarid areas, where rainfall and irrigation control soil water transmission properties; because this is a continuous remediation procedure, polysaccharide-induced soil changes are irreversible on a lifetime scale.

Alum Dispersivity of a soil is determined by the amount and type of soil clay and by the presence of Na as saturated cation. Dispersive clay soils exhibiting a low permeability and a low water velocity prevent transport of soil particles. However, once cracks develop, dispersed soil-clay particles are easily carried with water flowing through these cracks. In general, soil dispersivity in agricultural lands is controlled by gypsum product additives. However, for specific geotechnical projects, alum or aluminum sulfate ($Al_2(SO_4)_3{\cdot}18H_2O$) is also used. Ouhadi and Goodarzi (2006) tested the stability of dispersive soil clays following alum treatment (Fig. 18.13). It can be observed that the adding of 1.5 % alum, for example, caused the percent dispersion of the soil bentonite to drop to 40 % and then stabilize. With the application of alum, the soil structure changed from dispersive to more flocculated and nondispersive. The effect of aluminum sulfate on the dispersivity of clay soils could be due to a pH effect, which favors the interaction between clay particles and electrolyte, causing a decrease in soil dispersivity.

Alum treatment also affects the physicomechanical properties of a dispersive soil. Ouhadi and Goodarzi (2006) showed that under constant stress, the density of dispersed soil bentonite increased as the percentage of alum increased (Fig. 18.13b). This behavior can be attributed to a decrease in the double-layer thickness with an increase in added alum, due to the replacement of sodium ions by aluminum ions with a higher valence.

18.1.4 Pesticides

Pesticides (recall Sect. 4.1) comprise a large number of organic compounds, mainly with toxic properties, that may reach the land surface following agricultural practices or industrial accidents. According to their molecular properties, pesticides partition between solid and aqueous phases of the soil–subsurface geosystem (recall Part III), and under the influence of environmental conditions, pesticides may be redistributed among the land surface, vadose zone, and groundwater (recall Part IV). Pesticides may be retained on the soil and subsurface mineral phases by sorption. Sorption is a dynamic process for both the sorbing compound, which transfers from the fluid state to a solid state, and for the sorbent itself, which undergoes physical changes upon sorption (Pignatello 2012). When the sorption is irreversible, two interrelated processes are involved: desorption and binding. In general, cationic pesticides are adsorbed by clay minerals, while nonionic pesticides are adsorbed by the organic fraction of the soil–subsurface solid phase.

Adsorption of charged pesticides on the organic fraction of mineral phases (e.g., humic macromolecules) decreases according to the order Ca^{2+}-humate, H^{+}-humic acid, and Na^{+}-humate. This can be explained not only by the preference for exchange of ions of higher valence, but also by the effect of macromolecules, which is more important than consideration of valence. Adsorbed organo-cation complexes bridge negative charges with strands, and the interstrand binding causes macromolecule shrinkage, exclusion of water from the matrix, and precipitation of organoclay–humate complexes. Adsorption of nonionic pesticides on the soil–subsurface solid phase is controlled by enthalpy and entropy. Enthalpy-related adsorption forces include hydrogen bonding, ligand exchange processes, protonation mechanisms, π-π bonds, van der Waals forces, and chemisorption. Entropy-related adsorption forces, also referred to as hydrophobic sorption, involve the partitioning of nonpolar pesticides from the polar aqueous phase onto hydrophobic surfaces. Nonadsorptive retention of pesticides in the subsurface involves their accumulation to form a new bulk solid phase within pores and along solid phase boundaries. In some cases, bound pesticide residues may irreversibly alter the natural solid phase matrix (Yaron et al. 2010 and references therein).

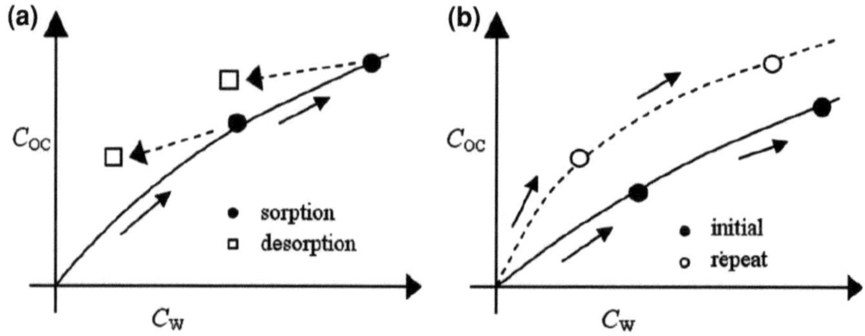

Fig. 18.14 Examples of irreversible adsorption: **a** hysteresis isotherm, immediately after first sorption; **b** conditioning effect, hysteresis after repetitive sorption–desorption cycles (Pignatello 2006)

Irreversible adsorption may occur as a result of hysteresis or by a conditional effect. *Hysteresis* is exhibited by nonuniqueness of adsorption–desorption curves within an adsorption isotherm (Sect. 5.8), and *conditioning* results from enhanced sorption following repeated experiments on the same sample after a previous adsorption–desorption cycle (Pignatello 2006, Fig. 18.14). In thermodynamic terms, a solute concentration cannot be matched to multiple sorbed concentrations, and as a consequence, irreversible sorption may arise following generation of metastable states.

Examples of irreversible retention of pesticides on geosorbents are illustrated below for both polar and nonionic compounds.

Cationic pesticides *Diquat* (1,1′-ethylene-2,2′-dipyridylium dibromide) and *Paraquat* (1,1′-dimethyl-4,4′-bipyridinium dichloride) are cationic herbicides used broadly for crop protection. They may contaminate the land surface and subsequently pollute the subsurface zone and groundwater after being leached by rain or irrigation water. Retention and release of these herbicides—charged organic molecules—on and from clay minerals have been the subject of many investigations during the second half of the last century. The complete adsorption of diquat and paraquat by bentonite and kaolinite, as measured by the absence of detectable bipyridyl in solutions equilibrated with these minerals, was observed by Weber et al. (1965). The extent of adsorption reaches the limit of the mineral cation exchange capacity (Knight and Tomlinson 1967). Weber and Weed (1968) confirmed that diquat and paraquat are adsorbed by montmorillonitic and kaolinitic clay minerals to approximately their cation exchange capacity. Adsorption–desorption experiments exhibit a hysteresis process. After desorption caused by leaching with 1 M$BaCl_2$, about 20 and 95 % of each of the herbicides was retained on kaolinite on montmorillonite, respectively. Knight and Denny (1970) explained the high retention of paraquat on dioctahedral montmorillonite by interlamellar absorption; the clay thickness of the silicate sheet expanded from 9.6 to 12.6 Å.

Hysteresis in the adsorption and desorption of diquat and paraquat may occur also when they are in competition with an additional ion for the adsorbing sites of

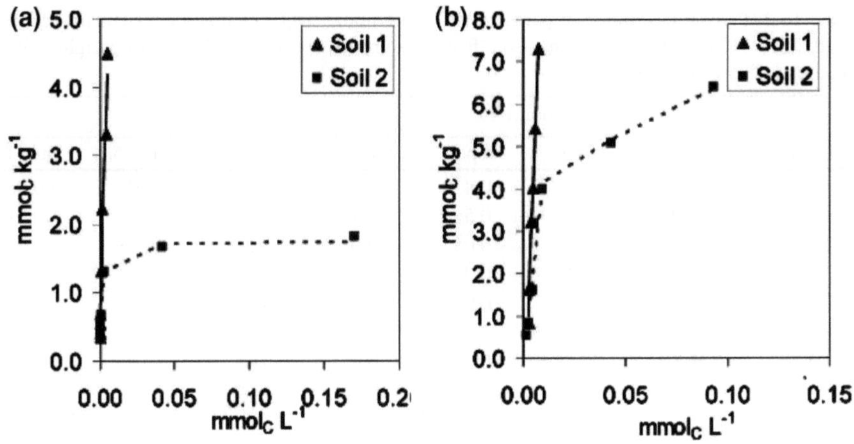

Fig. 18.15 Adsorption isotherms of **a** paraquat and **b** diquat in two acidic vineyard soils (1 and 2) (modified after Pateiro-Moure et al. 2007) reprinted with permission from Pateiro-Moure et al. (2007). Copyright 2007 American Chemical Society

the mineral fraction of soil. In vineyard soils, cationic herbicides may interact with inorganic cations such as copper, which is used as a fungicide, or from application of animal manure. Effects of aged copper on the adsorption–desorption of paraquat and diquat in polluted vineyard soils in Spain were studied by Pateiro-Moure et al. (2007). The two studied soils were acidic (pH 5.2–5.8) and characterized by a surface area of 66–78 m^2/g; the total Cu^{2+} contents of the two soils studied were 42 and 583 mg/kg. Both diquat and paraquat were adsorbed strongly by the soils (Fig. 18.15). However, the removal of Cu^{2+} from soils by EDTA extraction (37 and 14 %, respectively) led to small increases in the paraquat and diquat concentrations (1.7 and 0.3 %, respectively). The increase in herbicide retention was equivalent to the replacement of 6.4 and 1.6 % of the charges theoretically released by Cu^{2+} from the two studied soils.

Desorption tests showed that both diquat and paraquat are retained irreversibly on soil even after six sequential extractions (Table 18.2). One soil exhibited less than 1 % desorption of the two herbicides, while the second soil exhibited a slightly higher percentage. Paraquat was desorbed in proportions from 1 % in the first extraction step to 0.2 % in the sixth step. Diquat desorption never exceeded 4 % and tended to decrease as the desorption sequence progressed.

The adsorption–desorption of the cationic pesticide *chlordimeform* on montmorillonite clay was shown by Hermosin and Perez Rodriguez (1981) to be cation exchange coupled to coadsorption of neutral molecules and extraction of Al from the mineral. Adsorption of chlordimeform on multilayered montmorillomite is an irreversible process; in contrast, chlordimeform adsorbed on single layer kaolinite is easily removed by washing with aqueous solution. XRD data suggest that a single layer of chlordimeform cations adsorbs in the interlayer region of the formed chlordimeform–montmorillonite complex. The d-spacing values between

Table 18.2 Percent desorption of paraquat (PQ) and diquat (DQ) (modified after Pateiro-Moure et al. 2007) Reprinted with permission from Pateiro-Moure et al. (2007). Copyright 2007 American Chemical Society

		Desorption (%)					
Soil sample		Step 1	Step 2	Step 3	Step 4	Step 5	Step 6
1	PQ	1.0	0.4	0.3	0.2	<0.08	<0.08
	DQ	0.5	0.5	0.3	0.3	0.2	<0.01
EDTA-1	PQ	<0.08	<0.08	<0.08	<0.08	<0.08	<0.08
	DQ	<0.01	<0.01	<0.01	<0.01	<0.01	<0.01
2	PQ	1.5	0.9	0.7	0.7	0.5	0.2
	DQ	3.2	3.2	3.6	3.3	3.2	2.1
EDTA-2	PQ	0.3	0.2	0.4	0.4	0.3	0.2
	DQ	1.9	1.5	0.5	1.4	1.1	0.5

Standard deviations: lower than 3 %

air-dried and moist samples of chlordimeform–montmorillonite complex were practically equal (14.8 and 15.2 Å). Based on additional infrared spectra (Fig. 18.16), indicating that the adsorbed pesticide fixes the interlayer space, Hermosin and Perez Rodriguez (1981) considered that (1) water molecules associated inorganic exchangeable cations were displaced by chlordimeform, changing the natural clay matrix and forming a coordination complex $X(CfHCl)_m{}^{n+}$ in the interlayer spaces, or (2) inorganic exchangeable cations were displaced with their hydration spheres from the interlamellar spaces by nonhydrated chlordimeform molecules.

Glyphosate (N-phosphonomethylglycine)—a broad-spectrum herbicide—is a small molecule comprising three polar functional groups (phosphonic acid, carboxylic acid, and secondary amines), which is sorbed strongly by soil minerals. Glyphosate (GP) can only be sorbed onto variable-charge surfaces and not onto permanent negative sites of layer silicates, because it has anionic properties in a specific pH range (Borggaard and Gimsing 2008, and references therein). The most important soil and subsurface sorption sites for GP retention may be found on aluminum and iron oxides, defectively ordered aluminum silicates, and edges of layer silicates. Table 18.3 provides a view of ensemble GP sorption values on various synthetic earth minerals, as a function of their surface area. It can be seen that silicate clay (saturated with K^+) has a limited capacity to sorb GP; the amount appears to depend on the solution-saturating cation. Sorption of GP on montmorillonite, illite, and kaolinite decreases according to the following order of cation presence: $Al^{3+} > Fe^{3+} > Cu^{2+} > Mg^{2+} > Zn^{2+} > Mn^{2+} > Ca^{2+} > Na^+$ (Glass 1987). The increased sorption of divalent and trivalent cations in multilayered clays may be explained by the formation of complexes between interlayer cations and glyphosate. Borggaard and Gimsing (2008) suggest that the discrepancy between the two mineral groups (Table 18.3) can be attributed to the number and distribution of sorption sites. Oxides contain many adsorption sites; clay silicates offer available adsorption sites only on octahedral layers exposed to mineral edges.

Fig. 18.16 IR spectrum of (I) pure chlordimeform hydrochloride, and (II) **a** Mg–montmorillonite; **b** Mg–montmorillonite–chlordimeform complex; **c** Ca–montmorillonite; **d** Ca–montmorillonite–chlordimeform complex (modified after Hermosin and Perez Rodriguez 1981)

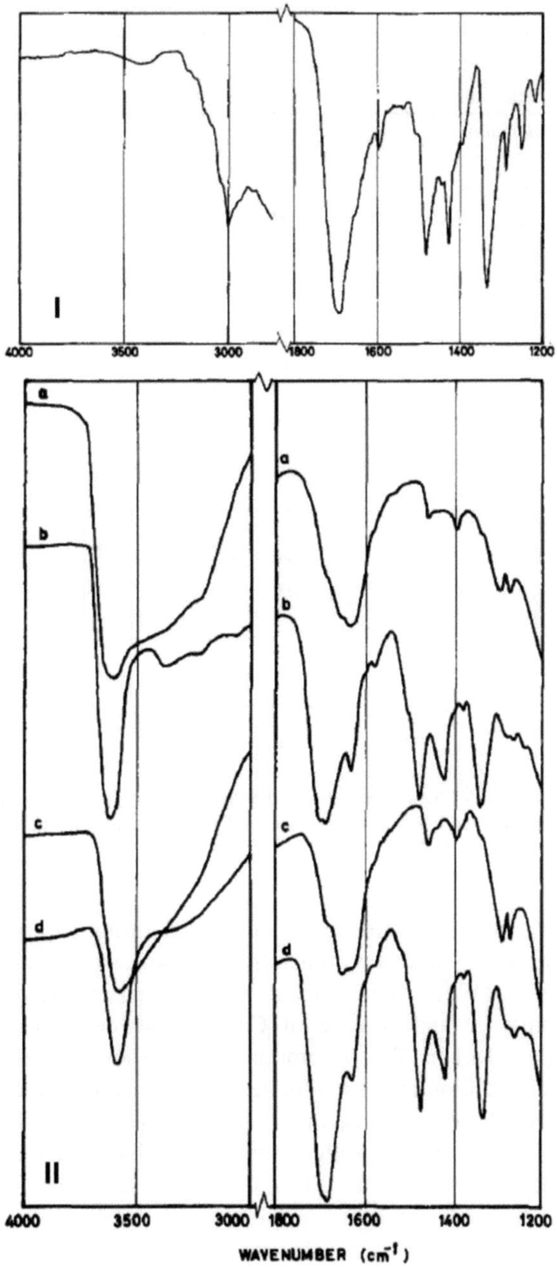

Table 18.3 Glyphosate sorption at pH 7 by synthetic aluminum and iron oxides, and pure K^+-saturated layer silicates. Reprinted from Borggaard and Gimsing (2008). Copyright 2008 with permission of John Wiley and Sons

Earth minerals	Specific surface area (m^2/g)	Glyphosate sorption (mmol/kg)
Aluminum and iron oxides		
Gibbsite	45	72
Ferrihydrite	343	635
Goethite	85	125
Hematite	33	86
Clay silicates		
K^+-Kaolinite 1	12	3.9
K^+-Kaolinite 2	22	6.9
K^+-Illite	43	5.2
K^+-Montmorillonite	32	6.5

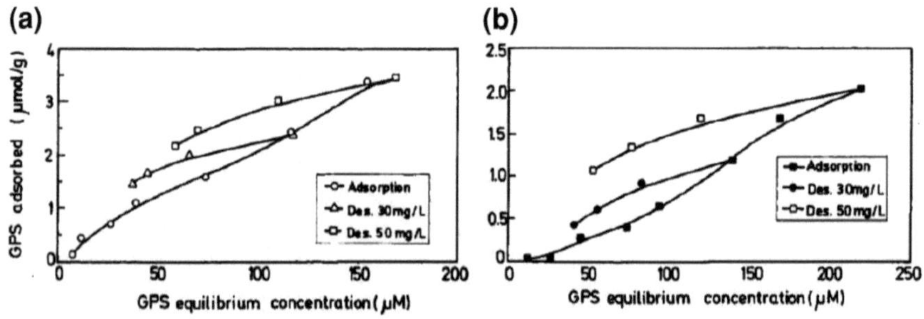

Fig. 18.17 Adsorption–desorption isotherms of glyphosate (*GP*) on montmorillonite **a** in the absence, and **b** in the presence of Cu. Reprinted with permission from Morillo et al. (1997). Copyright 1997 American Chemical Society

An hysteretic effect was observed during adsorption–desorption studies of GP with soils and clay minerals, suggesting potential irreversible retention. Morillo et al. (1997) examined GP retention and release on/from montmorillonite in the presence and absence of Cu^{2+} in the desorbing aqueous electrolyte solution (Fig. 18.17). The desorption curves showed that about 60 % of GP initially adsorbed on montmorillonite was retained on the clay surface after three successive desorption procedures. In this case, some GP initially adsorbed on broken borders of montmorillonite is replaced by hydroxyl groups or by water during the desorption process. When Cu^{2+} ions are added to the desorption solution, a reduction is observed in the amount of GP that is retained (Fig. 18.17b). Here, some interlayer Cu^{2+} desorbs by cation exchange with Na^+ ions from the electrolyte solution, favoring formation of GP complexes with Cu^{2+} in solution.

Nonionic pesticides *Chlorinated benzenes*—aromatic organic compounds with the chemical formula $C_6H_{6-n}Cl_n$—are common hydrophobic solvents widely used as intermediate materials in the manufacture of various nonionic pesticides. The adsorption of chlorinated benzenes on natural soil-sediment materials is illustrated below, with reference to the extent of retention reversibility and induced changes in the physical state of the natural sorbents.

Chen et al. (2000 and references therein) found that for most chemical-sediment interactions, a significant fraction of chlorobenzene is bound irreversibly to the sediment. The irreversible component has a finite maximum capacity in the range of 1/3–1/2 of the initially sorbed chemical. Results of experiments on irreversible adsorption of five chlorinated benzenes, differing in K_{ow} and solubility, and four natural sediments with organic matter concentration ranging from 0.27 to 4.1 %, are summarized in Table 18.4. Note that the irreversible partition coefficient (K_{oc}^{irr}) and maximum irreversible adsorption capacity (q_{max}^{irr}) are the most important parameters characterizing the irreversible adsorption component; K_{oc}^{irr} is defined as the ratio of the organic carbon normalized solid phase concentration to the solution phase concentration after extensive desorption, when the aqueous phase concentration has reached a significantly low and constant value.

Chen et al. (2000) proposed a model to describe the observed adsorption–desorption behavior. The labile fraction follows a linear isotherm, while the nonlinear behavior associated with the irreversible fraction is described by a Langmuir-type isotherm. Figure 18.18 compares the measured and calculated adsorption–desorption isotherms; the desorption behavior is described well by the isotherm. As the aqueous and solid phase concentrations decrease, the isotherm deviates increasingly from the linear model. Chen et al. (2000) suggest that irreversible adsorption is a consequence of interactions between hydrocarbon molecules and organic matter in sediments. The apparent organic carbon-based partition coefficient, K_{oc}, associated with the irreversible fraction is essentially constant for this chemical-sediment combination. As a result, irreversible retention of chlorinated benzene in specific sediments, given by the value of q_{max}^{irr}, is dependent upon the presence of organic matter in the sediment solid phase.

Natural organic matter (NOM) may be altered irreversibly by the influx of hydrophobic compounds into soils and sediments. More specifically, intermingling of sorbate molecules with natural organic matter macromolecules is considered to be a major factor in sorbent alteration, by formation of a three-dimensional solid phase of NOM (Pignatello 2012 and references therein). In addition, contaminant entrapment in NOM may also play a minor role in organic sorbent alteration. Here, we focus on sorption irreversibility of unlabeled (bulk) and labeled (tracer) carbon 1,4-dichlorobenzene (DCB) on peat (PP) and lignite (BZL), two natural geosorbents rich in organic matter (Sander and Pignatello 2009). The experiments were performed in duplicate on bulk and tracer solutions to confirm that the hysteresis effect is not an artifact. The surface areas of BZL and PP are 1.85 and 0.88 m^2/g, respectively. Hysteresis was quantified using the thermodynamic index of reversibility (TII), which on a normalized scale ranges from completely reversible sorption (0 value) to completely irreversible sorption (1 value). The experiments indicated that free exchange of sorbate DCB molecules (tracer and bulk) occurred between the solid and solution phases, under both sorption and desorption, at high and low concentrations. DCB hysteresis was found to be concentration- and

Table 18.4 Adsorption and desorption of chlorinated benzenes on sediments with different organic matter (OC) concentrations. Reprinted with permission from Chen et al. (2000). Copyright 2000 American Chemical Society

Exp	Sediment (% OC)	Compound	log K_{ow}	Adsorption		Desorption		
				$q_0^b (\mu g/g)$	log K_{oc} (L/kg of OC)	$q_{max}^{irr\ d}$ (μg/g)	$C_w^{max\ e}$ (μg/L)	log K_{oc}^{irr} (L/kg of OC)
1	Dickinson (1.5)	1,2-DCB	3.38	960[a]	3.05 ± 0.05[c]	6.37	0.81 ± 0.19	5.44 ± 0.10
2	Dickinson (1.5)	1,2,4-TCB	4.00	442[a]	3.42 ± 0.06[c]	13.6	1.3 ± 0.29	5.46 ± 0.11
3	Dickinson (1.5)	1,4-DCB	3.38	562	2.78	14.5	1.4 ± 0.55	5.38 ± 0.17
4	Dickinson (1.5)	1,2,3,4-TCB	4.55	688	3.87	15.3	2.0 ± 0.53	5.33 ± 0.10
5	Dickinson (1.5)	HCB	5.50	6.12	4.93	0.450	0.15 ± 0.02	5.31 ± 0.05
6	Lake Charles (4.1)	1,4-DCB	3.38	2,330	2.97	271	6.9 ± 1.1	5.32 ± 0.08
7	Lula (0.27)	1,4-DCB	3.38	116	2.86	0.970	0.92 ± 0.16	5.26 ± 0.08
8	Utica (2.8)	1,4-DCB	3.38	2,660	3.18	202	3.28 ± 0.16	5.81 ± 0.02

[a] Concentrations at the end of repetitive adsorptions, calculated from mass balance
[b] Measured by solid extraction after continuous adsorption
[c] Average values of individual steps in repetitive adsorptions
[d] Values fitted with the least-squares method using the irreversible isotherm and experimental adsorption and desorption data
[e] C_w^{max}, the average concentration observed in the last few desorption steps

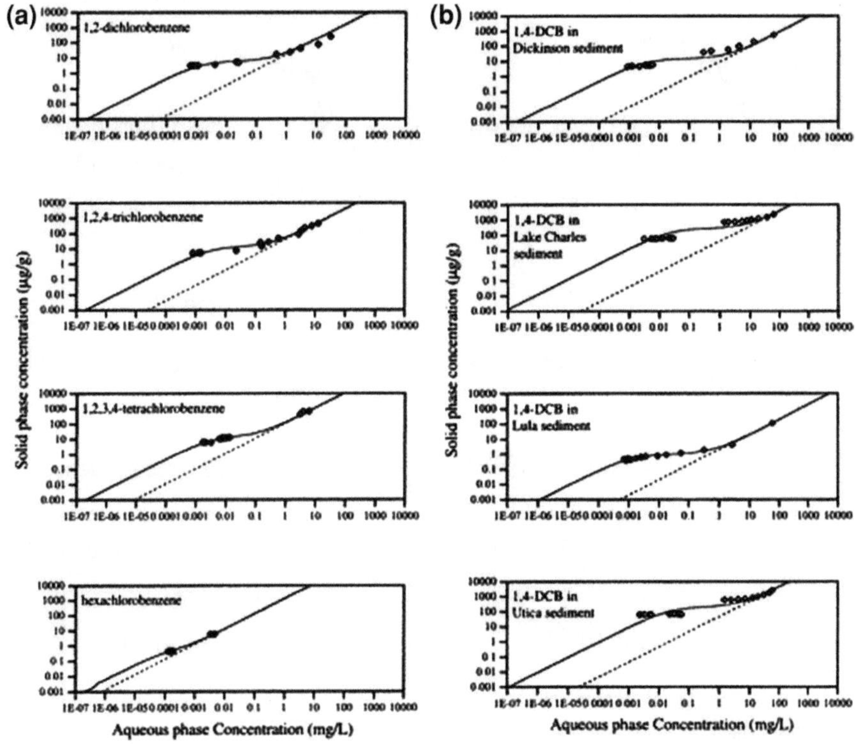

Fig. 18.18 **a** Plot of the model predicted versus experimentally observed isotherms of 1,2-dichlorobenzene, 1,2,4-trichlorobenzene, 1,2,3,4-tetrachlorobenzene, and hexachlorobenzene in Dickinson sediment. Solid lines are predicted isotherms using the irreversible adsorption model. Dotted lines are the linear isotherms plotted using the K_{oc} values obtained in adsorption experiments. The symbols are experimental observations. **b** Plot of the model predicted versus experimentally observed isotherms of 1,4-dichlorobenzene in Dickinson, Lake Charles, Lula, and Utica sediments. Solid lines are predicted isotherms using the irreversible adsorption model. Dotted lines are the linear isotherms plotted using the K_{oc} values obtained in adsorption experiments. The symbols are experimental observations. Reprinted with permission from Chen et al. (2000). Copyright 2000 American Chemical Society

sorbent-dependent and is due to matrix deformation. On peat, the TII of DCB hysteresis displayed a downward trend, from 0.6 to 0.3, while on lignite TII increased from 0 to 0.5 with increasing concentration.

Theoretically, self-diffusion in PP and BZL should be slower at lower concentrations, in both the uptake and release directions. In contrast, the experimental results show that uptake and release of bulk DCB are nearly the same as the tracer at the two concentrations (Fig. 18.19). In some cases, however, bulk DCB shows a biphasic rate while the tracer does not, which may be explained by sorbate-induced deformation of the matrix. Comparison between bulk sorption versus forward isotope exchange kinetics and bulk desorption versus reverse isotope exchange kinetics, at the sorption point (Fig. 18.20) provides further evidence of the matrix

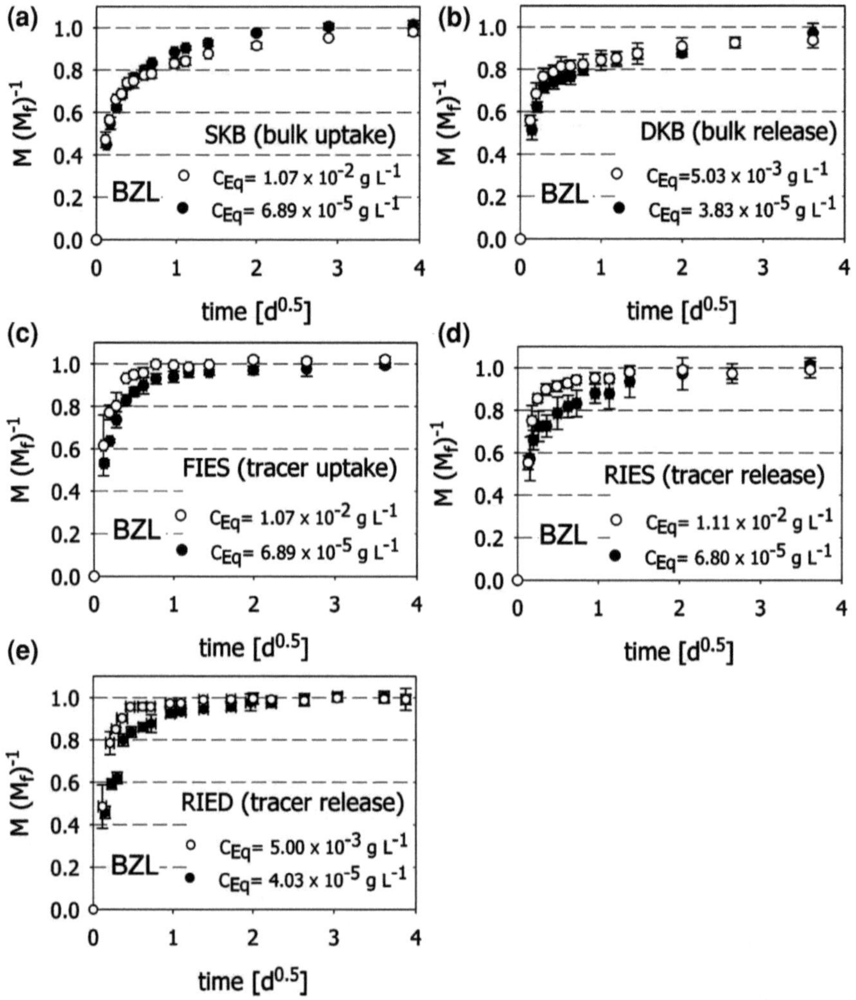

Fig. 18.19 Normalized uptake and release rates of bulk and tracer 1,4-dichlorobenzene (*DCB*) on Beulah-Zap (*BZL*) lignite during **a** sorption of bulk (SKB), **b** desorption of bulk (*DKB*), **c** forward isotope exchange at the sorption point (*FIES*), **d** reverse isotope exchange at the sorption point (*RIES*), and **e** reverse isotope exchange at the desorption point (*RIED*). M is the cumulative mass of DCB taken up or released, and M_f is the final mass taken up or released by the geosorbent. C_{Eq} is the final leveling-off concentration during SKB and DKB and the constant bulk concentration during isotope exchange steps. Reprinted from Sander and Pignatello (2009). Copyright 2009 with permission of John Wiley and Sons

effect. Based on these observations, Sander and Pignatello (2009) conclude that uptake and release of bulk DCB at the higher concentration are coupled to matrix exchange in peat and lignite.

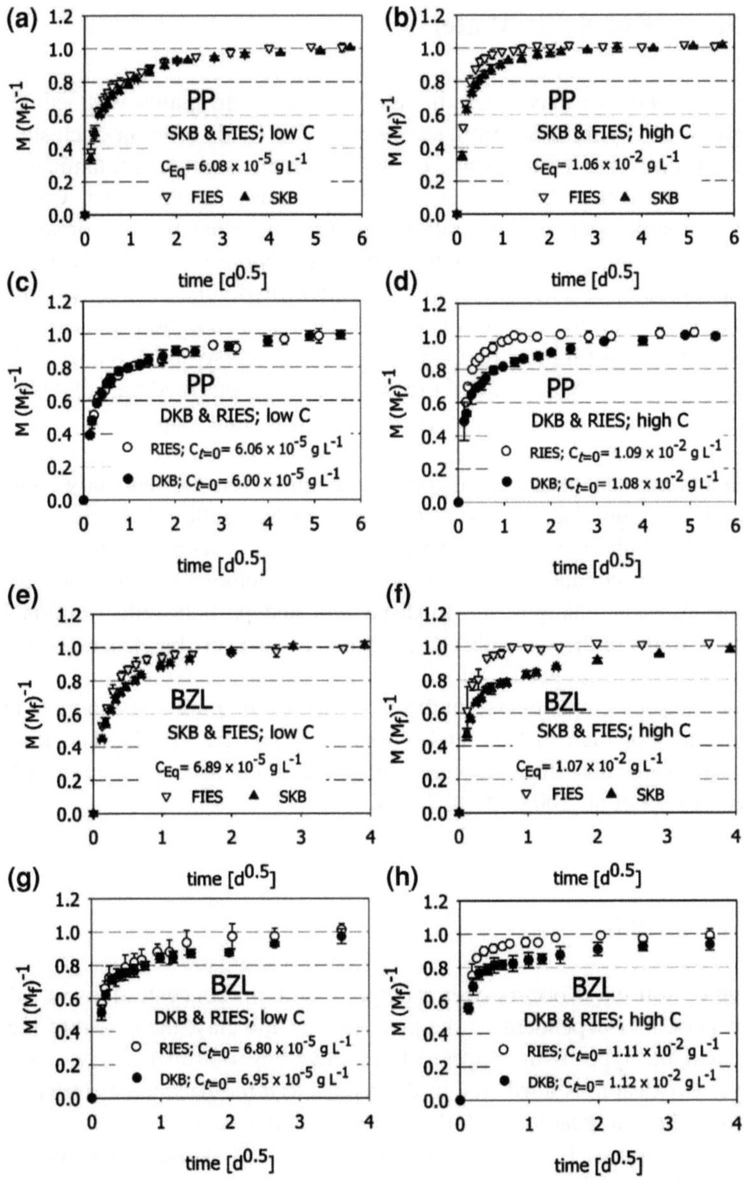

Fig. 18.20 Comparison of normalized uptake of 1,4-dichlorobenzene (*DCB*) to Pahokee peat (*PP*) soil (**a–d**) and to Beulah-Zap lignite (*BZL*) (**e–h**) during sorption kinetics of bulk (*SKB*) and forward isotope exchange at the sorption point (*FIES*) or during desorption kinetics of bulk (*DKB*) and reverse isotope exchange at the sorption point (*RIES*) at low and high concentrations indicated on the graphs. M is the cumulative mass of DCB taken up or released, and M_f is the final mass taken up or released by the geosorbent. C_{Eq} is the final leveling-off concentration, and CTO is the initial concentration before initiation of DKB and RIES, *left* lower concentration; *right* higher concentration filled symbol, bulk; open symbol tracer. Reprinted from Sander and Pignatello (2009). Copyright 2009 with permission of John Wiley and Sons

18.2 Impact of Sodic Water

Sodium contaminants may reach the land surface via irrigation with saline water and wastewater, by intrusion from seawater, and/or by dissolution of clustered salt crystals. Within the components of saline water, sodium contaminants may strongly affect soil and subsurface water transmission properties. Changes in hydraulic conductivity and infiltration rate are controlled by the porosity of the medium and by the clay content of the solid phase. Moreover, soil-clay mineralogy was determined to be a key factor in the extent of hydraulic conductivity decrease in soils leached with sodic water (Yaron and Thomas 1968). Soil–subsurface clays, with expanding layer mineralogy, may swell and disperse upon contact with sodic water, changing the architecture of the natural host porous medium solid phase and affecting its water transmission properties. Swelling and dispersion of clays have been proposed to explain the decrease in hydraulic conductivity of soils. Deflocculation and movement of clay in conducting pores were subsequently considered to be another major factor causing decreases in hydraulic conductivity (Frenkel et al. 1978; Shainberg and Letey 1984, and references therein). Whereas a change in soil hydraulic conductivity due to sodic water-induced swelling is a partially reversible process, changes due to dispersion and particle redistribution in the conducting pores of the soil–subsurface regime are irreversible. Changes occurring to various physical properties of soil and the subsurface regime under irrigation with sodic water were addressed in the early studies of Quirk and Schofield (1955), McNeal and Coleman (1966), and Russo and Bressler (1977). The impact of sodium on soil–subsurface contaminant-induced changes was revisited more recently (Yaron et al. 2008, 2010, 2012).

18.2.1 Swelling

Swelling of soil clays occurs over a range of sodium concentrations in saline or waste waters; it is dependent on the properties of the clay, the extent of the layer charge, and the type of interlayer cations. Swelling of soil clays, expressed by interlayer expansion, is attributed to the repulsive force between electrical double layers on adjacent platelets (McBride 1992), and based on the fact that layer silicate clays approach a maximum hydration state beyond which they do not extend. Farmer and Russell (1971) explained swelling qualitatively, as a simple model of H-bonding when interlayer expansion is driven initially by the energy of hydration of the exchangeable cations. Swelling was explained later by noting that the zeta potential of Na^+-clay is much lower than that anticipated from Na^+ ions fully dissociated from the surface in a suspension of layered silicates in water (Low 1981).

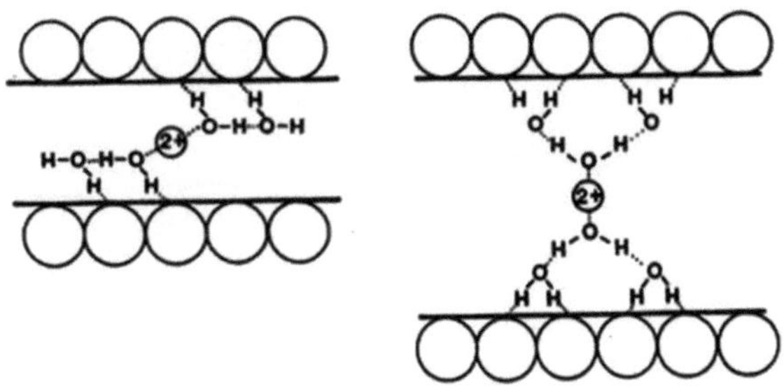

Fig. 18.21 Arrangement of H_2O chains linking a divalent exchange cation to the surfaces of a 2.0 nm hydrate of smectite (modified after Farmer 1978, in McBride 1989). Copyright 1989 Soil Science Society of America. Reprinted with permission

Interlayer expansion of layered clay is shown in Fig. 18.21, where the arrangements of H_2O chains linking a divalent cation to the surface of a 2.0 nm hydrate of smectite clay are characterized by 2:1 layers. The force-inducing clay expansion arises from ion–dipole interactions, while the force-opposing clay swelling emanates from electrostatic interactions. In general, the swelling of 2:1 silicates is considered to be induced by hydration of charge-compensating counter-ions in the interlayer space of clay.

Laird (2006) defined six separate processes driving the swelling of smectite clays saturated with alkali and alkaline earth cations in an aqueous system, as follows: crystalline swelling, double-layer swelling, quasi-crystal formation, cation demixing, covolume swelling, and Brownian swelling. In an aqueous system, some of these processes may act concurrently, affecting swelling dimension. An increase in layer charge leads to a decrease in crystalline swelling, but increases the size and stability of smectite clay quasicrystals. When two smectite quasi-crystals approach each other in an aqueous suspension, a portion of their double layer may fuse expelling anions, cations, and water from the contact surfaces. In dilute aqueous suspensions, Na-smectite can be almost completely delaminated, a diffuse double layer is formed between the individual layers, and each layer behaves as a stable individual colloid (Sposito 1992; Laird 2006). A schematic diagram depicting the breakup and formation of quasicrystals is given in Fig. 18.22, leading to an increase in soil-clay swelling.

In situ laser microscopy combined with digital image analysis was used by Suzuki et al. (2005) to investigate the swelling of bentonite aggregates in a NaCl solution of variable concentration. Changes in quasicrystal shape after contact with NaCl solutions are shown in Fig. 18.23. These results refer only to a simple NaCl water solution. However, in natural soil–subsurface conditions, saline irrigation water and groundwater contain also divalent cations, which can exchange with Na^+ ions in bentonite. As a consequence, the results of Suzuki et al. (2005) are significant only for understanding the mechanism of Na^+-controlled swelling of expandable soil clays.

Hysteresis was noted in investigations of clay swelling, beginning with the early pioneering studies of Mooney et al. (1952) and Norrish (1954). Swelling hysteresis was attributed to changes in the energy levels of expansion and

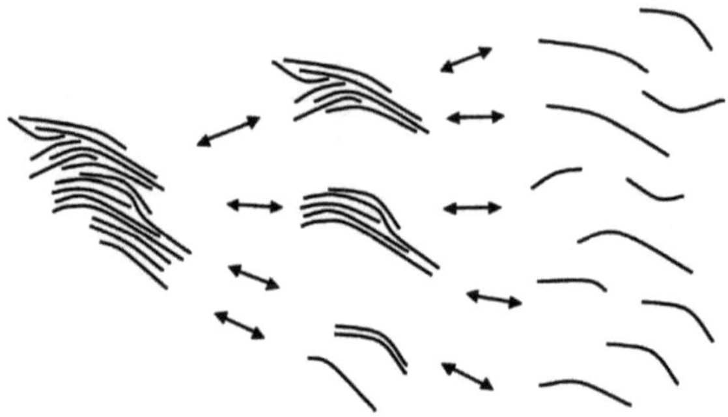

Fig. 18.22 Schematic diagram depicting the breakup and formation of quasicrystals. *On the left,* the layers are grouped together in a single quasi-crystal; in the middle the same layers are split into four quasi-crystals; *on the right,* the layers are completely delaminated (Laird 2006) Reprinted from Laird (2006). Copyright 2006 with permission from Elsevier

attraction (e.g., Wada et al. 1990) and to the rigidity of the unit layer (e.g., Norrish 1954) or to the clay-water system (e.g., Fu et al. 1990). Evidence of swelling hysteresis was observed in various experiments using gravimetric measurements and basal spacing determination, as well as various relative humidity and salt concentrations (Laird et al. 1995, and references therein). In general, swelling was explained in terms of restructured arrangements of the clay and changes in interactions between layers upon expansion or contraction. It was also considered that swelling and shrinking of clay minerals is a cause of hysteresis.

Correlation coefficients relating hysteresis magnitude to surface charge density, percentage of tetrahedral charge, and surface area are not significant, indicating that the magnitude of hysteresis is independent of the surface charge properties of the smectites. Based on these results, Laird et al. (1995) considered that crystalline swelling is an "irreversible thermodynamic process." In support of irreversibility, free energy changes occurring during swelling and collapse can be classified as free energy of expansion, free energy of attraction, and free energy of transition. Free energy of expansion is due to changes in the hydration states of the interlayer cations and negative charge sites, free energy of attraction is due to changes in the net electrostatic attraction between layers, and free energy of transition is due to the irreversible work needed to affect transfer of heat and mass during swelling and collapse. As a consequence, the cause of hysteresis in crystalline swelling is attributed to both intrinsic and extrinsic processes. Intrinsic hysteresis is a consequence of changes in expansion and attraction energy levels during swelling caused by the rigidity of the clay-water system. Extrinsic hysteresis is caused by factors that contribute to additional rigidity of smectite quasicrystals.

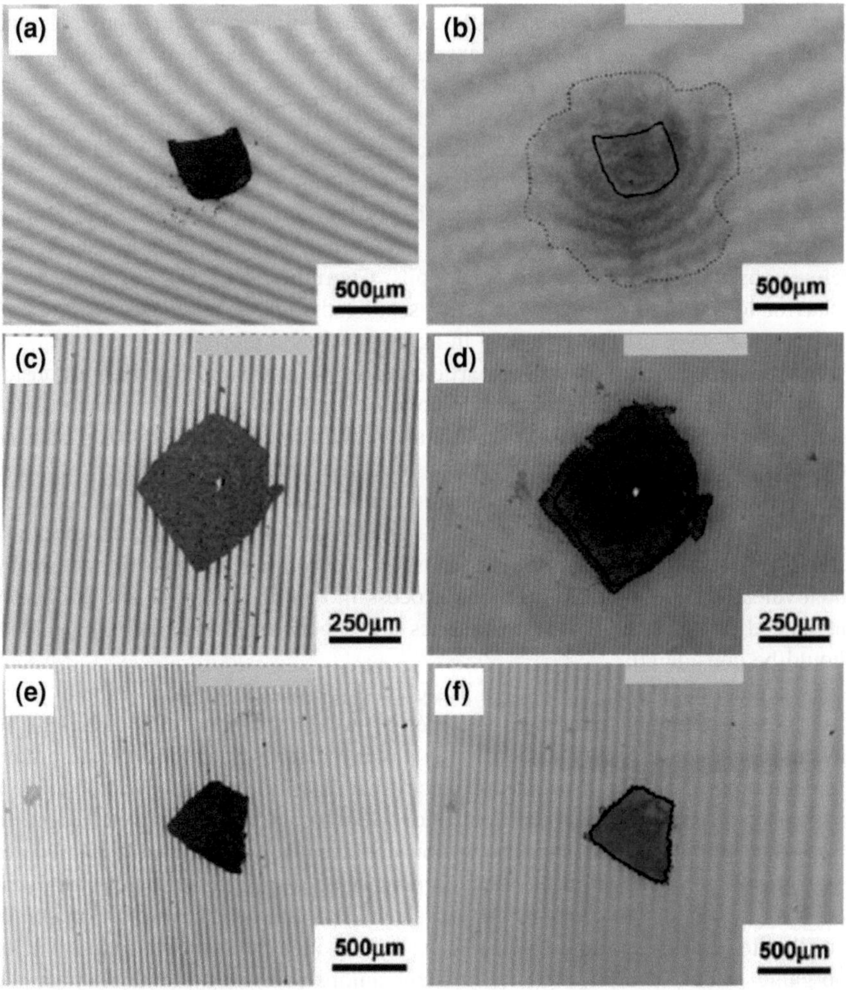

Fig. 18.23 Changes in a quasi-crystal shape before (**a**, **c**, **e**) and after (**b**, **d**, **f**) coming into contact with NaCl solutions with concentrations of **a**, **b** 0.01; **c**, **d** 0.30; **e**, **f** 1.00 mol/dm^3 NaCl. The solid line indicates the outline of a quasicrystal before coming into contact with the solution, and the dashed line indicates the outline of the quasicrystal 15 min after exposure to the solution (Suzuki et al. 2005) Reprinted from Suzuki et al. (2005). Copyright 2005 with permission from Elsevier

18.2.2 Dispersion and Flocculation

Dispersion and flocculation of colloidal earth material are other processes induced by sodium contamination, comprising both inorganic and organic compounds, which may affect water transfer properties of the soil–subsurface system. Principal factors related to clay dispersion–flocculation processes following contact of sodic water with a soil surface include the water electrolyte concentration, pH, and sodium adsorption ratio (SAR) expressed by the ratio between Na^+ and $Ca^{2+} + Mg^{2+}$ concentrations ($SAR = [Na^+] /[(Ca^{2+} + Mg^{2+})/2]^{1/2}$).

Clay dispersion is governed by the attractive and repulsive forces in the electrical double layer at the surface of charged colloids; the balance between the forces is determined by factors such as exchangeable cation concentration and ionic strength (Rengasamy and Olsson 1991). The critical ion concentration for clay dispersion from soil aggregates is lower than the flocculation values (see below) observed in the separated clay soils. Chorum et al. (1994) attributed this behavior to the high negative charge of separated soil clays, due to exposure of surfaces that were originally bound in aggregates.

Dispersion and release of clay particles from the soil solid matrix are considered to be the major source of mobile colloidal particles in the subsurface (Ryan and Elimelech 1996). Release of colloidal particles is controlled by their surface interaction with the solid matrix and by hydrodynamics of the kinetics of water flow. Summarizing the existing information in this field, Grolimund and Borkovec (2006) showed that (1) particle release is usually enhanced by lowering the ionic strength (2) the release process depends on the solution composition whereby monovalent counterions promote this process most effectively (3) the release is influenced by the pH (4) release kinetics are exponential, and (5) aging effects should be considered.

Flocculation, or coagulation, of Na-clay dispersion systems occurs in most cases between the negative edges and the negative faces of the clay particle. The flocculation value of a clay suspension in a sodic water is defined as the minimum concentration of Na^+ in the aqueous electrolyte solution necessary to flocculate a suspension of clay in a given time, under specific conditions of exchangeable cation concentration, pH, and suspension concentration. Based on a number of experiments on coagulation of Na-montmorillonite dispersions, Lagaly and Ziesmer (2003) found that the anisometric shape and charge distribution on montmorillonite particles caused very low critical coagulation concentration of inorganic salts. Adsorption of multivalent anions increased the density of the negative edge charges. As a consequence, the edge/face coagulation became a face/face coagulation with a high critical salt concentration. Lagaly and Ziesmer (2003) pointed out that in dispersion of high montmorillonite contents, the flocculation may be initiated by edge/edge aggregation.

Flocculation between the positive edge and the negative face occurs only at pH < 6. The critical flocculation concentrations of the clay fraction extracted from three strongly structured soils from Natal (two dominated by c and one by kaolinite) were compared with those of smectite from Wyoming and kaolinite from Georgia in Na-saturated forms (Frenkel et al. 1992). They found that the soil clays had much higher critical flocculation concentrations than their Na-saturated clay reference counterparts. This difference was attributed to the existence of humic acid coating on soil-clay surfaces.

18.2.3 Water Transmission Properties

Water transmission properties, expressed in terms of hydraulic conductivity (HC) and infiltration rate (IR), and solute movement in the soil and subsurface regions can be changed irreversibly as a result of sodic water disposal on the land surface. The hydraulic conductivity is a property that describes the ease with which water can move through pore spaces and/or fractures and depends on the intrinsic permeability of the material. The infiltration rate is the speed at which water enters the soil–subsurface domain. The extent of changes in transmission properties is controlled largely by the clay content and clay type present in the solid phase, and by the sodium concentration of the incoming water.

18.2.3.1 Hydraulic Conductivity

Swelling and dispersion are the two processes affecting the soil–subsurface hydraulic conductivity. The Na^+ cations from incoming contaminated water are adsorbed on clay minerals, inducing their swelling and/or dispersion. The extent of this process is governed by the concentration of the Na^+ cations in the incoming water. In their review on sodium and electrolyte concentration effects on soil hydraulic conductivity, Yaron and Shainberg (1973) noted that swelling is not the main cause for the decrease in HC of clayey soils. In contrast, dispersion of clay due to Na^+ is the main cause of irreversible reduction in HC of clays, because it results in rearrangements and blocking of the pores through which flow takes place.

It is clear that clay dispersion is a major factor that irreversibly affects the HC of soils irrigated with sodic water (Shainberg and Letey 1984, and references therein). Oster et al. (1980) found that clay dispersion is highly sensitive to low levels of exchangeable Na and increases sharply with very small increases in exchangeable sodium percentage (ESP). The movement of water through clay systems when HC was affected only by swelling and when the water movement was affected by both swelling and dispersion of particles is shown in Fig. 18.24.

Soil HC at equilibrium with salt-contaminated water is determined by the concentration of electrolytes, the ionic composition of the water, and the exchangeable cations in the soil complex (Quirk and Schofield 1955). In a mixed system, the ESP is the most important characteristic influencing HC. This was noted by Quirk (1957), who showed that soil permeability decreases with increasing percentage of exchangeable sodium. A quantitative relationship between solution concentration and ESP was expressed as a "threshold concentration." The threshold concept was proposed initially by Quirk and Schofield (1955) and is defined as the salt concentration needed to maintain the hydraulic conductivity in a Na-saturated system 20 % below that existing in a calcium-saturated system.

The relationship between soil hydraulic conductivity and electrolyte type, and concentration of the infiltrating water is affected by the soil–subsurface

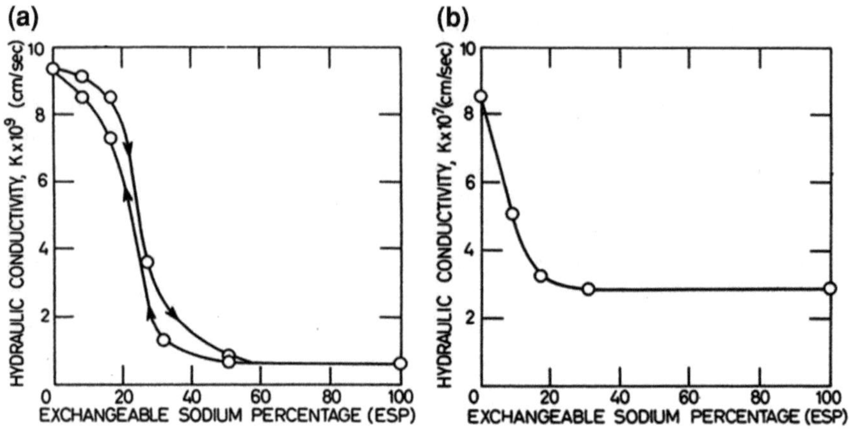

Fig. 18.24 The hydraulic conductivity of **a** the clay membrane, and **b** clay pastes, as a function of exchangeable sodium percentage, ESP (modified after Yaron and Shainberg 1973)

mineralogy. McNeal and Coleman (1966) indicated that in soils where the clay minerals are dominated by the extent of 2:1 layer silicates, and where there are moderate amounts of montmorillonite, ESP values in excess of 15 can often be maintained without causing a serious reduction in HC, provided that the concentration of the permeating solution is about 3 meq/L. Yaron and Thomas (1968) studied the effect of ESP on the hydraulic conductivity of various Texas soils, finding that a 20 % HC decrease was produced by an ESP of 10 in montmorillonite soils, by an ESP of 16 in a soil with mixed mineralogy, and by an ESP of 23 in a kaolinite soil.

The composition of the permeating water—in terms of the ratio between sodium and other cations—has a significant effect on the exchange process and on the hydraulic conductivity of soils. In dealing with the effect of saline or waste water on HC, one must consider the sodium adsorption ratio (SAR). Under dynamic conditions, when an initially nonsodic soil is irrigated with sodic water, the HC decreases continuously until the exchangeable sodium in the entire soil profile equilibrates with the permeating solution. Yaron and Thomas (1968) showed that this decrease in HC is characterized by a curve whose shape depends on the nature of the soil, the cationic composition, and the total salt concentration of the permeating solution. The hydraulic conductivity depends on the mean ESP of the soil profile and is independent of the volume of the effluent. Moreover, an increase in SAR results in a decrease in hydraulic conductivity. Figure 18.25 shows the reduction in HC when three electrolyte water solutions with one value of total concentration (11 meq/L) and three SAR values (7.5, 14.0, 28.0) were passed through columns of a Burleson loamy clay vertisol. At equilibrium, an SAR value of 7.5 resulted in a 15 % decrease in HC, whereas an SAR of 28 reduced initial HC by 84 %. These results are in agreement with those of other parallel studies (e.g., Reeve and Tamaddoni 1965; McNeal and Coleman 1966; McNeal 1968;

Fig. 18.25 Effect of sodium adsorption ratio (*SAR*) on soil hydraulic conductivity (soil Burleson; total electrolyte concentration of irrigation water = 11 meq/L. Reprinted from Yaron and Thomas (1968). Copyright 1968 with permission of John Wiley and Sons

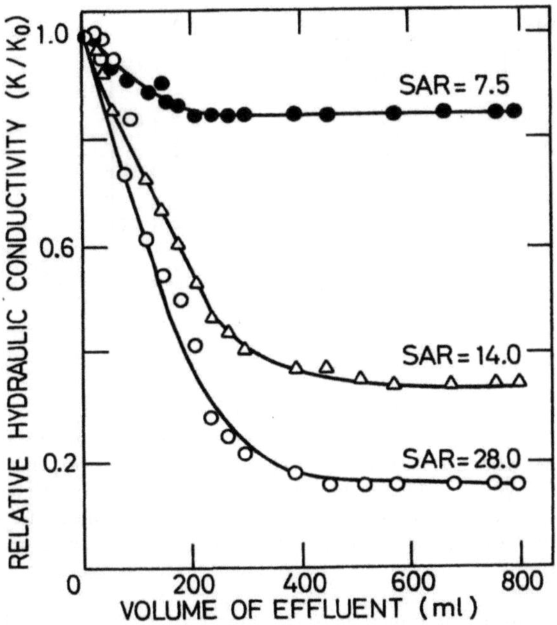

Lagerwerff et al. 1969), showing that total electrolyte concentration and SAR can be used to predict the decrease in saturated HC of a soil following percolation with sodic water.

The decrease in soil hydraulic conductivity can be attributed to changes occurring in the soil–subsurface matrix. Chen and Banin (1975) performed scanning electron microscopy measurements in a sandy soil (6 % clay) and a clayey soil (55 % clay), the clay fraction in both cases having similar mineralogy. With an increase in SAR, fine particles were detached from coarse sandy soil, creating a new matrix arrangement containing a continuous network in the voids. Although clay movement was more limited in the clayey soil, Chen and Banin (1975) considered that the same mechanism caused the changes in the solid matrix, indicating possible clay particle migration along the soil profile.

The irreversibility of the salt effect on soil hydraulic conductivity was emphasized by Dane and Klute (1977), who determined the effect of percolating solution composition on both HC and soil bulk density (Fig. 18.26). A clayey Weld soil was tested, with a composition of 24 % sand, 33 % silt, and 43 % clay. The clay mineralogy consisted of 60 % smectite, 20 % mica, and 20 % kaolin. When soil samples were subjected to percolation by NaCl solutions with decreasing concentration and a constant SAR of 40, it was observed that the decrease in total electrolyte concentration led to a decrease in HC. No changes were observed for a reference solution ($CaCl_2$) with a concentration of 1,000 meq/L and an SAR value of 0. The composition of the percolating solution affected also the soil bulk density. The data presented in Fig. 18.26b correspond to soil samples leached with

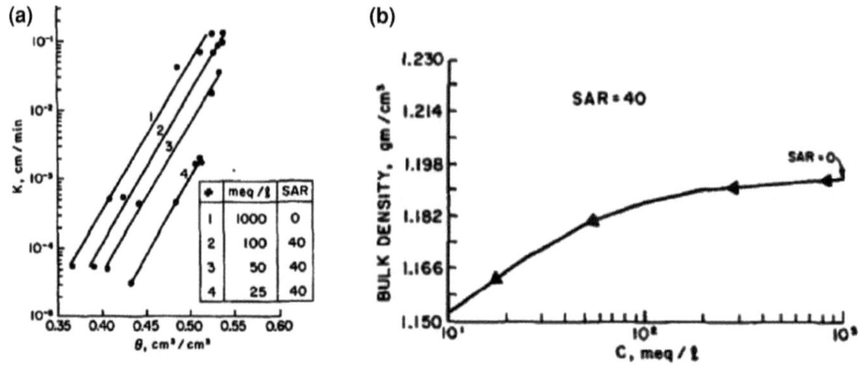

Fig. 18.26 **a** Hydraulic conductivity volumetric soil content relations for a soil sample (load 13.3 g/cm^2) subjected to a series of solutions of decreasing concentration at SAR = 40 (Dane and Klute 1977). **b** Bulk density during conditions of ponded solution for a soil sample (load 13.3 g/cm^2) subjected to series of solutions of varying concentration at SAR = 40. *Arrows* indicate the sequence of solution application (modified after Dane and Klute 1977). Copyright 1977 Soil Science Society of America. Reprinted with permission

solutions at constant SAR values of 0–40, respectively. When the SAR of the solutions was 40, the bulk density decreased with a decrease in the concentration of percolating solution.

Dane and Klute (1977) considered that a decrease in the electrolyte concentration of the surrounding solution caused clay platelets to increase their distance of separation. Under these conditions, the repulsive forces begin to dominate over the attractive forces, with the clay particles forming an unstable arrangement. A partial collapse of the clay particle configuration causes the bulk volume of the soil to decrease, explaining the increase in bulk density when solution concentration decreases. Repercolating the soil samples with an electrolyte solution of an increasing concentration does not enable a return to the initial state.

The effect of pH on clay dispersion in Na-saturated soils was reported by Schofield and Samson (1954), who showed that kaolinite edge faces become negatively charged at high pH and positively charged at low pH. An example of the effect of pH on kaolinitic shortlands soil clay in Na solution is found in Frenkel et al. (1992), where the critical flocculation value, defined as the NaCl concentration required to sediment 8 % of the clay after standing for 24 h, was measured. The critical flocculation values determined at pH values of 6.6, 7.3, and 8.5 are shown in Fig. 18.27.

18.2.3.2 Crust Formation and Infiltration Rate

Sealing the soil surface by crust formation, and subsequent decreases in the water infiltration rate, may be caused by the presence of sodium in the incoming water (see also Sect. 18.1.1). The extent of crust formation is affected by the texture and

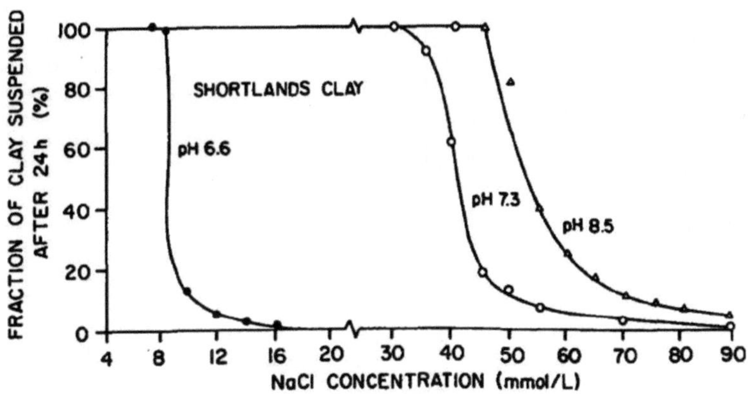

Fig. 18.27 Effect of pH on flocculation of shortlands clay in NaCl solution (Frenkel et al. 1992). Copyright 1992 Soil Science Society of America. Reprinted with permission

the clay mineralogy of the soils. The IR is much more sensitive than the HC to the exchangeable sodium percentage (ESP) of the soil. The higher susceptibility of the soil surface to low levels of ESP is explained by the mechanical impact of rain drops, which enhance dispersion, and by the solid phase matrix which slows clay dispersion and movement (Shainberg 1985). Figure 18.28 shows the effect of soil ESP on the water infiltration rate of sandy loam and silt loam soils. Clay dispersion during crust formation may also lead to irreversible transport of mineral colloids and to their redistribution with depth.

Crust formation can also occur under arid and semiarid climatic conditions, where high evaporation in the near-surface layer favors higher salt concentrations in the soil–subsurface residual water solutions. Such a pattern of salt accumulation may result from either evaporation of water infiltrating through the surface or by upwelling of solutes from subsurface sources. This phenomenon, which usually extends over long periods of time, is well known in the literature (e.g., Kijne 1973, Bresler et al. 1982).

18.3 Impact of Municipal and Industrial Contaminants

There are numerous types of compounds of industrial and urban origin that can reach the soil–subsurface system and irreversibly affect its original properties. We present here a number of examples that illustrate irreversible effects of acid rain, waste effluents and sludges, and compost, as well as petroleum products and other hydrocarbon products of urban origin, on soil–subsurface properties.

Acid rain is any form of atmospherically deposited acidic substance containing strong mineral acids, mainly of industrial and urban origin. Oil and hydrocarbon contamination of the soil–subsurface domain may be caused by accidental spillage

Fig. 18.28 The effect of soil exchangeable sodium percentage (*ESP*) on infiltration rates in two soils, **a** sandy loam (Hamra) and **b** silt loam (loess). Reprinted from Kazman et al. (1983). Copyright 1983 with permission of Wolters Kluwer Health

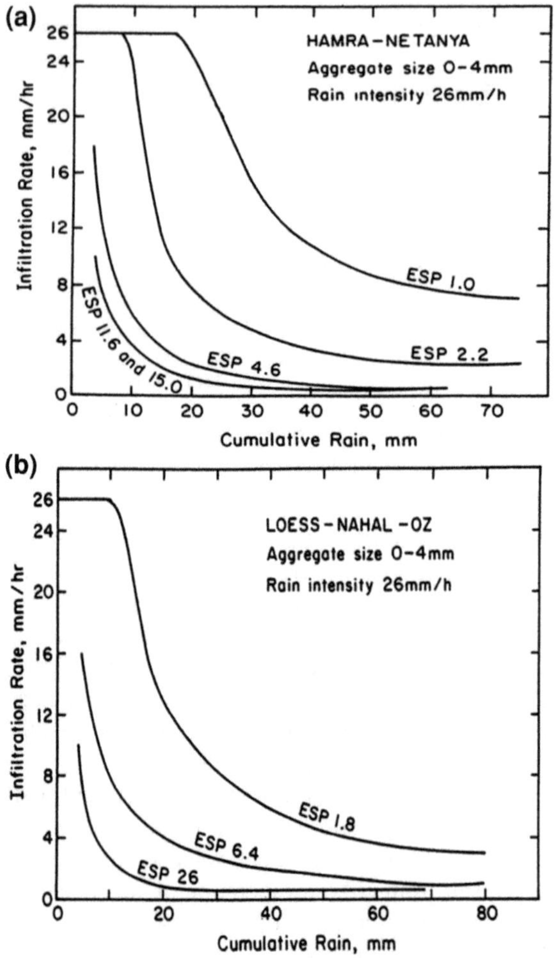

or leakage, or by intentional application, during urban or agricultural management. Long-term leakage of hydrocarbons can stimulate formation of oxidation–reduction zones in the soil–subsurface domain (Schumacher 1996).

18.3.1 Acid Rain Input

Acid rain represents an acidic deposition, which occurs in both wet and dry forms. The most common acidic substances are compounds containing hydrogen (H^+), sulfates (SO_4^{2-}), and nitrates (NO_3^-). The source of these compounds is from the combustion of fossil fuels such as coal, petroleum, and petroleum byproducts, primarily gasoline. Power plants that burn coal contribute over 50 % of atmospheric sulfates and 25 % of nitrates. While agriculture is also a major source of nitrates, we refer to acid rain as an inorganic contaminant of industrial and urban origin. Discussion of acid rain in the context of groundwater acidification is given in Sect. 17.2.

Rain and snow in equilibrium with carbon dioxide exhibit a pH of about 5.6. However, as result of a strong acid atmospheric pollution, the pH can decrease to 4. Records from a 30-year period (1963–1993) show, for example, that rain and snow in northeastern USA had an average annual pH of 4.05–4.3, and that sulfuric acid contributed 55–75 % of this measurable acidity (Likens and Bormann 1995). Another case is rapid industrialization in China, which has led to a significant increase in acid rains caused mainly by the extensive use of coal and higher emission of pollutants. Wet and dry deposition of acidifying compounds, including H_2SO_4 and HNO_3, occurs in most populated parts of China; the only exception is the desert areas, where alkaline dust largely neutralizes the acids in the deposition materials (Larssen et al. 2006).

The effect of acid deposition on soil depends on effects of both intensity and capacity (Reuss et al. 1987). Intensity effects include changes occurring in the soil solution as a response to strong acid anion concentration in the acid deposition, which is buffered by the properties of the natural soil. The main capacity effect occurs as a result of sulfate adsorption and the increased leaching of base cations associated with strong acid anion mobility. As a "mobile" anion from acidic deposition, SO_4^{2-} can be retained by soils from solution, regardless of whether or not this retention results in the stoichiometric displacement of another anion. In general, this retention is considered as "SO_4 adsorption." In many forest soils from the northeastern USA and some regions in Canada, it has been found that as a result of acid rain, the amount of SO_4^{2-} loading is presently higher compared with pre-industrial levels. Harrison et al. (1989) reviewed the degree of SO_4^{2-} adsorption reversibility following a decrease in anthropological inputs and suggested the pathway of soil solution sulfate concentration with and without adsorption (Fig. 18.29). It can be observed that an important portion of retained sulfate remains completely irreversible in the polluted soil.

Based on this observation, Harrison et al. (1989) examined, in a laboratory study, the relative SO_4^{2-} adsorption capacities and reversibility of a large number of soil samples. The samples were collected from a wide variety of forest sites in the USA, Canada, and Norway and were subject to different environmental conditions and past atmospheric deposition levels. The measure of reversibility of SO_4^{2-} adsorption was calculated from two estimates of irreversibility of adsorbed sulfate, by comparing differences in phosphate extractable fractions, and by comparing SO_4^{2-} desorption from treated and untreated soils. The results showed that $\sim$80 % of the tested soils irreversibly retained some of the adsorbed SO_4^{2-}. In general, soils that adsorbed higher amounts of SO_4^{2-} also showed higher amounts of irreversibly adsorbed sulfate (Fig. 18.30). On average, $\sim$36 % of the adsorbed SO_4^{2-} was adsorbed irreversibly, for all soils studied.

Dissolution of aluminum in acidic soils has an important irreversible effect on acid deposition. The depletion and redistribution of aluminum with depth have

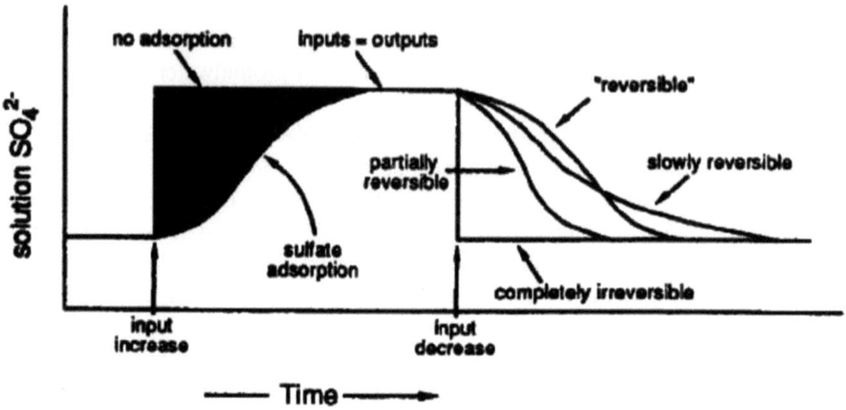

Fig. 18.29 Pathways of soil solution sulfate concentration with and without adsorption following deposition increase, and showing a variety of sulfate reversibility responses following deposition decreases (Harrison et al. 1989). Copyright 1989 American Society of Agronomy. Reprinted with permission

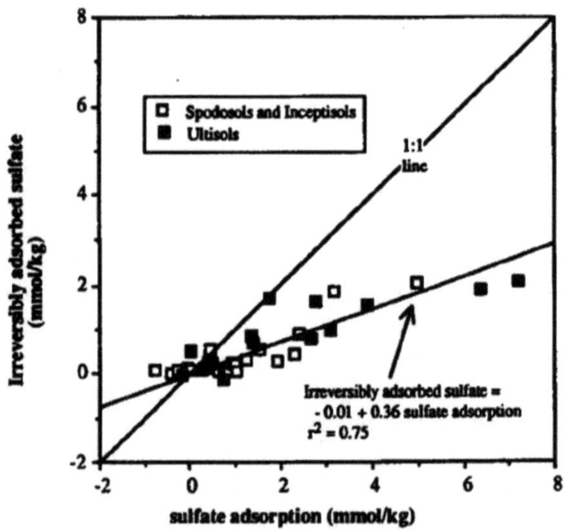

Fig. 18.30 Irreversibly adsorbed SO_4^{2-} versus total SO_4^{2-} adsorption from 0.25 mM $CaSO_4$ solution for soils (Harrison et al. 1989). Copyright 1989 American Society of Agronomy. Reprinted with permission

been studied in Dutch field conditions (van Bremen et al. 1984; Mulder et al. 1987, 1989). In these cases, it was observed that aluminum fractions are leached from acidic sandy soils, inducing depletion of certain aluminum fractions within several decades. Organic aluminum, for example, is removed from the surface layer at a rate of 1–2 % per year.

Aluminum is transferred to solution mainly from the organically complexed solid phase. Mulder et al. (1989) showed that depletion of alumino-organics in podzols was most pronounced in the upper layer of the soil profile, and that as a consequence, acid rain input produced a drastic change in the naturally occurring

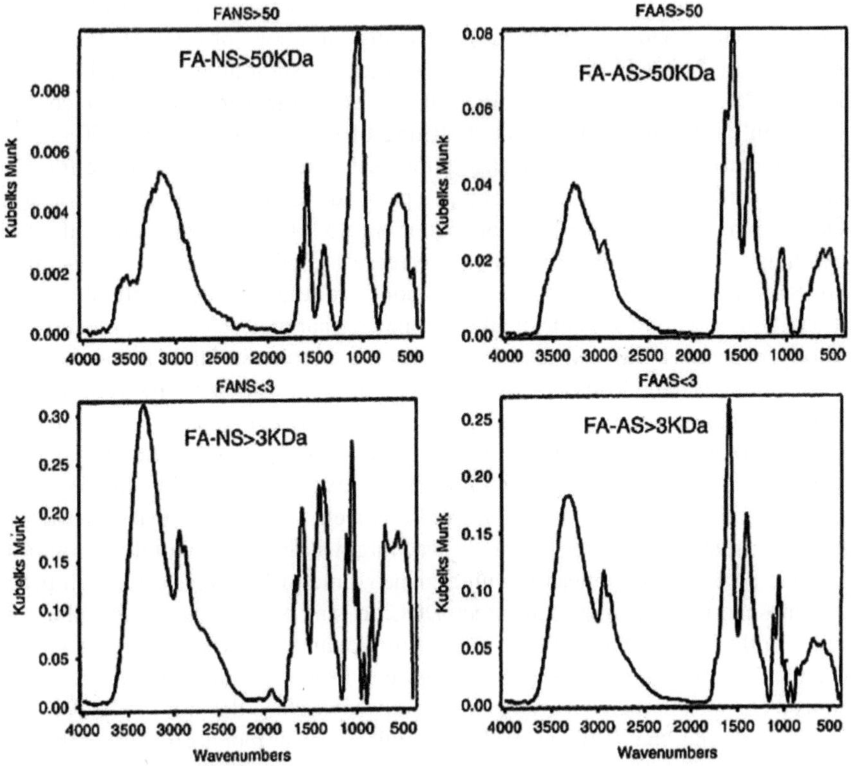

Fig. 18.31 FTIR spectra of fulvic acid (*FA*) indicating the structural differences induced by soil acidification. *NS* natural soil, *AS* acidified soil (modified after Calace et al. 2001). Reprinted from Calace et al. (2001). Copyright 2001 with permission from Elsevier

podzolization process. Mulder et al. (1989) concluded that "…the current rapid depletion of organic aluminum in the rooting zone of many acid sandy soils is irreversible on a time scale of decades or centuries."

Acid rains also may affect the molecular configuration and the properties of soil humic compounds. Calace et al. (2001) reported a laboratory experiment in which humic and fulvic acids extracted from a natural soil were subjected to simulated acid rains, until a soil pH of 4 was obtained. The effect of soil acidification on the humic and fulvic substances was determined in terms of humic/fulvic substance molecular weight and composition of the soil, before and after soil treatment. Neither molecular weight nor the FTIR spectrum of humic acid was influenced by the acidification process. In contrast, both the molecular weight and the FTIR spectrum of the fulvic acid were affected. The differences between FTIR spectra of pristine and acidified fulvic acids with molecular weights of 50 and 3 kDa, respectively, are shown in Fig. 18.31.

The fulvic acids extracted from the pristine soil (FA-NS) reveal a greater alcohol and carbohydrate content (1,050–1,250 cm^{-1}) and higher adsorption than the acidified samples (FA-AS). This is due to antisymmetrical vibration of $–COO^-$ groups (1,550–1,650 cm^{-1}); to –CO stretching vibrations of esters, ketones, and aldehydes; and to –CO absorption and $–NH_2$ deformation of amide I and II bands (1,650–1,750 cm^{-1}). In the acidified fulvic acid, the carboxyl group content (1,700 cm^{-1}) dominated the –CO stretching of esters (shoulder at 1,730 cm^{-1}). The differences between the two absorption peaks are more evident in the fraction smaller than 3 kDa, which shows only the peak due to $–COO^-$ groups. Fulvic acid acidification caused the peak corresponding to –OH deformation and –CO stretching of alcohols and phenols to decrease (1,050–1,350 cm^{-1}). These data show that the main effect of soil acidification is reflected in the structural composition of fulvic acids and in the breakup of the original molecules, due probably to a hydrolysis process. This suggests an irreversible increase in fulvic acid solubility.

Acid rain leaching leads also to changes in the composition of dissolved organic carbon (DOC). Release of different components of DOC obtained from a red clay soil (China), under leaching with simulated acid rain at pH of 2.5–5.6, was studied by Liu et al. (2009) using three-dimensional excitation-emission matrix spectroscopy (EEMS). Humic and fulvic acids, as well as microbial byproduct materials, were released at the beginning of leaching with low pH solution; in contrast, aromatic proteins were released with a greater time of leaching and a high pH.

Atmospheric deposition can affect DOC in soil by changing either the soil acidity or the ionic strength of the soil solution. An increase in soil acidity as a result of acid rain deposition may lead to a decrease in DOC. Potential anthropogenically induced changes in the DOC status in the soil–subsurface domain can be deduced by extrapolating the findings of large, multinational monitoring programs on atmospherically deposited sulfur and sea salts on continental surface waters. Monteith et al. (2007) used the results of 25 years of monitoring studies in Europe and North America, on DOC content in surface waters located on glaciated, acid-sensitive terrain, to define DOC trends resulting from changes in atmospheric deposition chemistry. Monitoring data showed that between 1990 and 2004, some of the sites measured a DOC increase; this finding was attributed to a gradual decline in the sulfate content of atmospheric deposition. A simple model that applied changes in deposition chemistry and catchment acid sensitivity were used to describe a partially buffered ecosystem where changes in organic acidity and rise in DOC were related to recovery from acidification. The effect on the ecosystem of a decrease in DOC due to acid deposition is irreversible. Even with global efforts to reduce sulfur emissions into the atmosphere, and a decrease in soil acidity, the soil dissolved organic carbon will never return to its initial concentration.

18.3.2 Impact of Effluents

Municipal and industrial effluents may contain dissolved organic and inorganic toxic chemicals, and suspended organic solids, which are introduced into the soil–subsurface system during land disposal of wastewater effluents and by the discharge of sludge during the effluent treatment process. Their effect on the properties of the soil and subsurface regime depends on their interaction with the system solid phase and their persistence and distribution with depth. As a result, physical properties of the pristine solid phase matrix, such as water retention capacity, bulk density, porosity, and structure, may be changed irreversibly. Both effluents and sludge are characterized by a high organic matter content which may vary from 1 to 70 %. As a function of their origin, effluents, treated wastewaters, and the resulting sludge may contain chemical contaminants that may cause irreversible changes to the soil–subsurface solid phase (Yaron et al. 1984; Feigin et al. 1991; Levy et al. 2010). Here, we consider surfactants that originate from effluents to illustrate irreversible changes in soil–subsurface properties.

Surfactants, or "surface active agents," are organic compounds of natural or anthropogenic origin that reduce the surface tension of water. Surfactants may interact with the porous medium solid phase, and as a consequence, they may affect the soil–subsurface water transmission properties. Surfactants are organic compounds comprising hydrophobic and hydrophilic components. Surfactants may be hydrophilic when the surfactant is ionic or highly polar. Surfactants may be anionic, cationic amphoteric, or nonionic. Due to an amphipathic structure, surfactant molecules exhibit a specific orientation at interfaces within multiphase systems where they concentrate. In these cases, the interfacial tension is reduced. Surfactants form micelles at critical solution concentrations.

When sewage water contaminated by surfactants reaches the land surface, e.g., in an irrigated field or in a disposal site, several physical interactions between the surfactant and the soil–subsurface solid phase occur (Tumeo 1997, and references therein). The surfactant induces changes in hydraulic conductivity, due to the alteration of the liquid phase properties and to a direct effect on the physical properties of the solid phase. While changes occurring in the liquid phase are reversible, surfactant-induced alteration of the solid phase becomes irreversible under specific conditions; this leads to an irreversible change in the soil–subsurface water transmission properties. Here, we present some examples from the literature that show the irreversibility of changes induced by anthropogenic surfactants on soil–subsurface water transmission properties.

The types of surfactant and soil control surfactant-induced effects on soil–subsurface water transmission properties. Allred and Brown (1994) performed laboratory tests on hydraulic conductivity of sandy and loam soils leached with nonionic, anionic, cationic, and amphoteric surfactants. It is apparent that the ionic surfactants caused a greater hydraulic conductivity decrease than the nonionic surfactants. Maximum hydraulic conductivity decreases were 47 % for the sandy soil and more than two orders of magnitude for the loamy soil (Fig. 18.32).

Similarly, Liu and Roy (1995) found that maximum adsorption and precipitation of the anionic surfactant sodium dodecylsulfate in a calcium-saturated subsurface soil (from Louisiana, USA) occurred when the surfactant concentration was close to the critical micelle concentration. Cation exchange between soil and this sodium surfactant led to the release of Ca^{2+}, subsequent precipitation of calcium, and irreversible decrease in hydraulic conductivity.

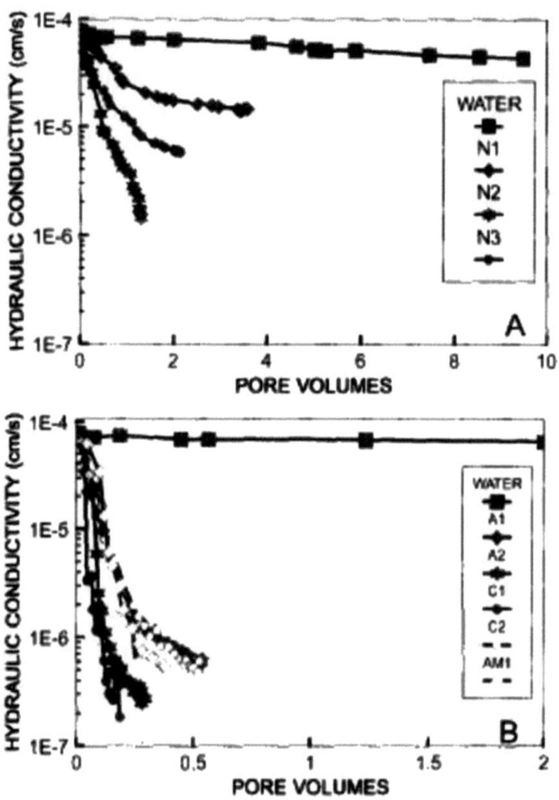

Fig. 18.32 Hydraulic conductivity versus pore volume (1 mol/kg solution concentration) for **a** nonionic surfactants, and **b** ionic surfactants. N1-3 nonionic; A1,2 anionic; C1.2 cationic; AM1-amphoteric surfactants. Reprinted from Allred and Brown (1994). Copyright 1994 with permission of John Wiley and Sons

Surfactant-induced deflocculation and dispersion of clay particles into the soil–subsurface matrix are other factors leading to irreversible changes in the soil hydraulic conductivity. The application of nonionic surfactants either before or during irrigation was found to increase the dispersion of hydrophobic sandy soils (Mustafa and Letey 1969), leading to a decreased flow rate (Miller et al. 1975).

Sodium dodecylbenzenesulfonate (SDBS) is a major constituent of synthetic detergents used in household products and industrial processes; it has been found in municipal sewage waters. Rao et al. (2006) used SDBS in a laboratory experiment to test the effect of anionic surfactants on dispersivity and hydraulic conductivity of sodium and calcium-saturated silt loam paddy soil. Dispersion of Na- and Ca-soil with SDBS in NaCl solutions was determined by measuring the relative absorbance (A_R) of the suspensions. Relative absorbance is defined as $A_R = A_x/A_0$, where A_x and A_0 are the absorbance of soil suspensions at increasing SDBS and NaCl concentrations, respectively.

The A_R of the suspension in the SDBS solution increased with SDBS concentration, indicating an increase in the stability of the suspensions (Fig. 18.33). To explain this behavior, Rao et al. (2006) suggested the following mechanism: (1) Adsorption of SDBS on the clay particles, which increases the repulsive forces and

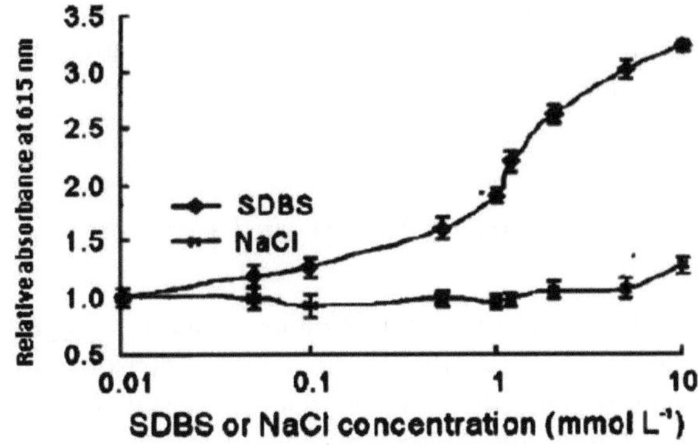

Fig. 18.33 Effect of sodium dodecylbenzenesulfonate (SDBS) and NaCl on the stability of soils saturated with calcium at 25 ± 1°C measured by relative absorbance (modified after Rao et al. 2006) Reprinted from Rao Rao et al. (2006). Copyright 2006 with permission from Elsevier

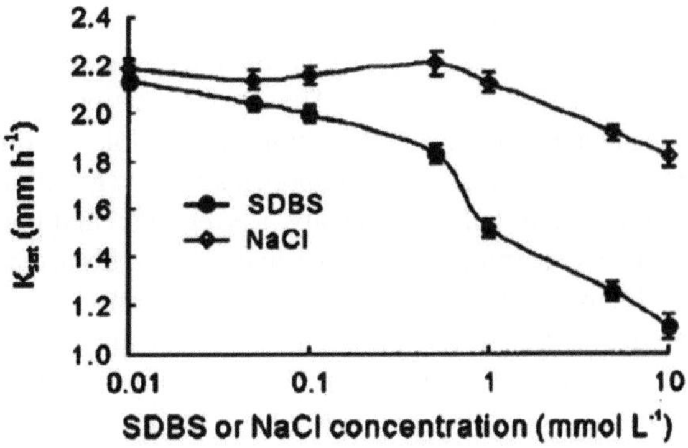

Fig. 18.34 Variations in saturated hydraulic conductivity, K_{sat}, with concentration of sodium dodecylbenzenesulfonate (SDBS) and NaCl injected into soils saturated with calcium at 25 ± 1°C (modified after Rao et al. 2006) Reprinted from Rao et al. (2006). Copyright 2006 with permission from Elsevier

causes dispersion of particles; (2) Adsorbed SDBS tends to decrease interface tension, supporting particle dispersion; and (3) increase of Na^+ in the Ca-soil suspension, promoting particle dispersion. Combination of these three processes results in an increase in the stability of the Ca-soil suspension. In contrast, the A_R of the suspension in NaCl solution varied little below 1.0 mmol/L NaCl. Above this value, the A_R increased slowly.

The dispersion of clay particles in Ca-soil columns leached by SDBS and NaCl-aqueous solution led to an irreversible decrease in soil hydraulic conductivity (Fig. 18.34). The higher the SDBS or NaCl concentration in the flushing solution, the greater the extent of the hydraulic conductivity decrease. However, as in the dispersion experiments, the effect of SDBS was much greater than that of NaCl.

These results show that the hydraulic conductivity of smectite clay-sand mixtures decreased following addition of various anions. Dispersed clay, however, appeared in the effluent only upon addition of citrate or hexametaphosphate (Frenkel et al. 1992). In the latter case, the hydraulic conductivity began to increase once maximum clay concentration reached the effluent. Dispersion of smectite was one order of magnitude lower than that of kaolinite; illite exhibits an intermediate value. Frenkel et al. (1992) attributed this behavior to kaolinite having the highest ratio of positively charged edge surfaces to negatively charged planar surfaces, which leads to a high adsorption capacity. In the case of smectite, hydraulic conductivity decreases through partial blocking of pores by short-distance migration of dispersed particles.

18.3.3 Impact of Sludge and Compost

Land disposal of untreated effluents and of sludge from wastewater purification plants, as well as urban waste compost, constitutes an additional factor leading to anthropogenically induced changes in soil and subsurface physical properties. Application of sludge obtained from wastewater purification treatments was considered, in the last half of the past century, to be a sound method for increasing soil fertility and improving soil physical properties.

Some of the organic carbon found in effluents, sludge, and urban waste compost is resistant to degradation. The organic matter content of the soil increases under repeated application; in this case, irreversible alteration of the physical properties is possible. The organic components of sewage effluents, sludge, and compost lead to changes in bulk soil density, aggregation, and porosity, affecting also the water transmission properties of the native soil–subsurface regime.

Many studies have shown an increase in the solid phase organic content after sludge application, which is usually explained by a parallel decrease in the bulk density and an increase in the total porosity. For soils with a wide range of textures, hydraulic conductivity in sludge-amended plots is usually higher than that in control plots, although a large quantity per unit area of sludge or compost additions is necessary to induce statistically significant changes in hydraulic conductivity (Metzger and Yaron 1987, and references therein).

Bulk density and porosity of the soil–subsurface solid phase is reduced after sludge application on land surface. Several studies have reported that the bulk density of a sludge-amended sandy soil is negatively correlated to the organic carbon content of the applied sludge, and to the application rate, and positively correlated to the soil porosity (Lindsay and Logan 1998; Bahremand et al. 2003). Based on analysis of 12 studies dealing with short- and long-term application of sludge and compost on bulk density, Khaleel et al. (1981) proposed an empirical relationship between the soil organic carbon content and the reduction in bulk density.

The effect of sludge application rate on the bulk density and porosity of a silt loam from Columbus, Ohio, USA, is shown in Fig. 18.35. The soil porosity increased approximately linearly ($r^2 = 0.728$; $P < 0.01$), while the bulk density decreased approximately linearly ($r^2 = 0.815$; $P < 0.01$) as a function of sludge application rate. Lindsay and Logan (1998) considered the effect of sludge on soil physical properties to depend largely on the rate of sludge decomposition and on the contribution to soil organic carbon. In the reported cases, these effects could still be seen four years after sludge application. This finding is in accord with the results of Paglia et al. (1981), who found that six months after sludge application

Fig. 18.35 Effect of sludge application rate on **a** total porosity, and **b** bulk density (modified after Lindsay and Logan 1998). Copyright 1998 American Society of Agronomy. Reprinted with permission

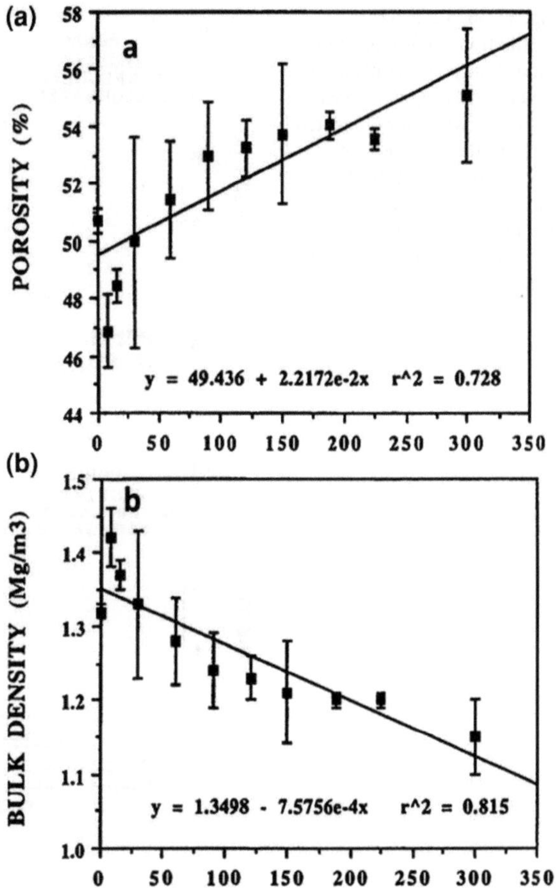

on a sandy loam soil, the total porosity reached values ranging between 24 and 31 %, as compared with 17.5 % for the control. Wei et al. (1985), too, showed that six years after application of a dewatered sludge to a clay soil, the "large pore" space increased on account of capillary pore space. These results suggest that despite the decomposition of organic matter, sludge-induced changes in soil porosity are, on a human lifetime scale, partially irreversible.

Urban waste compost disposal on a calcareous soil (Fluventic Xerochrept) was found to linearly increase the soil total porosity (pores > 5 μm) (Giusquiani et al. 1995). These authors found that pore shape and size distributions of amended soil were significantly different from those of the control plots (Fig. 18.36). In soils treated with a high rate of compost application, increases in elongated pores involved both transmission pores (50–500 μm) and those >500 μm. The increase in soil porosity in these soils also led to a decrease in bulk density. Compost addition also led to an increase in soil water retention, expressed by water potential measurements (Fig. 18.37).

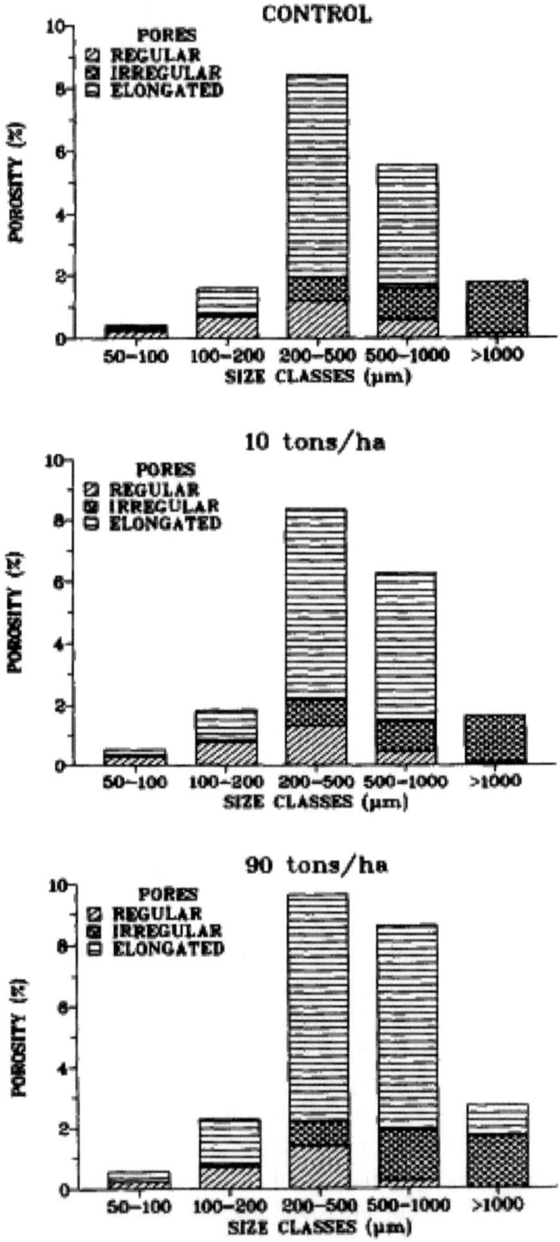

Fig. 18.36 Pore-size distribution according to the equivalent pore diameter for regular and irregular pores and to the width for elongated pores in the surface layer (0–10 cm) of control plots and in plots treated with low and high rates of compost application, at the June 1992 sampling time (Giusquiani et al. 1995)

Stability of aggregates is a function of how the cohesive forces between particles resist applied disruptive forces (Kemper and Rosenau 1986). Soil aggregation status is generally controlled by the organic carbon (OC) content, which may change following application of OC-rich sludge or compost. Indeed, laboratory

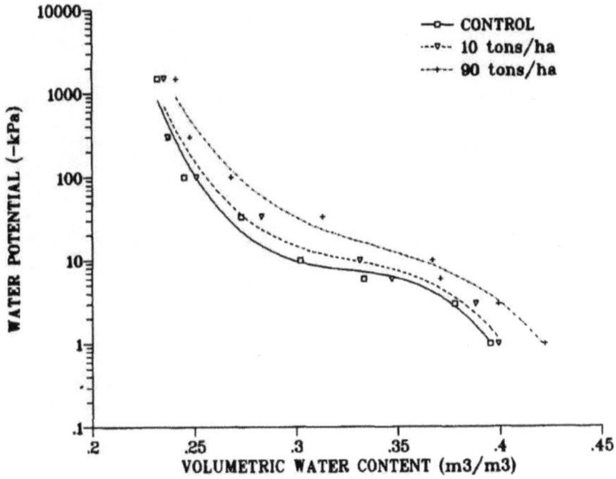

Fig. 18.37 Soil moisture retention curves in the surface layer (0–10 cm) of control and treated plots (Giusquiani et al. 1995)

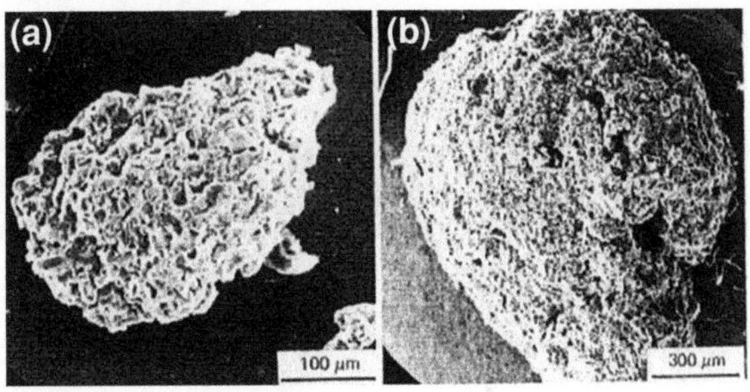

Fig. 18.38 SEM images showing general features of water-stable aggregates isolated from **a** a silt loam soil and **b** a silt loam sludge (5 %) mixture after 16 days of incubation at 25 °C (Metzger and Yaron 1987)

and field studies show that sludge application to soil induces an increase in the number and size of water-stable aggregates (WSA).

Figure 18.38 illustrates the typical morphology of WSA (200–2,000 μm in diameter) in a loamy soil following sludge application. Compared to control aggregates (Fig. 18.38a), WSA obtained in sludge-amended soil (Fig. 18.38b) are characterized by a large size and a loose and porous internal structure. From the aggregation patterns observed in Fig. 18.38, for three soils amended with sludge, three distinct phases can be identified: a short 10-day period characterized by a rapid rate of WSA formation, a period of similar duration displaying a moderate decrease in WSA content, and finally a longer period characterized by a slow secondary increase in WSA content over a scale of months (Metzger et al. 1986).

Experiments performed on various soils and under different climatic conditions confirmed that aggregate sizes in a sludge-amended soil decrease with time, over a scale of months (Metzger and Yaron 1987, and references therein). Moreover, long-lasting effects of sludge on the structural stability of modified aggregates are reported by Wei et al. (1985), who found aggregates in a modified silt clay loam soil, 6 years after 112 t/ha of sludge was added to a field soil in a single application.

Metzger et al. (1986) argue that the two main binding mechanisms responsible for aggregate formation are the cementation of primary particles by fungal

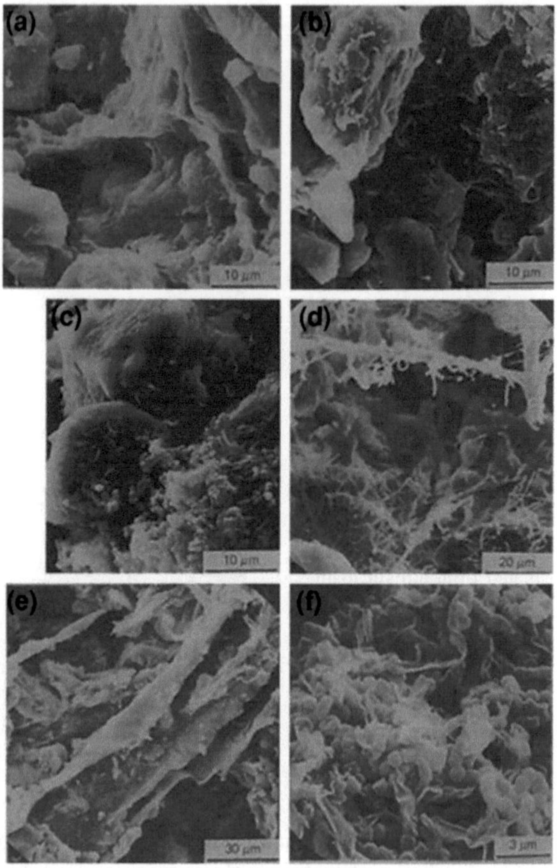

Fig. 18.39 SEM micrographs representative of internal structure of water-stable aggregates, in **a** a silt loam soil (control); and showing presence of bacteria, actinomycetes, and fungi in **b–f** a silt loam sludge (5 %) mixture after 16 days of incubation at 25 °C (Metzger and Yaron 1987)

polysaccharides, and the physical entanglement of primary particles in the fungal hyphae (Fig. 18.39). It was also observed that polysaccharides, as a byproduct of soil microbial activity, exhibit a positive correlation to WSA stability over time (e.g., Cheshire et al. 1984). Moreover, sludge water-soluble components were found to contribute to binding of well-defined soil constituents such as kaolinite or montmorillonite (Metzger and Robert 1985).

Soil aggregate stability following application of urban compost of different maturities is presented in Fig. 18.40, where effects of environmental factors such as mechanical breakdown or fast and slow wetting were considered by Annabi et al. (2007). These authors found that after addition of compost, the enhanced microbial activity improved aggregate stability mainly by increasing water repellency. The observed increase in interparticle cohesion was attributed to the inward diffusion of binding organic substances within the aggregates. The effect of microbial colonies growing on soil solid surfaces and forming polymer bridges that bind soil mineral particles was confirmed in a series of studies (Kimbursky et al. 1989; Chenu and Guerif 1991; Chenu 1993; Chenu and Tessier 1995). More

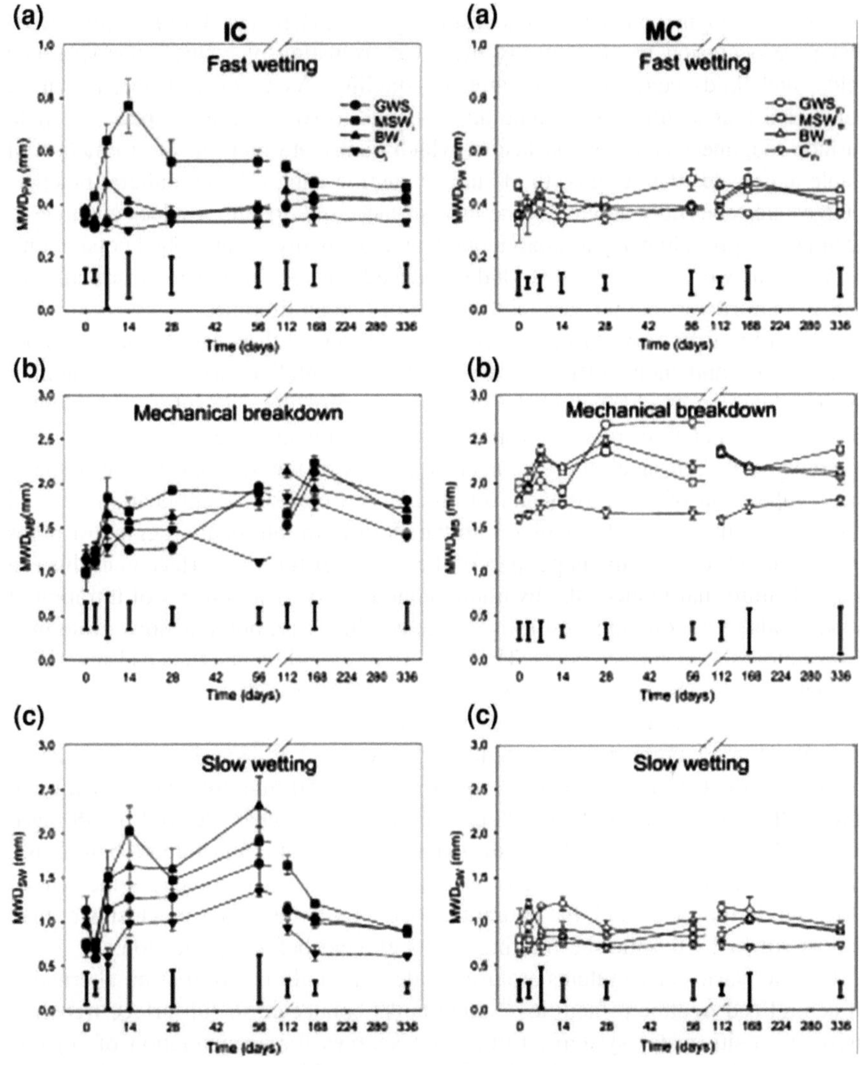

Fig. 18.40 Evolution of aggregate stability during the incubations of aggregate-compost mixtures: control (C), green waste and sewage sludge compost (GSW), municipal solid waste compost (MSW), and biowaste compost (BW) at two maturity levels: immature (IC and subscript *i*) and mature (MC and subscript *m*). Mean weight diameters (MWD) of three tests are presented: **a** fast wetting **b** mechanical breakdown, and **c** slow wetting. Note the different scales used on the *y*-axis (Annabi et al. 2007)

information on physical processes affecting microbial habitats in porous media appears in the review of Or et al. (2007).

The organic hydrophobic fraction of sludge has also been proposed as being responsible for aggregate stabilization. Based on an experiment on paper sludge amendments applied on clay loam and silt clay loam, Hafida et al. (2007) considered that pore occlusion by biologically mediated formation of neutral sugars, uronic acids, and lipids contributes to aggregate stability. As a consequence, it can be concluded that addition of organic amendments to soils causes increases in pore surface roughness and pore occlusion, which contribute more to the formation of stable aggregates than to growth of surface wetting angles. In agricultural practice, sludges are repeatedly applied to the land surface, and organic residues continue to accumulate, providing a quasicontinuous source of energy for microbial populations. As a result, we can expect that sludge-induced changes in the microstructure of WSA is an irreversible process on the scale of human lifetimes.

Clogging of pores by suspended particles, following disposal of sludges on the land surface and their vertical transport during rainfall or irrigation, is another process that causes changes in the microstructure of the soil–subsurface domain and on its water transmission properties. The dominant mechanism of clogging depends on the physical and chemical properties of the soil–subsurface domain, the composition of the wastewater or sludge, the method of application on land surface, and the timescale involved. In the short term, physical clogging of pores by suspended solids in the upper part of the soil layer forms a surface mat; this is a major feature that reduces the hydraulic conductivity. The process of filtration at depth leads to a reduction in soil hydraulic conductivity, but at a slower rate than that of clogging of surface pores. The stability of clogging, and thus its irreversible impact, should also be considered in terms of microbial habitats and activity during drying–wetting cycles in the unsaturated zone.

Soil column laboratory experiments by Vinten et al. (1983) showed the relative amounts of deposition of suspended solids in the top 50 mm, following leaching of three different soils with about 100 cm of wastewater (characterized by 98 mg/L suspended solids) (Fig. 18.41). In silt loam soil, the water was filtered before use, giving a concentration of 38 mg/L. The soil type strongly influences the rate of clogging. In the sandy soil, the hydraulic conductivity decreases much slower than that of silt loam, despite the filtering of the effluent before application.

On-site wastewater systems are used widely around the world as an alternative to centralized treatment facilities. Septic tanks connected to subsurface trenches, used in on-site water systems, form point sources for accumulation of organic matter components in the soil–subsurface domain. These components clog pores and cause a decrease in water transmission properties. The kinetics of suspended particle deposition in a porous medium significantly affect pore clogging, as shown by Amitay-Rosen et al. (2005) using noninvasive MRI measurements (Fig. 18.42).

Some studies have revealed biochemical processes under on-site wastewater disposal systems, with associated clogging and decrease in infiltration. McKinley and Siegrist (2010) analyzed the clogging layer (1.5 cm thickness) in an on-site wastewater disposal system in a sandy loam soil (Golden, Colorado), finding that most organic materials accumulated in the upper 0.7-cm segment. The concentrations of total organic carbon (TOC), humic acid (HA), and fulvic acid (FA), as well as of polysaccharides, found in the clogging layer are shown in Fig. 18.43. In this particular site, the less humified carbonaceous materials (polysaccharide and fulvic

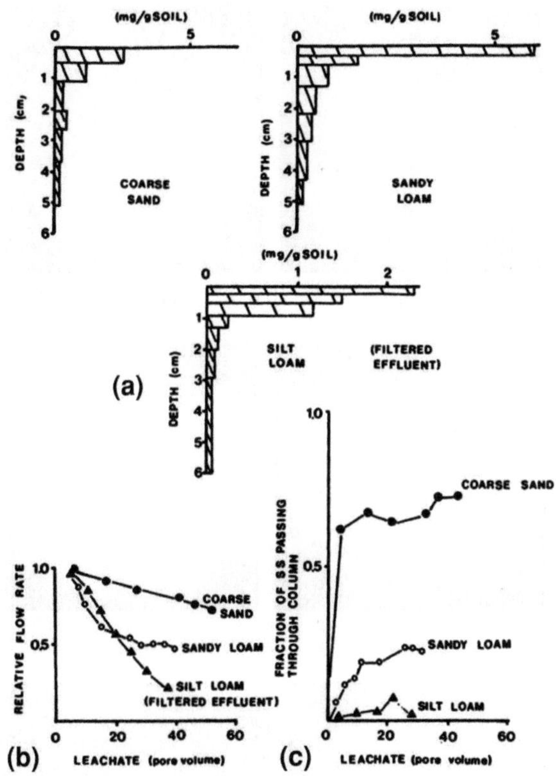

Fig. 18.41 **a** Deposited solids distribution in three soils leached with wastewater, showing **b** the effect on soil hydraulic conductivity, and **c** transport of suspended solids through the soils (Vinten et al. 1983) Reprinted with permission from Vinten et al. (1983). Copyright 1983 American Chemical Society

acids) were at low concentration compared with the more humified carbonaceous materials; the latter are less degradable because they contain a more complex molecular structure. As a result, the less degradable materials can accumulate in soil pores and become a key agent responsible for long term, irreversible clogging.

The microstructure of clogged soils in soil wastewater treatments under prolonged exposure was detected using fluorescence microscopy (Jiang and Matsumoto 1995). Detailed information on soil clogging by entrapment of suspended particles, microbial cells, and their metabolized products is shown in Fig. 18.44. For the investigated experimental conditions, Jiang and Matsumoto (1995) identified (1) physical clogging that occurs horizontally in the surface zone, due to trapping in pores of suspended and colloidal materials, and (2) biological clogging that occurs vertically along the soil profile, resulting from accumulation of microorganism products in soil pores under anaerobic conditions. Physical and biological clogging are interrelated and enhance each other. Under anaerobic conditions, decomposition of metabolic products is low, even tending to be negligible, when they are adsorbed on suspend solids and colloidal materials. This suggests irreversibility of the process.

Humic acid, originating from sewage sludge and composts, may lead to structural changes of soil–subsurface natural humic acid following long-term

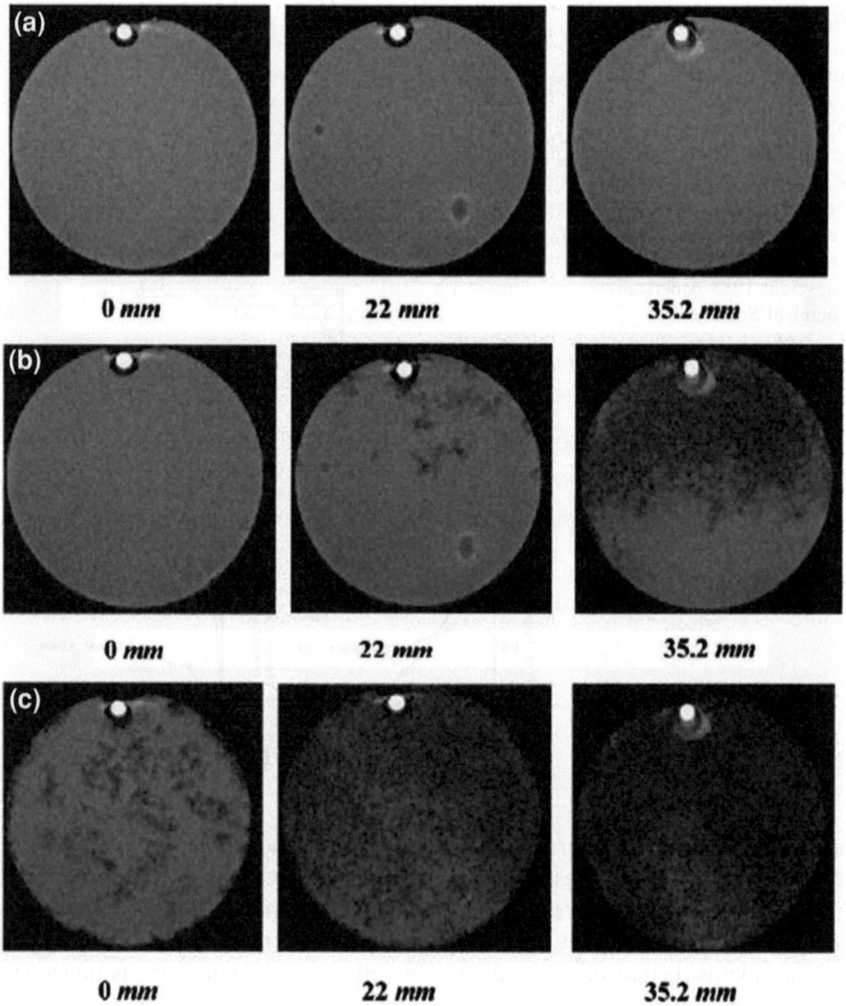

Fig. 18.42 Selected axial images of the porous medium at several distances from the inlet, for an experiment with 12 μm particles: **a** 0, **b** 120, and **c** 180 min. The column was oriented horizontally, with the reference tube (*white spot*) positioned at the top. Reprinted with permission from Amitay-Rosen et al. (2005). Copyright 2005 American Chemical Society

sludge and compost disposal on the land surface. Changes in the humic acid fraction (HA) of soil following disposal of sludge were shown by Boyd et al. (1980). These authors reported the association of proteinaceous and aliphatic compounds originating from disposed sludge with natural soil humic substances. Sludge-induced HA changes in the amended soil were confirmed by an increase in amide and C–H stretch infrared absorptions, as well by increased N and decreased C/H ratios. These results suggest that sludge-HA is incorporated into soil humic acid, thus modifying its original composition.

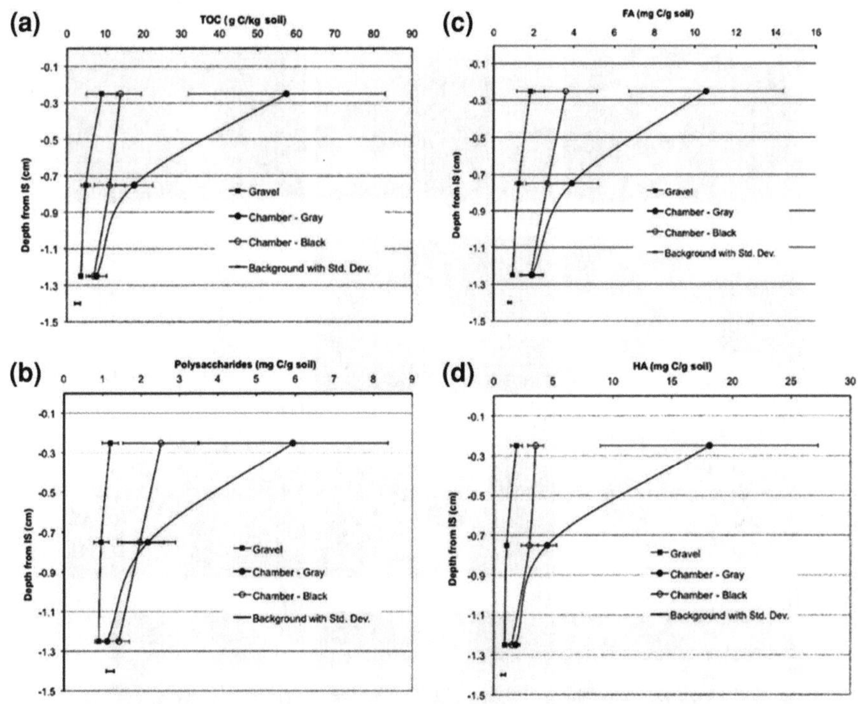

Fig. 18.43 Effect of sewage effluent organic matter on infiltration in a sandy loam soil. **a** Total organic carbon (TOC) versus penetration depth; **b** polysaccharides versus penetration depth, **c** fulvic acid (FA) versus penetration depth, and **d** humic acid (HA) versus penetration depth (modified after McKinley and Siegrist 2010). Copyright 2010 Soil Science Society of America. Reprinted with permission

The studies of Adani and Tambone (2005) and Adani et al. (2006) on sewage sludge and compost disposal on soils provide additional information on their effect on soil-HA composition. Comparison between diffuse reflectance infrared Fourier transform spectroscopy (DRIFT) spectra of HA extracted from sludge (Fig. 18.45a) and compost (Fig. 18.45b) contaminants, and from soils before and after their contamination, show changes in the soil HA after long-term contamination. Humic acid extracted from sludge-contaminated soil (ST-HA) showed a higher content of aliphatic fractions, a similar content of O-aliphatic-C and/or N-aliphatic-C, and a lower content of aromatic carbons than HA from untreated soil (SU).

Humic acid extracted from compost-contaminated soil (CM-HA) shows no relevant absorption bands at 830–1,100 cm^{-1} assigned to carbohydrates but exhibits bands at 1,038 and 1,231 cm^{-1} assigned to lignin. IR spectra show also the presence of aliphatic carboxylic acid and of amide groups. Most of the carboxyl groups in the CM–hHA are amide form, whereas free COOH groups prevailed in soil-HA.

The effect of annual addition of composed sewage sludge on the alteration of soil HA was studied by Fernandez et al. (2009). In a field experiment, the authors

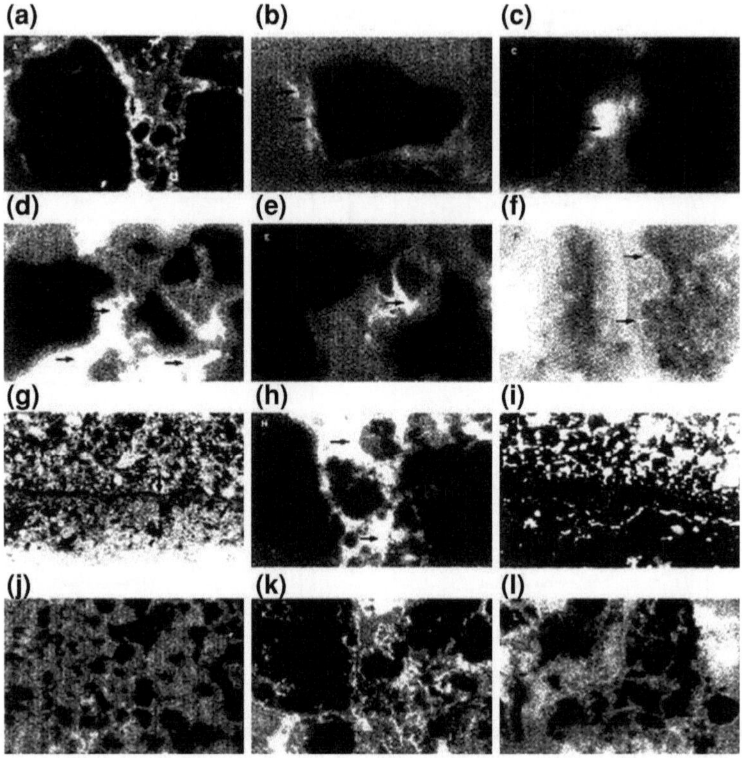

Fig. 18.44 Microstructure of clogged soil under anaerobic conditions. **a** Andosol after submergence for 4 months. Soil pores were filled with fine particles, suspended solid materials, bacteria cells, and bacterial synthesized products (frame length 750 μm). **b** Arenosol after submergence for 4 months. A bridge structure. Organic materials (*arrows*) clogged only a narrow portion between two sand particles (frame length 750 μm). **c** Andosol after submergence for 2 months. *Arrow* indicates the initial state of clogging in the portion connecting two grains. **d**, **e** Andosol after submergence for 4 months. Soil pores were fully sealed by polysaccharides. **f** Andosol after submergence for 4 months. Bacterial cells (*arrows*) in soil pores. **g** Fluvisol after submergence for 4 months. Compact grain structure. Dark girdle (*arrows*) consists of the layer accumulated materials. **h** Andosol after submergence for 4 months. Soil pores were sealed by suspended solid material, fine particles, and polysaccharides. **i** Andosol after submergence for 4 months. Structure of layer of accumulated materials. **j** Andosol after submergence for 4 months. Structure of layer of accumulated materials soil pores was completely filled with organic materials. Aggregates were destroyed to form fine particles. **k** Andosol after submergence for 4 months. An aggregate (*left*) and a destroyed aggregate (*right*). **l** As in (**k**), destroyed aggregate (Jiang and Matsumoto 1995). Reprinted with permission

found substantial differences in the chemical composition and structure between the HA of natural soil and that extracted from soils (CS-HA) after sewage sludge compost disposal. Compared to natural soil (SO-HA) humic acid, the soils contaminated by disposal of sewage sludge compost (CS-HA) display large S and N contents, which can be attributed to “the incorporation of protein decomposition

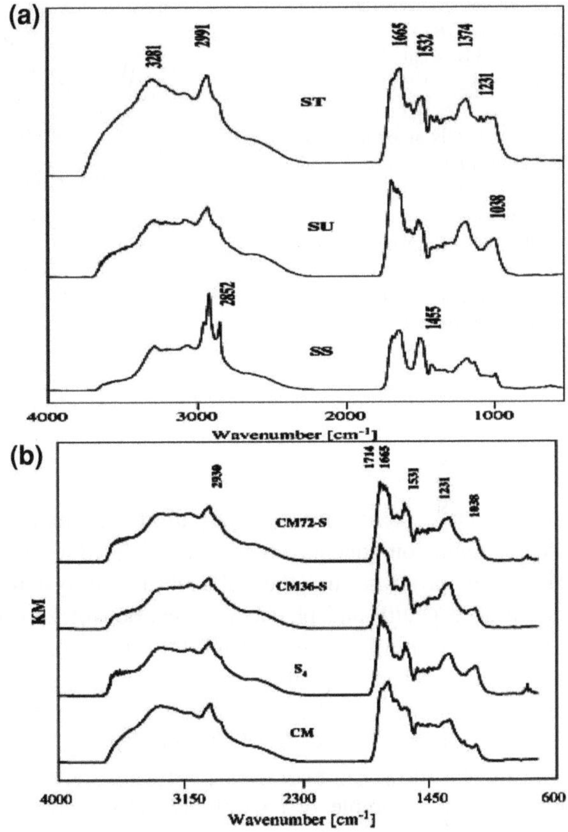

Fig. 18.45 DRIFT spectra of HA, **a** extracted from sludge (SS), untreated soil (SU) and treated soil (ST), and **b** extracted from compost (CM), control soil (S_4) and amended soil samples CM36-S and CM72-S (Adani and Tambone 2005 and Adani et al. 2006). Reprinted from Adani and Tambone (2005) and from Adani et al. (2006). Copyright 2006 with permission from Elsevier

products and S-containing surfactant residues in HA macromolecules" and a smaller C, O, COOH, and phenolic OH group. Senesi and Brunetti (1996) suggest that the low degree of humification of composted soils is related to the low amount of lignin structure present in sewage sludge. Changes in the FTIR spectra of HA isolated from sewage sludge compost and HA isolated from compost-contaminated soil may be seen in FTIR spectra (Fig. 18.46).

18.3.4 Petroleum Products

Petroleum products represent one of the main industrial pollutants spread widely on the land surface, around the world. Spilled petroleum products generally consist of complex hydrocarbon mixtures and may be found under gas stations, around pipelines or distillation plants, and along coastal seashores. These mixtures, which contain components with a large range of carbons (e.g., C_6–C_{30}) with different vapor pressures and solubilities may irreversibly affect the soil–subsurface regime.

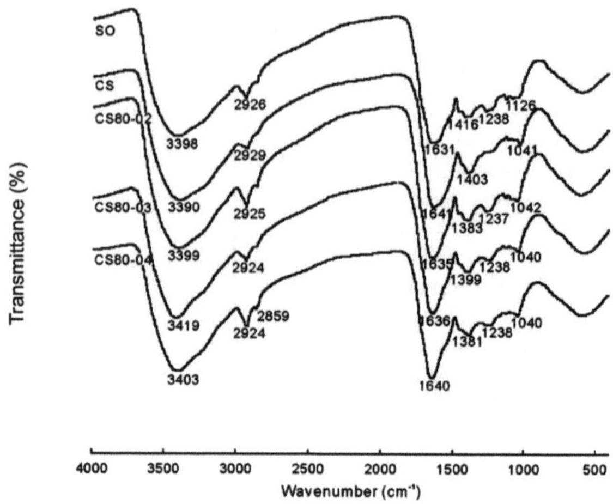

Fig. 18.46 Fourier transform infrared (*FTIR*) spectra of humic acids isolated from the unamended soil (*SO*), composted sewage sludge (*CS*), and soils amended with CS at 80 t/ha/yr for 1, 2, and 3 years 80-02, 80-03, and 80-04, respectively (Fernandez et al. 2009). Reprinted from Fernandez et al. (2009). Copyright 2009 with permission from Elsevier

The fate of kerosene—a widely used petroleum product containing about 100 hydrocarbon components—may be considered to illustrate the interaction between petroleum products and the soil–subsurface system (see also Sect. 8.2.3). The retention capacity and physical attenuation of kerosene in soil were studied by Fine et al. (1997), Yaron et al. (1998) and Dror et al. (2002). It was found that during disposal in the soil–subsurface domain, the light petroleum fractions volatilize or dissolve in the surrounding air and aqueous phases, leaving behind a nonaqueous liquid phase (NAPL) that contains a heavy hydrocarbon fraction. In the soil–subsurface domain, the petroleum hydrocarbon NAPL forms entrapped ganglia that resemble discontinuous blobs within large pores (Schwille 1984; Powers et al. 1994). These ganglia functionally alter the original porous medium architecture. In the case of kerosene contamination, for example, it was found that the kerosene residual content (KRC) in soil varies from soil to soil and is inversely linearly related to the moisture and linearly related to the soil texture as expressed by clay content (Fig. 18.47).

Under equilibrium conditions, multicomponent dissolution can be described in terms of a constant partition coefficient among all of the organic mixture components and the aqueous solution. An example of kerosene–water partitioning at equilibrium was presented in Chap. 8 (Fig. 8.18), where kerosene was in contact with an aqueous electrolyte solution (0.01 N NaCl) at a temperature of 22 °C for 100 h. From gas chromatogram analysis, it can be seen that aliphatic and branched aliphatic hydrocarbons in the range C_9–C_{16} comprise the major group of components in neat kerosene, with only a minor group of aromatic compounds. In the aqueous electrolyte solution, aromatic compounds, especially branched benzenes and naphthalene, make up the majority of the compounds in the aqueous phase. The C_9–C_{10} aliphatic and branched aliphatic components do not appear in the chromatogram of the aqueous kerosene, because aromatic components are several orders of magnitude more soluble than aliphatic components.

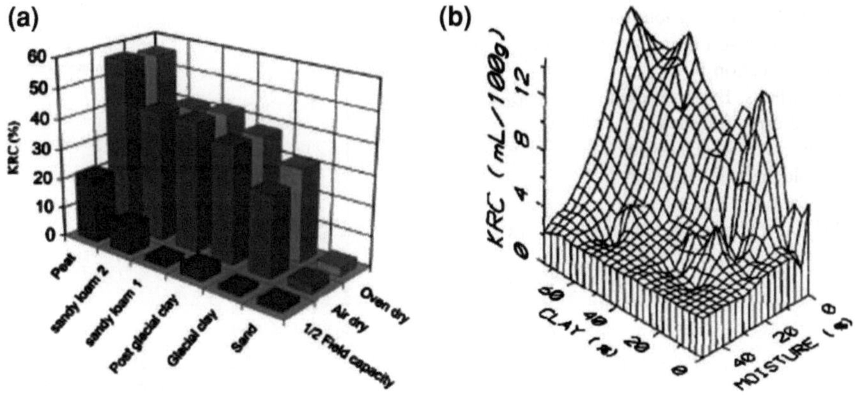

Fig. 18.47 Kerosene residual content (*KRC*) in soils **a** as affected by type of soil and moisture content (Jarsjö et al. 1994), and **b** as a function of clay and moisture content (Fine and Yaron 1993) Reprinted from Jarsjö et al. (1994). Copyright 1994 with permission from Elsevier, and from Fine and Yaron (1993). Copyright 1993 with permission from Elsevier

Fig. 18.48 Composition of kerosene as a function of viscosity (Gerstl et al. 1994). Copyright 1994 American Society of Agronomy. Reprinted with permission

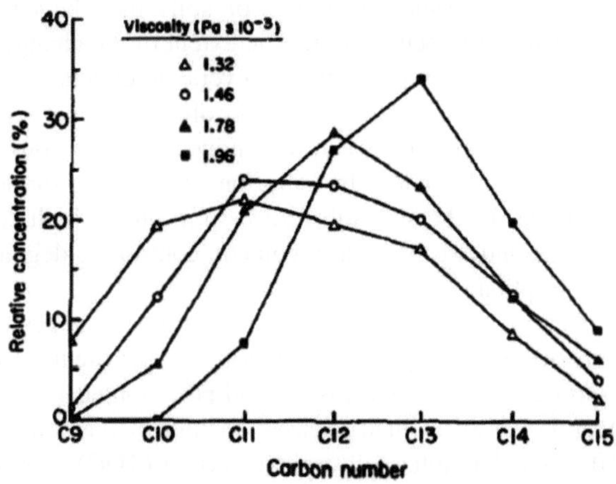

Loss of kerosene deposited on soil, through volatilization, is controlled by the vapor pressure of each of the kerosene components, concentration gradients in the subsurface, pore geometry, moisture content, temperature, and type of soil. The physical properties of the residual liquid also change as the composition of the initial mixture changes. Loss of the light fraction of kerosene by volatilization leads to an increase in the liquid viscosity. The relationship between the viscosity of kerosene after volatilization and the relative concentration of the major kerosene components is shown in Fig. 18.48. An increase in kerosene viscosity was accompanied by a relative increase in the heavy fraction (C_{13}–C_{15}) and a decrease in the light fraction (C_9–C_{11}). The increase in viscosity causes a change in soil–subsurface porosity, through the formation of ganglia and discontinuous blobs

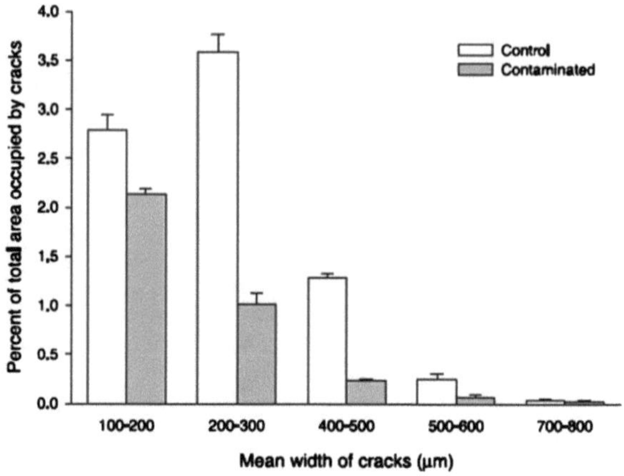

Fig. 18.49 Size distribution of cracks of a soil in response to contamination by hydrocarbons. Bars represent standard error for each measure ($n = 4$) (Caravaca and Roldán 2003) Reprinted from Caravaca and Roldán (2003). Copyright 2003 with permission from Elsevier

within large pores. Because these heavy kerosene residues are not easily biodegraded on a human lifetime scale, we can expect such porosity changes to be essentially irreversible.

Water transport properties of soils may be changed irreversibly following hydrocarbon contamination. The extent of the changes is controlled by the type of soil and pollutant properties. Irreversible effects on water transmission properties are related to the retention of petroleum products in the soil–subsurface matrix. Hydrocarbon products can be retained in the porous matrix by entrapment in the pore network and/or by sorption at particle surfaces. Physical and chemical properties of the soil–subsurface solid phase, including hydration status, mineral content, and organic matter content, control the degree of hydrocarbon entrapment and sorption.

Caravaca and Roldán (2003) examined changes occurring to soil texture and porosity following contamination by petroleum hydrocarbons. These authors considered an extreme case of soil pollution near an oil refinery in the province of Murcia (Spain), which released uncontrolled amounts of hydrocarbon over nearly 10 years. The total soil organic carbon (TOC) increased during this period from 7.2 to 73.8 g/kg. In contaminated soil samples, Caravaca and Roldán (2003) found a decrease in the clay faction from 33 to 21 %, relative to the natural soil, and an increase in sand fraction from 45 to 58 %. Soil macroporosity (defined as pores >50 μm) in the hydrocarbon-contaminated soil increased about 15-fold compared with that in the control soil, affecting water transport in the soil–subsurface system. This modification is reflected also in the pore size classes.

Shrinkage of the soil was determined by Caravaca and Roldán (2003) to be higher in the control plot than in the soil contaminated by hydrocarbons. Moreover, the size distribution of cracks in the control soil differed from the contaminated soil (Fig. 18.49).

Soil water-holding capacity and water repellency are additional factors that should be considered in relation to soil water transmission properties. The effect of varying petroleum hydrocarbon contamination on the hydraulic properties of

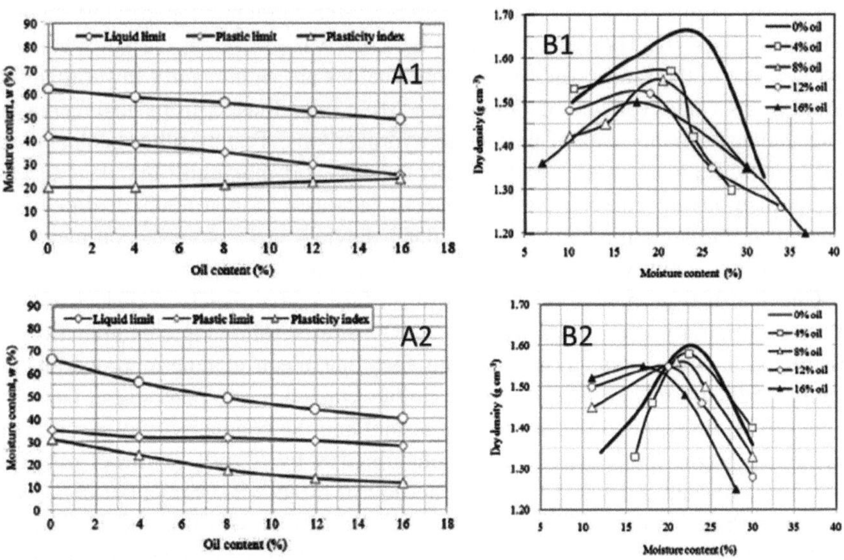

Fig. 18.50 Influence of oil contamination on **A** Atterberg limit, and **B** compaction curves for (1) sandy loam and (2) silt loam basaltic residual soil (modified after Rahman et al. 2010)

Ottawa sand was tested by Burckhard et al. (2000). In general, it was observed that soil water retention curves for hydrocarbon-contaminated sand were different from the retention curves of pristine Ottawa sand. At higher suction (i.e., lower pressure) values, the contaminated samples had a higher holding capacity, leading to a reduction in the soil hydraulic conductivity.

Soil water repellency caused by petroleum hydrocarbon contamination was reported by Roy et al. (2003), based on a large study of 12 hydrocarbon-contaminated soils from Alberta, Canada. Water repellency, defined by the molarity of ethanol droplet (MED) method, dichloromethane-extractable organics (DEO), and total organic carbon content (TOC), were examined. All three parameter values were significantly higher in hydrocarbon-contaminated repellent soils than in the adjacent pristine soils. The results indicated that, in the studied sites, petroleum hydrocarbon-induced repellency is predominantly a surface phenomenon. The fact that most crude oil naturally contains a small fraction of dichloromethane-insoluble material and that this fraction tends to substantially increase once reaching the soil (e.g., Connaughton et al. 1993; Reid et al. 2000), suggests that hydrocarbon-induced repellency of soils is an irreversible change at least over human lifetime scales.

Geotechnical and geoelectrical properties of the soil–subsurface system also may be affected irreversibly by oil contamination. A study of geotechnical properties of sandy loam and silt loam soils of basaltic origin (Rahman et al. 2010) showed changes in the soil Atterberg plastic and liquid limits following contamination by petroleum hydrocarbons (Fig. 18.50). It is clearly seen that both liquid and plastic limits decreased with increases in the amount of petroleum hydrocarbon in the soils. The decrease in liquid limit was greater in the silt loam soil (38 %) than in the sandy soil (16 %); these values are significantly lower than the liquid limits values (66 and 62 %) in the pristine soils. Plastic limits also increased with increases in amounts of petroleum hydrocarbons. Testing other geotechnical parameters, Rahman et al. (2010) found that the soil maximum dry density and optimum moisture content dropped due to increases in petroleum hydrocarbon content in the contaminated soils. Similar behaviors were observed on measurements of permeability and shear strength of the soils.

Geoelectrical properties of the soil–subsurface system, such as dielectric constant, conductivity, and resistivity, may also be altered following contamination by petroleum hydrocarbons. Based on the fact that electromagnetic properties of geomaterials are related directly to the liquid retained in the pore space, Darayan et al. (1998) determined the dielectric constant and the conductivity of a soil from the Texas Sugar Land area (USA), contaminated by a spill of diesel oil. The dielectric constant increased with the addition of diesel oil. Darayan et al. (1998) attributed these results to the replacement of water by diesel oil in the pore space.

Are changes to geotechnical and geoelectrical properties induced by petroleum hydrocarbon contamination of the soil–subsurface system irreversible? Given that a heavy fraction of the hydrocarbon remains in the soil–subsurface pore matrix, after volatilization of the light fraction, changes to these properties can be considered irreversible over human lifetime scales.

References

Adani F, Tambone F (2005) Lon-term effect of sewage sludge application on soil humic acids. Chemosphere 60:1214–1221

Adani F, Genevini P, Tambone F, Montoneri E (2006) Compost effect on soil humic acid: a NMR study. Chemosphere 65:1414–1418

Agassi M, Ben-Hur M (1991) Effect of slope length, aspect and phosphogypsum on runoff and erosion from steep slopes. Austral J Soil Res 29:197–203

Agassi M, Ben-Hur M (1992) Stabilizing steep slopes with soil conditioner and plants. Soil Technology 5:249–256

Agassi M, Shainberg I, Morin J (1990) Slope, aspect and phosphogypsum effects on runoff and erosion. Soil Sci Soc Am J 54:1102–1106

Ahuja LR, Cassel DK, Bruce RR, Barnes BB (1989) Evaluation of spatial distribution of hydraulic conductivity using effective porosity data. Soil Sci 148:404–411

Aldrich DG, Parker ER, Chapman HD (1945) Effects of several nitrogenous fertilizers and soil amendments on the physical and chemical properties of an irrigated soil. Soil Sci 59:299–312

Allred B, Brown GO (1994) Surfactant-induced reductions in soil hydraulic conductivity. Groundwater Monit Remed 14:174–184

Amitay-Rosen T, Cortis A, Berkowitz B (2005) Magnetic resonance imaging and quantitative analysis of particle deposition in porous media. Environ Sci Technol 39:7208–7216

Annabi M, Houot S, Francou C, Poitrenaud M, Le Bissonnais Y (2007) Soil aggregate stability improvement with urban composts of different maturities. Soil Sci Soc Am J 71:413–423

Bahremand MA, Afyuni M, Hajabbassi MA, Rezaeinejad Y (2003) Effects of sewage on soil physical properties. Agri Sci J Iran 6:1–9
Barak P, Jobe BO, Krueger AR, Peterson LA, Laird DA (1997) Effects of long-term soil acidification due to nitrogen fertilizer inputs in Wisconsin. Plant Soil 197:61–69
Bednarek W, Ereszka R (2007) Effect of liming and fertilization with nitrogen on the physicochemical properties of soil. Agrophysica 10:523–531
Ben-Hur M, Letey J (1989) Effect of polysaccharide, clay dispersion and impact energy on water infiltration. Soil Sci Soc Am J 53:233–238
Ben-Hur M, Faris J, Malik M, Letey J (1989) Polymers as soil conditioner under consecutive irrigations and rainfall. Soil Sci Soc Am J 5:1173–1177
Blevins RL, Thomas GW, Cornelius PL (1977) Influence of no-tillage and nitrogen fertilization on certain soil properties after 5 years of continuous corn. Agron J 69:383–386
Borggaard OK, Gimsing AL (2008) Fate of glyphosate in soil and possibility of leaching to ground and surface waters: a review. Pest Manage Sci 64:441–456
Bouman OT, Curtin D, Campbell CA, Biederbeck VO, Ukrainetz H (1995) Soil acidification from long-term use of anhydrous ammonia and urea. Soil Sci Soc Am J 59:1488–1494
Boyd SA, Sommers LE, Nelson W (1980) Change in the humic acid fraction of soil resulting from sludge application. Soil Sci Soc Am J 44:1179–1186
Bresler E, McNeal BL, Carter DL (1982) Saline and Sodic Soils. Springer-Verlag, Berlin, Germany 236 pp
Burckhard SR, Pirkl D, Schafer VR, Kulakow P, Leven BA (2000) A study of soil water-holding properties as affected by TPH contamination. In: Proceedings 2000 Conference Hazardous Waste Research, Denver, CO, 23–25 May 2000
Calace N, Fiorentini F, Petronio BM, Pietroletti M (2001) Effects of acid rain on soil humic compounds. Talanta 54:837–846
Caravaca F, Roldán A (2003) Assessing changes in physical and biological properties in a soil contaminated by oil sludge under semiarid Mediterranean conditions. Geoderma 117:53–61
Cavallaro N, McBride MB (1984) Zinc and copper sorption and fixation by an acid soil clay. Effect of selective dissolutions. Soil Sci Soc Am J 48:1050–1054
Chen Y, Banin A (1975) Scanning electron microscope (SEM) observations on soil structure changes induced by sodium calcium exchange in relation to hydraulic conductivity. Soil Sci 120:428–436
Chen W, Kan AT, Tomson MB (2000) Irreversible adsorption of chlorinated benzenes to natural sediments: implications for sediment quality criteria. Environ Sci Technol 34:385–392
Chenu C (1993) Clay- or sand-polysaccharide associations as models for the interface between micro-organisms and soil: water related properties and microstructure. Geoderma 56:143–156
Chenu C, Guerif J (1991) Mechanical strength of clay-minerals as influenced by an adsorbed polysaccharide. Soil Sci Soc Am J 55:1076–1080
Chenu C, Tessier D (1995) Low temperature scanning electron microscopy of clay and organic constituents and their relevance to soil microstructures. Scanning Microsc 9:989–1010
Cheshire MV, Sparling CP, Mundie CM (1984) Influence of soil type, crop and air drying on residual carbohydrate content and aggregate stability after treatment with periodate and tetraborate. Plant Soil 76:339–347
Chorum M, Murray RS, Osmund G, Rengassamy P (1994) Clay dispersion as influenced by pH and net particle charge of sodic soils. Austral J Soil Res 32:1243–1252
Connaughton DF, Stedinger JR, Lion LW, Shuler MKL (1993) Description of time-varying desorption kinetics. Environ Sci Technol 27:2397–2403
Dane JH, Klute A (1977) Salt effect on the hydraulic properties of a swelling soil. Soil Sci Soc Am J 41:1043–1049
Darayan S, Liu C, Shen LC, Shattuck D (1998) Measurement of electrical properties of contaminated soil. Geophys Prospect 46:477–488
Di Maio C (1996) Exposure of bentonite to salt solution: osmotic and mechanical effect. Geotechnique 46:695–707

Díaz-Raviña M, Calvo de Anta R, Bååth E (2007) Tolerance (PICT) of the bacterial communities to copper in vineyards soils from Spain. J Environ Qual 36:1760–1764
Dror I, Gerstl Z, Prost R, Yaron B (2002) Abiotic behavior of entrapped petroleum products in the subsurface during leaching. Chemosphere 49:1375–1388
Epstein L, Bassein S (2001) Pesticide applications of copper on perennial crops in California, 1993 to 1998. J Environ Qual 30:1844–1847
Farmer VC (1978) Water on particle surfaces. In: Greenland DJ, Hayes MHB (eds) The chemistry of soil constituents. Wiley, New York
Farmer VC, Russell JD (1971) Interlayer complexes in layer silicates: the structured water in lamellar ionic solutions. Trans Faraday Soc 67:2737–2749
Feigin A Ravina I, Shalhevet J (1991) Irrigation with treated sewage effluent. Advance series in agricultural science, vol 17. Springer, Heidelberg
Fernandez JM, Senesi N, Plaza C, Brunetti G, Polo A (2009) Effects of composed and thermally dried sewage sludges on soil and soil humic acid properties. Pedosphere 19:281–291
Fine P, Yaron B (1993) Outdoor experiments enhance volatilization by venting of kerosene from soil. J Contam Hydrol 12:355–374
Fine P, Graber ER, Yaron B (1997) Soil interactions with petroleum hydrocarbons: Abiotic processes. Soil Technol 10:133–153
Flanagan DC, Chaudhari K, Norton LD (2002) Polyacrylamide soil amendment effects on runoff and sediment yield on steep slopes II natural rainfall conditions. Trans Am Soc Agri Eng 45:1339–1351
Fox RL, Olson RA, Mazurak AP (1952) Persistence of ammonium ion and its effect upon physical and chemical properties of soil. Agron J 44:509–513
Frenkel H, Goertzen JO, Rohades JD (1978) Effects of clay type and content, exchangeable sodium percentage and electrolyte concentration on clay dispersion and soil hydraulic conductivity. Soil Sci Soc Am J 42:32–39
Frenkel H, Levy GJ, Fey MV (1992) Clay dispersion and hydraulic conductivity of clay-sand mixtures as affected by the addition of various anions. Clays Clay Miner 40:515–521
Fu MH, Zhang ZZ, Low PF (1990) Changes in the properties of a montmorillonite-water system during the adsorption and desorption of water hysteresis. Clays Clay Miner 38:485–492
Gerstl Z, Galin T, Yaron B (1994) Mass flow of a volatile organic liquid mixture (VOLM) in soils. J Environ Qual 37:487–493
Giusquiani PL, Pagliai M, Gigliotti G, Businelli D, Benetti A (1995) Urban waste compost: effects on physical, chemical, and biochemical soil properties. J Environ Qual 24:175–182
Glass RL (1987) Adsorption of glyphosate by soils and clay minerals. J Agri Food Chem 35:497–500
Greenland DJ, Gregory PJ, Nye PH (1997) Land resources: on the edge of Malthusian precipice? Philos Trans Royal Soc, Ser B-Biol Sci 352:861–867
Grolimund D, Borkovec M (2006) Release of colloidal particles in natural porous media by monovalent and divalent cations. J Contam Hydrol 87:155–175
Guidi G (1981) Relationship between organic matter of sewage sludge and physico-chemical properties of soil. In: L'Hermite PL, Ott E (eds) Treatment and use of sewage sludge. Reidel, Dordrecht, pp 530–544
Hafida Z, Caron J, Angers DA (2007) Pore occlusion by sugars and lipids as possible mechanism of aggregate stability in amended soils. Soil Sci Soc Am J 71:1831–1839
Harrison RB, Johnson DW, Todd DE (1989) Sulfate adsorption and desorption reversibility in a variety of forest soils. J Environ Qual 18:419–426
Haynes RJ (1984) Lime and phosphate in the soil-plant system. Adv Agron 37:249–315
Haynes RJ, Naidu R (1998) Influence of lime, fertilizer and manure applications on soil organic matter content and soil physical conditions: a review. Nutr Cycl Agroecosys 51:123–137
Hermosin MC, Perez Rodriguez JL (1981) Interaction of chlordimeform with clay minerals. Clays Clay Miner 29:141–152
Hilgard EW (1906) Soils-their formation, properties, composition and relation to climate and plant growth in humid and arid region. Macmillan, London

Hoyt PB (1981) Improvements in soil tilth and rapeseed emergence by lime applications on acid soils in the Peace River region. Can J Soil Sci 61:91–98

Jarsjö J, Destouni G, Yaron B (1994) Retention and volatilization of kerosene: laboratory experiments on glacial and post-glacial soils. J Contam Hydrol 17:167–185

Jiang Y, Matsumoto S (1995) Change in microstructure of clogged soil in soil wastewater treatment under prolonged submergence. Soil Sci Plant Nutr 41:207–213

Kazman SI, Shainberg I, Gal M (1983) Effect of low levels of exchangeable Na and applied gypsum in improving infiltration of a sodic soil. Soil Sci 135:184–192

Kelley WO, Arany A (1928) The chemical effect of gypsum, sulphur, iron sulphate and alum on alkali soils. Hilgardia 3:3939–3999

Kemper WD, Rosenau RC (1986) Aggregate stability and size distribution. SSSA Methods Soil Anal Agron Monogr 9:425

Kemper WD, Olsen J, De Mooy CJ (1975) Dissolution rate of gypsum in flowing water. Soil Sci Soc Am Proc 39:458–463

Khaleel R, Reddy KR, Overcash MR (1981) Changes in soil physical properties due to organic waste applications: a review. J Environ Qual 10:133–141

Kijne JW (1973) Evaporation from bare soils. In: Hadas A, Swartzendruber D, Rijtema PE, Fuchs M, Yaron B (eds) Physical aspects of soil water and salts in ecosystems, ecological studies, vol 4. Springer, Berlin

Kimbursky RS, Levanon D, Yaron B (1989) Role of fungi in stabilizing aggregates of sewage sludge amended soils. Soi Sci Soc Am J 53:1086–1091

Kiss S (1999) Enzymology of soils inoculated with microorganisms. Studia Universitatis Babes-Bolyai, Biologia 44:3–45

Knight BAG, Denny PJ (1970) The interaction of paraquat with soil: adsorption by expanding lattice mineral. Weed Res 10:40–48

Knight BAG, Tomlinson TE (1967) The interaction of paraquat with mineral soil. J Soil Sci 18:233–243

Lagaly G, Ziesmer S (2003) Colloid chemistry of clay minerals: the coagulation of montmorillonite dispersion. Adv Colloid Interface Sci 100–102:105–128

Lagerwerff JV, Nakayama FS, Frere HM (1969) Hydraulic conductivity related to porosity and swelling of soils. Soil Sci Soc Am Proc 33:3–12

Laird AD (2006) Influence of layer charge on swelling of smectites. Appl Clay Sci 34:74–87

Laird AD, Shang C, Thompson ML (1995) Hysteresis of crystalline swelling of smectites. J Colloid Interface Sci 171:240–245

Larssen T, Ludersen E, Tang DG, He Y, Gao JX, Liu HY, Duan L, Seip HM, Vogt RD, Mulder J, Shao M, Wang YH, Shang H, Zhang XS, Solberg S, Aas W, Okland T, Eilertsen O, Angell V, Liu QR, Zhao DW, Xiang RJ, Xiao JS, Luo JH (2006) Acid rain in China. Environ Sci Technol 40:418–425

Lebron I, Suarez DL, Yoshida T (2002) Gypsum effect on the aggregate size and geometry of three sodic soils under reclamation. Soil Sci Am J 66:92–98

Lentz RD, Sojka RE, Ross CW (2000) Polymer charge and molecular weight effects on treated irrigation furrow processes. Int J Sediments Res 15:17–30

Levy GJ, Sumner ME (1998) Mined and by-product gypsum as soil amendments and conditioners. In: Wallace A, Terry ER (eds) Handbook of soil conditioners. Marcel Dekker, New York, pp 445–462

Levy GJ, Fine P, Bar-Tal A (eds) (2010) Treated wastewater in agriculture: use and impacts on the soil environment and crops. Wiley-Blackwell, Oxford

Likens GE, Bormann FH (1995) Biogeochemistry of a Forested Ecosystem, 2nd edn. Springer, New York 159 pp

Lindsay BJ, Logan TL (1998) Field response of soil physical properties to sewage sludge. J Environ Qual 27:534–542

Liu M, Roy D (1995) Surfactant-induced interactions and hydraulic conductivity changes in soil. Waste Manag 15:463–470

Liu L, Song C, Yan Z, Li F (2009) Characterizing the release of different composition of dissolved organic matter in soil under acid rain leaching using three-dimensional excitation-emission matrix spectroscopy. Chemosphere 77:15–21

López-Periago JE, Arias-Estévez M, Nóvoa-Muñoz JC, Fernández-Calviño D, Soto B, Pèrez-Novo C, Simal-Gándara J (2008) Copper retention kinetics in acid soils. Soil Sci Soc Amer J 72:63–72

Low PF (1981) The swelling of clay III Dissociation of exchangeable cations. Soil Sci Soc Am J 45:1074–1078

Lutz JF, Pinto RA (1965) Effect of phosphorus on some physical properties of soils I Modulus of rupture. Soil Sci Soc Am Proc 29:458–460

Lutz JF, Garcia-Lagos R, Hilton HG (1966) Effect of phosphorus on some physical properties of soils II Water retention. Soil Sci Soc Am Proc 30:433–437

McBride MB (1989) Surface chemistry of soil minerals. In: Dixon JB, Weed SB (eds) Minerals in soil environments, 2nd edn. Soil Science Society of America Inc., Madison

McBride MB (1992) Surface chemistry of soil minerals. In: Dixon JB, Weed SB (eds) Minerals in soil environments, SSSA book series 1. Madison, Wisconsin, pp 35–91

McBride MB (2001) Cupric ion activity in peat soil as a toxicity indicator for maize. J Environ Qual 30:78–84

McKinley JW, Siegrist RL (2010) Accumulation of organic matter components in soil under conditions imposed by waste infiltration. Soil Sci Soc Amer J 74:1690–1700

McNeal BL (1968) Prediction of the effect of mixed salt solution on soil hydraulic conductivity. Proc Soil Sci Soc Am 32:190–193

McNeal BL, Coleman NT (1966) Effect of solution composition on soil hydraulic conductivity. Proc Soil Sci Soc Am 30:308–312

Metzger L, Robert M (1985) A scanning electron microscopy study of the interactions between sludge organic components and clay particles. Geoderma 35:159–167

Metzger L, Yaron B (1987) Influence of sludge organic matter on soil physical properties. Adv Soil Sci 7:141–173

Metzger L, Levanon D, Mingelgrin U (1986) Effect of sludge on soil structural stability: Microbiological aspects. Soil Sci Soc Am J 51:346–351

Miller WW, Valoras N, Letey J (1975) Movement of two nonionic surfactants in wettable and water repellent soils. Soil Sci Soc Am Proc 39:11–16

Monteith DT, Stoddard JL, Evans CD, de Wit HA, Forsius M, Høgåsen T, Wilander A, Skjelkvåle BL, Jeffries DS, Vuorenmaa J, Keller B, Kopácek J, Vesely J (2007) Dissolved organic carbon trends resulting from changes in atmospheric deposition chemistry. Nature 450:537–540

Mooney RW, Keenan AG, Wood LA (1952) Adsorption of water vapors by montmorillonite. J Am Chem Soc 74:1371–1374

Morel JL, Guckert A (1983) Influence of limed sludge on soil organic matter and soil physical properties In: Cartroux G, L'Hermite P, Suess E (eds) The influence of sewage sludge and activity in unsaturated porous media, vol 7. Agriculture Science, pp 39–45

Morillo E, Undabeytia T, Maqueda C (1997) Adsorption of glyphosate on the clay mineral montmorillonite: effect of Cu(II) in solution and adsorbed on the mineral. Environ Sci Technol 31:3588–3592

Mulder J, van Grinsven JJM, van Breemen N (1987) Impacts of acid atmospheric deposition on woodland soils in The Netherlands: III. Aluminum chemistry. Soil Sci Am J 51:1640–1646

Mulder J, van Breemen N, Eijck HC (1989) Depletion of soil aluminum by acid deposition and implication for acid neutralization. Nature 337:247–249

Mustafa MA, Letey J (1969) The effect of two nonionic surfactants on aggregate stability of soils. Soil Sci 107:343–347

Norrish K (1954) The swelling of montmorillonite. Disc Farad Soc 18:120–134

Or D, Smet BF, Wraith JM, Dechesne A, Friedman SP (2007) Physical constraints affecting bacterial habitats and activity in unsaturated porous media: a review. Adv Water Resour 30:1505–1527

Oster JD (1982) Gypsum usage in irrigated agriculture: a review. Fertilizer Res 3:73–89

Oster JD, Shainberg I, Wood JD (1980) Flocculation value and gel structure of sodium/calcium montmorillonite and illite suspensions. Soil Sci Soc Am J 44:955–959

Ouhadi VR, Goodarzi AR (2006) Assessment of the dispersive soil treated by alum. Eng Geol 85:91–101

Ouhadi VR, Yong RN, Sedighi M (2006) Influence of heavy metal contaminants at variable pH regimes on rheological behaviour of bentonite. Appl Clay Sci 32:217–231

Pagliai M, Giddi G, LaMarca M, Giachetti M, Lucamante G (1981) Effects of sewage sludge and compost on soil properties and aggregation. J Environ Qual 10:556–561

Pateiro-Moure M, Pérez-Novo C, Arias-Estévez M, López-Periago E, Martínez-Carballo E, Simal-Gándara J (2007) Influence of copper on the adsorption and desorption of paraquat, diquat and difenzoquat in vineyard acid soils. J Agri Food Chem 55:6219–6226

Pignatello JJ (2006) Fundamental issues in sorption related to physical and biological remediation of soils In: Twardowska K et al (eds) Soil and water pollution and remediation. Springer, Berlin, pp 3–23

Pignatello JJ (2012) Dynamic interactions of natural organic matter and organic compounds. J Soils Sediments doi:10.1007 /ss11368-012-0490-4

Pilsbury AF (1947) Factors influencing infiltration rates into Yolo loam. Soil Sci 64:177–181

Powers SE, Abriola LM, Dunkin JS, Weber WJ Jr (1994) Phenomenological models for transient water mass transfer processes. J Contam Hydrol 16:1–33

Quirk JP (1957) Effect of electrolyte concentration on soil permeability and water entry in irrigated soils. Int Conf Irri Drainage 8:115–123

Quirk JP, Schofield RK (1955) The effect of electrolyte concentration on soil permeability. J Soil Sci 6:251–259

Rahman AZ, Hamzah U, Taha MR (2010) Influence of oil contamination on geotechnical properties. Am J Appl Sci 7:954–961

Rao CH, Sridharan A (1984) Mechanism of sulfate adsorption by kaolinite. Clays Clay Miner 32:414–418

Rao PH, He M, Yang X, Zhang YC, Sun SQ, Wang JS (2006) Effect of an anionic surfactant on hydraulic conductivities of sodium- and calcium-saturated soils. Pedosphere 16:673–680

Reeve RC, Tamaddoni GH (1965) Effect of electrolyte concentration on laboratory permeability and field intake rate of a sodic soil. Soil Sci 99:261–266

Reid BJ, Jones KC, Semple KT (2000) Bioavailability of persistent organic pollutants in soils and sediments. Environ Pollut 108:103–122

Rengasamy P, Olsson KA (1991) Sodicity and soil structure. Austral J Soil Res 29:935–952

Reuss JO, Cosby BJ, Wright RF (1987) Chemical processes governing soil and water acidification. Nature 329:27–32

Roth CH, Pavan MA (1991) Effect of lime and gypsum on clay dispersion and infiltration in samples of a Brazilian Oxisol. Geodrma 48:351–361

Roy JL, McGill WB, Lowen HA, Johnson RL (2003) Relationship between water repellency and native and petroleum-derived organic carbon in soils. J Environ Qual 32:583–590

Russo D, Bresler E (1977) Analysis of the saturated-unsaturated hydraulic conductivity in a mixed sodium-calcium soil system. Soil Sci Soc Am J 41:706–710

Ryan JN, Elimelech M (1996) Colloid mobilization and transport in ground water. Colloids Surf A 107:1–56

U.S. Salinity Laboratory Staff (1954) Diagnosis and improvement of saline and alkali soils. In: USDA Agriculture Handbook, vol 60. United States Government Printing Office, Washington, DC

Sander M, Pignatello JJ (2009) sorption irreversibility of 1,4-dichlorobenzene in two natural organic matter-rich geosorbents. Environ Toxicol Chem 28:447–457

Schofield RK, Samson HR (1954) Flocculation of kaolinite due to the attraction of oppositely charge crystal. Discuss Faraday Soc 18:135–145

Schumacher D (1996) Hydrocarbons induce alteration of soils and sediments In: Schumacher D, Abrams MAS (eds) Hydrocarbon migration and its near surface expression, vol 66. AAPG Memoir, pp 71–89
Schwab AP, Ransom MD, Owensby CE (1989) Exchange properties of an argiustoll: effects of long-term ammonium nitrate fertilization. Soil Sci Soc Am J 53:1412–1417
Schwab AP, Owensby CE, Kulyingyong S (1990) Changes in soil chemical properties due to 40 years of fertilization. Soil Sci 149:35–43
Schwille F (1984) Migration of organic fluids immiscible with water in the unsaturated zone. In: Yaron GD, Heppie P (eds) Pollutants in porous media: the unsaturated zone between the soil surface and ground water, ecological studies, vol 47. Springer, New York, pp 27–48
Senesi N, Brunetti G (1996) Chemical and physicochemical parameters for quality evaluation of humic substances produced during composting. In: de Bertoldi M, Sequi P, Lemmes B, Papi T (eds) The science of composting. Blackie Academic and Professional, Glasgow, pp 195–212
Shainberg I (1985) The effect of exchangeable sodium and electrolyte concentration on crust formation. Adv Soil Sci 1:101–112
Shainberg I, Letey J (1984) Response of soils to sodic and saline conditions. Hilgardia 52:1–57
Shainberg I, Keren R, Frenkel H (1982) Response of sodic soils to gypsum and calcium chloride application. Soil Sci Soc Am J 46:113–119
Sposito G (1992) Characterization of particle surface charge. In: Buffle J, van Leeuwen (eds) Environmental particles. Lewis Publication, Chelsea, pp 291–314
Suzuki S, Prayongphan S, Ichikawa Y, Chae BG (2005) In situ observations of the swelling of bentonite aggregates in NaCl solution. Appl Clay Sci 29:89–98
Tumeo MA (1997) A survey of the causes of surfactant-induced changes in hydraulic conductivity. Ground Water Monit Rev 17:138–144
Uehara G, Gillman G (1981) The mineralogy, chemistry and physics of tropical soils with variable charge, Westview tropical agricultural series, vol 4. Boulder (Chap 3)
van Breemen N, Driscoll CT, Mulder J (1984) Depletion of soil aluminium by acid deposition and implications for acid neutralization. Nature 307:599–604
Vinten AJA, Yaron B, Nye PH (1983) Vertical transport of pesticides into soil when adsorbed on suspended particles. J Agri Food Chem 31:662–664
Wada N, Hines DR, Ahrenkiel SP (1990) X-ray-diffraction studies of hydration transitions in Na vermiculite. Phys Rev B 41:12895–12901
Warrington D, Shainberg I, Agassi M, Morin J (1989) Slope and phosphogypsum effects on runoff and erosion. Soil Sci Soc Am J 53:1201–1205
Warrington D, Shainberg I, Levy GJ, Bar-Or Y (1991) Polysaccharide and salt effects on infiltration and soil erosion: a rainfall simulation study. Soil Technol 4:79–91
Weber JB, Weed SB (1968) Adsorption and desorption of diquat and paraquat by montmorillonite and kaolinite clay minerals. Proc Soil Sci Soc Am 32:485–487
Weber JB, Perry PW, Upchurch RP (1965) The influence of temperature and time on the adsorption of paraquat, diquat, 1,4D and prometone by clays, charcoal and an anion exchange resin. Proc Soil Sci Soc Am 29:678–688
Wei QR, Lowery B, Peterson AE (1985) Effect of sludge application on physical properties of a silty clay soil. J Environ Qual 14:178–180
White RE (2008) Soil acidification: an overview for Australia and New Zealand. In: Proceedings of the conference Soil 2008—the living skin of planet earth, pp 218–234
White RE, Helyar KR, Ridley AM, Chen D, Heng LK, Evans J, Fisher R, Hirth JR, Mele PM, Morrison GR, Cresswell HP, Paydar Z, Dunin FX, Dove H, Simpson RJ (2000) Soil factors affecting the sustainability and productivity of perennial and annual pastures in the high rainfall zone of south-eastern Australia. Aust J Exp Agri 40:267–283
Wyszkowska J, Kucharski J, Lajszner W (2005) Enzymatic activities in different soils contaminated with copper. Polish J Environ Stud 14:659–664
Yaron B, Shainberg I (1973) Electrolytes and soil hydraulic conductivity. In: Yaron B, Danfors E, Vaadia Y (eds) Arid zone irrigation. Springer, Heidelberg, pp 189–199

Yaron B, Thomas GW (1968) Soil hydraulic conductivity as affected by sodic water. Water Resour Res 4:545–552

Yaron B, Vinten AJ, Fine P, Metzger L, Mingelgrin U (1984) Effect of solid organic components of sewage on some properties of the unsaturated zone. In: Yaron B, Dagan G, Goldshmid J (eds) Pollutants in porous media. Springer, Heidelberg, pp 162–184

Yaron B, Dror I, Graber E, Jarsjo E, Fine P, Gerstl Z (1998) Behavior of volatile organic mixtures in the soil environment. In: Rubin H, Narkis N, Carberry J (eds) Soil and pollution. Springer, Berlin, pp 37–56

Yaron B, Dror I, Berkowitz B (2008) Contaminant induced irreversible changes in properties of the soil-vadose-aquifer zone: an overview. Chemosphere 71:1321–1409

Yaron B, Dror I, Berkowitz B (2010) Contaminant geochemistry—a new perspective. Naturwiessenschaften 97:1–17

Yaron B, Dror I, Berkowitz B (2012) Soil subsurface change—chemical pollutant impacts. Springer, Heidelberg

Index

A
Acid base equilibria, 148
Acid dissociation constant, 126, 130
Acidification, 12, 185, 465–467, 469, 502, 503, 507, 508, 539, 541, 542
Acidity alkalinity, 21
Acid rain, 185, 465–467, 469, 508, 537–542
Acrolein, 397–399
Activity, 3, 4, 13, 18, 21, 22, 33, 34, 37, 38, 40, 42, 44, 46, 47, 56, 65, 67, 71, 81, 83, 120, 124, 128, 130, 131, 148, 150, 151, 154–156, 158, 164, 166, 177, 188, 222, 230–232, 280, 330, 351, 354, 362, 364, 381, 382, 384–386, 390, 412, 475, 477, 492, 506–509, 550, 552
Adsorption, 17, 20, 21, 29, 45–49, 97, 107–118, 120–129, 131, 133, 135–138, 148, 157, 160, 164, 169, 170, 173, 198, 200–218, 225–230, 232, 235–237, 256, 259, 268, 270, 277, 288, 296, 323–327, 330, 331, 358–361, 378, 391, 394, 397, 399, 401, 404, 405, 412, 416, 423, 473, 477, 492, 494, 510, 515, 517–520, 522–525, 532, 539–541, 543–546, 524
Adsorption desorption, 423
Adsorption desorption hysteresis, 229
Adsorption hysteresis, 135
Adsorption isotherm, 46, 109, 112, 128, 136, 169, 204, 206, 210, 215–218, 227, 236, 323, 519
Advection, 59, 255–261, 263, 264, 267, 269, 271, 277, 326, 329, 509
Advection dispersion, 257
Aerobic, 175, 176, 350, 354, 362, 365–368, 382, 383, 388, 389, 437, 438, 439, 442, 446, 449, 494, 496
Aerobic anaerobic, 389
Aerobically, 335, 336
Aggregate, 3, 10, 22, 115, 129, 137, 140, 155, 176, 180, 182, 188, 207, 220, 229, 247, 259, 302, 303, 305, 307, 312–314, 316, 331, 352, 409, 505–508, 511–513, 515, 529, 532, 548–552, 552, 556
Aggregation, 15, 23, 70, 180, 181, 302–304, 505, 510, 511, 532, 546, 549
Alkali, 121
Alkaline, 14, 15, 21, 47, 67, 121, 187, 207, 224, 350, 351, 360, 363, 379, 412, 501, 529, 539
Alkalinity, 21, 65, 165, 185, 187, 465–467
Alkylbenzene, 488, 481
Alkylphenols, 196, 426, 428, 430
Alum, 516, 517
Alumina, 418, 419
Aluminates, 423
Alumino, 540
Aluminosilicate, 7, 4
Aluminum, 4, 5, 13, 20, 26, 128, 185, 420, 421, 466, 509, 516, 517, 520, 522, 539–541
Amendment, 295, 512, 335, 501, 507, 510, 515, 552
Aminoparathion, 444
Ammonia, 175, 176, 350, 351, 357, 390, 501
Ammonia-N, 175
Ammonia-N volatilization, 175
Anaerobic, 22, 23, 175, 350, 354–356, 362, 365–368, 382, 383, 389, 393, 437, 438, 440–442, 444, 446, 449, 494–496, 553, 556
Anaerobic activity, 22
Anaerobic sludge, 176
Antibacterial, 97
Antimicrobial, 79
Apparent enhanced solubility, 160
Apparent hysteresis, 137
Apparent rate law, 116
Apparent solubility, 147, 159, 189–193, 197, 198
Apparently soluble, 193

DOI: 10.1007/978-3-642-54777-5, © Springer-Verlag Berlin Heidelberg 2014

Aquifer, 23, 24, 57, 59, 60, 69, 161, 225, 249, 251, 252, 257, 259, 262, 270, 338, 341, 342, 355, 362, 430, 434, 437, 447–451, 449, 451, 457–459, 461–467, 469, 470, 477, 478, 480, 481, 485, 488, 490–492, 494, 495, 510
Aquitard, 23, 59, 400
Arsenic, 26, 58, 63–65, 79, 234, 235, 296, 298, 301, 402, 404, 476–480
Arsenicals, 65
Atrazine, 84, 128, 129, 177, 209, 210, 230–232, 321, 325, 332, 337, 338, 350, 375–378, 385, 386, 442, 446–451, 480, 481, 486
Atrazine, deethylatrazine, 483
Atterberg limit, 561, 505, 506
Autooxidation, 357

B
Bacteria, 79, 97, 233, 381, 382, 390, 391, 393, 438, 442, 443, 449, 473, 481, 550
Bacterial, 54, 270, 354, 505, 506, 556
Benzene, 86, 92, 93, 157, 167, 189, 367, 388, 198–200, 430, 438, 482, 488, 490, 491, 494–496, 522–524, 558
Benzeneacetaldehyde, 483
Benzene, toluene, ethylbenzene, xylenes, 490
Beta-blockers, 98
Bicarbonate, 26, 360, 440, 501
Biocolloids, 270, 334
Biodiversity, 56, 57, 60, 81
Bound residues, 138–140, 230, 232, 233
Bromacil, 177, 321–332
Bromobenzene, 482
Brunauer-Emmet-Teller (BET), 46, 112, 113, 115, 409

C
Calcium, 13
Capillary fringe, 22, 91, 247, 250, 256, 273
Carbonate, 3, 5, 13, 24, 26, 37, 40, 56, 68, 133, 136, 296, 350, 351, 360, 362, 394, 465
Carbonate minerals, 40, 65
Carbonate rocks, 24
Carbon tetrachloride, 86
Carboxyl, 20, 124, 127, 208, 360, 378, 386, 447, 542
Carboxylate, 208
Carboxylic, 127, 207, 352, 360, 365, 375, 438, 386, 480, 555
Catalysis, 150, 188, 316, 351, 352, 373–375, 377, 378, 413, 414, 480
Cation exchange capacity, 8, 9, 108, 175, 502, 518, 290
Cation exchange, 8, 9, 46, 49, 108, 121, 127, 187, 198, 203, 216, 290, 522, 543
Cesium, 214–217
Chelates, 151
Chemical potential, 32–34, 122, 168
Chemisorption, 49, 116, 124, 399, 517
Chloride, 26, 65, 69, 86, 87, 196, 265, 295, 315, 321, 330, 360, 368, 394, 400, 458–461, 463, 477, 478
Chlordimeform, 205, 206, 519–521
Chlorobenzene, 482, 523
Chloromethylbenzene, 482
Chloronitrobenzene, 89
Clay, 4, 6–9, 11–13, 17, 20, 22–24, 39, 47–49, 58, 64, 65, 107, 108, 111, 117, 119–124, 126, 128, 133, 136–138, 140, 176, 178, 181, 198, 200, 203–206, 209–211, 213–216, 222, 223, 227–230, 249, 251, 259, 279, 280, 290, 295, 299, 316, 317, 330, 335, 339–341, 359, 373–376, 401, 402, 412, 414–416, 420, 425, 445, 470, 492–494, 504, 505, 509–513, 515–520, 522, 528–537, 542, 544, 545, 547, 552, 558–560
Clay dispersion, 509
Clayey, 223, 533, 535
Clay-loam, 317
Clay minerals, 416, 421, 517, 518, 522
Clay silicates, 520, 522
Clayton, 116
Clogging, 273, 502, 511, 552, 553
Coadsorption, 519
Colloid, 4, 17, 21, 108, 120, 127, 133, 134, 193, 213, 267, 269, 287, 288, 290–292, 333–337, 359, 374, 407, 469, 471, 502, 510, 532, 537
Colloidal materials, 25, 133, 285, 335, 553
Complex formation, 151, 377
Composite nonaqueous phase liquids (CNAPLs), 150
Compost, 537, 546–548, 550, 551, 553–557, 558
Copper, 11, 26, 63–65, 69, 217, 227, 304, 316, 319, 465, 474, 505–507, 519
Cosolvent, 153–155, 191, 192, 361, 492, 153, 155, 156, 160, 184, 190, 191
Crust, 61, 63, 68, 279, 280, 502, 509, 510, 536, 537
Crusting, 502
Cyclohexane, 172, 482

D

Decarboxylation, 385, 386
Deethylatrazine, 447, 486
Deflocculation, 544
Deisopropylatrazine, 447
Denitrification, 383, 389, 390
Dense NAPLs (DNAPLs), 273, 342
Desethylatrazine, 480, 481
Desorption, 17, 109, 113, 115, 128, 133, 135–138, 140, 141, 201, 203, 214, 224–230, 268, 270, 285, 323, 325, 326, 361, 363, 364, 391, 473, 517–520, 522–527, 539
Dichloride, 84, 87, 518
Dichlorobenzene, 89, 481, 482, 488, 523, 525–527
Diethylthiophosphate, 443
Differential rate law, 116
Diffuse double-layer model, 108
Diffusion, 18, 23, 45, 46, 49, 50, 59, 85, 116–118, 122, 129, 133, 135, 137, 140, 141, 163, 164–166, 168, 169, 173, 201, 202, 225, 255–258, 261, 267, 269, 271, 280, 312, 313, 328, 330, 331, 362, 364, 383, 437, 508, 509, 525, 550
Diffusionally, 19
Diquat, 84, 85, 203–206, 230, 518–520
Dispersion, 49, 124, 163, 164, 211, 255–261, 263, 264, 267, 269, 271, 277, 326, 329, 502, 505, 508–513, 515, 516, 528, 531–533, 536, 537, 544, 545
Dissolved, 15
Dissolved organic matter, 160, 188, 291, 304
Distribution coefficient, 115, 164, 209, 268, 159, 193, 214, 218, 219, 423
Dodecylbenzenesulfonate, 544, 545
Drainage, 13, 58, 59, 66, 187, 247, 248, 250, 252, 277, 278, 399, 510

E

Effluent, 19, 21, 56, 59, 133, 154, 155, 175, 176, 186, 187, 198, 214, 220, 221, 225, 294, 296, 297, 299, 301, 302, 307, 309–311, 335–338, 391, 432, 433, 458, 460, 481, 488, 489, 510, 511, 534, 537, 542, 543, 545, 546, 552, 555
Effluent irrigation, 176
Elimelech, 133
Elovich equation, 116, 118
Emerging contaminants, 486
Endocrine disruptor compounds, 214
Enthalpy, 29–31, 48, 123, 124, 154, 203, 517
Entropy, 29–31, 48, 49, 123, 125, 168, 517
Epoxidation, 385, 387
Equilibrium, 21, 23, 29, 31–33, 35, 37, 38, 40, 43, 46, 50, 57, 109, 112, 114–117, 123, 130, 131, 135–137, 147–151, 164–166, 168, 173, 175, 185, 187, 188, 191, 200, 211, 215–217, 223, 255, 258, 272, 273, 326, 330, 331, 352, 354, 355, 363, 373, 378, 391, 394, 409, 424–426, 430, 437, 458, 464, 468, 469, 476, 506, 533, 534, 539, 558
Erosion, 58, 69, 279, 280, 418, 502, 512–515
Ethylbenzene, 92, 172, 367, 490

F

Ferredoxin, 355
Field-scale, 248, 251, 255, 256, 259, 325–327, 432
Fingering, 247, 250, 273, 275, 277, 278
Flocculation, 108, 335, 505, 508, 509, 531, 532, 536, 537
Fluoride, 26, 214, 215, 360
Fluorophosphate, 57
Freundlich, 46, 112–114, 211, 268
Fringe, 275, 276
Fugacity, 114, 115, 150, 151, 168
Fulvic acid, 14, 15, 40, 108, 127, 153, 192, 285, 349–351, 378, 399, 473, 541, 542, 552, 555
Fulvic acid solubility, 542

G

Glyphosate, 520, 522
Glyphosphate, 139
Gouy-Chapman theory, 46
Gypsum, 4, 40, 59, 186, 187, 271, 501, 507, 509–512, 516
Gypsum solubility, 509

H

Heavy metal, 21, 47, 63, 66, 121, 130, 133, 188, 193, 201, 202, 205, 220, 227, 234, 235, 281, 285–289, 291, 295–297, 335, 360, 423, 469, 471, 473–475, 506
Heavy metal adsorption, 47
Heavy metal speciation, 360
Hemangiosarcomas, 87
Henry's law, 164–166
Herbicide, 57, 81, 84, 85, 89, 112, 114, 128, 136, 139, 177, 178, 203–205, 207, 209, 210, 230, 280, 321, 323–325, 330, 331, 333, 335, 337,

359, 375, 378, 386, 387, 442, 481, 484–486, 518–520
Hexachlorobenzene, 525
Hexachlorocyclohexane, 81, 128
Hexahydrobenzene, 392
Hexametaphosphate, 545
Hexane, 93, 111, 112, 167, 211, 212, 419, 421, 443
Hexaphosphate, 392
Homoionic, 203, 204, 229, 376
Human activity, 233
Humic acid, 14–16, 44, 127, 160, 206–209, 229, 230, 304, 306, 308, 309, 311, 314–316, 350, 352, 360, 378, 423, 424, 440, 441, 471, 532, 541, 552, 553, 555, 556, 558
Hydrazobenzene, 356
Hydrochloride, 521
Hydrogen bonding, 20, 48, 123, 154, 517
Hydrolysis, 211, 349–352, 362–364, 374–378, 388, 389, 397, 400, 405, 412–416, 442–445, 447, 480, 505, 542
Hydroxides, 3, 20, 39, 108, 128, 136, 207, 301, 509
Hysteresis, 46, 135, 137–140, 202, 225–227, 229, 230, 518, 523, 525, 529, 530
Hysteresis in the adsorption, 137
Hysteresis isotherm, 518

I

Imbibition, 247, 248, 250, 252
Infiltration, 26, 71, 98, 200, 217, 247–252, 255, 256, 258, 259, 274–280, 324, 338, 339, 481, 488, 502, 509–515, 528, 533, 536–538, 552, 555
Inorganic fractions, 503
Inorganic matter, 133
Insecticide, 57, 81, 82, 128, 139, 177, 211, 321, 328, 335, 364, 375, 442, 485
Intraorganic matter, 140
Iodobenzene, 482
Ion activity product, 130
Irradiation, 61, 359, 419–422
Irreversible, 31, 35, 47, 129, 135, 138, 140, 141, 201, 214, 457–459, 461, 464–466, 469, 471, 473, 474, 477, 478, 480, 486, 492, 495, 497, 501, 502, 505, 507–510, 512–519, 522, 523, 525, 528, 530, 533, 537, 539, 541–547, 552, 553, 560–562
Irreversible adsorption, 492
Irreversible isotherm, 524
Irrigation, 19, 21, 54, 58, 61, 133, 187, 189, 198, 200, 201, 210, 279, 291, 320–322, 324, 325, 327, 330, 332–334, 337, 458, 459, 461, 477, 501, 502, 510, 513, 515, 516, 518, 528, 529, 535, 544, 552
Isotherm, 46, 110–112, 114, 115, 136, 138, 165, 166, 214, 216, 268, 450, 523
Isothermal, 165, 166, 450
Isotherm model, 115
Isotherms, 109–115, 128, 135, 136, 169, 203, 204, 206–208, 210, 213–218, 227, 229, 236, 323, 525
Isotherms for the adsorption, 208

K

Kaolinite, 4, 7, 8, 12, 13, 20, 107, 108, 111, 112, 116, 126, 200–204, 211, 216, 290, 317, 318, 375, 376, 412–414, 415, 421, 518–520, 522, 532, 534, 545, 550
Kaolinitic, 536
Kerogen, 207, 229, 230
Kerosene, 157, 160, 180–185, 189, 190, 195, 197, 222, 223, 339–341, 361, 430, 431, 434–437, 492, 558–560

L

Landfill, 97, 474, 487, 488
Langmuir, 46, 112–115, 203, 215, 216, 236, 523
Lewis acid, 152, 374, 377, 480
Lewis acidity, 374
Ligand, 21, 25, 26, 39, 40, 48, 124, 130, 148, 151–153, 158, 159, 190, 192, 214, 217, 218, 220, 226, 227, 285, 349, 359, 360, 374, 377, 379, 382, 391, 394, 397, 404, 422, 423, 477, 510, 517
Light nonaqueous-phase liquids (LNAPLs), 91, 273
Lime, 176, 335, 336, 501, 507–509
Limed, 508
Lindane, 81, 82, 128, 167, 177, 230, 231

M

Macroorganic matter, 14
Metalloid, 63, 285–287, 296, 234, 285, 286, 296, 373, 476
Metal oxides, 13, 39, 202, 207, 373, 416, 420
Metals, 21, 47, 63, 121, 130, 133, 152, 158, 189, 192, 193, 201, 204, 220, 227, 234–236, 285–289, 291, 297, 352, 359, 360, 375, 379, 394, 399, 422
Methyl tertiary-butyl ether (MTBE), 92, 94, 490

Microbes, 79, 232, 252, 507
Microbial, 3, 13, 14, 22, 64, 133, 140, 150, 222, 232, 233, 251, 296, 298, 330, 354, 362, 367, 368, 381–386, 388, 389, 440, 447, 472, 492, 507, 542, 550–553
Microbial activity, 492
Microbially, 298, 355, 356, 364, 367, 381, 383, 384, 388, 389, 393, 440, 477
Micronutrient, 56, 505, 506
Minerals, 4, 6–8, 13, 20, 39, 47, 49, 56, 107, 121–123, 127, 133, 140, 202, 206, 228, 298, 316, 334, 335, 373–375, 388, 404, 409, 420, 421, 520, 530, 533, 534
Monochlorobenzene, 89
Montmorillonite clay, 9–12, 108, 111, 112, 116, 128, 200–204, 209, 211–215, 226, 227, 229, 317, 335, 337, 375, 376, 420–422, 493, 494, 518–522, 532, 534, 550

N
Napropamide, 321, 330–332
Natural attenuation, 367, 381, 457, 474, 492, 495, 497
Near solid phase water, 18–20
Netalloids, 469
Nickel, 68
Nitrification, 382, 383, 389, 502
Nitrification-denitrification, 389
Nitroaromatics, 407
Nitrobenzene, 357
Nitrogen, 42, 53, 54, 84, 89, 280, 378, 390, 425, 438, 447, 465, 483, 501, 502, 508
Nitrogen fertilizer, 53, 502, 503
Nitroreduction, 354
Nitrosobenzene, 409
n-Octane, 93, 167
Nonaqueous phase liquids (NAPLs), 25, 32, 91, 150, 160, 191, 247, 256, 273, 337, 338, 341, 491
Nonequilibrium, 307
Nonionic polyacrylamides, 513
Non metals, 21
Nonpoint source, 177, 281

O
Organic fraction, 3, 114, 127, 228, 229, 378, 394, 517
Organic matter, 12–17, 64, 69, 112, 121–123, 127–129, 136, 140, 141, 150, 159, 160, 178, 188, 191, 207, 213, 291, 303, 304, 312, 313, 316, 320, 323, 325, 337, 341, 354, 362, 378, 379, 382, 393, 399–402, 415–417, 420, 423, 444, 445, 447, 523, 524, 543, 546, 547, 555, 560
Organoclay, 517
Organo-mineral association, 108
Organophosphate, 83, 139, 350, 352, 375, 376, 384, 414, 415, 442
Organophosphate pesticides, 84
Organophosphorus, 57, 210, 211, 213, 399, 412
Organothiophosphates, 83
Orthophosphate, 391
Osmotic pressure, 506
Overland flow, 56, 280
Oxidation, 40, 42, 44, 45, 53, 64, 66, 86, 127, 148, 149, 152, 188, 213, 285, 296, 298, 350, 352, 354, 357, 359, 363–365, 367, 368, 373, 374, 382, 384–386, 389, 393, 401–407, 424, 438–440, 470, 473, 474, 477–480, 538
Oxidation-reduction equilibria, 149
Oxyhydroxides, 477

P
Paraquat, 84, 85, 203–209, 230, 231, 335, 337, 518–520
Parathion, 83, 111, 127, 129, 177, 178, 211–213, 227–229, 321, 328, 329, 375, 397, 399, 400, 412–417, 442–446
Partition coefficient, 114, 154, 164, 171, 178, 196, 197, 214, 215, 425, 426, 428, 429, 523, 558
Pentachloronitrobenzene, 405
Pentaphosphates, 392
Perchloroethylene (PCE), 87, 88, 342, 368, 430–432, 433, 437, 488
Peroxidation, 85
Personal care products, 94, 197, 316, 486, 489
Pesticides, 57, 79–84, 86, 89, 90, 138, 139, 153, 167, 168, 177–179, 203, 207, 209–211, 230, 231, 281, 285, 320, 321, 328, 335, 350, 352, 354, 363, 375, 383–385, 387–390, 412, 437, 442, 449, 450, 481, 483–486, 506, 517, 518, 522
Petroleum hydrocarbons, 25, 91, 93, 185, 234, 285, 365, 437, 438, 488, 490, 491, 494, 495, 560–562
Pharmaceuticals, 86, 89, 94, 153, 197, 316, 486, 489
Phenanthrene, 140, 158, 191, 207, 229, 232, 233, 237, 367

Phenylmethylbenzene, 482
Phosphate, 26, 56, 57, 62, 65, 89, 118, 136, 295, 335, 350, 351, 360, 377, 389, 392, 393, 413, 414, 416, 477, 440, 441, 480, 487, 503–505, 539
Phosphogypsum, 509, 511–513, 515, 516
Phosphorus, 56, 57, 89, 280, 334, 390, 391, 412, 501, 503, 504
Photolysis, 349, 357–359, 365, 416, 417, 419–421
Phyllosilicate, 4, 7, 20, 216, 202, 423
Plutonium, 404–406, 423, 470
Point source, 177, 237, 328, 330, 331, 446, 457, 458, 474–476, 552
Polycyclic aromatic hydrocarbons (PAHs), 150, 158, 191, 193, 194, 237, 238, 367, 419
Polysaccharide, 127, 160, 379, 515, 516, 550, 552, 555, 556
Preadsorption, 206
Preferential flow, 56, 248–250, 285, 313, 330–333, 335, 457
Primary, 39
Primary minerals, 4
Prometryn, 84, 209, 321, 332
Protonation/deprotonation, 17, 48, 124, 127, 375

Q
Quartz, 4–6, 24, 39, 200, 216, 290, 307–311, 314, 316–318, 423

R
Radiation, 61–63, 110, 359
Radionuclide, 61–63, 133, 215, 216, 404, 469–471
Rate laws, 34, 116
Reactivity, 4, 159, 409
Redox, 13, 42–45, 63, 64, 149, 150, 187, 188, 286, 298, 349, 350, 352, 354, 355, 357, 363–365, 368, 374, 376, 383–385, 393, 394, 397, 402, 405, 472, 477
Reduction, 29, 42–44, 66, 148, 149, 158, 161, 188, 189, 211, 225, 232, 270, 273, 285, 298, 325, 350, 352, 354–357, 362–365, 368, 382, 388–390, 393, 397–402, 404–411, 426, 440, 441, 448, 450, 464, 466, 504, 515, 516, 533, 534, 546, 552, 561
Repellency, 550, 560, 561
Residual organic liquid saturation, 132
Residues, 46
Retention hysteresis, 135
Reversible, 30, 31, 83, 135, 214, 227–229, 356, 357, 523, 528
Richards' equation, 250
Runoff, 54, 56, 59, 68, 94, 188, 220, 265, 279–281, 302, 333–335, 399, 509, 512–516

S
Salinity, 21, 47, 58, 60, 61, 156, 158, 195, 196, 214, 218, 219, 425, 426, 428–430, 458–461, 464, 510
Salinization, 57–61, 458–461, 463–465
Salting-out effect, 156–158, 195–198, 430
Secondary minerals, 4, 56
Sewage, 53, 94, 97, 98, 133, 155, 175, 198, 220, 289, 293, 295, 316, 335, 337, 458, 459, 543, 544, 546, 551, 553, 555–558
Sewage effluent, 98, 480, 481, 510
Sewage sludge, 175, 506
Shrink-swell property, 12
Silica, 4–7, 13, 56, 128, 418, 419, 421
Silicate, 3, 4, 7, 11, 12, 20, 56, 107, 108, 127, 350, 351, 505, 518, 520, 522, 528, 529, 534
Silicate minerals, 4
Sludge, 53, 175, 176, 189, 190, 192, 214, 220, 289, 293–297, 391, 507, 543, 546–558
Sludge-amended, 295
Sludges, 220, 537, 552
Smectite, 4, 7–13, 20, 107, 116, 120, 122, 123, 126, 137, 207, 208, 216, 290, 415, 416, 423, 529, 530, 532, 535, 545, 546
Sodic water, 198, 201, 510, 528, 532–535
Sodium adsorption ratio (SAR), 47, 157, 200, 201, 361, 532, 534–536
Soil, 13–15
Soil organic matter, 13, 127, 323, 325, 423
Solubility, 14, 15, 22, 23, 25, 32, 40, 66, 70, 86, 92–94, 126, 128–132, 147, 148, 150, 152–161, 164, 166, 167, 173, 184, 187–193, 195–198, 200, 209, 271, 273, 288, 307, 321, 325, 335, 361, 385, 394, 428, 430, 447, 474, 490–492, 503, 505, 523
Solubility product, 130, 131, 273
Sorption isotherms, 203
Spatial variability, 109, 210, 250, 260, 279, 280, 321, 323–325
Speciation, 21, 158, 192, 218, 352, 359, 360, 393, 394, 404, 405, 422, 423, 477–480
Steroids, 96
Straining, 132, 133, 269, 270

Sulfoxidation, 385, 388
Surfactant, 26, 84, 153, 155, 156, 190–192, 281, 314, 390, 488, 543, 544, 557
Suspended particles, 107, 132, 133, 147, 220, 280, 281, 301, 337, 362, 473, 480, 513, 552, 553
Swelling, 12, 67, 85, 137, 229, 528–530, 533

T
Tectosilicates, 4
Tetrachloride, 273
Tetrachlorobenzene, 525
Thiphosphate, 415
Toluene, 92, 93, 150, 167, 198–200, 367, 421, 437, 438, 440, 441, 490, 491, 493
Trace, 422
Trace elements (micronutrients), 5, 21, 113, 153, 189, 220, 223, 383, 394, 480, 501
Trace metals, 158, 188, 466
Transition, 21, 192, 379, 399
Transition metals, 121, 360
Tribromoneopentyl alcohol (TBNPA), 397, 400
Trichlorobenzene, 482, 525
Trichloroethylene (TCE), 87, 88, 198–200, 273, 342, 368, 430–433
Trichloroethylene petroleum hydrocarbons, 491
Trichloromethane (TCM), 433
Triphosphate, 56

U
Uranium, 26, 66, 67, 423, 470, 471
Urban runoff, 281

V
Vadose zone, 3, 14, 23, 26, 131–133, 173, 247, 248, 250–252, 255, 277, 302, 333, 423, 457, 458, 466, 481, 488, 517
van der Waals forces, 8, 49, 124, 141, 198, 269, 517
Vineyard, 506, 519
Volatilization, 88, 132, 137, 148, 163–166, 168, 169, 171, 173–185, 255, 267, 271, 274, 277, 278, 321, 339, 349, 361, 365, 367, 425, 434–437, 488, 492, 559
Volatilization, adsorption, 492

W
Wastewater, 57, 59, 200, 295, 296, 302, 423, 486, 488, 528, 542, 543, 546, 552, 553
Water transmission, 510–512, 516, 528, 533, 543, 552, 560
Weathering, 3, 5, 6, 12, 39, 40, 131, 137, 402, 465, 507

X
Xylene, 92, 157, 183, 184, 198, 199, 367, 490, 495, 496

MIX
Papier aus verantwortungsvollen Quellen
Paper from responsible sources
FSC® C105338

If you have any concerns about our products,
you can contact us on
ProductSafety@springernature.com

In case Publisher is established outside the EU,
the EU authorized representative is:
Springer Nature Customer Service Center GmbH
Europaplatz 3, 69115 Heidelberg, Germany

Printed by Libri Plureos GmbH
in Hamburg, Germany